Bacterial Membranes

Structural and Molecular Biology

Edited by

Han Remaut

Structural Biology Brussels
Vrije Universiteit Brussel and VIB
Brussels
Belgium

and

Rémi Fronzes

Groupe à 5 ans 'Biologie Structurale de la Sécretion Bactérienne'
Institut Pasteur
Paris
France

Caister Academic Press

Caister Academic Press
Norfolk, UK

www.caister.com

British Library Cataloguing-in-Publication Data
A catalogue record for this book is available from the British Library

ISBN: 978-1-908230-27-0

Description or mention of instrumentation, software, or other products in this book does not imply endorsement by the author or publisher. The author and publisher do not assume responsibility for the validity of any products or procedures mentioned or described in this book or for the consequences of their use.

Cover design adapted from Figures 8.1 and 8.3.

Printed and bound in Great Britain

Contents

Contributors

Pedro M. Alzari
Unite de Microbiologie Structurale and CNRS UMR 3528
Institut Pasteur
Paris
France

pedro.alzari@pasteur.fr

Gwénaëlle André-Leroux
Unite de Microbiologie Structurale and CNRS UMR 3528
Institut Pasteur
Paris
France

gwenaelle.andre@pasteur.fr

Jesús Arenas
Department of Molecular Microbiology and Institute of Biomembranes
Utrecht University
Utrecht
The Netherlands

j.a.arenasbusto@uu.nl

Suraaj Aulakh
Department of Molecular Biology and Biochemistry
Simon Fraser University
Burnaby
Canada

ska16@sfu.ca

Tanneke den Blaauwen
Bacterial Cell Biology
Swammerdam Institute for Life Sciences
University of Amsterdam
Amsterdam
The Netherlands

t.denblaauwen@uva.nl

Martine P. Bos
Department of Molecular Microbiology and Institute of Biomembranes
Utrecht University
Utrecht
The Netherlands

m.bos@uu.nl

Volkmar Braun
Max Planck Institute for Developmental Biology
Tübingen
Germany

volkmar.braun@tuebingen.mpg.de

Juliane K. Brülle
Institut für Medizinische Mikrobiologie
Universität Zürich
Zurich
Switzerland

jbruelle@imm.uzh.ch

Matthias J. Brunner
Research Institute of Molecular Biotechnology (IMBA)
Austrian Academy of Sciences;
Research Institute of Molecular Pathology (IMP)
Vienna
Austria

matthias.brunner@imba.oeaw.ac.at

Gaetano Castaldo
Structural and Molecular Microbiology;
Structural Biology Brussels
VIB Department of Structural Biology
Vrije Universiteit Brussel
Brussels
Belgium

gaetano.castaldo@vub.ac.be

Mamadou Daffé
Department of Molecular Mechanisms of Mycobacterial Infections
Institut de Pharmacologie et Biologie Structurale du Centre National de la Recherche Scientifique (CNRS)
Université Paul Sabatier
Toulouse
France

mamadou.daffe@ipbs.fr

Vincent van Dam
Department of Molecular Microbiology and Institute of Biomembranes
Utrecht University
Utrecht
The Netherlands

vandam.vincent@gmail.com

Fabien Debailleul
Department of Structure and Function of Biological Membranes
Structural Biology and Bioinformatics Center
Université Libre de Bruxelles
Brussels
Belgium

fabien.debailleul@ulb.ac.be

Gianni Dehò
Dipartimento di Scienze Biomolecolari e Biotecnologie
Università di Milano
Milan
Italy

gianni.deho@unimib.it

Adeline Derouaux
Centre d´ingénierie des protéines
Institut de Chimie
Université de Liège
Sart-Tilman
Belgium

adeline.derouaux@ulg.ac.be

Rémi Fronzes
Groupe à 5 ans 'Biologie Structurale de la Sécretion Bactérienne'
Institut Pasteur
Paris
France

remi.fronzes@pasteur.fr

Anthony M. George
School of Medical and Molecular Biosciences and iThree Institute
University of Technology Sydney
Broadway
Australia

tony.george@uts.edu.au

Cédric Govaerts
Department of Structure and Function of Biological Membranes
Structural Biology and Bioinformatics Center
Université Libre de Bruxelles
Brussels
Belgium

cgovaert@ulb.ac.be

Thomas Grau
Roche Diagnostics Ltd
Rotkreuz
Switzerland

thomas.grau@roche.com

Vinciane Grimard
Department of Structure and Function of Biological Membranes
Structural Biology and Bioinformatics Center
Université Libre de Bruxelles
Brussels
Belgium

vgrimard@ulb.ac.be

Peter M. Jones
School of Medical and Molecular Biosciences and iThree Institute
University of Technology Sydney
Broadway
Australia

peter.jones-2@uts.edu.au

Kelly H. Kim
Department of Molecular Biology and Biochemistry
Simon Fraser University
Burnaby
Canada

kkim@crystal.harvard.edu

Ulrich Kleinekathöfer
School of Engineering and Science
Jacobs University Bremen
Bremen
Germany

u.kleinekathoefer@jacobs-university.de

Hans-Georg Koch
Institut für Biochemie und Molekularbiologie
ZBMZ
Albert-Ludwigs-Universität Freiburg
Freiburg
Germany

Hans-Georg.Koch@biochemie.uni-freiburg.de

Renuka Kudva
Institut für Biochemie und Molekularbiologie
ZBMZ
Albert-Ludwigs-Universität Freiburg
Freiburg
Germany

renuka.kudva@biochemie.uni-freiburg.de

Patrick Kuhn
Institut für Biochemie und Molekularbiologie
ZBMZ
Albert-Ludwigs-Universität Freiburg
Freiburg
Germany

patrick.kuhn@biochemie.uni-freiburg.de

Kenji Kurokawa
Division of Phamaceutical Cell Biology
Faculty of Pharmaceutical Sciences
Nagasaki International University
Nagasaki
Japan

kurokawa@niu.ac.jp

Bok Luel Lee
National Research Laboratory of Defense Proteins
College of Pharmacy
Pusan National University
Busan
Korea

brlee@pusan.ac.kr

Marc Lensink
Interdisciplinary Research Institute (IRI) – Computational Biology
USR3078 CNRS
Villeneuve d'Ascq
France

marc.lensink@iri.univ-lillel.fr

Alvin Lo
Structural and Molecular Microbiology,
Structural Biology Brussels
VIB Department of Structural Biology
Vrije Universiteit Brussel
Brussels
Belgium

alvilo@vub.ac.be

Kozhinjampara R. Mahendran
Jacobs University Bremen
School of Engineering and Science
Bremen
Germany

mahendran.kozhinjampararadhakrishnan@chem.ox.ac.uk

Thomas C. Marlovits
Research Institute of Molecular Biotechnology (IMBA)
Austrian Academy of Sciences
Research Institute of Molecular Pathology (IMP)
Vienna
Austria

thomas.marlovits@imba.oeaw.ac.at

Mariano Martinez
Unite de Microbiologie Structurale and CNRS UMR 3528
Institut Pasteur
Paris
France

mariano.martinez@pasteur.fr

Hiroshi Nakayama
Biomolecular Characterization Team
RIKEN
Wako
Japan

knife@riken.jp

Suguru Okuda
Department of Chemistry and Chemical Biology
Harvard University
Cambridge, MA
USA

6kuda@fas.harvard.edu

Mark Paetzel
Department of Molecular Biology and Biochemistry
Simon Fraser University
Burnaby
British Columbia
Canada

mpaetzel@sfu.ca

Alessandra Polissi
Dipartimento di Biotecnologie e Bioscienze
Università di Milano-Bicocca
Milan
Italy

allessandra.polissi@unimib.it

Han Remaut
Structural Biology Brussels
Vrije Universiteit Brussel and VIB
Brussels
Belgium

han.remaut@vib-vub.be

Virginie Roussel-Jazédé
Department of Molecular Microbiology and Institute of Biomembranes
Utrecht University
Utrecht
The Netherlands

virginiejazede@free.fr

Jean-Marie Ruysschaert
Department of Structure and Function of Biological Membranes
Structural Biology and Bioinformatics Center
Université Libre de Bruxelles
Brussels
Belgium

jmruyss@ulb.ac.be

Peter Sander
Institut für Medizinische Mikrobiologie
Universität Zürich
Zurich
Switzerland

psander@imm.uzh.ch

Robert Schulz
Jacobs University Bremen
School of Engineering and Science
Bremen
Germany

r.schulz@jacobs-alumni.de

Paola Sperandeo
Dipartimento di Biotecnologie e Bioscienze
Università di Milano-Bicocca
Milan
Italy

paola.sperandeo@unimib.it

Lukas Sturm
Institut für Biochemie und Molekularbiologie
ZBMZ
Albert-Ludwigs-Universität Freiburg
Freiburg
Germany

sturm.lukas@gmail.com

Mohammed Terrak
Centre d'ingénierie des protéines
Institut de Chimie
Université de Liège
Sart-Tilman
Belgium

mterrak@ulg.ac.be

Hajime Tokuda
Faculty of Nutritional Sciences
University of Morioka
Takizawa
Iwate
Japan

htokuda@morioka-4.ac.jp

Jan Tommassen
Department of Molecular Microbiology and Institute of Biomembranes
Utrecht University
Utrecht
The Netherlands

j.p.m.tommassen@uu.nl

Andreas Tschumi
Institut für Medizinische Mikrobiologie
Universität Zürich
Zurich
Switzerland

atschumi@imm.uzh.ch

Riccardo Villa
Dipartimento di Biotecnologie e Bioscienze
Università di Milano-Bicocca
Milan
Italy

riccardo.villa@unimib.it

Waldemar Vollmer
Centre for Bacterial Cell Biology
Institute for Cell and Molecular Biosciences
Newcastle University
Newcastle upon Tyne
UK

w.vollmer@ncl.ac.uk

Helge Weingart
School of Engineering and Science
Jacobs University Bremen
Bremen
Germany

h.weingart@jacobs-university.de

Thomas Welte
Institut für Biochemie und Molekularbiologie
ZBMZ
Albert-Ludwigs-Universität Freiburg
Freiburg
Germany

thomas.welte@gmail.com

Mathias Winterhalter
School of Engineering and Science
Jacobs University Bremen
Bremen
Germany

m.winterhalter@jacobs-university.de

Benoît Zuber
Institute of Anatomy
University of Bern
Bern
Switzerland

benoit.zuber@ana.unibe.ch

Preface

Membranes are pivotal components of life. In an aqueous environment, these lipid bilayers form formidable insulators that demarcate the contained environment that forms a living cell, where biological processes can occur under controlled conditions. Membranes are such good insulators that they are a central player in biological energy generation in the form of ion gradients. As a further consequence, passage of ions, proteins, nucleic acids, nutrients and metabolites across membranes needs to be facilitated and requires an arsenal of dedicated channels and transporters. So does the transduction of signals and cues from the outside world into the cell and vice versa. In combination with supporting protein and glycan networks, membranes also provide shape and structure to cells and are important in cell motility. They further fulfil a scaffolding function for proteins and organelles exposed to the outside world and destined to interact with the extracellular environment. Clearly, a plethora of cellular processes evolve at and near cell membranes. In this book, the structural and molecular biology of these processes will be closer examined from a bacterial perspective.

In bacteria, two large groups can be identified based on the presence of either one or two lipid bilayers surrounding the cell. Historically these two groups have been referred to as, respectively, Gram-positive and Gram-negative bacteria based on their susceptibility or resistance to Gram staining. This staining indirectly reflects two different organizations of the cell envelope. In monoderm, or Gram-positive, bacteria the cell membrane is encapsulated by a thick layer of peptidoglycan, a network of cross-linked peptide and glycan units that provide shape and physical strength to the cell. In diderm or Gram-negative bacteria, a thin layer of peptidoglycan surrounds the cell membrane and is itself enclosed by a second lipid bilayer, the outer membrane. Both membranes have different protein and lipid composition and play specific roles. They enclose a cellular compartment referred to as the periplasmic space, that holds a dedicated repertoire of proteins but is devoid of nucleic acids or many of the metabolites associated with life. Additional layers such as two-dimensional protein arrays (or S-layers) and polysaccharide capsules are also present at the surface of many bacteria.

Many crucial processes are located at the cell envelope. Primarily, bacteria use this envelope as a protective layer from their hostile environment. S-layers and capsules protect bacteria from desiccation, low-pH, extracellular enzymes or virus infection. In both Gram-negative and Gram-positive bacteria, the rest of the cell envelope also plays a crucial role as a barrier around the cell. In both bacterial groups, the peptidoglycan layer shapes the cell and contributes to mechanical resistance of the cells. It is also directly involved in cell division. Because it is in contact with the outside world, this barrier is also an important interface. It selectively mediates the transport of molecules (solute and ions) and macromolecules (protein and DNA) in and out of the bacterial cell. Depending on the structure of their cell envelope, bacteria developed various systems to import and secrete molecules. These systems vary from individual channel or transporter proteins found in the bacterial inner- or outer membrane to complicated multiprotein complexes that span the entire cell envelope. Many surface appendages and adhesion proteins are also found at the surface of the bacteria. They are essential for bacterial survival by mediating cell motility, adhesion or various enzymatic activities. Their assembly or transport involves specialized systems, often related to secretion systems, located in bacterial cell envelope. Finally, bacteria developed sophisticated signal transduction systems at their surface to sense their environment or to communicate with their peers.

All these systems are essential for bacterial survival because they provide essential nutrients, help to bacteria to adapt quickly to changes in their environment or to live within bacterial communities. In addition, many pathogenic bacteria rely on these systems to infect their host. For example, many adhesion proteins or complexes that are displayed at the surface of the bacterial cell promote adhesion to and invasion of the host cell. The polysaccharide capsule or some enzymes exposed at the surface impair phagocytosis by the host during infection. Furthermore, given their central role in sustaining life, it should not surprise that membranes and membrane-associated processes form the subject of a series of antibacterial agents and molecules.

An impressive effort has been made by the microbiology community to understand the molecular details of the cell envelope, its biogenesis and function. In the recent years, some tremendous progresses have been obtained by a combination of molecular and structural biology approaches. In this book, specialists in the field present a selection of these recent progresses across the spectrum of the versatile functions and involvements of membranes in bacterial physiology. We are grateful to them for their valuable contributions.

Han Remaut and Rémi Fronzes

Part I

Membrane Composition and Synthesis

Bacterial Cell Wall Growth, Shape and Division

1

Adeline Derouaux, Mohammed Terrak, Tanneke den Blaauwen and Waldemar Vollmer

Abstract

The shape of a bacterial cell is maintained by its peptidoglycan sacculus that completely surrounds the cytoplasmic membrane. During growth the sacculus is enlarged by peptidoglycan synthesis complexes that are controlled by components linked to the cytoskeleton and, in Gram-negative bacteria, by outer membrane regulators of peptidoglycan synthases. Cell division is achieved by a large assembly of essential cell division proteins, the divisome, that coordinates the synthesis and hydrolysis of peptidoglycan during septation. Coccal species such as *Staphylococcus aureus* grow exclusively by synthesis and cleavage of a cross-wall. Ovococci like *Streptococcus pneumoniae* elongate at a central growth zone resulting in their lancet-shape. Rod-shaped species elongate either at the side-wall coordinated by the MreB cytoskeleton, like *Escherichia coli* or *Bacillus subtilis,* or at the poles like *Corynebacterium glutamicum*. Bacteria have different mechanisms to achieve bent or helical cell shape, involving cytoskeletal proteins, periplasmic flagella or peptidoglycan hydrolases, and to form branched, filamentous cell chains. Peptidoglycan enzymes and cytoskeletal proteins are validated targets for antimicrobial compounds. Recent approaches applying structure-based inhibitor design, high-throughput screening assays and whole cell assays have identified a large number of novel inhibitors of cytoskeletal proteins and enzymes of the peptidoglycan biosynthesis pathway.

Introduction

Bacteria come in an amazing variety of shapes and their cell size varies by more than six orders of magnitude (Young, 2006). Next to the rod-shaped model bacteria *Escherichia coli* or *Bacillus subtilis* round, bent, curved and helical species exist. Some bacteria even adopt exotic shapes such as lemon, teardrop, triangle, square or others. Moreover, many bacteria grow as linear or branched chains of cells or they form various types of multicellular aggregates. Despite the wide variety of bacterial morphology, cells of a given species are remarkably homogenous in shape and size and bacteria are capable to propagate with their inherited shape from one generation to the next indicating that cell shape is maintained by robust mechanisms (Young, 2006; Philippe *et al.*, 2009). A particular cell shape might have evolved to optimize bacteria's motility, nutrient uptake and reproduction, or to avoid predation in its environmental niche. However, in many cases we can only speculate as to why a bacterial species has adopted its specific cell shape, and the selective pressures underlying cell shape evolution are not known (Young, 2006; Margolin, 2009). Based on recent research we are beginning to understand how bacterial cell shape is achieved and maintained during growth and cell division.

The cell envelope peptidoglycan (PG; also called murein) is essential to maintain cell shape in most bacteria. PG forms an exoskeleton (sacculus) made of glycan strands that are cross-linked by short peptides. As a result, isolated sacculi retain the size and shape of the bacterial cell (Weidel and Pelzer, 1964; Vollmer *et al.*, 2008a). PG is a specific component of the bacterial cell envelope and is required in most bacteria to withstand the internal cytoplasmic turgor pressure. Inhibition of PG synthesis or its uncontrolled degradation leads to cell lysis making the PG biosynthetic pathway a prime target for antibacterial compounds. During growth and cell division the sacculus is enlarged by incorporation of PG precursor molecules. This process involves different PG synthases and hydrolases and their regulators. The molecular details of PG growth are not known. According to the current understanding PG enzymes form multiprotein complexes for PG synthesis and hydrolysis during sacculus growth (Höltje, 1998), and these complexes are spatiotemporally controlled from inside the cell by components linked to the bacterial cytoskeleton (Margolin, 2009). In Gram-negative bacteria, sacculus growth is also regulated from the outside by the recently identified outer membrane lipoproteins (Paradis-Bleau *et al.*, 2010; Typas *et al.*, 2010).

In this chapter, we will discuss PG structure and its variation among different bacteria species. We will then present the current knowledge about how PG is synthesized by coordinated activities of PG enzymes to enlarge the sacculus and how bacteria divide. We will also summarize the growth of bacteria with particular cell shapes such as spherical, lancet-shaped, rod-shaped, bent or helical rod-shaped, and branched filament shape. Finally, we will outline the current research on antimicrobials targeting PG growth and cell morphogenesis followed by our view on the future perspectives of this research field.

Peptidoglycan structure and architecture

PG has a net-like structure in which glycan chains are connected by short peptides (Vollmer *et al.*, 2008a). Despite this common overall PG structure and the presence of conserved components, such as muramic acid and D-alanine, there is considerable species- and strain-specific structural variation and heterogeneity as outlined in the following sections.

The glycan chains in peptidoglycan

The glycan chains are linear and made of alternating, β-1,4-linked *N*-acetylglucosamine (GlcNAc) and *N*-acetylmuramic acid (MurNAc) residues (Fig. 1.1) (Vollmer, 2008). In all Gram-negative and some Gram-positive bacteria the MurNAc residue at the chain end carries a 1,6-anhydro ring (MurNAcAnh). The mean length of the glycan chains varies between <10 and >500 disaccharide units depending on species and growth conditions (summarized in Vollmer and Seligman, 2010). Species with short glycan chains are *Helicobacter pylori* and *Staphylococcus aureus*, whereas *Bacillus subtilis* has remarkably long glycan chains. The longest individual glycan chains isolated from *B. subtilis* were ~5000 disaccharide units (~5 μm) long and, hence, longer than the bacterial cell itself (Hayhurst *et al.*, 2008). *E. coli* PG has glycan chains ranging from 1 to ~80 disaccharide units (Harz *et al.*, 1990). Newly synthesized glycan chains are 60–80 disaccharides long, and they are processed within minutes after synthesis to their final length (average 20–40 disaccharide units), whereby chains of 8–14 disaccharide units are the most abundant (Glauner and Höltje, 1990; Harz *et al.*,

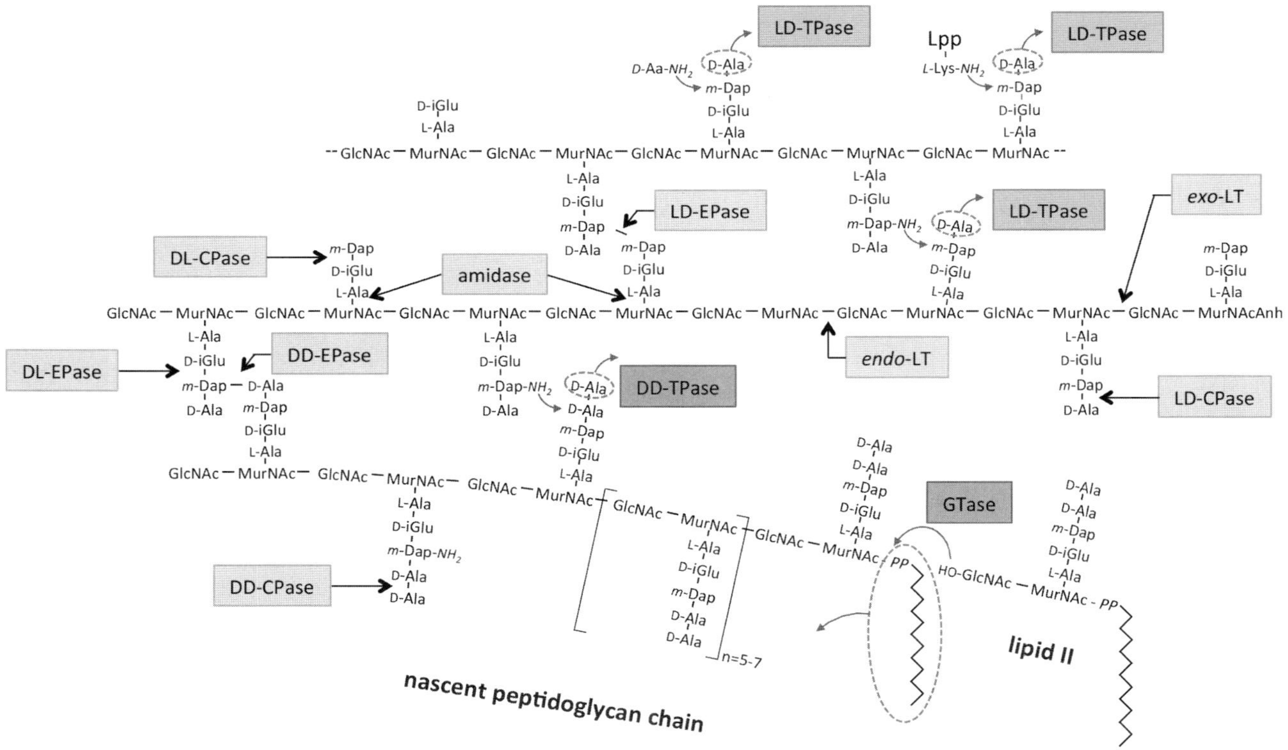

Figure 1.1 Synthesis and hydrolysis of *E. coli* peptidoglycan. The bottom part shows the polymerization of a nascent PG chain by a glycosyltransferase (GTase) and its attachment to mature PG by a DD-transpeptidases (DD-TPase). The mature PG (top) contains different peptides and has a MurNAcAnh residue at the glycan chain ends. The cleavage sites of DD-, DL- and LD-carboxypeptidase (DD-CPase, DL-CPase, LD-CPase), DD- and LD-endopeptidases (DD-EPase, LD-EPase), *exo*- and *endo*-lytic transglycosylase (*exo*-LT, *endo*-LT) and MurNAc-L-Ala amidase (amidase) are shown. LD-transpeptidases (LD-TPase) form 3–3 cross-links, attach the outer membrane lipoprotein Lpp and/or attach unusual D-amino acids (D-Aa-NH_2) to the PG. GlcNAc, *N*-acetylglucosamine; MurNAc, *N*-acetylmuramic acid; MurNAcAnh, 1,6-anhydro-*N*-acetylmuramic acid; zigzag line, undecaprenol; *PP*, pyrophosphate.

1990). In several species it has been observed that glycan chains of stationary phase cells are shorter than those of exponentially growing cells (Vollmer and Seligman, 2010). The implications of glycan chain length on PG stability and architecture are not well understood.

In most bacterial species the glycan chains become modified and this can occur at different stages of PG synthesis (Vollmer, 2008). The glycan chains of a group of closely related species of actinomycetales (for example, *Mycobacterium* species) contain a variable fraction, depending on growth conditions, of glycolylated muramic acid residues (MurN-Glyc) instead of the normal acetylated MurNAc. The glycolyl modification is introduced at the stage of cytoplasmic precursor synthesis by the NamH monooxygenase (Uchida *et al.*, 1999; Raymond *et al.*, 2005).

Other modifications are more abundant and occur during or shortly after glycan chain synthesis. Glycan chains are subject to N-deacetylation and/or O-acetylation in many Gram-positive pathogens (Vollmer, 2008). These secondary modifications confer resistance to lysozyme, an important enzyme of the innate immune system capable of lysing invading bacteria by cleaving PG glycan chains. N-deacetylation can occur at GlcNAc, MurNAc or both sugars to a variable extent by the activity of PG deacetylase PgdA (Vollmer and Tomasz, 2000). This enzyme was identified in *S. pneumoniae* and has since then been found in many other species containing deacetylated PG (Vollmer, 2008; Davis and Weiser, 2011). PgdA is a member of the carbohydrate esterase family 4 (PFAM01522) and related to chitin deacetylases and the rhizobial nodulation factor deacetylase NodB (Blair *et al.*, 2005). Purified PgdA is active against pneumococcal PG and artificial substrates. Inhibitory compounds have recently been identified by virtual high-throughput screening and experimentally confirmed (Bui *et al.*, 2011).

O-acetylated PG contains glycan chains with extra acetyl groups linked to the C6-OH group of MurNAc, GlcNAc or both (Clarke and Dupont, 1992). However, the O-acetylation of MurNAc yielding 2,6-*N,O*-diacetylmuramic acid is the most common. This PG modification occurs in a large number of Gram-positive and Gram-negative species (Vollmer, 2008). The process of O-acetylation is not well understood. Two types of *O*-acetyltransferases have been described. The first type of O-acetyltransferases are encoded by genes similar to *oatA*, the first described PG-O-acetyltransferase gene identified in *S. aureus* (Bera *et al.*, 2005). Subsequently, homologues of *oatA* have been found to be required for PG O-acetylation in *S. pneumoniae* and other Gram-positive bacteria (Crisostomo *et al.*, 2006; Vollmer, 2008). Interestingly, *Lactobacillus plantarum* contains two *oatA* homologues with different specificity (Turk *et al.*, 2011). While *oatA* encodes a MurNAc-O-acetyltransferase, *oatB* is the first gene known to encode a GlcNAc-O-acetyltransferase. The second type of O-actyltransferases, called Pat, is found in some Gram-negative bacteria and *B. anthracis* (Weadge *et al.*, 2005; Laaberki *et al.*, 2011). In many bacteria the *pat O*-acetyltransferase gene is often adjacent to an *ape* gene encoding a PG O-acetylesterase, suggesting that addition and removal of O-acetyl groups might be coordinated (Weadge and Clarke, 2006, 2007). In addition to conferring resistance to lysozyme, PG N-deacetylation and O-acetylation have been implicated in modulating the activity of endogenous PG hydrolases (autolysins). Lytic transglycosylases require a free C6-OH group for activity and are unable to cleave O-acetylated PG and, hence, the abundance and sites of O-acetylated PG glycan chains could be important in regulating the activity of lytic transglycosylases (Clarke and Dupont, 1992).

In Gram-positive bacteria the glycan chains serve as attachment sites of secondary cell wall polymers to PG. Most of these polymers, such as wall teichoic acid, teichuronic acid, capsule polysaccharides and arabinogalactan, are linked to the C6-OH of MurNAc via a phosphodiester bond and an oligosaccharide linkage unit (Vollmer, 2008). By contrast, the type III capsule polysaccharide of *Streptococcus agalactiae* is linked to GlcNAc residues. The mechanism of attachment of these secondary cell wall polymers to PG in Gram-positive bacteria and the enzymes catalysing the attachment reactions are largely unknown.

The peptides in peptidoglycan

The peptides are amide-linked to the lactoyl group of MurNAc and contain amino acids that are not present in proteins, such as D-alanine, D-glutamic acid and *meso*-diaminopimelic acid (*m*-Dap) (Schleifer and Kandler, 1972). The prevailing peptide structure in the PG of Gram-negative bacteria is L-Ala-γ-D-Glu-*m*-Dap-D-Ala-D-Ala (Fig. 1.1). Gram-positive bacteria show a much greater diversity with respect to the composition and sequence of their peptides (Schleifer and Kandler, 1972). Their peptides are modified by amidation of carboxylic groups (at position 2 and/or 3) and they often contain a branch of 1–6 amino acids linked to the diamino acid at position 3, which generally shows the greatest diversity amongst the species (Schleifer and Kandler, 1972). Other modifications of the peptides include hydroxylation, acetylation and the covalent attachment of cell wall proteins catalysed by sortases and LD-transpeptidases (Dramsi *et al.*, 2008). In many species the peptides are trimmed by PG carboxypeptidases shortly after their synthesis. For example, shortening of the pentapeptides in the PG of *E. coli* leads to tetrapeptides, which are highly abundant, some of which are further trimmed to tri- and dipeptides (Glauner *et al.*, 1988; Glauner and Höltje, 1990).

Peptides can be modified by incorporation of unusual D-amino acids, for example D-Cys, D-Met or D-Trp, by what was long thought to be a side-reaction of an LD-transpeptidase that normally forms 3–3 cross-links or attaches lipoprotein to PG (Fig. 1.1) (Caparros *et al.*, 1991; Caparros *et al.*, 1992). This view has changed with the discovery in *Vibrio cholerae* of amino acid racemases that produce unusual D-amino acids, which accumulate in the growth medium at millimolar concentration (Lam *et al.*, 2009). These D-amino acids were incorporated into the PG in stationary phase cells of different bacteria. Although the physiological role of these modifications are not known, D-amino acids might alter the activities of PG enzymes in stationary phase or serve as signals for developmental decisions (Cava *et al.*, 2011).

Peptides of neighbouring glycan chains can be connected by the formation of cross-links (Fig. 1.1). The PG of Gram-negative bacteria contains mainly 4–3 (or DD-) cross-links, i.e. the D-Ala at position 4 of one peptide is amide-linked to the *m*-Dap at position 3 of another peptide (Glauner *et al.*, 1988; Quintela *et al.*, 1995). These cross-links are formed by a DD-transpeptidases (penicillin-binding proteins; see below). *E. coli* PG has also low proportion of 3–3 (or LD-) cross-links that are formed by LD-transpeptidases (Glauner and Höltje, 1990) (Fig. 1.1). 3–3 cross-links are abundant in β-lactam resistant enterococci, *M. tuberculosis* and *Clostridium difficile* (Mainardi *et al.*, 2005; Mahapatra *et al.*, 2008; Mainardi *et al.*, 2008; Peltier *et al.*, 2011). The corynebacteria utilize a 2–4 type of PG cross-linkage in which the glutamic acid at position 2 is linked via a single diamino acid, often D-ornithine, to the D-Ala residue at position 4 of another peptide (Schleifer and Kandler, 1972).

More than two peptides can be connected to form trimeric, tetrameric and higher oligomeric peptide structures, which are rare in the PG of Gram-negative bacteria but can be very abundant in Gram-positive bacteria such as *Staphylococcus aureus* (Snowden and

Perkins, 1990). Thus, the degree of peptide cross-linkage is very variable but appears generally higher in Gram-positive bacteria (Vollmer and Seligman, 2010). According to data from solid-state NMR spectroscopy, the degree of cross-linkage, and not so much the length of the glycan chains or the thickness of the PG, correlates with the overall structural flexibility of the PG net (Kern *et al.*, 2010).

Peptidoglycan architecture

PG sacculi from Gram-negative bacteria have remarkable biophysical properties (Vollmer and Höltje, 2004). They are very thin with a thickness of ~2.5–7 nm in the hydrated state (Labischinski *et al.*, 1991; Yao *et al.*, 1999; Gan *et al.*, 2008). Sacculi are also highly elastic, allowing the surface area to reversibly expand up to 3-fold without rupture of covalent bonds (Koch and Woeste, 1992; Yao *et al.*, 1999). Isolated sacculi are more deformable in the direction of the long axis, which is consistent with PG-models aligning the flexible peptide cross-links in this direction, and the more rigid glycan chains in the direction perpendicular to the long axis (Boulbitch *et al.*, 2000). In addition, isolated sacculi have pores with a mean radius of about 2.1 nm (Demchick and Koch, 1996). This study estimated that stretched sacculi, as they exist in the cell due to the turgor, are penetrable by globular proteins of up to 50 kDa, consistent with the molecular sieving effect of PG in osmotically shocked cells (Vazquez-Laslop *et al.*, 2001).

There is currently no method available that can resolve the molecular details of glycan chains and peptides in PG sacculi. The models on PG architecture differ with respect to the orientation of the glycan chains relative to the bacterial membrane and to the long axis of the cell (Labischinski *et al.*, 1983; Dmitriev *et al.*, 1999; Koch, 2000). Recent cryo-electron microscopy data suggest that isolated sacculi from the Gram-negative *E. coli* and *Caulobacter crescentus* are made of a single layer in which glycan chains are somewhat disordered but running mainly and on average in the direction of the short axis of the cell (Gan *et al.*, 2008). This conclusion fits to earlier data on the thickness and elasticity of the PG layer and the amount of PG per cell (Vollmer and Höltje, 2004).

Atomic force microscopy (AFM) on PG sacculi from the Gram-positive *B. subtilis* and *S. aureus* suggest a more complex architecture (Hayhurst *et al.*, 2008; Turner *et al.*, 2010). These data suggest that in *B. subtilis* PG the glycan chains form bundles that coil into 50-nm-wide cables, which run roughly perpendicular to the long axis of the cell. Interestingly, in both *B. subtilis* and *S. aureus* the septal PG region showed cables that spiral towards the centre of the closing septum. More experiments with high-resolution techniques will be required to decipher the molecular architecture of PG and associated cell wall polymers in Gram-positive bacteria.

Peptidoglycan synthesis

The synthesis, membrane translocation and polymerization of the PG precursor take place within and at the interfaces of the cytoplasmic membrane (Typas *et al.*, 2012). Most of the proteins involved are membrane-bound with one (bitopic) or several (polytopic) membrane spanning segments. Table 1.1 contains the PG synthases, PG hydrolases and their regulators present in *E. coli*.

The last monomeric PG precursor before the formation of the PG polymer is lipid II, a disaccharide pentapeptide consisting of β-1,4-linked GlcNAc and MurNAc attached

Table 1.1 Peptidoglycan synthases and hydrolases and their regulators in *E. coli*

Protein/category	Gene	Function/remarks
PG precursor synthesis		
MurA, MurB	*murA*, *murB*	Phosphoenoltransferase (MurA) and reductase (MurB) for the synthesis of UDP-MurNAc
DadX, Alr, MurI	*dadX*, *alr*, *murI*	Amino acid racemases for the formation of D-Ala (DadX, Alr) and D-Glu (MurI)
MurC, MurD, MurE, MurF, Ddl	*murC*, *murD*, *murE*, *murF*, *ddl*	ATP-dependent amino acid ligases for the formation of D-Ala-D-Ala (Ddl) and UDP-MurNAc-pentapeptide
MraY, MurG	*mraY*, *murG*	Synthesis of lipid I (MraY) and lipid II (MurG) at the cytoplasmic membrane
PG synthases		
PBP1A	*mrcA* (*ponA*)	Major GTase and DD-TPase; class A PBP; mainly involved in cell elongation
PBP1B	*mrcB* (*ponB*)	Major GTase and DD-TPase; class A PBP; mainly involved in cell division
PBP1C	*pbpC*	GTase and DD-TPase; class A PBP; unknown role
PBP2	*mrdA* (*pbpA*)	DD-TPase; class B PBP; essential for cell elongation
PBP3	*ftsI*	DD-TPase; class B PBP; essential for cell division
MtgA	*mtgA* (*mgt*)	GTase; unknown function
Activators of PG synthases		
LpoA	*lpoA*	Outer membrane lipoprotein; interacts with PBP1A and activates TPase activity
LpoB	*lpoB*	Outer membrane lipoprotein; interacts with PBP1B and activates TPase activity
PG hydrolases		
Slt70	*slt*	Periplasmic lytic transglycosylase involved in septum cleavage
MltA, MltB, MltC, MltD, MltE (EmtA), MltF	*mltA*, *mltB*, *mltC*, *mltD*, *mltE* (*emtA*), *mltF*	Outer membrane-anchored lytic transglycosylases; MltA, MltB, MltC and MltD participate in septum cleavage during cell division
AmiA, AmiB, AmiC	*amiA*, *amiB*, *amiC*	Periplasmic amidases for septum cleavage during cell division; AmiB and AmiC localize to the division site
AmiD	*amiD*	Outer membrane lipoprotein; amidase
PBP5, PBP6, PBP6B	*dacA*, *dacC*, *dacD*	DD-CPases; class C PBPs; removal of the terminal D-Ala in newly made PG
PBP4	*dacB*	DD-EPase and DD-CPase; class C PBP; involved in septum cleavage
PBP7	*pbpG*	DD-EPase; class C PBP; involved in septum cleavage
MepA	*mepA*	DD/LD-EPase; unknown function
Spr, YdhO	*spr*, *ydhO*	DD-EPase; CHAP family of amidohydrolase/peptidases; involved in cell elongation
YebA	*yebA*	DD-EPase; LytM family of metallopeptidases; involved in cell elongation
Activators of PG hydrolases		
EnvC	*envC*	LytM family outer membrane protein; activates AmiA and AmiB

Table 1.1 (continued)

Protein/category	Gene	Function/remarks
NlpD	*nlpD*	LytM family outer membrane protein; activates AmiC
LD-Transpeptidases		
YnhG, YcbB	*ynhG, ycbB*	Periplasmic LD-TPases that synthesize 3–3 cross-links
ErfK, YbiS, YcfS	*erfK, ybiS, ycfS*	Periplasmic LD-TPases that attach Lpp to PG
Braun's lipoprotein		
Lpp	*lpp*	Outer membrane lipoprotein; covalently attached to PG by LD-TPases

to the membrane-bound undecaprenyl carrier via a pyrophosphate group (Fig. 1.1). The synthesis of lipid II and the cytoplasmic intermediates of PG have recently been reviewed (Barreteau *et al.*, 2008; Bouhss *et al.*, 2008). Lipid II is assembled on the cytoplasmic side of the plasma membrane from the soluble intermediates. The translocase MraY (Fig. 1.3) transfers the phospho-MurNAc-pentapeptide moiety of the nucleotide substrate UDP-MurNAc-pentapeptide onto the undecaprenyl phosphate carrier to form the lipid I, the undecaprenyl pyrophosphoryl-MurNAc-pentapeptide. The MurG transferase (Fig. 1.3) catalyses the addition of GlcNAc to lipid I to form lipid II. Initially anchored in the inner leaflet of the cytoplasmic membrane by the undecaprenyl moiety, lipid II is flipped to the outer leaflet of the cytoplasmic membrane by proteins of the SEDS (shape, elongation, division and sporulation) family, RodA and FtsW (Fig. 1.3) (Mohammadi *et al.*, 2011). The flipping of the precursor and its utilization by PG synthases are coupled processes (van Dam *et al.*, 2007; Mohammadi *et al.*, 2011). PG synthases polymerize the glycan chains by glycosyltransferase (GTase) reactions and form the peptide cross-links by transpeptidase (TPase) reactions (Fig. 1.1) (Sauvage *et al.*, 2008). During GTase reaction the catalytic glutamate (E233 in *E. coli* PBP1B) catalyses the deprotonation of the GlcNAc 4-OH of lipid II, the activated nucleophile then directly attacks the C1 of the lipid-linked MurNAc of the growing polysaccharide chain leading to the formation of a β-1,4-glycosidic bond. The undecaprenyl pyrophosphate anchor is released and is then recycled by dephosphorylation to the monophosphate form and flipped to the inner leaflet of the plasma membrane (a process perhaps coupled to the translocation of lipid II by FtsW) for new rounds of lipid II transport (Bouhss *et al.*, 2008). The TPase reaction involves a pentapeptide donor, which loses the terminal D-Ala residue during the reaction, and an acceptor that can be a tri-, tetra- or pentapeptide, whereby the carboxylic group of the donor is linked to the amino group of the acceptor to form the new amide bond (Fig. 1.1). The TPase domain of the PG synthases covalently bind penicillin and other β-lactams, which mimic the terminal D-Ala-D-Ala of the lipid II peptide stem, via a serine residue of the active site and therefore are commonly known as penicillin-binding proteins (PBPs).

Peptidoglycan synthases

Based on sequence analysis and activities the PG synthases can be divided into three major groups: the bifunctional GTase/TPase class A PBPs, the monofunctional TPase class B PBPs and the monofunctional GTase (Mtg or Mgt) (Sauvage *et al.*, 2008). In addition, many species have LD-transpeptidases that are penicillin-insensitive [but sensitive to the β-lactam

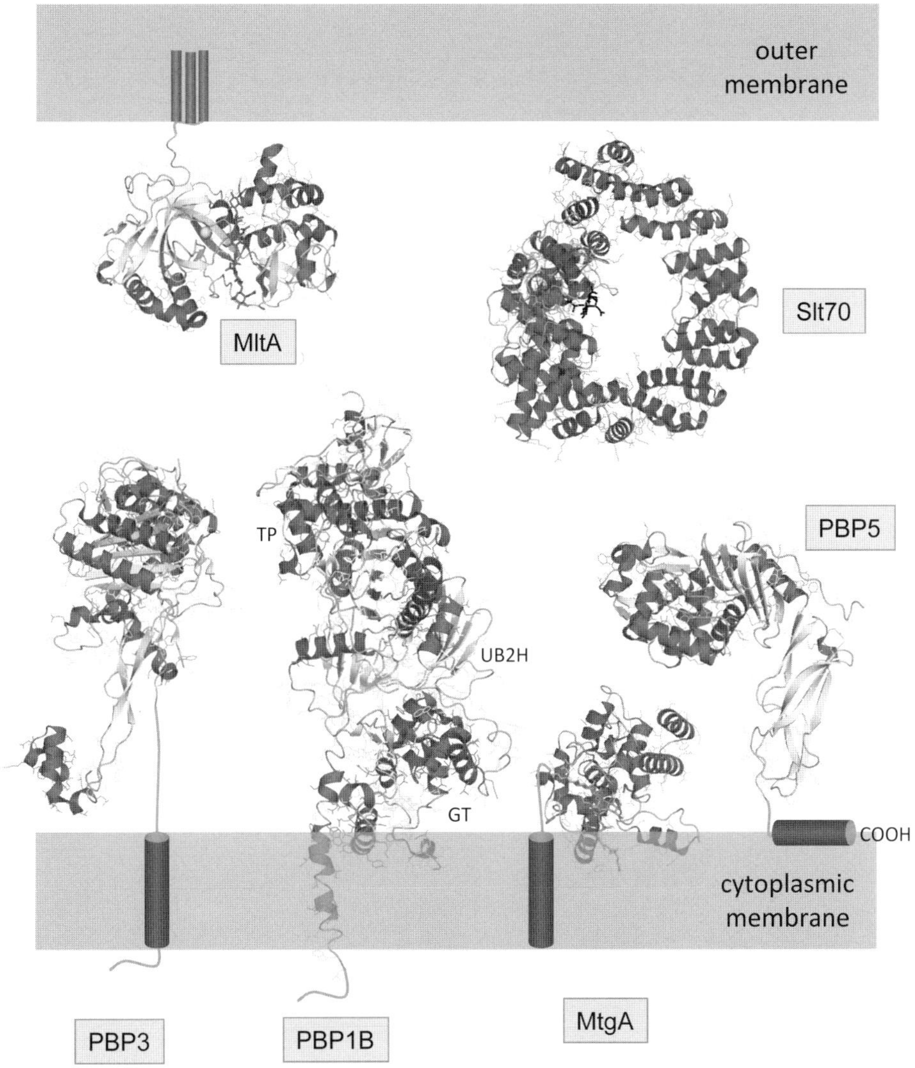

Figure 1.2 Crystal structures of PG synthases and hydrolases. The synthases PBP3, PBP1B and MtgA are anchored to the cytoplasmic membrane by a hydrophobic helix near the N-terminus (grey rod). In the case of PBP1B this membrane anchor is present in the crystal structure. PBP5 attaches to the membrane surface by a C-terminal amphipathic helix (grey rod with COOH). Slt70 is a soluble, periplasmic enzyme; MltA is attached to the outer membrane by an N-terminal lipid modification typical for bacterial lipoproteins (small grey rods). Helices are shown in dark grey, β-sheets are shown in grey and loops in light grey. PBP3 from *Pseudomonas aeruginosa*, amino acid residues 60–561 (amino acids 116–140, 188–216, 491–500 are not visible), PDB accession number 3PBN (Han *et al.*, 2010); PBP1B from *E. coli*, amino acid residues 66–799 (amino acids 249–267 are not visible), PDB accession number 3FWL (Sung *et al.*, 2009); MtgA from *Staphylococcus aureus* in complex with moenomycin, amino acid residues 60–268 (amino acids 60–67 are from the tag), PDB accession number 3HZS (Heaslet *et al.*, 2009); PBP5 from *E. coli*, amino acid residues 33–384, PDB accession number 1NZO (Nicholas *et al.*, 2003); Slt70 from *E. coli*, amino acid residues 28–645, PDB accession number 1QTE (van Asselt *et al.*, 1999b); MltA from *E. coli*, amino acid residues 23–357, PDB accession number 2PI8 (van Straaten *et al.*, 2007).

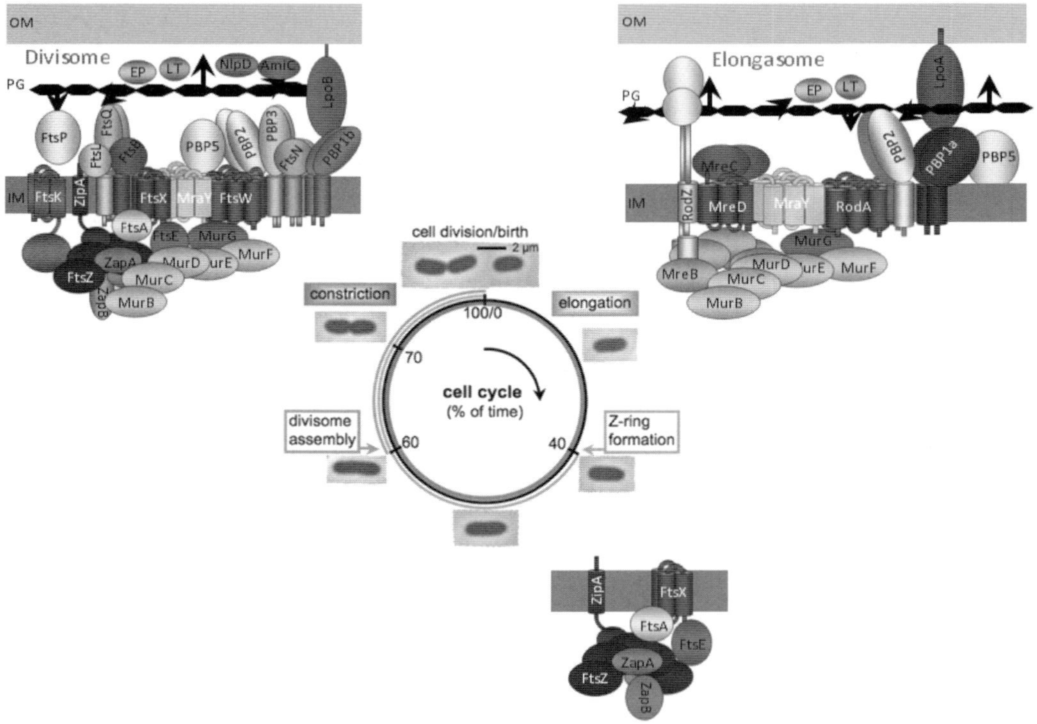

Figure 1.3 Cell cycle of *E. coli* and peptidoglycan synthase complexes active during elongation and division. The middle part shows the cell cycle of *E. coli* and the timing of Z-ring formation, divisome assembly and constriction, with representative micrographs of cells at the various stages. The elongasome (upper right side) assembles at MreB filaments and is responsible for the synthesis of PG at the cylindrical part of the cell. The divisome (upper left side) assembles at FtsZ-polymers to synthesize the new poles during cell division. The lower right part shows the assembly of the early cell division proteins at midcell; these drive the pre-septal phase of cell elongation. IM, inner membrane; OM, outer membrane; EP, endopeptidase; LT, lytic transglycosylase; MurB-F, enzymes for PG precursor synthesis. Note that the proteins and membranes are not drawn to scale.

imipenem (Mainardi *et al.*, 2007)] and catalyse different reactions, including the formation of 3–3-cross-links, the covalent attachment of the outer membrane lipoprotein Lpp to PG (in Gram-negative bacteria) and, presumably, the incorporation of D-amino acids (Fig. 1.1) (Caparros *et al.*, 1992; Mainardi *et al.*, 2000, 2008; Magnet *et al.*, 2007). PG synthases are all anchored to the cytoplasmic membrane by an N-terminal hydrophobic helix following a short cytoplasmic sequence while the catalytic domains are located in the periplasm (Figs. 1.2 and 1.3).

The number of class A and B PBPs is correlated with the life cycle and the shape of the bacteria (Zapun *et al.*, 2008). The coccus *Neisseria gonorrhoeae* has one representative of each class while the rod-shaped and sporulating *B. subtilis* contains four PBPs of class A and six of class B. In *E. coli*, the essential set of PG synthases consist of the bi-functional PBP1B (or PBP1A) and the monofunctional TPases PBP2 and PBP3 (Denome *et al.*, 1999). The bifunctional PBP1C and the GTase MtgA are non-essential and may play an accessory role. PBP2 and PBP3 have specific functions in cell elongation and division, respectively, as demonstrated by the phenotypes of thermosensitive mutants or upon inhibition by specific β-lactams (Spratt, 1975). In other cases such as the homologous *E. coli* class A PBP1A and PBP1B the phenotypic differences of mutants are more subtle. The single mutants are viable while the absence of both PBPs is lethal. The single mutants have similar phenotypes unless other mutations are added (Garcia del Portillo and de Pedro, 1990; Typas *et al.*, 2010) and they differ in their sensitivity to β-lactams (Yousif *et al.*, 1985). Cellular localization and protein interactions identified suggest that PBP1B is more active in cell division whereas PBP1A is more involved in cell elongation (Bertsche *et al.*, 2006; Typas *et al.*, 2010; van der Ploeg *et al.*, 2013). *E. coli* PBP1A and PBP1B are active in polymerizing lipid II to glycan chains and to perform peptide cross-linking reactions *in vitro*, and PBP1B is more active at conditions that favour dimerization (Bertsche *et al.*, 2005; Born *et al.*, 2006). Both enzymes perform GTase reactions in the absence of TPase reactions, but they require ongoing GTase reactions for expressing efficient TPase activity, suggesting that both activities are coupled (den Blaauwen *et al.*, 1990). Based on the crystal structure of PBP1B (Fig. 1.2), it has been suggested that the TPase domain utilizes peptides of the growing glycan chain (produced by the GTase domain) as donors for TPase reactions (Sung *et al.*, 2009). Interestingly, PBP1A was capable of attaching by transpeptidation a fraction of the newly synthesized PG to sacculi (Born *et al.*, 2006), a reaction that occurs during PG growth in the cell (Burman and Park, 1984; de Jonge *et al.*, 1989; Glauner and Höltje, 1990).

The bifunctional GTase/TPase enzymes are the main PG synthases and at least one is required for viability in *E. coli*, *S. aureus* and *S. pneumoniae*. However, the essentiality of class A PBPs, in laboratory conditions, has been questioned as a general rule in *B. subtilis* and *Enterococcus faecalis* (McPherson and Popham, 2003; Arbeloa *et al.*, 2004). These two Gram-positive bacteria possess four and three class A PBPs, respectively, and they lack *mtgA* genes in their genomes. Yet, all known genes encoding class A PBPs could be deleted without a loss in viability, and so it was hypothesized that a yet unknown GTase is able to compensate for the absence of the classical PG GTase.

The monofunctional TPases consist of a C-terminal TPase domain associated with an N-terminal domain of unknown function; they generally play a specific role in relation with cell morphology (i.e. PBP2 and PBP3 in *E. coli*).

The monofunctional GTases (Mtg) are homologues of the GTase domain of class A PBPs and belong to family 51 of GT (GT51) in the CAZy database (Cantarel *et al.*, 2009). Mtg is not present in all PG containing bacteria and, if present (1 or 2 genes per genome), they are not essential (Denome *et al.*, 1999; Reed *et al.*, 2011). Hence, Mtg probably has an accessory role in cell wall synthesis. *S. aureus* contains two Mtg enzymes named MtgA (also referred to as MGT) (Terrak and Nguyen-Disteche, 2006) and SgtA; deletion of both show that they are not essential for growth in laboratory conditions. The two Mtg enzymes catalyse PG polymerization *in vitro* from lipid II and the activity of MtgA was about ten-fold higher than that of SgtA. MtgA but not SgtA was able to replace the GT activity of the class A PBP2 of *S. aureus* showing that the MtgA is functional *in vivo* and that the activities observed *in vitro* for the two Mtg enzymes may also reflect their activities *in vivo* (Reed *et al.*, 2011).

Structure of peptidoglycan synthases

Fig. 1.2 shows the crystal structures of three representative PG synthases. Many crystal structures of penicillin-binding (PB) domains and associated domains of class B and C PBPs have been determined with a variety of β-lactams and other ligands, but there are only few structures of GTase domains (Macheboeuf *et al.*, 2006; Sauvage *et al.*, 2008; Mattei *et al.*, 2010). The general fold of all PB domains is similar in all PBPs but there are many local structural variations and differences in the specificities of ligands. The TPase domain is composed of an α- and α/β-subdomain, with the catalytic cavity harbouring the active serine located at the interface of the two subdomains. Several crystal structures of class B PBPs with low affinity to penicillin have been determined: PBP2x of *Streptococcus pneumoniae* (Pares *et al.*, 1996), PBP5 of *Enterococcus faecium* (Sauvage *et al.*, 2002), PBP2a from a methicillin-resistant *S. aureus* (Lim and Strynadka, 2002) and PBP2 from *Neisseria gonorrhoeae* (Powell *et al.*, 2009). The general fold of the N-terminal, non-catalytic domain of class B PBPs has an elongated shape composed mainly by long and shorter β-strands and few small helices (see PBP3, Fig. 1.2).

The first structures of bifunctional class A PBPs and GTase domains were solved only recently: (i) *S. aureus* PBP2 (apo and with the GTase inhibitor moenomycin bound) (Lovering *et al.*, 2007; Lovering *et al.*, 2008a,b), (ii) *E. coli* PBP1B in complex with moenomycin (Sung *et al.*, 2009), (iii) *Aquifex aeolicus* PBP1a GTase domain (Yuan *et al.*, 2007) and its complexes with neryl-moenomycin (Yuan *et al.*, 2008; Fuse *et al.*, 2010) and (iv) the *S. aureus* monofunctional glycosyltransferase mutant MtgA(E100Q) in complex with moenomycin (Fig. 1.2) (Heaslet *et al.*, 2009). The N-terminal GTase domain of *S. aureus* PBP2 is linked to the C-terminal TPase domain via a linker domain composed a small β-sheet and one α-helix. The different crystal structures of *S. aureus* PBP2 show large GTase-TPase interdomain movement and raise the question on the significance of these conformational changes in term of catalytic activities, protein–protein interactions and regulation (Lovering *et al.*, 2008a).

Structure of the GTase domain

The structures of GTase domains show high degree of similarity (RMSD Cα 1.45–1.7 Å). They are composed mainly of α-helices organized in two lobes (head and jaw domains) between which the catalytic cleft is located. Interestingly, the fold of the head domain of the GTase and of the active site shows structural similarities with that of the phage λ lysozyme (λL), a muramidase like T4 lysozyme and Slt70 (Evrard *et al.*, 1998). Glu19 of λL superposes not only with the corresponding Glu residues in other lysozymes but also with the catalytic Glu residue in family GT51 GTases. Type C lysozymes and GT51 GTases have a second carboxylic residue in the active site that is absent in λL (see below). The jaw domain is specific to GT51 family and has a hydrophobic region partly embedded in the cytoplasmic membrane (see PBP1B, Fig. 1.2).

The GTase domain contains five highly conserved motifs that characterize the GT51 family. In *E. coli* PBP1B the glutamate residues of motif 1 (Glu233) and motif 3 (Glu290) are important for catalysis whereby the Glu233 is the catalytic base (Terrak *et al.*, 1999). In addition, the residues Asp234, Phe237 and His240 of motif 1 and Thr267, Gln271 and Lys274 of motif 2 play a critical role in the polymerase activity of PBP1B (Terrak *et al.*, 2008). Mutation of the corresponding residues in other GTases, with few exceptions, gave comparable results (Yuan *et al.*, 2007; Heaslet *et al.*, 2009). Most of the GTase structures solved have a disordered region located between the essential motifs 1 and 2. This highly mobile region was visible in the newly solved structures of PBP2 and MtgA. It occupies a central position in the enzymatic cleft and was suggested to play an important role in the processive catalysis mechanism of glycan chain elongation through folding unfolding movements (Lovering *et al.*, 2008b). This region is partly buried in the membrane and separates the two substrate-binding pockets, the growing chain donor site and lipid II acceptor site.

The crystal structure of *E. coli* PBP1B has been determined with its transmembrane (TM) helix (66–96), showing the TM residues 83–88 interacting with residues 292–296 of the GT domain (Fig. 1.2) (Sung *et al.*, 2009). The majority of the PG enzymes and associated cell division and elongation proteins have TM segments embedding them partly or completely in the cytoplasmic membrane (Tables 1.1 and 1.2). Hence, interactions within the membrane should play an important role in the coordination between proteins locating on both sides of the membrane and in the control of PG synthesis. In addition to the classical GTase/TPase catalytic domains of bifunctional PBPs, *E. coli* PBP1B contains an additional domain called UB2H (UvrB domain 2 homologue), composed of ~100 amino acid residues (109–200) located between the TM and GTase domain in the protein sequence (between GTase and TPase domain in the structure) and which is restricted to PBPs of subclass A2 (Goffin and Ghuysen, 1998; Sauvage *et al.*, 2008). Homologue domains of UB2H of the DNA repair system (UvrA and TRCF) are involved in protein–protein interactions. The UB2H domain was found to mediate the interaction of PBP1B with the outer membrane-anchored lipoprotein LpoB resulting in an increase of the TPase activity (see below) (Typas *et al.*, 2010). UB2H folds as five antiparallel β-strands (β2–β6) and one α-helix and has more interactions with the TPase than the GTase domain in agreement with the activation of TPase by LpoB (Typas *et al.*, 2010).

PG hydrolases

Numerous PG hydrolases have been identified in bacteria and eukaryotes. Collectively, they are able to cleave almost every chemical bond in the PG (Vollmer *et al.*, 2008b). Owing to their enzymatic specificity, some PG hydrolases, like muramidases, endopeptidases and *N*-acetylmuramoyl-L-alanine amidases, have the capability to completely degrade the PG sacculus. At certain conditions these enzymes cause lysis of their own cell and are therefore called 'autolysins'. Other PG hydrolases, such as carboxypeptidases, are unable to degrade the sacculus but have important regulatory roles in PG growth. PG hydrolases are thought to be essential for cell wall growth, because the net-like PG structure requires cleavage of covalent bonds to allow the insertion of new PG for enlargement of the surface area. Indeed,

Table 1.2 Cell division and cell elongation proteins in *E. coli*

Protein/category	Gene	Function/remarks
Cell division		
1. Inhibition of Z-ring-localization		
MinC, MinD, MinE	*minC*, *minD*, *minE*	Inhibition of FtsZ polymerization at the poles
SlmA	*slmA*	Inhibition of FtsZ polymerization in the vicinity of the nucleoid (nucleoid occlusion)
2. Early cell division proteins		
FtsZ	*ftsZ*	Cytoskeletal protein essential for cell division; polymerizes in a ring at mid-cell; tubulin structural homologue
FtsA	*ftsA*	Cytoplasmic actin homologue; binds to the cytoplasmic membrane via an amphipathic helix, stabilizes the FtsZ-ring and attaches it to the membrane
ZipA	*zipA*	Bitopic membrane protein; attaches the FtsZ-ring to the cytoplasmic membrane
ZapA, ZapB, ZapC, ZapD	*zapA*, *zapB*, *zapC*, *zapD*	Cytoplasmic proteins, whose absence affects the stability of the Z-ring during its formation. ZapB is bound to the Z-ring by ZapA and interacts with the ter macro domain organizing protein MatP and might sense the presence of the nucleoids during division
FtsE	*ftsE*	Cytoplasmic ATPase essential for cell division
FtsX	*ftsX*	Integral cytoplasmic membrane protein that recruits FtsE; essential for cell division
FtsK	*ftsK*	Integral cytoplasmic membrane protein essential for cell division involved in DNA transport and in the recruitment of cell division proteins of the late stage
3. Late cell division proteins		
FtsQ, FtsL, FtsB	*ftsQ*, *ftsL*, *ftsB*	Cytoplasmic proteins essential for cell division; they form a pre-complex
FtsW	*ftsW*	Lipid II flippase; essential for cell division
PBP3	*ftsI*	Class B PBP; essential for cell division
FtsN	*ftsN*	Bitopic cytoplasmic membrane protein essential for cell division; interacts with PBP1B and PBP3; binds PG via its C-terminal SPOR domain
DamX, DedD	*damX*, *dedD*	Bitopic cytoplasmic SPOR domain proteins that localize at the division site
RlpA	*rlpA*	Outer membrane lipoprotein with SPOR domain; localizes at the division site
FtsP	*ftsP* (*sufI*)	Periplasmic protein that stabilizes the divisome at stress conditions
Cell elongation		
MreB	*mreB*	Cytoskeletal protein essential for cell elongation; structural homologue of actin; forms filaments attached to the cytoplasmic membrane
MreC	*mreC*	Bitopic membrane protein essential for cell elongation; interacts with MreB
MreD	*mreD*	Integral membrane protein essential for cell elongation; interacts with MreC

Table 1.2 (continued)

Protein/category	Gene	Function/remarks
RodZ	*rodZ*	Bitopic membrane protein essential for cell elongation; interacts with MreB
RodA	*rodA* (*mrdB*)	Lipid II flippase essential for cell elongation
PBP2	*mrdA*	Class B PBP; essential for cell elongation

E. coli. requires at least one of the DD-endopeptidases Spr, YdhO and YebA for growth (Singh *et al.*, 2012), and *B. subtilis* cells cannot grow upon depletion of two DD-endopeptidases, CwlO and LytE (Bisicchia *et al.*, 2007). Moreover, interactions between PG synthases and hydrolases support a model of multienzyme complexes for PG synthesis that combine all the synthetic and hydrolytic activities required to enlarge the sacculus (see below) (Höltje, 1993, 1998). PG hydrolases are active in growing *E. coli* cells: They cause the release of as much as 40–50% of the total PG from the sacculus per generation in a process called PG turnover (Goodell and Schwarz, 1985). The turnover material (muropeptides) is efficiently recycled (Goodell, 1985; Jacobs *et al.*, 1994; Uehara and Park, 2008). In *Citrobacter freundii* and *Enterobacter cloacae* PG turnover is coupled with the induction of β-lactamase via the sensing of the levels of PG intermediates from *de novo* synthesis and muropeptides from PG turnover (Jacobs *et al.*, 1997). Some PG hydrolases have established roles in cleavage of the septum during or after cell division, like the *E. coli* amidases AmiA, AmiB and AmiC, and the *S. pneumoniae* glucosaminidase LytB. In both cases and many other examples, deletion of the PG hydrolase gene(s) results in the formation of long chains of unseparated cells. In the next section we summarize the different PG hydrolase specificities. The *E. coli* PG hydrolases are listed in Table 1.1. The interactions of PG hydrolases and their roles during cell elongation and division will be described later.

Lytic transglycosylases

The lytic transglycosylases are exo- or endo-muramidases catalysing the intramolecular transglycosylation reaction combining the hydrolysis of the β-1,4 glycosidic bond between MurNAc and GlcNAc and the concomitant formation of a 1,6–anhydro bond at MurNAc (Fig. 1.1) (Höltje *et al.*, 1975; Scheurwater *et al.*, 2008). Catalysis of this reaction requires only one catalytic glutamate residue, which deprotonates the hydroxyl group at the C-6 of MurNAc, allowing direct nucleophilic attack at C-1 of the same amino sugar and the formation of the 1,6-anhydro ring (Thunnissen *et al.*, 1994). The mechanism and the reaction product of the lytic transglycosylases are different from those of the hydrolytic lysozymes, which catalyse attack of carbon C1 of the intermediate by a water molecule, thus completing the hydrolysis of the glycosidic bond and the release of reducing MurNAc (Jolles, 1996).

E. coli contains seven lytic transglycosylases, six of them are lipoproteins bound to the outer membrane [MltA, MltB, MltC, MltD, MltE (EmtA) and MltF] and one is soluble in the periplasm (Slt70) (Table 1.1) (Romeis *et al.*, 1993; Heidrich *et al.*, 2002; Scheurwater and Clarke, 2008). Differences in the substrate specificity are observed between these proteins. For example, Slt70 only cleaves PG chains carrying peptides and is inactive on PG lacking the stem peptides while MltA can cleave either form (Romeis *et al.*, 1993; Ursinus and Höltje, 1994).

Representative crystal structures of lytic transglycosylases include those of the *E. coli* enzymes Slt70, MltA (Fig. 1.2) and the soluble fragment Slt35 from MltB. The crystal structures of Slt70 have been solved in apo-form and bound to PG fragment (van Asselt *et al.*, 1999a), or the inhibitor bulgecin (Thunnissen *et al.*, 1995b). Slt70 is a multimodular protein. The N-terminal domain (1–448) is composed of α-helices that fold in a ring-shaped structure. The catalytic domain (residues 449–618) is at the C-terminus, which itself is divided into two lobes with the catalytic centre located in the groove containing the catalytic acid/base Glu478 (Thunnissen *et al.*, 1994, 1995a; van Asselt *et al.*, 1999b). MltA structures are available with and without chitohexaose (van Straaten *et al.*, 2005, 2007). MltA has a different fold than the other known lytic transglycosylases. It contains two domains having two distinct β-barrel topologies separated by a deep groove. Asp308 is the single catalytic residue in this enzyme. The structure of the proteolytic product of MltB, Slt35 was solved in apo form and with PG fragments or bulgecin (van Asselt *et al.*, 1999a, 2000). The structure of Slt35 reveals an ellipsoid molecule with three domains called alpha, beta and core domains. The fold of the core domain (catalytic domain) resembles that of lysozyme with an EF-hand calcium-binding motif that was shown to be important for the stability of the protein (van Asselt and Dijkstra, 1999). The catalytic residue Glu162 of Slt35 is at equivalent position to Glu478 of Slt70.

N-acetylmuramyl-L-alanine amidases

N-acetylmuramyl-L-alanine amidases cleave between the L-Ala of the stem peptide and the MurNAc (Fig. 1.1). *E. coli* has five amidases; AmiA, AmiB and AmiC (LytC-type) are located in the periplasm, the periplasmic AmiD (fold as T7-lysozyme) is attached to the outer membrane and AmpD is cytoplasmic (Table 1.1). AmiA, AmiB and AmiC contribute to septum cleavage during cell division (see below). *In vitro* experiments confirmed that AmiA is a zinc metalloenzyme active on polymeric PG and that it requires at least tetrasaccharide PG fragments as substrate (Lupoli *et al.*, 2009). EnvC was shown to activate the PG hydrolysis activities of AmiA and AmiB *in vitro* whereas NlpD specifically activated that of AmiC (Uehara *et al.*, 2010). The activation of AmiB involves a conformational change that moves an α-helix away from the active site to allow the access of the peptidoglcyan substrate (Yang *et al.*, 2012). Total amidase activity seems to be greater during septation than during elongation because dividing cells release more PG than elongating cells (Uehara and Park, 2008). This is consistent with the localization of AmiB and AmiC, and of their activators EnvC and NlpD at the division site. However, the periplasmic amidases were also shown to release 6-fold less PG material than lytic transglycosylases during septation (Uehara and Park, 2008).

Carboxypeptidases and endopeptidases

DD-carboxypeptidases (*E. coli* PBP5, PBP6 and PBP6B) belong to the class C PBPs also known as the low molecular mass PBPs (LMM-PBPs) (Sauvage *et al.*, 2008). They remove the C-terminal D-Ala residue from pentapeptides to perhaps regulate the degree of cross-linking in PG and the activities of amidases. The DD-endopeptidases PBP4, PBP7 and MepA hydrolyse D-Ala–*m*-DAP cross-bridges. MepA has an additional LD-endopeptidase activity enabling it to hydrolyse *m*-DAP–*m*-DAP cross-bridges (Fig. 1.1 and Table 1.1).

PBP5 is the major DD-carboxypeptidase in *E. coli*. In addition to the catalytic domain, PBP5 has a C-terminal domain ending with an amphipathic helix that associates the protein to the membrane (Fig. 1.2) (Nicholas *et al.*, 2003). Both the membrane bound and soluble forms of PBP5 converted pentapeptides to tetrapeptides *in vitro* and *in vivo*, and the enzymes accepted a range of pentapeptide-containing substrates, including PG sacculi, lipid II and muropeptides, and artificial substrates (Potluri *et al.*, 2010). PBP5 contributes to maintaining normal cell shape and diameter in *E. coli* (Nelson and Young, 2000, 2001). PBP5 localizes in the lateral wall and at the division site and its septal localization depends on the membrane-anchoring domain. Its localization at sites of ongoing PG synthesis requires enzyme activity showing that its localization is substrate dependent (Potluri *et al.*, 2010).

The penicillin-insensitive MepA belongs to the LytM-type zinc metallopeptidase family and has structural similarity to lysostaphin that cleaves in the pentaglycine interpeptide bridge of staphylococcal PG. PBP4 and PBP7 seem to play an auxiliary role in cell morphogenesis (Meberg *et al.*, 2004; Priyadarshini *et al.*, 2006) and deletion of the *pbpG* (PBP7) gene together with the *dacA* (PBP5) gene causes morphological defects (Meberg *et al.*, 2004). In addition to the catalytic domain, PBP4 has two domains of unknown function. The protein is loosely associated with the membrane and could be released in the periplasm to interact with the PG (Harris *et al.*, 1998; Kishida *et al.*, 2006). The DD-endopeptidase PBP7 lacks DD-carboxypeptidase activity and is active on high molecular weight sacculi but not on soluble muropeptides (Romeis and Höltje, 1994a).

Interactions of PG synthases and hydrolases

The available data support a model according which the activities of PG enzymes are coordinated and regulated by multiple protein–protein interactions. Some PG enzymes are preferentially active in cell division or elongation, respectively (Fig. 1.3). This view is in accordance with the model proposed by Höltje who suggested that PG synthases and hydrolases form multienzyme complexes that enlarge the PG layer during cell elongation and division by a defined growth mechanism (Höltje, 1993, 1996, 1998). The hypothetical complexes contain the synthetic enzymes anchored in the cytoplasmic membrane and the hydrolases anchored to the outer membrane. The formation of such trans-periplasmic complexes has recently been supported by the identification of outer membrane lipoprotein activators of PG synthases (Paradis-Bleau *et al.*, 2010; Typas *et al.*, 2010). While to date it has not been possible to isolate intact complexes, possibly because they dissociate when cells are broken, a number of interactions between PG enzymes, their regulators and cell elongation/division proteins have been reported, lending further support for the existence of PG synthesis complexes (summarized for *E. coli* in Vollmer and Bertsche, 2008; Typas *et al.*, 2012). Here we mention these interactions in the following respective sections on cell division and cell elongation.

Cell division

Cell division in Gram-negative bacteria requires the timed redirection of length growth of the cylinder to invagination of the three layered cell envelope and synthesis of two new cell poles. Table 1.2 contains the *E. coli* cell division proteins. Several stages can be discriminated during this process (Fig. 1.3). First, FtsZ needs to polymerize into a ring at midcell at the correct moment in the cell cycle to initiate cell division. The second stage prepares the mid cell position for cell pole synthesis and consists of recruitment of the division machinery and the redirection of PG synthesis. The last stage is the synthesis of the new cell poles by the division machinery.

First stage: localization of the Z-ring

The cytoskeletal protein FtsZ is a tubulin homologue that is present in about 5000 copies in an average *E. coli* cell grown with a mass doubling time of 80 min. It polymerizes by binding GTP at the interface between two monomers (see crystal structure of an FtsZ dimer in Fig. 1.4) (Scheffers *et al.*, 2002) and its critical concentration for polymerization is well below its cellular concentration. Consequently, the polymerization of FtsZ has to be continuously inhibited in the cytoplasmic compartment to prevent premature cell division. The nucleoid occupies a large part of the cylindrical cell and is actively transcribed into RNA and about 30% of all proteins to be translated are membrane proteins that are inserted into the membrane by ribosomes that dock onto the SecYEG protein translocase in the cytoplasmic membrane (for a review see Driessen and Nouwen, 2008). This is thought to create a continuum between the nucleoid and the membrane where FtsZ cannot form stable polymers due to the ongoing protein translocation activity. In addition, the DNA binding protein SlmA has a high affinity for FtsZ when it is DNA bound. The SlmA-bound FtsZ is not able to polymerize (Bernhardt and de Boer, 2005; Cho *et al.*, 2011; Tonthat *et al.*, 2011).

The inhibition of FtsZ in the cylindrical part of the cell due to the presence of the nucleoids (nucleoid occlusion) leaves only the cell poles free to polymerize. However, this is prevented by the min system [see for a review (Lutkenhaus, 2008)], which consists of the MinD protein that in its ATP bound state is attached to the cytoplasmic membrane of one cell pole with its C-terminal 10 residues amphipathic helix. MinD-ATP binds MinC that inhibits FtsZ polymerization. The third protein of the system, MinE, competes with MinC for MinD-ATP binding and stimulates the ATPase activity of MinD. In the ADP bound state MinD loses its affinity for the membrane and for MinE and MinC. It is released in the cytoplasm, where it will exchange the ADP for ATP. The release of the Min proteins at one pole creates a concentration gradient that causes a flux from the initially occupied pole to the non-occupied pole, where MinD-ATP now inserts and recruits MinC after which the competition with MinE starts all over again. This process results in a continuous oscillation of the Min proteins from one pole to the opposite pole and consequently the inhibition of FtsZ polymerization at both poles (Raskin and de Boer, 1999). *B. subtilis*, *S. aureus* and *C. crescentus* have slightly different solutions to prevent FtsZ polymerization (Wu and Errington, 2004; Thanbichler and Shapiro, 2006; Wu *et al.*, 2009; Schofield *et al.*, 2010; Veiga *et al.*, 2011).

Segregation of the replicating genome and length growth coincide. Consequently due to the absence of sufficient DNA and SlmA at mid cell at a specific length of the bacterium, FtsZ is allowed to polymerize at mid cell at about 40% of the bacterial cell cycle (Fig. 1.3) (den Blaauwen *et al.*, 1999). Whether the Min system and the nucleoid occlusion are sufficient to determine the timing and position of FtsZ polymerization is not known. During its attempts to form a ring at mid cell FtsZ is aided by at least ZapA, the protofilament cross-linking tetrameric protein (Fig. 1.4) (Mohammadi *et al.*, 2009; Monahan *et al.*, 2009). FtsZ protofilaments are tethered to the membrane by the essential bitopic membrane protein ZipA and the essential protein FtsA that both bind to the same C-terminal region of FtsZ (Fig. 1.4) (Haney *et al.*, 2001; Pichoff and Lutkenhaus, 2005). FtsA binds with its C-terminal amphipathic helix to the cytoplasmic membrane (Pichoff and Lutkenhaus, 2005) and loss of this

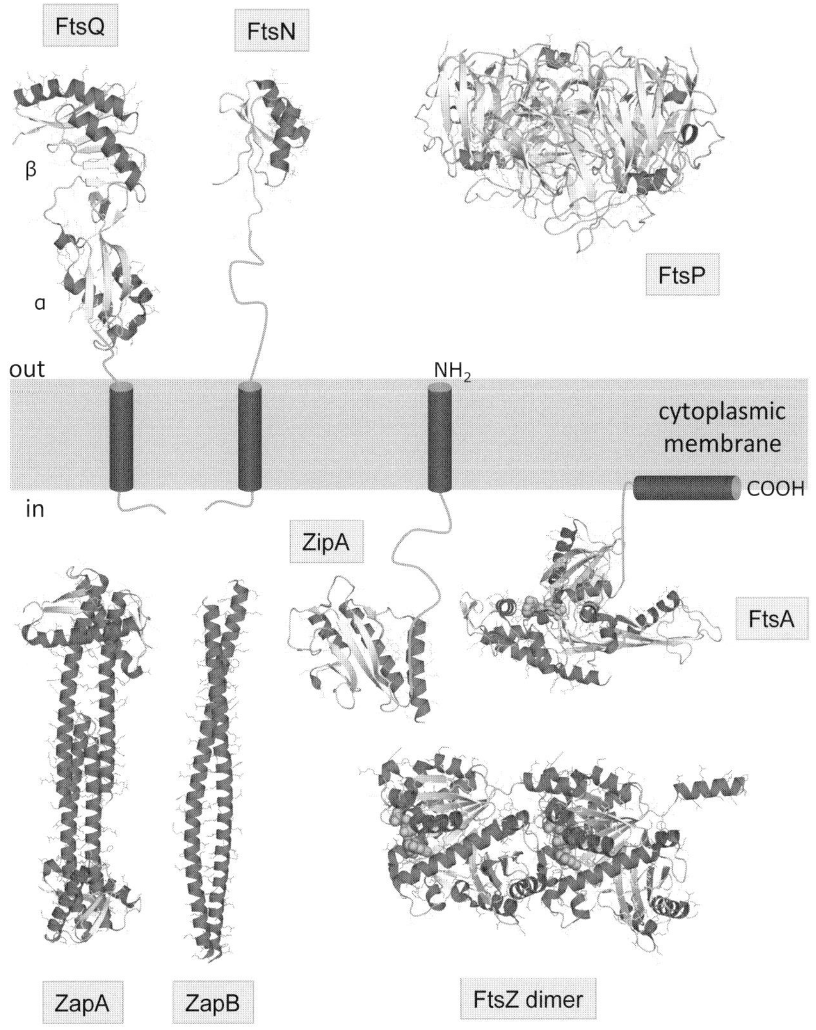

Figure 1.4 Crystal structures of cell division proteins. FtsZ, FtsA, ZapA, ZapB and ZipA are cytoplasmic proteins. FtsZ binds a GTP in its dimer interface (spheres). ZapA and ZapB also form dimers and are mainly composed of α-helixes forming coiled-coils. FtsA attaches to the membrane by a C-terminal amphipathic helix (grey rod). ZipA, FtsQ and FtsN are membrane anchored by an N-terminal hydrophobic helix (grey rod). FtsN has a long, unstructured region in the periplasm. FtsP is a soluble, periplasmic protein. Helices are shown in dark grey, β-sheets are shown in grey and loops in light grey. FtsZ dimer from *Methanocaldococcus jannaschii*, amino acid residues 22–355, PDB accession number 1W5B (Löwe and Amos, 1998). ZapA from *Pseudomonas aeruginosa*, amino acid residues 6–97, PDB accession number 1T3U (Low *et al.*, 2004); ZapB from *E. coli*, amino acids 3–81, PDB accession number 2JEE (Ebersbach *et al.*, 2008): FtsA from *Thermotoga maritima*, amino acids 6–392 (amino acids 320 to 326 are not visible) and with ATP in sphere representation, PDB accession number 1E4G (van den Ent and Löwe, 2000); ZipA with the FtsZ-binding domain from *E. coli*, amino acid residues 190–328, PDB accession number 1F46 (Mosyak *et al.*, 2000); FtsQ from *E. coli*, amino acid residues from 58–260, PDB accession number 2VH1 (van den Ent *et al.*, 2008); FtsN C-terminal SPOR domain from *E. coli*, amino acid residues 243–319, PDB accession number 1UTA (Yang *et al.*, 2004); FtsP from *E. coli*, amino acid residues 30–469 (amino acids 296 to 313 are not visible), PDB accession number 2UXV (Tarry *et al.*, 2009).

helix causes FtsA to form rod-like aggregates in the cytoplasm that cause the cells to grow in C-shape (Gayda *et al.*, 1992). A mutant FtsA*, in which arginine 268 has been replaced by tryptophan, completely bypasses the need for ZipA (Geissler *et al.*, 2003) suggesting that the universally conserved protein FtsA is more important than ZipA for the organization of the Z-ring. During its attempts to form a ring at mid cell FtsZ is aided by at least ZapA, the proto filament cross-linking tetrameric protein (Fig. 1.4) (Mohammadi *et al.*, 2009; Monahan *et al.*, 2009) and perhaps also by ZapC and ZapD (Durand-Heredia *et al.*, 2011, 2012; Hale *et al.*, 2011). Different combinations of stabilizing proteins can be found in bacteria other than *E. coli* (Adams and Errington, 2009). ZapA also binds the dimeric ZapB (Fig. 1.4) and links it to FtsZ. ZapB interacts with the DNA binding protein MatP (Espeli *et al.*, 2012) that organizes and compacts the ter macro domain of the genome (Dupagne *et al.*, 2012; Espeli *et al.*, 2012). Since the termini of the genomes are in the vicinity of the mid-cell or even at mid-cell in the case of unsuccessful segregation, the ZapB-MatP interaction might function as a sensor for DNA segregation defects.

Second stage: redirecting cell envelope synthesis

In *E. coli* PG synthesis occurs by insertion of building units in the cylindrical part of the cell for length growth and completely new synthesis of the new cell poles during cell division (de Pedro *et al.*, 1997). For some time it has been known that PG synthesis at mid cell increases before the synthesis of the new cell poles or the appearance of any invagination could be observed by phase contrast or electron microscopy (Woldringh *et al.*, 1987; de Pedro *et al.*, 1997). This pre-septal mode of PG synthesis produces a small band of new PG and its synthesis is dependent on the Z-ring formation at mid cell (Fig. 1.3). In *C. crescentus* this band is much broader than in *E. coli* and so pre-septal PG synthesis contributes to a large portion of side-wall synthesis (elongation) (Fig. 1.5) (Aaron *et al.*, 2007). The localization of cell division proteins is complete after ~60% of the bacterial cell cycle, which is ~20% later than the formation of the Z-ring. This so-called divisome maturation coincides with the appearance of a new constriction or cell pole synthesis (Fig. 1.3) (Aarsman *et al.*, 2005). The DD-TPases PBP2 and PBP3 are essential for length growth and cell division, respectively. Increasing evidence exists that PBP2 is temporarily involved in mid cell PG synthesis (den Blaauwen *et al.*, 2003; Varma and Young, 2004; Vats and Rothfield, 2007; Vats *et al.*, 2009) and that from the time point of FtsZ-ring formation to full divisome formation a slow build up of PBP3 at mid cell occurs (Costa *et al.*, 2008; van der Ploeg *et al.*, 2013). The function of the band of pre-septal PG is not yet known, but it might be used by the cells to define the middle of the cell and to assist in recruitment of other proteins involved in cell division such as PBP1B, FtsN and other SPOR domain containing proteins (see below), LpoB and PBP5 as mid-cell localization of these proteins seem to depend on substrate recognition rather than, or in addition to the recruitment by other proteins (Bertsche *et al.*, 2006; Möll and Thanbichler, 2009; Möll *et al.*, 2010; Poggio *et al.*, 2010; Potluri *et al.*, 2010; Typas *et al.*, 2010).

Third stage: divisome maturation and synthesis of the new poles

After ~60% of the cell cycle in cells grown with a mass doubling of ~80 min the division machinery seems to be complete and starts to synthesize the new cell poles (Fig. 1.3) (Aarsman *et al.*, 2005). More than 20 proteins, many of which are not essential, have now been recognized to be involved in cell division (Gueiros-Filho and Losick, 2002; Bernhardt and de Boer, 2003, 2004; Bertsche *et al.*, 2006; Gerding *et al.*, 2007, 2009; Samaluru *et al.*, 2007; Derouaux *et al.*, 2008; Ebersbach *et al.*, 2008; Karimova *et al.*, 2009; Uehara *et al.*, 2009; Arends *et al.*, 2010; Paradis-Bleau *et al.*, 2010; Potluri *et al.*, 2010; Typas *et al.*, 2010; Hale *et al.*, 2011). There is a striking pattern of dependency in mid-cell localization of cell division proteins in mutants lacking others. For example, no cell division protein localizes to mid-cell upon depletion of *ftsZ*, whereas FtsZ, FtsA, ZipA and ZapA but not FtsQ, FtsL, FtsB, FtsW, FtsI (PBP3) and FtsN localize at mid-cell upon depletion of *ftsK*. Testing the ability for mid-cell localization of all cell division proteins in mutants lacking other cell division genes gave an almost linear hierarchy of interdependence for mid-cell localization (Chen and Beckwith, 2001). Thus, for a long time it was thought that the assembly of the divisome required for each protein to localize at mid-cell the presence of its upstream partner protein. Presently, it seems more likely that physiological conditions such as a stable FtsZ-ring (Goehring and Beckwith, 2005; Goehring *et al.*, 2006; Rico *et al.*, 2010) or the presence of substrate (see above) determine or enhances the ability of proteins to localize. In the next sections we summarize the role of late cell division proteins.

FtsE, FtsX and FtsK

FtsX is an integral membrane protein with four membrane-spanning helices (Arends *et al.*, 2009). It recruits the cytoplasmic ATPase FtsE to the septal ring (Schmidt *et al.*, 2004), which in turn interacts with FtsZ (Fig. 1.3) (Corbin *et al.*, 2007). FtsEX, the binding of ATP and probably its hydrolysis by FtsE are essential for the recruitment of late localizing proteins and cell division at growth in medium at low osmolarity (Reddy, 2007; Arends *et al.*, 2009). Despite its similarity to ABC transporters FtsEX does not seem to be involved in transport (Arends *et al.*, 2009). Instead, FtsX interacts via a periplasmic loop with EnvC and recruits it to the division site to activate two septum-splitting amidases, AmiA and AmiB (Uehara *et al.*, 2010; Yang *et al.*, 2011). The ATPase activity of FtsEX is required for septum cleavage indicating that the periplasmic cleavage of peptidoglycan is regulated by cytoplasmic hydrolysis of ATP (Yang *et al.*, 2011). FtsEX is also required for the recruitment of FtsK, which is involved in cell division as well as in DNA segregation (Yu *et al.*, 1998). FtsK is an integral membrane protein with four membrane-spanning helices (Dorazi and Dewar, 2000) essential for cell division (Yu *et al.*, 1998) followed by a long linker domain and a RecA-type ATPase domain at its C-terminus that constitutes a motor domain for DNA translocation (Dubarry and Barre, 2010). In the case that DNA replication leads to concatenated chromosomes, the motor domain consisting of a hexamer of FtsK molecules (Massey *et al.*, 2006) transports the DNA until it has positioned the *dif* sites near the terminus of DNA replication. Two tyrosine recombinases, XerC and XerD that interact with FtsK and recognize the *dif* sites then deconcatenate the chromosomes (Aussel *et al.*, 2002).

FtsQ, FlsL and FtsB

The essential protein FtsQ (31 kDa) forms a precomplex with FtsL and FtsB (Fig. 1.3) (Buddelmeijer and Beckwith, 2004), the mid-cell localization of which is dependent on the presence of FtsK (Chen and Beckwith, 2001; Chen *et al.*, 2002). FtsQ is bitopic membrane protein. The periplasmic region of FtsQ consists of three domains: the α-domain that is homologous to polypeptide transport associated domains (POTRA) and is required for the localization of FtsQ, the β-domain that interacts with FtsL and FtsB (van den Ent *et al.*, 2008) and the γ-domain, whose structure is not known (Fig. 1.4). The function of FtsQ is not known but with only 25 copies (Carson *et al.*, 1991) it might determine the number of proteins complexes that effectively synthesize the septum. FtsL and FtsB are thought to interact through a coiled coil that ends in C-terminal globular domains that interacts with the β-domain of FtsQ (Masson *et al.*, 2009; Villanelo *et al.*, 2011). In a bacterial two-hybrid system, the β-domain of FtsQ was reported to be essential for the interaction with FtsW, FtsI and FtsN as well (D'Ulisse *et al.*, 2007). The abundance and stability of FtsL varies considerably in the different bacteria species investigated (Daniel and Errington, 2000; Noirclerc-Savoye *et al.*, 2005) pointing again to a role for the FtsQLB complex in regulation of divisome assembly.

FtsW, PBP3 (FtsI), PBP1B and FtsN

The lipid II PG precursor translocase FtsW (Mohammadi *et al.*, 2011) is an integral membrane protein with 10 membrane spanning sequences and a large periplasmic domain that forms a precomplex with the cell division-specific PG TPase PBP3 (also called FtsI) (Fraipont *et al.*, 2011). Their localization is dependent on the presence of FtsQLB (Wang *et al.*, 1998; Mercer and Weiss, 2002) suggesting that FtsW-PBP3 and FtsQLB are preassembled independently and interact with each other (Goehring *et al.*, 2006). The periplasmic loop between the 9th and the 10th transmembrane segment of FtsW appears to be involved in the interaction with both PBP3 and the bifunctional PG synthase PBP1B. This loop may thus play an important role in the positioning of these synthases within the divisome (Fraipont *et al.*, 2011). The two non-essential PG synthases PBP1C and MtgA might also participate in septal PG synthesis. For example, PBP1C interacts with PBP1B and PBP3 (Schiffer and Höltje, 1999) and MtgA interacts with PBP3, FtsW and FtsN in bacterial two-hybrid assays (Derouaux *et al.*, 2008).

PBP3 was found to interact with several proteins of the divisome: FtsA, FtsK, FtsQ, FtsL, FtsB, FtsW and FtsN (Di Lallo *et al.*, 2003; Karimova *et al.*, 2005; D'Ulisse *et al.*, 2007; Müller *et al.*, 2007; Alexeeva *et al.*, 2010; Fraipont *et al.*, 2011). PBP3 requires FtsW for its localization (Mercer and Weiss, 2002) and the direct interaction between purified PBP3 and FtsW was demonstrated by co-immunoprecipitation experiments (Fraipont *et al.*, 2011). *In vitro* studies and *in vivo* cross-linking showed that PBP3 also interacts with PBP1B and MtgA (Bertsche *et al.*, 2006; Derouaux *et al.*, 2008). PBP3 is required for the midcell localization of PBP1B (Bertsche *et al.*, 2006). The first 70 residues of PBP3 are sufficient for the interaction with FtsW, FtsQ and PBP1B (Bertsche *et al.*, 2006; D'Ulisse *et al.*, 2007; Fraipont *et al.*, 2011) while the first 250 residues are needed for the interaction with FtsL (Karimova *et al.*, 2005). PBP3 mutated in residue G57, S61, L62 or R210 failed to recruit FtsN and thus these residues may be involved in the interaction between PBP3 and FtsN (Wissel and Weiss, 2004). The 1–56 peptide of PBP3 containing the membrane-spanning segment is sufficient for the interaction with FtsW and PBP3 dimerization and is essential for its localization at the division site (Weiss *et al.*, 1999; Piette *et al.*, 2004; Wissel and Weiss, 2004; Fraipont *et al.*, 2011).

FtsN was shown to interact with PBP1B (Müller *et al.*, 2007) and PBP3 (Bertsche *et al.*, 2006) and in bacterial two-hybrid assays also with FtsA and FtsQ (Di Lallo *et al.*, 2003; Karimova *et al.*, 2005). It is a bitopic membrane protein whose periplasmic region contains three short helices followed by a long flexible linker and a C-terminal PG binding SPOR domain that is not essential (Fig. 1.4) (Ursinus *et al.*, 2004; Yang *et al.*, 2004). The function of three other SPOR domain proteins (DamX, DedD and RlpA) is unknown, but they are also septally localized and mutants lacking multiple SPOR domain proteins have defects in cell division (Arends *et al.*, 2010). When targeted to the periplasm the SPOR domain is sufficient for septal localization presumably because it recognizes a specific, yet unknown, septal PG structure, or a cell division protein (Gerding *et al.*, 2009; Möll and Thanbichler, 2009; Arends *et al.*, 2010). FtsN is relatively abundant in *E. coli* with 1000 molecules per average cell grown in minimal glucose medium (Aarsman *et al.*, 2005) and >4000 molecules in cells grown in rich medium (Ursinus *et al.*, 2004). In the absence of FtsN, the division machine readily dissociates, suggesting that FtsN affects the stability and dynamics of the divisome (Rico *et al.*, 2010). Interestingly, FtsN was found to stimulate the polymerase activity of PBP1B presumably by stabilizing the dimeric form of PBP1B (Müller *et al.*, 2007). Since PBP1B is present in the division site and interacts with PBP3 (Bertsche *et al.*, 2006), FtsN may activate or regulate the concerted activities of the two PBPs.

Peptidoglycan synthesis during cell division

Synthesis of lipid II at the cytoplasmic membrane by MraY and MurG occurs in close association with the division machinery (Fig. 1.3) (Mohammadi *et al.*, 2007; Barreteau *et al.*, 2008; White *et al.*, 2010; den Blaauwen, unpublished results). This may then allow FtsW to directly translocate the newly synthesized and relatively rare lipid II molecules for immediate insertion into the growing PG layer by the PG synthases. PBP1B and MtgA have their lipid II-binding GTase domain close to the surface of the cytoplasmic membrane (between 25 and 30 Å) (Heaslet *et al.*, 2009; Sung *et al.*, 2009), whereas the peptide binding TPase sites of *E. coli* PBP1B (Sung *et al.*, 2009) and *Pseudomonas aeruginosa* PBP3 (Han *et al.*, 2010) are on top of the molecules about 90 Å away from the membrane. It is possible that PBP1B and MtgA compete for lipid II and PBP1B and PBP3 compete for the peptide side chains of the PG building blocks and that this serves some regulatory function. The DD-carboxypeptidase PBP5 with an active site 70 Å above the membrane is present at the division site to remove the terminal amino acid from pentapeptides of the growing glycan strands. The resulting tetrapeptides are lost as donors for the DD-transpeptidase reaction but can still serve as acceptors. The recently identified outer membrane lipoprotein LpoB is essential for the *in vivo* function of PBP1B (Typas *et al.*, 2010). LpoB stimulates *in vitro* the TPase activity of PBP1B by interacting with the UB2H domain of PBP1B at maximally 60 Å above the cytoplasmic membrane (Typas *et al.*, 2010). This accumulation of different proteins involved in the stimulation or prevention of cross-links in the septal PG suggests a subtle fine-tuning of the density of the PG network in the course of cell division.

The role of peptidoglycan hydrolases in cell division

PG hydrolases are required for cell separation during or after cell division, as demonstrated by many examples of PG hydrolase mutant strains that form chains of non-separated cells (Vollmer *et al.*, 2008b). Gram-positive and Gram-negative bacteria differ in the timing of septum cleavage. The former first divide to synthesize a closed cross-wall between the daughter cells, followed by cleavage of the cross-wall to separate the cells. In most Gram-negative species synthesis and cleavage of the septum occur simultaneously, leading to the constrictive mode of cell division typical for these species (compare *E. coli* and *B. subtilis*, Fig. 1.5).

Originally thought to consist of a single cell wall layer, recent cryo electron microscopy images of frozen hydrated sections revealed a multilayered appearance of septa in the Gram-positive *S. aureus* and *B. subtilis* (Matias and Beveridge, 2005, 2006, 2007). Outside the cytoplasmic membrane there is an inner wall zone of low electron density (called IWZ) that has been suggested to have similar properties as the periplasmic space in Gram-negative species (Matias and Beveridge, 2006, 2008). The next layer is the outer wall zone (OWZ) that, in Gram-positive bacteria, contains the PG with associated anionic polymers like wall teichoic acid. In the septum the two OWZ layers sandwich another layer of low density called middle low-density zone (MLZ). Cleavage of PG hydrolases through the MLZ results in the separation of the daughter cells and is accompanied by the 'inflation' of the disc-shaped septal walls to the hemispherical poles of the daughter cells (Vollmer and Seligman, 2010). The *S. aureus* Atl autolysin, a bifunctional glucosaminidase-amidase, localizes outside the cell wall in a ring at the future division site (Yamada *et al.*, 1996). Interestingly, localization of Atl is strongly impaired in mutants lacking wall teichoic acids (Schlag *et al.*, 2010). Wall teichoic acids are absent at new septal regions and, hence, it has been proposed that Atl localizes to the septum and is restricted to cleave septal PG because the wall teichoic acid prevents its action at other sites of old cell wall (Schlag *et al.*, 2010).

In *E. coli* about 30% of newly made septum PG is removed shortly after its synthesis by the hydrolases (Uehara and Park, 2008). In this species mainly the amidases AmiA, AmiB and AmiC but also lytic transglycosylases and endopeptidases contribute to septum cleavage during cell division (Heidrich *et al.*, 2001, 2002; Priyadarshini *et al.*, 2006). In agreement with the participation of lytic transglycosylases and endopeptidases in septum cleavage (Heidrich *et al.*, 2002) Slt70, MltA or MltB coupled to a Sepharose column have been shown to retain the PBP1B, PBP2 (except for MltB) and PBP3 and the non essential PBP1C (von Rechenberg *et al.*, 1996; Schiffer and Höltje, 1999; Vollmer *et al.*, 1999; Vollmer and Bertsche, 2008). In addition, Slt70 retains PBP7/8 (PBP8 is a degradation product of PBP7) (Romeis and Höltje, 1994b) and MltA retains MipA that mediates interaction between PBP1B and MltA (Vollmer *et al.*, 1999). The interaction between PBP1B and MltA was shown to be mediated by the UB2H domain of PBP1B (Sung *et al.*, 2009).

Triple deletion of *amiA*, *amiB* and *amiC* result in the formation of long cell chains (90% of the population form chains of 6–24 cells) and this phenotype is less pronounced in the single mutants (Heidrich *et al.*, 2001). AmiB and AmiC localize to the division site while AmiA does not (Bernhardt and de Boer, 2003). AmiC contains two domains, an N-terminal non catalytic domain that mediates its localization to the septal ring and a C-terminal catalytic domain (LytC) that removes preferentially tetrapeptides from PG (Priyadarshini *et al.*, 2006). Interestingly, the amidases are regulated in the cell by members of the LytM endopeptidase family, EnvC and NlpD, that appear not to have an own enzymatic activity. EnvC stimulates AmiA and AmiB and NlpD stimulates AmiC *in vitro* (Uehara *et al.*, 2010). Consistent with their role as amidase activators, mutants lacking these LytM proteins cannot separate the daughter cells and form cell chains (Uehara *et al.*, 2009). In contrast to EnvC and NlpD, which are inactive, other LytM proteins with PG hydrolase activity have been identified in two other Gram-negative species. DipM localizes to mid-cell during septation, is essential for septum cleavage and is required to maintain normal PG thickness in *Caulobacter crescentus* (Collier, 2010; Möll *et al.*, 2010; Poggio *et al.*, 2010). The three recently identified LytM-type endopeptidases Csd1, Csd2 and Csd3 are required for helical cell shape in *Helicobacter pylori* (see below) (Sycuro *et al.*, 2010).

Membrane lipid composition affects cell division

The major phospholipids building the *E. coli* cytoplasmic membrane are the zwitterionic phosphatidylethanolamine (PE, ~70% of total lipids) and the negatively charged phosphatidylglycerol (PG, ~20%) and 10% cardiolipin (CL, ~10%). *E. coli* cells lacking the *pss* gene required for PE synthesis are viable when grown in the presence of high concentrations of divalent cations to stabilize the negative charge of the remaining lipids (DeChavigny *et al.*, 1991). The cells filament in the absence of the cations and constriction but not the localization of the early cell division proteins FtsZ, FtsA and ZipA is inhibited (Mileykovskaya *et al.*, 1998). The membrane association and the membrane insertion of many proteins are dependent on negatively charged phospholipids because of their interaction with the protein translocation machinery (SecAYEG) and with the membrane-associating amphipathic helices of proteins such as FtsA and MinD. In addition the lipid composition of the membrane affects the structure of integral membrane proteins. Minicells produced by polar division are enriched in CL (Koppelman *et al.*, 2001) suggesting that CL is either needed for the curvature of the membrane or is involved in division itself. An *E. coli psgA* null mutant that lacks the major anionic phospholipids PG and CL seems to divide normally because it has replaced these phospholipids by a new negatively charged lipid, N-acylphosphatidylethanolamine that accumulates at the poles and division site and in minicells of this strain (Mileykovskaya *et al.*, 2009). These results indicate that the lipid composition of the membrane affects cell division. However, it is not known if certain phospholipids are specifically required for division and if so, what their role would be.

The membrane potential is required for membrane attachment and correct localization of the cell division protein FtsA and of the FtsZ-affector protein MinD via their amphipathic helix, explaining the requirement of a membrane potential for cell division (Strahl and Hamoen, 2010).

Growth with different cell shapes

Deciphering the molecular mechanism allowing bacteria to grow with a large diversity of cell shapes is an important research area in microbiology. Bacterial cell shape appears to be determined by the coordinated activities of PG synthesis complexes that drive cell elongation, division and bacterial cytoskeletal elements (Margolin, 2009). Here, we will first describe how cells grow with spherical or ovoid shape. We then present two mechanisms by which rod-shaped bacteria elongate: (i) by incorporation of new PG at the side-wall, a process dependent on the bacterial actin-like protein MreB, or (ii) by an MreB-independent mode of elongation at the tip of the cell. Finally, we provide recent data on the growth of cells with more elaborate elongated shapes, such as curved, helical and branched filamentous shapes (Fig. 1.5).

Growth with spherical shape

Spherical bacteria, or cocci, like the Gram-positive *S. aureus*, lack MreB and grow exclusively by FtsZ-dependent septum synthesis, which accounts for the synthesis of the entire new hemisphere of each daughter cell (Fig. 1.5) (Pinho and Errington, 2003; Zapun *et al.*, 2008). After septum formation PG hydrolases, such as Atl, cleave the septum for the separation of the daughter cells (Yamada *et al.*, 1996). During this process, the planar septal PG disc is transformed into the half-spheres of the daughter cells. The mechanisms regulating

PG hydrolase activity and the details of changes in PG architecture during the 'inflation' of the septal PG are poorly understood (Vollmer and Seligman, 2010). Phylogenetic studies suggest that cocci evolved from rods by loss of function of the machinery required for cylindrical elongation (Siefert and Fox, 1998).

Growth with ovoid shape

While spherical cocci exclusively grow by forming and cleavage of septa, ovococci additionally employ an MreB-independent peripheral growth, which occurs mainly at mid cell before cell division (Daniel and Errington, 2003) and which is responsible for the longitudinal expansion that contributes to their ovoid shape (Fig. 1.5). In the Gram-positive *S. pneumoniae*, PBP2x and PBP2b incorporate new PG at the septum and sidewall (peripheral growth), respectively (Morlot *et al.*, 2003; Zapun *et al.*, 2008). Both septal and peripheral PG synthesis enzymes or complexes are associated to FtsZ, as MreB is absent (Zapun *et al.*, 2008). *S. thermophilus* requires RodA, a PG precursor flippase essential for rod-shape in many species (see below), to maintain its ovoid shape (Thibessard *et al.*, 2002). Interestingly, filamentation of ovococci can be induced by methicillin, a beta-lactam antibiotic that inhibits cell division (Lleo *et al.*, 1990) and *Lactococcus lactis* can elongate to form long filaments by synthesizing PG at the FtsZ ring positions when septation is inhibited (Perez-Nunez *et al.*, 2011).

Growth as a straight rod

The elongation of a rod-shaped *E. coli* cell occurs by the insertion of new PG into the cylindrical part of the cell at a limited number (~50) of PG growth sites (den Blaauwen *et al.*, 2008). Fluorescent vancomycin labelling in *B. subtilis* showed that the incorporation of this nascent PG occurs in a helical pattern (Daniel and Errington, 2003; Tiyanont *et al.*, 2006).

Rod-shaped cells of *E. coli* can be converted into spherical cells by treatment with certain β-lactam antibiotics such as mecillinam, which inhibits PBP2 (Waxman and Strominger, 1983), or by inactivation of genes encoding MreB, MreC, RodA, RodZ and PBP2 (Table 1.2) (Normark *et al.*, 1969; Matsuzawa *et al.*, 1989; Wachi *et al.*, 1989; Kruse *et al.*, 2005; Shiomi *et al.*, 2008). All these proteins are associated to or inserted in the inner membrane where they form the elongasome multi-protein complex that performs cell elongation (Fig. 1.3).

In *E. coli*, the *rodA* gene forms an operon with the *pbpA* and *dacA* genes encoding PBP2 and PBP5, respectively (Begg *et al.*, 1986; Matsuzawa *et al.*, 1989). RodA is part of the SEDS (shape, elongation, division and sporulation) family of proteins that include the cell division protein FtsW and the sporulation protein SpoVE of *B. subtilis*. FtsW has recently been identified as lipid II flippase and, hence, RodA likely translocates lipid II across the cytoplasmic membrane during cell elongation (Mohammadi *et al.*, 2011). RodA is also required for rod-shape in Gram-positive bacteria like *B. subtilis* (Henriques *et al.*, 1998).

E. coli PBP2 is a class B PBP present in ~100 copies per cell (Dougherty *et al.*, 1996). Its inhibition by the β-lactam antibiotic mecillinam induces the formation of spheres that expand and lyse because there is not enough FtsZ molecules to allow division. Overexpression of FtsZ restores cell growth in strains with inactivated PBP2 (Vinella *et al.*, 1993). PBP2 localizes in a patchy pattern along the cytoplasmic membrane and at mid-cell during early phase of cell division (den Blaauwen *et al.*, 2003; van der Plaeg *et al.*, 2013). PBP2 interacts with the class A PBP1A and both synthases cooperate in the synthesis of new PG and its

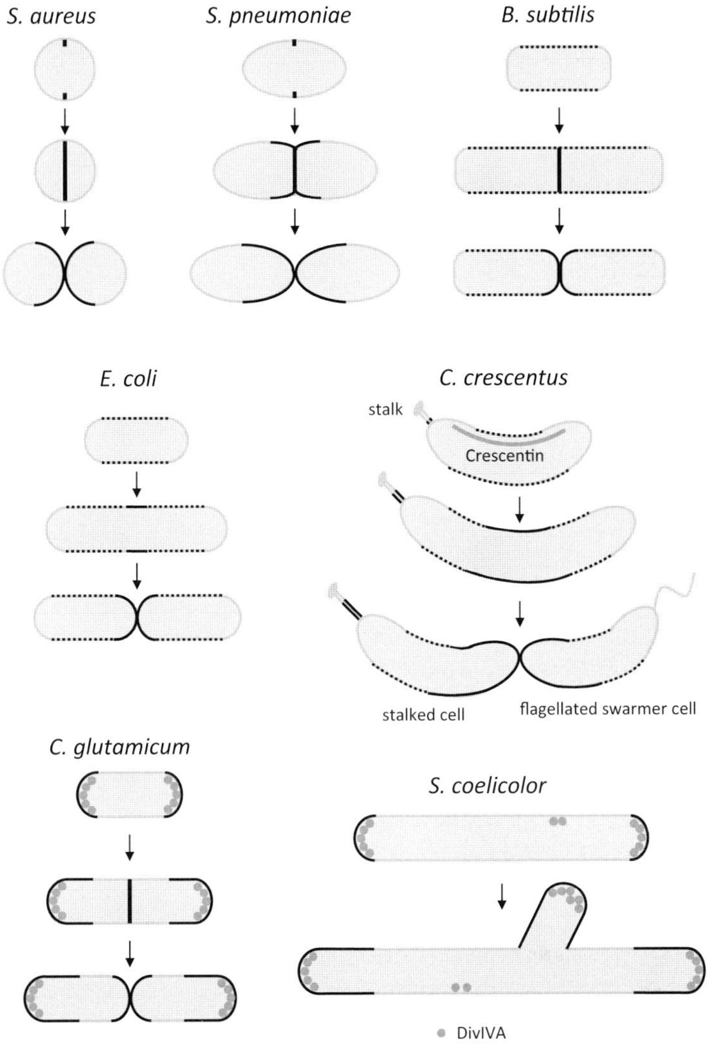

Figure 1.5 Cell shape and sites of peptidoglycan growth during the cell cycle in representative bacteria. Old PG present in newborn cells (top pictures) is shown as grey line; new PG is shown as black dots or lines. Spherical cells of *S. aureus* grow by septum synthesis and cleavage in an FtsZ-dependent manner. In the ovococcus *S. pneumoniae* PG is synthesized at the septum and sidewall in an FtsZ-dependent manner. The rod-shaped *B. subtilis*, *E. coli* and *C. crescentus* elongate by PG incorporation of new PG at the sidewall in an MreB-dependent manner, and then switch to FtsZ-dependent septal PG synthesis. In the Gram-positive *B. subtilis* synthesis and cleavage of the septum are separate events, whereas both processes occur simultaneously in the Gram-negative *E. coli* and *C. crescentus* resulting in constrictive mode of division. In *E. coli* and, more pronounced, in *C. crescentus* there is also a pre-septal phase of PG synthesis, during which the cell elongates in an FtsZ-dependent manner at mid-cell. The crescentin filament causes *C. crescentus* to grow with a curved mode. This bacterium also divides asymmetrically to produce a stalked and a swarmer cell. Growth of the stalk during the cell cycle is indicated. *C. glutamicum* does not have MreB and grows at the poles in a DivIVA-dependent manner, and for cell division at the septum in an FtsZ-dependent manner. Also *S. coelicolor* filaments grow at the poles, and this species also form branches in a DivIVA-dependent manner. Note that the bacterial species are not drawn to scale.

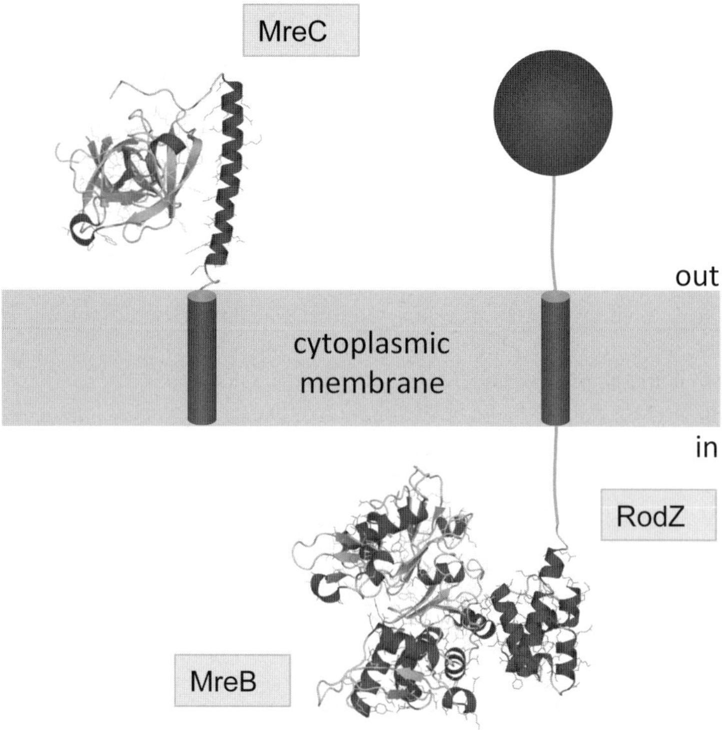

Figure 1.6 Crystal structures of cell elongation proteins. The cytoplasmic MreB is linked to the membrane by interaction with the bitopic RodZ. MreC has an N-terminal membrane anchor helix (dark grey rod) and locates mainly in the periplasm. Helices are shown in dark grey, β-sheets are shown in grey and loops in light grey. Complex of MreB (amino acid residues 2–336) and RodZ (amino acid residues 2–88) from *Thermotoga maritima*, PDB accession number 2WUS (van den Ent *et al.*, 2010); MreC from *Listeria monocytogenes*, amino acid residues 74–284, PDB accession number 2J5U (van den Ent *et al.*, 2006).

attachment to sacculi in PG synthesis assays. They likely perform most of the side-wall PG synthesis during elongation (Banzhaf *et al.*, 2012).

MreB, MreC, MreD and RodZ have no enzymatic role in PG synthesis but are essential for rod-shape (Figs. 1.3 and 1.6 and Table 1.2). MreB is a cytoskeletal protein present in most rod-shaped bacteria and absent in cocci and ovococci (Jones *et al.*, 2001). The crystal structure of MreB is similar to that of eukaryotic actin (Fig. 1.6) (van den Ent *et al.*, 2001; Amos *et al.*, 2004). MreB forms dynamic and flexible filaments in an ATP-dependent manner. A22 and related compounds rapidly disrupt MreB filaments in the cell resulting in a block in cell elongation and growth into spherical cells (Iwai *et al.*, 2002; Gitai *et al.*, 2005; Takacs *et al.*, 2010) GFP-MreB localized in a helical pattern just under the cytoplasmic membrane (Jones *et al.*, 2001; Shih *et al.*, 2003; Figge *et al.*, 2004; Karczmarek *et al.*, 2007; Vats and Rothfield, 2007). Recent studies with high-resolution fluorescence microscopy in *B. subtilis* challenge the localization of MreB in a continuous helix. The images indicate that MreB filaments consist of small patches that move underneath the cytoplasmic membrane perpendicular to the cell length in a direction, which is roughly that of the glycan chains in the sacculus (Dominguez-Escobar *et al.*, 2011; Garner *et al.*, 2011). Interestingly, both studies also demonstrate that MreB-movement depends on ongoing PG synthesis, indicating that cytoskeletal dynamics and PG synthesis during cell elongation are interdependent processes. Some species possess several MreB homologues. The three variants in *B. subtilis* (MreB, Mesh and Mbl) interact with each other and show a partial redundancy in directing rod-shaped growth (Kawai *et al.*, 2009), and they colocalize and move perpendicular to the length of the cell (Dominguez-Escobar *et al.*, 2011; Garner *et al.*, 2011). MreBH recruits a cell wall hydrolase, LytE, and thus might regulate cell wall turnover (Carballido-Lopez *et al.*, 2006).

E. coli MreC is a bitopic membrane protein with a large periplasmic domain. The crystal structures of the periplasmic domain from *Listeria monocytogenes* (van den Ent *et al.*, 2006) and from *Streptococcus pneumoniae* (Lovering and Strynadka, 2007) suggest that the protein is capable to form dimers mediated by its N-terminal helix or oligomeric filaments (Fig. 1.6) (van den Ent *et al.*, 2006). Fluorescence-tagged MreC appears to localize in a helix in the periplasm (Dye *et al.*, 2005; Gitai *et al.*, 2005). *E. coli* MreD is an integral membrane protein. Bacterial two-hybrid experiments showed that MreB interacts with MreC but not with MreD. MreC interacts with MreB, MreD and PBP2. In *E. coli*, the association between these proteins is essential for elongation as the depletion of any one of them results in spherical growth (Kruse *et al.*, 2005). The MreBCD complex seems to position and/or recruit the lipid II and PG synthetic machinery in order to assure the disperse insertion of nascent PG into the existing cell wall allowing the maintenance of a rod shape (Daniel and Errington, 2003; Mohammadi *et al.*, 2007; White *et al.*, 2010). Remarkably, the MreB cytoskeleton affects the mechanical properties of the cell, i.e. disruption of MreB filaments leads to a reduction in bending stiffness (Wang *et al.*, 2010).

Recently, another conserved bacterial shape protein, called RodZ, has been identified (Shiomi *et al.*, 2008; Alyahya *et al.*, 2009; Bendezu *et al.*, 2009). RodZ has one transmembrane segment. Its cytoplasmic domain interacts with MreB via a helix–turn–helix motif as shown in the crystal structure of the complex (Fig. 1.6) (van den Ent *et al.*, 2010). RodZ co-localizes with MreB and is important for the formation of MreB filaments (Shiomi *et al.*, 2008; Alyahya *et al.*, 2009; Bendezu *et al.*, 2009). An *E. coli rodZ* mutant is wider than the wild type. Overproduction of MreB causes *E. coli* cells to become wider, but when RodZ is overexpressed simultaneously, the cells are able to maintain their normal diameter (Bendezu *et al.*, 2009), indicating that being membrane-bound RodZ may assist membrane anchoring or may alter the dynamics of MreB filaments (van den Ent *et al.*, 2010).

In *E. coli*, the dynamic cell elongation complex includes MreB, MreC, MreD, RodZ, the MraY-MurG enzymes for lipid II synthesis, the lipid II flippase RodA, the PG synthases PBP2 and PBP1A, the outer membrane activator of PBP1A, LpoA, and, presumably, PG hydrolases (Fig. 1.3). LpoA is essential for the *in vivo* function of PBP1A, interacts with a certain region in PBP1A, called ODD, and stimulates the PBP1A TPase activity *in vitro* (like LpoB interacts with and stimulates PBP1B; see above) (Typas *et al.*, 2010). The outer membrane-localized Lpo proteins must reach through the pores in the PG net in order to interact with their cognate PG synthase. Hence, it has been suggested that Lpo-mediated activation of PG synthesis is responsive to the state of the PG sacculus and is more efficient through a more stretched PG with larger pores, and less efficient through a relaxed PG with smaller pores. Hence, the activation of PBPs by outer membrane proteins could be a homeostatic mechanism to achieve constant and homogeneous PG thickness and surface density, and to adjust the PG synthesis rate to the overall cellular growth rate (Typas *et al.*, 2010, 2012). In addition, PG hydrolases were suggested to be involved for cleaving covalent bonds in the PG to allow the insertion of the newly made PG into the sacculus (see above and Höltje, 1996).

Rod-shape without MreB

Some rod-shaped bacteria, such as corynebacteria, lack MreB, MreC and MreD. These species elongate from the cell poles as observed by vancomycin staining of the nascent PG (Fig. 1.5) (Daniel and Errington, 2003). In *Corynebacterium glutamicum* DivIVA, a protein conserved in many Gram-positive bacteria, localizes to the cell poles and triggers tip growth together with PBP1A, PBP1B and RodA (Valbuena *et al.*, 2007). DivIVA has a carboxy-terminal coiled-coil domain with homology to eukaryotic tropomyosins. It is involved in tip growth and in septum synthesis probably by recruiting PG synthases (Margolin, 2009). Depletion of DivIVA results in spherical cells, whereas overproduction induces swollen poles (Letek *et al.*, 2009). Purified DivIVA forms 'doggy bone' structures *in vitro* that could form *in vivo* a lattice associated to the membrane (Stahlberg *et al.*, 2004). DivIVA is also present in *B. subtilis* where it is involved in spatial control of cell division via MinJ and MinCD (Bramkamp *et al.*, 2008; Patrick and Kearns, 2008), and in attachment of chromosomes to cell poles via an interaction with RacA (Lenarcic *et al.*, 2009). DivIVA localizes to cell poles by recognizing negative membrane curvature (Lenarcic *et al.*, 2009; Ramamurthi and Losick, 2009). The crystal structure of the *B. subtilis* DivIVA N-terminal domain suggests that DivIVA interacts with the membrane via hydrophobic residues inserted into the membrane and positively charged residues bound to the membrane surface (Oliva *et al.*, 2010). *S. aureus* and *S. pneumoniae* also possess DivIVA. *C. glutamicum* has in addition a cytoskeletal protein called RsmP that affects cell shape by an unknown mechanism and that is required for polar growth (Fiuza *et al.*, 2010).

Other bacteria including many α-proteobacteria and the γ-proteobacteria *Francisella tularensis* are rod-shaped and lack MreB, MreC, MreD and DivIVA. It is not known how these Gram-negative species elongate. As these bacteria have a tendency to form branches, they might grow at the tips like corynebacteria but with the help of another protein than DivIVA (Margolin, 2009).

Growth as a curved rod

C. crescentus cells have curved rod-shape and tapered poles. An asymmetric cell division gives rise to two different daughter cells, a mobile swarmer cell that carries a polar flagellum, and a stalked cell that has a thin polar extension, called the stalk, which carries an adhesive organelle at its tip (Fig. 1.5). Mutant cells lacking the *creS* gene, encoding crescentin, have straight rod-shape (Ausmees *et al.*, 2003). Crescentin has a similar, coiled-coil-rich structure as that of eukaryotic intermediate filaments and purified crescentin can spontaneously assemble into filaments without the need of any nucleotide cofactors (Ausmees *et al.*, 2003). One large cytoskeletal filamentous structure of crescentin spans the cytoplasmic membrane at the inner curvature of the cells parallel to the length axes of the cell (Fig. 1.5). The crescentin filament acts like an expanded spring, exerting a force on the cell envelope that reduces the strain on the PG at the inner curvature of the cell (Cabeen *et al.*, 2009). This causes the cell to grow slower at the inner curvature and faster at the outer curvature to maintain its bent cell shape. Interestingly, crescentin is also able to cause curved growth in *E. coli* cells (Cabeen *et al.*, 2009). FRAP (fluorescence recovery after photobleaching) microscopy experiments showed that the crescentin filament is extremely stable inside the cell as are eukaryotic intermediate filaments unlike FtsZ or MreB filaments, which are much more dynamic (Charbon *et al.*, 2009). The interaction of crescentin with the membrane requires the presence of MreB filaments. Inhibition of MreB polymerization causes crescentin filaments to mislocalize and become non-functional. Consistent with MreB dependency, crescentin was not able to induce cell curvature in *Agrobacterium tumefaciens*, which is rod-shaped but lacks MreB (Charbon *et al.*, 2009)

The *C. crescentus* stalk contains all envelope layers including the PG. In addition to mediating adherence to a solid surface the stalk provides an advantage to the cell in case of nutrient limitation; the elongation of that thin part of the cell increases the surface area available for nutrient absorption while decreasing the surface-to-volume ratio (Wagner *et al.*, 2006). Decreasing phosphate concentrations induce the growth of a longer stalk (Gonin *et al.*, 2000). The stalk is newly synthesized at the same pole after a swarmer cell ejects its flagellum, and is elongating from the base in each generation (Fig. 1.5). Stalk synthesis appears to be a specialized form of cell elongation. Inactivation of PBP2 or depletion of MreB or RodA results in a defect in stalk elongation (Wagner *et al.*, 2005). A new class of cytoskeletal proteins, the bactofilins BacA and BacB, localize in proximity of the stalked pole where they form a sheet-like structure interacting with the membrane. The bactofilin structure recruits the PG synthase PBPC involved in stalk morphogenesis in an MreB-independent manner (Kühn *et al.*, 2010).

Growth with helical shape

Although helical cell shape is abundant in bacteria the components and mechanisms to achieve helical growth are not well understood. Gram-negative spirochaetes such as *Borrelia burgdorferi*, which is the causative agent of Lyme disease, have periplasmic flagella essential not only for motility but also for helical cell shape (Goldstein *et al.*, 1994). The periplasmic flagella might generate the force to bend the cell into a helix (Wolgemuth *et al.*, 2006). Cells with non-rotating flagella or fixed cells lose the regular helix-shape but nevertheless maintain some residual curvature indicating that the rotating flagellum is not the sole determinant of cell curvature and twist. In addition, some spirochaetes keep their helical shape even without flagella, but the machinery involved in the synthesis of a helical cell envelope has yet to be identified (Bromley and Charon, 1979; Ruby *et al.*, 1997).

The Gram-negative ε-proteobacterium *Helicobacter pylori* colonizes the human stomach and is capable of inducing chronic gastric inflammation. Its helical cell shape is important for pathogenesis presumably by allowing movement through the viscous epithelial mucus layer perhaps like a corkscrew (Hazell *et al.*, 1986). Helical shape can be thought of as the sum of three morphogenic processes: cell elongation, curvature and twist. How *H. pylori* elongates is not known. It has the typical cell elongation genes *mreB* and *mreC* but lacks *mreD* and *rodZ* (Margolin, 2009). Moreover, MreB is not essential for viability and its absence does not affect cell shape but cell length, chromosome segregation and pathogenicity (Waidner *et al.*, 2009). *H. pylori* has two coiled-coil rich proteins (Ccrp) that are required for normal cell shape by unknown mechanism (Waidner *et al.*, 2009). Recently, four proteins involved in helical cell shape maintenance have been identified, the bactofilin Ccm, the PG DD-Endopeptidases Csd1 and Csd2, and the DD-endopeptidase/DD-carboxypeptidase Csd3 (Bonis *et al.*, 2010; Sycuro *et al.*, 2010). Csd1, Csd2 and Csd3 belong to the LytM-type peptidase family. Deletion of one or more of the corresponding genes results in loss of helical cell shape and the formation of more or less curved rods (Sycuro *et al.*, 2010). Although the molecular mechanisms have to be determined it has been proposed that cell curvature and twist is generated and maintained by local relaxation of cross-links in the PG, catalysed by Csd1–3 and regulated by Ccm, resulting in helical shape of *H. pylori* (Sycuro *et al.*, 2010). Such a mechanism would be consistent with modelling results of PG cross-linking and cell shape (Huang *et al.*, 2008).

Plant-pathogenic spiroplasma are minimal sized, helical-shaped bacteria that lack a PG cell wall (Bove *et al.*, 2003). The helical shape of *Spiroplasma melliferum* is maintained by a cytoskeletal apparatus composed mainly of Fib that lacks similarity to any eukaryotic or prokaryotic cytoskeletal protein. Fib forms filament bundles in a ribbon-like helix, which attaches to the inner surface of the membrane along the shortest (inner) helical line and twists the cell into a helix. Interestingly, MreB has been co-purified with the cytoskeletal ribbon (Trachtenberg, 2006; Trachtenberg *et al.*, 2008).

Growth with branched cell chains

Some Gram-positive species of the actinobacteria phylum, including the streptomycetes, form branches or hyphae analogous to filamentous fungi. The hyphae elongate at their tips without the need for MreB, like the single-cell rod-shaped corynebacteria (see above) and mycobacteria. *Streptomyces* species germinate from a spore and grow first as a vegetative branched mycelium. It then forms an aerial branched mycelium, which may differentiate in chains of spores. Branching requires a new cell pole to be created in the cylindrical part of the filament. DivIVA localization triggers the new pole formation (Fig. 1.5) (Hempel *et al.*, 2008). The membrane curvature at the new pole can be important for the stabilization of the DivIVA cluster, as it is in *B. subtilis*, but it probably does not determine the selection of the site where the new pole will be created (Flardh, 2010). Additional coiled-coil proteins have been identified with a role in cell shape of *Streptomyces*. Scy is a long coiled-coil protein involved in controlling filamentous growth (Walshaw *et al.*, 2010) whereas FilP assembles in filaments that localize at the poles and near branch points (Bagchi *et al.*, 2008; Flardh, 2010). Atomic force microscopy (AFM) showed that a *S. coelicolor* FilP-null mutant has more deformable hyphae than wild type (Bagchi *et al.*, 2008). RodA is essential for polar cell wall synthesis during hyphae growth, whereas FtsW is required for septum formation during sporulation (Mistry *et al.*, 2008). *S. coelicolor* possesses three MreB-like proteins. None of these are essential for cell elongation but MreB and Mbl co-localize at the sporulation septum and are required for spore cell wall synthesis (Heichlinger *et al.*, 2011). The role of the third MreB-like protein (SCO6166) is not known.

Shape of cyanobacteria

Cyanobacteria come in a large diversity of shapes. They have cell walls similar to Gram-negative species, except that the PG is thicker (Hoiczyk and Hansel, 2000). They can adopt different morphology depending on environmental factors like nutrient limitation, temperature, desiccation or light (Singh and Montgomery, 2011). How they build those shapes is mainly unknown. The *mreBCD* operon is often found in cyanobacteria and a mutation in mreB produce spherical cells in Anabaena sp. PCC 7120, which is normally rod-shaped (B. Hu *et al.*, 2007). Most of the spherical cyanobacteria lack MreB.

Antibiotic inhibition of peptidoglycan growth and cell division

Inhibitors of PG synthesis

PG inhibiting antibiotics (β-lactams, glycopeptides) are still widely used, but their efficacy is continuously decreasing with the emergence of new multidrug-resistant strains. Therefore the search for new antibiotic classes, with completely different mode of action, is urgently needed. In this respect several groups have investigated new strategies to find new antibacterial scaffolds and old cell wall inhibitors have been revisited.

Non-β-lactam transpeptidase inhibitors

Lactivicin (LTV) is a natural γ-lactam antibiotic that inhibits the transpeptidase activity of PBPs and also of β-lactamases. LTV reacts with the nucleophilic serine of these enzymes, which leads to the opening of the cycloserine and lactone rings of LTV as was shown in the crystal structures of *S. pneumoniae* PBP1b and *Bacillus licheniformis* BS3 β-lactamase complexes (Macheboeuf *et al.*, 2007; Brown *et al.*, 2010).

Boronic acids with appropriate side chains have been shown to be potent inhibitors of DD-peptidases. The crystal structures show that they usually adopt a tetrahedral conformation, bound to the nucleophilic serine of the active site and mimic the transition state of the enzymatic reaction (Nicola *et al.*, 2005; Inglis *et al.*, 2009; Adediran *et al.*, 2010; Dzhekieva *et al.*, 2010). Moreover, peculiar binding was observed in the crystal structures of the DD-peptidase R39 with amidomethylboronic acids complexes. The boron forms a tricovalent adduct with the γ-O of Ser49, Ser298, and the terminal amine group of Lys410, three key residues involved in the catalytic mechanism of PBPs (Zervosen *et al.*, 2011).

Structure-based high-throughput screening and direct enzymatic screening of compounds have been used to search for non-β-lactam inhibitors of transpeptidases. These lead to compounds of variable scaffolds and some of them had promising inhibitory activities for low affinity PBPs such as PBP2x from *S. pneumoniae,* and also exhibit good antibacterial activities against a panel of Gram-positive bacteria (Turk *et al.*, 2011).

GT inhibitors, moenomycin and small molecules

There is no clinically used antibiotic against the PG GTases. Yet, these enzymes represent a validated target of high interest for the development of new antibacterials. The only well-characterized inhibitor of the GTases is the natural product moenomycin, a potent antibacterial against Gram-positives that unfortunately suffers from poor physicochemical properties (Welzel, 2005). Its structure consists of a pentasaccharide linked to moenocinyl lipid by a phosphoglycerate that mimics the undecaprenyl-linked growing chain of the PG and binds to the donor site of the enzyme (Fig. 1.2). The minimal structure that retains antibacterial activity is a trisaccharide linked to the moenocinyl by a phosphoglycerate (Welzel, 2005). The crystal structures of GTase with moenomycin show that most interactions are mediated by the trisaccharide units (CEF) (Lovering *et al.*, 2008b). The two negatively charged groups of the phosphoglycerate moiety, a phosphoryl group and a carboxylate, makes a critical contact with conserved residues of the GTase active site (Lovering *et al.*, 2008b; Fuse *et al.*, 2010). These interactions were suggested to help orient the lipid chain for binding along the hydrophobic groove facing the cytoplasmic membrane (Lovering *et al.*, 2008b; Fuse *et al.*, 2010). The lipid chain also plays an important role in binding and its removal abolishes both antibacterial and *in vitro* GTase activity (Welzel, 2005).

Recent research aimed to identify new GTase inhibitors with scaffolds different from moenomycin from chemical libraries. Structure-based approaches and direct high-throughput screening assays have been performed. Substrate analogues and compounds with moenomycin and substrate features have been synthesized and evaluated (Dumbre *et al.* 2012; Derouaux *et al.*, 2013). Using structure-based virtual screening, tryptamine-based molecules were identified and shown to exhibit antibacterial activity against several Gram-positive bacteria (MIC 4–8 μg/ml) and to inhibit the polymerase activity from lipid II of several GTases (IC_{50} 30–60 μM) (Derouaux *et al.*, 2011). These compounds have been confirmed to specifically target cell wall synthesis. The most active compound was further characterized by NMR and found to interact with the lipid II substrate via the pyrophosphate motif.

High-throughput screening assays have been performed on the 2M compound library to search for methicillin-resistant *S. aureus* (MRSA) inhibitors. The positive hits (252) were tested using the fluorescence anisotropy (FA) assay for competition with moenomycin binding to GTase, which led to the discovery of 16 non-carbohydrate molecules able to interfere with moenomycin binding. One identified compound, a salicylanilide-based molecule inhibits PG polymerization from lipid II *in vitro* (Cheng *et al.*, 2010). Other analogues of the same scaffold also showed transglycosylase inhibition activity *in vitro*. Cheng and colleagues identified three compounds that interfere with moenomycin binding using the FA-assay but their effect on the transglycosylase activity with lipid II was not tested (Cheng *et al.*, 2008). A small-molecules library of iminocyclitol-based compounds was synthesized using a combinatorial approach. 32 compounds were found to interfere with moenomycin binding to GTase in FA assay, of which two active inhibitors were confirmed by lipid II-based activity assay (Shih *et al.*, 2010). The new molecules described in these studies show different structural scaffold and could be promising leads for the development of novel antibiotics directed against the multidrug-resistant bacteria.

Inhibitors of cell division – FtsZ as novel antibiotic target

The bacterial cytoskeleton has been recognized as potential new target for the development of novel antimicrobials (Vollmer, 2006). FtsZ is essential in all bacteria. Antibiotics that inhibit FtsZ polymerization or its interaction with some of its associated proteins could either be of the broad spectrum or narrow spectrum type. A potential draw back could be its homology to tubulin. However, FtsZ inhibitors that appear to inhibit tubulin as well can be further investigated as anti-cancer drugs. Many compounds that inhibit the interaction of FtsZ with other essential cell division protein should not inhibit the partners of tubulin because only the polymerization mechanism and the protein structure of FtsZ and tubulin are similar but not their amino acid sequences. Even the GTP binding pocket of both proteins is sufficiently different to be discriminated by inhibitors (Läppchen *et al.*, 2008). FtsZ has been included in many high-throughput screening assays of synthetic compounds [4-aminofurazan derivative-A189 (Ito *et al.*, 2006), OBTA (Beuria *et al.*, 2009), trisubstituted benzimidazoles (Kumar *et al.*, 2011), Zantrins (Margalit *et al.*, 2004), 3-methoxybenzamine derivatives (Haydon *et al.*, 2008; Haydon *et al.*, 2010; Adams *et al.*, 2011)], anti-*Mycobacterium tuberculosis* screens (White *et al.*, 2002; Reynolds *et al.*, 2004), natural compounds [(berberine (Domadia *et al.*, 2008; Boberek *et al.*, 2010), chrysophaentins (Plaza *et al.*, 2010), cinnamaldehyde (Domadia *et al.*, 2007), curcumin (Rai *et al.*, 2008), sanguinarine (Beuria *et al.*, 2005), sulfoalkylresorcinol (Kanoh *et al.*, 2008), viriditoxin

(Wang *et al.*, 2003) and dichamanetin (Urgaonkar *et al.*, 2005)], anti-tubulin compounds [5,5-bis-8-anilino-1-naphthalenesulfonate (Yu and Margolin, 1998), totarol (Jaiswal *et al.*, 2007) and taxanes (Huang *et al.*, 2006)] or rational drug design of GTP analogues (Läppchen *et al.*, 2005, 2008; Paradis-Bleau *et al.*, 2007).

The inhibitory action of the compounds can be divided into molecules that prevent the association of FtsZ monomers that enhance or inhibit the GTPase activity of FtsZ and thereby prolong the lifetime of the protofilaments, or that cause extensive bundling, which also prolongs the lifetime of the filaments. The dynamic nature of the FtsZ polymers is essential for the correct timing and progression of cell division in bacteria, therefore compounds that stabilize or destabilize the protofilaments can be equally potential as new antibiotics. Compounds that specifically inhibit the interaction of FtsZ with one of its partner proteins have not been reported as far as we know, but this is also not tested in the majority of the cases. Many of these lead molecules inhibit FtsZ *in vitro* and cell division *in vivo* in various bacterial species including the pathogenic *M. tuberculosis*, methicillin-resistant *S. aureus* (MRSA) and vancomycin-resistant enterococci (VRE). Some of the compounds have been patented (Awasthi *et al.*, 2011) but very few have been tested in animal models. One exception is the compound PC190723, which was able to cure mice from *S. aureus* infection (Haydon *et al.*, 2008). This compound has been identified in an *in vivo* screen using two fluorescent reporters, one to detect the inhibition of cell division and the other to simultaneously report the continuation of normal growth ensuring de-selection of compounds that were not specific for division inhibition (Stokes *et al.*, 2005).

Overall, many compounds have been found to specifically inhibit FtsZ one way or another and we expect that the most promising compounds are included in lead optimization and will be funnelled in clinical trials in the near future.

Inhibitors of cell elongation

In *E. coli* and *C. crescentus* MreB filaments are disrupted by the small molecule *S*-(3,4-dichlorbenzyl) isothiourea (also called A22) resulting inhibition of cell elongation, the formation of spherical or lemon-shaped cells and lysis (Iwai *et al.*, 2002; Gitai *et al.*, 2005). More recently, a series of derivatives of A22 has been synthesized and shown to be active against different Gram-negative bacteria (Takacs *et al.*, 2010). Although the application of MreB-inhibitors is so far limited as research tools the MreB cytoskeleton will in the near future probably prove to have a high potential as target for new antimicrobial drugs.

Future trends

Over the last decades we have made significant progress in understanding bacterial growth, cell division and cell shape. We begin to understand how bacteria enlarge their cell wall and how the process is regulated by cytoskeletal elements, without knowing the precise molecular details. Novel components of the cell wall synthetic machinery have been identified only recently, for example the new regulators of PG synthases and hydrolases in Gram-negative bacteria, and it is almost certain that more components are to be discovered in model bacteria that are already intensively studied and in more 'exotic', less studied species. Advances in technology are likely to drive new discoveries. For example, biophysical methods such as solid-state NMR spectroscopy, electron cryotomography and atomic force microscopy will provide more detailed insights into PG architecture. High-resolution fluorescence microscopy and X-ray crystallography of membrane proteins will revolutionize imaging the cellular localization, dynamics, and interactions of PG enzymes and their regulators. The *in vitro* systems available for PG synthesis and cell division will advance towards the goal to reconstitute the processes in the test tube to decipher the molecular mechanisms of PG growth and cell division. All these new technological developments will likely help us to discover the common principles and mechanisms in bacterial cell wall architecture and synthesis, and to pinpoint distinct mechanisms used by a subset of species to grow with a particular morphology.

Acknowledgements

We thank Phillip Aldridge, Newcastle University, for critical reading of the manuscript. This work was supported by an EMBO long-term fellowship (to A.D.), the European Commission within the DIVINOCELL HEALTH-F3–2009-223431 and EUR-INTAFAR LSHM-CT-2004-512138 projects (to T.d.B. and W.V.), the BBSRC (to W.V.), by the Belgian program of Interuniversity Attraction Poles initiated by the Federal Office for Scientific Technical and Cultural Affairs (PAI number P6/19). M.T. is a Research Associate of the National Fond for Scientific Research (FRS-FNRS, Belgium) and supported by the Fonds de la Recherche Fondamentale Collective (contract number 2.4543.12).

References

Aaron, M., Charbon, G., Lam, H., Schwarz, H., Vollmer, W., and Jacobs-Wagner, C. (2007). The tubulin homologue FtsZ contributes to cell elongation by guiding cell wall precursor synthesis in *Caulobacter crescentus*. Mol. Microbiol. *64*, 938–952.

Aarsman, M.E., Piette, A., Fraipont, C., Vinkenvleugel, T.M., Nguyen-Distèche, M., and den Blaauwen, T. (2005). Maturation of the *Escherichia coli* divisome occurs in two steps. Mol. Microbiol. *55*, 1631–1645.

Adams, D.W., and Errington, J. (2009). Bacterial cell division: assembly, maintenance and disassembly of the Z ring. Nat. Rev. Microbiol. *7*, 642–653.

Adams, D.W., Wu, L.J., Czaplewski, L.G., and Errington, J. (2011). Multiple effects of benzamide antibiotics on FtsZ function. Mol. Microbiol. *80*, 68–84.

Adediran, S.A., Kumar, I., Nagarajan, R., Sauvage, E., and Pratt, R.F. (2010). Kinetics of reactions of the *Actinomadura* R39 DD-peptidase with specific substrates. Biochem. *50*, 376–387.

Alexeeva, S., Gadella, T.W. Jr., Verheul, J., Verhoeven, G.S., and den Blaauwen, T. (2010). Direct interactions of early and late assembling division proteins in *Escherichia coli* cells resolved by FRET. Mol. Microbiol. *77*, 384–398.

Alyahya, S.A., Alexander, R., Costa, T., Henriques, A.O., Emonet, T., and Jacobs-Wagner, C. (2009). RodZ, a component of the bacterial core morphogenic apparatus. Proc. Natl. Acad. Sci. U.S.A. *106*, 1239–1244.

Amos, L.A., van den Ent, F., and Löwe, J. (2004). Structural/functional homology between the bacterial and eukaryotic cytoskeletons. Curr. Opin. Cell Biol. *16*, 24–31.

Arbeloa, A., Segal, H., Hugonnet, J.E., Josseaume, N., Dubost, L., Brouard, J.P., Gutmann, L., Mengin-Lecreulx, D., and Arthur, M. (2004). Role of class A penicillin-binding proteins in PBP5-mediated beta-lactam resistance in *Enterococcus faecalis*. J. Bacteriol. *186*, 1221–1228.

Arends, S.J., Kustusch, R.J., and Weiss, D.S. (2009). ATP-binding site lesions in FtsE impair cell division. J. Bacteriol. *191*, 3772–3784.

Arends, S.J., Williams, K., Scott, R.J., Rolong, S., Popham, D.L., and Weiss, D.S. (2010). Discovery and characterization of three new *Escherichia coli* septal ring proteins that contain a SPOR domain: DamX, DedD, and RlpA. J. Bacteriol. *192*, 242–255.

Ausmees, N., Kuhn, J.R., and Jacobs-Wagner, C. (2003). The bacterial cytoskeleton: an intermediate filament-like function in cell shape. Cell *115*, 705–713.

Aussel, L., Barre, F.X., Aroyo, M., Stasiak, A., Stasiak, A.Z., and Sherratt, D. (2002). FtsK Is a DNA motor protein that activates chromosome dimer resolution by switching the catalytic state of the XerC and XerD recombinases. Cell *108*, 195–205.

Awasthi, D., Kumar, K., and Ojima, I. (2011). Therapeutic potential of FtsZ inhibition: a patent perspective. Expert Opin. Ther. Pat. *21*, 657–679.

Bagchi, S., Tomenius, H., Belova, L.M., and Ausmees, N. (2008). Intermediate filament-like proteins in bacteria and a cytoskeletal function in *Streptomyces*. Mol. Microbiol. *70*, 1037–1050.

Banzhaf, M., van den Berg van Saparoea, B., Terrak, M., Fraipont, C., Egan, A., Philippe, J., Zapun, A., Breukink, E., Nguyen-Distèche, M., den Blaauwen, T., *et al.* (2012). Cooperativity of peptidoglycan synthases active in bacterial cell elongation. Mol. Microbiol. *85*, 179–194.

Barreteau, H., Kovac, A., Boniface, A., Sova, M., Gobec, S., and Blanot, D. (2008). Cytoplasmic steps of peptidoglycan biosynthesis. FEMS Microbiol. Rev. 32, 168–207.

Begg, K.J., Spratt, B.G., and Donachie, W.D. (1986). Interaction between membrane proteins PBP3 and RodA is required for normal cell shape and division in *Escherichia coli*. J. Bacteriol. *167*, 1004–1008.

Bendezu, F.O., Hale, C.A., Bernhardt, T.G., and de Boer, P.A. (2009). RodZ (YfgA) is required for proper assembly of the MreB actin cytoskeleton and cell shape in *E. coli*. EMBO J. *28*, 193–204.

Bera, A., Herbert, S., Jakob, A., Vollmer, W., and Götz, F. (2005). Why are pathogenic staphylococci so lysozyme resistant? The peptidoglycan O-acetyltransferase OatA is the major determinant for lysozyme resistance of *Staphylococcus aureus*. Mol. Microbiol. *55*, 778–787.

Bernhardt, T.G., and de Boer, P.A. (2003). The *Escherichia coli* amidase AmiC is a periplasmic septal ring component exported via the twin-arginine transport pathway. Mol. Microbiol. *48*, 1171–1182.

Bernhardt, T.G., and de Boer, P.A. (2004). Screening for synthetic lethal mutants in *Escherichia coli* and identification of EnvC (YibP) as a periplasmic septal ring factor with murein hydrolase activity. Mol. Microbiol. *52*, 1255–1269.

Bernhardt, T.G., and de Boer, P.A. (2005). SlmA, a nucleoid-associated, FtsZ binding protein required for blocking septal ring assembly over chromosomes in *E. coli*. Mol. Cell *18*, 555–564.

Bertsche, U., Breukink, E., Kast, T., and Vollmer, W. (2005). *In vitro* murein peptidoglycan synthesis by dimers of the bifunctional transglycosylase-transpeptidase PBP1B from *Escherichia coli*. J. Biol. Chem. *280*, 38096–38101.

Bertsche, U., Kast, T., Wolf, B., Fraipont, C., Aarsman, M.E., Kannenberg, K., von Rechenberg, M., Nguyen-Distèche, M., den Blaauwen, T., Höltje, J.-V., *et al.* (2006). Interaction between two murein (peptidoglycan) synthases, PBP3 and PBP1B, in *Escherichia coli*. Mol. Microbiol. *61*, 675–690.

Beuria, T.K., Santra, M.K., and Panda, D. (2005). Sanguinarine blocks cytokinesis in bacteria by inhibiting FtsZ assembly and bundling. Biochemistry *44*, 16584–16593.

Beuria, T.K., Singh, P., Surolia, A., and Panda, D. (2009). Promoting assembly and bundling of FtsZ as a strategy to inhibit bacterial cell division: a new approach for developing novel antibacterial drugs. Biochem. J. *423*, 61–69.

Bisicchia, P., Noone, D., Lioliou, E., Howell, A., Quigley, S., Jensen, T., Jarmer, H., and Devine, K.M. (2007). The essential YycFG two-component system controls cell wall metabolism in *Bacillus subtilis*. Mol. Microbiol. 65, 180–200.

Blair, D.E., Schuttelkopf, A.W., MacRae, J.I., and van Aalten, D.M. (2005). Structure and metal-dependent mechanism of peptidoglycan deacetylase, a streptococcal virulence factor. Proc. Natl. Acad. Sci. U.S.A. *102*, 15429–15434.

Boberek, J.M., Stach, J., and Good, L. (2010). Genetic evidence for inhibition of bacterial division protein FtsZ by berberine. PLoS One 5, e13745.

Bonis, M., Ecobichon, C., Guadagnini, S., Prevost, M.C., and Boneca, I.G. (2010). A M23B family metallopeptidase of *Helicobacter pylori* required for cell shape, pole formation and virulence. Mol. Microbiol. *78*, 809–819.

Born, P., Breukink, E., and Vollmer, W. (2006). *In vitro* synthesis of cross-linked murein and its attachment to sacculi by PBP1A from *Escherichia coli*. J. Biol. Chem. *281*, 26985–26993.

Bouhss, A., Trunkfield, A.E., Bugg, T.D., and Mengin-Lecreulx, D. (2008). The biosynthesis of peptidoglycan lipid-linked intermediates. FEMS Microbiol. Rev. *32*, 208–233.

Boulbitch, A., Quinn, B., and Pink, D. (2000). Elasticity of the rod-shaped Gram-negative eubacteria. Phys. Rev. Lett. *85*, 5246–5249.

Bove, J.M., Renaudin, J., Saillard, C., Foissac, X., and Garnier, M. (2003). *Spiroplasma citri*, a plant pathogenic molligute: relationships with its two hosts, the plant and the leafhopper vector. Annu. Rev. Phytopathol. *41*, 483–500.

Bramkamp, M., Emmins, R., Weston, L., Donovan, C., Daniel, R.A., and Errington, J. (2008). A novel component of the division-site selection system of *Bacillus subtilis* and a new mode of action for the division inhibitor MinCD. Mol. Microbiol. *70*, 1556–1569.

Bromley, D.B., and Charon, N.W. (1979). Axial filament involvement in the motility of *Leptospira interrogans.* J. Bacteriol. *137,* 1406–1412.
Brown, T. Jr., Charlier, P., Herman, R., Schofield, C.J., and Sauvage, E. (2010). Structural basis for the interaction of lactivicins with serine beta-lactamases. J. Med. Chem. *53,* 5890–5894.
Buddelmeijer, N., and Beckwith, J. (2004). A complex of the *Escherichia coli* cell division proteins FtsL, FtsB and FtsQ forms independently of its localization to the septal region. Mol. Microbiol. *52,* 1315–1327.
Bui, N.K., Turk, S., Buckenmaier, S., Stevenson-Jones, F., Zeuch, B., Gobec, S., and Vollmer, W. (2011). Development of screening assays and discovery of initial inhibitors of pneumococcal peptidoglycan deacetylase PgdA. Biochem. Pharmacol. *82,* 43–52.
Burman, L.G., and Park, J.T. (1984). Molecular model for elongation of the murein sacculus of *Escherichia coli.* Proc. Natl. Acad. Sci. U.S.A. *81,* 1844–1848.
Cabeen, M.T., Charbon, G., Vollmer, W., Born, P., Ausmees, N., Weibel, D.B., and Jacobs-Wagner, C. (2009). Bacterial cell curvature through mechanical control of cell growth. EMBO J. *28,* 1208–1219.
Cantarel, B.L., Coutinho, P.M., Rancurel, C., Bernard, T., Lombard, V., and Henrissat, B. (2009). The Carbohydrate-Active EnZymes database (CAZy): an expert resource for Glycogenomics. Nucleic Acids Res. *37,* D233–238.
Caparros, M., Torrecuadrada, J.L., and de Pedro, M.A. (1991). Effect of D-amino acids on *Escherichia coli* strains with impaired penicillin-binding proteins. Res. Microbiol. *142,* 345–350.
Caparros, M., Pisabarro, A.G., and de Pedro, M.A. (1992). Effect of D-amino acids on structure and synthesis of peptidoglycan in *Escherichia coli.* J. Bacteriol. *174,* 5549–5559.
Carballido-Lopez, R., Formstone, A., Li, Y., Ehrlich, S.D., Noirot, P., and Errington, J. (2006). Actin homolog MreBH governs cell morphogenesis by localization of the cell wall hydrolase LytE. Dev. Cell *11,* 399–409.
Carson, M.J., Barondess, J., and Beckwith, J. (1991). The FtsQ protein of *Escherichia coli*: membrane topology, abundance, and cell division phenotypes due to overproduction and insertion mutations. J. Bacteriol. *173,* 2187–2195.
Cava, F., Lam, H., de Pedro, M.A., and Waldor, M.K. (2011). Emerging knowledge of regulatory roles of D-amino acids in bacteria. Cell Mol. Life Sci. *68,* 817–831.
Charbon, G., Cabeen, M.T., and Jacobs-Wagner, C. (2009). Bacterial intermediate filaments: *in vivo* assembly, organization, and dynamics of crescentin. Genes Dev. *23,* 1131–1144.
Chen, J.C., and Beckwith, J. (2001). FtsQ, FtsL and FtsI require FtsK, but not FtsN, for co-localization with FtsZ during *Escherichia coli* cell division. Mol. Microbiol. *42,* 395–413.
Chen, J.C., Minev, M., and Beckwith, J. (2002). Analysis of *ftsQ* mutant alleles in *Escherichia coli*: complementation, septal localization, and recruitment of downstream cell division proteins. J. Bacteriol. *184,* 695–705.
Cheng, T.J., Sung, M.T., Liao, H.Y., Chang, Y.F., Chen, C.W., Huang, C.Y., Chou, L.Y., Wu, Y.D., Chen, Y.H., Cheng, Y.S., *et al.* (2008). Domain requirement of moenomycin binding to bifunctional transglycosylases and development of high-throughput discovery of antibiotics. Proc. Natl. Acad. Sci. U.S.A. *105,* 431–436.
Cheng, T.J., Wu, Y.T., Yang, S.T., Lo, K.H., Chen, S.K., Chen, Y.H., Huang, W.I., Yuan, C.H., Guo, C.W., Huang, L.Y., *et al.* (2010). High-throughput identification of antibacterials against methicillin-resistant *Staphylococcus aureus* (MRSA) and the transglycosylase. Bioorg. Med. Chem. *18,* 8512–8529.
Cho, H., McManus, H.R., Dove, S.L., and Bernhardt, T.G. (2011). Nucleoid occlusion factor SlmA is a DNA-activated FtsZ polymerization antagonist. Proc. Natl. Acad. Sci. U.S.A. *108,* 3773–3778.
Clarke, A.J., and Dupont, C. (1992). O-acetylated peptidoglycan: its occurrence, pathobiological significance, and biosynthesis. Can. J. Microbiol. *38,* 85–91.
Collier, J. (2010). A new factor stimulating peptidoglycan hydrolysis to separate daughter cells in *Caulobacter crescentus.* Mol. Microbiol. *77,* 11–14.
Corbin, B.D., Wang, Y., Beuria, T.K., and Margolin, W. (2007). Interaction between cell division proteins FtsE and FtsZ. J. Bacteriol. *189,* 3026–3035.
Costa, T., Priyadarshini, R., and Jacobs-Wagner, C. (2008). Localization of PBP3 in *Caulobacter crescentus* is highly dynamic and largely relies on its functional transpeptidase domain. Mol. Microbiol. *70,* 634–651.
Crisostomo, M.I., Vollmer, W., Kharat, A.S., Inhülsen, S., Gehre, F., Buckenmaier, S., and Tomasz, A. (2006). Attenuation of penicillin resistance in a peptidoglycan O-acetyl transferase mutant of *Streptococcus pneumoniae.* Mol. Microbiol. *61,* 1497–1509.
D'Ulisse, V., Fagioli, M., Ghelardini, P., and Paolozzi, L. (2007). Three functional subdomains of the *Escherichia coli* FtsQ protein are involved in its interaction with the other division proteins. Microbiology *153,* 124–138.

Daniel, R.A., and Errington, J. (2000). Intrinsic instability of the essential cell division protein FtsL of *Bacillus subtilis* and a role for DivIB protein in FtsL turnover. Mol. Microbiol. *36,* 278–289.

Daniel, R.A., and Errington, J. (2003). Control of cell morphogenesis in bacteria: two distinct ways to make a rod-shaped cell. Cell *113,* 767–776.

Davis, K.M., and Weiser, J.N. (2011). Modifications to the peptidoglycan backbone help bacteria to establish infection. Infect. Immun. *79,* 562–570.

de Jonge, B.L., Wientjes, F.B., Jurida, I., Driehuis, F., Wouters, J.T., and Nanninga, N. (1989). Peptidoglycan synthesis during the cell cycle of *Escherichia coli*: composition and mode of insertion. J. Bacteriol. *171,* 5783–5794.

de Pedro, M.A., Quintela, J.C., Höltje, J.-V., and Schwarz, H. (1997). Murein segregation in *Escherichia coli.* J. Bacteriol. *179,* 2823–2834.

DeChavigny, A., Heacock, P.N., and Dowhan, W. (1991). Sequence and inactivation of the pss gene of *Escherichia coli.* Phosphatidylethanolamine may not be essential for cell viability. J. Biol. Chem. *266,* 5323–5332.

Demchick, P., and Koch, A.L. (1996). The permeability of the wall fabric of *Escherichia coli* and *Bacillus subtilis.* J. Bacteriol. *178,* 768–773.

den Blaauwen, T., Aarsman, M., and Nanninga, N. (1990). Interaction of monoclonal antibodies with the enzymatic domains of penicillin-binding protein 1b of *Escherichia coli.* J. Bacteriol. *172,* 63–70.

den Blaauwen, T., Buddelmeijer, N., Aarsman, M.E., Hameete, C.M., and Nanninga, N. (1999). Timing of FtsZ assembly in *Escherichia coli.* J. Bacteriol. *181,* 5167–5175.

den Blaauwen, T., Aarsman, M.E., Vischer, N.O., and Nanninga, N. (2003). Penicillin-binding protein PBP2 of *Escherichia coli* localizes preferentially in the lateral wall and at mid-cell in comparison with the old cell pole. Mol. Microbiol. *47,* 539–547.

den Blaauwen, T., de Pedro, M.A., Nguyen-Distèche, M., and Ayala, J.A. (2008). Morphogenesis of rod-shaped sacculi. FEMS Microbiol. Rev. *32,* 321–344.

Denome, S.A., Elf, P.K., Henderson, T.A., Nelson, D.E., and Young, K.D. (1999). *Escherichia coli* mutants lacking all possible combinations of eight penicillin binding proteins: viability, characteristics, and implications for peptidoglycan synthesis. J. Bacteriol. *181,* 3981–3993.

Derouaux, A., Wolf, B., Fraipont, C., Breukink, E., Nguyen-Distèche, M., and Terrak, M. (2008). The monofunctional glycosyltransferase of *Escherichia coli* localizes to the cell division site and interacts with penicillin-binding protein 3, FtsW, and FtsN. J. Bacteriol. *190,* 1831–1834.

Derouaux, A., Turk, S., Olrichs, N.K., Gobec, S., Breukink, E., Amoroso, A., Offant, J., Bostock, J., Mariner, K., Chopra, I., *et al.* (2011). Small molecule inhibitors of peptidoglycan synthesis targeting the lipid II precursor. Biochem. Pharmacol. *81,* 1098–1105.

Derouaux, A., Sauvage, E., and Terrak, M. (2013). Peptidoglycan glycosyltransferase substrate mimics as templates for the design of new antibacterial drugs. Front. Immunol. *4,* 78.

Di Lallo, G., Fagioli, M., Barionovi, D., Ghelardini, P., and Paolozzi, L. (2003). Use of a two-hybrid assay to study the assembly of a complex multicomponent protein machinery: bacterial septosome differentiation. Microbiology *149,* 3353–3359.

Dmitriev, B.A., Ehlers, S., and Rietschel, E.T. (1999). Layered murein revisited: a fundamentally new concept of bacterial cell wall structure, biogenesis and function. Med. Microbiol. Immunol. *187,* 173–181.

Domadia, P., Swarup, S., Bhunia, A., Sivaraman, J., and Dasgupta, D. (2007). Inhibition of bacterial cell division protein FtsZ by cinnamaldehyde. Biochem. Pharmacol. *74,* 831–840.

Domadia, P.N., Bhunia, A., Sivaraman, J., Swarup, S., and Dasgupta, D. (2008). Berberine targets assembly of *Escherichia coli* cell division protein FtsZ. Biochemistry *47,* 3225–3234.

Dominguez-Escobar, J., Chastanet, A., Crevenna, A.H., Fromion, V., Wedlich-Soldner, R., and Carballido-Lopez, R. (2011). Processive movement of MreB-associated cell wall biosynthetic complexes in bacteria. Science *333,* 225–258.

Dorazi, R., and Dewar, S.J. (2000). Membrane topology of the N-terminus of the *Escherichia coli* FtsK division protein. FEBS Lett. *478,* 13–18.

Dougherty, T.J., Kennedy, K., Kessler, R.E., and Pucci, M.J. (1996). Direct quantitation of the number of individual penicillin-binding proteins per cell in *Escherichia coli.* J. Bacteriol. *178,* 6110–6115.

Dramsi, S., Magnet, S., Davison, S., and Arthur, M. (2008). Covalent attachment of proteins to peptidoglycan. FEMS Microbiol. Rev. *32,* 307–320.

Driessen, A.J., and Nouwen, N. (2008). Protein translocation across the bacterial cytoplasmic membrane. Annu. Rev. Biochem. *77,* 643–667.

Dubarry, N., and Barre, F.X. (2010). Fully efficient chromosome dimer resolution in *Escherichia coli* cells lacking the integral membrane domain of FtsK. EMBO J. *29,* 597–605.

Dumbre, S., Derouaux, A., Lescrinier, E., Piette, A., Joris, B., Terrak, M., and Herdewijn, P. (2012). Synthesis of modified peptidoglycan precursor analogues for the inhibition of glycosyltransferase. J. Am. Chem. Soc. *134,* 9343–9351.

Dupaigne, P., Tonthat, N.K., Espeli, O., Whitfill, T., Boccard, F., and Schumacher, M.A. (2012). Molecular basis for a protein-mediated DNA-bridging mechanism that functions in condensation of the *E. coli* chromosome. Mol. Cell *48,* 560–571.

Durand-Heredia, J.M., Yu, H.H., De Carlo, S., Lesser, C.F., and Janakiraman, A. (2011). Identification and characterization of ZapC, a stabilizer of the FtsZ ring in *Escherichia coli*. J. Bacteriol. *193,* 1405–1413.

Durand-Heredia, J., Rivkin, E., Fan, G., Morales, J., and Janakiraman, A. (2012). Identification of ZapD as a cell division factor that promotes the assembly of FtsZ in *Escherichia coli*. J. Bacteriol. *194,* 3189–3198.

Dye, N.A., Pincus, Z., Theriot, J.A., Shapiro, L., and Gitai, Z. (2005). Two independent spiral structures control cell shape in *Caulobacter*. Proc. Natl. Acad. Sci. U.S.A. *102,* 18608–18613.

Dzhekieva, L., Rocaboy, M., Kerff, F., Charlier, P., Sauvage, E., and Pratt, R.F. (2010). Crystal structure of a complex between the *Actinomadura* R39 DD-peptidase and a peptidoglycan-mimetic boronate inhibitor: interpretation of a transition state analogue in terms of catalytic mechanism. Biochemistry *49,* 6411–6419.

Ebersbach, G., Galli, E., Moller-Jensen, J., Löwe, J., and Gerdes, K. (2008). Novel coiled-coil cell division factor ZapB stimulates Z ring assembly and cell division. Mol. Microbiol. *68,* 720–735.

Espeli, O., Borne, R., Dupaigne, P., Thiel, A., Gigant, E., Mercier, R., and Boccard, F. (2012). A MatP–divisome interaction coordinates chromosome segregation with cell division in *E. coli*. EMBO J. *31,* 3198–3211.

Evrard, C., Fastrez, J., and Declercq, J.P. (1998). Crystal structure of the lysozyme from bacteriophage lambda and its relationship with V and C-type lysozymes. J. Mol. Biol. *276,* 151–164.

Figge, R.M., Divakaruni, A.V., and Gober, J.W. (2004). MreB, the cell shape-determining bacterial actin homologue, co-ordinates cell wall morphogenesis in *Caulobacter crescentus*. Mol. Microbiol. *51,* 1321–1332.

Fiuza, M., Letek, M., Leiba, J., Villadangos, A.F., Vaquera, J., Zanella-Cleon, I., Mateos, L.M., Molle, V., and Gil, J.A. (2010). Phosphorylation of a novel cytoskeletal protein (RsmP) regulates rod-shaped morphology in *Corynebacterium glutamicum*. J. Biol. Chem. *285,* 29387–29397.

Flardh, K. (2010). Cell polarity and the control of apical growth in *Streptomyces*. Curr. Opin. Microbiol. *13,* 758–765.

Fraipont, C., Alexeeva, S., Wolf, B., van der Ploeg, R., Schloesser, M., den Blaauwen, T., and Nguyen-Distèche, M. (2011). The integral membrane FtsW protein and peptidoglycan synthase PBP3 form a subcomplex in *Escherichia coli*. Microbiology *157,* 251–259.

Fuse, S., Tsukamoto, H., Yuan, Y., Wang, T.S., Zhang, Y., Bolla, M., Walker, S., Sliz, P., and Kahne, D. (2010). Functional and structural analysis of a key region of the cell wall inhibitor moenomycin. ACS Chem. Biol. *5,* 701–711.

Gan, L., Chen, S., and Jensen, G.J. (2008). Molecular organization of Gram-negative peptidoglycan. Proc. Natl. Acad. Sci. U.S.A. *105,* 18953–18957.

Garcia del Portillo, F., and de Pedro, M.A. (1990). Differential effect of mutational impairment of penicillin-binding proteins 1A and 1B on *Escherichia coli* strains harboring thermosensitive mutations in the cell division genes *ftsA, ftsQ, ftsZ,* and *pbpB*. J. Bacteriol. *172,* 5863–5870.

Garner, E.C., Bernard, R., Wang, W., Zhuang, X., Rudner, D.Z., and Mitchison, T. (2011). Coupled, circumferential motions of the cell wall synthesis machinery and MreB filaments in *B. subtilis*. Science *333,* 222–225.

Gayda, R.C., Henk, M.C., and Leong, D. (1992). C-shaped cells caused by expression of an *ftsA* mutation in *Escherichia coli*. J. Bacteriol. *174,* 5362–5370.

Geissler, B., Elraheb, D., and Margolin, W. (2003). A gain-of-function mutation in *ftsA* bypasses the requirement for the essential cell division gene *zipA* in *Escherichia coli*. Proc. Natl. Acad. Sci. U.S.A. *100,* 4197–4202.

Gerding, M.A., Ogata, Y., Pecora, N.D., Niki, H., and de Boer, P.A. (2007). The trans-envelope Tol–Pal complex is part of the cell division machinery and required for proper outer-membrane invagination during cell constriction in *E. coli*. Mol. Microbiol. *63,* 1008–1025.

Gerding, M.A., Liu, B., Bendezu, F.O., Hale, C.A., Bernhardt, T.G., and de Boer, P.A. (2009). Self-enhanced accumulation of FtsN at division sites and roles for other proteins with a SPOR domain (DamX, DedD, and RlpA) in *Escherichia coli* cell constriction. J. Bacteriol. *191,* 7383–7401.

Gitai, Z., Dye, N.A., Reisenauer, A., Wachi, M., and Shapiro, L. (2005). MreB actin-mediated segregation of a specific region of a bacterial chromosome. Cell *120*, 329–341.
Glauner, B., and Höltje, J.-V. (1990). Growth pattern of the murein sacculus of *Escherichia coli*. J. Biol. Chem. *265*, 18988–18996.
Glauner, B., Höltje, J.-V., and Schwarz, U. (1988). The composition of the murein of *Escherichia coli*. J. Biol. Chem. *263*, 10088–10095.
Goehring, N.W., and Beckwith, J. (2005). Diverse paths to midcell: assembly of the bacterial cell division machinery. Curr. Biol. *15*, R514–526.
Goehring, N.W., Gonzalez, M.D., and Beckwith, J. (2006). Premature targeting of cell division proteins to midcell reveals hierarchies of protein interactions involved in divisome assembly. Mol. Microbiol. *61*, 33–45.
Goffin, C., and Ghuysen, J.M. (1998). Multimodular penicillin-binding proteins: an enigmatic family of orthologs and paralogs. Microbiol. Mol. Biol. Rev. *62*, 1079–1093.
Goldstein, S.F., Charon, N.W., and Kreiling, J.A. (1994). *Borrelia burgdorferi* swims with a planar waveform similar to that of eukaryotic flagella. Proc. Natl. Acad. Sci. U.S.A. *91*, 3433–3437.
Gonin, M., Quardokus, E.M., O'Donnol, D., Maddock, J., and Brun, Y.V. (2000). Regulation of stalk elongation by phosphate in *Caulobacter crescentus*. J. Bacteriol. *182*, 337–347.
Goodell, E.W. (1985). Recycling of murein by *Escherichia coli*. J. Bacteriol. *163*, 305–310.
Goodell, E.W., and Schwarz, U. (1985). Release of cell wall peptides into culture medium by exponentially growing *Escherichia coli*. J. Bacteriol. *162*, 391–397.
Gueiros-Filho, F.J., and Losick, R. (2002). A widely conserved bacterial cell division protein that promotes assembly of the tubulin-like protein FtsZ. Genes Dev. *16*, 2544–2556.
Hale, C.A., Shiomi, D., Liu, B., Bernhardt, T.G., Margolin, W., Niki, H., and de Boer, P.A. (2011). Identification of *Escherichia coli* ZapC (YcbW) as a component of the division apparatus that binds and bundles FtsZ polymers. J. Bacteriol. *193*, 1393–1404.
Han, S., Zaniewski, R.P., Marr, E.S., Lacey, B.M., Tomaras, A.P., Evdokimov, A., Miller, J.R., and Shanmugasundaram, V. (2010). Structural basis for effectiveness of siderophore-conjugated monocarbams against clinically relevant strains of *Pseudomonas aeruginosa*. Proc. Natl. Acad. Sci. U.S.A. *107*, 22002–22007.
Haney, S.A., Glasfeld, E., Hale, C., Keeney, D., He, Z., and de Boer, P. (2001). Genetic analysis of the *Escherichia coli* FtsZ–ZipA interaction in the yeast two-hybrid system. Characterization of FtsZ residues essential for the interactions with ZipA and with FtsA. J. Biol. Chem. *276*, 11980–11987.
Harris, F., Demel, R., de Kruijff, B., and Phoenix, D.A. (1998). An investigation into the lipid interactions of peptides corresponding to the C-terminal anchoring domains of *Escherichia coli* penicillin-binding proteins 4, 5 and 6. Biochim. Biophys. Acta *1415*, 10–22.
Harz, H., Burgdorf, K., and Höltje, J.-V. (1990). Isolation and separation of the glycan strands from murein of *Escherichia coli* by reversed-phase high-performance liquid chromatography. Anal Biochem. *190*, 120–128.
Haydon, D.J., Stokes, N.R., Ure, R., Galbraith, G., Bennett, J.M., Brown, D.R., Baker, P.J., Barynin, V.V., Rice, D.W., Sedelnikova, S.E., *et al.* (2008). An inhibitor of FtsZ with potent and selective anti-staphylococcal activity. Science *321*, 1673–1675.
Haydon, D.J., Bennett, J.M., Brown, D., Collins, I., Galbraith, G., Lancett, P., Macdonald, R., Stokes, N.R., Chauhan, P.K., Sutariya, J.K., *et al.* (2010). Creating an antibacterial with *in vivo* efficacy: synthesis and characterization of potent inhibitors of the bacterial cell division protein FtsZ with improved pharmaceutical properties. J. Med. Chem. *53*, 3927–3936.
Hayhurst, E.J., Kailas, L., Hobbs, J.K., and Foster, S.J. (2008). Cell wall peptidoglycan architecture in *Bacillus subtilis*. Proc. Natl. Acad. Sci. U.S.A. *105*, 14603–14608.
Hazell, S.L., Lee, A., Brady, L., and Hennessy, W. (1986). *Campylobacter pyloridis* and gastritis: association with intercellular spaces and adaptation to an environment of mucus as important factors in colonization of the gastric epithelium. J. Infect. Dis. *153*, 658–663.
Heaslet, H., Shaw, B., Mistry, A., and Miller, A.A. (2009). Characterization of the active site of *S., aureus* monofunctional glycosyltransferase (Mtg) by site-directed mutation and structural analysis of the protein complexed with moenomycin. J. Struct. Biol. *167*, 129–135.
Heichlinger, A., Ammelburg, M., Kleinschnitz, E.M., Latus, A., Maldener, I., Flardh, K., Wohlleben, W., and Muth, G. (2011). The MreB-like protein Mbl of *Streptomyces coelicolor* A3(2) depends on MreB for proper localization and contributes to spore wall synthesis. J. Bacteriol. *193*, 1533–1542.

Heidrich, C., Templin, M.F., Ursinus, A., Merdanovic, M., Berger, J., Schwarz, H., de Pedro, M.A., and Höltje, J.V. (2001). Involvement of N-acetylmuramyl-L-alanine amidases in cell separation and antibiotic-induced autolysis of *Escherichia coli*. Mol. Microbiol. *41*, 167–178.

Heidrich, C., Ursinus, A., Berger, J., Schwarz, H., and Höltje, J.-V. (2002). Effects of multiple deletions of murein hydrolases on viability, septum cleavage, and sensitivity to large toxic molecules in *Escherichia coli*. J. Bacteriol. *184*, 6093–6099.

Hempel, A.M., Wang, S.B., Letek, M., Gil, J.A., and Flardh, K. (2008). Assemblies of DivIVA mark sites for hyphal branching and can establish new zones of cell wall growth in *Streptomyces coelicolor*. J. Bacteriol. *190*, 7579–7583.

Henriques, A.O., Glaser, P., Piggot, P.J., and Moran, C.P. Jr. (1998). Control of cell shape and elongation by the rodA gene in *Bacillus subtilis*. Mol. Microbiol. *28*, 235–247.

Hoiczyk, E., and Hansel, A. (2000). Cyanobacterial cell walls: news from an unusual prokaryotic envelope. J. Bacteriol. *182*, 1191–1199.

Höltje, J.-V. (1993). 'Three for one'- A simple growth mechanism that guarantees a precise copy of the thin, rod-shaped murein sacculus of *Escherichia coli*. In Bacterial Growth and Lysis – Metabolism and Structure of the Bacterial Sacculus, de Pedro, M.A., Höltje, J.V., and Löffelhardt, W., eds. (Plenum Press, New York), pp. 419–426.

Höltje, J.-V. (1996). A hypothetical holoenzyme involved in the replication of the murein sacculus of *Escherichia coli*. Microbiology *142*, 1911–1918.

Höltje, J.-V. (1998). Growth of the stress-bearing and shape-maintaining murein sacculus of *Escherichia coli*. Microbiol. Mol. Biol. Rev. *62*, 181–203.

Höltje, J.-V., Mirelman, D., Sharon, N., and Schwarz, U. (1975). Novel type of murein transglycosylase in *Escherichia coli*. J. Bacteriol. *124*, 1067–1076.

Hu, B., Yang, G., Zhao, W., Zhang, Y., and Zhao, J. (2007). MreB is important for cell shape but not for chromosome segregation of the filamentous cyanobacterium *Anabaena* sp. PCC 7120. Mol. Microbiol. *63*, 1640–1652.

Huang, Q., Kirikae, F., Kirikae, T., Pepe, A., Amin, A., Respicio, L., Slayden, R.A., Tonge, P.J., and Ojima, I. (2006). Targeting FtsZ for antituberculosis drug discovery: noncytotoxic taxanes as novel antituberculosis agents. J. Med. Chem. *49*, 463–466.

Huang, K.C., Mukhopadhyay, R., Wen, B., Gitai, Z., and Wingreen, N.S. (2008). Cell shape and cell-wall organization in Gram-negative bacteria. Proc. Natl. Acad. Sci. U.S.A. *105*, 19282–19287.

Inglis, S.R., Zervosen, A., Woon, E.C., Gerards, T., Teller, N., Fischer, D.S., Luxen, A., and Schofield, C.J. (2009). Synthesis and evaluation of 3-(dihydroxyboryl)benzoic acids as D,D-carboxypeptidase R39 inhibitors. J. Med. Chem. *52*, 6097–6106.

Ito, H., Ura, A., Oyamada, Y., Tanitame, A., Yoshida, H., Yamada, S., Wachi, M., and Yamagishi, J. (2006). A 4-aminofurazan derivative-A189-inhibits assembly of bacterial cell division protein FtsZ *in vitro* and *in vivo*. Microbiol. Immunol. *50*, 759–764.

Iwai, N., Nagai, K., and Wachi, M. (2002). Novel S-bezylsiothiourea compound that induces spherical cells in *Escherichia coli* probably by acting on a rod-shaped-determining protein(s) other than penicillin-binding protein 2. Biosci. Biotechnol. Biochem. *66*, 2658–2662.

Jacobs, C., Huang, L.J., Bartowsky, E., Normark, S., and Park, J.T. (1994). Bacterial cell wall recycling provides cytosolic muropeptides as effectors for beta-lactamase induction. EMBO J. *13*, 4684–4694.

Jacobs, C., Frère, J.M., and Normark, S. (1997). Cytosolic intermediates for cell wall biosynthesis and degradation control inducible beta-lactam resistance in Gram-negative bacteria. Cell *88*, 823–832.

Jaiswal, R., Beuria, T.K., Mohan, R., Mahajan, S.K., and Panda, D. (2007). Totarol inhibits bacterial cytokinesis by perturbing the assembly dynamics of FtsZ. Biochemistry *46*, 4211–4220.

Jolles, P., ed. (1996). Lysozymes: model enzymes in BioChem. Biol. (Basel, Birkhäuser).

Jones, L.J., Carballido-Lopez, R., and Errington, J. (2001). Control of cell shape in bacteria: helical, actin-like filaments in *Bacillus subtilis*. Cell *104*, 913–922.

Kanoh, K., Adachi, K., Matsuda, S., Shizuri, Y., Yasumoto, K., Kusumi, T., Okumura, K., and Kirikae, T. (2008). New sulfoalkylresorcinol from marine-derived fungus, Zygosporium sp. KNC52. J. Antibiot. (Tokyo) *61*, 192–194.

Karczmarek, A., Martinez-Arteaga, R., Alexeeva, S., Hansen, F.G., Vicente, M., Nanninga, N., and den Blaauwen, T. (2007). DNA and origin region segregation are not affected by the transition from rod to sphere after inhibition of *Escherichia coli* MreB by A22. Mol. Microbiol. *65*, 51–63.

Karimova, G., Dautin, N., and Ladant, D. (2005). Interaction network among *Escherichia coli* membrane proteins involved in cell division as revealed by bacterial two-hybrid analysis. J. Bacteriol. *187*, 2233–2243.

Karimova, G., Robichon, C., and Ladant, D. (2009). Characterization of YmgF, a 72-residue inner membrane protein that associates with the *Escherichia coli* cell division machinery. J. Bacteriol. *191*, 333–346.

Kawai, Y., Asai, K., and Errington, J. (2009). Partial functional redundancy of MreB isoforms, MreB, Mbl and MreBH, in cell morphogenesis of *Bacillus subtilis*. Mol. Microbiol. *73*, 719–731.

Kern, T., Giffard, M., Hediger, S., Amoroso, A., Giustini, C., Bui, N.K., Joris, B., Bougault, C., Vollmer, W., and Simorre, J.P. (2010). Dynamics characterization of fully hydrated bacterial cell walls by solid-state NMR: evidence for cooperative binding of metal ions. J. Am. Chem. Soc. *132*, 10911–10919.

Kishida, H., Unzai, S., Roper, D.I., Lloyd, A., Park, S.Y., and Tame, J.R. (2006). Crystal structure of penicillin binding protein 4 (dacB) from *Escherichia coli*, both in the native form and covalently linked to various antibiotics. Biochemistry *45*, 783–792.

Koch, A.L. (2000). Simulation of the conformation of the murein fabric: the oligoglycan, penta-muropeptide, and cross-linked nona-muropeptide. Arch. Microbiol. *174*, 429–439.

Koch, A.L., and Woeste, S. (1992). Elasticity of the sacculus of *Escherichia coli*. J. Bacteriol. *174*, 4811–4819.

Koppelman, C.M., Den Blaauwen, T., Duursma, M.C., Heeren, R.M., and Nanninga, N. (2001). *Escherichia coli* minicell membranes are enriched in cardiolipin. J. Bacteriol. *183*, 6144–6147.

Kruse, T., Bork-Jensen, J., and Gerdes, K. (2005). The morphogenetic MreBCD proteins of *Escherichia coli* form an essential membrane–bound complex. Mol. Microbiol. *55*, 78–89.

Kühn, J., Briegel, A., Morschel, E., Kahnt, J., Leser, K., Wick, S., Jensen, G.J., and Thanbichler, M. (2010). Bactofilins, a ubiquitous class of cytoskeletal proteins mediating polar localization of a cell wall synthase in *Caulobacter crescentus*. EMBO J. *29*, 327–339.

Kumar, K., Awasthi, D., Lee, S.Y., Zanardi, I., Ruzsicska, B., Knudson, S., Tonge, P.J., Slayden, R.A., and Ojima, I. (2011). Novel Trisubstituted Benzimidazoles, Targeting Mtb FtsZ, as a New Class of Antitubercular Agents. J. Med. Chem. *54*, 374–381.

Laaberki, M.H., Pfeffer, J., Clarke, A.J., and Dworkin, J. (2011). O-Acetylation of peptidoglycan is required for proper cell separation and S-layer anchoring in *Bacillus anthracis*. J. Biol. Chem. *286*, 5278–5288.

Labischinski, H., Barnikel, G., and Naumann, D. (1983). The state of order of bacterial peptidoglycan. In The target of penicillin: the murein sacculus of bacterial cell walls architecture and growth, Hakenbeck, R., Höltje, J.-V., and Labischinski, H., eds. (de Gruyter, Berlin/New York), pp. 49–54.

Labischinski, H., Goodell, E.W., Goodell, A., and Hochberg, M.L. (1991). Direct proof of a 'more-than-single-layered' peptidoglycan architecture of *Escherichia coli* W7: a neutron small-angle scattering study. J. Bacteriol. *173*, 751–756.

Lam, H., Oh, D.C., Cava, F., Takacs, C.N., Clardy, J., de Pedro, M.A., and Waldor, M.K. (2009). D-amino acids govern stationary phase cell wall remodeling in bacteria. Science *325*, 1552–1555.

Läppchen, T., Hartog, A.F., Pinas, V.A., Koomen, G.J., and den Blaauwen, T. (2005). GTP analogue inhibits polymerization and GTPase activity of the bacterial protein FtsZ without affecting its eukaryotic homologue tubulin. Biochemistry *44*, 7879–7884.

Läppchen, T., Pinas, V.A., Hartog, A.F., Koomen, G.J., Schaffner-Barbero, C., Andreu, J.M., Trambaiolo, D., Lowe, J., Juhem, A., Popov, A.V., *et al.* (2008). Probing FtsZ and tubulin with C8-substituted GTP analoques reveals differences in their nucleotide binding sites. Chem. Biol. *15*, 189–199.

Lenarcic, R., Halbedel, S., Visser, L., Shaw, M., Wu, L.J., Errington, J., Marenduzzo, D., and Hamoen, L.W. (2009). Localisation of DivIVA by targeting to negatively curved membranes. EMBO J. *28*, 2272–2282.

Letek, M., Fiuza, M., Ordonez, E., Villadangos, A.F., Flardh, K., Mateos, L.M., and Gil, J.A. (2009). DivIVA uses an N-terminal conserved region and two coiled-coil domains to localize and sustain the polar growth in *Corynebacterium glutamicum*. FEMS Microbiol. Lett. *297*, 110–116.

Lim, D., and Strynadka, N.C. (2002). Structural basis for the beta lactam resistance of PBP2a from methicillin-resistant *Staphylococcus aureus*. Nat. Struct. Biol. *9*, 870–876.

Lleo, M.M., Canepari, P., and Satta, G. (1990). Bacterial cell shape regulation: testing of additional predictions unique to the two-competing-sites model for peptidoglycan assembly and isolation of conditional rod-shaped mutants from some wild-type cocci. J. Bacteriol. *172*, 3758–3771.

Lovering, A.L., and Strynadka, N.C. (2007). High-resolution structure of the major periplasmic domain from the cell shape-determining filament MreC. J. Mol. Biol. *372*, 1034–1044.

Lovering, A.L., de Castro, L.H., Lim, D., and Strynadka, N.C. (2007). Structural insight into the transglycosylation step of bacterial cell-wall biosynthesis. Science *315*, 1402–1405.

Lovering, A.L., De Castro, L., and Strynadka, N.C. (2008a). Identification of dynamic structural motifs involved in peptidoglycan glycosyltransfer. J. Mol. Biol. *383*, 167–177.

Lovering, A.L., Gretes, M., and Strynadka, N.C. (2008b). Structural details of the glycosyltransferase step of peptidoglycan assembly. Curr. Opin. Struct. Biol. *18*, 534–543.

Low, H.H., Moncrieffe, M.C., and Löwe, J. (2004). The crystal structure of ZapA and its modulation of FtsZ polymerisation. J. Mol. Biol. *341*, 839–852.

Löwe, J., and Amos, L.A. (1998). Crystal structure of the bacterial cell-division protein FtsZ. Nature *391*, 203–206.

Lupoli, T.J., Taniguchi, T., Wang, T.S., Perlstein, D.L., Walker, S., and Kahne, D.E. (2009). Studying a cell division amidase using defined peptidoglycan substrates. J. Am. Chem. Soc. *131*, 18230–18231.

Lutkenhaus, J. (2008). Min oscillation in bacteria. Adv. Exp. Med. Biol. *641*, 49–61.

Macheboeuf, P., Contreras-Martel, C., Job, V., Dideberg, O., and Dessen, A. (2006). Penicillin binding proteins: key players in bacterial cell cycle and drug resistance processes. FEMS Microbiol. Rev. *30*, 673–691.

Macheboeuf, P., Fischer, D.S., Brown, T. Jr., Zervosen, A., Luxen, A., Joris, B., Dessen, A., and Schofield, C.J. (2007). Structural and mechanistic basis of penicillin-binding protein inhibition by lactivicins. Nat. Chem. Biol. 3, 565–569.

Magnet, S., Bellais, S., Dubost, L., Fourgeaud, M., Mainardi, J.L., Petit-Frère, S., Marie, A., Mengin-Lecreulx, D., Arthur, M., and Gutmann, L. (2007). Identification of the L,D-transpeptidases responsible for attachment of the Braun lipoprotein to *Escherichia coli* peptidoglycan. J. Bacteriol., 3927–3931.

Mahapatra, S., Crick, D.C., McNeil, M.R., and Brennan, P.J. (2008). Unique structural features of the peptidoglycan of *Mycobacterium leprae*. J. Bacteriol. *190*, 655–661.

Mainardi, J.L., Legrand, R., Arthur, M., Schoot, B., van Heijenoort, J., and Gutmann, L. (2000). Novel mechanism of beta-lactam resistance due to bypass of DD-transpeptidation in *Enterococcus faecium*. J. Biol. Chem. *275*, 16490–16496.

Mainardi, J.L., Fourgeaud, M., Hugonnet, J.E., Dubost, L., Brouard, J.P., Ouazzani, J., Rice, L.B., Gutmann, L., and Arthur, M. (2005). A novel peptidoglycan cross-linking enzyme for a beta-lactam-resistant transpeptidation pathway. J. Biol. Chem. *280*, 38146–38152.

Mainardi, J.L., Hugonnet, J.E., Rusconi, F., Fourgeaud, M., Dubost, L., Moumi, A.N., Delfosse, V., Mayer, C., Gutmann, L., Rice, L.B., *et al.* (2007). Unexpected inhibition of peptidoglycan LD-transpeptidase from *Enterococcus faecium* by the beta-lactam imipenem. J. Biol. Chem. *282*, 30414–30422.

Mainardi, J.L., Villet, R., Bugg, T.D., Mayer, C., and Arthur, M. (2008). Evolution of peptidoglycan biosynthesis under the selective pressure of antibiotics in Gram-positive bacteria. FEMS Microbiol. Rev. *32*, 386–408.

Margalit, D.N., Romberg, L., Mets, R.B., Hebert, A.M., Mitchison, T.J., Kirschner, M.W., and RayChaudhuri, D. (2004). Targeting cell division: small-molecule inhibitors of FtsZ GTPase perturb cytokinetic ring assembly and induce bacterial lethality. Proc. Natl. Acad. Sci. U.S.A. *101*, 11821–11826.

Margolin, W. (2009). Sculpting the bacterial cell. Curr. Biol. *19*, R812–822.

Massey, T.H., Mercogliano, C.P., Yates, J., Sherratt, D.J., and Löwe, J. (2006). Double-stranded DNA translocation: structure and mechanism of hexameric FtsK. Mol. Cell *23*, 457–469.

Masson, S., Kern, T., Le Gouellec, A., Giustini, C., Simorre, J.P., Callow, P., Vernet, T., Gabel, F., and Zapun, A. (2009). Central domain of DivIB caps the C-terminal regions of the FtsL/DivIC coiled-coil rod. J. Biol. Chem. *284*, 27687–27700.

Matias, V.R., and Beveridge, T.J. (2005). Cryo-electron microscopy reveals native polymeric cell wall structure in *Bacillus subtilis* 168 and the existence of a periplasmic space. Mol. Microbiol. *56*, 240–251.

Matias, V.R., and Beveridge, T.J. (2006). Native cell wall organization shown by cryo-electron microscopy confirms the existence of a periplasmic space in *Staphylococcus aureus*. J. Bacteriol. *188*, 1011–1021.

Matias, V.R., and Beveridge, T.J. (2007). Cryo-electron microscopy of cell division in *Staphylococcus aureus* reveals a mid-zone between nascent cross walls. Mol. Microbiol. *64*, 195–206.

Matias, V.R., and Beveridge, T.J. (2008). Lipoteichoic acid is a major component of the *Bacillus subtilis* periplasm. J. Bacteriol. *190*, 7414–7418.

Matsuzawa, H., Asoh, S., Kunai, K., Muraiso, K., Takasuga, A., and Ohta, T. (1989). Nucleotide sequence of the *rodA* gene, responsible for the rod shape of *Escherichia coli*: *rodA* and the *pbpA* gene, encoding penicillin-binding protein 2, constitute the *rodA* operon. J. Bacteriol. *171*, 558–560.

Mattei, P.J., Neves, D., and Dessen, A. (2010). Bridging cell wall biosynthesis and bacterial morphogenesis. Curr. Opin. Struct. Biol. *20*, 749–755.

McPherson, D.C., and Popham, D.L. (2003). Peptidoglycan synthesis in the absence of class A penicillin-binding proteins in *Bacillus subtilis*. J. Bacteriol. *185*, 1423–1431.

Meberg, B.M., Paulson, A.L., Priyadarshini, R., and Young, K.D. (2004). Endopeptidase penicillin-binding proteins 4 and 7 play auxiliary roles in determining uniform morphology of *Escherichia coli*. J. Bacteriol. *186*, 8326–8336.

Mercer, K.L., and Weiss, D.S. (2002). The *Escherichia coli* cell division protein FtsW is required to recruit its cognate transpeptidase, FtsI (PBP3), to the division site. J. Bacteriol. *184*, 904–912.

Mileykovskaya, E., Sun, Q., Margolin, W., and Dowhan, W. (1998). Localization and function of early cell division proteins in filamentous *Escherichia coli* cells lacking phosphatidylethanolamine. J. Bacteriol. *180*, 4252–4257.

Mileykovskaya, E., Ryan, A.C., Mo, X., Lin, C.C., Khalaf, K.I., Dowhan, W., and Garrett, T.A. (2009). Phosphatidic acid and N-acylphosphatidylethanolamine form membrane domains in *Escherichia coli* mutant lacking cardiolipin and phosphatidylglycerol. J. Biol. Chem. *284*, 2990–3000.

Mistry, B.V., Del Sol, R., Wright, C., Findlay, K., and Dyson, P. (2008). FtsW is a dispensable cell division protein required for Z-ring stabilization during sporulation septation in *Streptomyces coelicolor*. J. Bacteriol. *190*, 5555–5566.

Mohammadi, T., Karczmarek, A., Crouvoisier, M., Bouhss, A., Mengin-Lecreulx, D., and den Blaauwen, T. (2007). The essential peptidoglycan glycosyltransferase MurG forms a complex with proteins involved in lateral envelope growth as well as with proteins involved in cell division in *Escherichia coli*. Mol. Microbiol. *65*, 1106–1121.

Mohammadi, T., Ploeger, G.E., Verheul, J., Comvalius, A.D., Martos, A., Alfonso, C., van Marle, J., Rivas, G., and den Blaauwen, T. (2009). The GTPase activity of *Escherichia coli* FtsZ determines the magnitude of the FtsZ polymer bundling by ZapA *in vitro*. Biochemistry *48*, 11056–11066.

Mohammadi, T., van Dam, V., Sijbrandi, R., Vernet, T., Zapun, A., Bouhss, A., Diepeveen-de Bruin, M., Nguyen-Distèche, M., de Kruijff, B., and Breukink, E. (2011). Identification of FtsW as a transporter of lipid-linked cell wall precursors across the membrane. EMBO J. *30*, 1425–1432.

Möll, A., and Thanbichler, M. (2009). FtsN-like proteins are conserved components of the cell division machinery in proteobacteria. Mol. Microbiol. *72*, 1037–1053.

Möll, A., Schlimpert, S., Briegel, A., Jensen, G.J., and Thanbichler, M. (2010). DipM, a new factor required for peptidoglycan remodelling during cell division in *Caulobacter crescentus*. Mol. Microbiol. *77*, 90–107.

Monahan, L.G., Robinson, A., and Harry, E.J. (2009). Lateral FtsZ association and the assembly of the cytokinetic Z ring in bacteria. Mol. Microbiol. *74*, 1004–1017.

Morlot, C., Zapun, A., Dideberg, O., and Vernet, T. (2003). Growth and division of *Streptococcus pneumoniae*: localization of the high molecular weight penicillin-binding proteins during the cell cycle. Mol. Microbiol. *50*, 845–855.

Mosyak, L., Zhang, Y., Glasfeld, E., Haney, S., Stahl, M., Seehra, J., and Somers, W.S. (2000). The bacterial cell-division protein ZipA and its interaction with an FtsZ fragment revealed by X-ray crystallography. EMBO J. *19*, 3179–3191.

Müller, P., Ewers, C., Bertsche, U., Anstett, M., Kallis, T., Breukink, E., Fraipont, C., Terrak, M., Nguyen-Distèche, M., and Vollmer, W. (2007). The essential cell division protein FtsN interacts with the murein (peptidoglycan) synthase PBP1B in *Escherichia coli*. J. Biol. Chem. *282*, 36394–36402.

Nelson, D.E., and Young, K.D. (2000). Penicillin binding protein 5 affects cell diameter, contour, and morphology of *Escherichia coli*. J. Bacteriol. *182*, 1714–1721.

Nelson, D.E., and Young, K.D. (2001). Contributions of PBP 5 and DD-carboxypeptidase penicillin binding proteins to maintenance of cell shape in *Escherichia coli*. J. Bacteriol. *183*, 3055–3064.

Nicholas, R.A., Krings, S., Tomberg, J., Nicola, G., and Davies, C. (2003). Crystal structure of wild-type penicillin-binding protein 5 from *Escherichia coli*: implications for deacylation of the acyl–enzyme complex. J. Biol. Chem. *278*, 52826–52833.

Nicola, G., Peddi, S., Stefanova, M., Nicholas, R.A., Gutheil, W.G., and Davies, C. (2005). Crystal structure of *Escherichia coli* penicillin-binding protein 5 bound to a tripeptide boronic acid inhibitor: a role for Ser-110 in deacylation. Biochemistry *44*, 8207–8217.

Noirclerc-Savoye, M., Le Gouellec, A., Morlot, C., Dideberg, O., Vernet, T., and Zapun, A. (2005). *In vitro* reconstitution of a trimeric complex of DivIB, DivIC and FtsL, and their transient co-localization at the division site in *Streptococcus pneumoniae*. Mol. Microbiol. *55*, 413–424.

Normark, S., Boman, H.G., and Matsson, E. (1969). Mutant of *Escherichia coli* with anomalous cell division and ability to decrease episomally and chromosomally mediated resistance to ampicillin and several other antibiotics. J. Bacteriol. *97*, 1334–1342.

Oliva, M.A., Halbedel, S., Freund, S.M., Dutow, P., Leonard, T.A., Veprintsev, D.B., Hamoen, L.W., and Löwe, J. (2010). Features critical for membrane binding revealed by DivIVA crystal Structure EMBO J. *29*, 1988–2001.

Paradis-Bleau, C., Beaumont, M., Sanschagrin, F., Voyer, N., and Levesque, R.C. (2007). Parallel solid synthesis of inhibitors of the essential cell division FtsZ enzyme as a new potential class of antibacterials. Bioorg. Med. Chem. *15*, 1330–1340.

Paradis-Bleau, C., Markovski, M., Uehara, T., Lupoli, T.J., Walker, S., Kahne, D.E., and Bernhardt, T.G. (2010). Lipoprotein cofactors located in the outer membrane activate bacterial cell wall polymerases. Cell *143*, 1110–1120.

Pares, S., Mouz, N., Petillot, Y., Hakenbeck, R., and Dideberg, O. (1996). X-ray structure of *Streptococcus pneumoniae* PBP2x, a primary penicillin target enzyme. Nat. Struct. Biol. *3*, 284–289.

Patrick, J.E., and Kearns, D.B. (2008). MinJ (YvjD) is a topological determinant of cell division in *Bacillus subtilis*. Mol. Microbiol. *70*, 1166–1179.

Peltier, J., Courtin, P., El Meouche, I., Lemee, L., Chapot-Chartier, M.P., and Pons, J.L. (2011). *Clostridium difficile* has an original peptidoglycan structure with high level of N-acetylglucosamine deacetylation and mainly 3–3 cross-links. J. Biol. Chem. *286*, 29053-29. 62.

Perez-Nunez, D., Briandet, R., David, B., Gautier, C., Renault, P., Hallet, B., Hols, P., Carballido-Lopez, R., and Guedon, E. (2011). A new morphogenesis pathway in bacteria: unbalanced activity of cell wall synthesis machineries leads to coccus-to-rod transition and filamentation in ovococci. Mol. Microbiol. *79*, 759–771.

Philippe, N., Pelosi, L., Lenski, R.E., and Schneider, D. (2009). Evolution of penicillin-binding protein 2 concentration and cell shape during a long-term experiment with *Escherichia coli*. J. Bacteriol. *191*, 909–921.

Pichoff, S., and Lutkenhaus, J. (2005). Tethering the Z ring to the membrane through a conserved membrane targeting sequence in FtsA. Mol. Microbiol. *55*, 1722–1734.

Piette, A., Fraipont, C., den Blaauwen, T., Aarsman, M.E., Pastoret, S., and Nguyen-Distèche, M. (2004). Structural determinants required to target penicillin-binding protein 3 to the septum of *Escherichia coli*. J. Bacteriol. *186*, 6110–6117.

Pinho, M.G., and Errington, J. (2003). Dispersed mode of *Staphylococcus aureus* cell wall synthesis in the absence of the division machinery. Mol. Microbiol. *50*, 871–881.

Plaza, A., Keffer, J.L., Bifulco, G., Lloyd, J.R., and Bewley, C.A. (2010). Chrysophaentins A-H, antibacterial bisdiarylbutene macrocycles that inhibit the bacterial cell division protein FtsZ. J. Am. Chem. Soc. *132*, 9069–9077.

van der Ploeg, R., Verheul, J., Vischer, N.O., Alexeeva, S., Hoogendoorn, E., Postma, M., Banzhaf, M., Vollmer, W., and den Blaauwen, T. (2013). Colocalization and interaction between elongasome and divisome during a preparative cell division phase in *Escherichia coli*. Mol. Microbiol. *87*, 1074–1087.

Poggio, S., Takacs, C.N., Vollmer, W., and Jacobs-Wagner, C. (2010). A protein critical for cell constriction in the Gram-negative bacterium *Caulobacter crescentus* localizes at the division site through its peptidoglycan-binding LysM domains. Mol. Microbiol. *77*, 74–89.

Potluri, L., Karczmarek, A., Verheul, J., Piette, A., Wilkin, J.M., Werth, N., Banzhaf, M., Vollmer, W., Young, K.D., Nguyen-Distèche, M., *et al.* (2010). Septal and lateral wall localization of PBP5, the major D,D-carboxypeptidase of *Escherichia coli*, requires substrate recognition and membrane attachment. Mol. Microbiol. *77*, 300–323.

Powell, A.J., Tomberg, J., Deacon, A.M., Nicholas, R.A., and Davies, C. (2009). Crystal structures of penicillin-binding protein 2 from penicillin-susceptible and -resistant strains of *Neisseria gonorrhoeae* reveal an unexpectedly subtle mechanism for antibiotic resistance. J. Biol. Chem. *284*, 1202–1212.

Priyadarshini, R., Popham, D.L., and Young, K.D. (2006). Daughter cell separation by penicillin-binding proteins and peptidoglycan amidases in *Escherichia coli*. J. Bacteriol. *188*, 5345–5355.

Quintela, J.C., Caparros, M., and de Pedro, M.A. (1995). Variability of peptidoglycan structural parameters in Gram-negative bacteria. FEMS Microbiol. Lett. *125*, 95–100.

Rai, D., Singh, J.K., Roy, N., and Panda, D. (2008). Curcumin inhibits FtsZ assembly: an attractive mechanism for its antibacterial activity. Biochem. J. *410*, 147–155.

Ramamurthi, K.S., and Losick, R. (2009). Negative membrane curvature as a cue for subcellular localization of a bacterial protein. Proc. Natl. Acad. Sci. U.S.A. *106*, 13541–13545.

Raskin, D.M., and de Boer, P.A. (1999). Rapid pole-to-pole oscillation of a protein required for directing division to the middle of *Escherichia coli*. Proc. Natl. Acad. Sci. U.S.A. *96*, 4971–4976.

Raymond, J.B., Mahapatra, S., Crick, D.C., and Pavelka, M.S. Jr. (2005). Identification of the *namH* gene, encoding the hydroxylase responsible for the N-glycolylation of the mycobacterial peptidoglycan. J. Biol. Chem. *280*, 326–333.

von Rechenberg, M., Ursinus, A., and Höltje, J.-V. (1996). Affinity chromatography as a means to study multienzyme complexes involved in murein synthesis. Microb. Drug Resist. *2*, 155–157.

Reddy, M. (2007). Role of FtsEX in cell division of *Escherichia coli*: viability of *ftsEX* mutants is dependent on functional SufI or high osmotic strength. J. Bacteriol. *189*, 98–108.

Reed, P., Veiga, H., Jorge, A.M., Terrak, M., and Pinho, M.G. (2011). Monofunctional transglycosylases are not essential for *Staphylococcus aureus* cell wall synthesis. J. Bacteriol. *193*, 2549–2556.
Reynolds, R.C., Srivastava, S., Ross, L.J., Suling, W.J., and White, E.L. (2004). A new 2-carbamoyl pteridine that inhibits mycobacterial FtsZ. Bioorg. Med. Chem. Lett. *14*, 3161–3164.
Rico, A.I., Garcia-Ovalle, M., Palacios, P., Casanova, M., and Vicente, M. (2010). Role of *Escherichia coli* FtsN protein in the assembly and stability of the cell division ring. Mol. Microbiol. *76*, 760–771.
Romeis, T., Vollmer, W., and Höltje, J.-V. (1993). Characterization of three different lytic transglycosylases in *Escherichia coli*. FEMS Microbiol. Lett. *111*, 141–146.
Romeis, T., and Höltje, J.-V. (1994a). Penicillin-binding protein 7/8 of *Escherichia coli* is a DD-endopeptidase. Eur. J. Biochem. *224*, 597–604.
Romeis, T., and Höltje, J.-V. (1994b). Specific interaction of penicillin-binding proteins 3 and 7/8 with soluble lytic transglycosylase in *Escherichia coli*. J. Biol. Chem. *269*, 21603–21607.
Ruby, J.D., Li, H., Kuramitsu, H., Norris, S.J., Goldstein, S.F., Buttle, K.F., and Charon, N.W. (1997). Relationship of *Treponema denticola* periplasmic flagella to irregular cell morphology. J. Bacteriol. *179*, 1628–1635.
Samaluru, H., SaiSree, L., and Reddy, M. (2007). Role of SufI (FtsP) in cell division of *Escherichia coli*: evidence for its involvement in stabilizing the assembly of the divisome. J. Bacteriol. *189*, 8044–8052.
Sauvage, E., Kerff, F., Fonze, E., Herman, R., Schoot, B., Marquette, J.P., Taburet, Y., Prevost, D., Dumas, J., Leonard, G., *et al.* (2002). The 2.4-A crystal structure of the penicillin-resistant penicillin-binding protein PBP5fm from *Enterococcus faecium* in complex with benzylpenicillin. Cell Mol. Life Sci. *59*, 1223–1232.
Sauvage, E., Kerff, F., Terrak, M., Ayala, J.A., and Charlier, P. (2008). The penicillin-binding proteins: structure and role in peptidoglycan biosynthesis. FEMS Microbiol. Rev. *32*, 234–258.
Scheffers, D.J., de Wit, J.G., den Blaauwen, T., and Driessen, A.J. (2002). GTP hydrolysis of cell division protein FtsZ: evidence that the active site is formed by the association of monomers. Biochemistry *41*, 521–529.
Scheurwater, E.M., and Clarke, A.J. (2008). The C-terminal domain of *Escherichia coli* YfhD functions as a lytic transglycosylase. J. Biol. Chem. *283*, 8363–8373.
Scheurwater, E., Reid, C.W., and Clarke, A.J. (2008). Lytic transglycosylases: bacterial space-making autolysins. Int. J. Biochem. Cell Biol. *40*, 586–591.
Schiffer, G., and Höltje, J.-V. (1999). Cloning and characterization of PBP 1C, a third member of the multimodular class A penicillin-binding proteins of *Escherichia coli*. J. Biol. Chem. *274*, 32031–32039.
Schlag, M., Biswas, R., Krismer, B., Kohler, T., Zoll, S., Yu, W., Schwarz, H., Peschel, A., and Götz, F. (2010). Role of staphylococcal wall teichoic acid in targeting the major autolysin Atl. Mol. Microbiol. *75*, 864–873.
Schleifer, K.H., and Kandler, O. (1972). Peptidoglycan types of bacterial cell walls and their taxonomic implications. Bacteriol. Rev. *36*, 407–477.
Schmidt, K.L., Peterson, N.D., Kustusch, R.J., Wissel, M.C., Graham, B., Phillips, G.J., and Weiss, D.S. (2004). A predicted ABC transporter, FtsEX, is needed for cell division in *Escherichia coli*. J. Bacteriol. *186*, 785–793.
Schofield, W.B., Lim, H.C., and Jacobs-Wagner, C. (2010). Cell cycle coordination and regulation of bacterial chromosome segregation dynamics by polarly localized proteins. EMBO J. *29*, 3068–3081.
Shih, Y.L., Le, T., and Rothfield, L. (2003). Division site selection in *Escherichia coli* involves dynamic redistribution of Min proteins within coiled structures that extend between the two cell poles. Proc. Natl. Acad. Sci. U.S.A. *100*, 7865–7870.
Shih, H.W., Chen, K.T., Chen, S.K., Huang, C.Y., Cheng, T.J., Ma, C., Wong, C.H., and Cheng, W.C. (2010). Combinatorial approach toward synthesis of small molecule libraries as bacterial transglycosylase inhibitors. Org. Biomol. Chem. *8*, 2586–2593.
Shiomi, D., Sakai, M., and Niki, H. (2008). Determination of bacterial rod shape by a novel cytoskeletal membrane protein. EMBO J. *27*, 3081–3091.
Siefert, J.L., and Fox, G.E. (1998). Phylogenetic mapping of bacterial morphology. Microbiology *144*, 2803–2808.
Singh, S.P., and Montgomery, B.L. (2011). Determining cell shape: adaptive regulation of cyanobacterial cellular differentiation and morphology. Trends Microbiol. *19*, 278–285.
Singh S.K., SaiSree, L., Amrutha, R.N., and Reddy, M. (2012). Three redundant murein endopeptidases catalyse an essential cleavage step in peptidoglycan synthesis of *Escherichia coli* K12. Mol. Microbiol. *86*, 1036–1051.

Snowden, M.A., and Perkins, H.R. (1990). Peptidoglycan cross-linking in *Staphylococcus aureus*. An apparent random polymerisation process. Eur. J. Biochem. *191*, 373–377.

Spratt, B.G. (1975). Distinct penicillin binding proteins involved in the division, elongation, and shape of *Escherichia coli* K12. Proc. Natl. Acad. Sci. U.S.A. *72*, 2999–3003.

Stahlberg, H., Kutejova, E., Muchova, K., Gregorini, M., Lustig, A., Muller, S.A., Olivieri, V., Engel, A., Wilkinson, A.J., and Barak, I. (2004). Oligomeric structure of the *Bacillus subtilis* cell division protein DivIVA determined by transmission electron microscopy. Mol. Microbiol. *52*, 1281–1290.

Stokes, N.R., Sievers, J., Barker, S., Bennett, J.M., Brown, D.R., Collins, I., Errington, V.M., Foulger, D., Hall, M., Halsey, R., *et al.* (2005). Novel inhibitors of bacterial cytokinesis identified by a cell-based antibiotic screening assay. J. Biol. Chem. *280*, 39709–39715.

Strahl, H., and Hamoen, L.W. (2010). Membrane potential is important for bacterial cell division. Proc. Natl. Acad. Sci. U.S.A. *107*, 12281–12286.

Sung, M.T., Lai, Y.T., Huang, C.Y., Chou, L.Y., Shih, H.W., Cheng, W.C., Wong, C.H., and Ma, C. (2009). Crystal structure of the membrane-bound bifunctional transglycosylase PBP1b from *Escherichia coli*. Proc. Natl. Acad. Sci. U.S.A. *106*, 8824–8829.

Sycuro, L.K., Pincus, Z., Gutierrez, K.D., Biboy, J., Stern, C.A., Vollmer, W., and Salama, N.R. (2010). Peptidoglycan crosslinking relaxation promotes *Helicobacter pylori's* helical shape and stomach colonization. Cell *141*, 822–833.

Takacs, C.N., Poggio, S., Charbon, G., Pucheault, M., Vollmer, W., and Jacobs-Wagner, C. (2010). MreB drives *de novo* rod morphogenesis in *Caulobacter crescentus* via remodeling of the cell wall. J. Bacteriol. *192*, 1671–1684.

Tarry, M., Arends, S.J., Roversi, P., Piette, E., Sargent, F., Berks, B.C., Weiss, D.S., and Lea, S.M. (2009). The *Escherichia coli* cell division protein and model Tat substrate SufI (FtsP) localizes to the septal ring and has a multicopper oxidase-like Structure J. Mol. Biol. *386*, 504–519.

Terrak, M., Ghosh, T.K., van Heijenoort, J., Van Beeumen, J., Lampilas, M., Aszodi, J., Ayala, J.A., Ghuysen, J.M., and Nguyen-Distèche, M. (1999). The catalytic, glycosyl transferase and acyl transferase modules of the cell wall peptidoglycan-polymerizing penicillin-binding protein 1b of *Escherichia coli*. Mol. Microbiol. *34*, 350–364.

Terrak, M., and Nguyen-Distèche, M. (2006). Kinetic characterization of the monofunctional glycosyltransferase from *Staphylococcus aureus*. J. Bacteriol. *188*, 2528–2532.

Terrak, M., Sauvage, E., Derouaux, A., Dehareng, D., Bouhss, A., Breukink, E., Jeanjean, S., and Nguyen-Distèche, M. (2008). Importance of the conserved residues in the peptidoglycan glycosyltransferase module of the class A penicillin-binding protein 1b of *Escherichia coli*. J. Biol. Chem. *283*, 28464–28470.

Thanbichler, M., and Shapiro, L. (2006). MipZ, a spatial regulator coordinating chromosome segregation with cell division in *Caulobacter*. Cell *126*, 147–162.

Thibessard, A., Fernandez, A., Gintz, B., Leblond-Bourget, N., and Decaris, B. (2002). Effects of *rodA* and *pbp2b* disruption on cell morphology and oxidative stress response of *Streptococcus thermophilus* CNRZ368. J. Bacteriol. *184*, 2821–2826.

Thunnissen, A.M., Dijkstra, A.J., Kalk, K.H., Rozeboom, H.J., Engel, H., Keck, W., and Dijkstra, B.W. (1994). Doughnut-shaped structure of a bacterial muramidase revealed by X-ray crystallography. Nature *367*, 750–753.

Thunnissen, A.M., Isaacs, N.W., and Dijkstra, B.W. (1995a). The catalytic domain of a bacterial lytic transglycosylase defines a novel class of lysozymes. Proteins *22*, 245–258.

Thunnissen, A.M., Rozeboom, H.J., Kalk, K.H., and Dijkstra, B.W. (1995b). Structure of the 70-kDa soluble lytic transglycosylase complexed with bulgecin A. Implications for the enzymatic mechanism. Biochemistry *34*, 12729–12737.

Tiyanont, K., Doan, T., Lazarus, M.B., Fang, X., Rudner, D.Z., and Walker, S. (2006). Imaging peptidoglycan biosynthesis in *Bacillus subtilis* with fluorescent antibiotics. Proc. Natl. Acad. Sci. U.S.A. *103*, 11033–11038.

Tonthat, N.K., Arold, S.T., Pickering, B.F., Van Dyke, M.W., Liang, S., Lu, Y., Beuria, T.K., Margolin, W., and Schumacher, M.A. (2011). Molecular mechanism by which the nucleoid occlusion factor, SlmA, keeps cytokinesis in check. EMBO J. *30*, 154–164.

Trachtenberg, S. (2006). The cytoskeleton of spiroplasma: a complex linear motor. J. Mol. Microbiol. Biotechnol. *11*, 265–283.

Trachtenberg, S., Dorward, L.M., Speransky, V.V., Jaffe, H., Andrews, S.B., and Leapman, R.D. (2008). Structure of the cytoskeleton of *Spiroplasma melliferum* BC3 and its interactions with the cell membrane. J. Mol. Biol. *378*, 778–789.

Turk, S., Verlaine, O., Gerards, T., Zivec, M., Humljan, J., Sosic, I., Amoroso, A., Zervosen, A., Luxen, A., Joris, B., *et al.* (2011). New noncovalent inhibitors of penicillin-binding proteins from penicillin-resistant bacteria. PLoS One *6*, e19418.

Turner, R.D., Ratcliffe, E.C., Wheeler, R., Golestanian, R., Hobbs, J.K., and Foster, S.J. (2010). Peptidoglycan architecture can specify division planes in *Staphylococcus aureus*. Nat. Commun. *1*, 1–9.

Typas, A., Banzhaf, M., van Saparoea, V.B., Verheul, J., Biboy, J., Nichols, R.J., Zietek, M., Beilharz, K., Kannenberg, K., von Rechenberg, M., *et al.* (2010). Regulation of peptidoglycan synthesis by outer membrane proteins Cell *143*, 1097–1109.

Typas, A., Banzhaf, M., Gross, C.A., and Vollmer, W. (2012). From the regulation of peptidoglycan synthesis to bacterial growth and morphology. Nat. Rev. Microbiol. *10*, 123–136.

Uchida, K., Kudo, T., Suzuki, K.I., and Nakase, T. (1999). A new rapid method of glycolate test by diethyl ether extraction, which is applicable to a small amount of bacterial cells of less than one milligram. J. Gen. Appl. Microbiol. *45*, 49–56.

Uehara, T., and Park, J.T. (2008). Growth of *Escherichia coli*: significance of peptidoglycan degradation during elongation and septation. J. Bacteriol. *190*, 3914–3922.

Uehara, T., Dinh, T., and Bernhardt, T.G. (2009). LytM-domain factors are required for daughter cell separation and rapid ampicillin-induced lysis in *Escherichia coli*. J. Bacteriol. *191*, 5094–5107.

Uehara, T., Parzych, K.R., Dinh, T., and Bernhardt, T.G. (2010). Daughter cell separation is controlled by cytokinetic ring-activated cell wall hydrolysis. EMBO J. *29*, 1412–1422.

Urgaonkar, S., La Pierre, H.S., Meir, I., Lund, H., RayChaudhuri, D., and Shaw, J.T. (2005). Synthesis of antimicrobial natural products targeting FtsZ: (±)-dichamanetin and (±)-2′ "-hydroxy-5"benzylisouvarinol-B. Org. Lett. *7*, 5609–5612.

Ursinus, A., and Höltje, J.-V. (1994). Purification and properties of a membrane-bound lytic transglycosylase from *Escherichia coli*. J. Bacteriol. *176*, 338–343.

Ursinus, A., van den Ent, F., Brechtel, S., de Pedro, M., Höltje, J.-V., Löwe, J., and Vollmer, W. (2004). Murein (peptidoglycan) binding property of the essential cell division protein FtsN from *Escherichia coli*. J. Bacteriol. *186*, 6728–6737.

Valbuena, N., Letek, M., Ordonez, E., Ayala, J., Daniel, R.A., Gil, J.A., and Mateos, L.M. (2007). Characterization of HMW-PBPs from the rod-shaped actinomycete *Corynebacterium glutamicum*: peptidoglycan synthesis in cells lacking actin-like cytoskeletal structures. Mol. Microbiol. *66*, 643–657.

van Asselt, E.J., and Dijkstra, B.W. (1999). Binding of calcium in the EF-hand of *Escherichia coli* lytic transglycosylase Slt35 is important for stability. FEBS Lett. *458*, 429–435.

van Asselt, E.J., Dijkstra, A.J., Kalk, K.H., Takacs, B., Keck, W., and Dijkstra, B.W. (1999a). Crystal structure of *Escherichia coli* lytic transglycosylase Slt35 reveals a lysozyme-like catalytic domain with an EF-hand. Structure Fold. Des. *7*, 1167–1180.

van Asselt, E.J., Thunnissen, A.M., and Dijkstra, B.W. (1999b). High resolution crystal structures of the *Escherichia coli* lytic transglycosylase Slt70 and its complex with a peptidoglycan fragment. J. Mol. Biol. *291*, 877–898.

van Asselt, E.J., Kalk, K.H., and Dijkstra, B.W. (2000). Crystallographic studies of the interactions of *Escherichia coli* lytic transglycosylase Slt35 with peptidoglycan. Biochemistry *39*, 1924–1934.

van Dam, V., Sijbrandi, R., Kol, M., Swiezewska, E., de Kruijff, B., and Breukink, E. (2007). Transmembrane transport of peptidoglycan precursors across model and bacterial membranes. Mol. Microbiol. *64*, 1105–1114.

van den Ent, F., and Löwe, J. (2000). Crystal structure of the cell division protein FtsA from *Thermotoga maritima*. EMBO J. *19*, 5300–5307.

van den Ent, F., Amos, L.A., and Löwe, J. (2001). Prokaryotic origin of the actin cytoskeleton. Nature *413*, 39–44.

van den Ent, F., Leaver, M., Bendezu, F., Errington, J., de Boer, P., and Löwe, J. (2006). Dimeric structure of the cell shape protein MreC and its functional implications. Mol. Microbiol. *62*, 1631–1642.

van den Ent, F., Vinkenvleugel, T.M., Ind, A., West, P., Veprintsev, D., Nanninga, N., den Blaauwen, T., and Löwe, J. (2008). Structural and mutational analysis of the cell division protein FtsQ. Mol. Microbiol. *68*, 110–123.

van den Ent, F., Johnson, C.M., Persons, L., de Boer, P., and Löwe, J. (2010). Bacterial actin MreB assembles in complex with cell shape protein RodZ. EMBO J. *29*, 1081–1090.

van Straaten, K.E., Dijkstra, B.W., Vollmer, W., and Thunnissen, A.M. (2005). Crystal structure of MltA from *Escherichia coli* reveals a unique lytic transglycosylase fold. J. Mol. Biol. *352*, 1068–1080.

van Straaten, K.E., Barends, T.R., Dijkstra, B.W., and Thunnissen, A.M. (2007). Structure of *Escherichia coli* lytic transglycosylase MltA with bound chitohexaose: implications for peptidoglycan binding and cleavage. J. Biol. Chem. *282*, 21197–21205.

Varma, A., and Young, K.D. (2004). FtsZ collaborates with penicillin binding proteins to generate bacterial cell shape in *Escherichia coli*. J. Bacteriol. *186*, 6768–6774.

Vats, P., and Rothfield, L. (2007). Duplication and segregation of the actin (MreB) cytoskeleton during the prokaryotic cell cycle. Proc. Natl. Acad. Sci. U.S.A. *104*, 17795–17800.

Vats, P., Shih, Y.L., and Rothfield, L. (2009). Assembly of the MreB-associated cytoskeletal ring of *Escherichia coli*. Mol. Microbiol. *72*, 170–182.

Vazquez-Laslop, N., Lee, H., Hu, R., and Neyfakh, A.A. (2001). Molecular sieve mechanism of selective release of cytoplasmic proteins by osmotically shocked *Escherichia coli*. J. Bacteriol. *183*, 2399–2404.

Veiga, H., Jorge, A.M., and Pinho, M.G. (2011). Absence of nucleoid occlusion effector Noc impairs formation of orthogonal FtsZ rings during *Staphylococcus aureus* cell division. Mol. Microbiol. *80*, 1366–1380.

Villanelo, F., Ordenes, A., Brunet, J., Lagos, R., and Monasterio, O. (2011). A model for the *Escherichia coli* FtsB/FtsL/FtsQ cell division complex. BMC Struct. Biol. *11*, 28.

Vinella, D., Joseleau-Petit, D., Thevenet, D., Bouloc, P., and D'Ari, R. (1993). Penicillin-binding protein 2 inactivation in *Escherichia coli* results in cell division inhibition, which is relieved by FtsZ overexpression. J. Bacteriol. *175*, 6704–6710.

Vollmer, W. (2006). The prokaryotic cytoskeleton: a putative target for inhibitors and antibiotics? Appl. Microbiol. Biotechnol. *73*, 37–47.

Vollmer, W. (2008). Structural variation in the glycan strands of bacterial peptidoglycan. FEMS Microbiol. Rev. *32*, 287–306.

Vollmer, W., and Bertsche, U. (2008). Murein (peptidoglycan) structure, architecture and biosynthesis in *Escherichia coli*. Biochim. Biophys. Acta *1778*, 1714–1734.

Vollmer, W., and Höltje, J.-V. (2004). The architecture of the murein (peptidoglycan) in Gram-negative bacteria: vertical scaffold or horizontal layer(s)? J. Bacteriol. *186*, 5978–5987.

Vollmer, W., and Seligman, S.J. (2010). Architecture of peptidoglycan: more data and more models. Trends Microbiol. *18*, 59–66.

Vollmer, W., and Tomasz, A. (2000). The *pgdA* gene encodes for a peptidoglycan N-acetylglucosamine deacetylase in *Streptococcus pneumoniae*. J. Biol. Chem. *275*, 20496–20501.

Vollmer, W., von Rechenberg, M., and Höltje, J.-V. (1999). Demonstration of molecular interactions between the murein polymerase PBP1B, the lytic transglycosylase MltA, and the scaffolding protein MipA of *Escherichia coli*. J. Biol. Chem. *274*, 6726–6734.

Vollmer, W., Blanot, D., and de Pedro, M.A. (2008a). Peptidoglycan structure and architecture. FEMS Microbiol. Rev. *32*, 149–167.

Vollmer, W., Joris, B., Charlier, P., and Foster, S. (2008b). Bacterial peptidoglycan (murein) hydrolases. FEMS Microbiol. Rev. *32*, 259–286.

Wachi, M., Doi, M., Okada, Y., and Matsuhashi, M. (1989). New *mre* genes *mreC* and *mreD*, responsible for formation of the rod shape of *Escherichia coli* cells. J. Bacteriol. *171*, 6511–6516.

Wagner, J.K., Galvani, C.D., and Brun, Y.V. (2005). *Caulobacter crescentus* requires RodA and MreB for stalk synthesis and prevention of ectopic pole formation. J. Bacteriol. *187*, 544–553.

Wagner, J.K., Setayeshgar, S., Sharon, L.A., Reilly, J.P., and Brun, Y.V. (2006). A nutrient uptake role for bacterial cell envelope extensions. Proc. Natl. Acad. Sci. U.S.A. *103*, 11772–11777.

Waidner, B., Specht, M., Dempwolff, F., Haeberer, K., Schaetzle, S., Speth, V., Kist, M., and Graumann, P.L. (2009). A novel system of cytoskeletal elements in the human pathogen *Helicobacter pylori*. PLoS Pathog. *5*, e1000669.

Walshaw, J., Gillespie, M.D., and Kelemen, G.H. (2010). A novel coiled-coil repeat variant in a class of bacterial cytoskeletal proteins. J. Struct. Biol. *170*, 202–215.

Wang, J., Galgoci, A., Kodali, S., Herath, K.B., Jayasuriya, H., Dorso, K., Vicente, F., Gonzalez, A., Cully, D., Bramhill, D., *et al.* (2003). Discovery of a small molecule that inhibits cell division by blocking FtsZ, a novel therapeutic target of antibiotics. J. Biol. Chem. *278*, 44424–44428.

Wang, L., Khattar, M.K., Donachie, W.D., and Lutkenhaus, J. (1998). FtsI and FtsW are localized to the septum in *Escherichia coli*. J. Bacteriol. *180*, 2810–2816.

Wang, S., Arellano-Santoyo, H., Combs, P.A., and Shaevitz, J.W. (2010). Actin-like cytoskeleton filaments contribute to cell mechanics in bacteria. Proc. Natl. Acad. Sci. U.S.A. *107*, 9182–9185.

Waxman, D.J., and Strominger, J.L. (1983). Penicillin-binding proteins and the mechanism of action of beta-lactam antibiotics. Annu. Rev. Biochem. *52*, 825–869.

Weadge, J.T., and Clarke, A.J. (2006). Identification and characterization of O-acetylpeptidoglycan esterase: a novel enzyme discovered in *Neisseria gonorrhoeae*. Biochemistry *45*, 839–851.
Weadge, J.T., and Clarke, A.J. (2007). *Neisseria gonorrheae* O-acetylpeptidoglycan esterase, a serine esterase with a Ser-His-Asp catalytic triad. Biochemistry *46*, 4932–4941.
Weadge, J.T., Pfeffer, J.M., and Clarke, A.J. (2005). Identification of a new family of enzymes with potential O-acetylpeptidoglycan esterase activity in both Gram-positive and Gram-negative bacteria. BMC Microbiol. *5*, 49.
Weidel, W., and Pelzer, H. (1964). Bagshaped macromolecules – a new outlook on bacterial cell walls. Adv. Enzymol. *26*, 193–232.
Weiss, D.S., Chen, J.C., Ghigo, J.M., Boyd, D., and Beckwith, J. (1999). Localization of FtsI (PBP3) to the septal ring requires its membrane anchor, the Z ring, FtsA, FtsQ, and FtsL. J. Bacteriol. *181*, 508–520.
Welzel, P. (2005). Syntheses around the transglycosylation step in peptidoglycan biosynthesis. Chem. Rev. *105*, 4610–4660.
White, C.L., Kitich, A., and Gober, J.W. (2010). Positioning cell wall synthetic complexes by the bacterial morphogenetic proteins MreB and MreD. Mol. Microbiol. *76*, 616–633.
White, E.L., Suling, W.J., Ross, L.J., Seitz, L.E., and Reynolds, R.C. (2002). 2-Alkoxycarbonylaminopyridines: inhibitors of *Mycobacterium tuberculosis* FtsZ. J. Antimicrob. Chemother. *50*, 111–114.
Wissel, M.C., and Weiss, D.S. (2004). Genetic analysis of the cell division protein FtsI (PBP3): amino acid substitutions that impair septal localization of FtsI and recruitment of FtsN. J. Bacteriol. *186*, 490–502.
Woldringh, C., Huls, P., Pas, E., Brakenhoff, G.H., and Nanninga, N. (1987). Topography of peptidoglycan synthesis during elongation and polar cap formation in a cell division mutant of *Escherichia coli* MC43100. J. Gen. Microbiol. *133*, 575–586.
Wolgemuth, C.W., Charon, N.W., Goldstein, S.F., and Goldstein, R.E. (2006). The flagellar cytoskeleton of the spirochetes. J. Mol. Microbiol. Biotechnol. *11*, 221–227.
Wu, L.J., and Errington, J. (2004). Coordination of cell division and chromosome segregation by a nucleoid occlusion protein in *Bacillus subtilis*. Cell *117*, 915–925.
Wu, L.J., Ishikawa, S., Kawai, Y., Oshima, T., Ogasawara, N., and Errington, J. (2009). Noc protein binds to specific DNA sequences to coordinate cell division with chromosome segregation. EMBO J. *28*, 1940–1952.
Yamada, S., Sugai, M., Komatsuzawa, H., Nakashima, S., Oshida, T., Matsumoto, A., and Suginaka, H. (1996). An autolysin ring associated with cell separation of *Staphylococcus aureus*. J. Bacteriol. *178*, 1565–1571.
Yang, D.C., Peters, N.T., Parzych, K.R., Uehara, T., Markovski, M., and Bernhardt, T.G. (2011). An ATP-binding cassette transporter-like complex governs cell-wall hydrolysis at the bacterial cytokinetic ring. Proc. Natl. Acad. Sci. U.S.A. *108*, 1052–1060.
Yang, D.C., Tan, K., Joachimiak, A., and Bernhardt, T.G. (2012). A conformational switch controls cell-wall-remodelling enzymes required for bacterial cell division. Mol. Microbiol. *85*, 768–781.
Yang, J.C., Van Den Ent, F., Neuhaus, D., Brevier, J., and Löwe, J. (2004). Solution structure and domain architecture of the divisome protein FtsN. Mol. Microbiol. *52*, 651–660.
Yao, X., Jericho, M., Pink, D., and Beveridge, T. (1999). Thickness and elasticity of Gram-negative murein sacculi measured by atomic force microscopy. J. Bacteriol. *181*, 6865–6875.
Young, K.D. (2006). The selective value of bacterial shape. Microbiol. Mol. Biol. Rev. *70*, 660–703.
Yousif, S.Y., Broome-Smith, J.K., and Spratt, B.G. (1985). Lysis of *Escherichia coli* by beta-lactam antibiotics: deletion analysis of the role of penicillin-binding proteins 1A and 1B. J. Gen. Microbiol. *131*, 2839–2845.
Yu, X.C., and Margolin, W. (1998). Inhibition of assembly of bacterial cell division protein FtsZ by the hydrophobic dye 5,5'-bis-(8-anilino-1-naphthalenesulfonate). J. Biol. Chem. *273*, 10216–10222.
Yu, X.C., Tran, A.H., Sun, Q., and Margolin, W. (1998). Localization of cell division protein FtsK to the *Escherichia coli* septum and identification of a potential N-terminal targeting domain. J. Bacteriol. *180*, 1296–1304.
Yuan, Y., Barrett, D., Zhang, Y., Kahne, D., Sliz, P., and Walker, S. (2007). Crystal structure of a peptidoglycan glycosyltransferase suggests a model for processive glycan chain synthesis. Proc. Natl. Acad. Sci. U.S.A. *104*, 5348–5353.
Yuan, Y., Fuse, S., Ostash, B., Sliz, P., Kahne, D., and Walker, S. (2008). Structural analysis of the contacts anchoring moenomycin to peptidoglycan glycosyltransferases and implications for antibiotic design. ACS Chem. Biol. *3*, 429–436.
Zapun, A., Vernet, T., and Pinho, M.G. (2008). The different shapes of cocci. FEMS Microbiol. Rev. *32*, 345–360.

Zervosen, A., Herman, R., Kerff, F., Herman, A., Bouillez, A., Prati, F., Pratt, R.F., Frere, J.M., Joris, B., Luxen, A., *et al.* (2011). Unexpected tricovalent binding mode of boronic acids within the active site of a penicillin binding protein. J. Am. Chem. Soc. *133*, 10839–10848.

The Outer Membrane of Gram-negative Bacteria: Lipopolysaccharide Biogenesis and Transport

2

Paola Sperandeo, Riccardo Villa, Gianni Dehò and Alessandra Polissi

Abstract

The cell envelope of Gram-negative bacteria consists of two distinct membranes, the inner (IM) and the outer membrane (OM), separated by an aqueous compartment, the periplasm. The OM contains in the outer leaflet the lipopolysaccharide (LPS), a complex glycolipid with important biological functions. In the host, it elicits the innate immune response whereas in the bacterium it is responsible for the peculiar permeability barrier properties exhibited by the OM. LPS is synthesized in the cytoplasm and at the inner leaflet of the IM. It needs to cross two different compartments, the IM and the periplasm, to reach its final destination at the cell surface. In this chapter we will first summarize LPS structure, functions and biosynthetic pathway and then review in more details the studies that have led in the last decade to elucidate the protein machinery that ferries LPS from the IM to its final destination in the OM.

Introduction

All living cells are surrounded by the cytoplasmic membrane, a unit membrane whose overall architecture (a fluid lipid bilayer with integral and peripheral membrane proteins) is conserved among the three domains of life (Chapter 1). Nevertheless, the chemical composition of the lipid bilayer poses a divide between *Archaea*, whose membrane lipids consist of isoprenoid hydrocarbon chains linked to glycerol-1-phosphate through an ether linkage, and both *Bacteria* and *Eukarya*, which contain glycerol-3-phosphate diesters of linear fatty acids (De Rosa *et al.*, 1991; Wachtershauser, 2003; see also Chapter 1 and references therein).

Outside of the universally conserved cytoplasmic membrane, most prokaryotes have developed complex and varied peripheral architectures, collectively named the cell wall, that provide additional strength and protection against environmental assaults and contribute to the cell shape determination (Beveridge, 1999; Ellen *et al.*, 2010; Silhavy *et al.*, 2010). The great majority of *Bacteria* are surrounded by an additional lipid bilayer, the outer membrane (OM), and are thus described as diderm bacteria; the OM is not present in monoderm bacteria, which possess the cytoplasmic membrane as the unique lipid membrane (Gupta, 1998; Desvaux *et al.*, 2009; Sutcliffe, 2011).

The prototypical OM has been characterized in great detail over the last half-century in *Proteobacteria*, particularly in *Enterobacteriaceae*, and it is characterized by a peculiar glycolipid, the lipopolysaccharide (LPS), that forms the outer leaflet of the lipid bilayer, whereas the inner leaflet is composed of phospholipids. OM proteins (OMPs) and lipoproteins are also embedded and anchored, respectively, in the OM.

The architecture of *Proteobacteria* cell envelope that emerged from these studies has long since been considered the standard for all Gram-negative bacteria. It consists of the inner (cytoplasmic) membrane (IM) and the LPS-containing OM that delimit a periplasmic space with a thin layer of murein. Conversely, both low and high G + C% Gram-positive bacteria (*Firmicutes* and *Actinobacteria*, respectively) have been traditionally considered as monoderms. However, it is now recognized that different non-LPS OM architectures can be found in both Gram-negative and Gram-positive bacteria (Sutcliffe, 2010). For example, the Gram-negative *Thermotogae* appear to be surrounded by an OM not containing LPS (Plotz *et al.*, 2000; Sutcliffe, 2010), whereas a mycolic acid-based OM is present in the *Corynebacterineae*, a suborder of Gram-positive bacteria that comprises mycobacteria and other genera such as *Corynebacterium*, and *Nocardia* (Minnikin, 1991; Zuber *et al.*, 2008; Niederweis *et al.*, 2010). This latter example, in particular, suggests that functionally analogous OM architectures may have independently evolved in bacteria, thus highlighting the functional relevance of an additional outer lipid bilayer for bacterial adaptation. On the other hand, the Gram-negative *Chloroflexi* appear to be monoderms (Sutcliffe, 2011); thus, the Gram-positive versus Gram-negative classification of *Bacteria* does not coincide with the monoderm versus diderm grouping and neither criterion should be taken as a discriminating phylogenetic character (Fig. 2.1).

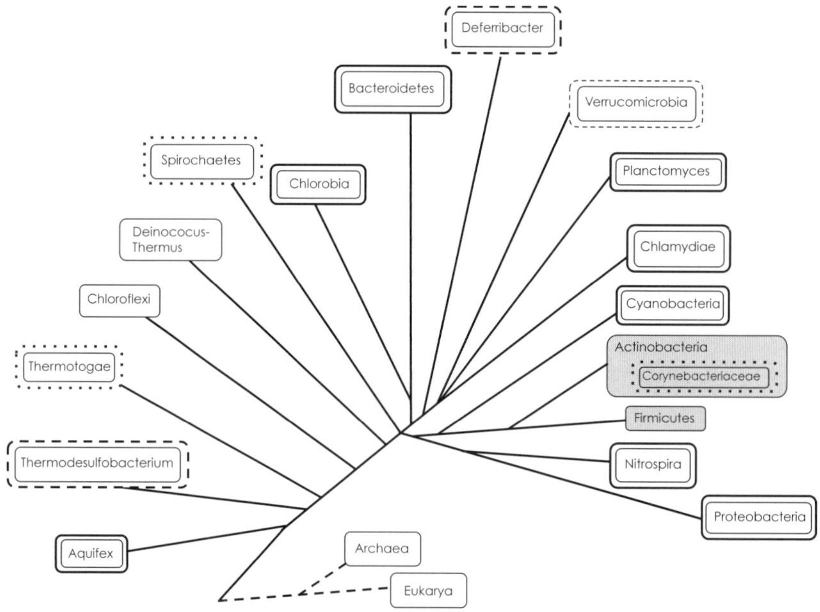

Figure 2.1 Distribution of monoderm an diderm cells within the universal phylogenetic tree. Major lineages of Bacteria are shown. A grey box indicates Gram-positive bacteria. Monoderm cells are surrounded by a single line, whereas an additional outer line identifies the OM of diderm bacteria as follows: thick continuous line, OM with LPS; thick dashed line, LPS predicted but not experimentally demonstrated; thick dotted line, OM with different lipid composition; thin dashed line, data not available. Data from Sutcliffe (2010).

The presence of a highly structured OM poses several problems as of its biogenesis. Both lipid and protein components not only must be synthesized in the cytoplasm and/or at the IM level and translocated across the IM lipid bilayer, but also must traverse the aqueous periplasmic space and be assembled at the amphipathic final destination. The cell compartments external to the IM are devoid of ATP and other high-energy carriers. As a consequence the energy to build up periplasmic and OM structures is either provided by exergonic reactions (thus involving substrates that have been energized before their translocation across the IM) or transduced by devices (usually protein machines) connected to the IM and capable of exploiting the energy released by ATP hydrolysis in the cytoplasm or the proton motive force.

In this chapter, after a short overview on the OM structure and functions as they emerge from intensive studies on (mainly) *Proteobacteria*, we will concentrate on its peculiar component, the LPS. We will first summarize its structure and biosynthetic pathway and then review in more detail the studies that have led in the last decade to elucidate the machinery that ferries LPS from the IM to its final destination in the OM.

Whereas our present knowledge of the LPS transport mechanism leaves many open questions and poses new challenges for future research, the identification and characterization of the seven essential components of the machine devoted to LPS transport (Lpt complex) has provided a solid base to further dissect the transport mechanism as well as new potential targets for antibacterial drugs design.

An overview of OM structure, functions and evolution

The topography of the bacterial cell wall was made possible by the development of electron microscopy techniques. Early studies in the 1960s convincingly showed that the basic structure surrounding Gram-negative bacteria is composed by an inner and an outer membrane, both seemingly 'unit membranes', separated by intermediated layers, including the peptidoglycan layer (Bladen and Mergenhagen, 1964). In Gram-positive bacteria, on the other hand, the OM was missing and a thicker peptidoglycan layer was present (reviewed by Glauert and Thornley, 1969). The striking correlation between cell wall structure and Gram staining was rationalized much later when it was shown that upon ethanol treatment the crystal violet–potassium iodide precipitate is retained within the cell by the thick peptidoglycan layer of Gram-positive bacteria, whereas it is washed away through the thinner murein sacculus of Gram-negative bacteria that are thus decolorized (Beveridge and Davies, 1983).

Assessing a detailed chemical composition of the OM was facilitated by the different buoyant densities of the two membranes (approximately 1.22 and 1.15 for the OM and IM, respectively), which could therefore be fractionated by equilibrium centrifugation in sucrose density gradients and analysed separately. Our present understanding of the OM structure and composition is diagrammed in Fig. 2.2, in the context of the Gram-negative cell envelope. This picture emerges from studies mainly performed on model Gram-negative bacteria, especially *Proteobacteria*, such as *Escherichia coli*, *Salmonella enterica* serovar Typhimurium and other *Enterobacteriaceae*, *Pseudomonas aeruginosa*, *Neisseria meningitidis* and others, whereas for other phyla information is less complete.

Several structural and functional aspects differentiate the OM from the plasma membrane. The most striking structural difference is the asymmetry of the OM bilayer. Whereas

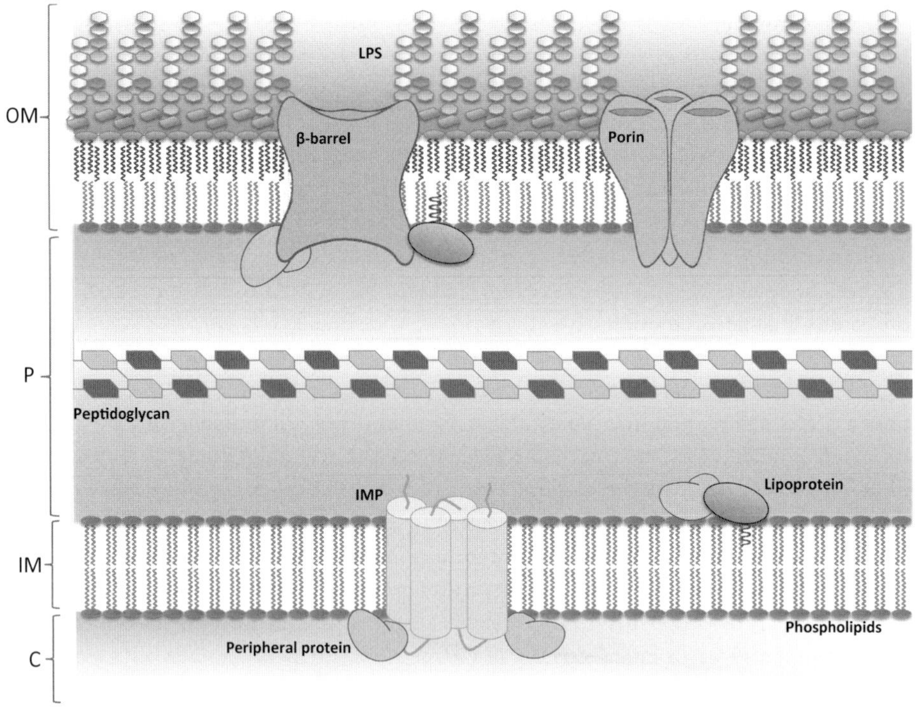

Figure 2.2 The Gram-negative cell envelope. The envelope of Gram-negative bacteria is composed of an inner membrane (IM), the periplasm (P) and an outer membrane (OM). The IM is a symmetric lipid bilayer composed of phospholipids, integral proteins (IMP) that span the membrane by α-helical transmembrane domains, and peripheral proteins associated to the inner leaflet of the IM. The periplasm (P) is hydrophilic gel-like compartment located between IM and OM and containing a layer of peptidoglycan. The OM is an asymmetric bilayer composed of phospholipid in the inner leaflet and Lipopolysaccharide (LPS) towards the outside. The OM also contains integral proteins folded in β-barrel conformation and trimeric proteins forming channels known as porins. Both IM and OM contain lipoproteins anchored to their periplasmic faces. Transenvelope secretory machines (see Figure 2.5 as an example) are not shown.

the periplasmic side is made by a layer of the same type of phospholipids that compose both leaflets of the IM, LPS paves the OM layer facing the environment outside the cell. This was first observed by immuno-electron microscopy (Muhlradt and Golecki, 1975) and then demonstrated by Kamio and Nikaido (1976) who showed that in intact cells of *S. typhimurium* (now renamed *S. enterica* serovar Typhimurium) phospholipids could not be chemically modified by an OM-impermeable macromolecular reagent. The unique chemical structure and properties of LPS, as discussed in 'Lipopolysaccharide structure and biosynthesis' below, are mainly responsible of the peculiar properties of the OM. This second lipid bilayer with an additional external hydrophilic region of long polysaccharide chains, endows Gram-negative bacteria with a strong additional diffusion barrier, which accounts for the generally higher resistance of Gram-negative bacteria, as compared with most Gram-positives, to many toxic chemicals such as antibiotics and detergents (e.g. bile salts) and to survive hostile environments such as the gastrointestinal tracts of mammals, encountered during host colonization or infection (Gunn, 2000; Nikaido, 2003).

Cell-environment exchanges across the OM are thus ensured by OM proteins, that are implicated in several functions such as nutrients uptake, transport and secretion of various molecules (proteins, polysaccharides, DNA or drugs), assembly of proteins or proteinaceous structures at the OM, and other types of interactions with the external environment and the underlying cell compartments. As discussed elsewhere in this book (Chapters 3 and 4) typical OM integral proteins (OMPs) are β-barrel proteins, whereas OM-associated proteins are generally lipoproteins that are anchored to the periplasmic side of the OM via a lipid tail attached to an N-terminal cysteine residue (Sankaran and Wu, 1994). Bacterial lipoproteins are mostly OM associated, but IM lipoproteins are also known. The role of lipoproteins is little understood; the best known is Lpp (or Braun's protein), the most abundant protein in *E. coli*, that anchors the peptidoglycan layer to the OM (Braun, 1975).

Many OMPs associate as trimeric pores or channels that allow passive diffusion across the OM of small hydrophilic molecules such as mono or oligosaccharides, amino acids, ions and/or catabolism waste products, with various degrees of specificity, whereas other proteins are part of energy consuming active transport systems, especially for the transport of larger molecules (efflux pumps, TonB-dependent high-affinity receptors, ABC transporters) that are connected to IM proteins to form transenvelope machines energized by ATP hydrolysis in the cytoplasm or by the IM proton gradient. Other OM proteins are devoted to secretion of proteins (with their final destination outside the OM or in the OM itself), either in concert with or independently of the main SecA-dependent secretory system. For this energy costly secretion process, proteins may be first translocated in the periplasm in an energized form and then pass through the specific OMP transporter consuming the accumulated energy, or may be transported by transenvelope machines as mentioned above for other large molecules (Nikaido, 2003; Knowles *et al.*, 2009; Karuppiah *et al.*, 2011; this chapter; this book, Chapters 7, 8 and 11; see also Fig. 2.5). The panoply of secretion systems that have evolved in bacteria meets the needs of a vast variety protein structures and final destinations, including the intracellular milieu of eukaryotic cells (Holland, 2010).

OM creates the unique organelle of diderm bacteria, the periplasmic space, a viscous (periplasm proteins concentration is higher than in the cytoplasm) hydrophilic compartment lying between the IM and OM. Several processes that are vital to growth and viability of the cell occur in this compartment. Proteins residing in the periplasmic space fulfil important functions in the detection, processing and transport of nutrients into the cell, including breakdown by nucleases, peptidases, and phosphatases of large or charged molecules that cannot be transported through the IM; periplasmic chaperones (including proteins involved in disulfide bond formation) promote the biogenesis of periplasmic, outer membrane, and external appendages proteins such as pili and fimbriae; detoxifying enzymes (such as β-lactamases) preserve the cell from obnoxious chemicals (Oliver, 1996).

One of the major cell processes occurring in the periplasm is the synthesis of the peptidoglycan layer, the largest cell polymer that surrounds the bacterial cell forming the seamless murein sacculus (Gan *et al.*, 2008; Vollmer and Seligman, 2010). This cellular exoskeleton is the main structure responsible for the cell shape and the mechanical strength and elasticity of the bacterial envelope, which can withstand turgor pressure up to three atmospheres (Koch, 1998). Biosynthesis of the murein sacculus must be very carefully coordinated with that of IM and OM during cell growth and division. Understanding formation of the cell septum that separates two newborn bacterial cells at the biochemical, structural and topological levels remains one of the unresolved problems of the bacterial cell biology, although impressive advances have been obtained in this field in the last years (Margolin, 2009). Likewise, biogenesis of IM and OM is an exciting and rapidly advancing field that will be reviewed in this and other chapters of this book.

Electron microscopy observations of adhesion regions between OM and EM known as Bayer's bridges (Bayer, 1968) have been thought of for some time as potential sites for lipid trafficking and possibly protein transport between the two membranes. The idea of intermembrane adhesion zones was later considered an artefact and abandoned (Kellenberger, 1990). However, as mentioned above, protein machines that cross the periplasmic space and the murein layer are now well documented and, as we will see in the next paragraphs, appear to be implicated in LPS transport. Thus Bayer's bridges could be re-evaluated as proteinaceous structure connecting the two envelope membranes.

Outer membrane is an essential organelle of most *Bacteria* (Fig. 2.1) that creates the periplasmic space, an intermediate region between plasma membrane and the environment in which many vital functions are expressed. It may be proposed that the LPS-containing OM is a primary feature of *Bacteria* and that a modified non-LPS OM may have evolved in some diderm phyla. Alternatively, an ancestral OM might not have contained LPS, which could thus be a subsequent specialization of the OM. In this scenario, monoderm bacteria such as Gram-positive bacteria could have lost OM as a secondary adaptation, compensating the lack of OM with a more complex murein wall. Ironically, a subgroup of Gram-positives (*Corynebacteriaceae*) reinvented an outer lipid bilayer unrelated to LPS.

Lipopolysaccharide structure and biosynthesis

Lipopolysaccharide is a unique glycolipid present in Gram-negative bacteria. Immunological, genetics and biochemical studies have contributed to the determination of LPS chemical structure. The availability of several so called 'rough (R) mutants' in *Salmonella* that showed a typical distinct colony morphology compared to the wild type 'smooth (S)' strain provided a powerful tool for initial LPS structural analysis. In fact compared to the wild type S strain, R mutants were sensitive to infection by the P22 phage, showed different serological properties and did not contain rhamnose and mannose residues (Beckmann *et al.*, 1964; Nikaido *et al.*, 1964; Subbaiah and Stocker, 1964) later shown to be specific of the O-antigen portion (Luederitz *et al.*, 1965). More recently biochemical and genetic approaches have fully elucidated the biosynthesis of this complex molecule. Also in the last few years a lot of progress has been made in determining the exact chemical structure not only for *Enterobacteria* but for an increasing number of *Proteobacteria*.

Structure and functions

LPS is typically organized into three structural domains: lipid A, a core oligosaccharide and a highly variable O-antigen constituted of repeating oligosaccharide units (Fig. 2.3) (Raetz and Whitfield, 2002). Lipid A is a unique glycolipid that forms the outer hydrophobic leaflet of the OM (Raetz *et al.*, 2007). The core is covalently linked to lipid A and can be further divided into inner (lipid A proximal) and outer core. The chemical structure of the outer core is variable, whereas the inner core region tends to be quite conserved within a genus or a family. In all species so far analysed, 3-deoxy-D-*manno*-oct-2 ulosonic acid (Kdo) is the first residue linking the inner core to lipid A, and thus Kdo is a chemical hallmark of LPS and a marker of Gram-negative bacteria (Holst, 2007). The O-antigen is the distal, surface

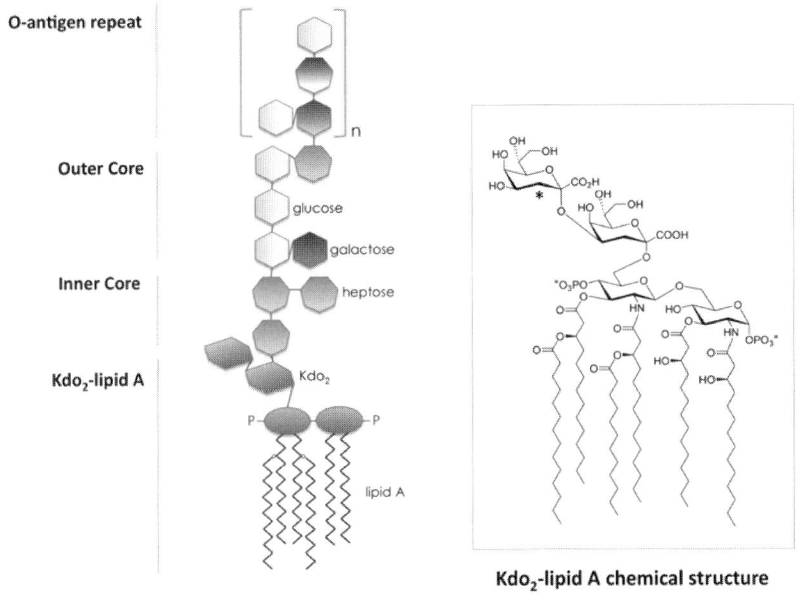

Figure 2.3 Structure of LPS. A. General structure of LPS in Gram-negative bacteria. The Kdo_2 Lipid A moiety, the inner and outer core and the O-antigen repeat are shown. Glucosamine residues are indicated as ovals, Kdo residues as slant hexagons, heptose as heptagon, galactose and glucose as light and dark grey hexagons, respectively. A single repeating unit composing the O-antigen polysaccharide is shown. B. Chemical structure of Kdo_2-lipid A. The glucosamine disaccharide backbone and the Kdo disaccharide are shown. The asterisk indicates the position modified by Kdo dioxygenase to generate the Ko moiety (see text for details).

exposed LPS moiety and responsible of the immunogenic properties of this macromolecule; it is the most variable portion, a feature used as a tool for strains classification based on the different serological properties (Raetz and Whitfield, 2002). In many pathogenic Gram-negative bacteria the O-antigen is a virulence factor that enables the bacterium to escape killing by phagocytosis and serum complement (Raetz and Whitfield, 2002).

Although differing in sugar composition, the common structure of LPS may be seen as the core oligosaccharide and the lipid A; indeed, these two structural domains are present in all LPS analysed so far (Holst, 2007). LPS is essential in most Gram-negative bacteria with the notable exception of *Neisseria meningitidis* (Steeghs *et al.*, 1998); however, the LPS structural requirements for bacterial viability may vary across genera/species. In *E. coli* the minimal LPS structure required for growth has been defined as Kdo_2-lipidA (Raetz and Whitfield, 2002), although the lethal phenotype of Kdo-deficient mutants may be overcome by several suppressor mutations (Meredith *et al.*, 2006). In contrast, to be viable, *Pseudomonas aeruginosa* requires the full inner core and at least part of the outer core in addition to lipid A (Rahim *et al.*, 2000; Walsh *et al.*, 2000).

The structural complexity of LPS reflects the multiple functions displayed by this macromolecule. The outer hydrophilic layer of LPS leaflet in the OM represents a very effective barrier for the spontaneous diffusion of lipophilic compounds, whereas the core, together with the phospholipids of the internal leaflet, forms a hydrophobic barrier. LPS is also a potent activator of the innate immune response and lipid A (also known as endotoxin) represents the conserved molecular pattern recognized by innate immune receptors (Miller *et al.*, 2005).

Several LPS features contribute to the peculiar permeability properties exhibited by the OM: (i) in enterobacteria grown under usual laboratory conditions, the LPS fatty acids substituents are saturated and are thus thought to form a low fluidity gel-like layer (Nikaido, 2003); (ii) the core region is negatively charged due to phosphoryl substituents and sugar acids such as Kdo; in addition, a strong lateral interaction between LPS molecules occurs by the bridging action of Mg^{2+} and Ca^{2+} divalent cations that counteract the negative repulsive charges and stabilize the structure (Nikaido, 2003; Holst, 2007); (iii) finally, the strong association of LPS to OMPs such as FhuA, a ferric hydroxamate uptake receptor, offers an additional mode of interaction between neighbouring LPS molecules (Ferguson *et al.*, 2000). LPS organization is disrupted by defects in assembly of OM components (Ruiz *et al.*, 2006), in mutants producing LPS severely truncated in sugar chains ('deep rough' mutants) (Young and Silver, 1991) or by exposure to antimicrobial peptides and chelating agents such as EDTA, which displace the divalent cations that shield the repulsive charges between LPS molecules (Nikaido, 2003). In all these cases, the consequence is that much of the LPS layer is shed and phospholipids from the inner leaflet migrate into the breached areas of the outer leaflet. These locally symmetrical bilayer rafts are more permeable to hydrophobic molecules, which can thus gain access to the periplasm while the OM continues to retain the more polar periplasmic contents (Nikaido, 2005). Therefore, appreciable levels of phospholipids in the outer leaflet of the OM are detrimental to the cell and thus, not surprisingly, cells have evolved systems to monitor the asymmetry of the OM and to respond either by removing phospholipids from the outer leaflet or by modifying LPS.

Two main mechanisms have been described that restore OM asymmetry by acting on phospholipids migrated into the outer leaflet: the phospholipase PldA (also known as OMPLA), which degrades the invading lipid molecules, and the Mla pathway, which removes phospholipids from the outer leaflet.

PldA normally exists as an inactive monomer in the OM. PldA phospholipase activity is modulated by a reversible dimerization mechanism triggered by events that promote the migration of glycerophospholipids into the OM outer leaflet (Dekker, 2000). Activated PldA sequesters and destroys the invading lipid substrates, thus the enzyme proposed function is to maintain lipid asymmetry of the OM under stress conditions provided that enough substrate is available to promote dimerization (Dekker, 2000).

The Mla (maintenance of OM lipid asymmetry) proteins constitute a highly conserved ABC transport system that prevents phospholipids accumulation in the outer leaflet of the OM under non stress conditions. Mla operates in the absence of PldA but the converse is not true; indeed, the Mla pathway inhibits the activation of phospholipases in non-stressful conditions (Malinverni and Silhavy, 2009). Based on these observations it has been proposed that the Mla proteins constitute a bacterial intermembrane phospholipid trafficking system (Malinverni and Silhavy, 2009). In agreement with the proposed function of the Mla system, we recently found that in cells depleted for LptC, an IM protein implicated in LPS transport to the OM (see below), MlaD is up-regulated, thus suggesting that LPS export to the cell surface and phospholipids removal from the OM are functionally interconnected pathways (P.L. Mauri and A. Polissi, unpublished results).

An alternative response to OM asymmetry perturbation consists in LPS modification. LPS can be palmitoylated at the position 2 of lipid A by PagP, an OM β-barrel acyltransferase that utilizes phospholipids migrated in the OM as the substrate (Bishop *et al.*, 2000). The product of the PagP reaction is an hepta-acylated LPS which possesses increased hydrophobicity (Bishop, 2008) and therefore contributes to restore the permeability barrier function of the OM. The active site residues of PagP map to the extracellular surface of the outer membrane, and thus the reaction can proceed only upon phospholipids migration to the outer leaflet (Hwang *et al.*, 2002). In *Salmonella pagP* is regulated by the PhoP/PhoQ regulatory system, which senses low Mg^{2+} concentration, a condition encountered during infection (Guo *et al.*, 1998). Moreover, it appears that Lipid A palmitoylation is also a regulated process in other human, insect, and plant pathogens (Fukuoka *et al.*, 2001; Derzelle *et al.*, 2004; Rebeil *et al.*, 2004), thus supporting the hypothesis that such LPS modification contributes to adaptation to the host.

Lipid A is the LPS conserved structure that is recognized by specific receptors on cells of the innate immune system. Innate immune receptors recognize microorganisms specific motifs named PAMPs (Pathogens Associated Molecular Patterns) to signal and activate complex signalling cascades that lead to the release of pro-inflammatory cytokines (Miller *et al.*, 2005). Recognition of lipid A requires the TLR4–MD2 complex (Medzhitov *et al.*, 1997; Shimazu *et al.*, 1999) and the accessory protein CD14 and LPB (LPS binding protein) (Miyake, 2006). LPB converts oligomeric micelles of LPS to a monomer for delivery to CD14 which in turn transfers LPS to the TLR4-MD2 receptors complex (Miyake, 2006). The structures of TLR4, MD2 and CD14 have been solved both alone and in complex with either lipid A or lipid A antagonists (Kim *et al.*, 2005, 2007; Park *et al.*, 2009). TLR4 and MD2 form a heterodimer and LPS binding induces the dimerization of the TLR4/MD2 complex to form the activated heterotetrameric complex that initiates signal transduction. MD2, which belongs to a small family of lipid binding proteins, plays a key role in initial lipid A recognition by accommodating the lipid acyl chains into its large hydrophobic pocket. The two phosphate groups of lipid A bind to the TLR4/MD2 complex by interaction with positively charged residues located on both proteins (Park *et al.*, 2009). These structural studies greatly contributed to our understanding of how lipid A is recognized and how it induces the innate immune response. This information may be extremely useful to understand the molecular basis of recognition and binding of other LPS binding proteins including the Lpt proteins (see below) implicated in LPS export to the cell surface.

Overview of LPS biosynthesis

The biosynthesis of LPS is a complex process requiring spatial and temporal coordination of several independent pathways that converge in an ordered assembly line to give the mature molecule (Fig. 2.4) (Raetz and Whitfield, 2002; Samuel and Reeves, 2003; Valvano, 2003). The lipid A-core domain is synthesized in the cytoplasm and at the inner leaflet of the IM and requires the convergent synthetic pathways of the lipid A moiety, the Kdo residue and the oligosaccharide core. Then the assembled lipid A-core moiety is flipped over the IM by the ABC transporter MsbA and becomes exposed in the periplasm (Polissi and Georgopoulos, 1996; Zhou *et al.*, 1998). The biosynthesis of the O-antigen repeating units occurs at the cytoplasmic face of the IM where it is assembled on a lipid carrier and then translocated across the IM. The lipid A-core and O-antigen biosynthetic pathways converge with the

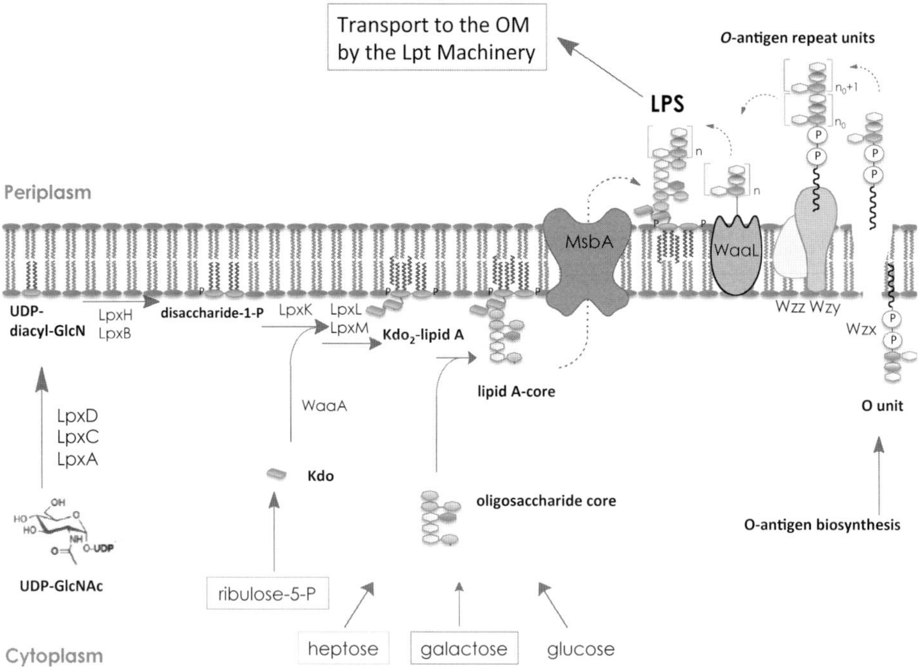

Figure 2.4 Convergent pathways for LPS biosynthesis in *E. coli*. Starting from bottom left (cytoplasm and inner leaflet of IM): UDP-2,3-diacylglucosamine (UDP-diacyl-GlcN) is synthesized in the cytoplasm by the action of LpxA, LpxC and LpxD enzymes. The synthesis of β–1′–6 linked disaccharide (Disaccharide-1-P) requires LpxH and LpxB. The Kdo_2-lipid A is synthesized from the sequential action of LpxK, WaaA, which transfers two molecules of Kdo, and the late acyltransferases LpxL and LpxM. Core oligosaccharide is assembled on Kdo_2-lipid A via sequential glycosyl transfer of sugar precursors. Lipid A-core is translocated across the IM by the ABC transporter MsbA. Starting from bottom right (cytoplasm and inner leaflet of IM): *O*-antigen repeat units are synthesized in the cytoplasm and at the IM; they are then transported and (outer leaflet of IM) polymerized via a separated pathway (Wzx–Wzy dependent pathway). Lipid A-core ligation to O-antigen polysaccharide occurs at the periplasmic face of the IM by the action of WaaL ligase. Symbols are as shown in Figure 2.2. LPS is then delivered to the Lpt machinery (see text for details).

ligation of O-antigen to the lipid A-core moiety at the periplasmic side of the IM mediated by the WaaL ligase to form the mature LPS molecule. The O-antigen domain is not essential and is missing in common laboratory *E. coli* K12 strains due to mutations in *wbbL*, a gene implicated in O-antigen repeat biosynthesis (Reeves *et al.*, 1996; Rubires *et al.*, 1997).

Biosynthesis of lipid A core

The enzymes for lipid A biosynthesis are constitutively expressed and are located in the cytoplasm or at the inner leaflet of the IM (Raetz and Whitfield, 2002; Raetz *et al.*, 2009). The first step of lipid A biosynthesis is the acylation of UDP-N-acetylglucosamine (UDP-GlcNAc) catalysed by LpxA. The enzyme has a strict dependence for a β-hydroxymyristoyl acyl carrier protein and functions as an accurate hydrocarbon ruler that incorporates a β-hydroxymyristoyl chain two orders of magnitude faster than β-hydroxylauroyl or β-hydroxypalmitoyl chain (Anderson and Raetz, 1987; Wyckoff *et al.*, 1998). The next step

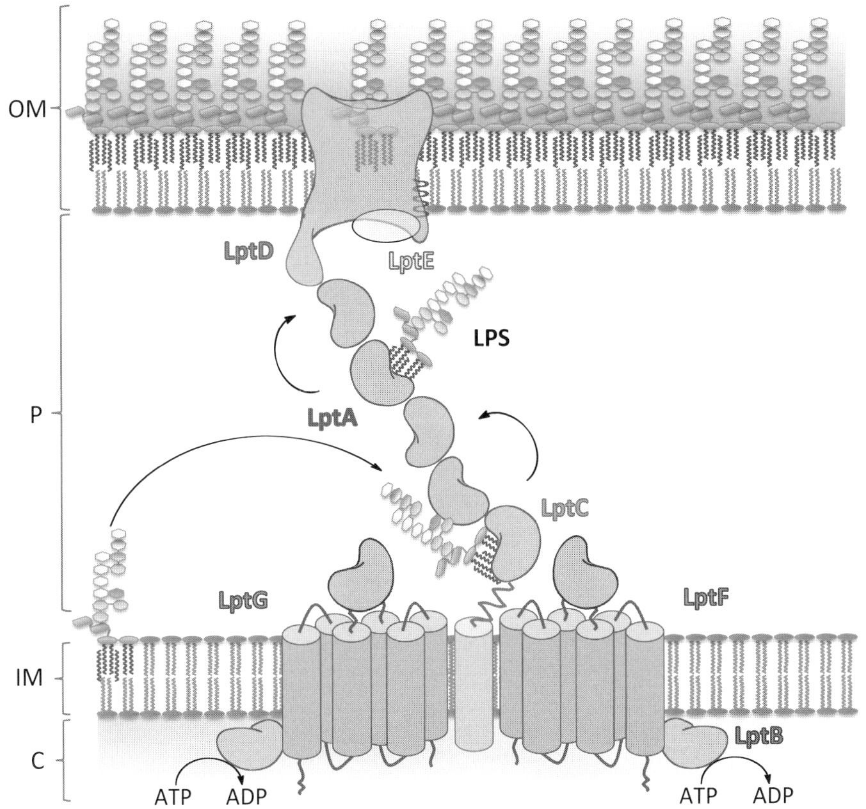

Figure 2.5 LPS transport through the cell envelope. The flipped LPS molecule is extracted from the IM by the ABC transporter LptBCFG. According to the 'trans–envelope complex' model LptA, LptE and LptD constitute a multiprotein complex with LptBCFG which spans the cell envelope by bridging IM and OM components (see text for details). The IM and OM components are indicated as in Figure 2.2. The name of the specific Lpt components is indicated; the bean shape indicates the OstA_N domains present in different proteins of the complex (see Figure 2.6 and text for details). A colour version of this figure is located in the plate section at the back of the book.

involves the deacetylation of UDP-3-O-acyl-GlcNAc catalysed by LpxC, a Zn^{2+}-dependent enzyme (Jackman *et al.*, 1999). LpxC is an attractive target for developing antibiotics inhibiting lipid A biosynthesis as it is well conserved in diverse Gram-negative bacteria and does not possess sequence similarity to other deacetylases or amidases (Onishi *et al.*, 1996). Following deacetylation a second β-hydroxymyristoyl chain is added by LpxD to form UDP-2,3-diacylglucosamine (Bartling and Raetz, 2008) which is cleaved by the pyrophosphatase LpxH to produce UMP and 2,3-diacylglucosamine-1-phosphate also known as lipid X (Babinski *et al.*, 2002). The disaccharide synthase LpxB catalyses the condensation of one molecule of UDP-2,3-diacylglucosamine with one molecule of lipid X to form a β-1′-6-linked disaccharide (Radika and Raetz, 1988).

LpxA, LpxC and LpxD are soluble proteins whereas LpxH and LpxB are peripheral membrane proteins (Raetz *et al.*, 2009); the homologues of such genes have been identified in other Gram-negative bacteria by sequence comparison, with the exception of LpxH, which appears to be missing in all *α-Proteobacteria* and many *δ-Proteobacteria,* although they use a similar lipid A biosynthetic pathway (Price *et al.*, 1994; Gonzalez *et al.*, 2006). Based on the notion that genes involved in the same pathway are often clustered, a gene (named LpxI) located between the *lpxA* and *lpxB* in *Caulobacter crescentus* has been shown to encode an alternative pyrophosphatase and to rescue the conditional lethal phenotype of an *lpxH*-deficient *E. coli* mutant. LpxH from *C. crescentus* catalyses *in vitro* the hydrolysis of UDP-2,3-diacylglucosamine thus indicating that LpxH and LpxI are functional homologues (Metzger and Raetz, 2010).

An integral IM protein, LpxK, catalyses the addition of a phosphate group to the 4′-position of the tetra-acylated disaccharide 1-phosphate thus producing lipid IV_A (Garrett *et al.*, 1997). The reaction catalysed by LpxK precedes the addition of the Kdo residues by the bifunctional enzyme WaaA. Kdo is synthesized by a separate pathway and requires four sequentially acting enzymes (Cipolla *et al.*, 2009). WaaA is a CMP-Kdo dependent transferase that catalyses the sequential incorporation of two activated CMP-Kdo residues in *E. coli* (Belunis and Raetz, 1992). WaaA homologues from different bacterial species can transfer up to four Kdo residues, thus accounting for the differences observed in the structure of lipid A-core moieties across species (Raetz and Whitfield, 2002; Holst, 2007). However, in few species such as *Yersinia pestis* and *Burkholderia cepacia* the outer Kdo residue may be replaced by the stereochemically similar sugar Ko (D-glycero-D-*talo*-oct 2-ulosonic acid) in which the axial hydrogen atom at the 3-position is replaced by an OH group (Fig. 2.3) (Vinogradov *et al.*, 2002; Isshiki *et al.*, 2003). The biosynthesis of Ko has not been elucidated yet; however, a recent study reports the identification of a unique Kdo hydrolase (KdoO) that is present in *Burkholderia ambifaria* and in *Y. pestis* that catalyses the hydroxylation of the deoxy-sugar residue Kdo (Chung and Raetz, 2011). The biological function of Ko is not known; it has been speculated that the extra OH group in Ko may facilitate hydrogen bonding between adjacent LPS molecules and therefore provide an advantage under stress conditions (Chung and Raetz, 2011).

In *E. coli* and *Salmonella* the synthesis of the final hexa-acylated lipid A (Kdo_2-lipid A) requires the two late acyltransferases LpxL and LpxM that catalyse the addition of secondary acyl chains to the distal glucosamine (Clementz *et al.*, 1996; Clementz *et al.*, 1997). However, *Pseudomonas aeruginosa* LPS biosynthesis differs in that fully acylated lipid A is required before Kdo residues addition (King *et al.*, 2009). The additional sugars composing the oligosaccharide core are then added to Kdo_2-lipid A by specific glycosyl-transferases to generate the lipid A-core structure (Raetz and Whitfield, 2002).

The enzymes for the biosynthesis of Kdo_2 -lipid A are constitutively expressed. However, in *E. coli* the production of Kdo_2 -lipid A is post-transcriptionally regulated by FtsH, an essential membrane bound protease belonging to the AAA family (ATPase associated with various cellular activities) that controls the turnover of LpxC (Ogura *et al.*, 1999). Mutations in *ftsH* lead to increased cellular levels of LpxC and are lethal (Ogura *et al.*, 1999). This can be explained by the fact that both lipid A and phospholipid biosynthetic pathways largely depend on the same precursor molecule, R-3-hydroxymyristoyl ACP. The increased level of LpxC may thus effectively deplete the R-3-hydroxymyristoyl ACP pool, thus leading to an imbalanced phospholipid/LPS ratio in the OM (Ogura *et al.*, 1999), which, as mentioned above, is crucial for survival of most Gram-negative bacteria. More recently, it has been shown that FtsH also controls the turnover of WaaA, the CMP-Kdo-dependent transferase that catalyses incorporation of Kdo residues (Katz and Ron, 2008). Therefore

FtsH dependent proteolysis seems to be essential for balancing the levels of two key components of the lipid A-core moiety. Control of LPS biosynthesis by FtsH-mediated proteolysis, however, is not a widespread mechanism across Gram-negative bacteria but seems restricted to *Enterobacteria*; indeed the C-terminus of LpxC (where sequence specific degradation signals are located) differs significantly among species whereas the overall sequence of LpxC is highly conserved (Langklotz *et al.*, 2011). Interestingly, the turnover of LpxC in some *α-Proteobacteria* such as *Agrobacterium tumefaciens* and *Rhodobacter capsulatus* depends on the Lon protease, whereas in *P. aeruginosa* the control of LPS biosynthesis seems to be independent of proteolysis and lipid A core biosynthesis might thus be regulated by a yet unknown mechanism (Langklotz *et al.*, 2011).

Assembly of mature LPS at the outer leaflet of IM

Translocation of lipid A core across the IM

After biosynthesis, the lipid A-core is anchored to the IM with its hydrophilic moiety exposed to the cytoplasm and is then flipped across the IM by the essential ABC (ATP binding cassette) transporter MsbA (Fig. 2.4) (Table 2.1) (Davidson *et al.*, 2008). The 64.3 kDa peptide encoded by *msbA* is described as a 'half-transporter' and the functional MsbA protein is presumed to be a homodimer. MsbA belongs to a class of ABC transporters with the transmembrane domain (composed by 6 membrane-spanning helices believed to contain the substrate-binding site) is fused to the nucleotide-binding domain (NBD) (Davidson *et al.*, 2008). Substrate transport is driven by the energy provided by ATP hydrolysis.

MsbA was originally identified in *E. coli* as a multicopy suppressor of the thermosensitive phenotype of a *htrB* deletion mutant (Karow and Georgopoulos, 1993). *htrB* (now renamed *lpxL*) encodes one of the two late Kdo-dependent acyltransferases responsible for the addition of lauryl moieties to the tetra-acylated Kdo_2-lipid IVA, thus forming the penta-acylated Kdo_2-lipid A (Clementz *et al.*, 1996). Mutants in *htrB/lpxL* are not viable at temperatures above 33°C and produce underacylated LPS that is not efficiently transported to the OM (Zhou *et al.*, 1998). Under non-permissive conditions, the *htrB/lpxL* null mutant shows alterations in cell morphology (such as formation of bulges and filaments), accumulates phospholipids (Karow *et al.*, 1992) and the tetra-acylated LPS precursor in the IM (Polissi and Georgopoulos, 1996; Zhou *et al.*, 1998). Expression of *msbA* from a plasmid vector in the *htrB/lpxL* null mutant suppresses the thermosensitive growth defect and the abnormal phospholipids overproduction, and restores tetra-acylated LPS precursor translocation and transport to the OM. Therefore, the higher expression of MsbA at higher temperature does not restore lipid IVA acylation to give lipid A but seems to facilitate the transport of the immature LPS form to the OM (Zhou *et al.*, 1998). By contrast, MsbA depleted cells accumulate hexa-acylated lipid A at the IM (Zhou *et al.*, 1998), thus further implicating MsbA in LPS transport. The key role of MsbA in lipid trafficking was proposed by Doerrler and co-workers (Doerrler *et al.*, 2001), who showed that in an *E. coli* thermosensitive *msbA* mutant carrying a single amino acid substitution (A270T) in a transmembrane region of the protein, the transport of both LPS and phospholipids to the OM was inhibited at the non permissive temperature, thus suggesting that *E. coli* MsbA is needed to export both major membrane lipids (Doerrler *et al.*, 2001).

In *N. meningitidis* the *msbA* gene is not essential for cell viability as this bacterium can survive without LPS (Steeghs *et al.*, 1998). *N. meningitidis msbA* mutants produce reduced

Table 2.1 Components of the LPS transport machine.

Gene name	Synonyms	Chromosomal location	Protein MW (kDa)	Protein localization	Protein properties/function	References
lptA	*yhbN*	*yrbG–lptB* locus	18.6	Periplasm/associates to IM and OM Lpt components	Binds LPS	Serina et al. (2004), Sperandeo *et al.* (2006, 2007, 2008), Tran et al. (2008), Chng et al. (2010a)
lptB	*yhbG*	*yrbG–lptB* locus	26.7	Cytoplasm/IM-associated	ABC protein; component of IM ABC transporter	Stenberg *et al*. (2005), Sperandeo *et al*. (2006, 2007, 2008)
lptC	*yrbK*	*yrbG–lptB* locus	21.6	IM/periplasm exposed	Component of IM ABC transporter; binds LPS	Sperandeo *et al*. (2006, 2008), Narita *et al*. (2009), Tran *et al*. (2010)
lptD	*imp; ostA*	*lptD–apaH* operon	87	OM	β-barrel component of OM complex for LPS assembly	Sampson *et al*. (1989), Aono *et al*. (1994), Braun and Silhavy (2002), Bos *et al*. (2004), Chng *et al*. (2010b)
lptE	*rlpB*	*leuS–cobC* operon	21.2	OM	LptD-associated lipoprotein; binds LPS	Takase *et al*. (1987), Wu *et al*. (2006), Chng *et al*. (2010b)
lptF	*yjgP*	*lptF–lptG* operon	40.2	IM	Component of IM ABC transporter	Ruiz *et al*. (2008), Narita *et al*. (2009)
lptG	*yjgQ*	*lptF–lptG* operon	39.5	IM	Component of IM ABC transporter	Ruiz *et al*. (2008), Narita *et al*. (2009)
msbA		*ycaI–ycaQ* operon	64.3	IM	ABC transporter; LPS flipping over IM	Karow and Georgopoulos (1993), Doerrler *et al*. (2001), Doerrler *et al*. (2004), Eckford and Sharom (2010)

amounts of LPS, a feature typical of mutants in LPS transport in this organism, but possess an OM mostly composed of phospholipids, indicating that phospholipid transport to the OM is not impaired and suggesting a difference in general lipid transport with respect to *E. coli* (Tefsen *et al.*, 2005a). In *P. aeruginosa* MsbA is essential as expected for an organism that requires the lipid A with at least part of the outer core to be viable (see above). However, *msbA* from *E. coli* cannot cross-complement *msbA* merodiploid cells of *P. aeruginosa*. Moreover, differences between the corresponding gene products are remarked by the observation that the kinetic parameters of purified and reconstituted *P. aeruginosa* MsbA considerably differ from those of *E. coli* MsbA (Ghanei *et al.*, 2007).

A topological analysis of lipids *in vivo* can be performed using as markers covalent modifications catalysed by compartment-specific enzymes. The topology of newly synthesized lipid A in the temperature sensitive $msbA_{A270T}$ mutant was assessed in a polymyxin-resistant genetic background (Doerrler *et al.*, 2004). In *E. coli* and *Salmonella* polymyxin resistance depends on enzymes acting at the periplasmic side of the IM that covalently modify lipid A with cationic substituents (Raetz *et al.*, 2007). Upon MsbA inactivation at high temperature, newly synthesized lipid A was not modified, suggesting that the molecule accumulates in the IM facing the cytoplasm (Doerrler *et al.*, 2004). This is consistent with a model of MsbA-mediated LPS flipping over the membrane leaflets, rather than translocation and ejection from the bilayer.

Very recently several mutations in the NBD of *E. coli* MsbA have been characterized *in vitro* by fluorescent ATP binding, radioactive ATP hydrolysis assays, and Electron Paramagnetic Resonance (EPR) spectroscopy (Schultz *et al.*, 2011a,b). Analysis of two loss-of-function mutations, L511P and D512G, originally identified by Polissi and Georgopoulos as unable to support cell growth *in vivo* but still able to bind ATP (Polissi and Georgopoulos, 1996), revealed that the L511P substitution prevents effective ATP hydrolysis whereas the D512G mutant enzyme is still proficient in ATP hydrolysis but does not undergo to the conformational rearrangement required for flipping lipid A, as assessed by EPR spectroscopy (Schultz *et al.*, 2011b). The characterization of E506Q and H537A amino acid substitutions revealed that the corresponding proteins are still able to bind ATP but exhibit a severely reduced rate of ATP hydrolysis (Schultz *et al.*, 2011a). Spectroscopy data show that the mutated proteins are locked in a closed dimer conformation even when the hydrolysed nucleotide is released (Schultz *et al.*, 2011a). Collectively these studies have identified residues within the NBD domain that are either necessary for efficient ATP hydrolysis (L511) or for the conformational rearrangements required during flipping (E506, D512, H537).

Several *in vitro* studies have been performed to evaluate MsbA substrate specificity. The basal ATPase activity of purified MsbA reconstituted into liposomes is stimulated by hexa-acylated lipid A, Kdo_2-lipidA, or LPS but not by underacylated lipid A precursors, suggesting that hexa-acylated LPS is the substrate required for the transport (Doerrler and Raetz, 2002), in line with previous genetic and biochemical evidence (Zhou *et al.*, 1998). This work was further expanded by functional reconstitution of the protein into proteoliposomes of *E. coli* lipids to estimate MsbA binding affinities for nucleotides and putative transport substrates (Eckford and Sharom, 2008). Using purified labelled MsbA, simultaneous high-affinity binding of lipid A and daunorubicin was demonstrated (Siarheyeva and Sharom, 2009). These results indicate that MsbA contains two substrate binding sites that communicate with both the nucleotide-binding domain and with each other. One is a high-affinity binding site for the physiological substrate, lipid A, and the other site interacts with drugs with comparable affinity. Thus, MsbA may function as both a lipid flippase and a multidrug transporter (Siarheyeva and Sharom, 2009). Early attempts to demonstrate MsbA-mediated lipid flipping *in vitro* failed (Kol *et al.*, 2003). However, a direct measurement of the lipid flippase activity of purified MsbA in a reconstituted system has been recently reported (Eckford and Sharom, 2010).

The X-ray crystal structures of MsbA from the three closely related orthologues from *E. coli*, *Vibrio cholerae*, and *S. enterica* (serovar Typhimurium) in different conformations were recently reported (Ward *et al.*, 2007), after the original MsbA structures were withdrawn due to the discovery of a flaw in the software used to solve them (Chang *et al.*, 2006). The overall shape and domain organization of MsbA resemble that of the 3.0-Å structure of the putative bacterial multidrug transporter Sav1866 (Dawson and Locher, 2006) and the 8-Å cryo-EM structure of Pgp (Rosenberg *et al.*, 2005). The analyses of crystal structures of MsbA trapped in different conformations indicate that this molecule may undergo large ranges of motion that may be required for substrate transport (Ward *et al.*, 2007). Collectively, these results show that MsbA has the potential, at least *in vitro*, to handle a variety of substrates as expected from a protein belonging to the subfamily of drug-efflux transporters. However, *in vivo* MsbA displays a remarkable selectivity towards the LPS substrates, being capable of translocating only hexa-acylated but not penta- or tetra-acylated LPS. This observation, together with data that will be discussed in the following paragraphs, suggests that MsbA may play the role of 'quality control system' for LPS export to the OM.

Lipid-A core modification systems

Following MsbA mediated translocation the nascent core-lipid A moiety may undergo diverse covalent modifications during its transit from the outer surface of the IM to the OM. These modifications are not essential for growth but confer an advantage to the bacterium in evading the innate immune system (Raetz *et al.*, 2007). The majority of such modifications are regulated and, in most cases, relevant only during specific phases of the bacteria life cycle. Regulation of LPS modifications has been extensively studied in *Salmonella*, where it occurs via the PhoP/PhoQ and PmrA/PmrB two-component systems (Gunn, 2008).

Several bacteria such as *Rhizobium* and *Francisella* can remove the phosphate moieties from positions 1 and 4′ of lipid A by two distinct inner membrane phosphatases designated LpxE and LpxF, respectively (Raetz *et al.*, 2007). Interestingly, lipid A cannot be dephosphorylated when LpxE and LpxF are expressed in a conditional *E. coli* MsbA mutant unable to transport the core-lipid A across the IM (Wang *et al.*, 2004, 2006); this is consistent with the proposed localization of their active sites at the periplasmic side of the IM.

Decoration of phosphate groups may occur in both *E. coli* and *Salmonella* by addition of L-Ara4N (4-amino-4-deoxy-L-arabinose) and PEtN (phosphoethanolamine) catalysed by the ArnT (Trent *et al.*, 2001b) and EptA (Lee *et al.*, 2004) enzymes, respectively. These modifications mask phosphate groups with positively charged moieties and, when present in LPS, confer resistance to antimicrobial peptides. Expression of both enzymes is regulated by the PmrA/PmrB bacterial two-component system (Gunn, 2008).

The number of acyl chains in core-lipid A may also be modulated. Three different enzymes a have been implicated in such modifications: PagP, PagL and LpxR. The OMP PagP, mentioned above, catalyses the addition of palmitate residue at position 2 of lipid A acyl chains. According to both X-ray and NMR structures of PagP from *E. coli*, the active site of the enzyme faces the exterior of the cell (Hwang *et al.*, 2002). PagL is an OM lipase regulated by the PhoP/PhoQ two component system responsible of 3-O-deacylation of lipid A (Trent *et al.*, 2001a) whereas LpxR is a distinct OM lipase cleaving 3′-O-linked acyl chains (Reynolds *et al.*, 2006).

Based on their subcellular localization and mechanism of action, lipid A modification enzymes have been extremely useful as reporters for LPS trafficking within the bacterial envelope (see below).

O-antigen biosynthesis, transport and ligation to the lipid A core

The structural diversity of the O-antigens stems from variation in sugar composition and the sequence of sugars and linkages. The oligosaccharide units (O-units) composing the O-antigens are synthesized as lipid-linked intermediates and then assembled (Fig. 2.4). The lipid component is undecaprenyl phosphate (Und-P), a C_{55} polyisoprenoid derivative (Whitfield, 1995; Raetz and Whitfield, 2002). The enzymes implicated in the synthesis of the O-unit are either integral membrane proteins or associated with the cytoplasmic site of the IM by ionic interaction (Raetz and Whitfield, 2002). Most of the O-units is exported and assembled by the so-called Wzx/Wzy-dependent pathway (Raetz and Whitfield, 2002). At least three proteins, Wzx, Wzy and Wzz, are involved in this export pathway. Wzx is an integral membrane protein postulated as a candidate for the O-unit flippase across IM (Marolda *et al.*, 2010). The Wzy protein is required for the polymerization of Und-PP-linked O-units at the periplasmic face of the IM (McConnell *et al.*, 2001). Chain length distribution of *O*-antigen polysaccharide depends on Wzz that belongs to a family of protein called 'polysaccharide co-polymerases' (Morona *et al.*, 2000). An ABC transporter-dependent pathway may represent an alternative export mechanism. The most significant features of this pathway are that the completion of the O-specific polysaccharide occurs at the cytosolic side of the IM and the export of the polymer across IM requires an ABC transporter (Zhang *et al.*, 1993).

Irrespective of the export and polymerization mode, the assembly of the mature LPS molecule occurs at the periplasmic face of the IM where ligation of assembled Und-PP linked O-antigens to the lipid A-core moiety takes place. This reaction is catalysed by a specific glycosyltransferase, an integral membrane protein encoded by the *waaL* gene (Raetz and Whitfield, 2002). Mutant strains devoid of *waaL* are viable but cannot ligate O-antigen molecules to lipid A core and thus produce LPS lacking O-antigen polysaccharide and accumulate membrane bound Und-PP linked O-antigen molecules (McGrath and Osborn, 1991). WaaL displays relaxed substrate specificity, as donor Und-PP linked glycans for the ligation reaction can originate from various biosynthesis pathways. For example colanic acid, a cell surface capsular material that is produced upon cold shock or other stress conditions and that is usually loosely associated with the bacterial cell, can be covalently linked to the lipid A-core by WaaL at the same position as the O-antigen (Meredith *et al.*, 2007; Sperandeo *et al.*, 2008).

Finally, the mature LPS molecule is transported to the cell surface. In the following paragraphs the advances made in the last decade in understanding the LPS export pathway downstream of MsbA-mediated translocation across the IM will be reviewed.

LPS transport to the outer membrane

The Lpt machinery: identification of the genes, structure and organization of the components across IM and OM

The mature LPS molecule assembled at the periplasmic face of the IM must then traverse the aqueous periplasmic compartment before being inserted and correctly assembled at the OM. As mentioned above the periplasm is devoid of high-energy phosphate bound molecules as ATP (Oliver, 1996), therefore the transport across the periplasm occurs in absence of an obvious energy source.

In 1972, exploiting sucrose density gradient ultracentrifugation to separate IM and OM from *S. enterica (serovar* Typhimurium), Osborn and collaborators demonstrated for the first time that LPS transport from the site of synthesis at the IM to the OM is unidirectional (Osborn *et al.*, 1972). However, it took several decades to unravel the first molecular details of this process. Unlike MsbA, whose role in LPS flipping across the IM has been clearly established during the last two decades (Doerrler *et al.*, 2001, 2004; Doerrler and Raetz, 2002) (see above), most of the factors involved in LPS transport downstream of MsbA have been identified only in the past 5 years. Nevertheless, during these few years, the combined contribution of genetic, biochemical and bioinformatic approaches has led different laboratories to discover the Lpt complex required for LPS transport downstream of MsbA.

The *E. coli* Lpt (lipopolysaccharide transport) complex is composed of seven essential and variably conserved proteins (LptABCDEFG) that are located in every cellular compartment: cytoplasm, IM, periplasm and OM (Fig. 2.5) (Table 2.1). The Lpt complex provides energy for LPS extraction from the IM and mediates its transport across the aqueous periplasm and its insertion and assembly at the OM (Ruiz *et al.*, 2009; Sperandeo *et al.*, 2009).

This complex may be divided in three subassemblies: LptBCFG, LptA, and LptDE which are located at the IM, in the periplasm, and at the OM, respectively. LptBFG is an IM-associated ABC transporter that harbours an atypical subunit constituted by the bitopic IM protein LptC, whose function in the ABC transporter has not yet been clarified (Narita and Tokuda, 2009). LptB is the ATP binding domain of this transporter and is phylogenetically related to ABC proteins of hydrophobic amino acid uptake systems (Saurin *et al.*, 1999). LptF and LptG are the transmembrane subunits of this ABC transporter. LptA is a periplasmic protein and is reminiscent of the substrate binding proteins often related to importers in *E. coli*. At the OM resides the LptDE complex, which is composed by the β-barrel protein LptD and the lipoprotein LptE.

The reason why this field has been left unexplored for so many years is that the identification of OM biogenesis factors has been challenged by the lack of specific phenotypes in LPS transport mutants that often made difficult the design of genetic selections.

The long journey of LPS from the IM to the OM has been unveiled starting from its end. *lptD* (formerly designated *imp* for increased membrane permeability or *ostA* for organic solvent tolerance) was the first gene isolated in a genetic selection designed with the aim of obtaining mutations affecting OM permeability (Sampson *et al.*, 1989). In that pioneering work, a mutant lacking the maltodextrins-specific channel LamB was grown in the presence of maltodextrins larger than maltotriose as a sole energy and carbon source and mutants with altered OM permeability, which allowed the entry of these large molecules through the OM, were isolated. Two such mutants bore mutations that not only allowed growth on maltodextrins but also conferred sensitivity to several hydrophobic and hydrophilic antibiotics, thus suggesting that OM barrier integrity was impaired. These two mutations mapped into the *lptD* gene that was shown in the same work to be essential.

Interestingly, *lptD* was identified again five years later in an independent genetic screen as the responsible of increased resistance to organic solvent and designated *ostA* (Aono *et al.*, 1994), thus confirming that alterations of LptD functionality actually result in permeability defects.

LptD is an 87 kDa OM protein characterized by a C-terminal β-barrel domain (aa 203–784) and a periplasmic N-terminal domain (a.a. 25–202). This protein is conserved among the major classes of *Proteobacteria* and its presence in different genomes correlates with the presence of the second lipid A biosynthesis enzyme LpxC (Table 2.2).

Initially, a role for LptD in OMPs biogenesis was proposed, based on the observation that LptD depletion results in accumulation of newly synthesized proteins and lipids in a membrane fraction with higher density than the OM in sucrose density gradient centrifugation (Braun and Silhavy, 2002). The appearance of this novel membrane fraction was attributed to the unbalanced protein/lipid ratio resulting from OMPs mislocalization. This hypothesis was further supported by the finding that *lptD* belongs to the σ^E regulon, which controls envelope biogenesis genes in response to extracytoplasmic stresses (Dartigalongue *et al.*, 2001) and by the observation that it is genetically linked to *surA*, a gene coding for the primary chaperon involved in transit of the bulk mass of OMPs through the periplasm (Missiakas *et al.*, 1996; Sklar *et al.*, 2007).

However, the function of LptD was clearly demonstrated two years later by Tommassen's group by exploiting the ability of *N. meningitidis* to survive without LPS (Steeghs *et al.*, 1998). The authors demonstrated that in mutants lacking the neisserial *lptD* orthologue, which are viable, LPS is not accessible to extracellularly added neuraminidase, an enzyme that modifies LPS by adding sialic acid residues, and its lipid A moiety is not deacylated by the ectopically expressed OM deacylase PagL, thus suggesting that these *lptD* mutants are unable to transport LPS to the cell surface (Bos *et al.*, 2004). In addition in these mutants the LPS total content in the cell is dramatically decreased, as previously observed in an *msbA* knockout mutant that in this organism is viable (Tefsen *et al.*, 2005a).

The role of *E. coli* LptD in LPS assembly to the cell surface was further confirmed by two different works showing that LptD depletion prevents newly synthesized LPS from reaching the OM (Wu *et al.*, 2006; Sperandeo *et al.*, 2008).

As early observations suggested that LptD exists as a high molecular weight complex in the OM (Braun and Silhavy, 2002), Kahne and co-workers searched for additional Lpt factors by affinity purification. Using a His-tagged version of LptD the authors enriched on a Ni-NTA column the LptD-containing protein complex from solubilized OM extracts of *E. coli*. The LptD-interacting protein was subjected to tandem mass spectrometry and was identified as the essential 21.2 kDa rare lipoprotein formerly known as RlpB, and now renamed LptE (Wu *et al.*, 2006). The role of LptE in LPS transport to the OM was demonstrated by assessing the occurrence of PagP-mediated lipid A modification in newly synthesized LPS extracted from LptE depleted cells. In both LptE and LptD depleted cells newly synthesized LPS fails to be modified by PagP, thus proving that both proteins are implicated in the LPS transport pathway (Wu *et al.*, 2006).

Recent work by the same group has provided new insights into the structure and biogenesis of the LptDE complex. Using proteolysis experiments coupled to size exclusion chromatography, they demonstrated that the C-terminal domain of LptD strongly interacts

Table 2.2 Distribution of protein domains functioning in LPS biogenesis[1]

Phylum	Order	LpxC[2]	LpxK[2]	WaaA[2]	OstA_N[3]	LptC[3]	LptE[3]	YjgP_YjgQ[3]
		PF03331[3]	PF02606[3]	PF04413.10[3]	PF03968[3]	PF06835[3]	PF04390[3]	PF03739[3]
α-Proteobacteria	Rickettsiales	19	21	17	15	18	0	19
	Rhizobiales	87	88	89	89	82	0	89
	Rhodospirillales	17	17	19	16	16	0	17
β-Proteobacteria	Neisseriales	32	32	32	31	32	32	32
	Burkholderiales	88	85	89	88	88	87	87
	Methylophilales	3	3	3	3	3	3	3
γ-Proteobacteria	Enterobacteriales	162	169	178	173	164	172	174
	Pasteurellales	33	32	34	33	32	33	33
	Pseudomonadales	42	41	42	43	42	42	41
δ-Proteobacteria	Desulfovibrionales	11	13	11	10	0	0	11
	Desulfuromonadales	11	11	11	11	0	3	11
	Myxococcales	6	6	6	6	0	0	6
ε-Proteobacteria	Campylobacterales	46	46	46	46	0	0	46
Acidobacteria		3	4	3	2	3	0	3
Aquificae		9	10	10	10	7	0	10
Cyanobacteria		57	3	1	38	54	0	56
Fusobacteria		15	15	15	13	10	0	15
Thermotogae		0	0	0	7	2	0	12

[1]Several diderm bacteria phyla are selected. For α-, β-, γ-, δ-, ε-proteobacteria the presence of the specific functional domains are also shown in the most representative genera. Data reported are the numbers of proteins in each phylum/genus matching the PFAM entries characteristic of the selected proteins. Data were taken from the PFAM entries for each domain, accessed at http.//pfam.sanger.ac.uk
[2]Protein domains selected (LpxC, Lpxk WaaA) belong to enzymes involved in the LPS biosynthetic pathway to identify LPS producing bacteria.
[3]OstA_N, LptC, LptE, YjgP_YjgQ are selected domains of proteins involved in the LPS transport to the OM.

with LptE and that the stoichiometric ratio of the two proteins in the heterodimeric complex is 1:1. The stable association revealed by proteolysis experiments suggests that LptE may be important to correctly fold the C-terminal domain of LptD, which appears to be unstable when overexpressed without LptE (Chng *et al.*, 2010b). The interaction with LptE seems to be required for the formation of the disulfide bonds of LptD, which is essential for its function (Ruiz *et al.*, 2010). Indeed it has been shown that LptE forms a plug buried in the lumen of the mature β-barrel formed by the C-terminal domain of LptD and that the two proteins associate via an extensive interface which involves a predicted extracellularly exposed loop of LptD (Freinkman *et al.*, 2011). This strong interaction may also explain the previous observation that LptE is functional even without its N-terminal lipid anchor (Chng *et al.*, 2010b). LptE does not seem to simply play a structural role in LPS biogenesis, as it has been demonstrated to specifically bind LPS (Chng *et al.*, 2010b). Finally, in a screening for suppressor mutants of a two-codon *lptE* deletion that altered LptE interaction with LptD, suppressors were isolated that mapped not only in *lptD* but also in *bamA* (belonging to the Bam complex that assemble OMP at the OM; see also Chapter 3), revealing that LptE association has a role in LptD assembly by the Bam β-barrel assembly machinery (Chimalakonda *et al.*, 2011).

The remaining Lpt components, LptABCFG, are inserted in or associated to the IM. These factors were discovered by different approaches. *lptA*, *lptB* and *lptC* (formerly *yhbN*, *yhbG* and *yrbK* respectively) were identified by Polissi and co-workers using a genetic screen designed to identify novel essential functions in *E. coli* (Serina *et al.*, 2004). In this work a Tn5-derived mini-transposon carrying the inducible *araBp* arabinose promoter oriented outward at one end was used to generate mutants that were subsequently assayed for conditional lethal phenotypes. This genetic selection led to the identification of a chromosomal locus containing novel essential functions. Along with *lptA*, *lptB* and *lptC*, this locus contains two LPS biosynthesis genes (*kdsD* and *kdsC* coding for two enzymes involved in Kdo biosynthesis (Meredith and Woodard, 2003; Wu and Woodard, 2003)); the sequence and chromosomal organization of the genes at this locus are conserved among Gram-negative bacteria, especially in the *Enterobacteriaceae*. In *E. coli*, the entire locus is transcribed from a single upstream promoter, but at least two complex internal promoter regions may allow differential expression of the different genes (Sperandeo *et al.*, 2007; Martorana *et al.*, 2011).

All the three *lpt* genes turned out to be essential in subsequent studies and this feature together with the high degree of conservation and the genetic linkage with LPS biosynthesis genes strongly suggested a role in OM biogenesis and possibly in LPS transport (Sperandeo *et al.*, 2006).

The analysis of conditional mutants in each gene allowed Polissi and collaborators to validate this hypothesis. Membrane fractionation experiments using sucrose density gradient centrifugation revealed that depletion of LptA, LptB and LptC leads to (i) arrest of cell growth after few generations; (ii) accumulation of abnormal membrane structure in the periplasm; (iii) appearance of an anomalous LPS form (visible in tricine SDS-PAGE as a ladder-like banding of high molecular weight species); and, more importantly, (iv) block of the transport to the OM of *de novo* synthesized LPS, which accumulated in a novel membrane fraction with intermediate density between IM and OM (Sperandeo *et al.*, 2007; Sperandeo *et al.*, 2008). A closer inspection of the modified LPS extracted from LptA-LptB and LptC depleted cells revealed that repeated units of colanic acid were ligated to the inner core of LPS by the O-antigen ligase WaaL (Sperandeo *et al.*, 2008). As the active site of this enzyme is located at the periplasmic side of the IM, it was postulated that when LPS transport is impaired, newly synthesized LPS stacked in the IM could become substrate of WaaL, which has relaxed substrate specificity. For this reason, this LPS modification was suggested to be diagnostic of LPS transport impairment (Sperandeo *et al.*, 2008). Raetz and co-worker made similar observations by exploiting the ectopic expression of lipid A 3-O-deacylase PagL from *Salmonella* and the lipid A 1-phosphatase LpxE from *Francisella* as OM and periplasmic markers, respectively, of LPS topology in a novel temperature sensitive LptA mutant. They demonstrated that at the non-permissive temperature LptA inactivation leads to lipid A-core arrest at the outer side of the IM where it becomes substrate of LpxE. Interestingly, the newly synthesized lipid A-core extracted from the LptA-inactivated mutant cells is not modified by PagL, whose active site is localized at the OM. These observations confirmed that LptA is required to transfer LPS from the periplasmic side of the IM to the OM (Ma *et al.*, 2008).

LptB is a 26.7 kDa protein possessing the nucleotide binding domain typical of ABC transporters. Initial evidence revealed that LptB was associated to the IM in a high molecular complex of approximately 140 kDa, although the interacting partners of LptB were not identified (Stenberg *et al.*, 2005).

LptA is a 18.6 kDa periplasmic protein, with an N-terminal signal sequence that is processed in the mature form (Tran *et al.*, 2008). In early works, *E. coli* LptA versions fused to a C-terminal His tag and overexpressed from a plasmid were reported to have a periplasmic localization (Sperandeo *et al.*, 2007; Tran *et al.*, 2008); however, in a recent paper by Kahne and co-workers it has been demonstrated that physiologically expressed LptA is able to associate with both IM and OM (Chng *et al.*, 2010a). Similar observations had been already made for the neisserial LptA homologue LptH (Bos *et al.*, 2007). *lptA* and *lptB* are co-transcribed in a dicistronic operon belonging to the σ^E regulon, which is implicated in envelope stress response (Dartigalongue *et al.*, 2001). Interestingly, the σ^E-dependent *lptAp* promoter seems to be exclusively activated by an LPS specific stress, but the fine regulation of this promoter is still unknown (Martorana *et al.*, 2011).

LptC is a small bitopic 21.1-kDa protein that is anchored to the IM by an uncleaved signal sequence. This protein possesses an N-terminal transmembrane segment and a large soluble C-terminal domain exposed to the periplasm (Tran *et al.*, 2010).

In Gram-negative bacteria transmembrane components of ABC transporters are constituted either by one protein with 12 transmembrane segments or two proteins with six transmembrane segments each (Davidson *et al.*, 2008); for this reason it was immediately clear that LptC could not be the transmembrane partner of LptB and LptA and that some components of the Lpt transporter were missing. These were identified by Ruiz and collaborators using a bioinformatic approach exploiting the high degree of conservation of OM biogenesis proteins among Gram-negative bacteria, including endosymbionts whose genome is dramatically reduced. To search for novel Lpt factors, the authors selected as a model organism the endosymbiont *Blochmannia floridanus*, an Enterobacteriaceae with a reduced proteome (14% the *E. coli* proteome; Gil *et al.*, 2003) but containing most of the OM biogenesis factors identified so far in *E. coli*. This approach led to the discovery of two essential six-transmembrane-domain IM proteins, LptF (40.4 kDa) and LptG (39.6 kDa) (formerly YjgP and YjgQ, respectively), as the transmembrane components of the novel Lpt ABC transporter. In *E. coli*, the genes encoding LptF and LptG belong to an operon unlinked to *lptB*. The involvement of LptF and LptG in LPS transport was demonstrated using conditional expression mutants and analysing the PagP-mediated modification of *de novo* synthesized LPS in LptF- or/and LptG-depleted cells. The lack of LPS modification and its accumulation at the IM upon depletion revealed that the two proteins are actually required for LPS transport downstream MsbA (Ruiz *et al.*, 2008). Recently, it has been confirmed that LptBCFG proteins physically interact and display ATPase activity (Narita and Tokuda, 2009).

Based on bioinformatic analysis it is reasonable to postulate that the proteins required for LPS transport so far identified and described in this paragraph represent the entire set of essential components of the LPS transport machine (Ruiz *et al.*, 2008). Genetic evidence suggests that the complex functions as a single device (Sperandeo *et al.*, 2008); however, the molecular mechanisms underlying the LPS transport still wait to be clarified and at the moment only models are available.

Mechanism of LPS transport: facts and models

The body of work available so far reveals that LPS transport requires an IM associated ABC transporter, composed by LptBFG and the atypical subunit LptC with the stoichiometric ratio of 2:1:1:1 (Narita and Tokuda, 2009), a periplasmic subunit, LptA, and an OM-inserted two-component complex LptDE. However, the molecular mechanism by which this complex achieves the unidirectional LPS transport from IM to OM is far from being understood.

The main obstacle encountered in dissecting the role of each Lpt component is that depletion of any Lpt factor leads to the same phenotypes ultimately resulting in LPS accumulation in the periplasmic leaflet of the IM and no intermediate stages have been so far identified. This fact, on one hand, makes it impossible to perform epistasis experiments; on the other hand, it provides strong evidence that the Lpt machinery operates as a single device in a step downstream of the MsbA-mediated LPS translocation across the IM (Sperandeo *et al.*, 2008).

Owing to its amphipathic nature, LPS transport cannot occur by simple diffusion and needs an energy transducing device to cross the aqueous periplasmic space; such a device is expected to cross the cell envelope spanning each cell compartment from the cytoplasm to the OM.

Three main models have been proposed to account for the transport mechanism: the vesicle-mediated movement, the chaperone-mediated transit across the periplasm, and the transport at IM–OM fusion sites (compatible with the so-called 'Bayer bridges').

A model based on transport mediated by membrane vesicles was abandoned early because of the short space between IM and OM and the observation that the peptidoglycan layer could represent a barrier for such a bulky vehicle (Dijkstra and Keck, 1996). Moreover, vesicles have never been documented within the periplasm.

By analogy with the OM lipoprotein transport mechanism (see also Chapter 4), the chaperone-mediated transport model implies a soluble periplasmic protein that binds LPS and shields its lipid portion, thus allowing its diffusion across the periplasm. In the lipoprotein transport system, the periplasmic protein LolA receives its substrate from the IM-associated ABC transporter LolCDE, which provides energy for the conformational changes required by LolA to accommodate the lipid moiety of lipoproteins in its cavity. LolA is then responsible to deliver its cargo to an OM associated receptor lipoprotein, LolB, which ultimately inserts it into the OM (Tokuda, 2009). According to the chaperone-mediated model, LptA could be the soluble carrier that receives LPS from the IM ABC transporter LptBFGC, diffuses across the periplasm and delivers it to the OM complex LptDE. Consistent with this model, LptA binds LPS *in vitro* (Tran *et al.*, 2008) and, interestingly, also LptC is able to bind LPS *in vitro*; moreover, LptA can displace LPS from LptC in line with their location and their proposed placement in a unidirectional export pathway (Tran *et al.*, 2010).

However, some substantial differences exist between Lol and Lpt transporters. First of all, LptC is an atypical subunit that has no counterpart in the Lol transporter. The ATPase activity displayed by LptBFG and LptBCFG exhibit the same K_m and V_{max} values, suggesting that LptC does not affect the kinetic parameter of the ATPase activity (Narita and Tokuda, 2009). Moreover, it has been demonstrated that LPS transport can occur in spheroplasts, where the periplasmic soluble content has been effectively drained, indicating that all the components required for LPS transport remain stably associated to the spheroplast; finally, no LPS carrier has been identified in periplasmic extracts using the same approach that allowed the isolation of LolA (Tefsen *et al.*, 2005b).

The third model suggests the existence of bridges connecting IM and OM and was proposed more than 40 years ago by Manfred E. Bayer (Bayer, 1968, 1991). Whatever the nature of the bridges (proteinaceous or lipidic), it was postulated that they could facilitate the transit of hydrophobic molecules through the periplasm. Some initial evidence in *S. typhimurium* supported this model. First of all, in 1973 it was reported that newly synthesized LPS appears in zones of adhesion between IM and OM (Muhlradt *et al.*, 1973). In line with this observation, Ishidate and co-workers, using sucrose density gradient centrifugation, identified a lighter OM domain (OM_L fraction) where newly synthesized LPS transiently accumulates and demonstrated that in OM_L IM and OM components were present along with murein, evoking the existence of bridges between IM and OM (Ishidate *et al.*, 1986). In a very recent work by Kahne's group, the OM_L fraction was isolated and it was demonstrated that all the Lpt proteins co-fractionate in this membrane fraction. In the same paper evidences supporting the physical interaction between the seven Lpt proteins were provided, supporting the idea that Bayer bridges actually exist and are constituted by a transenvelope protein complex (Chng *et al.*, 2010a). Genetic evidence also support the transenvelope model: first of all, the observation mentioned above in this chapter that depletion of any component of the Lpt machine results in a similar phenotype (i.e. LPS accumulation at the periplasmic side of the IM) (Sperandeo *et al.*, 2008); second, a recent paper by Polissi and collaborators revealed that any mutation impairing Lpt complex assembly results in LptA degradation, suggesting that LptA could be a sensor of properly bridged IM and OM (Sperandeo *et al.*, 2011).

Some open questions remain to be solved: what is the mode of interaction between the different Lpt proteins? Which are the LPS binding determinants in LptA, LptC and LptE? How does the Lpt complex accomplish the LPS transport?

Very recently, LptA–LptC interaction has been demonstrated *in vivo* and *in vitro* (Bowyer *et al.*, 2011; Sperandeo *et al.*, 2011), suggesting that LptC may function as the IM docking site at the IM. The notion that LptA binds LPS with higher affinity than LptC suggests that LptC might use the energy provided by the ATP hydrolysis to extract LPS from the IM and then deliver it to LptA (Tran *et al.*, 2010). As previously mentioned LptC does not affect the kinetic parameter of the ATPase activity of the LptBCFG complex (Narita and Tokuda, 2009). As it has been shown that neither LPS nor lipid A are able to stimulate the ATPase

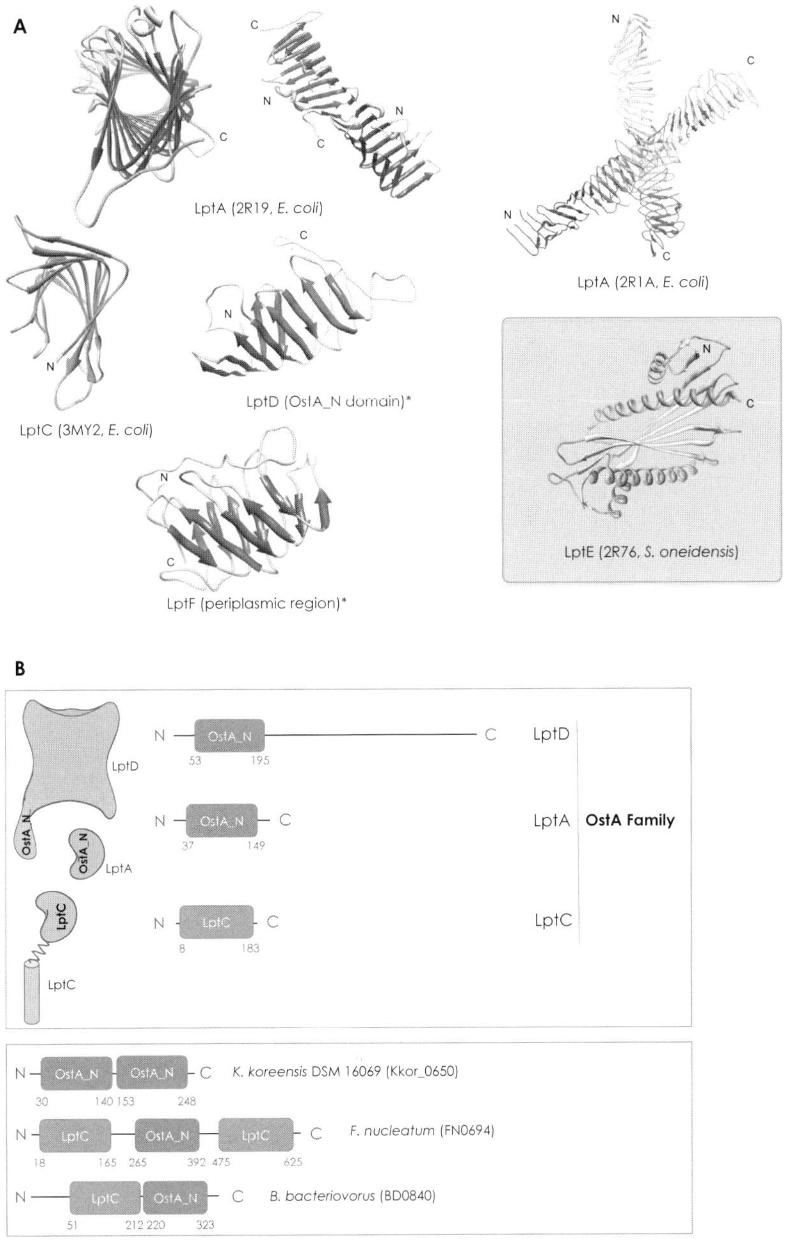

Figure 2.6 Crystal structures and structure prediction of Lpt proteins. (A) The crystal structure of LptA, LptC and LptE are reported together with the respective PDB codes. LptA structures obtained in the presence (PDB: 2R1A) or in the absence of LPS (PDB: 2R19) are shown. The structure of LptE has been solved from three different *E. coli* orthologues, only the structure of LptE from *Shewanella oneidensis* is shown. Structure predictions are marked with an asterisk: OstA_N domain of *E. coli* LptD and periplasmic region of *E. coli* LptF. (B) Upper panel: conserved functional domains in Lpt proteins. The OstA domain family includes the OstA_N signature in both LptD and LptA and LptC signature in LptC. Lower panel: domains organization of PF03968 (OstA_N) and PF036835 (LptC) in selected genomes. A colour version of this figure is located in the plate section at the back of the book.

activity of either LptBFG or LptBCFG (Narita and Tokuda, 2009), it is likely that in this *in vitro* assay some component, as LptA for example, was missing.

The crystal structure of LptA has been solved in the presence and absence of LPS (Fig. 2.6A). LptA presents a novel fold consisting of 16 antiparallel β-strands folded to resemble a semiclosed β-jellyroll; the structure in not completely symmetrical and it opens slightly at the N- and C-termini. In the presence of LPS, LptA molecules associate in a head-to-tail fashion forming fibrils containing a hydrophobic groove. According to the hypothesis that LptA physically connects IM and OM, the interior cavity of LptA fibres could ultimately accommodate LPS (Suits *et al.*, 2008).

Interestingly, the recently solved crystal structure of LptC revealed a similar fold to LptA with 15 antiparallel β-strands, although the two proteins share very low sequence similarity (Tran *et al.*, 2010). LptA and LptC belong to the same OstA family of the N-terminal domain of LptD, which has been recently demonstrated to be essential for LptD function *in vivo* (Bos *et al.*, 2007; Chng *et al.*, 2010b) (Fig. 2.6A). It may be postulated that LptA bridges the membranes by interacting with LptC at the IM and the N-terminal domain of LptD at the OM. In line with this hypothesis is the observation that *in vitro* LptC may form dimers (Sperandeo *et al.*, 2011; Bowyer *et al.*, 2011). Moreover, structure prediction of the periplasmic region of LptF reveals a striking similarity to LptC structure (Fig. 2.6A). Therefore it seems that the OstA family domain would be the determinant required for Lpt proteins interaction in creating the transenvelope machinery (Fig. 2.6A and B).

LptA, LptC and LptE have been demonstrated to bind specifically LPS (Tran *et al.*, 2008, 2010; Chng *et al.*, 2010b); however, the determinants for LPS binding are unknown. One possibility is that LPS exploits the hydrophobic cavity formed by the OstA-family domains of LptA, LptC and, possibly, the N-terminal domain of LptD to traverse the periplasm. At the OM, binding of LPS with LptE might trigger a conformational change that could be ultimately transmitted to the C-terminal domain of LptD allowing LPS insertion in the OM (Chng *et al.*, 2010b).

LPS binding analysis performed on LptA has revealed that the Lpt complex has relaxed substrate specificity as this protein is able to bind hexa and tetra-acylated lipid A *in vitro* (Tran *et al.*, 2008). However, it is well established that the minimal essential portion of LPS required to sustain cell viability is composed of lipid A and two molecules of Kdo (Raetz and Whitfield, 2002). This implies that MsbA must perform the quality control step in LPS transport. Several lines of evidence support this notion. For example, MsbA overexpression from a multicopy plasmid can rescue mutants lacking *htrB/lpxL* acyltransferase (Karow and Georgopoulos, 1993) or defective in the first Kdo biosynthetic enzyme (KdsD and its paralogue GutQ) (Meredith *et al.*, 2006) and thus unable to synthesize a complete Kdo_2-lipid A. This evidence implies that *in vivo* MsbA can flip under-acylated and non-glycosylated lipid A precursors, although with low efficiency, and that these molecules can be efficiently transported by the downstream Lpt machinery. This hypothesis has been further supported by the isolation of two classes of suppressor mutations allowing growth of a *waaA* deletion mutant unable to ligate Kdo to lipid A and accumulating lipid IV_A (Mamat *et al.*, 2008). The first class of suppressor mutations carries a single amino acid substitution in MsbA, resulting in more relaxed substrate specificity. Indeed, those mutants are viable and possess an OM composed by lipid IV_A The second class of suppressor mutations mapped in *yhjD*, a gene coding for a conserved integral IM protein whose function is not known. The suppressor allele (*yhjD400*) consists in a single amino acid substitution in YhjD that seems to activate an alternative transport pathway, independently by MsbA (Mamat *et al.*, 2008). Indeed, in *yhjD400* genetic background *msbA* turns out to be dispensable. Finally, in line with these overall observations, Raina and co-worker have recently isolated a suppressor-free *waaC lpxL lpxM lpxP* mutant defective in heptosyltransferase I and late acyltransferase genes. This mutant is viable under slow growth condition at low temperatures, although producing a Kdo_2-lipid IV_A LPS precursor, and shows a constitutive envelope stress response. Interestingly, *waaC lpxL lpxM lpxP* growth at normal temperature can be rescued by chromosomal D498V suppressor mutation in MsbA or by wild-type MsbA overexpression (Klein *et al.*, 2009).

Phylogeny of *lpt* genes

As discussed above, the pathway of LPS translocation from IM to the OM has been characterized mainly using the *E. coli* and *N. meningitidis* model systems. A phylum level analysis of Lpt proteins conservation shows that not all the LPS-producing bacteria contain a complete set of *E. coli* Lpt proteins homologues (Sutcliffe, 2010).

However, searching Lpt orthologous proteins in LPS producing diderm bacteria by a standard BLAST analysis might be misleading. For example, Haarmaan and co-workers (2010) showed that the proteins of the Lpt complex localized within the IM (LptF and LptG) are generally present in Gram-negative bacteria, whereas the periplasmic proteins LptA, LptC and LptE result hard to be detected by BLAST analysis. This is not surprising considering that the homologues of these proteins in rather closely related strains within *γ-Proteobacteria* share a low identity level (e.g. the identity of *P. aeruginosa* and *E. coli* LptC and LptE is 19% and 21%, respectively).

For this reason, given the essential role played by the Lpt protein machinery in *E. coli*, alternative search methods are desirable to explore in greater depth the available genomic and proteomic information.

A more stringent approach could be to use the functional domains found in Lpt proteins (Fig. 2.6A and B) to search for homologues in PFAM database (pfam.sanger.ac.uk/).

By this approach, we found that the Lpt proteins are globally present in β- and *γ-Proteobacteria* (Table 2.2). A closer examination of the distribution at phylum level of PFAM domain PF03968 (OstA-N domain, present in both LptA and the N-terminus of LptD) in diderm bacteria reveals that LptA and LptD homologues are widely distributed, as it is possible to find PF03968 domain even in *Thermotogae*, which do not possess LPS biosynthesis genes (Plotz *et al.*, 2000; Sutcliffe, 2010). Also PF03739 domain (YjgP_YjgQ domain present in both LptF and LptG) results widely distributed, as expected for an ABC transporter subunit.

On the contrary, the examination of the PF06835 and PF04390 domains (LptC and LptE, respectively) reveals a narrower distribution. LptC homologues are apparently missing in δ- and ε-*Proteobacteria*, whereas the presence of LptE homologues appears to be restricted to the β- and *γ-Proteobacteria*, with the only exception of some *Desulfuromonadales* within δ-*Proteobacteria*. Accordingly, in a recent work it has been reported that in some phyla known to produce LPS the Lpt pathway is either completely missing (e.g. in chlamydiae) or lacks some components (e.g. in *Bacteroidetes*, *Chlorobi* and *Cyanobacteria*) (Sutcliffe, 2010).

However, the absence of LptC or LptE homologues in LPS producing bacteria is at odds with the essential role of these proteins in *E. coli*. Therefore, other identification criteria in addition to significant PFAM hits need to be exploited to find potential LptE and LptC-like proteins. In Gram-negative bacteria, the clusters of *lpt* genes are generally conserved. Therefore, a possible strategy to detect the missing Lpt components not identified by other approaches based on sequence similarity of proteins or protein domains could be to inspect more closely the sequences flanking identified conserved homologues.

For example, in ε-*Proteobacteria* several proteins implicated in LPS synthesis and transport can be detected by the above bioinformatic methods, including the LpxC, LpxK and WaaA enzymes, the OstA_N-like proteins, and the IM protein complex components LptFG, whereas LptC and LptE appear to be missing (Table 2.2). In *H. pylori* strain 26695, a gene belonging to a locus composed of two ORFs encodes the putative periplasmic LptA homologue (HP1569), whereas the upstream gene (HP1569) encodes a putative protein with no significant similarity with proteins of known function. However, its structural prediction (performed with I-TASSER prediction server available at http://zhanglab.ccmb.med.umich.edu/I-TASSER) indicates that HP1569 is a putative IM bitopic conserved protein of 197 residues that shows a structure similar to that of *E. coli* LptC (unpublished data).

It thus appears that the genetic organization of *H. pylori* putative *lptC-lptA-lptB* genes resembles the organization found in the *E. coli* genome. A similar observation can be done for LptE, which in *E. coli* is located between the housekeeping genes *holA* and *leuS*. In *H. pylori*, ORF HP1546, which is flanked by a *leuS* homologue, codes for a putative lipoprotein as suggested by *in silico* analysis using LipoP 1.0 server (http://www.cbs.dtu.dk/services/LipoP/). HP1569 and HP1546 might thus represent highly diverged LptC and LptE orthologues.

Curiously, in some species a different organization of conserved domains is observed. As an example, in *Kangiella koreensis* two PF03968 (OstA_N) domains are present in the LptA homologue whereas in *Fusobacterium nucleatum* and in *Bdellovibrio bacteriovorus* the PF03968 (OstA_N) domain is fused in a single polypeptide with one or two PF036835 (LptC) domains, thus highlighting the importance of this interaction in LPS transport pathway (Fig. 2.6B).

The lack of amino acid sequence conservation in proteins belonging to the Lpt pathway may reflect the mode of interaction between Lpt proteins and LPS. It is possible that Lpt binding to LPS (and/or lipid A) may not implicate a few specific amino acids, rather peculiar chemico-physical features extended to the whole three dimensional structure.

A notable example supporting this idea is the human protein MD-2, an LPS-binding protein involved in the innate immune response to bacterial infections (Miyake, 2004). As reported by Park and co-workers (Park *et al.*, 2009), the molecular determinants of human MD-2 implicated in lipid A recognition are not conserved in homologous genes of other mammals, although the overall three-dimensional structure is highly conserved.

Intriguingly, in the non-LPS diderm phylum of *Thermotogae* (Plotz *et al.*, 2000), despite the lack of LPS biosynthetic pathway, LptA and LptC homologues have been identified (Table 2.2), suggesting that the structural motifs described for Lpt proteins are not exclusively correlated to lipid A transport and might have evolved in some genomes to transport other lipids.

Conclusions and future trends

LPS is an essential glycolipid of the OM of Gram-negative bacteria endowed of important biological functions. LPS largely contributes to the permeability barrier properties exhibited by the OM that protects cells from the entry of toxic compounds. Moreover, the lipid A moiety of LPS is the conserved structure recognized by the TLR4/MD2 receptors responsible for activation of the host innate immune system. The LPS biogenetic pathway is a complex process requiring the coordinate action of over 30 genes in every cellular compartment, namely cytoplasm, IM, periplasm and OM. The structure and the biosynthesis of LPS have been known for many years; however, only recent advances have led to the identification of the LPS transport protein machine that extracts LPS from the intracellular site of synthesis to the environment-exposed final destination. Very strong genetic and biochemical evidences indicate that the Lpt machinery functions as a single device and that the seven Lpt proteins composing the system physically interact to form a transenvelope complex. However, neither the detailed mechanisms of LPS translocation across the periplasm and its insertion at the OM nor the molecular requirements for LPS binding to LptA, LptC and LptE are known. The Lpt transport system consists of a single apparatus spanning IM and OM and the structural complexity of LPS, the transported molecule, together with the composite organization of the protein machinery poses a major challenge for studying the mechanism of LPS transport. The crystal structure of Lpt proteins in complex with their ligand would give insight in the mechanistic aspects of the LPS export. Moreover, the development of an *in vitro* transport assay is needed to dissect the molecular mechanism of the process and to define the individual role played by each of these seven proteins. The Lpt machinery represents a composite cellular target to design/identify molecules able to inhibit not only the function of any single protein but also the assembly of the complex. The identification of inhibitors that specifically target LPS transport *in vitro* and more importantly *in vivo* may represent additional tools to dissect the transport pathway and may open the door to novel antibiotic strategies targeting OM biogenesis.

References

Anderson, M.S., and Raetz, C.R. (1987). Biosynthesis of lipid A precursors in *Escherichia coli*. A cytoplasmic acyltransferase that converts UDP-N-acetylglucosamine to UDP-3-O-(R-3 hydroxymyristoyl)-N-acetylglucosamine. J. Biol. Chem. *262*, 5159–5169.

Aono, R., Negishi, T., Aibe, K., Inoue, A., and Horikoshi, K. (1994). Mapping of organic solvent tolerance gene *ostA* in *Escherichia coli* K-12. Biosci. Biotechnol. Biochem. *58*, 1231–1235.

Babinski, K.J., Ribeiro, A.A., and Raetz, C.R. (2002). The *Escherichia coli* gene encoding the UDP-2,3-diacylglucosamine pyrophosphatase of lipid A biosynthesis. J. Biol. Chem. *277*, 25937–25946.

Bartling, C.M., and Raetz, C.R. (2008). Steady-state kinetics and mechanism of LpxD, the N-acyltransferase of lipid A biosynthesis. Biochemistry *47*, 5290–5302.

Bayer, M.E. (1968). Areas of adhesion between wall and membrane of *Escherichia coli*. J. Gen. Microbiol. *53*, 395–404.

Bayer, M.E. (1991). Zones of membrane adhesion in the cryofixed envelope of *Escherichia coli*. J. Struct. Biol. *107*, 268–280.

Beckmann, I., Subbaiah, T.V., and Stocker, B.A. (1964). Rough mutants of *Salmonella typhimurium*. II. Serological and chemical investigations. Nature *201*, 1299–1301.

Belunis, C.J., and Raetz, C.R. (1992). Biosynthesis of endotoxins. Purification and catalytic properties of 3-deoxy-D-manno-octulosonic acid transferase from *Escherichia coli*. J. Biol. Chem. *267*, 9988–9997.

Beveridge, T.J. (1999). Structures of gram-negative cell walls and their derived membrane vesicles. J. Bacteriol. *181*, 4725–4733.

Beveridge, T.J., and Davies, J.A. (1983). Cellular responses of *Bacillus subtilis* and *Escherichia coli* to the Gram stain. J. Bacteriol. *156*, 846–858.

Bishop, R.E. (2008). Structural biology of membrane-intrinsic beta-barrel enzymes: sentinels of the bacterial outer membrane. Biochim. Biophys. Acta *1778*, 1881–1896.

Bishop, R.E., Gibbons, H.S., Guina, T., Trent, M.S., Miller, S.I., and Raetz, C.R. (2000). Transfer of palmitate from phospholipids to lipid A in outer membranes of gram-negative bacteria. EMBO J. *19*, 5071–5080.

Bladen, H.A., and Mergenhagen, S.E. (1964). Ultrastructure of *Veillonella* and morphological correlation of an outer membrane with particles associated with endotoxic activity. J. Bacteriol. *88*, 1482–1492.

Bos, M.P., Tefsen, B., Geurtsen, J., and Tommassen, J. (2004). Identification of an outer membrane protein required for the transport of lipopolysaccharide to the bacterial cell surface. Proc. Natl. Acad. Sci. U.S.A. *101*, 9417–9422.

Bos, M.P., Robert, V., and Tommassen, J. (2007). Biogenesis of the Gram-negative bacterial outer membrane. Annu. Rev. Microbiol. *61*, 191–214.

Bowyer, A., Baardsnes, J., Ajamian, E., Zhang, L., and Cygler, M. (2011). Characterization of interactions between LPS transport proteins of the Lpt system. Biochem. Biophys. Res. Commun. *404*, 1093–1098.

Braun, M., and Silhavy, T.J. (2002). Imp/OstA is required for cell envelope biogenesis in *Escherichia coli*. Mol. Microbiol. *45*, 1289–1302.

Braun, V. (1975). Covalent lipoprotein from the outer membrane of *Escherichia coli*. Biochim. Biophys. Acta *415*, 335–377.

Chang, G., Roth, C.B., Reyes, C.L., Pornillos, O., Chen, Y.J., and Chen, A.P. (2006). Retraction. Science *314*, 1875.

Chimalakonda, G., Ruiz, N., Chng, S.S., Garner, R.A., Kahne, D., and Silhavy, T.J. (2011). Lipoprotein LptE is required for the assembly of LptD by the beta-barrel assembly machine in the outer membrane of *Escherichia coli*. Proc. Natl. Acad. Sci. U.S.A. *108*, 2492–2497.

Chng, S.S., Gronenberg, L.S., and Kahne, D. (2010a). Proteins required for lipopolysaccharide assembly in *Escherichia coli* form a transenvelope complex. Biochemistry *49*, 4565–4567.

Chng, S.S., Ruiz, N., Chimalakonda, G., Silhavy, T.J., and Kahne, D. (2010b). Characterization of the two-protein complex in *Escherichia coli* responsible for lipopolysaccharide assembly at the outer membrane. Proc. Natl. Acad. Sci. U.S.A. *107*, 5363–5368.

Chung, H.S., and Raetz, C.R. (2011). Dioxygenases in *Burkholderia ambifaria* and *Yersinia pestis* that hydroxylate the outer Kdo unit of lipopolysaccharide. Proc. Natl. Acad. Sci. U.S.A. *108*, 510–515.

Cipolla, L., Polissi, A., Airoldi, C., Galliani, P., Sperandeo, P., and Nicotra, F. (2009). The Kdo biosynthetic pathway toward OM biogenesis as target in antibacterial drug design and development. Curr. Drug Discov. Technol. *6*, 19–33.

Clementz, T., Bednarski, J.J., and Raetz, C.R. (1996). Function of the *htrB* high temperature requirement gene of *Escherchia coli* in the acylation of lipid A: HtrB catalyzed incorporation of laurate. J. Biol. Chem. *271*, 12095–12102.

Clementz, T., Zhou, Z., and Raetz, C.R. (1997). Function of the *Escherichia coli* msbB gene, a multicopy suppressor of htrB knockouts, in the acylation of lipid A. Acylation by MsbB follows laurate incorporation by HtrB. J. Biol. Chem. *272*, 10353–10360.

Dartigalongue, C., Missiakas, D., and Raina, S. (2001). Characterization of the *Escherichia coli* σ^E regulon. J. Biol. Chem. *276*, 20866–20875.

Davidson, A.L., Dassa, E., Orelle, C., and Chen, J. (2008). Structure, function, and evolution of bacterial ATP-binding cassette systems. Microbiol. Mol. Biol. Rev. *72*, 317–64.

Dawson, R.J., and Locher, K.P. (2006). Structure of a bacterial multidrug ABC transporter. Nature *443*, 180–185.

De Rosa, M., Trincone, A., Nicolaus, B., Gambacorta, A., and di Prisco, G. (1991). Archeabacteria: lipids, membrane structures, and adaptation to environmental stresses. In Life under Extreme Conditions, Prisco, G., ed. (Springer, Berlin, Germany), pp. 61–87.

Dekker, N. (2000). Outer-membrane phospholipase A: known structure, unknown biological function. Mol. Microbiol. *35*, 711–717.

Derzelle, S., Turlin, E., Duchaud, E., Pages, S., Kunst, F., Givaudan, A., and Danchin, A. (2004). The PhoP-PhoQ two-component regulatory system of *Photorhabdus luminescens* is essential for virulence in insects. J. Bacteriol. *186*, 1270–1279.

Desvaux, M., Hebraud, M., Talon, R., and Henderson, I.R. (2009). Secretion and subcellular localizations of bacterial proteins: a semantic awareness issue. Trends Microbiol. *17*, 139–145.

Dijkstra, A.J., and Keck, W. (1996). Peptidoglycan as a barrier to transenvelope transport. J. Bacteriol. *178*, 5555–5562.

Doerrler, W.T., and Raetz, C.R. (2002). ATPase activity of the MsbA lipid flippase of *Escherichia coli*. J. Biol. Chem. *277*, 36697–36705.

Doerrler, W.T., Reedy, M.C., and Raetz, C.R. (2001). An *Escherichia coli* mutant defective in lipid export. J. Biol. Chem. *276*, 11461–11464.
Doerrler, W.T., Gibbons, H.S., and Raetz, C.R. (2004). MsbA-dependent translocation of lipids across the inner membrane of *Escherichia coli*. J. Biol. Chem. *279*, 45102–45109.
Eckford, P.D., and Sharom, F.J. (2008). Functional characterization of *Escherichia coli* MsbA: interaction with nucleotides and substrates. J. Biol. Chem. *283*, 12840–12850.
Eckford, P.D., and Sharom, F.J. (2010). The reconstituted *Escherichia coli* MsbA protein displays lipid flippase activity. Biochem. J. *429*, 195–203.
Ellen, A.F., Zolghadr, B., Driessen, A.M., and Albers, S.V. (2010). Shaping the archaeal cell envelope. Archaea. *2010*, 608243.
Ferguson, A.D., Welte, W., Hofmann, E., Lindner, B., Holst, O., Coulton, J.W., and Diederichs, K. (2000). A conserved structural motif for lipopolysaccharide recognition by procaryotic and eucaryotic proteins. Structure *8*, 585–592.
Freinkman, E., Chng, S.S., and Kahne, D. (2011). The complex that inserts lipopolysaccharide into the bacterial outer membrane forms a two-protein plug-and-barrel. Proc. Natl. Acad. Sci. U.S.A. *108*, 2486–2491.
Fukuoka, S., Brandenburg, K., Muller, M., Lindner, B., Koch, M.H., and Seydel, U. (2001). Physico-chemical analysis of lipid A fractions of lipopolysaccharide from *Erwinia carotovora* in relation to bioactivity. Biochim. Biophys. Acta *1510*, 185–197.
Gan, L., Chen, S., and Jensen, G.J. (2008). Molecular organization of Gram-negative peptidoglycan. Proc. Natl. Acad. Sci. U.S.A. *105*, 18953–18957.
Garrett, T.A., Kadrmas, J.L., and Raetz, C.R. (1997). Identification of the gene encoding the *Escherichia coli* lipid A 4′-kinase. Facile phosphorylation of endotoxin analogs with recombinant LpxK. J. Biol. Chem. *272*, 21855–21864.
Ghanei, H., Abeyrathne, P.D., and Lam, J.S. (2007). Biochemical characterization of MsbA from *Pseudomonas aeruginosa*. J. Biol. Chem. *282*, 26939–26947.
Gil, R., Silva, F.J., Zientz, E., Delmotte, F., Gonzalez-Candelas, F., Latorre, A., Rausell, C., Kamerbeek, J., Gadau, J., Holldobler, B., *et al.* (2003). The genome sequence of *Blochmannia floridanus*: comparative analysis of reduced genomes. Proc. Natl. Acad. Sci. U.S.A. *100*, 9388–9393.
Glauert, A.M., and Thornley, M.J. (1969). The topography of the bacterial cell wall. Annu. Rev. Microbiol. *23*, 159–198.
Gonzalez, V., Santamaria, R.I., Bustos, P., Hernandez-Gonzalez, I., Medrano-Soto, A., Moreno-Hagelsieb, G., Janga, S.C., Ramirez, M.A., Jimenez-Jacinto, V., Collado-Vides, J., *et al.* (2006). The partitioned *Rhizobium etli* genome: genetic and metabolic redundancy in seven interacting replicons. Proc. Natl. Acad. Sci. U.S.A. *103*, 3834–3839.
Gunn, J.S. (2000). Mechanisms of bacterial resistance and response to bile. Microbes Infect. *2*, 907–913.
Gunn, J.S. (2008). The *Salmonella* PmrAB regulon: lipopolysaccharide modifications, antimicrobial peptide resistance and more. Trends Microbiol. *16*, 284–290.
Guo, L., Lim, K.B., Poduje, C.M., Daniel, M., Gunn, J.S., Hackett, M., and Miller, S.I. (1998). Lipid A acylation and bacterial resistance against vertebrate antimicrobial peptides. Cell *95*, 189–198.
Gupta, R.S. (1998). What are archaebacteria: life's third domain or monoderm prokaryotes related to gram-positive bacteria? A new proposal for the classification of prokaryotic organisms. Mol. Microbiol. *29*, 695–707.
Haarmann, R., Ibrahim, M., Stevanovic, M., Bredemeier, R., and Schleiff, E. (2010). The properties of the outer membrane localized Lipid A transporter LptD. J. Phys. Condens. Matter 22 454124.
Holland, I.B. (2010). The extraordinary diversity of bacterial protein secretion mechanisms. Methods Mol. Biol. *619*, 1–20.
Holst, O. (2007). The structures of core regions from enterobacterial lipopolysaccharides – an update. FEMS Microbiol. Lett. *271*, 3–11.
Hwang, P.M., Choy, W.Y., Lo, E.I., Chen, L., Forman-Kay, J.D., Raetz, C.R., Prive, G.G., Bishop, R.E., and Kay, L.E. (2002). Solution structure and dynamics of the outer membrane enzyme PagP by NMR. Proc. Natl. Acad. Sci. U.S.A. *99*, 13560–13565.
Ishidate, K., Creeger, E.S., Zrike, J., Deb, S., Glauner, B., MacAlister, T.J., and Rothfield, L.I. (1986). Isolation of differentiated membrane domains from *Escherichia coli* and *Salmonella typhimurium*, including a fraction containing attachment sites between the inner and outer membranes and the murein skeleton of the cell envelope. J. Biol. Chem. *261*, 428–443.
Isshiki, Y., Zahringer, U., and Kawahara, K. (2003). Structure of the core-oligosaccharide with a characteristic D-glycero-alpha-D-talo-oct-2-ulosylonate-(2→4)-3-deoxy-D-manno-oct-2-ulo sonate

[alpha-Ko-(2→4)-Kdo] disaccharide in the lipopolysaccharide from *Burkholderia cepacia*. Carbohydr. Res. *338*, 2659–2666.

Jackman, J.E., Raetz, C.R., and Fierke, C.A. (1999). UDP-3-O-(R-3-hydroxymyristoyl)-N-acetylglucosamine deacetylase of *Escherichia coli* is a zinc metalloenzyme. Biochemistry *38*, 1902–1911.

Kamio, Y., and Nikaido, H. (1976). Outer membrane of *Salmonella typhimurium*: accessibility of phospholipid head groups to phospholipase c and cyanogen bromide activated dextran in the external medium. Biochemistry *15*, 2561–2570.

Karow, M., Fayet, O., and Georgopoulos, C. (1992). The lethal phenotype caused by null mutations in the *Escherichia coli htrB* gene is suppressed by mutations in the *accBC* operon, encoding two subunits of acetyl coenzyme A carboxylase. J. Bacteriol. *174*, 7407–7418.

Karow, M., and Georgopoulos, C. (1993). The essential *Escherichia coli msbA* gene, a multicopy suppressor of null mutations in the *htrB* gene, is related to the universally conserved family of ATP-dependent translocators. Mol. Microbiol. *7*, 69–79.

Karuppiah, V., Berry, J.L., and Derrick, J.P. (2011). Outer membrane translocons: structural insights into channel formation. Trends Microbiol. *19*, 40–48.

Katz, C., and Ron, E.Z. (2008). Dual role of FtsH in regulating lipopolysaccharide biosynthesis in *Escherichia coli*. J. Bacteriol. *190*, 7117–7122.

Kellenberger, E. (1990). The 'Bayer bridges' confronted with results from improved electron microscopy methods. Mol. Microbiol. *4*, 697–705.

Kim, H.M., Park, B.S., Kim, J.I., Kim, S.E., Lee, J., Oh, S.C., Enkhbayar, P., Matsushima, N., Lee, H., Yoo, O.J., *et al.* (2007). Crystal structure of the TLR4–MD-2 complex with bound endotoxin antagonist Eritoran. Cell *130*, 906–917.

Kim, J.I., Lee, C.J., Jin, M.S., Lee, C.H., Paik, S.G., Lee, H., and Lee, J.O. (2005). Crystal structure of CD14 and its implications for lipopolysaccharide signaling. J. Biol. Chem. *280*, 11347–11351.

King, J.D., Kocincova, D., Westman, E.L., and Lam, J.S. (2009). Review: lipopolysaccharide biosynthesis in *Pseudomonas aeruginosa*. Innate. Immun. *15*, 261–312.

Klein, G., Lindner, B., Brabetz, W., Brade, H., and Raina, S. (2009). *Escherichia coli* K-12 Suppressor-free Mutants Lacking Early Glycosyltransferases and Late Acyltransferases: minimal lipopolysaccharide structure and induction of envelope stress response. J. Biol. Chem. *284*, 15369–15389.

Knowles, T.J., Scott-Tucker, A., Overduin, M., and Henderson, I.R. (2009). Membrane protein architects: the role of the BAM complex in outer membrane protein assembly. Nat. Rev. Microbiol. *7*, 206–214.

Koch, A.L. (1998). The biophysics of the gram-negative periplasmic space. Crit. Rev. Microbiol. *24*, 23–59.

Kol, M.A., van Dalen, A., de Kroon, A.I., and de Kruijff, B. (2003). Translocation of phospholipids is facilitated by a subset of membrane-spanning proteins of the bacterial cytoplasmic membrane. J. Biol. Chem. *278*, 24586–24593.

Langklotz, S., Schakermann, M., and Narberhaus, F. (2011). Control of lipopolysaccharide biosynthesis by FtsH-mediated proteolysis of LpxC is conserved in enterobacteria but not in all gram-negative bacteria. J. Bacteriol. *193*, 1090–1097.

Lee, H., Hsu, F.F., Turk, J., and Groisman, E.A. (2004). The PmrA-regulated pmrC gene mediates phosphoethanolamine modification of lipid A and polymyxin resistance in *Salmonella enterica*. J. Bacteriol. *186*, 4124–4133.

Luederitz, O., Risse, H.J., Schulte-Holthausen, H., Strominger, J.L., Sutherland, I.W., and Westphal, O. (1965). Biochemical studies of the smooth-rough mutation in *Salmonella minneso*ta. J. Bacteriol. *89*, 343–354.

Ma, B., Reynolds, C.M., and Raetz, C.R. (2008). Periplasmic orientation of nascent lipid A in the inner membrane of an *Escherichia coli* LptA mutant. Proc. Natl. Acad. Sci. U.S.A. *105*, 13823–13828.

Malinverni, J.C., and Silhavy, T.J. (2009). An ABC transport system that maintains lipid asymmetry in the gram-negative outer membrane. Proc. Natl. Acad. Sci. U.S.A. *106*, 8009–8014.

Mamat, U., Meredith, T.C., Aggarwal, P., Kuhl, A., Kirchhoff, P., Lindner, B., Hanuszkiewicz, A., Sun, J., Holst, O., and Woodard, R.W. (2008). Single amino acid substitutions in either YhjD or MsbA confer viability to 3-deoxy-d-manno-oct-2-ulosonic acid-depleted *Escherichia coli*. Mol. Microbiol. *67*, 633–648.

Margolin, W. (2009). Sculpting the bacterial cell. Curr. Biol. *19*, R812-R822.

Marolda, C.L., Li, B., Lung, M., Yang, M., Hanuszkiewicz, A., Rosales, A.R., and Valvano, M.A. (2010). Membrane topology and identification of critical amino acid residues in the Wzx O-antigen translocase from *Escherichia coli* O157:H4. J. Bacteriol. *192*, 6160–6171.

Martorana, A.M., Sperandeo, P., Polissi, A., and Deho, G. (2011). Complex transcriptional organization regulates an *Escherichia coli* locus implicated in lipopolysaccharide biogenesis. Res. Microbiol. *162*, 470–482.

McConnell, M.R., Oakes, K.R., Patrick, A.N., and Mills, D.M. (2001). Two functional O-polysaccharide polymerase wzy (rfc) genes are present in the rfb gene cluster of Group E1 *Salmonella enterica* serovar Anatum. FEMS Microbiol. Lett. *199*, 235–240.

McGrath, B.C., and Osborn, M.J. (1991). Localization of the terminal steps of O-antigen synthesis in Salmonella typhimurium. J. Bacteriol. *173*, 649–654.

Medzhitov, R., Preston-Hurlburt, P., and Janeway, C.A. Jr. (1997). A human homologue of the Drosophila Toll protein signals activation of adaptive immunity. Nature *388*, 394–397.

Meredith, T.C., and Woodard, R.W. (2003). *Escherichia coli* YrbH is a D-arabinose 5-phosphate isomerase. J. Biol. Chem. *278*, 32771–32777.

Meredith, T.C., Aggarwal, P., Mamat, U., Lindner, B., and Woodard, R.W. (2006). Redefining the requisite lipopolysaccharide structure in *Escherichia coli*. ACS Chem. Biol. *1*, 33–42.

Meredith, T.C., Mamat, U., Kaczynski, Z., Lindner, B., Holst, O., and Woodard, R.W. (2007). Modification of lipopolysaccharide with colanic acid (M-antigen) repeats in *Escherichia coli*. J. Biol. Chem. *282*, 7790–7798.

Metzger, L.E., and Raetz, C.R. (2010). An alternative route for UDP-diacylglucosamine hydrolysis in bacterial lipid A biosynthesis. Biochemistry *49*, 6715–6726.

Miller, S.I., Ernst, R.K., and Bader, M.W. (2005). LPS, TLR4 and infectious disease diversity. Nat. Rev. Microbiol. *3*, 36–46.

Minnikin, D.E. (1991). Chemical principles in the organization of lipid components in the mycobacterial cell envelope. Res. Microbiol. *142*, 423–427.

Missiakas, D., Betton, J.M., and Raina, S. (1996). New components of protein folding in extracytoplasmic compartments of *Escherichia coli* SurA, FkpA and Skp/OmpH. Mol. Microbiol. *21*, 871–884.

Miyake, K. (2004). Innate recognition of lipopolysaccharide by Toll-like receptor 4-MD-2. Trends Microbiol. *12*, 186–192.

Miyake, K. (2006). Roles for accessory molecules in microbial recognition by Toll-like receptors. J. Endotoxin. Res. *12*, 195–204.

Morona, R., Van Den, B.L., and Daniels, C. (2000). Evaluation of Wzz/MPA1/MPA2 proteins based on the presence of coiled-coil regions. Microbiology *146*, 1–4.

Muhlradt, P.F., and Golecki, J.R. (1975). Asymmetrical distribution and artifactual reorientation of lipopolysaccharide in the outer membrane bilayer of *Salmonella typhimurium*. Eur. J. Biochem. *51*, 343–352.

Muhlradt, P.F., Menzel, J., Golecki, J.R., and Speth, V. (1973). Outer membrane of *Salmonella*. Sites of export of newly synthesised lipopolysaccharide on the bacterial surface. Eur. J. Biochem. *35*, 471–481.

Narita, S., and Tokuda, H. (2009). Biochemical characterization of an ABC transporter LptBFGC complex required for the outer membrane sorting of lipopolysaccharides. FEBS Lett. *583*, 2160–2164.

Niederweis, M., Danilchanka, O., Huff, J., Hoffmann, C., and Engelhardt, H. (2010). Mycobacterial outer membranes: in search of proteins. Trends Microbiol. *18*, 109–116.

Nikaido, H. (2003). Molecular basis of bacterial outer membrane permeability revisited. Microbiol. Mol. Biol. Rev. *67*, 593–656.

Nikaido, H. (2005). Restoring permeability barrier function to outer membrane. Chem. Biol. *12*, 507–509.

Nikaido, H., Mikaido, K., Subbaiah, T.V., and Stocker, B.A. (1964). Rough mutants of *Salmonella typhimurium*. III. Enzymatic synthesis of nucleotide-sugar compounds. Nature *201*, 1301–1302.

Ogura, T., Inoue, K., Tatsuta, T., Suzaki, T., Karata, K., Young, K., Su, L.H., Fierke, C.A., Jackman, J.E., Raetz, C.R., *et al.* (1999). Balanced biosynthesis of major membrane components through regulated degradation of the committed enzyme of lipid A biosynthesis by the AAA protease FtsH (HflB) in *Escherichia coli*. Mol. Microbiol. *31*, 833–844.

Oliver, D.B. (1996). Periplasm. In *Escherichia coli* and *Salmonella* Cell Mol. Biol., Neidhardt, F.C., ed. (ASM press, Washington DC), pp. 88–103.

Onishi, H.R., Pelak, B.A., Gerckens, L.S., Silver, L.L., Kahan, F.M., Chen, M.H., Patchett, A.A., Galloway, S.M., Hyland, S.A., Anderson, M.S., *et al.* (1996). Antibacterial agents that inhibit lipid A biosynthesis. Science *274*, 980–982.

Osborn, M.J., Gander, J.E., and Parisi, E. (1972). Mechanism of assembly of the outer membrane of Salmonella typhimurium. Site of synthesis of lipopolysaccharide. J. Biol. Chem. *247*, 3973–3986.

Park, B.S., Song, D.H., Kim, H.M., Choi, B.S., Lee, H., and Lee, J.O. (2009). The structural basis of lipopolysaccharide recognition by the TLR4–MD-2 complex. Nature *458*, 1191–1195.

Plotz, B.M., Lindner, B., Stetter, K.O., and Holst, O. (2000). Characterization of a novel lipid A containing D-galacturonic acid that replaces phosphate residues. The structure of the lipid a of the lipopolysaccharide from the hyperthermophilic bacterium Aquifex pyrophilus. J. Biol. Chem. *275*, 11222–11228.

Polissi, A., and Georgopoulos, C. (1996). Mutational analysis and properties of the *msbA* gene of *Escherichia coli*, coding for an essential ABC family transporter. Mol. Microbiol. *20*, 1221–1233.

Price, N.P., Kelly, T.M., Raetz, C.R., and Carlson, R.W. (1994). Biosynthesis of a structurally novel lipid A in Rhizobium leguminosarum: identification and characterization of six metabolic steps leading from UDP-GlcNAc to 3-deoxy-D-manno-2-octulosonic acid2-lipid IVA. J. Bacteriol. *176*, 4646–4655.

Radika, K., and Raetz, C.R. (1988). Purification and properties of lipid A disaccharide synthase of *Escherichia coli*. J. Biol. Chem. *263*, 14859–14867.

Raetz, C.R., and Whitfield, C. (2002). Lipopolysaccharide endotoxins. Annu. Rev. Biochem. *71*, 635–700.

Raetz, C.R., Reynolds, C.M., Trent, M.S., and Bishop, R.E. (2007). Lipid A modification systems in Gram-negative bacteria. Annu. Rev. Biochem. *76*, 295–329.

Raetz, C.R., Guan, Z., Ingram, B.O., Six, D.A., Song, F., Wang, X., and Zhao, J. (2009). Discovery of new biosynthetic pathways: the lipid A story. J. Lipid Res. *50* (Suppl), S103-S108.

Rahim, R., Burrows, L.L., Monteiro, M.A., Perry, M.B., and Lam, J.S. (2000). Involvement of the rml locus in core oligosaccharide and O polysaccharide assembly in *Pseudomonas aeruginosa*. Microbiology *146*, 2803–2814.

Rebeil, R., Ernst, R.K., Gowen, B.B., Miller, S.I., and Hinnebusch, B.J. (2004). Variation in lipid A structure in the pathogenic yersiniae. Mol. Microbiol. *52*, 1363–1373.

Reeves, P.R., Hobbs, M., Valvano, M.A., Skurnik, M., Whitfield, C., Coplin, D., Kido, N., Klena, J., Maskell, D., Raetz, C.R., *et al.* (1996). Bacterial polysaccharide synthesis and gene nomenclature. Trends Microbiol. *4*, 495–503.

Reynolds, C.M., Ribeiro, A.A., McGrath, S.C., Cotter, R.J., Raetz, C.R., and Trent, M.S. (2006). An outer membrane enzyme encoded by *Salmonella typhimurium* lpxR that removes the 3′-acyloxyacyl moiety of lipid A. J. Biol. Chem. *281*, 21974–21987.

Rosenberg, M.F., Callaghan, R., Modok, S., Higgins, C.F., and Ford, R.C. (2005). Three-dimensional structure of P-glycoprotein: the transmembrane regions adopt an asymmetric configuration in the nucleotide-bound state. J. Biol. Chem. *280*, 2857–2862.

Rubires, X., Saigi, F., Pique, N., Climent, N., Merino, S., Alberti, S., Tomas, J.M., and Regue, M. (1997). A gene (wbbL) *from Serratia marcescens* N28b (O4) complements the rfb-50 mutation of *Escherichia coli* K-12 derivatives. J. Bacteriol. *179*, 7581–7586.

Ruiz, N., Kahne, D., and Silhavy, T.J. (2006). Advances in understanding bacterial outer-membrane biogenesis. Nat. Rev. Microbiol. *4*, 57–66.

Ruiz, N., Gronenberg, L.S., Kahne, D., and Silhavy, T.J. (2008). Identification of two inner-membrane proteins required for the transport of lipopolysaccharide to the outer membrane of *Escherichia coli*. Proc. Natl. Acad. Sci. U.S.A. *105*, 5537–5542.

Ruiz, N., Kahne, D., and Silhavy, T.J. (2009). Transport of lipopolysaccharide across the cell envelope: the long road of discovery. Nat. Rev. Microbiol. *7*, 677–683.

Ruiz, N., Chng, S.S., Hiniker, A., Kahne, D., and Silhavy, T.J. (2010). Nonconsecutive disulfide bond formation in an essential integral outer membrane protein. Proc. Natl. Acad. Sci. U.S.A. *107*, 12245–12250.

Sampson, B.A., Misra, R., and Benson, S.A. (1989). Identification and characterization of a new gene of *Escherichia coli* K-12 involved in outer membrane permeability. Genetics *122*, 491–501.

Samuel, G., and Reeves, P. (2003). Biosynthesis of O-antigens: genes and pathways involved in nucleotide sugar precursor synthesis and O-antigen assembly. Carbohydr. Res. *338*, 2503–2519.

Sankaran, K., and Wu, H.C. (1994). Lipid modification of bacterial prolipoprotein. Transfer of diacylglyceryl moiety from phosphatidylglycerol. J. Biol. Chem. *269*, 19701–19706.

Saurin, W., Hofnung, M., and Dassa, E. (1999). Getting in or out: early segregation between importers and exporters in the evolution of ATP-binding cassette (ABC) transporters. J. Mol. Evol. *48*, 22–41.

Schultz, K.M., Merten, J.A., and Klug, C.S. (2011a). Characterization of the E506Q and H537A Dysfunctional Mutants in the *E. coli* ABC Transporter MsbA. Biochemistry *50*, 3599–3608.

Schultz, K.M., Merten, J.A., and Klug, C.S. (2011b). Effects of the L511P and D512G mutations on the *Escherichia coli* ABC transporter MsbA. Biochemistry *50*, 2594–2602.

Serina, S., Nozza, F., Nicastro, G., Faggioni, F., Mottl, H., Dehò, G., and Polissi, A. (2004). Scanning the *Escherichia coli* chromosome by random transposon mutagenesis and multiple phenotypic screening. Res. Microbiol. *155*, 692–701.

Shimazu, R., Akashi, S., Ogata, H., Nagai, Y., Fukudome, K., Miyake, K., and Kimoto, M. (1999). MD-2, a molecule that confers lipopolysaccharide responsiveness on Toll-like receptor 4. J. Exp. Med. *189*, 1777–1782.

Siarheyeva, A., and Sharom, F.J. (2009). The ABC transporter MsbA interacts with lipid A and amphipathic drugs at different sites. Biochem. J. *419*, 317–328.

Silhavy, T.J., Kahne, D., and Walker, S. (2010). The bacterial cell envelope. Cold Spring Harb. Perspect. Biol. *2*, a000414.

Sklar, J.G., Wu, T., Kahne, D., and Silhavy, T.J. (2007). Defining the roles of the periplasmic chaperones SurA, Skp, and DegP in *Escherichia coli*. Genes Dev. *21*, 2473–2484.

Sperandeo, P., Pozzi, C., Dehò, G., and Polissi, A. (2006). Non-essential KDO biosynthesis and new essential cell envelope biogenesis genes in the *Escherichia coli yrbG-yhbG* locus. Res. Microbiol. *157*, 547–558.

Sperandeo, P., Cescutti, R., Villa, R., Di Benedetto, C., Candia, D., Dehò, G., and Polissi, A. (2007). Characterization of *lptA* and *lptB*, two essential genes implicated in lipopolysaccharide transport to the outer membrane of *Escherichia coli*. J. Bacteriol. *189*, 244–253.

Sperandeo, P., Lau, F.K., Carpentieri, A., De Castro, C., Molinaro, A., Dehò, G., Silhavy, T.J., and Polissi, A. (2008). Functional analysis of the protein machinery required for transport of lipopolysaccharide to the outer membrane of *Escherichia coli*. J. Bacteriol. *190*, 4460–4469.

Sperandeo, P., Dehò, G., and Polissi, A. (2009). The lipopolysaccharide transport system of Gram-negative bacteria. Biochim. Biophys. Acta *1791*, 594–602.

Sperandeo, P., Villa, R., Martorana, A.M., Samalikova, M., Grandori, R., Deho, G., and Polissi, A. (2011). New insights into the Lpt machinery for lipopolysaccharide transport to the cell surface: LptA–LptC interaction and LptA stability as sensors of a properly assembled transenvelope complex. J. Bacteriol. *193*, 1042–1053.

Steeghs, L., den Hartog, R., den Boer, A., Zomer, B., Roholl, P., and van derLey, P. (1998). Meningitis bacterium is viable without endotoxin. Nature *392*, 449–450.

Stenberg, F., Chovanec, P., Maslen, S.L., Robinson, C.V., Ilag, L.L., von Heijne, G., and Daley, D.O. (2005). Protein complexes of the *Escherichia coli* cell envelope. J. Biol. Chem. *280*, 34409–34419.

Subbaiah, T.V., and Stocker, B.A. (1964). Rough mutants of *Salmonella typhimurium*. I. Genetics. Nature *201*, 1298–1299.

Suits, M.D., Sperandeo, P., Dehò, G., Polissi, A., and Jia, Z. (2008). Novel structure of the conserved Gram-negative lipopolysaccharide transport protein A and mutagenesis analysis. J. Mol. Biol. *380*, 476–488.

Sutcliffe, I.C. (2010). A phylum level perspective on bacterial cell envelope architecture. Trends Microbiol. *18*, 464–470.

Sutcliffe, I.C. (2011). Cell envelope architecture in the Chloroflexi: a shifting frontline in a phylogenetic turf war. Environ. Microbiol. *13*, 279–282.

Takase, I., Ishino, F., Wachi, M., Kamata, H., Doi, N., Asoh, S., Matsuzawa, H., Ohta, T., and Matsuhashi, M. (1987). Genes encoding two lipoproteins in the leuS-dacA region of the *Escherichia coli* chromosome. J. Bacteriol. *169*, 5692–5699.

Tefsen, B., Bos, M.P., Beckers, F., Tommassen, J., and de Cock, H. (2005a). MsbA is not required for phospholipid transport in *Neisseria meningitidis*. J. Biol. Chem. *280*, 35961–35966.

Tefsen, B., Geurtsen, J., Beckers, F., Tommassen, J., and de Cock, H. (2005b). Lipopolysaccharide transport to the bacterial outer membrane in spheroplasts. J. Biol. Chem. *280*, 4504–4509.

Tokuda, H. (2009). Biogenesis of outer membranes in Gram-negative bacteria. Biosci. Biotechnol. Biochem. 73, 465–473.

Tran, A.X., Trent, M.S., and Whitfield, C. (2008). The LptA protein of *Escherichia coli* is a periplasmic lipid A-binding protein involved in the lipopolysaccharide export pathway. J. Biol. Chem. *283*, 20342–20349.

Tran, A.X., Dong, C., and Whitfield, C. (2010). Structure and functional analysis of LptC, a conserved membrane protein involved in the lipopolysaccharide export pathway in *Escherichia coli*. J. Biol. Chem. *285*, 33529–33539.

Trent, M.S., Pabich, W., Raetz, C.R., and Miller, S.I. (2001a). A PhoP/PhoQ-induced Lipase (PagL) that catalyzes 3-O-deacylation of lipid A precursors in membranes of *Salmonella typhimurium*. J. Biol. Chem. *276*, 9083–9092.

Trent, M.S., Ribeiro, A.A., Doerrler, W.T., Lin, S., Cotter, R.J., and Raetz, C.R. (2001b). Accumulation of a polyisoprene-linked amino sugar in polymyxin-resistant Salmonella typhimurium and *Escherichia coli*: structural characterization and transfer to lipid A in the periplasm. J. Biol. Chem. *276*, 43132–43144.

Valvano, M.A. (2003). Export of O-specific lipopolysaccharide. Front Biosci. *8*, s452-s471.

Vinogradov, E.V., Lindner, B., Kocharova, N.A., Senchenkova, S.N., Shashkov, A.S., Knirel, Y.A., Holst, O., Gremyakova, T.A., Shaikhutdinova, R.Z., and Anisimov, A.P. (2002). The core structure of the lipopolysaccharide from the causative agent of plague, *Yersinia pestis*. Carbohydr. Res. *337*, 775–777.

Vollmer, W., and Seligman, S.J. (2010). Architecture of peptidoglycan: more data and more models. Trends Microbiol. *18*, 59–66.

Wachtershauser, G. (2003). From pre-cells to Eukarya – a tale of two lipids. Mol. Microbiol. *47*, 13–22.

Walsh, A.G., Matewish, M.J., Burrows, L.L., Monteiro, M.A., Perry, M.B., and Lam, J.S. (2000). Lipopolysaccharide core phosphates are required for viability and intrinsic drug resistance in *Pseudomonas aeruginosa*. Mol. Microbiol. 35, 718–727.

Wang, X., Karbarz, M.J., McGrath, S.C., Cotter, R.J., and Raetz, C.R. (2004). MsbA transporter-dependent lipid A 1-dephosphorylation on the periplasmic surface of the inner membrane: topography of *Francisella novicida* LpxE expressed in *Escherichia coli*. J. Biol. Chem. *279*, 49470–49478.

Wang, X., McGrath, S.C., Cotter, R.J., and Raetz, C.R. (2006). Expression cloning and periplasmic orientation of the *Francisella novicida* lipid A 4′-phosphatase LpxF. J. Biol. Chem. *281*, 9321–9330.

Ward, A., Reyes, C.L., Yu, J., Roth, C.B., and Chang, G. (2007). Flexibility in the ABC transporter MsbA: alternating access with a twist. Proc. Natl. Acad. Sci. U.S.A. *104*, 19005–19010.

Whitfield, C. (1995). Biosynthesis of lipopolysaccharide O antigens. Trends Microbiol. 3, 178–185.

Wu, J., and Woodard, R.W. (2003). *Escherichia coli* YrbI is 3-deoxy-D-manno-octulosonate 8-phosphate phosphatase. J. Biol. Chem. *278*, 18117–18123.

Wu, T., McCandlish, A.C., Gronenberg, L.S., Chng, S.S., Silhavy, T.J., and Kahne, D. (2006). Identification of a protein complex that assembles lipopolysaccharide in the outer membrane of *Escherichia coli*. Proc. Natl. Acad. Sci. U.S.A. *103*, 11754–11759.

Wyckoff, T.J., Lin, S., Cotter, R.J., Dotson, G.D., and Raetz, C.R. (1998). Hydrocarbon rulers in UDP-N-acetylglucosamine acyltransferases. J. Biol. Chem. *273*, 32369–32372.

Young, K., and Silver, L.L. (1991). Leakage of periplasmic enzymes from envA1 strains of *Escherichia coli*. J. Bacteriol. *173*, 3609–3614.

Zhang, L., al Hendy, A., Toivanen, P., and Skurnik, M. (1993). Genetic organization and sequence of the rfb gene cluster of Yersinia enterocolitica serotype O:3: similarities to the dTDP-L-rhamnose biosynthesis pathway of *Salmonella* and to the bacterial polysaccharide transport systems. Mol. Microbiol. *9*, 309–321.

Zhou, Z., White, K.A., Polissi, A., Georgopoulos, C., and Raetz, C.R. (1998). Function of *Escherichia coli MsbA*, an essential ABC family transporter, in lipid A and phospholipid biosynthesis. J. Biol. Chem. *273*, 12466–12475.

Zuber, B., Chami, M., Houssin, C., Dubochet, J., Griffiths, G., and Daffe, M. (2008). Direct visualization of the outer membrane of mycobacteria and corynebacteria in their native state. J. Bacteriol. *190*, 5672–5680.

Outer Membrane Protein Biosynthesis: Transport and Incorporation of Outer Membrane Proteins (in)to the Outer Membrane Bilayer

3

Kelly H. Kim, Suraaj Aulakh and Mark Paetzel

Abstract

The outer membrane is a unique structural feature of Gram-negative bacteria. Within the outer membrane reside β-barrel outer membrane proteins that serve many important functions such as nutrient uptake, virulence, and cell signalling. Proper folding and assembly of these proteins are therefore essential for cell viability. Gram-negative bacteria possess a specialized proteinaceous machine known as the β-barrel assembly machinery (BAM) complex that is responsible for the proper assembly of β-barrel proteins into the outer membrane. This chapter summarizes the current status of knowledge about outer membrane protein biosynthesis, and the significant progress that has been made towards understanding the structure and function of the bacterial BAM complex.

Introduction

One of the most recognizable features of Gram-negative bacteria is the outer membrane. Whereas the cell wall of Gram-positive bacteria consists of a plasma membrane and a thick peptidoglycan layer, Gram-negative bacteria have an inner membrane (IM), an outer membrane (OM), and a thin peptidoglycan layer in the periplasmic space between the IM and the OM (Fig. 3.1) (Silhavy *et al.*, 2010). The most obvious advantage of the OM is its ability to serve as a physical barrier against the extracellular environment. The asymmetric bilayer of the OM is different from the IM phospholipid bilayer. The OM outer leaflet is composed of glycolipids, mainly lipopolysaccharide (LPS), which is a known endotoxin that triggers an immune response in humans (Raetz and Whitfield, 2002). The inner leaflet of the OM contains the phospholipids, primarily phosphatidylethanolamine and phosphatidylglycerol (Morein *et al.*, 1996).

Aside from lipids, about half of the OM's mass consists of proteins (Koebnik *et al.*, 2000). These proteins generally fall into two categories: lipoproteins, which are found in the periplasmic space and covalently anchored by an amino-terminal cysteine residue to the inner leaflet phospholipids; and transmembrane proteins, which span the entire width of the bilayer (Koebnik *et al.*, 2000; Bos and Tommassen, 2004). The latter category is generally referred to as outer membrane proteins (OMPs), and current structural information shows

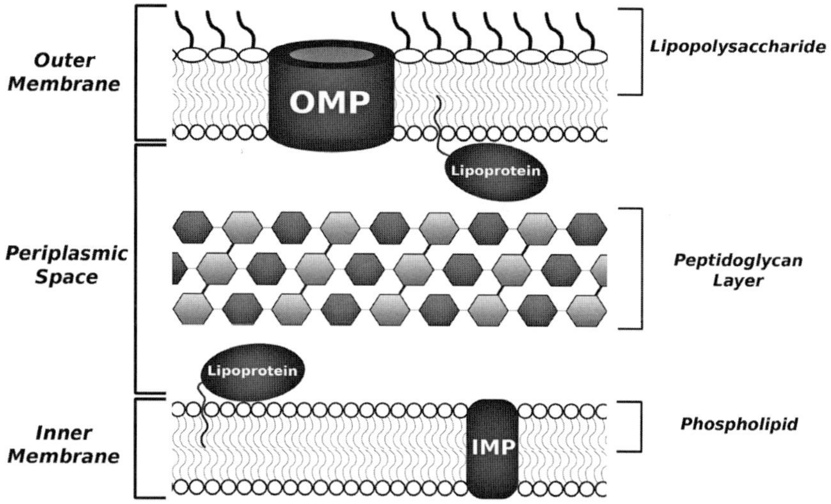

Figure 3.1 Gram-negative bacterial cell envelope. The cell envelope of a Gram-negative bacteria consists of the periplasm with a peptidoglycan layer surrounded by two membranes (inner and outer membranes). Unlike the IM, which is mostly made up of phospholipids, the OM is an asymmetric lipid-bilayer containing only lipopolysaccharides (LPS) in its outer leaflet. The membrane associated proteins of the Gram-negative bacterial cell envelope can be largely divided into three categories – outer membrane proteins (OMPs), inner membrane proteins (IMPs), and lipid-attached lipoproteins (found on the periplasmic leaflets of the IM and the OM).

most OMPs to have a β-barrel fold (Fairman *et al.*, 2011). The only exceptions found so far are Wza from *Escherichia coli*, PorB from *Corynebacterium glutamicum*, and the Type IV outer membrane secretion complex found throughout Gram-negative bacteria (Dong *et al.*, 2006; Collins and Derrick, 2007; Ziegler *et al.*, 2008; Chandran *et al.*, 2009). For the purpose of this chapter, the term OMP will refer specifically to integral β-barrel proteins found in the OM.

Despite the structural similarities of these β-barrel OMPs, the functions are quite diverse. Some OMPs are involved in the flow of nutrients across the OM, some function as exporters of virulence factors, and some help remove toxins, including antibiotics, out of the cell (Koebnik *et al.*, 2000). The first section of this chapter takes a closer look into the different types of OMPs and their respective roles. However, in order for these proteins to be functional, they must first be exported from the cytosol, across the IM and periplasmic space, and correctly assembled in the OM. The final folding and assembly step is now believed to be carried out by a multi-protein system known as the β-barrel assembly machinery (BAM) complex. Improper assembly by or absence of the BAM complex can lead to defects in the OM, eventually leading to cell death (Gentle *et al.*, 2004). Homologues of the BAM complex can also be found in mitochondria and chloroplasts.

With antibiotic resistance on the rise, researchers are in pursuit of novel drug targets, especially in the case of Gram-negative bacteria. Most antibiotics today are directed to cellular processes in the periplasmic space or further inside the cell, for which they must first pass the OM (Delcour, 2009). Thus, understanding the pathway of OMP biogenesis not only answers the question of how OMPs get into the OM, but also provides a potential new target for antibiotics. This chapter reviews the current state of knowledge in the field of OMP biogenesis: how a diverse group of β-barrel proteins travel from the cytosol to the OM, and finally become assembled by the BAM complex.

Outer membrane proteins (OMPs)

OMPs are β-barrel proteins found in the outer membrane of Gram-negative bacteria, mitochondria and chloroplasts. They usually have an even number of β-strands ranging from 8 to 24, arranged in an anti-parallel fashion, and each OMP prefers a particular oligomeric state (Fig. 3.2) (Koebnik *et al.*, 2000; Schulz, 2000; Wimley, 2003; Meng *et al.*, 2009). Although the overall β-barrel scaffold of the OMPs may be consistent, differences in the number of strands, the length and properties of the loops, and oligomeric state add up to a diverse group of proteins with distinct functions (Fig. 3.3 and Table 3.1). To demonstrate the range of diversity, six general categories of OMPs are provided below, with specific examples from *E. coli*.

Non-specific porins

Despite its general use today, the term 'porin' was originally coined to describe a class of OMP channels involved in non-specific diffusion of solutes (Nakae, 1976; Nikaido, 1994, 2003). Porins are usually found as homotrimers of 16-stranded β-barrel subunits, and the hydrophilic interior allows the transport of hydrophilic molecules smaller than 600 Da into the cell (Nikaido, 1994; Delcour, 2003). While defined as 'non-specific,' porins can be selective in terms of the size and charge of the molecule. Examples of porins include OmpC (preference for small, positively charged solutes), OmpF (preference for large, positively charged solutes), and PhoE (preference for negatively charged solutes) (Nikaido, 2003).

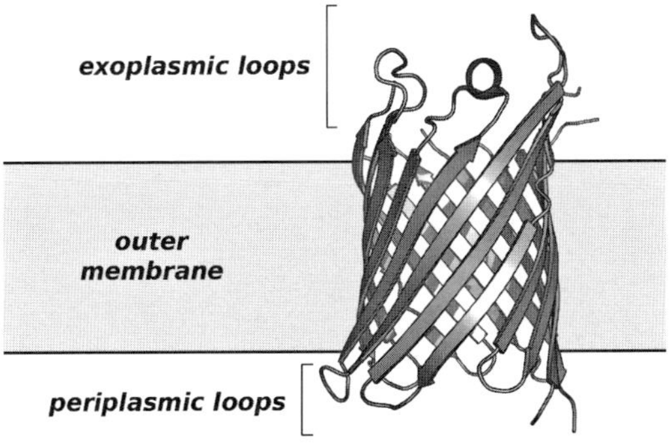

Figure 3.2 Outer membrane proteins (OMPs). OMPs are found predominantly as β-barrel proteins. The β-strands in OMPs are typically arranged in an anti-parallel fashion, but they vary in number depending on the protein. OMPs are usually asymmetric in shape, with the exoplasmic loops being larger than the periplasmic loops. In this figure, the structure of an *E. coli* OMP, OmpG (PDB: 2F1C), is shown.

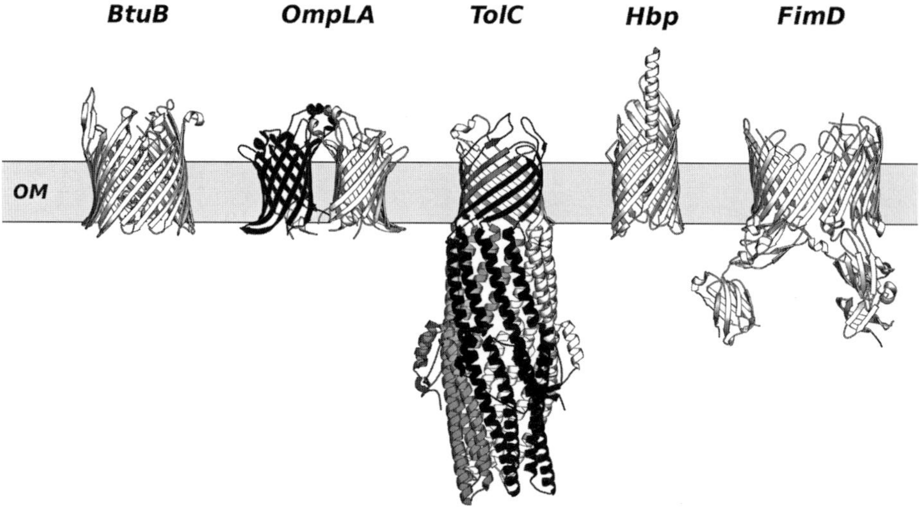

Figure 3.3 Diversity of bacterial OMP structure. Some examples of *E. coli* OMPs are shown. OMPs can be found either as a monomer (ex. BtuB, Hbp and FimD; PDB: 2GUF, 3AEH and 3RFZ), an oligomer where each subunit creates its own β-barrel (ex. OmpLA; PDB: 1QD6), or an oligomer where the multiple subunits come together to form one β-barrel (ex. TolC; PDB: 1EK9).

Substrate specific channels

Aside from porins, there are other OMPs that also act as pores for larger solutes to pass through. However, these pores are specific for their substrates, and can be referred to as channels rather than porins (Nikaido, 2003). For example, BtuB (Fig. 3.3) is specific for the uptake of vitamin B12, Fhua for iron, LamB for maltose and other sugars, and Tsx for nucleosides. Some of these channels (ex. LamB and Tsx) are independent and allow passive diffusion of solutes upon contact, while others (ex. BtuB and FhuA) are TonB-dependent channels as they require assistance for the active transport of substrates (Nikaido, 1994). The TonB complex is found in the IM and spans the periplasmic space to interact with these latter channels and provide the energy required for uptake of the substrate (Postle and Kadner, 2003; Noinaj *et al.*, 2010; Krewulak and Vogel, 2011). LamB is a homotrimer of 16-stranded β-barrel subunits, and thus it has often been classified as a substrate specific porin and has even been given the name maltoporin. However, to avoid confusion with non-specific porins and to stay consistent with the original definition, here it is classified as a substrate specific channel (Nakae, 1976; Nikaido, 2003).

Translocons for the export of substrates

In addition to the import of molecules, the OM allows certain molecules to travel outside the cell as well. For example, many proteins are synthesized in the cytosol, but need to be secreted outside the cell for proper function. In a Gram-negative bacterium, many OMPs can be found playing a role in the different export pathways. TolC is an OMP involved in the Type I secretion pathway that allows the export of proteins without the use of the SecYEG translocon at the IM. TolC forms a unique trimer structure, with three protomers contributing four strands each to form a single 12-stranded β-barrel (Fig. 3.3). In addition

Table 3.1 List of currently reviewed *Escherichia coli* OMPs in the UniProt database

Protein	Length in amino acids	Proposed function/category	UniProt ID	Evidence as BAM substrate
AfaC	859	Usher	P53517	
Ag43	1039	Autotransporter	P39180	Rossiter *et al*. (2011)
AggC	842	Usher	P46005	
AIDA-I	1286	Autotransporter	Q03155	Jain and Goldberg (2007)
BglH	538	Sugar transport	P26218	
BtuB	614	Vitamin B12 transport	P06129	
CirA	663	Iron transport	P17315	
CS3–2	937	Usher	P15484	
CssD	819	Usher	P53512	
EaE (Intimin)	934	Autotransporter	P43261	Bodelón *et al*. (2009)
EatA	1364	Autotransporter	Q84GK0	
ElfC	866	Usher	P75857	
EspC	1305	Autotransporter	Q9EZE7	
EspP	1300	Autotransporter	Q7BSW5	
FadL	446	Long-chain fatty acid transport	P10384	
FaeD	812	Usher	P06970	
FanD	783	Usher	P12050	
FasD	835	Usher	P46000	
FecA	774	Iron transport	P13036	
FepA	746	Iron transport	P05825	
FhuA	747	Iron transport	P06971	
FhuE	729	Iron transport	P16869	
FimD	878	Usher	P30130	Palomino *et al*. (2011)
Fiu	760	Iron transport	P75780	
FocD	875	Usher	P46009	
Hbp	1377	Autotransporter	O88093	Sauri *et al*. (2009)
HtrE	865	Usher	P33129	
IutA	732	Iron transporter	P14542	
LamB	446	Maltose transport	P02943	Malinverni *et al*. (2006)
LptD	784	LPS assembly protein	P31554	
MipA	248	Peptidoglycan synthesis scaffold protein	P0A908	
NanC	238	N-acetylneuraminic acid transport	P69856	
NfrA	990	N4 bacteriophage receptor	P31600	
NmpC	365	Porin	P21420	
OmpA	346	OM stability, bacterial conjugation	P0A910	Malinverni *et al*. (2006)
OmpC	367	Porin	P06996	Malinverni *et al*. (2006)

Table 3.1 (continued)

Protein	Length in amino acids	Proposed function/category	UniProt ID	Evidence as BAM substrate
OmpF	362	Porin	P02931	Malinverni *et al.* (2006)
OmpG	301	Sugar transport	P76045	
OmpL	230	Sugar transport	P76773	
OmpLA	289	Phospholipase	P0A921	
OmpN	377	Porin	P77747	
OmpP	315	Protease	P34210	
OmpT	317	Protease	P09169	Hagan *et al.* (2010)
OmpW	212	Colicin S4 receptor	P0A915	
OmpX	171	Adhesin	P0A917	
PapC	836	Usher	P07110	
PagP	186	Lipid A palmitoyltransferase	P37001	
PcoB	296	Copper resistance	Q47453	
Pet	1295	Autotransporter	O68900	Rossiter *et al.* (2011)
PgaA	807	Poly-beta-1,6-*N*-acetyl-D-glucosamine transport	P69434	
PhoE	351	Porin	P02932	Robert *et al.* (2006)
Pic	1371	Autotransporter	Q8CWC7	
Sat	1295	Autotransporter	Q8FDW4	
SfmD	867	Usher	P77468	
TibA	989	Autotransporter	Q9XD84	
TolC	493	OM export protein	P02930	Malinverni *et al.* (2006)
TraN	602	Transfer of F plasmid during conjugation	P24082	
Tsh	1377	Autotransporter	Q47692	
Tsx	294	Nucleoside specific channel	P0A927	
UidC	421	Involved in glucuronide transport	Q47706	
YaeT (BamA)	810	Assembly of OMPs	P0A940	
YbgQ	815	Usher	P75750	
YedS	397	Porin	P76335	
YehB	826	Usher	P33341	
YejO	863	Autotransporter	P33924	
YfcU	881	Usher	P77196	
YhcD	793	Usher	P45420	
YiaT	246	Peptidoglycan synthesis scaffold protein	P37681	
YncD	700	Channel (TonB-dependent)	P76115	
YpjA	1526	Autotransporter	P52143	
YqiG	821	Usher	P76655	

Table 3.1 (continued)

Protein	Length in amino acids	Proposed function/category	UniProt ID	Evidence as BAM substrate
YraJ	838	Usher	P42915	
YuaO	1758	Autotransporter	Q9JMS5	
YuaQ	1371	Autotransporter	Q9JMS3	

to the barrel, a long periplasmic domain is present that forms an α-tunnel, which allows interaction with other components of the pathway found in the periplasm and IM (Koronakis *et al.*, 2000). Aside from protein export, TolC is also known to export small molecules and drugs, contributing to antibiotic resistance (Zgurskaya *et al.*, 2011). Other examples of OMPs acting as translocons are those involved in the two-partner secretion pathway (a component of the Type V secretion pathway). In this pathway, the secreted protein requires a specific OMP translocon for export. A famous example is FhaC from *Bordetella pertussis*, which is involved in the export of filamentous haemagglutinin, an adhesin secreted during infection. FhaC has a 16-stranded β-barrel at the C-terminus, with a periplasmic region at the N-terminus (Clantin *et al.*, 2007).

Autotransporters

Some proteins destined for transport outside the cell can be secreted without an additional OMP channel. In these cases, the proteins contain their own C-terminal β-barrel domain that acts as a transporter for the N-terminal passenger protein that is to be secreted. Because of their ability to transport themselves, these proteins are known as autotransporters. The C-terminal transporter domain is generally a 12-stranded barrel, and in many cases found as a monomer (Oomen *et al.*, 2004; van Ulsen, 2011). The passenger proteins are usually virulence factors secreted by pathogenic strains of *E. coli*. Examples include adhesins such as AIDA-I and Ag43, and proteases such as Hbp (Fig. 3.3) and Pet (van Ulsen, 2011). Intimin, another example, is an attaching and effacing protein. Interestingly, intimin does not fall into the classical definition of autotransporters as its passenger domain is at the C-terminus, while the β-barrel transporter is at the N-terminus. The structure of this β-barrel has yet to be solved, and some experiments suggest possible homodimer formation (Bodelon *et al.*, 2009).

Enzymes

Aside from transport, there are other functions required to take place at the OM, some of which are carried out by enzymatic OMPs. To date, only three enzymatic OMPs have been discovered in *E. coli*: OmpLA, OmpT, and PagP (Bishop, 2008). OmpLA (phospholipase A) is a 12-stranded β-barrel that hydrolyses phospholipids in the OM (Fig. 3.3). Its active site is located in the LPS-containing outer leaflet, where it can detect the presence of phospholipids that disrupt the asymmetry of the OM. OmpLA is found as a monomer, and dimerizes upon substrate presentation to become active (Dekker *et al.*, 1997). OmpT is a protease that has a 10-stranded β-barrel fold that specifically cleaves between two basic residues of a protein, with substrates shown to include antimicrobial peptides released by host immune responses (Sugimura and Nishihara, 1988; Stumpe *et al.*, 1998; Vandeputte-Rutten *et al.*, 2001). The oligomeric state of OmpT is unknown, but early gel-filtration studies suggest a possible pentamer formation (Sugimura and Nishihara, 1988). Interestingly, some strains of *E. coli* have another protease in the OM known as OmpP, which is a homologue of OmpT found encoded on the F-plasmid (Bishop, 2008). Finally, PagP is an eight-stranded β-barrel that transfers a palmitate chain from a phospholipid in the inner leaflet to the Lipid A component of a LPS molecule in the outer leaflet. Because of its role in maintaining this essential piece of the OM, PagP is being studied as a potential drug target (Bishop, 2005).

Structural OMPs

This final category classifies OMPs that contribute to the formation and integrity of the cell wall structure. For example, Mipa is an OMP that acts as a scaffolding protein to mediate the interaction between specific enzymes involved in peptidoglycan synthesis in the periplasm (Vollmer *et al.*, 1999; Vollmer and Bertsche, 2008). BamA is a component of the BAM complex which catalyses the insertion of OMPs into the OM. It has an N-terminal periplasmic region and a C-terminal β-barrel (more discussion on this protein in later sections of this chapter). LptD is an OMP involved in the final stages of LPS assembly in the OM (Okuda and Tokuda, 2011). OmpX is a part of a family of OMPs that are involved in adhesion and entry into host cells. Its structure has been solved and shows the presence of an 8-stranded β-barrel (Vogt and Schulz, 1999). The final family of proteins that can be classified here are the fimbrial usher proteins, which are involved in the formation of the pili subunits found on the exterior of the bacterium. For example, FimD is a 24-stranded β-barrel (Fig. 3.3) that serves as an usher to transport and polymerize subunits of the Type I pili, with the help of the periplasmic chaperone FimC (Phan *et al.*, 2011).

General overview of OMP biosynthesis

In eukaryotes and prokaryotes, most proteins are synthesized in the cytosol, with almost half of them requiring membrane targeting for proper function (Schatz and Dobberstein, 1996). In eukaryotes, these proteins could be destined for an organelle membrane or lumen, the cytoplasmic membrane, or secretion outside the cell. For prokaryotes, the options are limited to the IM, periplasm, OM, or extracellular space. In order for these proteins to function correctly, they must reach their target membrane or compartment efficiently. The information for their delivery is encoded in the primary sequence of the protein, usually found at the N- or C-terminus. The overall OMP biogenesis process is summarized below and also in Fig. 3.4.

Synthesis and transport to the inner membrane

In prokaryotes, to ensure proper targeting of proteins destined for the IM or further, an N-terminal signal directs them to the IM. The signal sequence of transmembrane IM proteins (IMPs) is co-translationally recognized and directed to the IM by the signal recognition particle (SRP). In many cases, this signal is encoded within the N-terminal transmembrane region of the IMP, and thus a separate cleavable signal sequence is not needed (Dalbey *et al.*, 2011).

In contrast, targeting of precursor OMPs (subsequently referred to as pre-OMPs) in Gram-negative bacteria occurs post-translationally, involving a cleavable N-terminal signal

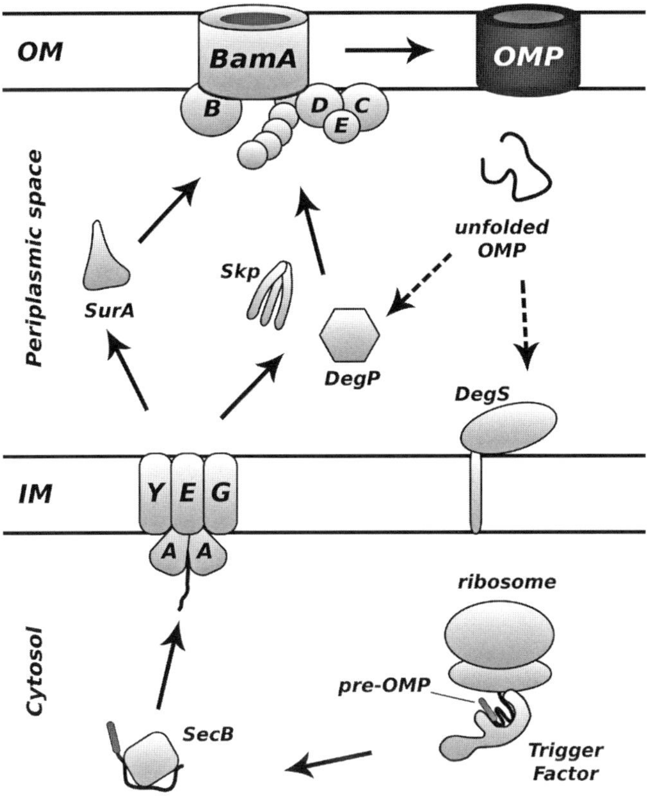

Figure 3.4 OMP biogenesis and degradation. OMP synthesis takes place in the cytosol. The newly synthesized OMP precursor (pre-OMP) contains an N-terminal signal sequence that targets the protein to the IM. The N-terminal signal of the pre-OMP is first bound by a trigger factor as it is being synthesized by the ribosome, and subsequently transferred to a cytoplasmic chaperone SecB. The pre-OMP is then released from SecB to an ATP-powered motor protein SecA, which facilitates the translocation of the pre-OMP across the IM via the SecYEG channel. At the IM, the N-terminal signal sequence is removed by signal peptidase I, and the pre-OMP (still in an unfolded form) is released into the periplasm. The protein then takes either the SurA or the Skp/DegP pathway to travel to the BAM complex of the OM. By an unknown mechanism, the BAM complex mediates the folding and membrane insertion of OMPs. Misfolded or aggregated OMPs in the periplasm are recognized and degraded by DegP and DegS.

sequence with the following features: a positively charged N-terminal region, a hydrophobic region, and a polar C-terminal region containing the cleavage site (Gierasch, 1989; von Heijne, 1990). SRP has a preference for hydrophobic regions as its association with the signal sequence strengthens with increased hydrophobicity (Valent *et al.*, 1995). Thus, SRP preferentially binds to the IMP signal allowing another protein, trigger factor (TF), to bind to the pre-OMP as it emerges from the ribosome (Ferbitz *et al.*, 2004; Driessen and Nouwen, 2008; Hoffmann *et al.*, 2010). TF then passes the pre-OMP over to the cytoplasmic chaperone SecB, which associates with regions of the protein that are believed to be buried in the natively folded form (Knoblauch *et al.*, 1999; Bechtluft *et al.*, 2010). This ensures that the pre-OMP remains in a stable unfolded state, which is required for translocation through the SecYEG complex at the IM (Driessen and Nouwen, 2008).

Translocation at the inner membrane

Once at the IM, SecB transfers the pre-OMP to SecA, a homodimeric ATPase bound to the SecYEG translocation complex. ATP binding is required to allow SecB release from this complex, and then SecA begins to thread the pre-OMP through the SecYEG channel. The mechanism by which SecA does this is not known; however, one proposed model suggests an ATP-dependent process in which the two-helix-finger domain of SecA acts as a piston to push the pre-OMP through (Zimmer *et al.*, 2008; Cross *et al.*, 2009; Kusters and Driessen, 2011).

The SecYEG translocon (also known as the Sec translocon) is a well conserved complex whose homologues are also found in the endoplasmic reticulum (ER) membrane of eukaryotes, where it is known as the Sec61 translocon (Pohlschroder *et al.*, 1997; Cross *et al.*, 2009). As the name suggests, SecYEG is a heterotrimer of the IMPs SecY, SecE, and SecG, which come together to form a central pore. The diameter of this pore is too small for the pre-OMP polypeptide chain to enter, and thus it must widen and form an open channel for translocation to occur. This could occur simply by conformational changes within one protomer of SecYEG, or by further assembly of the complex into a dimeric or tetrameric form to create a larger channel (Driessen and Nouwen, 2008; du Plessis *et al.*, 2011). Thus, the oligomeric state required for proper *in vivo* translocation is still under debate. The SecYEG translocon is also found associated with another heterotrimeric complex, SecDFYajC, which is not essential for translocation, but has been shown to enhance the activity of SecYEG through an unknown mechanism (Driessen and Nouwen, 2008; du Plessis *et al.*, 2011).

Transport to the outer membrane

As the pre-OMP begins to emerge into the periplasmic space, the signal peptide is recognized and cleaved by signal peptidase I (SPaseI). SPaseI specifically cleaves after the conserved Ala-X-Ala sequence in the C-terminal region of the signal peptide, releasing the mature OMP into the periplasmic space (Paetzel *et al.*, 2002). From here, the OMP must travel through the dense periplasm to the OM before being folded. Periplasmic chaperones SurA, Skp, and DegP have been shown to associate with OMPs, and are believed to be the major players involved in stabilizing these unfolded proteins during their transport. SurA recognizes unfolded proteins by aromatic residues, a common feature of OMPs, especially at their C-terminus (Bitto and McKay, 2003). It has been shown that, aside from the N-terminal signal sequence for IM targeting, OMPs carry another signal on the C-terminus which is recognized by the BAM complex for assembly and insertion at the OM (Robert *et al.*, 2006). On the other hand, Skp recognizes unfolded proteins by their exposed hydrophobic regions (Qu *et al.*, 2007). The trimeric structure of Skp mimics the shape of a jellyfish with three tentacles, and forms a hydrophobic interior serving as a protective cavity for the OMPs (Walton and Sousa, 2004). Similarly, DegP oligomerizes to form large cages that can also protect the OMPs from aggregation or degradation by proteases (Merdanovic *et al.*, 2011). Unlike SurA and Skp, substrate specificity of DegP is broad as it can recognize unfolded, misfolded and mislocalized proteins in the periplasm. In addition to its function as a chaperone, DegP also shows a proteolytic activity at high temperatures (more detailed discussion in the 'Degradation of OMPs' section). Gene knockout studies suggest that Skp and DegP may function in a separate pathway than SurA, as individual gene absences are viable, but a double *surA*⁻ and *degP*⁻ or a double *skp*⁻ and *surA*⁻ mutant is synthetically lethal. This also suggests that both pathways play a redundant role, and that at least one pathway must always be functional for cell survival (Rizzitello *et al.*, 2001).

Folding and insertion into the outer membrane

After being transported to the OM, the OMPs are then folded and inserted into the membrane by the BAM complex. In *E. coli*, this multi-protein machinery consists of an OMP known as BamA (previously known as YaeT) and four OM lipoproteins: BamB, BamC, BamD, and BamE (previously YfgL, NlpB, YfiO, and SmpA) (Hagan *et al.*, 2011; Ricci and Silhavy, 2011). BamA has a C-terminal β-barrel domain spanning the membrane and an N-terminal periplasmic region composed of five polypeptide transport associated (POTRA) domains. The absence of BamA results in an accumulation of unfolded proteins in the periplasm, a deformed OM, as well as cell death; hence it is an essential protein found in all Gram-negative bacteria, with homologues also in mitochondria and chloroplasts. Out of the four lipoproteins, only BamD is essential and conserved. When a newly synthesized OMP reaches the OM, it is proposed that its C-terminal targeting sequence is recognized by BamD, directing the OMP towards the BAM complex. The exact mechanism of OMP folding and insertion is not well understood yet, and a detailed discussion of current models is provided in the 'Proposed mechanisms of the BAM complex' section of this chapter.

Degradation of OMPs

Protein degradation is an essential component of quality control. Stress on the OMP synthesis pathway, such as overproduction, can cause OMPs to become misfolded, aggregated, or mislocalized. Fortunately, there are systems in place to remove the defective OMPs from the synthesis pathway. When OMPs are mislocalized and not correctly targeted to the BAM complex, their C-termini activate the DegS protease which initiates a cascade of events in the SigmaE pathway that eventually leads to a decrease in the expression level of OMPs. This lowers the stress put on the OMP synthesis pathway, preventing further mislocalization. Similarly, any damaged or misfolded OMPs are recognized by DegP, which can use its proteolytic function to initiate degradation (Merdanovic *et al.*, 2011).

In vitro studies of OMP folding and membrane insertion

Prior to the discovery of the BAM complex, OMP folding studies were mainly carried out *in vitro*. Compared with α-helical membrane proteins, β-barrel membrane proteins such as OMPs tend to be less hydrophobic. This is because β-barrels have solvent-exposed polar residues facing the interior of the barrel, in addition to hydrophobic residues on the lipid bilayer-facing surfaces (Koebnik *et al.*, 2000; Fairman *et al.*, 2011). β-Barrels are therefore less likely to aggregate in solution, and have made studying and monitoring their folding behaviours *in vitro* simpler. In contrast, the generally more hydrophobic α-helical membrane proteins have been more difficult to work with and to examine their *in vitro* folding behaviours (Tamm *et al.*, 2004). Studying protein folding *in vitro* has advantages of being able to perform controlled experiments on isolated proteins, and to more easily test how various factors influence the folding behaviour of a protein. In the case of OMPs, several *in vitro* thermodynamics and kinetics studies were carried out even before the discovery of the BAM complex, and have provided valuable insights into biochemical properties and folding mechanisms of OMPs (Kleinschmidt, 2006; Burgess *et al.*, 2008).

Various studies have demonstrated that some OMPs can spontaneously fold and insert themselves into a lipid bilayer in the absence of an energy source or a folding factor. For example, an eight-stranded β-barrel protein OmpA has been shown to denature in 6–8 M urea and subsequently refold into a lipid bilayer when urea concentration is reduced via rapid dilution (Surrey and Jahnig, 1992). OmpA has since been frequently used as a model to study the kinetics of folding and insertion into lipid bilayers *in vitro*. The kinetics studies of OmpA folding and insertion into dioleoylphosphatidylcholine (DOPC) bilayers at various temperatures, coupled with the time-resolved tryptophan fluorescence quenching (TDFQ) technique, has suggested that OmpA refolding exhibits three distinct kinetic phases: (1) the fastest and temperature-independent phase attributed to the initial binding of unfolded OmpA to the membrane surface, (2) the slower and strongly temperature-dependent phase that corresponds to a deeper but partial insertion of OmpA into the membrane, and (3) the slowest phase of β-barrel maturation that was only observed to occur at temperatures greater than 30°C (Kleinschmidt, 2003; Tamm *et al.*, 2004; Kleinschmidt, 2006). These and other *in vitro* studies suggest that OMP folding (i.e. acquiring inter-strand hydrogen bonds and forming β-hairpins) and membrane insertion are coupled and that both occur in a synchronized manner.

In vitro studies of OMP folding has also allowed direct examination of how lipid bilayer composition influences the folding behaviour and stability of OMPs. As OMPs are membrane proteins and their folding takes place as they translocate into the OM, the chemical and physical properties of the lipid bilayer must have a significant influence on the folding kinetics of OMPs. Both the IM and the inner leaflet of the OM in Gram-negative bacteria are composed of ~75% phosphatidylethanolamine (PE), ~20% phosphatidylglycerol (PG), and ~5% cardiolipin (CL) (Diedrich and Cota-Robles, 1974; Morein *et al.*, 1996). Calculated free energy of urea-induced unfolding of OmpA has been shown to rise as an increasing amount of 1-palmitoyl-2-oleoyl-*sn*-phosphatidylethanolamine (POPE) was included in a bilayer composed of POPC (PC is not found in bacterial membranes). Higher PG and CL content was also observed to increase OmpA stability in the membrane (Tamm *et al.*, 2004). One may wonder whether LPS, which is only present on the outer leaflet of OM in bacteria, has any effect on the OmpA folding. Surprisingly, inclusion of LPS into the membrane bilayer has been reported to inhibit OmpA folding and insertion (Tamm *et al.*, 2004); however, this result is not conclusive because LPS was incorporated in both leaflets of the membrane in the reported experiment, whereas the OM is asymmetrical and contains LPS only on the outer leaflet.

Assisted folding of OMPs *in vivo* by the BAM complex

How OMPs are assembled inside living cells is only now starting to be understood. Uncertainty had lingered for a long time regarding the existence of a proteinaceous machine in the OM, analogous to the Sec system of the IM. After all, it had been observed that OMPs could spontaneously fold and insert into a lipid bilayer *in vitro* (Kleinschmidt, 2003; Tamm *et al.*, 2004; Kleinschmidt, 2006). That led many scientists to ask the question, 'Does *in vivo* OMP assembly require assistance?' There are two major reasons as to why cells would require a protein complex that would facilitate the folding and insertion of OMPs: (1) The kinetics of folding and insertion of OMPs *in vitro* is too slow to be biologically plausible, and (2) the spontaneous insertion of OMPs *in vitro* does not explain the exclusive insertion of these proteins into the OM, and not the IM (Tamm *et al.*, 2004). In other words, a proteinaceous machine is required in order to increase the kinetics and specificity of the process.

BamA (formerly known as Omp85 or YaeT) is the first OMP assembly factor that was discovered in 2003. BamA was first identified in *Neisseria meningitidis*, but its homologues are found in all Gram-negative bacteria, as well as in endosymbiotically derived eukaryotic organelles, namely the mitochondria and chloroplasts (Voulhoux *et al.*, 2003; Gentle *et al.*, 2005). The strong conservation of the gene encoding BamA, together with the observation that the gene is essential for the viability of the cells, reinforces the fundamental importance of BamA, and its involvement in the OMP biogenesis is well supported by various experimental studies.

The first non-experimental evidence that BamA is involved in OMP biogenesis is the location of its gene within the bacterial genome. The *omp85* gene coding for BamA is located immediately adjacent to the *skp* gene, which encodes the periplasmic chaperone Skp involved in OMP biogenesis (Gentle *et al.*, 2004). Furthermore, the *omp85* gene is found in a cluster with other genes that are involved in the biosynthesis of LPS, which is exclusively found on the outer leaflet of the OM in Gram-negative bacteria (Genevrois *et al.*, 2003). The involvement of BamA in OMP biogenesis became clearer when a BamA conditional mutant study observed a periplasmic accumulation of unfolded OMPs in a BamA-depleted strain (Voulhoux *et al.*, 2003). Earlier studies suggested that the observed effect of BamA on OMPs could be an indirect one, and that BamA could actually be involved in LPS transport with its absence leading to the depletion of LPS, which could affect the OMP assembly (Genevrois *et al.*, 2003). However, later studies showed that LPS presence in the OM is not required for the correct assembly of the OMPs (Doerrler and Raetz, 2005). A direct involvement of BamA in OMP biogenesis was further supported by a protein interaction study, which demonstrated that BamA binds to unfolded OMPs *in vitro* (Voulhoux *et al.*, 2003; Knowles *et al.*, 2008).

Following the discovery of BamA and its involvement in an unknown OMP assembly pathway, it was later discovered that *E. coli* BamA exists in a larger protein complex with four other lipoproteins, namely BamB, C, D, and E (Hagan *et al.*, 2011; Ricci and Silhavy, 2011). BamA along with the lipoproteins are now known as the BAM complex (Fig. 3.4), and its structure and mechanism is just starting to be understood. The rest of this book chapter summarizes the current literature and understanding of the BAM complex.

The components of the BAM complex

In *E. coli*, five different proteins assemble to form the BAM complex, and together they ensure the proper incorporation of β-barrel proteins into the OM. The components of the BAM complex are named alphabetically from BamA to BamE in order of decreasing molecular weight (Table 3.2). BamA, the largest and first discovered component, is an OMP itself that adopts a β-barrel fold to span the OM lipid bilayer (Gentle *et al.*, 2005; Ricci and Silhavy, 2011). BamB, C, D and E on the other hand are lipoproteins, meaning they are soluble proteins anchored to the periplasmic surface of the OM by a diacylglycerol linked to the N-terminal cysteine (Hayashi and Wu, 1990; Hagan *et al.*, 2011).

Table 3.2 The components of the BAM complex

	Protein	MW[1](kDa)	pI[1]	Essential?	Knockout phenotype[2]	Conserved in[3]
Membrane protein	BamA (YaeT/ Omp85)	88.4	4.9	Yes	Defective in assembly of all OMPs studied	All Gram-negative bacteria and eukaryotes
Lipoproteins	BamB (YfgL)	39.9	4.6	No	Increased OM permeability Significant defects in OMP assembly (especially larger OMPs)	All proteobacteria (except δ- and ε-proteobacteria)
	BamC (NLpB)	34.4	5.0	No	Increased OM permeability Mild defects in OMP assembly	β- and γ-proteobacteria
	BamD (YfiO)	25.8	5.5	Yes	Defective in assembly of all OMPs studied	All Gram-negative bacteria
	BamE (SmpA)	10.4	6.8	No	Increased OM permeability Mild defects in OMP assembly	All proteobacteria (except δ- and ε-proteobacteria)

[1]These values are calculated based on the *E. coli* Bam component sequences (UniProt IDs: P0A940, P77774, P0A903, P0AC02 and P0A937).
[2]Voulhoux *et al.* (2003), Doerrler and Raetz (2005), Onufryk *et al.* (2005), Werner and Misra (2005), Charlson *et al.* (2006), Malinverni *et al.* (2006), Ruiz *et al.* (2006), Sklar *et al.* (2007a).
[3]Anwari *et al.* (2012).

The association of the lipoproteins with BamA has been shown by co-immunoprecipitation analysis, as well as by size-exclusion chromatography (Wu *et al.*, 2005; Malinverni *et al.*, 2006; Sklar *et al.*, 2007a; Hagan *et al.*, 2010; Kim *et al.*, 2011a). However, the exact stoichiometry of a functional unit of the BAM complex is not clear. Size-exclusion chromatographic and native gel analyses of BamA (both when extracted from OM and when refolded from inclusion bodies) suggest that BamA may exist as a tetramer (Robert *et al.*, 2006; Hritonenko, 2011). However, a recent *in vitro* study showed that a functional unit of the BAM complex reconstituted into a proteoliposome has a BamA:B:C:D:E ratio of 1:1:1:1:1(or 2) (Hagan *et al.*, 2010).

Of the five proteins making up the BAM complex, only BamA and BamD are essential for cell viability (Knowles *et al.*, 2009b; Tommassen, 2010). Neither the gene encoding BamA or BamD can be deleted without causing cell death, suggesting that the functions of both proteins are absolutely critical for the cell survival (Voulhoux *et al.*, 2003; Onufryk *et al.*, 2005). Both BamA and BamD depletion strains display severe defects in OMP assembly, as indicated by reduced levels of OMPs in the OM and accumulation of unfolded OMPs in the periplasm (Voulhoux *et al.*, 2003; Doerrler and Raetz, 2005; Onufryk *et al.*, 2005; Werner and Misra, 2005; Malinverni *et al.*, 2006). Deletion of the gene encoding BamB also results in significant defects in OMP assembly, although it is not lethal (Onufryk *et al.*, 2005; Charlson *et al.*, 2006; Hagan *et al.*, 2010). Cells lacking BamB show increased membrane permeability as indicated by higher susceptibility to antibiotics, and cells are unable to assemble many OMPs, especially those forming larger β-barrels (Charlson *et al.*, 2006; Ruiz *et al.*, 2006). Absence of BamC or BamE causes mild defects in OMP assembly and increased membrane permeability (Onufryk *et al.*, 2005; Sklar *et al.*, 2007a). Based on the observed knockout phenotypes, it seems that BamA and BamD function in the most critical steps of OMP assembly process, while BamB, C, and E play more important roles in improving the efficiency. Or, it is possible that the non-essential lipoproteins BamB, C, and E are mainly involved in helping the assembly of OMPs that are less important for cell survival.

It should be noted that different protein subunits of the BAM complex show different degrees of conservation across different species of Gram-negative bacteria. In proteobacteria, for example, BamA and BamD are found ubiquitously, while BamB and BamE are absent in δ-proteobacteria and ε-proteobacteria. BamC, which is the least conserved component of the BAM complex, is only found in β-proteobacteria and γ-proteobacteria (Anwari *et al.*, 2012). It should also be noted that a recent study has identified yet another lipoprotein, BamF, as a part of the BAM complex. The N-terminal region of BamF shares some sequence homology to BamC, and interestingly it is only found in α-proteobacteria in which BamC is absent (Anwari *et al.*, 2012). As not much is known about the structure and function of BamF at present, the main focus of this review hereafter will be on BamA, B, C, D, and E.

Structural studies of the BAM complex

Numerous crystal and NMR structures of the proteins of the BAM complex have become available in recent years (Kim *et al.*, 2012). We now have basic structural information on all individual components of the complex, which is summarized below for each protein.

BamA

BamA is the largest (88.4 kDa) and the best conserved component of the BAM complex (Gentle *et al.*, 2005). It is also the only protein of the complex that is an integral membrane protein (Knowles *et al.*, 2009b). That is, BamA is an OMP that resides in the OM by forming a β-barrel structure (Gentle *et al.*, 2005). Extending away from the β-barrel of BamA into the periplasmic space are the five polypeptide transport-associated (POTRA) domains, numbered POTRA1 to POTRA5 in the N- to C-terminal direction (Fig. 3.5A) (Misra, 2007). The C-terminal β-barrel domain and a varying number of the POTRA domains at the N-terminus are the two main structural features that distinguish proteins in the Omp85 superfamily (Sanchez-Pulido *et al.*, 2003; Gentle *et al.*, 2005; Knowles *et al.*, 2009b). In addition to BamA, the Omp85 superfamily in Gram-negative bacteria includes the following two groups of proteins: (1) the transporter proteins involved in the two-partner secretion systems that serve to translocate specific substrates across the OM (Sanchez-Pulido *et al.*, 2003), and (2) the recently discovered TamA, which assists the assembly of autotransporter passenger domains (the N-terminal region of autotransporters secreted through the C-terminal β-barrel domain of the same protein, which is assembled by the BAM complex) (Selkrig *et al.*, 2012).

The first structural insight into BamA came from the crystal structure of the POTRA domains. In 2007, Kim and his colleagues successfully crystallized and solved the structure of the four N-terminal POTRA domains of *E. coli* BamA (i.e. POTRA1–4) (Kim *et al.*, 2007). Their initial attempt to crystallize POTRA1–5 was unsuccessful due to the poor stability of the protein; nevertheless, this was the first glimpse of the BAM complex structure. The crystal structure solved by Kim *et al.* (2007) revealed that the four POTRA domains of BamA have similar protein folds despite being sequentially diverse overall. Each POTRA domain has a β-α-α-β-β topology, which folds into a compact structure comprising a three-stranded mixed β-sheet overlaid with a pair of anti-parallel helices (Fig. 3.5B) (Kim *et al.*, 2007; Misra, 2007). The overall structure of POTRA1–4 has a twisted fish-hook shape, resulting from right-handed rotation of successive POTRA domains (Kim *et al.*, 2007). Many conserved residues are found at the interfaces between neighbouring POTRA domains, and it was initially thought that the interdomain contacts between the successive POTRA domains keep POTRA1–4 in the rigid fish-hook conformation (Kim *et al.*, 2007). In the following year, however, Knowles *et al.* (2008) showed by a small-angle X-ray scattering (SAXS) analysis that the POTRA domains can exist in a different, more extended conformation in solution. This observation was confirmed later by another crystal structure of POTRA1–4 that was solved by Gatzeva-Topalova *et al.* (2008).

The POTRA1–4 structure solved by Gatzeva-Topalova *et al.* (2008) is drastically different from the one solved by Kim *et al.* (2007) in that the conformation of the protein is much more extended (Fig. 3.5B and C). Superposition of the two structures reveals that the conformations of POTRA1–2 and POTRA3–4, but not POTRA2–3, remain the same in both structures. The NMR and the crystal structures of POTRA4–5 reported later showed that POTRA4–5 forms a rigid, non-flexible structure (Gatzeva-Topalova *et al.*, 2010; Zhang *et al.*, 2011). Taken together, these results indicate that the interdomain interaction between POTRA2 and POTRA3 is not as extensive as those found in POTRA1–2, or POTRA3–4–5 (Kim *et al.*, 2007; Gatzeva-Topalova *et al.*, 2008; Knowles *et al.*, 2008; Gatzeva-Topalova *et al.*, 2010). The flexibility of the linker between POTRA2 and POTRA3 therefore acts as a hinge region, allowing the POTRA domains to adopt either the compact (earlier structure) or the extended (later structure) conformation (Fig. 3.5C) (Gatzeva-Topalova *et al.*, 2008). Although the SAXS data suggest that the extended form of the POTRA domains is found in solution, it is possible that the function of the POTRA domains require them to undergo a conformational change to a bent form during the catalysis of OMP folding and insertion.

While a number of POTRA domain structures of BamA have been reported, the exact structure of the C-terminal β-barrel domain still remains unknown. However, the β-barrel domain of BamA is expected to share close structural similarity with that of FhaC (14% sequence identity), a protein transporter in *Bordetella pertussis* that mediates the secretion of filamentous haemagglutinin (FHA). FhaC, like BamA, belongs to the Omp85 superfamily of proteins, but differs from BamA in that it has only one specific substrate (i.e. FHA), has only two POTRA domains, and does not mediate the membrane insertion of its substrate (Sanchez-Pulido *et al.*, 2003; Jacob-Dubuisson *et al.*, 2009). As a member of the Omp85 superfamily, FhaC has the characteristic POTRA-β-barrel domain architecture (Gentle *et al.*, 2005; Jacob-Dubuisson *et al.*, 2009). In 2007, the crystal structure of the full length FhaC protein was solved by Clantin *et al.*, and showed FhaC to be a 16-stranded β-barrel with two POTRA domains (Fig. 3.5D) (Clantin *et al.*, 2007). Based on the secondary structure prediction, BamA is also expected to form a 16-stranded β-barrel. The channel formed by the β-barrel in FhaC is partially obstructed by a conserved extracellular loop of the barrel that folds over into the pore, as well as by an N-terminal α-helix preceding the POTRA domains (Clantin *et al.*, 2007). The diameter of the FhaC channel partially obstructed by

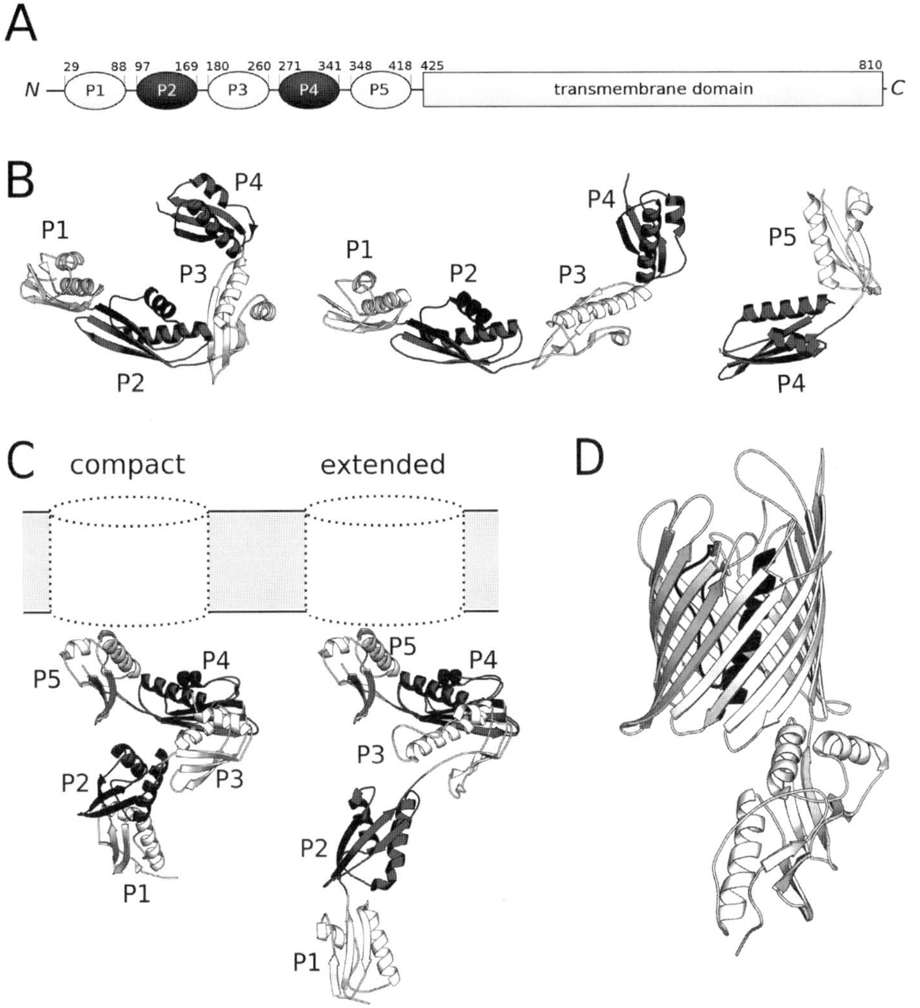

Figure 3.5 Structural features of BamA. (A) The domain structure of BamA shows an N-terminal periplasmic region that contains five POTRA domains along with a C-terminal transmembrane domain. (B) Ribbon diagrams of the POTRA1–4 and POTRA4–5 domains (PDB: 3Q6B) are shown. The structures for POTRA1–4 show the possibility for bent (PDB: 2QDF) and extended (PDB: 3EFC) conformations. (C) Models of full length POTRA domains of BamA in bent and extended conformations are shown. The models were prepared by extending the POTRA1–4 structures by a structure alignment of the overlapping POTRA4 domain of POTRA4–5 structure. (D) The structure of FhaC (2QDZ) from *Bordetella pertussis* (BamA homologue) shows the presence of only two POTRA domains, as well as a helix and a conserved exoplasmic loop (both shown in black) that can insert themselves inside the β-barrel.

the two structural elements is approximately 3 Å (Clantin *et al.*, 2007). Sequence alignment with FhaC shows that the conserved loop is found in BamA as well, but not the N-terminal α-helix. The pore diameter of BamA may therefore resemble that of FhaC without the N-terminal α-helix, which is estimated to be approximately 8 Å (Clantin *et al.*, 2007). However, the calculated channel diameter of BamA based on diffusion rates of solutes of different sizes is 25Å, suggesting an oligomeric state (Robert *et al.*, 2006). Whether BamA exists as a monomer or a multimer remains a subject of further investigation.

The crystal structure of FhaC not only helps predict the structure of the BamA β-barrel domain, but also the orientation of the BamA POTRA domains relative to the β-barrel domain. Although FhaC differs from BamA in that it has only two POTRA domains instead of five, the arrangement of the β-barrel and the POTRA domains in BamA is believed to be similar to that observed in the FhaC crystal structure. From the crystal structure, it can be seen that the C-terminal POTRA domain of FhaC is situated very close to the membrane, and that it extends away from the centre of the β-barrel (Clantin *et al.*, 2007). Superposing the POTRA5 of a spliced model of BamA POTRA1–5 onto the C-terminal POTRA domain of the FhaC structure assists visualizing what the entire BamA molecule might look like. Still, despite predicted structural similarity, FhaC and BamA are not expected to share the same functional mechanism; this is because the main function of FhaC is the secretion of its substrate, while BamA is involved in catalysing the concerted events of substrate folding and membrane insertion. Therefore, structures of the β-barrel domain and the full-length BamA will need to be solved in the future to provide insights into the mechanistic details of BamA.

BamB

BamB is the largest (39.9 kDa) and also the first lipoprotein of the BAM complex to have had its structure solved following the POTRA domains of BamA. In 2011, various crystal structures of BamB were solved independently by four different groups. The structures revealed that BamB has an architecture that resembles a short cylinder with a narrow funnel shaped channel running down the cylindrical axis (Kim and Paetzel, 2011). The BamB structure could also be described as having a ring-like shape, which arises from the eight bladed β-propeller fold of the protein (Albrecht and Zeth, 2011; Heuck *et al.*, 2011; Kim and Paetzel, 2011; Noinaj *et al.*, 2011) (Fig. 3.6). The propeller has eight 'blades' arranged toroidally around a pseudo-8-fold axis (Fig. 3.6B). Each blade is a twisted β-sheet consisting of four anti-parallel β-strands, and long loops (ranging from 5 to 21 amino acid residues in length) that link the neighbouring blades together. Similar to many other proteins with β-propeller folds, the ring structure of BamB is stabilized by having one N-terminal and three C-terminal β-strands form one of the blades (known as the '1+3 Velcro closure' arrangement). Analysis of the electrostatic properties of BamB reveals that the protein has a predominantly negatively charged surface, especially at the solvent-filled central channel of the protein (Kim and Paetzel, 2011; Noinaj *et al.*, 2011). It is not known whether the negatively charged pore of BamB has any functional implications.

Mapping conserved residues onto the protein structure often helps in the identification of functionally important residues. In the BamB structure, there are several conserved residues buried inside the protein (Kim and Paetzel, 2011). Many of these conserved residues are found repeatedly in each blade, and the hydrophobic nature of these residues appears to play a structurally important role of promoting inter-blade contacts by forming interlocking hydrophobic interactions (Kim and Paetzel, 2011). Interestingly, many conserved solvent-exposed residues are localized onto one surface of the β-propeller structure, which is formed by the loops connecting neighbouring blades (Heuck *et al.*, 2011; Kim and Paetzel, 2011; Noinaj *et al.*, 2011). Also found on the same side of the propeller, and within close proximity of this region, are the residues (Leu192, Leu194, Arg195, Asp246 and Asp248) that were

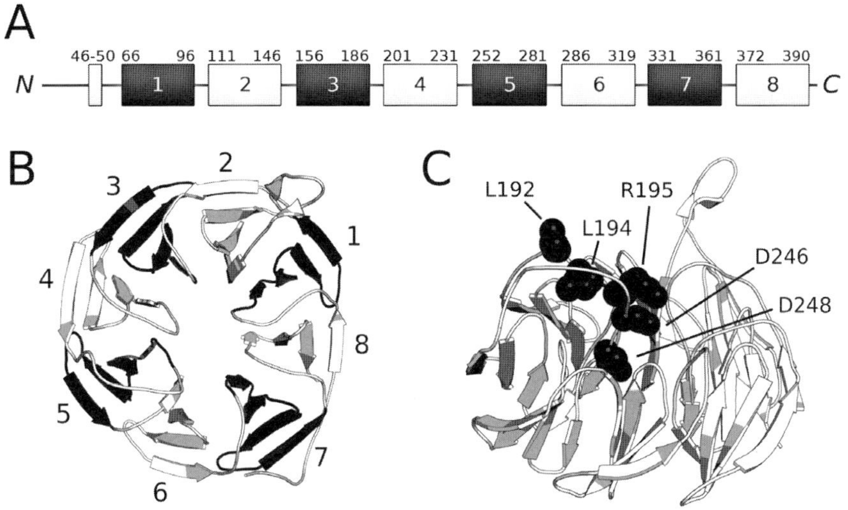

Figure 3.6 Structural features of BamB. (A) The domain structure of BamB shows the presence of eight domains that together form the β-propeller structure. Note that residues 46–50 form a strand that is a part of blade 8. (B) A ribbon diagram of BamB (PDB: 3P1L) shows the eight-bladed β-propeller structure, with each blade numbered as in A. (C) Conserved residues of BamB are shown as black spheres, and are found clustered on one face of the β-propeller structure. These residues are believed to be important for interaction with BamA.

previously determined by mutagenesis to be important for interaction with BamA (Fig. 3.6C) (Vuong *et al.*, 2008; Albrecht and Zeth, 2011; Heuck *et al.*, 2011; Kim and Paetzel, 2011; Noinaj *et al.*, 2011).

The β-propeller fold is a common protein fold found in proteins with diverse functions. The BamB structure shows a significant degree of similarity in protein topology and architecture to several other proteins with the β-propeller fold. These include *Pseudomonas putida* ADH IIB (an alcohol dehydrogenase; PDB: 1KV9), *Homo sapiens* Fbw7 (part of an ubiquitin ligase complex; PDB: 2QVR), *Paracoccus pantotrophus* cytochrome cd1 (nitrite reductase; PDB: 1HJ5), and *Saccharomyces cerevisiae* Sif2p (a transcriptional co-repressor of meiotic genes; PDB: 1R5M) (Kim and Paetzel, 2011). All of these proteins have an eight-bladed β-propeller structure, and most of them have additional domains positioned on either side of the β-propeller domain. In all four proteins, protein or ligand binding sites are found in the β-propeller domain, more specifically on the surface that is equivalent to the surface of BamB where the conserved residues and the five residues important for BamA interaction are located (Kim and Paetzel, 2011). The conserved molecular surface of BamB therefore appears to be the site of BamA interaction (Albrecht and Zeth, 2011; Heuck *et al.*, 2011; Kim and Paetzel, 2011).

BamC

BamC is a 34.4 kDa lipoprotein with a unique modular structure. The first structural study of BamC was carried out by Knowles and his colleagues who used NMR to predict the secondary structure of the protein (Knowles *et al.*, 2009a). Their NMR data predicted that BamC has a mixture of α- and β-secondary structure elements, and that approximately 70 residues at the N-terminus of BamC is unstructured (Knowles *et al.*, 2009a). Subjecting BamC to a small amount of a broad-range protease such as subtilisin and chymotrypsin results in the degradation of the unstructured N-terminal tail and produces two protease-resistant fragments (12.2 kDa and 14.5 kDa) (Albrecht and Zeth, 2010; Kim *et al.*, 2011b; Warner *et al.*, 2011). Taken together, these results suggested that BamC has two independently folding domains (the N- and the C-terminal domains) following the unstructured N-terminal region (Fig. 3.7A). This prediction of the BamC domain architecture was found to be in close agreement with later NMR and crystal structure data of BamC that became available later.

As yet, efforts to crystallize the full-length BamC protein have been unsuccessful; however, separate structures of the individual domains have been solved successfully. The first structures of BamC to be reported were those of the globular domains. These were the separate structures of the N- and the C-terminal domains of BamC that were solved in 2011 by both NMR and X-ray crystallography (Albrecht and Zeth, 2011; Warner *et al.*, 2011). Surprisingly, the structures show that the two domains of BamC have the same 'helix-grip fold' despite sharing low sequence identity with each other (12%) (Fig. 3.7B) (Albrecht and Zeth, 2011; Warner *et al.*, 2011). Both the N- and the C-terminal domains of BamC consist of a central 6-stranded anti-parallel β-sheet with two helical units packing tightly against the sheet (Albrecht and Zeth, 2011; Kim *et al.*, 2011b; Warner *et al.*, 2011). The C-terminal domain has an extra 3_{10}-helix in one of the loops connecting the neighbouring β-strands (Kim *et al.*, 2011b). The NMR structure and backbone amide dynamics studies by Warner *et al.* (2011) showed that the two globular domains of BamC are joined by a highly flexible α-helical linker.

BamC has an unusually long (~70 residues) unstructured region following the lipid-attached cysteine at its N-terminus (Fig. 3.7B) (Knowles *et al.*, 2009a). Full-length BamC protein with the unstructured N-terminal tail was reported to be difficult to crystallize, probably due to the disordered conformation of the N-terminal tail (Albrecht and Zeth, 2010; Kim *et al.*, 2011b); however, the N-terminal domain of BamC with its unstructured region could be successfully crystallized in complex with BamD (Fig. 3.7C) (Kim *et al.*, 2011a). In this BamCD subcomplex structure, the N-terminal region lacks secondary structure as previously predicted (Kim *et al.*, 2011a). Instead, the unstructured region bends into a lasso-shape and makes an extensive protein–protein interaction with BamD (Kim *et al.*, 2011a). In fact, domain truncation and protein interaction analysis suggests that the unstructured N-terminal region is essential for the BamCD complex formation (Kim *et al.*, 2011a).

Sequence comparisons of BamC from several different species of Gram-negative bacteria indicate that the majority of the conserved blocks of sequence reside in the unstructured N-terminal region and within the C-terminal domain (Albrecht and Zeth, 2011; Kim *et al.*, 2011a,b; Warner *et al.*, 2011). Despite having the same protein fold as the C-terminal domain, the N-terminal domain of BamC does not comprise many conserved residues (Warner *et al.*, 2011). The conserved residues of BamC map to the BamD interaction interface in the N-terminal tail, and a negatively charged concave surface of the C-terminal domain (Kim *et al.*, 2011a,b). The conserved groove of the C-terminal domain is also the site of a crystal contact where an α-helix from the neighbouring BamC molecule binds (Kim *et al.*, 2011b). The conserved surface of the C-terminal domain (Fig. 3.7D) therefore has been proposed to be a site of protein–protein interaction, although this will need to be verified experimentally in the future.

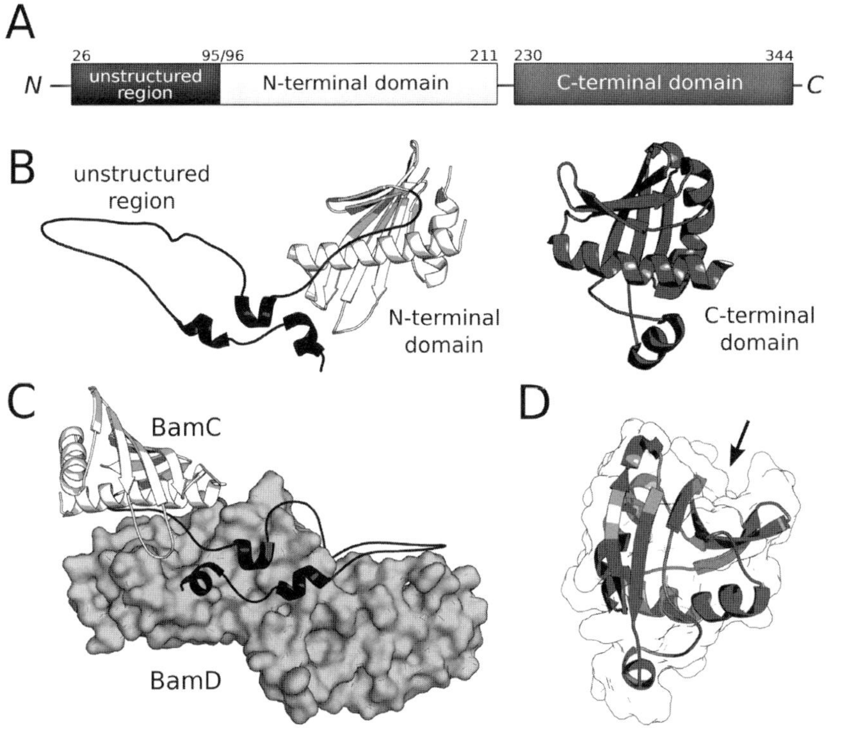

Figure 3.7 Structural features of BamC. (A) The domain structure of BamC shows the presence of three domains: an unstructured region at the N-terminus followed by two domains known as the N-terminal domain and the C-terminal domain. (B) The C-terminal domain (PDB: 2YH5) was solved separately from the N-terminal domain with the unstructured region (PDB: 3TGO), but both globular domains have a similar helix-grip fold. (C) The BamC unstructured region (black) and N-terminal domain (white) was co-crystallized with BamD (PDB: 3TGO). The resulting structure shows the unstructured region to form a long loop that interacts with BamD (grey). (D) The C-terminal domain of BamC (PDB: 3SNS) is shown with the conserved groove that may serve as potential protein–protein interaction surface indicated by an arrow.

BamD

BamD is a 25.8 kDa lipoprotein that is not only essential for proper function of the BAM complex, but also for the viability of the bacteria (Onufryk *et al.*, 2005). The first structure of BamD was solved by Sandoval *et al.* (2011), who successfully crystallized BamD from the thermophilic bacteria *Rhodothermus marinus*. The crystal structure of *E. coli* BamD soon followed, and it showed only minor differences from the earlier structure. The *E. coli* BamD structure consists of 10 α-helices, whereas *R. marinus* BamD contains one extra helix at the C-terminus (Albrecht and Zeth, 2011; Kim *et al.*, 2011a; Sandoval *et al.*, 2011). In both structures, BamD is an entirely α-helical protein containing five tetratricopeptide repeats (TPRs), which are helix–turn–helix structural motifs found in a wide range of proteins (Fig. 3.8) (Albrecht and Zeth, 2011; Kim *et al.*, 2011a; Sandoval *et al.*, 2011). The five TPR motifs of BamD are found in tandem repeats and are arranged in a parallel fashion, which results in BamD having the elongated rod-like shape.

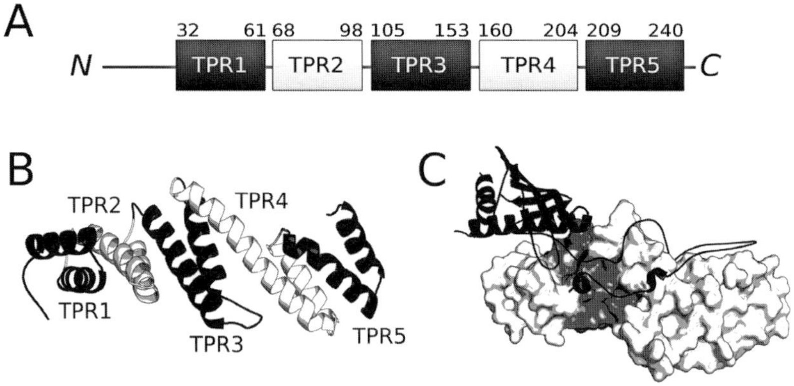

Figure 3.8 Structural features of BamD. (A) The domain structure of BamD shows the presence of five tetratricopeptide repeat (TPR) motifs. (B) The ribbon diagram of the BamD structure (PDB: 3TGO) is shown. Ten helices form the five TPR motifs which are numbered as in A. (C) The BamCD structure (PDB: 3TGO) shows that BamC (black) binds to a proposed substrate binding pocket of BamD (white; the pocket is shown in dark grey).

Structural analysis of the molecular surface properties reveals that there are three main patches of conserved surfaces on BamD (Kim *et al.*, 2011a; Sandoval *et al.*, 2011). The first group of solvent-exposed conserved residues are found on the N-terminal half (TPR 1–3) of BamD that forms a groove similar to the targeting signal binding pockets of other proteins such as PEX5 (peroxisomal targeting signal receptor) and Tom70 (mitochondrial import receptor) (Albrecht and Zeth, 2011; Sandoval *et al.*, 2011). This structural similarity has led to the hypothesis that BamD may function as a targeting signal receptor that recognizes the C-terminal OM targeting sequence of OMP substrates. However, in the BamCD complex structure, the same conserved pocket of BamD is occluded by the unstructured N-terminal region of BamC (Fig. 3.8C) (Kim *et al.*, 2011a). The second patch of conserved residues are also found on the N-terminal TPR motifs, on the opposite side from the binding groove, and the last cluster of conserved residues are found on the C-terminal half of the BamD molecule (Kim *et al.*, 2011a; Sandoval *et al.*, 2011). These conserved regions may serve as additional protein binding surfaces for BamD to interact with other components of the BAM complex or with the substrates (Kim *et al.*, 2011a).

BamE

BamE is the smallest (10.4 kDa) component of the BAM complex, and unlike other lipoproteins of the complex, it is observed to exist in both monomeric and dimeric forms in solution (Albrecht and Zeth, 2011; Kim *et al.*, 2011c; Knowles *et al.*, 2011). The first structural insight of BamE came from the determination of the NMR structure of the protein in the monomeric form. The BamE monomer has long unstructured N- and C-termini, and a globular domain that folds as two α-helices packed against a three-stranded anti-parallel β-sheet (Fig. 3.9A) (Kim *et al.*, 2011c; Knowles *et al.*, 2011). NMR-based amide backbone dynamics analysis shows that the unstructured N- and the C-termini, as well as one of the loops connecting the β-strands, are highly flexible on the subnanosecond timescale (Kim *et al.*, 2011c). Interestingly, BamE shares significant structural similarity with β-lactamase

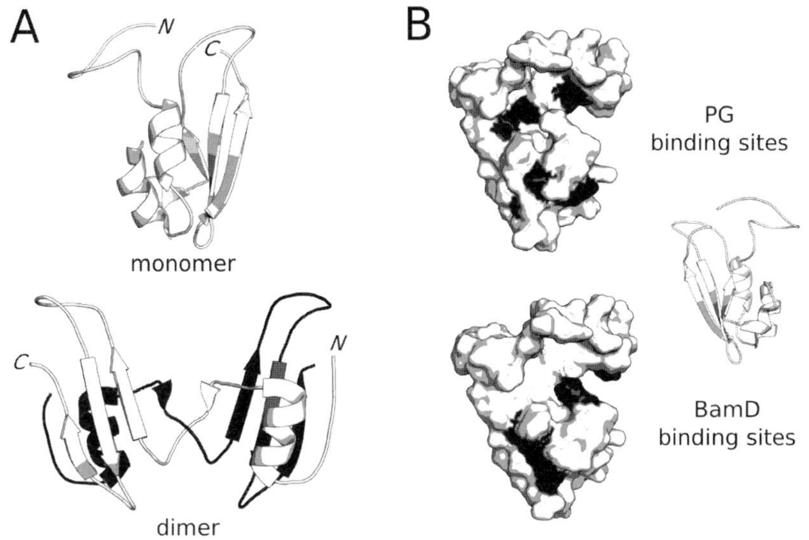

Figure 3.9 Structural features of BamE. (A) The ribbon diagrams of a BamE monomer (PDB: 2KXX) and a dimer (PDB: 2YH9) are shown. The dimer structure shows how the two monomers (shown in white and black) exchange N-terminal α-helices to form a domain-swapped dimer. (B) The residues previously identified to be important for PG and BamD interactions are shown in black on the surface diagrams of the BamE monomer. Both surface diagrams are shown in the same orientations as the ribbon diagram shown on the right.

inhibitor protein (BLIP), a protein that inhibits a variety of class A β-lactamase enzymes (Vanini *et al.*, 2008; Kim *et al.*, 2011c). It is also interesting to note that the flexible loop of BamE is topologically equivalent to the loop of BLIP that binds and inhibits the active site of β-lactamase enzymes (Kim *et al.*, 2011c). Therefore, it can be postulated that the flexible loop of BamE may serve a similar function as a protein binding motif.

BamE expressed and purified from *E. coli* has been shown to exist in a kinetically trapped dimeric state (Kim *et al.*, 2011c). The dimer formation of BamE appears to be irreversible under various pH and salt concentration conditions tested *in vitro* (Kim *et al.*, 2011c). The crystal structure of BamE dimer was solved by Albrecht and Zeth (2011), and it explained why the BamE dimer seems so stable under varying conditions. The crystal structure revealed that BamE forms a domain-swapped dimer, in which the α-helices of the two monomers are exchanged such that the helical units from one monomer pack against the β-sheet of another monomer (Fig. 3.9B) (Albrecht and Zeth, 2011). Many aliphatic and aromatic residues found at the dimer interface are conserved among different bacterial species, suggesting that dimerization could be functionally important for BamE. However, currently there are conflicting data on which oligomeric form of BamE is biologically relevant. While one study reported that the formation of the BamE dimer is a result of protein misfolding under temperature stress, another study reported that BamE purified from a native outer membrane exhibits a dimeric state (Albrecht and Zeth, 2011; Knowles *et al.*, 2011). Further experiments are required to identify which oligomeric form of BamE is found *in vivo*, or if both forms are functional in the BAM complex.

Assembly of the BAM complex

As described above, efforts by various groups have led to the structure determination of all individual components of the BAM complex. While this is a great step forward, full understanding of the BAM complex structure and function requires knowledge of how the different components of the complex interact with each other. In this respect, several mutagenesis studies combined with protein–protein interaction analysis, as well as structural studies, have yielded results that provide insights into the oligomeric organization of the BAM complex.

BamA was first identified as an essential OMP assembly factor in 2003 (Voulhoux *et al.*, 2003). In separate studies, BamB (formerly known as YfgL) had also been implicated in OMP assembly owing to the observation that deleting the gene encoding BamB results in lower levels of OMPs such as OmpA and LamB (Ruiz *et al.*, 2006). In 2005, Wu and his colleagues published their results that showed BamB and BamA, as well as BamC (formerly known as NlpB), can be co-immunoprecipitated together in *E. coli* (Wu *et al.*, 2005). In the following year, BamD (formerly known as YfiO) was identified as another member of the complex, also via co-immunoprecipitation (Malinverni *et al.*, 2006). The interaction of BamA/B/C/D was further dissected by performing pull-down experiments in strains defective in BamB, BamC or BamD (Malinverni *et al.*, 2006). In the absence of BamB and BamC, BamD co-purified with BamA (Malinverni *et al.*, 2006). While the absence of BamC still allowed co-purification of BamA/B/D, only BamA/B (and not BamC) could co-purify in cells depleted of BamD. (Malinverni *et al.*, 2006). Taken together, these results indicate that BamB and BamD interact directly, but independently, with BamA. Furthermore, the inability of BamC to co-purify with the rest of the proteins in the absence of BamD suggests that BamC binds to BamD directly but not to BamA. The last component of the BAM complex, BamE (formerly known as SmpA), was identified in 2007 (Sklar *et al.*, 2007a). Using a co-immunoprecipitation technique similar to the one mentioned above, BamE was found to interact with BamC and BamD, which appears to stabilize the direct interaction between BamA and BamD (Sklar *et al.*, 2007a). BamE has also recently been shown to modulate a conformational change of BamA, although neither the structural details of this conformational change nor the mechanism of how BamE causes the change are currently understood (Rigel *et al.*, 2012).

The interaction between BamA and the lipoproteins was explored further in the study published by Kim and his colleagues in 2007. They created a series of BamA mutants that have POTRA domain deletions, and tested for their ability to co-purify with the lipoproteins in *E. coli*. The results from their experiments showed that BamB cannot co-purify with BamA when any one of the POTRA 2, 3, 4 or 5 domains is missing (Kim *et al.*, 2007). On the other hand, deletion of POTRA5 led to the inability of BamC/D/E to associate with BamA (Kim *et al.*, 2007). It therefore seems that while BamB interacts with POTRA2–5, BamC/D/E interacts with POTRA5. Based on these data, it was confirmed that the POTRA domains of BamA serve as docking sites for the lipoproteins BamB/C/D/E.

At present, there is not much known about the nature and mode of interaction in the BamA–BamD association. For the BamA–BamB interaction, however, mutagenesis studies have identified residues that are critical for the association of the two proteins. On BamB, as described earlier, the conserved residues (Leu192, Leu194, Arg195, Asp246, and Asp248) found in loops on one face of its β-propeller structure are important for BamA association (Vuong *et al.*, 2008; Heuck *et al.*, 2011; Kim and Paetzel, 2011; Noinaj *et al.*, 2011). In

BamA, the residues important for BamB association are located on a β-bulge in the second β-strand of POTRA3 (Ile240 and Asp241) (Kim *et al.*, 2007). This bulge forms one edge of the POTRA3 β-sheet, and interestingly it is observed to participate in crystal packing via β-augmentation (a mode of protein interaction in which a strand from one protein is added to an existing β-sheet of another) in two independently solved crystal structures (Kim *et al.*, 2007; Gatzeva-Topalova *et al.*, 2008). Analysis of crystal packings of BamB has also revealed that the outermost β-strands of the β-propeller blades can mediate protein–protein interaction by β-augmentation (Heuck *et al.*, 2011). Although the outer edges of the blades are not where the residues previously determined to be important for BamA interaction are found, they could be additional contact points between BamA and BamB. As POTRA2–5 have been shown to interact with BamB, it is likely that more residues than currently identified are found in the BamA–BamB interaction interface.

Within the BAM complex, there are lipoprotein–lipoprotein interactions in addition to the BamA–lipoprotein interactions. The lipoproteins BamC, BamD, and BamE have been shown to interact with each other via co-immunoprecipitation study, but the exact mode of interaction is not fully understood yet. A recent co-crystal structure of the essential lipoprotein BamD with BamC has given the first clue as to how these two components of the BAM complex interact with each other. In this structure, the unstructured ~70-residues-long N-terminal region of BamC is shown to make an extensive interaction along the longitudinal axis of the BamD molecule (Kim *et al.*, 2011a). Surprisingly, domain truncation studies revealed that only the N-terminal region of BamC is absolutely required for the BamC–BamD association (Kim *et al.*, 2011a). A BamC mutant that has intact N- and the C-terminal globular domains but is missing the unstructured region fails to form a heterodimeric complex with BamD (Kim *et al.*, 2011a). The globular domains of BamC therefore may be important for interaction with BamE, but further experiments are necessary to validate if that is the case. A map of interactions between the various components of the BAM complex is summarized in Fig. 3.10.

Proposed BAM mechanisms

As structural information about the BAM complex is emerging, so is knowledge about the functional aspects of the complex. Genetics, mutagenesis and *in vitro* analyses, together with the structural data, allow us to make educated guesses as to how the BAM complex carries out its function. Although the exact mechanism is not clearly understood yet, the process of OMP assembly by the BAM complex could be broken down into three major steps: (1) substrate recognition (selection), (2) substrate binding, and (3) catalysis of the folding and insertion of the substrates into the OM. In this section, we summarize the current literature with regards to each of the three steps of the OMP assembly.

Substrate selection

Despite their ability to insert into different lipid bilayers of varying composition *in vitro*, OMPs only assemble at the OM and not at the IM (Patel *et al.*, 2009). This implies that the primary amino acid sequence of OMPs must contain information that targets them to the OM and to the BAM complex, but where does this information reside within the OMP sequences?

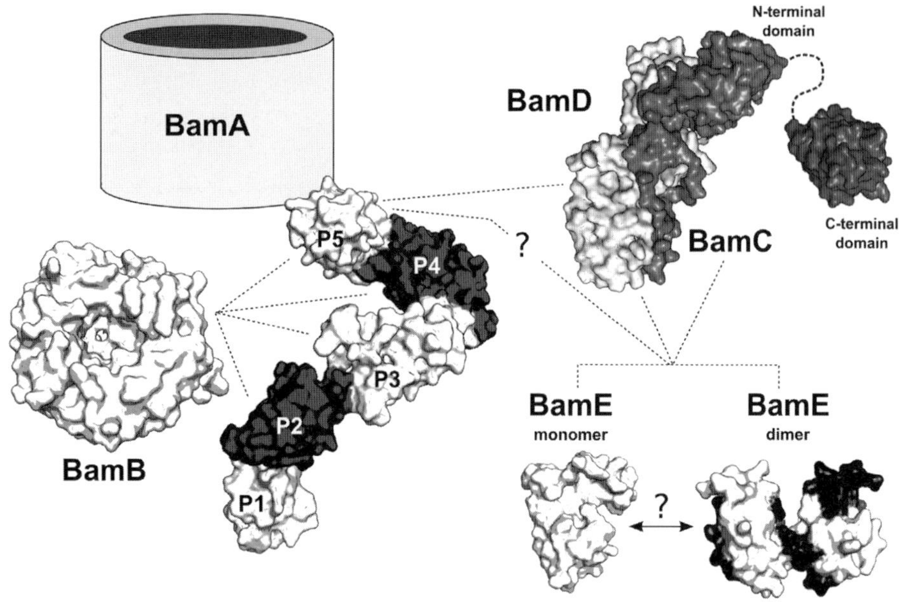

Figure 3.10 Interactions within the BAM complex. This figure summarizes all currently known and suspected interactions between the components of the BAM complex. Structures of all proteins (except for the β-barrel domain of BamA) are shown as surface diagrams, and associations between two proteins are indicated by dotted lines. While BamB interacts with POTRA 2, 3, 4 and 5, BamD interacts with POTRA 5. BamB–POTRA association is independent from that of BamD–POTRA association. BamC binds BamD, but not the POTRA domains. BamE has been shown to bind BamD and BamC, but it is not clear if it also interacts with any of the POTRA domains. It is also not clear which form of BamE (monomer, dimer or both) is found in the BAM complex.

In 1991, Struyve and her colleagues first noticed that deletion of the C-terminal segment of PhoE (phosphate porin) prevents the assembly of the protein into the OM (Struyve *et al.*, 1991). Multiple sequence alignments of bacterial OMPs subsequently revealed that the C-termini of the vast majority of OMPs consists of a phenylalanine (or tryptophan) at the C-terminal position, and hydrophobic residues at positions 3 (mostly tyrosine), 5, 7 and 9 from the C-terminus (Fig. 3.11A) (Struyve *et al.*, 1991). The C-terminal phenylalanine is strongly conserved, and this led Struyve and her colleagues to examine the possible role of the conserved C-terminal residue on the folding behaviour of PhoE. They reported that the level of PhoE detected in the OM decreases significantly when the C-terminal phenylalanine is deleted or mutated, suggesting that the conserved C-terminal residue is essential for correct assembly of PhoE. A few years later, the same group performed a follow-up study using immunocytochemical labelling, which revealed accumulation of PhoE in the periplasm when the C-terminal phenylalanine of PhE was mutated (de Cock *et al.*, 1997). Taken together, these results suggest an important role for the conserved C-terminal residue in OM targeting.

Following the discovery of the BAM complex and its role in OMP assembly, a hypothesis that the BAM complex might recognize its OMP substrates via their C-terminal signature sequence emerged. As BamA is essential and its N-terminal POTRA domain (POTRA 1)

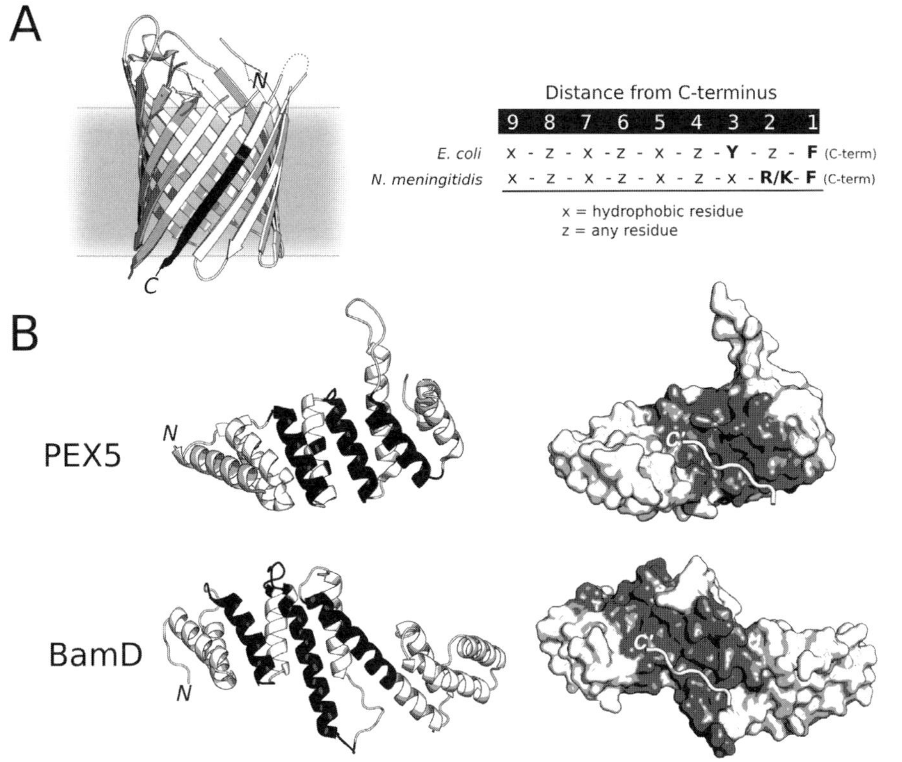

Figure 3.11 Substrate recognition by the BAM complex. (A) An OM targeting signal resides at the C-terminus of an OMP sequence. The location of the targeting signal within a folded OmpG β-barrel (PDB: 2F1C) is shown (black). The targeting signal sequences of OMPs seem to be species specific. For comparison, trends in the C-terminal sequences of *E. coli* and *N. meningitidis* OMPs are shown (right). (B) BamD of the BAM complex has a predicted function of OMP targeting signal recognition, in a similar fashion to how PEX5 recognizes the peroxisomal targeting signals of its substrates. BamD (PDB: 3TGO) and PEX5 (PDB: 2C0L) show significant structural similarity, especially in the region that forms the targeting signal binding site (shown in black in the ribbon diagram and dark grey in the surface diagrams). As shown in the surface diagram, the peroxisomal targeting signal (white) binds to PEX5 in an extended form. BamD is predicted to bind the C-terminal targeting sequences of OMPs in a similar manner.

has been shown to bind the periplasmic chaperone SurA (Bennion *et al.*, 2010), BamA was initially considered by many to be the most likely candidate for performing the substrate recognition function. The idea that the C-terminal signature sequence of OMPs is recognized by BamA was first investigated by Robert *et al.* (2006), who showed that various OMPs, as well as a peptide mimicking the C-terminal signature sequence of PhoE alone, can bind and modulate BamA channel activity. In the same study, Robert *et al.* (2006) observed that PorA from *N. meningitidis* is unable to induce the channel opening of *E. coli* BamA, despite having the conserved C-terminal phenylalanine and hydrophobic residues at positions 3, 5, 7 and 9 from the C-terminus. Subsequent comparison of the C-terminal sequences of *E. coli* and *N. meningitidis* OMPs show that *N. meningitidis* OMPs differ from the *E. coli* OMPs in that a positively charged residue (predominantly lysine or arginine) is found at position

2 (Robert *et al.*, 2006). In *E. coli* OMPs, the amino acid found at position 2 is usually glutamine. Interestingly, PorA can be assembled correctly into the *E. coli* OM when the lysine residue at position 2 of PorA is mutated to glutamine (Robert *et al.*, 2006). Based on this result, Robert *et al.* (2006) suggested that BamA recognizes its OMP substrates by a species-specific C-terminal motif.

While it has been established that denatured OMP substrates bind directly and modulate channel conductance of BamA (Voulhoux *et al.*, 2003; Robert *et al.*, 2006), recently emerging experimental data suggest that BamD may also play an important role in the initial substrate selection. When Sandoval *et al.* (2011) solved the first crystal structure of BamD from *R. marinus*, they showed through structure comparison that the N-terminal half (TPR1–3) of BamD forms a pocket that superimposes very closely with the binding pockets of the other proteins where protein substrates in extended conformations bind (Fig. 3.11B) (Albrecht and Zeth, 2011; Kim *et al.*, 2011a; Sandoval *et al.*, 2011). These proteins include peroxisomal targeting signal receptor (PEX5) and Hsp-organizing protein (HOP) (Sandoval *et al.*, 2011). Both proteins recognize the C-termini of their substrates, and this has led to the speculation that BamD might serve as an OMP targeting signal receptor for the BAM complex by recognizing the C-terminal signature sequences of OMPs (Sandoval *et al.*, 2011). Shortly after the crystal structure of *R. marinus* BamD was published, Albrecht and Zeth reported the crystal structure of *E. coli* BamD, along with the finding that a truncated form of BamD consisting only of TPR1–3 that harbours the proposed binding pocket crosslinks with synthetic peptides harbouring the OMP C-terminal targeting sequence (Albrecht and Zeth, 2011). Interestingly, however, the recently solved BamCD subcomplex crystal structure revealed that the proposed binding pocket of BamD is bound to the unstructured N-terminal region of BamC (Fig. 3.8C) (Kim *et al.*, 2011a). It is clear from the BamCD structure that the proposed binding pocket of BamD will not be able to bind the C-terminal signature sequence of OMPs, as it is completely occluded by BamC. Nevertheless, the structural similarity between BamD and those of other targeting signal recognition proteins is conspicuous, which gives rise to a question of whether the substrate binding activity of BamD is regulated by BamC interaction.

So which protein, BamA or BamD, initially recognizes the OMP substrates via the C-terminal targeting signal sequence? While the C-terminal fragment of a canonical OMP substrate PhoE has been shown to bind BamA and modulate its channel activity (Robert *et al.*, 2006), the crosslinking data and structural resemblance of BamD to other targeting signal receptor proteins favour the hypothesis that the N-terminal binding pocket of BamD functions in the initial substrate recognition and selection (Albrecht and Zeth, 2011). Future studies of substrate binding affinity to BamA and BamD, as well as further biochemical characterization of the interactions between the OM targeting signal and BamA/BamD, will shed light on which protein acts as the main OM targeting signal receptor.

Substrate binding

Prior to incorporation into the OM, OMPs must be kept in unfolded forms to prevent misfolding and aggregation. Chaperones such as SurA and Skp play important roles in this process by binding OMPs while they traverse through the dense periplasmic space to reach the OM (Sklar *et al.*, 2007b; Volokhina *et al.*, 2011). Once at the OM, it is currently not clear how the OMPs are passed on to the BAM complex for subsequent assembly process. Does the substrate get progressively transferred from the chaperone to the BAM complex as the folding and membrane insertion take place? Or does the chaperone unload the substrate all at once, implying that the BAM complex needs to keep the substrate in a non-aggregated form before the assembly process begins? In either case, the proteins of the BAM complex must be able to bind OMP substrates in unfolded forms. The ability of BamA to bind OMPs has already been established, but the location of the binding sites and the mode of interaction are still subjects of further study. In addition to BamA, BamB has also been suggested to be involved in substrate binding (Voulhoux *et al.*, 2003; Heuck *et al.*, 2011).

As described earlier, the POTRA domains of BamA and BamB have been shown to be capable of participating in protein–protein interaction via β-augmentation (Kim *et al.*, 2007; Gatzeva-Topalova *et al.*, 2008; Knowles *et al.*, 2008; Heuck *et al.*, 2011). Both the POTRA and the BamB structures exhibit β-sheets with the edges of the sheets exposed and available for hydrogen bonds (Fig. 3.12; Hagan *et al.*, 2011). In the case of the POTRA domains, Knowles *et al.* (2008) showed by NMR titration experiment that addition of various nascent β-strand peptides derived from PhoE induced chemical shift changes in residues found on the outer edges of the β-sheets in POTRA 1 and POTRA 2. Furthermore, POTRA 3 was shown to be involved in crystal packing by β-augmentation in two separate crystal structures despite different crystallization conditions (Fig. 3.12A) (Kim *et al.*, 2007; Gatzeva-Topalova *et al.*, 2008). Since all POTRA domains share the same basic structure, it seems reasonable to predict that POTRA 4 and POTRA 5 may also be able to participate in protein–protein interaction via β-augmentation. Similarly, β-augmentation has been observed in BamB

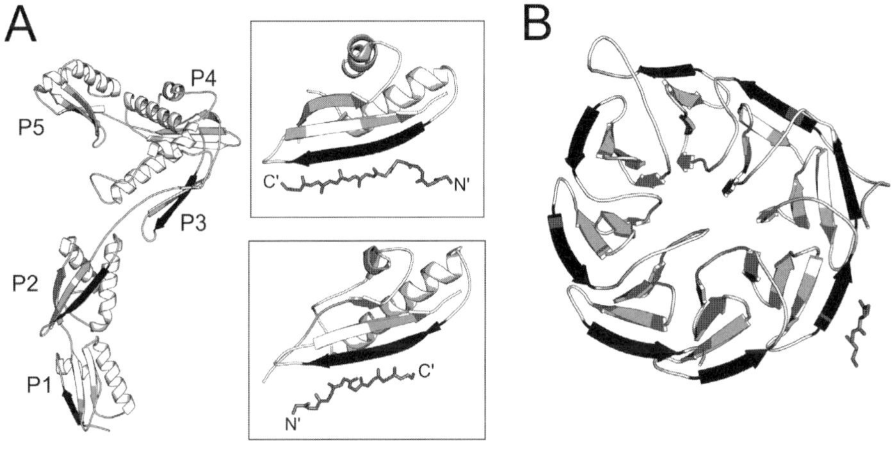

Figure 3.12 Potential substrate binding sites within the BAM complex. (A) A model of the full length BamA POTRA domains (POTRA 1–5) was built by superposing the POTRA1–4 structure (PDB: 3EFC) with the POTRA4–5 structure (PDB: 3Q6B). Coloured black are the regions that have been shown experimentally to be the sites of protein–protein interaction. Crystal packing of POTRA domains revealed that they can interact with other proteins via β-augmentation. Close-up views of β-augmentation of POTRA seen in two separate crystal structures are shown. One of the β-strands of POTRA 3 has been observed to interact with a strand from another protein in both parallel (top; PDB: 2QDF) and anti-parallel (bottom; PDB: 3EFC) fashion. (B) Similar to the POTRA domains, a BamB crystal structure (PDB: 3PRW) shows that the outermost strand in one of the β-propeller blades can interact with another strand (the main chain of which is shown as a stick model) via β-augmentation. All the solvent exposed β-strands that could potentially serve as β-augmentation sites are shown in black.

crystals as well, in which the outermost β-strand of one of the blades in the BamB structure is seen bound to a strand of a neighbouring BamB molecule (Heuck *et al.*, 2011). As the BamB structure consists of eight blades that are very close to each other in topology, BamB could provide eight potential substrate binding sites (Fig. 3.12B).

If both the POTRA domains of BamA and BamB can bind substrates, how are their roles distinguished? Although OMP assembly is significantly reduced in its absence, BamB is not an essential component of the BAM complex. On the other hand, at least one of the POTRA domains, POTRA 5 (the closest to the membrane), is required for proper function. Considering the essential nature and close proximity to the β-barrel domain of BamA (where the actual catalysis of OMP folding and membrane insertion is thought to take place), the POTRA domains likely serve as the major substrate binding sites and as the passage that leads substrates towards the β-barrel domain. The role of BamB may then be more of a supportive one, since its absence significantly decreases the efficiency but does not halt the proper functioning of the BAM complex. As the OMPs that are most affected by BamB deletion are relatively large (16- to 24-stranded β-barrels), it has been suggested that BamB could aid BamA function by increasing the substrate binding capacity (Heuck *et al.*, 2011). It is also possible that BamB functions as a reservoir of substrates when the BAM complex function is in high demand, binding substrates and preventing them from aggregating until they can be delivered to BamA for subsequent assembly process.

Protein folding and membrane insertion

The least understood aspect of the BAM complex function is the process by which OMPs are folded and inserted into the OM to adopt the final β-barrel structure. Although controversial, it has been suggested that OMPs exist in a partially folded state in the periplasm. Aside from the periplasmic chaperones, the conformational change of the POTRA domains from the extended to the bent state (when bound to an unfolded OMP) has been speculated to facilitate formation of β-hairpins in the substrate prior to membrane insertion (Gatzeva-Topalova *et al.*, 2008). *In vitro* β-barrel folding studies suggest the presence of membrane bound folding intermediates, and that completion of folding and membrane insertion of OMPs take place in a concerted manner (Kleinschmidt, 2003; Burgess *et al.*, 2008). As OMPs can spontaneously fold and insert into membranes *in vitro* and there is no ATP source in the periplasm, the BAM complex function is thought to be mainly associated with increasing the kinetics of the natural folding process of OMPs. Several different models of OMP assembly by the BAM complex have been proposed to date (Fig. 3.13), and they are described below.

The earliest model for the BAM complex assembly mechanism proposed that a substrate folds within the pore formed by the β-barrel domain of BamA, and then is laterally released into the OM. If this model is correct, the channel formed by BamA must be large enough to accommodate a folded substrate. Currently available data suggest that a channel formed by a single BamA molecule is not large enough to contain a fully folded OMP. Whereas the pore diameter of BamA estimated from electrophysiological data is 25Å, the structure of a BamA homologue, FhaC, has a measured diameter of 16Å even under the assumption that the components partially blocking the pore of FhaC are relocated out of the channel (Clantin *et al.*, 2007). However, studies have shown that BamA in monomeric form can properly assemble OmpT (approximately 25Å in diameter) (Hagan *et al.*, 2011). Also, BamA has been shown to be responsible for the folding of large OMPs such as FimD, which is more

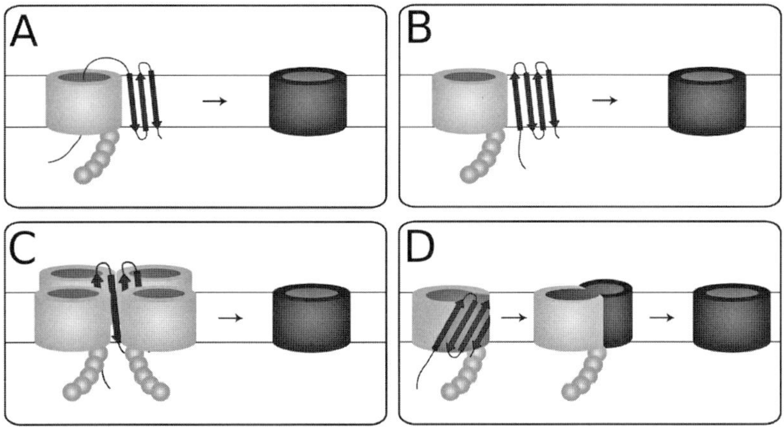

Figure 3.13 Different models of OMP assembly. Four different models of how the BAM complex may facilitate the folding and insertion of OMPs are shown. BamA is shown in light grey, and the substrate protein in black. The lipoproteins BamB/C/D/E are not shown in these models for clarity. The outer membrane is outlined with black lines, with the extracellular space above and the periplasmic space below. (A) In the first model, the substrate protein is first translocated across the outer membrane through the channel formed by the β-barrel domain of BamA. The substrate then inserts and folds into the outer membrane lipid bilayer from outside the cell. (B) In the second model, the substrate inserts into the lipid bilayer from the periplasmic face of the outer membrane. Instead of using the channel of BamA, the insertion and folding of OMPs occur at the BamA–lipid interface. In this model, the outer wall of the β-barrel of BamA provides a scaffold for the substrate folding. (C) This model is similar to the second model, but assumes that BamA forms an oligomeric structure. The coordinated events of substrate folding and membrane insertion are contained within the space formed by the BamA tetramer. The mature, folded OMP substrate is then released laterally into the lipid bilayer of the outer membrane. (D) In the last model, the OMP substrate uses the N- and the C-terminal β-strands of BamA as folding templates. The hydrogen bonds between the two terminal β-strands within the barrel of BamA are displaced by the incoming substrate, in a way that the first and the last β-strands of the substrate form hydrogen bonds with the terminal strands of BamA. The substrate therefore becomes a part of the BamA structure, and the substrate folds and inserts into the OM as a β-sheet held between the terminal strands of BamA. The β-sheet of the substrate is then closed to form a β-barrel, and the new mature folded and assembled OMP is released into the lipid bilayer.

than twice as large as OmpT (Palomino *et al.*, 2011). The channel formed by BamA may be large enough if BamA oligomerizes to form a single β-barrel; but, as mentioned above, it seems that monomeric BamA is fully capable of carrying out its function. Furthermore, lateral release of a folded substrate implies breaking several hydrogen bonds of the BamA β-barrel domain, which would be energetically costly.

The difficulty of biochemically detecting and characterizing folding intermediates, as well as lack of a BamA-substrate structure, presents a large challenge to the understanding of the folding and insertion mechanism of the BAM complex. The only structure available of a protein that is closely related to BamA is that of a distant homologue, FhaC. The function of FhaC is different from BamA in that it is responsible for translocating one specific substrate across the OM, rather than folding and inserting the substrate into the OM bilayer (Clantin *et al.*, 2007; Delattre *et al.*, 2010). However, its structure still provides valuable insights into the BamA function, as BamA is predicted to have the similar overall structure as FhaC (except that BamA has five and FhaC has only two POTRA domains). The channel formed by the β-barrel of FhaC is approximately 3 Å in diameter, which is too small for accommodating even an unfolded polypeptide (Clantin *et al.*, 2007). However, studies suggest binding of the substrate (FHA) causes a conformational change in FhaC, which could subsequently increase the diameter of the pore from 3 Å to 16 Å (Clantin *et al.*, 2007; Delattre *et al.*, 2010). This increase in the pore size could make enough room for the substrate to enter the channel within the barrel in an unfolded extended form (Clantin *et al.*, 2007). As it has been shown that substrate binding increases the channel conductivity of BamA (Stegmeier and Andersen, 2006), it is possible that BamA also undergoes a similar conformational change as in FhaC. If this is the case, the β-barrel lumen of BamA could serve as a conduit for a substrate to be translocated across the OM in an unfolded form. This model implies that the substrate would emerge into the extracellular space, and that substrate folding/insertion would take place on the outside of the cell (Fig. 3.13A). However, it is difficult to imagine OMPs assembling efficiently without extra folding factors on the extracellular surface of the outer membrane. Perhaps it is possible that LPS, which is only present on the outer leaflet of the OM, promotes substrate folding; but currently there isn't sufficient experimental evidence to associate LPS with OMP folding efficiency.

The two models described above assume that the β-barrel lumen of BamA plays the critical role by either providing an isolated environment for OMP folding or providing a passage for substrate to cross the OM. However, a more recently proposed model predicts that OMP substrates use the outer wall of the BamA β-barrel as a scaffold for folding and membrane insertion (Fig. 3.13B) (Hagan *et al.*, 2011). In this model, an unfolded or partially folded substrate will start to insert between the BamA–lipid interfaces as it starts acquiring tertiary structure. Alternatively, if BamA forms a tetramer *in vivo*, the substrate folding could be contained in the space formed by the four BamA subunits. The limited folding space may facilitate the closing of the β-sheet into a β-barrel, and the β-barrel would be released laterally into the lipid bilayer (Fig. 3.13c). This model requires that BamA subunits within the proposed tetramer be able to associate and dissociate with each other to allow substrate release. This model does not require BamA to have a large channel nor does it require breaking inter-strand hydrogen bonds for substrate release.

The mechanism of folding and membrane insertion by the BAM complex remains largely speculative to date. Here, we propose yet another model that combines and modifies the ideas from each of the models described earlier (Fig. 3.13D). As mentioned already, the structural and functional information now tells us that the channel formed by BamA is nowhere close to being large enough to hold a folded OMP, and breaking inter-strand hydrogen bonds of a β-barrel to release the substrate seems energetically very costly. To take into account the new structural data and to minimize the energy cost, we suggest the model be modified as follows. Instead of the unfolded substrate folding within the β-barrel of BamA, the N- and the C-terminal β-strands of BamA (the two strands that hydrogen bond with each other to close the β-sheet into a β-barrel) could serve as folding templates for the substrate. More specifically, the hydrogen bonds between the two terminal β-strands of BamA would be interrupted and replaced by an incoming substrate, which would form new hydrogen bonds with the terminal strands of BamA. In other words, the β-barrel domain of BamA would be opened and augmented by addition of strands from the substrate to form a larger temporary chimeric β-barrel. As the substrate adopts a β-sheet structure held between the two terminal strands of BamA, it will close into a β-barrel and 'bud off' BamA into the lipid bilayer. Although this model requires the β-barrel of BamA to open up, the energy cost of breaking hydrogen bonds is compensated for by having the substrate forming new hydrogen bonds with BamA via β-augmentation. Also, this model keeps the hydrophobic residues of a substrate always facing the membrane, and hydrophilic residues always facing away from the OM bilayer. However, like with other models, many aspects of this model are speculative and need to be experimentally validated in the future.

Do lipoproteins BamB-E play any role in these last steps of OMP assembly? Ieva *et al.* (2011) used site-specific cross-linking in combination with pulse-chase labelling to show that BamB and BamD remain bound to an OMP substrate longer than BamA, and suggested that BamB and BamD may function at a later stage of assembly such as substrate release (Ieva *et al.*, 2011). BamE, on the other hand, has been shown to bind specifically to phosphatidylglycerol (Knowles *et al.*, 2011), which has previously been shown to enhance the insertion of OMPs into liposomes although the reason for this is unclear. Based on this observation, it has been hypothesized that the function of BamE may be to recruit phosphatidylglycerol to enhance OMP membrane insertion (Endo *et al.*, 2011). How the functions of the lipoproteins are coordinated with that of BamA remains largely enigmatic. Biochemical and kinetics studies of how substrates interact with each of the BAM components may help identify steps in the process of OMP assembly by the BAM complex.

Eukaryotic homologues

Homologues of the BAM complex also exist in the outer membranes of mitochondria and chloroplasts (Fig. 3.14). Similar to the bacterial OMPs, mitochondrial and chloroplastic OMPs are also synthesized in the cytosol prior to being targeted. However, for these eukaryotic proteins, the signal sequence directs the OMP to the organelle membrane (mitochondrion/chloroplast) rather than the plasma membrane of the cell.

The SAM complex in mitochondria

In the mitochondrial system, before being inserted into the mitochondrial outer membrane (MOM), the substrate proteins are first imported into the mitochondrion via the translocase of outer mitochondrial membrane (TOM). After entering the intermembrane space (IMS), Tim chaperones transport the OMPs back to the MOM for assembly by the sorting and assembly machinery (SAM) complex (Fig. 3.14B). The primary component of this complex is Sam50 (the BamA homologue) which contains only one POTRA domain facing the IMS. It appears that the POTRA domain plays an important role in substrate release as this function is hindered when the domain is absent (Stroud *et al.*, 2011). Instead of lipoproteins, two cytosolic proteins, Sam35 and Sam37, have been identified as the main accessory proteins, with Sam35 being essential for cell survival (Milenkovic *et al.*, 2004; Paschen *et al.*, 2005). Current research suggests Sam35 to be involved in substrate recognition and Sam37 involved in substrate release (Paschen *et al.*, 2005; Chacinska *et al.*, 2009).

The TOC complex in chloroplasts

For chloroplasts, protein import from the cytosol into the stroma involves passing the translocons at the outer and inner envelopes of chloroplasts (the TOC and TIC complexes) (Oreb *et al.*, 2008). In the case of chloroplastic OMPs found in the outer envelope

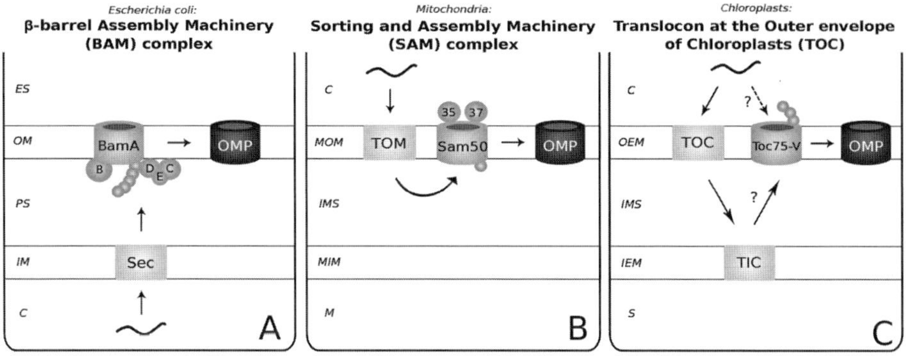

Figure 3.14 OMP assembly systems in eukaryotes. In both Gram-negative bacteria and eukaryotes, outer membrane β-barrel proteins are first synthesized in the cytosol of the cell and then targeted to either the inner membrane (bacteria) or the proper organelle (mitochondria or chloroplasts). This figure compares the three pathways as the unfolded substrate protein (black curved line) is directed by associated translocons to the assembly complex consisting of the core BamA homologue and accessory proteins, to form the final folded β-barrel (black cylinder). For simplicity, other proteins and chaperones involved in the pathways are not shown. (A) The *E. coli* β-barrel Assembly Machinery (BAM) complex consists of membrane embedded BamA, and four accessory lipoproteins: BamB, C, D, and E. Substrate proteins cross the inner membrane via the Sec translocase, and travel through the periplasmic space before being assembled by the BAM complex at the outer membrane. (B) In the mitochondrial system, the substrate proteins enter via the translocase of outer mitochondrial membrane (TOM) and are assembled by the sorting and assembly machinery (SAM) complex. The BamA homologue is Sam50, which works together with cytosolic proteins Sam35 and Sam37 for insertion of OMPs into the outer mitochondrial membrane. (C) In chloroplasts, the translocons at the outer and inner envelopes of chloroplasts (TOC/TIC complexes) are believed to be involved in assembly of OMPs. The BamA homologue is Toc75-V, with accessory proteins yet to be identified. It is unclear if the substrate proteins travel to the stroma prior to being assembled or if they are directly assembled into the outer envelope membrane from the cytosol. ES=extracellular space; OM=outer membrane; PS=periplasmic space; IM=inner membrane space; C=cytosol; MOM=mitochondrial outer membrane; IMS=intermembrane space; MIM=mitochondrial inner membrane; M=matrix; OEM=outer envelope membrane of chloroplast; IEM=inner envelope membrane of chloroplast; S=stroma.

membrane (OEM), it was previously proposed that the OMPs travel into the stroma using the TOC/TIC complexes, and then travel back to the OEM for assembly by Toc75-V (the BamA homologue) (Fig. 3.14C). This was based on the assumption that the three POTRA domains of Toc75-V face the IMS similar to Sam50 of the mitochondrial system. However, a recent study has shown the POTRA domains to exist in the opposite orientation, with the POTRA domains facing towards the cytosol (Sommer *et al.*, 2011). This new finding suggests the possibility of OMPs to be imported directly by Toc75-V and immediately inserted into the OEM, without the use of the TOC/TIC pathway. However, with the exact pathway unknown, and essential accessory proteins yet to be identified, the mechanism of chloroplastic OMP assembly is less understood and requires more research (Fairman *et al.*, 2011; Schleiff *et al.*, 2011).

Discussion

The field of OMP biogenesis research has made exciting progress in recent years. Unlike α-helical membrane proteins of the IM, assembly of β-barrel OMPs into the OM had been relatively poorly understood. Although *in vitro* studies gave important insights and possible mechanisms of OMP folding into lipid bilayers, how the β-barrel maturation process takes place inside cells remained uncertain until genetics and protein–protein interaction studies identified the BAM complex and its components as the OMP assembly factors. The BAM complex catalyses the essential function of OMP folding and membrane insertion, and newly emerging data are starting to reveal the structure and mechanism of the BAM complex.

So what do the currently available structural and functional data tell us about how the BAM complex works? The current consensus is that BamA and BamD perform the essential function of substrate recognition and assembly while BamB, BamC, and BamE increase the efficiency of the process. Based on its structural similarity with other targeting signal recognition proteins, BamD likely recognizes the β-barrel precursors (OMP substrates) via their C-terminal targeting sequence. Since SurA is known to bind the POTRA1 domain of BamA, it is currently not known if the POTRA domains are also involved in substrate recognition. Or, it is possible that the chaperone–substrate complex needs to be recognized and selected by BamD first, prior to docking onto the POTRA domains for subsequent assembly steps. In either case, the substrate is thought to eventually bind the POTRA domains via β-augmentation. Conformational flexibility of the POTRA domains may perhaps promote initial β-hairpin formation, as the substrate is guided towards the BamA β-barrel domain. BamA then catalyses synchronized folding and membrane insertion of the substrate by a mechanism that is still poorly understood. During the entire OMP assembly process, BamB could provide extra substrate binding surfaces to aid the POTRA domains (again, via β-augmentation), and BamE may recruit PG to enhance the folding efficiency. Finally, BamC could use its unstructured N-terminus to regulate the targeting signal binding activity of BamD.

Comparing the eukaryotic and bacterial OMP assembly systems yields some interesting similarities and differences between them. For example, unlike BamA, homologues of the lipoproteins BamB-E are not found in eukaryotes (Tommassen, 2010; Ricci and Silhavy, 2011). However, the SAM complex of mitochondria is known to consist of Sam50 (BamA homologue) along with accessory proteins. One of these accessory proteins, Sam35, is essential and has been shown to function in substrate recognition (Milenkovic *et al.*, 2004; Paschen *et al.*, 2005; Chacinska *et al.*, 2009). In bacteria, the essential lipoprotein BamD is predicted to serve a similar role, although Sam35 and BamD do not seem to share any sequence homology. Understanding the roles of the accessory proteins in the eukaryotic OMP assembly systems may therefore shed some light on the exact functions of the BAM lipoproteins, and vice versa.

Future trends

Although great progress has been made in recent years, many aspects of the BAM complex function still remain to be elucidated in order to learn the molecular mechanism of β-barrel assembly. The major challenge of studying the BAM complex is that currently there is no probe for detecting OMP folding intermediates *in vivo*, which makes it difficult to study exactly what stage of the OMP assembly process is affected when a certain mutation is introduced into the system. Developing a technique that can track the folding status of a substrate would tremendously help dissecting the role of each components of the BAM complex. Time-resolved tryptophan fluorescence quenching techniques have been reported to be useful to track the position of substrate undergoing folding relative to the membrane (i.e. how deep the substrate is inserted into the membrane) (Kleinschmidt, 2003). Perhaps this technique can be incorporated into the *in vitro* BAM complex system reconstituted into a proteoliposome that has proven to be very helpful in studying the BAM complex in isolation and in a controlled environment.

Structural studies of the BAM complex have not only shown us what each component of the BAM complex looks like, but have also provided clues to the functional roles of each protein. Future research efforts will need to address the questions posed by structural analysis of the BAM proteins. For instance, more experimental evidence is needed to validate the role of BamD as an OMP targeting signal receptor and the ability of BamB to bind substrates. In terms of structural research, future structural investigation of the BAM complex should focus on determining how the BAM components are arranged within the complex, in addition to determining the structure of the BamA β-barrel domain. Co-crystal structures of a substrate bound to BamA POTRA domains, BamD or BamB would also provide a great deal of insight into the BAM-OMP specificity. These structural studies should be accompanied by binding kinetics studies to characterize how the BAM proteins interact with each other and substrates.

Could better understanding of the BAM complex be utilized for medical applications? The BAM complex has been suggested to be a suitable drug target for novel antibiotics and vaccine development. The BAM complex is not only essential for the survival of Gram-negative bacteria, but it is functionally non-redundant (i.e. there is no other back-up system in the cell that can perform the same function). Furthermore, the BAM complex is found in the outer membrane, implying uncomplicated drug delivery strategy. In order to realize its potential as a drug target, however, continuing research efforts are required to elucidate the molecular mechanism of the BAM complex.

References

Albrecht, R., and Zeth, K. (2010). Crystallization and preliminary X-ray data collection of the *Escherichia coli* lipoproteins BamC, BamD and BamE. Acta Crystallogr. Sect. F. Struct. Biol. Cryst. Commun. *66*, 1586–1590.

Albrecht, R., and Zeth, K. (2011). Structural basis of outer membrane protein biogenesis in bacteria. J. Biol. Chem. *286*, 27792–27803.

Anwari, K., Webb, C.T., Poggio, S., Perry, A.J., Belousoff, M., Celik, N., Ramm, G., Lovering, A., Sockett, R.E., Smit, J., Jacobs-Wagner, C., and Lithgow, T. (2012). The evolution of new lipoprotein subunits of the bacterial outer membrane BAM complex. Mol. Microbiol. *84*, 832–844.

Bechtluft, P., Nouwen, N., Tans, S.J., and Driessen, A.J. (2010). SecB – a chaperone dedicated to protein translocation. Mol. Biosyst *6*, 620–627.

Bennion, D., Charlson, E.S., Coon, E., and Misra, R. (2010). Dissection of beta-barrel outer membrane protein assembly pathways through characterizing BamA POTRA 1 mutants of *Escherichia coli*. Mol. Microbiol. *77*, 1153–1171.

Bishop, R.E. (2005). The lipid A palmitoyltransferase PagP: molecular mechanisms and role in bacterial pathogenesis. Mol. Microbiol. *57*, 900–912.

Bishop, R.E. (2008). Structural biology of membrane-intrinsic beta-barrel enzymes: sentinels of the bacterial outer membrane. Biochim. Biophys. Acta *1778*, 1881–1896.

Bitto, E., and McKay, D.B. (2003). The periplasmic molecular chaperone protein SurA binds a peptide motif that is characteristic of integral outer membrane proteins. J. Biol. Chem. *278*, 49316–49322.

Bodelon, G., Marin, E., and Fernandez, L.A. (2009). Role of periplasmic chaperones and BamA (YaeT/Omp85) in folding and secretion of intimin from enteropathogenic *Escherichia coli* strains. J. Bacteriol. *191*, 5169–5179.

Bos, M.P., and Tommassen, J. (2004). Biogenesis of the Gram-negative bacterial outer membrane. Curr. Opin. Microbiol. *7*, 610–616.

Burgess, N.K., Dao, T.P., Stanley, A.M., and Fleming, K.G. (2008). Beta-barrel proteins that reside in the *Escherichia coli* outer membrane *in vivo* demonstrate varied folding behavior *in vitro.* J. Biol. Chem. *283*, 26748–26758.

Chacinska, A., Koehler, C.M., Milenkovic, D., Lithgow, T., and Pfanner, N. (2009). Importing mitochondrial proteins: machineries and mechanisms. Cell *138*, 628–644.

Chandran, V., Fronzes, R., Duquerroy, S., Cronin, N., Navaza, J., and Waksman, G. (2009). Structure of the outer membrane complex of a type IV secretion system. Nature *462*, 1011–1015.

Charlson, E.S., Werner, J.N., and Misra, R. (2006). Differential effects of yfgL mutation on *Escherichia coli* outer membrane proteins and lipopolysaccharide. J. Bacteriol. *188*, 7186–7194.

Clantin, B., Delattre, A.S., Rucktooa, P., Saint, N., Meli, A.C., Locht, C., Jacob-Dubuisson, F., and Villeret, V. (2007). Structure of the membrane protein FhaC: a member of the Omp85-TpsB transporter superfamily. Science *317*, 957–961.

Collins, R.F., and Derrick, J.P. (2007). Wza: a new structural paradigm for outer membrane secretory proteins? Trends Microbiol. *15*, 96–100.

Cross, B.C., Sinning, I., Luirink, J., and High, S. (2009). Delivering proteins for export from the cytosol. Nat. Rev. Mol. Cell Biol. *10*, 255–264.

Dalbey, R.E., Wang, P., and Kuhn, A. (2011). Assembly of bacterial inner membrane proteins. Annu. Rev. Biochem. *80*, 161–187.

de Cock, H., Struyve, M., Kleerebezem, M., van der Krift, T., and Tommassen, J. (1997). Role of the carboxy-terminal phenylalanine in the biogenesis of outer membrane protein PhoE of *Escherichia coli* K-12. J. Mol. Biol. *269*, 473–478.

Dekker, N., Tommassen, J., Lustig, A., Rosenbusch, J.P., and Verheij, H.M. (1997). Dimerization regulates the enzymatic activity of *Escherichia coli* outer membrane phospholipase A. J. Biol. Chem. *272*, 3179–3184.

Delattre, A.S., Clantin, B., Saint, N., Locht, C., Villeret, V., and Jacob-Dubuisson, F. (2010). Functional importance of a conserved sequence motif in FhaC, a prototypic member of the TpsB/Omp85 superfamily. FEBS J. *277*, 4755–4765.

Delcour, A.H. (2003). Solute uptake through general porins. Front. Biosci. *8*, d1055–1071.

Delcour, A.H. (2009). Outer membrane permeability and antibiotic resistance. Biochim. Biophys. Acta *1794*, 808–816.

Diedrich, D.L., and Cota-Robles, E.H. (1974). Heterogeneity in lipid composition of the outer membrane and cytoplasmic membrane and cytoplasmic membrane of Pseudomonas BAL-31. J. Bacteriol. *119*, 1006–1018.

Doerrler, W.T., and Raetz, C.R. (2005). Loss of outer membrane proteins without inhibition of lipid export in an *Escherichia coli* YaeT mutant. J. Biol. Chem. *280*, 27679–27687.

Dong, C., Beis, K., Nesper, J., Brunkan-Lamontagne, A.L., Clarke, B.R., Whitfield, C., and Naismith, J.H. (2006). Wza the translocon for *E. coli* capsular polysaccharides defines a new class of membrane protein. Nature *444*, 226–229.

Driessen, A.J., and Nouwen, N. (2008). Protein translocation across the bacterial cytoplasmic membrane. Annu. Rev. Biochem. *77*, 643–667.

du Plessis, D.J., Nouwen, N., and Driessen, A.J. (2011). The Sec translocase. Biochim. Biophys. Acta *1808*, 851–865.

Endo, T., Kawano, S., and Yamano, K. (2011). BamE structure: the assembly of beta-barrel proteins in the outer membranes of bacteria and mitochondria. EMBO Rep. *12*, 94–95.

Fairman, J.W., Noinaj, N., and Buchanan, S.K. (2011). The structural biology of beta-barrel membrane proteins: a summary of recent reports. Curr. Opin. Struct. Biol. *21*, 523–531.

Ferbitz, L., Maier, T., Patzelt, H., Bukau, B., Deuerling, E., and Ban, N. (2004). Trigger factor in complex with the ribosome forms a molecular cradle for nascent proteins. Nature *431*, 590–596.

Gatzeva-Topalova, P.Z., Walton, T.A., and Sousa, M.C. (2008). Crystal structure of YaeT: conformational flexibility and substrate recognition. Structure *16*, 1873–1881.

Gatzeva-Topalova, P.Z., Warner, L.R., Pardi, A., and Sousa, M.C. (2010). Structure and flexibility of the complete periplasmic domain of BamA: the protein insertion machine of the outer membrane. Structure *18*, 1492–1501.

Genevrois, S., Steeghs, L., Roholl, P., Letesson, J.J., and van der Ley, P. (2003). The Omp85 protein of *Neisseria meningitidis* is required for lipid export to the outer membrane. EMBO J. *22*, 1780–1789.
Gentle, I., Gabriel, K., Beech, P., Waller, R., and Lithgow, T. (2004). The Omp85 family of proteins is essential for outer membrane biogenesis in mitochondria and bacteria. J. Cell Biol. *164*, 19–24.
Gentle, I.E., Burri, L., and Lithgow, T. (2005). Molecular architecture and function of the Omp85 family of proteins. Mol. Microbiol. *58*, 1216–1225.
Gierasch, L.M. (1989). Signal sequences. Biochemistry *28*, 923–930.
Hagan, C.L., Kim, S., and Kahne, D. (2010). Reconstitution of outer membrane protein assembly from purified components. Science *328*, 890–892.
Hagan, C.L., Silhavy, T.J., and Kahne, D. (2011). Beta-barrel membrane protein assembly by the Bam complex. Annu. Rev. Biochem. *80*, 189–210.
Hayashi, S., and Wu, H.C. (1990). Lipoproteins in bacteria. J. Bioenerg. Biomembr. *22*, 451–471.
Heuck, A., Schleiffer, A., and Clausen, T. (2011). Augmenting beta-augmentation: structural basis of how BamB binds BamA and may support folding of outer membrane proteins. J. Mol. Biol. *406*, 659–666.
Hoffmann, A., Bukau, B., and Kramer, G. (2010). Structure and function of the molecular chaperone trigger factor. Biochim. Biophys. Acta *1803*, 650–661.
Hritonenko, V. (2011). Quaternary structure of Omp85/YaeT/BamA of *Yersinia pestis*. Am. J. Biochem. Mol. Biol. *1*, 185–199.
Ieva, R., Tian, P., Peterson, J.H., and Bernstein, H.D. (2011). Sequential and spatially restricted interactions of assembly factors with an autotransporter beta domain. Proc. Natl. Acad. Sci. U.S.A. *108*, E383–391.
Jacob-Dubuisson, F., Villeret, V., Clantin, B., Delattre, A.S., and Saint, N. (2009). First structural insights into the TpsB/Omp85 superfamily. Biol. Chem. *390*, 675–684.
Jain, S., and Goldberg, M.B. (2007). Requirement for YaeT in the outer membrane assembly of autotransporter proteins. J. Bacteriol. *189*, 5393–5398.
Kim, K.H., and Paetzel, M. (2011). Crystal structure of *Escherichia coli* BamB, a lipoprotein component of the beta-barrel assembly machinery complex. J. Mol. Biol. *406*, 667–678.
Kim, K.H., Aulakh, S., and Paetzel, M. (2011a). Crystal structure of the {beta}-barrel assembly machinery BamCD complex. J. Biol. Chem. *286*, 39116–39121.
Kim, K.H., Aulakh, S., Tan, W., and Paetzel, M. (2011b). Crystallographic analysis of the C-terminal domain of the *Escherichia coli* lipoprotein BamC. Acta Crystallogr. Sect. F. Struct. Biol. Cryst. Commun. *67*, 1350–1358.
Kim, K.H., Kang, H.S., Okon, M., Escobar-Cabrera, E., McIntosh, L.P., and Paetzel, M. (2011c). Structural characterization of *Escherichia coli* BamE, a lipoprotein component of the beta-barrel assembly machinery complex. Biochemistry *50*, 1081–1090.
Kim, K.H., Aulakh, S., and Paetzel, M. (2012). The bacterial outer membrane beta-barrel assembly machinery. Protein Sci. *21*, 751–768.
Kim, S., Malinverni, J.C., Sliz, P., Silhavy, T.J., Harrison, S.C., and Kahne, D. (2007). Structure and function of an essential component of the outer membrane protein assembly machine. Science *317*, 961–964.
Kleinschmidt, J.H. (2003). Membrane protein folding on the example of outer membrane protein A of *Escherichia coli*. Cell Mol. Life Sci. *60*, 1547–1558.
Kleinschmidt, J.H. (2006). Folding kinetics of the outer membrane proteins OmpA and FomA into phospholipid bilayers. Chem. Phys. Lipids *141*, 30–47.
Knoblauch, N.T., Rudiger, S., Schonfeld, H.J., Driessen, A.J., Schneider-Mergener, J., and Bukau, B. (1999). Substrate specificity of the SecB chaperone. J. Biol. Chem. *274*, 34219–34225.
Knowles, T.J., Jeeves, M., Bobat, S., Dancea, F., McClelland, D., Palmer, T., Overduin, M., and Henderson, I.R. (2008). Fold and function of polypeptide transport-associated domains responsible for delivering unfolded proteins to membranes. Mol. Microbiol. *68*, 1216–1227.
Knowles, T.J., McClelland, D.M., Rajesh, S., Henderson, I.R., and Overduin, M. (2009a). Secondary structure and (1)H, (13)C and (15)N backbone resonance assignments of BamC, a component of the outer membrane protein assembly machinery in *Escherichia coli*. Biomol. NMR Assign *3*, 203–206.
Knowles, T.J., Scott-Tucker, A., Overduin, M., and Henderson, I.R. (2009b). Membrane protein architects: the role of the BAM complex in outer membrane protein assembly. Nat. Rev. Microbiol. *7*, 206–214.
Knowles, T.J., Browning, D.F., Jeeves, M., Maderbocus, R., Rajesh, S., Sridhar, P., Manoli, E., Emery, D., Sommer, U., Spencer, A., *et al.* (2011). Structure and function of BamE within the outer membrane and the beta-barrel assembly machine. EMBO Rep. *12*, 123–128.
Koebnik, R., Locher, K.P., and Van Gelder, P. (2000). Structure and function of bacterial outer membrane proteins: barrels in a nutshell. Mol. Microbiol. *37*, 239–253.

Koronakis, V., Sharff, A., Koronakis, E., Luisi, B., and Hughes, C. (2000). Crystal structure of the bacterial membrane protein TolC central to multidrug efflux and protein export. Nature *405*, 914–919.

Krewulak, K.D., and Vogel, H.J. (2011). TonB or not TonB: is that the question? Biochem. Cell Biol. *89*, 87–97.

Kusters, I., and Driessen, A.J. (2011). SecA, a remarkable nanomachine. Cell Mol. Life Sci. *68*, 2053–2066.

Malinverni, J.C., Werner, J., Kim, S., Sklar, J.G., Kahne, D., Misra, R., and Silhavy, T.J. (2006). YfiO stabilizes the YaeT complex and is essential for outer membrane protein assembly in *Escherichia coli*. Mol. Microbiol. *61*, 151–164.

Meng, G., Fronzes, R., Chandran, V., Remaut, H., and Waksman, G. (2009). Protein oligomerization in the bacterial outer membrane (Review). Mol. Membr. Biol. *26*, 136–145.

Merdanovic, M., Clausen, T., Kaiser, M., Huber, R., and Ehrmann, M. (2011). Protein quality control in the bacterial periplasm. Annu. Rev. Microbiol. *65*, 149–168.

Milenkovic, D., Kozjak, V., Wiedemann, N., Lohaus, C., Meyer, H.E., Guiard, B., Pfanner, N., and Meisinger, C. (2004). Sam35 of the mitochondrial protein sorting and assembly machinery is a peripheral outer membrane protein essential for cell viability. J. Biol. Chem. *279*, 22781–22785.

Misra, R. (2007). First glimpse of the crystal structure of YaeT's POTRA domains. ACS Chem. Biol. *2*, 649–651.

Morein, S., Andersson, A., Rilfors, L., and Lindblom, G. (1996). Wild-type *Escherichia coli* cells regulate the membrane lipid composition in a 'window' between gel and non-lamellar structures. J. Biol. Chem. *271*, 6801–6809.

Nakae, T. (1976). Identification of the outer membrane protein of *E. coli* that produces transmembrane channels in reconstituted vesicle membranes. Biochem. Biophys. Res. Commun. *71*, 877–884.

Nikaido, H. (1994). Porins and specific diffusion channels in bacterial outer membranes. J. Biol. Chem. *269*, 3905–3908.

Nikaido, H. (2003). Molecular basis of bacterial outer membrane permeability revisited. Microbiol. Mol. Biol. Rev. *67*, 593–656.

Noinaj, N., Guillier, M., Barnard, T.J., and Buchanan, S.K. (2010). TonB-dependent transporters: regulation, structure, and function. Annu. Rev. Microbiol. *64*, 43–60.

Noinaj, N., Fairman, J.W., and Buchanan, S.K. (2011). The crystal structure of BamB suggests interactions with BamA and its role within the BAM complex. J. Mol. Biol. *407*, 248–260.

Okuda, S., and Tokuda, H. (2011). Lipoprotein sorting in bacteria. Annu. Rev. Microbiol. *65*, 239–259.

Onufryk, C., Crouch, M.L., Fang, F.C., and Gross, C.A. (2005). Characterization of six lipoproteins in the sigmaE regulon. J. Bacteriol. *187*, 4552–4561.

Oomen, C.J., van Ulsen, P., van Gelder, P., Feijen, M., Tommassen, J., and Gros, P. (2004). Structure of the translocator domain of a bacterial autotransporter. EMBO J. *23*, 1257–1266.

Oreb, M., Tews, I., and Schleiff, E. (2008). Policing Tic 'n' Toc, the doorway to chloroplasts. Trends Cell Biol. *18*, 19–27.

Paetzel, M., Karla, A., Strynadka, N.C., and Dalbey, R.E. (2002). Signal peptidases. Chem. Rev. *102*, 4549–4580.

Palomino, C., Marin, E., and Fernandez, L.A. (2011). The fimbrial usher FimD follows the SurA-BamB pathway for its assembly in the outer membrane of *Escherichia coli*. J. Bacteriol. *193*, 5222–5230.

Paschen, S.A., Neupert, W., and Rapaport, D. (2005). Biogenesis of beta-barrel membrane proteins of mitochondria. Trends Biochem. Sci. *30*, 575–582.

Patel, G.J., Behrens-Kneip, S., Holst, O., and Kleinschmidt, J.H. (2009). The periplasmic chaperone Skp facilitates targeting, insertion, and folding of OmpA into lipid membranes with a negative membrane surface potential. Biochemistry *48*, 10235–10245.

Phan, G., Remaut, H., Wang, T., Allen, W.J., Pirker, K.F., Lebedev, A., Henderson, N.S., Geibel, S., Volkan, E., Yan, J., *et al.* (2011). Crystal structure of the FimD usher bound to its cognate FimC-FimH substrate. Nature *474*, 49–53.

Pohlschroder, M., Prinz, W.A., Hartmann, E., and Beckwith, J. (1997). Protein translocation in the three domains of life: variations on a theme. Cell *91*, 563–566.

Postle, K., and Kadner, R.J. (2003). Touch and go: tying TonB to transport. Mol. Microbiol. *49*, 869–882.

Qu, J., Mayer, C., Behrens, S., Holst, O., and Kleinschmidt, J.H. (2007). The trimeric periplasmic chaperone Skp of *Escherichia coli* forms 1:1 complexes with outer membrane proteins via hydrophobic and electrostatic interactions. J. Mol. Biol. *374*, 91–105.

Raetz, C.R., and Whitfield, C. (2002). Lipopolysaccharide endotoxins. Annu. Rev. Biochem. *71*, 635–700.

Ricci, D.P., and Silhavy, T.J. (2011). The Bam machine: a molecular cooper. Biochim. Biophys. Acta *1818*, 1067–1084.

Rigel, N.W., Schwalm, J., Ricci, D.P., and Silhavy, T.J. (2012). BamE modulates the *Escherichia coli* beta-barrel assembly machine component BamA. J. Bacteriol. *194*, 1002–1008.
Rizzitello, A.E., Harper, J.R., and Silhavy, T.J. (2001). Genetic evidence for parallel pathways of chaperone activity in the periplasm of *Escherichia coli*. J. Bacteriol. *183*, 6794–6800.
Robert, V., Volokhina, E.B., Senf, F., Bos, M.P., Van Gelder, P., and Tommassen, J. (2006). Assembly factor Omp85 recognizes its outer membrane protein substrates by a species-specific C-terminal motif. PLoS Biol. *4*, e377.
Rossiter, A.E., Leyton, D.L., Tveen-Jensen, K., Browning, D.F., Sevastsyanovich, Y., Knowles, T.J., Nichols, K.B., Cunningham, A.F., Overduin, M., Schembri, M.A., *et al.* (2011). The essential β-barrel assembly machinery complex components BamD and BamA are required for autotransporter biogenesis. J. Bacteriol. *193*, 4250–4253.
Ruiz, N., Wu, T., Kahne, D., and Silhavy, T.J. (2006). Probing the barrier function of the outer membrane with chemical conditionality. ACS Chem. Biol. *1*, 385–395.
Sanchez-Pulido, L., Devos, D., Genevrois, S., Vicente, M., and Valencia, A. (2003). POTRA: a conserved domain in the FtsQ family and a class of beta-barrel outer membrane proteins. Trends Biochem. Sci. *28*, 523–526.
Sandoval, C.M., Baker, S.L., Jansen, K., Metzner, S.I., and Sousa, M.C. (2011). Crystal structure of BamD: an essential component of the beta-Barrel assembly machinery of gram-negative bacteria. J. Mol. Biol. *409*, 348–357.
Sauri, A., Soprova, Z., Wickström, D., de Gier, J.W., Van der Schors, R.C., Smit, A.B., Jong, W.S., and Luirink, J. (2009). The Bam (Omp85) complex is involved in secretion of the autotransporter haemoglobin protease. Microbiology *155*, 3982–3991.
Schatz, G., and Dobberstein, B. (1996). Common principles of protein translocation across membranes. Science *271*, 1519–1526.
Schleiff, E., Maier, U.G., and Becker, T. (2011). Omp85 in eukaryotic systems: one protein family with distinct functions. Biol. Chem. *392*, 21–27.
Schulz, G.E. (2000). Beta-Barrel membrane proteins. Curr. Opin. Struct. Biol. *10*, 443–447.
Selkrig, J., Mosbahi, K., Webb, C.T., Belousoff, M.J., Perry, A.J., Wells, T.J., Morris, F., Leyton, D.L., Totsika, M., Phan, M.D., *et al.* (2012). Discovery of an archetypal protein transport system in bacterial outer membranes. Nat. Struct. Mol. Biol. *19*, 506–510.
Silhavy, T.J., Kahne, D., and Walker, S. (2010). The bacterial cell envelope. Cold Spring Harb Perspect. Biol. *2*, a000414.
Sklar, J.G., Wu, T., Gronenberg, L.S., Malinverni, J.C., Kahne, D., and Silhavy, T.J. (2007a). Lipoprotein SmpA is a component of the YaeT complex that assembles outer membrane proteins in *Escherichia coli*. Proc. Natl. Acad. Sci. U.S.A. *104*, 6400–6405.
Sklar, J.G., Wu, T., Kahne, D., and Silhavy, T.J. (2007b). Defining the roles of the periplasmic chaperones SurA, Skp, and DegP in *Escherichia coli*. Genes Dev. *21*, 2473–2484.
Sommer, M.S., Daum, B., Gross, L.E., Weis, B.L., Mirus, O., Abram, L., Maier, U.G., Kuhlbrandt, W., and Schleiff, E. (2011). Chloroplast Omp85 proteins change orientation during evolution. Proc. Natl. Acad. Sci. U.S.A. *108*, 13841–13846.
Stegmeier, J.F., and Andersen, C. (2006). Characterization of pores formed by YaeT (Omp85) from *Escherichia coli*. J. Biochem. *140*, 275–283.
Stroud, D.A., Becker, T., Qiu, J., Stojanovski, D., Pfannschmidt, S., Wirth, C., Hunte, C., Guiard, B., Meisinger, C., Pfanner, N., *et al.* (2011). Biogenesis of mitochondrial beta-barrel proteins: the POTRA domain is involved in precursor release from the SAM complex. Mol. Biol. Cell *22*, 2823–2833.
Struyve, M., Moons, M., and Tommassen, J. (1991). Carboxy-terminal phenylalanine is essential for the correct assembly of a bacterial outer membrane protein. J. Mol. Biol. *218*, 141–148.
Stumpe, S., Schmid, R., Stephens, D.L., Georgiou, G., and Bakker, E.P. (1998). Identification of OmpT as the protease that hydrolyzes the antimicrobial peptide protamine before it enters growing cells of *Escherichia coli*. J. Bacteriol. *180*, 4002–4006.
Sugimura, K., and Nishihara, T. (1988). Purification, characterization, and primary structure of *Escherichia coli* protease VII with specificity for paired basic residues: identity of protease VII and OmpT. J. Bacteriol. *170*, 5625–5632.
Surrey, T., and Jahnig, F. (1992). Refolding and oriented insertion of a membrane protein into a lipid bilayer. Proc. Natl. Acad. Sci. U.S.A. *89*, 7457–7461.
Tamm, L.K., Hong, H., and Liang, B. (2004). Folding and assembly of beta-barrel membrane proteins. Biochim. Biophys. Acta *1666*, 250–263.

Tommassen, J. (2010). Assembly of outer-membrane proteins in bacteria and mitochondria. Microbiology *156*, 2587–2596.

Valent, Q.A., Kendall, D.A., High, S., Kusters, R., Oudega, B., and Luirink, J. (1995). Early events in preprotein recognition in *E. coli*: interaction of SRP and trigger factor with nascent polypeptides. EMBO J. *14*, 5494–5505.

van Ulsen, P. (2011). Protein folding in bacterial adhesion: secretion and folding of classical monomeric autotransporters. Adv. Exp. Med. Biol. *715*, 125–142.

Vandeputte-Rutten, L., Kramer, R.A., Kroon, J., Dekker, N., Egmond, M.R., and Gros, P. (2001). Crystal structure of the outer membrane protease OmpT from *Escherichia coli* suggests a novel catalytic site. EMBO J. *20*, 5033–5039.

Vanini, M.M., Spisni, A., Sforca, M.L., Pertinhez, T.A., and Benedetti, C.E. (2008). The solution structure of the outer membrane lipoprotein OmlA from *Xanthomonas axonopodis* pv. citri reveals a protein fold implicated in protein–protein interaction. Proteins *71*, 2051–2064.

Vogt, J., and Schulz, G.E. (1999). The structure of the outer membrane protein OmpX from *Escherichia coli* reveals possible mechanisms of virulence. Structure *7*, 1301–1309.

Vollmer, W., and Bertsche, U. (2008). Murein (peptidoglycan) structure, architecture and biosynthesis in *Escherichia coli*. Biochim. Biophys. Acta *1778*, 1714–1734.

Vollmer, W., von Rechenberg, M., and Holtje, J.V. (1999). Demonstration of molecular interactions between the murein polymerase PBP1B, the lytic transglycosylase MltA, and the scaffolding protein MipA of *Escherichia coli*. J. Biol. Chem. *274*, 6726–6734.

Volokhina, E.B., Grijpstra, J., Stork, M., Schilders, I., Tommassen, J., and Bos, M.P. (2011). Role of the periplasmic chaperones Skp, SurA, and DegQ in outer membrane protein biogenesis in *Neisseria meningitidis*. J. Bacteriol. *193*, 1612–1621.

von Heijne, G. (1990). The signal peptide. J. Membr. Biol. *115*, 195–201.

Voulhoux, R., Bos, M.P., Geurtsen, J., Mols, M., and Tommassen, J. (2003). Role of a highly conserved bacterial protein in outer membrane protein assembly. Science *299*, 262–265.

Vuong, P., Bennion, D., Mantei, J., Frost, D., and Misra, R. (2008). Analysis of YfgL and YaeT interactions through bioinformatics, mutagenesis, and biochemistry. J. Bacteriol. *190*, 1507–1517.

Walton, T.A., and Sousa, M.C. (2004). Crystal structure of Skp, a prefoldin-like chaperone that protects soluble and membrane proteins from aggregation. Mol. Cell *15*, 367–374.

Warner, L.R., Varga, K., Lange, O.F., Baker, S.L., Baker, D., Sousa, M.C., and Pardi, A. (2011). Structure of the BamC two-domain protein obtained by Rosetta with a limited NMR data set. J. Mol. Biol. *411*, 83–95.

Werner, J., and Misra, R. (2005). YaeT (Omp85) affects the assembly of lipid-dependent and lipid-independent outer membrane proteins of *Escherichia coli*. Mol. Microbiol. *57*, 1450–1459.

Wimley, W.C. (2003). The versatile beta-barrel membrane protein. Curr. Opin. Struct. Biol. *13*, 404–411.

Wu, T., Malinverni, J., Ruiz, N., Kim, S., Silhavy, T.J., and Kahne, D. (2005). Identification of a multicomponent complex required for outer membrane biogenesis in *Escherichia coli*. Cell *121*, 235–245.

Zgurskaya, H.I., Krishnamoorthy, G., Ntreh, A., and Lu, S. (2011). Mechanism and function of the outer membrane channel TolC in multidrug resistance and physiology of enterobacteria. Front. Microbiol. *2*, 189.

Zhang, H., Gao, Z.Q., Hou, H.F., Xu, J.H., Li, L.F., Su, X.D., and Dong, Y.H. (2011). High-resolution structure of a new crystal form of BamA POTRA4–5 from *Escherichia coli*. Acta Crystallogr. Sect. F. Struct. Biol. Cryst. Commun. *67*, 734–738.

Ziegler, K., Benz, R., and Schulz, G.E. (2008). A putative alpha-helical porin from *Corynebacterium glutamicum*. J. Mol. Biol. *379*, 482–491.

Zimmer, J., Nam, Y., and Rapoport, T.A. (2008). Structure of a complex of the ATPase SecA and the protein-translocation channel. Nature *455*, 936–943.

Bacterial Lipoproteins: Biogenesis, Virulence/ Pathogenicity and Trafficking

4

Hajime Tokuda, Peter Sander, Bok Luel Lee, Suguru Okuda, Thomas Grau, Andreas Tschumi, Juliane K. Brülle, Kenji Kurokawa and Hiroshi Nakayama

Abstract

The mechanisms underlying the biogenesis and outer membrane sorting of lipoproteins have been mostly clarified in *Escherichia coli*. Three enzymes catalyse the post-translational modification of lipoproteins with a membrane anchor comprising a thioether-linked diacylglycerol and an amide-linked fatty acid. The Lol system, comprising five Lol proteins, mediates the sorting of lipoproteins to the outer membrane. The three enzymes and five proteins are essential for *E. coli* and are widely conserved in Gram-negative diderms having cytoplasmic and outer membranes. High G+C content Gram-positive bacteria such as those belonging to the genera *Mycobacterium* and *Corynebacterium* have an outer membrane-like structure. The structure and biogenesis of the cell envelope have been intensively studied in mycobacterial species including medically important *Mycobacterium tuberculosis*. Mycobacterial lipoproteins are triacylated, like those of Gram-negative diderms. The enzyme catalysing N-acylation was recently identified. The lipoprotein biosynthesis pathway is important for virulence of *M. tuberculosis*. The functions of individual mycobacterial lipoproteins are discussed in relation to envelope biogenesis, virulence and influence on immune systems. The N-terminal structures of lipoproteins of low G+C content Gram-positive monoderms are surprisingly diverse. An enzyme catalysing the *N*-acylation of the N-terminal Cys of lipoproteins has not been found in this class of bacteria. However, lipoproteins of *Staphylococcus* species are *N*-acylated. Moreover, three novel structures of lipidated N-terminal Cys were revealed on mass spectrometric analyses. A possible biosynthetic pathway generating these structures is discussed.

Introduction

Bacterial lipoproteins constitute a class of membrane proteins functioning outside of the cytoplasmic membrane. After lipoprotein precursors have been translocated across the cytoplasmic membrane, the conserved N-terminal Cys residue of lipoproteins is modified with lipids, which anchor the lipoproteins to membranes. The functions of lipoproteins are diverse; they are involved in drug export, substrate import, construction of extracytoplasmic structures, molecular chaperone activity in the extracytoplasm, activation of Toll-like receptor (TLR) of host cells, etc.

Gram staining is an empirical method for differentiating bacterial species into two large groups, Gram-positive and Gram-negative bacteria. It is almost always the first step in the identification of bacterial organisms, particularly in the medical microbiology laboratory. The Gram-reactivity of a bacterium is based on the ability of the cell to retain the purple primary staining agent crystal violet after undergoing Gram staining. This ability is not based on the actual structure of the cell envelope of the bacterium, although there is often a correlation. It was therefore proposed that bacteria should be classified whether they have one (monoderm) or two (diderm) membranes (Gupta, 1998; Desvaux *et al.*, 2009). Gram-negative bacteria are mostly diderms with a cytoplasmic membrane, a thin peptidoglycan layer and an outer membrane that acts as a permeability barrier to small hydrophilic molecules (Fig. 4.1). Typical members are Proteobacteria, e.g. *Escherichia coli*. High GC Gram-positive bacteria are generally diderms while low GC ones are monoderms. Low GC Gram positive bacteria are covered by multiple peptidoglycan layers (Fig. 4.1).

Lipoproteins in monoderm bacteria are only anchored to the outer surface of the cytoplasmic membrane. On the other hand, ones in a typical diderm, *Escherichia coli*, are anchored to the periplasmic surface of both the cytoplasmic (inner) and outer membranes. Lipoproteins anchored to the outer surface of the outer membrane are also known in Gram-negative diderms, although such lipoproteins are few in *E. coli*. The localization of lipoproteins in Gram-positive diderms remains largely unknown.

The pathways generating mature lipoproteins from their precursors have been extensively studied in *E. coli*. It has been established that lipoproteins in Gram-negative diderms are triacylated through three sequential modification reactions (Fig. 4.2). The first enzyme, Lgt, modifies the universally conserved Cys in the so-called 'lipoprotein box or lipobox' with a diacylglycerol moiety derived from phosphatidylglycerol (Sankaran and Wu, 1994) when lipoprotein precursors are translocated from the cytoplasm to the periplasmic side of the cytoplasmic membrane. The lipobox is a consensus sequence, (LVI)-(ASTVI)-(GAS)-C, comprising the C-terminal three residues of the signal peptide and the N-terminal Cys in the mature region (http://www.mrc-lmb.cam.ac.uk/genomes/dolop/). The second enzyme, LspA or signal peptidase II, cleaves the signal peptide of the diacylated lipoprotein precursors, rendering the Cys residue a new N-terminus. The last enzyme, Lnt, catalyses the

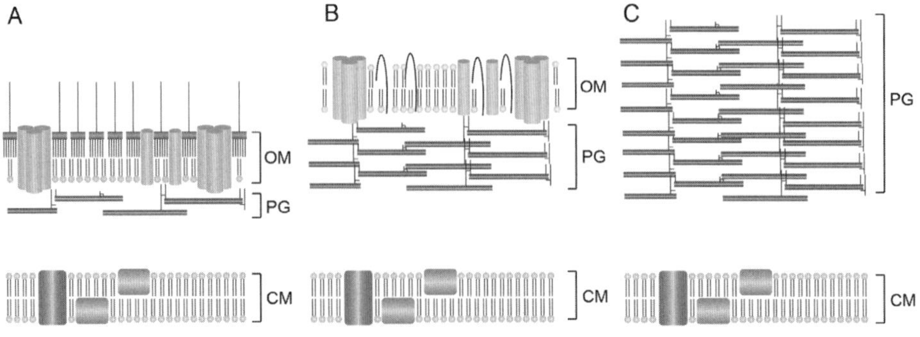

Figure 4.1 Schematic representation of the envelope structures of three bacteria, *E. coli* (A), *M. tuberculosis* (B), and *S. aureus* (C), discussed mainly in this chapter. CM, PG and OM represent the cytoplasmic membrane, peptidoglycan and outer membrane, respectively.

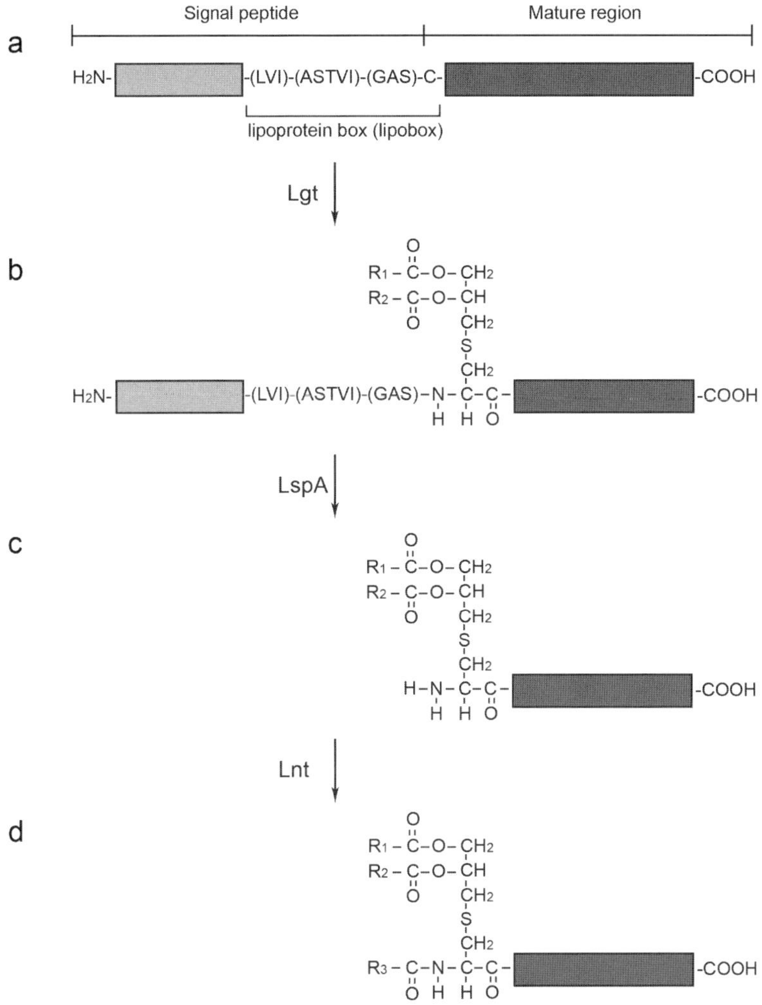

Figure 4.2 Lipoprotein processing in bacteria. The lipoprotein maturation pathway found in Gram-negative diderm is shown. Essentially the same pathway is present in Gram-positive diderms. A lipoprotein precursor [pre-prolipoprotein, (a)] is modified by phosphatidylglycerol/prolipoprotein diacylglyceryl transferase (Lgt), a thioether linkage being formed between the N-terminal Cys of the mature region and diacylglycerol. The signal sequence of diacylated prolipoprotein (b) is cleaved by prolipoprotein signal peptidase (LspA). The apolipoprotein (c) is N-acylated by phospholipid/apolipoprotein transacylase (Lnt), and becomes a triacylated mature lipoprotein (d). An Lnt homologue is not found in low GC content Gram-positive bacteria, although some of lipoproteins are triacylated.

aminoacylation of the N-terminal Cys. It has been thought that triacylated lipoproteins are limited to Gram-negative bacteria because Lnt homologues have not been found in Gram-positive bacteria. However, Lnt homologues were recently found in a Gram-positive diderm (Tschumi *et al.*, 2009). Moreover, it has been reported that the amino group of Cys are also diversely modified even in Gram-positive monoderms (Kurokawa *et al.*, 2009; Asanuma *et al.*, 2011; Kurokawa *et al.*, 2012a; Nakayama *et al.*, 2012).

The aims of this chapter are to discuss, based on recent findings, the mode of N-terminal modification, the enzymes involved in lipoprotein processing, the physiological functions of lipoproteins, and the sorting of lipoproteins in both monoderms and diderms. The Lyme disease spirochaete *Borrelia burgdorferi* is a Gram-negative diderm that possesses many lipoproteins (Fraser *et al.*, 1997), but has no lipopolysaccharides (LPS). Readers interested in spirochaete lipoproteins are referred to reports by Haake (2000) and Schulze *et al.* (2010).

Lipoproteins in Gram-negative diderm bacteria

The outer membrane is an essential organelle for most Gram-negative diderms, and comprises phospholipids, lipopolysaccharides (LPS), membrane spanning β-barrel proteins, and lipoproteins. These components are transported from the cytoplasmic to the outer membrane through the periplasm. Recent studies revealed that many lipoproteins are involved in the transport of and insertion of these components into the outer membrane. Among these four components, the sorting of lipoprotein to the outer membrane has been most extensively studied and clarified.

Lipoprotein-processing enzymes

Three well-conserved enzymes sequentially mediate the maturation reactions (Fig. 4.2) (Sankaran and Wu, 1994):

1. Phosphatidylglycerol/prolipoprotein diacylglyceryl transferase (Lgt) catalyses the transfer of the diacylglyceryl moiety from phosphatidylglycerol (PG) in the cytoplasmic membrane to the N-terminal Cys residue in the mature region through a thioether linkage.
2. Lipoprotein-specific signal peptidase (LspA or signal peptidase II) then cleaves the hydrophobic signal peptide.
3. The third enzyme, phospholipid/apolipoprotein transacylase (Lnt), attaches an acyl chain to the free amino group of the N-terminal Cys residue.

These three enzymes are essential for the growth of *E. coli* and *Salmonella enterica* serovar Typhimurium (Wu, 1966). The N-acylation of lipoproteins by Lnt has been shown to be critically important for their sorting *in vitro* (Fukuda *et al.*, 2002). An *E. coli* mutant, in which the chromosomal *lnt* gene was placed under an arabinose promoter, revealed *in vivo* that apolipoproteins were accumulated in the cytoplasmic membrane in the absence of arabinose (Robichon *et al.*, 2005). Lnt was found to have at least six transmembrane helices with a nitrilase-like periplasmic domain containing the catalytic triad Glu267–Lys335–Cys387 (Robichon *et al.*, 2005; Vidal-Ingigliardi *et al.*, 2007). It was proposed for the mechanism of the third lipid attachment that Lnt attacks the *sn*-1 acyl chain of a phospholipid and thereby forms a thioester acyl-enzyme intermediate, followed by the transfer of the acyl group to the α-amino group of the diacylglyceryl Cys (Buddelmeijer and Young, 2010). Recently, the *N*-acyltransferase activity of *E. coli* Lnt was found to strongly depend on phospholipid head group and acyl chain composition (Hillmann *et al.*, 2011). Lnt prefers phospholipids carrying a small head group, such as phosphatidylethanolamine (PE) or PG, as an acyl donor, and exhibits significant activity only when an unbranched acyl chain is present at position *sn*-1 with an unsaturated one at position *sn*-2 (Hillmann *et al.*, 2011). Finally, the *lnt* gene was found to be dispensable when an ABC transporter, LolCDE complex, is overproduced in *E. coli* cells lacking the major outer membrane lipoprotein Lpp (Narita and Tokuda, 2011).

The Lol system

In Gram-negative bacteria such as *E. coli*, triacylated mature lipoproteins are transported to the outer membrane by the Lol (localization of lipoproteins) system or are retained in the cytoplasmic membrane, depending on the residue at position 2, which is located adjacent to the triacylated N-terminal Cys. The Lol system is composed of five proteins, periplasmic carrier protein LolA, outer membrane lipoprotein receptor LolB, and cytoplasmic membrane ABC transporter LolCDE (Tokuda and Matsuyama, 2004). These five Lol proteins are conserved in Gram-negative bacteria, and are essential for the growth of *E. coli* and *S. enterica*, as the three processing enzymes are. Membrane subunits LolC and LolE of the LolCDE complex are only conserved in diderm bacteria. The conservation of LolB is much lower, and seems to be limited to β- and γ-proteobacteria among diderms. It is interesting that the proteins involved in the early step of lipoprotein processing or sorting are more strongly conserved than those involved in a later step (Okuda and Tokuda, 2011). The diderm bacteria, in which LolB is not found, might have some other outer membrane proteins or lipoproteins that can compensate for the LolB function.

Periplasmic carrier protein LolA

The finding of a lipoprotein-specific carrier protein, LolA, revealed how hydrophobic lipoproteins cross the hydrophilic periplasmic space and how the membrane specificity of lipoproteins is determined (Matsuyama *et al.*, 1995). Spheroplasts secrete various periplasmic and hydrophobic β-barrel outer membrane proteins into the medium, whereas outer membrane-specific lipoproteins, such as Lpp, are retained in the cytoplasmic membrane of spheroplasts. However, Lpp is secreted into the spheroplast medium when the periplasmic fraction is added externally. The Lpp-releasing activity in the periplasmic fraction has been purified, and a water-soluble protein named LolA has been identified. LolA also releases other outer membrane-specific lipoproteins such as Pal, BamC, Slp and RlpA, whereas cytoplasmic membrane-specific lipoproteins AcrA and NlpA remain in spheroplasts even when LolA is added. These observations suggest that the localization of lipoproteins in the outer membrane is determined by whether or not they are released from the cytoplasmic membrane. Lipoproteins released from spheroplasts were found to form a water-soluble complex with LolA in a stoichiometry of 1:1. Moreover, when LolA was genetically depleted, outer membrane-specific lipoproteins were accumulated in the cytoplasmic membrane (Tajima *et al.*, 1998). LolA was thus found to play essential roles in the release of lipoproteins from the cytoplasmic membrane and their transport through the hydrophilic periplasm.

Outer membrane receptor LolB

The LolA–lipoprotein complex described above has been isolated from a spheroplast supernatant, and then incubated with cytoplasmic and outer membranes. It was found that lipoproteins were specifically transferred to the outer membrane, but not to the cytoplasmic membrane (Matsuyama *et al.*, 1995). This observation suggested that an unidentified factor in the outer membrane accepts lipoproteins from LolA and then mediates their outer membrane localization. Solubilized outer membrane proteins have been fractionated and then reconstituted into proteoliposomes to identify this factor. The outer membrane lipoprotein named LolB was thus identified as a lipoprotein receptor anchored to the outer membrane (Matsuyama *et al.*, 1997). LolB depletion caused the accumulation of outer membrane-specific lipoproteins both in the periplasm as a complex with LolA and in the cytoplasmic membrane (Tanaka *et al.*, 2001). A LolB derivative, mLolB, lacking an N-terminal acyl chain is able to form a complex with lipoproteins as LolA does, indicating that LolA and LolB are functionally similar although their amino acid sequences are dissimilar. The modes of lipoprotein binding to LolA and LolB, and the importance of the lipid anchor of LolB are discussed in a later section.

Cytoplasmic membrane ABC transporter LolCDE

The outer membrane is the final destination of lipoproteins, and they are never released into the periplasm even in the presence of LolA. The release of outer membrane-specific lipoproteins from right-side-out membrane vesicles occurs in ATP- and LolA-dependent manners (Yakushi *et al.*, 1998). These observations indicate that there is an ATPase that catalyses the release of lipoproteins exclusively in the cytoplasmic membrane. Cytoplasmic membrane proteins have been solubilized, fractionated and reconstituted into liposomes together with outer membrane-specific lipoproteins to examine the lipoprotein-releasing activity of each fraction in the presence of ATP and LolA. The LolCDE complex, which belongs to the ABC (ATP-binding cassette) transporter super family, was thus identified (Yakushi *et al.*, 2000). This complex consists of LolC, LolD, and LolE subunits in a stoichiometry of 1:2:1. LolC and LolE are membrane subunits, each of which has four membrane-spanning regions and one large periplasmic loop (Yasuda *et al.*, 2009). LolD is a nucleotide-binding subunit with conserved ABC signature, Walker A, and Walker B motifs.

ABC transporters generally mediate the ATP-dependent transport of diverse substrates across the lipid bilayers of various organisms. They are composed of four domains; two transmembrane domains (TMDs), and two nucleotide-binding domains (NBDs). ABC transporters in eukaryotic cells mainly function as drug exporters, of which the TMDs and NBDs are expressed in a single polypeptide chain. On the other hand, most bacterial ABC transporters are involved in the uptake of nutrients with a specific substrate-binding protein, and have four domains frequently in separate polypeptide chains (Davidson and Chen, 2004; Dawson *et al.*, 2007). Some bacterial ABC transporters mediate the export of various drugs, inhibitors, polysaccharides and proteins.

LolCDE, having eight membrane-spanning regions in total, is assumed to be a derivative of a bacterial exporter, whereas exporters generally have at least twelve membrane-spanning regions. MacB, another bacterial ABC transporter, also has eight transmembrane regions in total, and exports enterotoxin II from the periplasm to the exterior in collaboration with a membrane fusion protein, MacA, and a multifunctional outer membrane channel, TolC (Kobayashi *et al.*, 2003; Yamanaka *et al.*, 2008). MacB is predicted to have a large periplasmic loop like LolC and LolE have. Both LolCDE and MacB release substrates after they have been translocated across the cytoplasmic membrane. Their characteristic structures might be related to their unique functions.

Sorting of lipoproteins to the outer membrane of Gram-negative bacteria

The molecular events occurring during the transport of lipoproteins from the cytoplasmic to the outer membrane through the periplasm have been extensively studied in *E. coli* (Fig. 4.3).

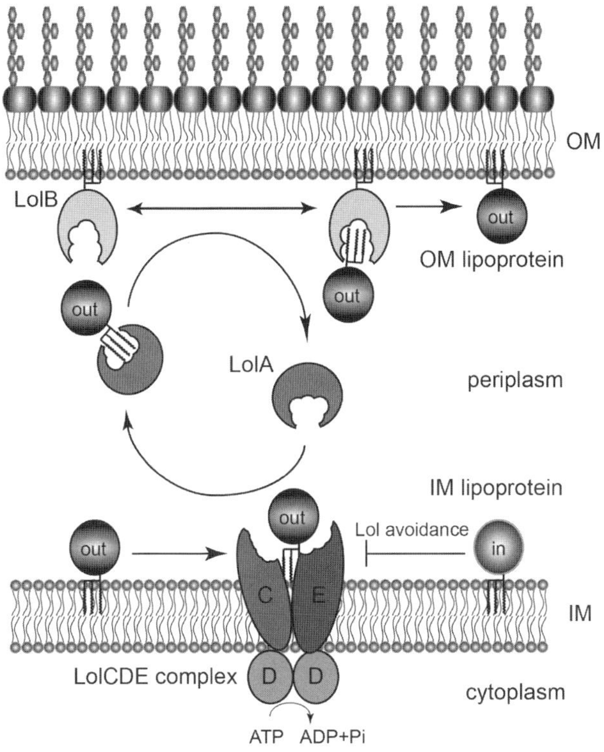

Figure 4.3 The Lol pathway. An ABC transporter, the LolCDE complex, recognizes an outer membrane-specific mature lipoprotein in the cytoplasmic membrane. The Asp residue at position 2 functions as an 'Lol avoidance signal' and prevents the recognition of lipoproteins by LolCDE. This causes the retention of lipoproteins in the cytoplasmic membrane. The released lipoprotein forms a complex with a periplasmic chaperone, LolA, in an ATP-dependent manner. The LolA–lipoprotein complex crosses the periplasm and reaches the outer membrane, where the lipoprotein is transferred from LolA to LolB. LolB is itself a lipoprotein anchored to the outer membrane and mediates the incorporation of lipoproteins into the periplasmic leaflet of the outer membrane.

Outer membrane-specific lipoproteins are recognized by LolCDE in the cytoplasmic membrane and then transferred to LolA in an ATP-dependent manner. Lipoproteins cross the periplasmic space as a complex with LolA and are transferred to LolB anchored to the outer membrane, followed by incorporation of the lipid moiety of the lipoproteins into the outer membrane. In this section, the mechanisms underlying the lipoprotein transfer reactions through five Lol proteins are discussed.

Sorting signals of lipoproteins

Yamaguchi *et al.* (Yamaguchi *et al.*, 1988) first indicated the importance of Asp at position 2 for the cytoplasmic membrane localization of lipoproteins in *E. coli* by demonstrating that the substitution of Ser with Asp at position 2 caused the retention of an outer membrane-specific lipoprotein in the cytoplasmic membrane. Moreover, substitution of Asp at position 2 of a cytoplasmic membrane-specific lipoprotein with another residue was shown to cause the outer membrane-localization of the lipoprotein. It was reported later that the residue at position 3 variously affects the Asp-dependent cytoplasmic membrane localization of lipoproteins (Gennity and Inouye, 1991). To examine this '+2 rule', the residues at positions 2 and 3 were systematically changed and the LolA-dependent release of lipoproteins from the cytoplasmic membrane was examined. It was found that only Asp at position 2 caused the retention of a lipoprotein in the cytoplasmic membrane when the residue at position 3 was Ser (Terada *et al.*, 2001). The residue at position 3 was then found to variously affect the cytoplasmic membrane retention of lipoproteins by Asp at position 2. Asp at position 2 was the strongest cytoplasmic membrane signal when Asp, Glu or Gln was at position 3 (Terada *et al.*, 2001). The combination of Asp and Asn or Asp and Arg at positions 2 and 3, respectively, also functioned as a relatively strong cytoplasmic membrane signal. Native cytoplasmic membrane-specific lipoproteins of *E. coli* have Asp at position 2 and Asp, Glu or Gln at position 3. Taken together, these results suggested that the negative charge or amide group at position 3 increases the potency of Asp at position 2 as a cytoplasmic membrane retention signal. Asn at position 2 exceptionally functions as a cytoplasmic membrane retention signal only when the residue at position 3 is Asp, as in the case of *E. coli* native lipoprotein AcrE (Klein *et al.*, 1991; Seiffer *et al.*, 1993). It remains to be elucidated why Lys or His with an amide group at position 3 decreases the retention of cytoplasmic membrane-specific lipoproteins (Terada *et al.*, 2001).

The bacterial genome carries a number of putative lipoprotein genes (Juncker *et al.*, 2003; Babu *et al.*, 2006). In the family Enterobacteriaceae, such as *S. enterica* serovar Typhimurium, *Shigella flexneri, Yersinia pseudotuberculosis, Erwinia carotovora,* and *Klebsiella oxytoca,* the '+2 rule' seems to be conserved, as judged on the direct visualization of fluorescence-labelled lipoproteins *in vivo* (Lewenza *et al.*, 2006). However, the sorting signal may vary in other bacteria even though they possess Lol protein homologues. For example, the residue at position 2 of MexA, a cytoplasmic membrane lipoprotein constituting the multidrug efflux pump of *Pseudomonas aeruginosa,* is Gly. Chimeric lipoproteins consisting of various regions of MexA and outer membrane-specific lipoprotein OprM revealed that Lys and Ser at positions 3 and 4, respectively, but not Gly at position 2, are the typical cytoplasmic membrane signals of *P. aeruginosa.* Although Asp at position 2 also caused the cytoplasmic membrane retention of lipoproteins, the residues at positions 3 and 4 were found to be the innate sorting signals for lipoproteins in *P. aeruginosa* (Narita and Tokuda, 2007). Indeed, the Lys-Ser signal functioned as the cytoplasmic membrane retention signal in proteoliposomes reconstituted with a LolCDE homologue purified from *P. aeruginosa,* while reconstitution of *E. coli* LolCDE caused the release of this lipoprotein. It should be noted that LolA purified from not only *E. coli* but also *P. aeruginosa* functions in this release from proteoliposomes (Tanaka *et al.*, 2007). Asp at position 2 functions as a cytoplasmic membrane retention signal whether LolCDE is derived from *E. coli* or *P. aeruginosa.* These findings suggest that the mechanisms underlying the Lol-dependent localization of lipoproteins are similar in *E. coli* and *P. aeruginosa,* although the cytoplasmic membrane retention signal of lipoproteins depends on the properties of LolCDE. The lipoprotein-sorting signals in Gram-negative bacteria seem to be more diverse than previously expected.

It was once thought to be possible that Asp at position 2 inhibits N-acylation and therefore causes retention of lipoproteins in the cytoplasmic membrane. However, this was found not to be the case, and cytoplasmic membrane-specific lipoproteins with Asp at position 2 were shown to be N-acylated (Fukuda *et al.*, 2002). It was also revealed that non-*N*-acylated lipoproteins (apolipoproteins) are hardly released from the cytoplasmic membrane, indicating that LolCDE recognizes triacylated lipoproteins. As mentioned above, the *lnt* gene can be deleted when cells overproduce LolCDE and lack either Lpp or the genes encoding the three transpeptidases, which form cross-links between Lpp and peptidoglycan (Narita and Tokuda, 2011). The diacylated lipoproteins in this *Δlnt* mutant were poor substrates for LolCDE, therefore overproduction of LolCDE was required for their correct sorting.

The Lol avoidance signal

The reason why Asp at position 2 causes the retention of lipoproteins in the cytoplasmic membrane has been studied by means of reconstitution experiments. Pal(S2D), a derivative of the outer membrane-specific lipoprotein Pal possessing Asp in place of Ser at position 2, was not released from the proteoliposomes reconstituted with LolCDE in the presence of LolA and ATP (Yakushi *et al.*, 2000). The ATPase activity of LolCDE was stimulated by Pal, but not by Pal(S2D) (Masuda *et al.*, 2002). Moreover, the release of an outer membrane-specific lipoprotein from proteoliposomes was completely inhibited by an excess amount of Pal in the same proteoliposomes, whereas an excess amount of Pal(S2D) had no effect (Masuda *et al.*, 2002). These observations suggested that Asp at position 2 functions as a 'Lol avoidance signal' and thereby prevents the recognition of lipoproteins by LolCDE. Indeed, various LolCDE mutants have been isolated that localize cytoplasmic membrane-specific lipoproteins to the outer membrane (Narita *et al.*, 2003; Sakamoto *et al.*, 2010). These findings indicate that both LolA and LolB recognize lipoproteins even though they have Asp at position 2 when the LolCDE mutants release those lipoproteins. An altered conformation of LolCDE most likely accounts for the suppression of the Lol avoidance function of Asp at position 2.

The mechanism underlying lipoprotein-recognition by LolCDE was further examined with proteoliposomes reconstituted with LolCDE and chemically modified lipoproteins (Hara *et al.*, 2003). A Pal derivative with Cys in place of Ser at position 2 was released from reconstituted proteoliposomes even after the Cys residue had been modified with SH-specific reagents. Since N-acylation is essential for recognition by LolCDE, it seemed likely that LolCDE recognizes the N-terminal Cys residue modified with three acyl chains, the sole common structure of lipoproteins. The oxidation of Cys to cysteic acid resulted in the generation of a Lol avoidance signal. The calculated distances between C*α* and the negative charges of Asp and cysteic acid are very similar. On the other hand, Glu at position 2 does not function as a Lol-avoidance signal although it is negatively charged and has similar properties to Asp. These results suggest that the distance between the negative charge and C*α* at position 2 is critical for the LolCDE avoidance function.

E. coli membranes contain 70–75% PE, 20–25% PG, and about 5% cardiolipin (CL). Examination of the release of lipoproteins from proteoliposomes reconstituted with various phospholipids revealed that Asp at position 2 functioned as the Lol avoidance signal in proteoliposomes reconstituted with phosphatidylcholine (PC) possessing a positive charge. In contrast, the Lol avoidance function of Asp was abolished if the proteoliposomes were reconstituted with *E. coli* phospholipids pretreated with an amine-specific reagent (Hara *et al.*, 2003). These observations suggested that the steric and electrostatic complementarity between Asp at position 2 and phospholipids such as PE having a positive charge is important for the Lol avoidance mechanism. However, an *E. coli* mutant unable to produce PE can grow on a medium supplemented with a high concentration of magnesium (DeChavigny *et al.*, 1991), and the sorting of lipoproteins was found to be normal in the mutant cells

(Miyamoto and Tokuda, 2007), indicating that the positive charge of phospholipids is not essential for the Lol avoidance signal. The phospholipids in this mutant were comprised 50% CL and 50% PG. CL is thought to exhibit a non-bilayer phospholipid property in the presence of a high concentration of magnesium (Rietveld *et al.*, 1995). Both the release of lipoproteins and the ATPase activity of LolCDE were stimulated by magnesium when proteoliposomes were reconstituted with CL alone (Miyamoto and Tokuda, 2007). However, the lipoproteins with Asp at position 2 were also released under these conditions. In contrast, PG added to CL-liposomes increasingly suppressed the release of lipoproteins with Asp and lowered the ATPase activity of LolCDE. Taken together, these findings indicate that phospholipids have diverse effects on lipoprotein sorting in *E. coli*. It remains unknown how PG negatively affects the release of lipoproteins with Asp at position 2.

Isolation of liganded LolCDE

As far as reported, no ABC transporter has been co-purified with its substrate. On the other hand, a liganded form of LolCDE could be purified if ATP was absent during its purification. Various lipoproteins co-purified with LolCDE were all outer membrane-specific, but not cytoplasmic membrane-specific (Ito *et al.*, 2006), indicating that the liganded LolCDE represents an intermediate of the lipoprotein release reaction in the cytoplasmic membrane. LolCDE liganded exclusively with Pal in a molar ratio of 1:1 was then purified. Taking advantage of this novel liganded LolCDE, the molecular events involved in the release of lipoproteins by LolCDE was divided into the following steps:

1. Outer membrane-specific lipoproteins are recognized by LolCDE in an ATP-independent manner unless the lipoprotein has Asp at position 2. The affinity of LolD for ATP increases upon lipoprotein binding.
2. ATP-binding to LolD decreases the strength of the hydrophobic interaction between lipoproteins and LolCDE through a conformational change of LolCDE.
3. Lipoproteins are transferred from LolCDE to LolA upon the hydrolysis of ATP. However, when LolA is not present, lipoproteins remain associated with LolCDE even if ATP is hydrolysed. This step involves a conformational change of LolA, as discussed later.

The crystal structure of the LolCDE complex has not been solved while a methanococcal LolD homologue exhibits a very similar tertiary fold to those of the ATPase subunits of other ABC transporters (Yuan *et al.*, 2001; Smith *et al.*, 2002). Since LolCDE can be assumed to be a variant of ABC exporter, the molecular mechanism of LolCDE might be similar to that proposed for crystallized ABC exporter Sav1866 of *Staphylococcus aureus* (Dawson and Locher, 2006, 2007).

Structures of LolA and LolB

Both LolA and LolB transiently bind lipoproteins and then transfer the bound lipoproteins to LolB and the outer membrane, respectively. Their functions are similar but their amino acid sequences exhibit no similarity. In contrast, the crystal structures of *E. coli* LolA and LolB solved at 1.65 and 1.9 Å resolution, respectively, exhibit remarkable similarity to each other (Takeda *et al.*, 2003). Both LolA and LolB form an incomplete β-barrel structure composed of 11 anti-parallel β-strands with a lid comprising three α-helices (Fig. 4.4). The

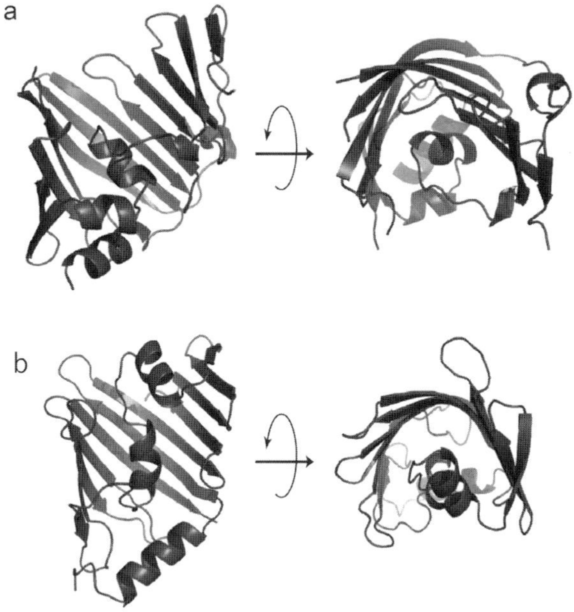

Figure 4.4 Crystal structures of (a) LolA and (b) LolB. The structures of LolA (PDB:1IWL) and LolB (1IWM) are strikingly similar, despite their dissimilar sequences. Both structures form a hydrophobic cavity consisting of 11 anti-parallel β-strand and three α-helices. These hydrophobic cavities bind the acyl chains of lipoproteins. The structures on the left were rotated by 90 degrees around the horizontal axis to give the structures on the right.

incomplete β-barrel and the lid form a hydrophobic cavity, which was shown to bind the acyl chains of lipoproteins, as discussed later. Their overall structures are similar, but two differences are important for their functional differentiation. An extra loop comprising a short helix and a twelfth β-strand at the C-terminus of LolA prevents retrograde transfer of lipoproteins to the cytoplasmic membrane by inhibiting the interaction between LolA and phospholipids (Okuda *et al.*, 2008). The Arg residue at position 43 in the β2-strand of LolA forms hydrogen bonds with some residues in the lid, and therefore closes the cavity, whereas the cavity of LolB is open. The LolA(R43L) mutant, in which Arg at position 43 is replaced by Leu, can accept lipoproteins from LolCDE, but cannot transfer them to LolB (Miyamoto *et al.*, 2001). Therefore, an unusually large amount of the LolA–lipoprotein complex is accumulated in the periplasm (Taniguchi *et al.*, 2005). The hydrophobic interaction of LolA(R43L) with lipoproteins is as strong as that of LolB (Taniguchi *et al.*, 2005). This is why lipoproteins are not transferred from LolA(R43L) to LolB. In contrast, the interaction between wild-type LolA and lipoproteins is the weakest, allowing efficient transfer of lipoproteins to LolB. It is also noteworthy that the hydrophobic cavity of LolA is formed from aromatic residues, whereas that of LolB consists mainly of Leu and Ile, whose hydrophobic side chains are more flexible than those of aromatic residues (Takeda *et al.*, 2003). These differences are critical for the efficient one-way transfer of lipoproteins in an energy-independent manner.

It remains unknown how the three acyl chains are localized in LolA and LolB of *E. coli*, because the hydrophobic cavities of LolA and LolB are only large enough to accommodate a single acyl chain, i.e. not more. The crystal structure of *P. aeruginosa* LolA was recently shown to be almost the same as that of *E. coli* (Remans *et al.*, 2010). Additional hydrophobic patches were found on the surface of *P. aeruginosa* LolA. These patches might be the binding sites for other acyl chains of lipoproteins. However, LolA(R43L) crystals exhibited two structures, open and closed conformations. The hydrophobic cavity of the open conformation is larger (~1700 Å^3) (Y. Oguchi and H. Tokuda, unpublished) than that of the closed form, which is essentially identical to the free form of wild-type LolA (Oguchi *et al.*, 2008). However, the hydrophobic cavity of wild-type LolA was found to undergo opening and closing upon the binding and release of lipoproteins, respectively (Oguchi *et al.*, 2008; Watanabe *et al.*, 2008). LprG, a lipoprotein of Gram-positive diderm *M. tuberculosis*, has a very similar structure to that of LolA. The hydrophobic cavity of LprG is large enough to accommodate three acyl chains (Drage *et al.*, 2010). Since its size was reported to be ~1500 Å^3, the open conformation of LolA seems to be able to accommodate all three acyl chains inside the hydrophobic cavity.

In addition to that of LprG, the structures of the N-terminal domains of RseB, LppX and VioE are similar to those of LolA and LolB. RseB is a member of the envelope stress response system leading to the induction of σ^E expression. RseB is proposed to bind an unfolded lipoprotein in the hydrophobic cavity (Kim *et al.*, 2007). LppX is a lipoprotein of *M. tuberculosis* required for the translocation of complex lipids, the phthiocerol dimycocerosates (DIM), to the outer layer of the cell. Its large hydrophobic cavity was thought to be sufficient to accommodate a single DIM molecule (Sulzenbacher *et al.*, 2006). These proteins seem to have similar function to LolA and LolB although the transported substrates differ. VioE is speculated to play a key role in the biosynthesis of violacein, a purple pigment with antibacterial and cytotoxic properties (Hirano *et al.*, 2008; Ryan *et al.*, 2008).

A soluble LolB derivative, mLolB, lacking an N-terminal lipid anchor is functional and able to replace LolB, although a higher amount of mLolB is required for normal growth (Tsukahara *et al.*, 2009). It was then found that mLolB expressed in the periplasm incorporates lipoproteins into not only the outer but also the cytoplasmic membrane or liposomes, indicating that mLolB has a lipid-targeting function, and does not distinguish the cytoplasmic and outer membranes. Therefore, the N-terminal membrane anchor of LolB is important for prevention of the mislocalization of lipoproteins to the cytoplasmic membrane. The Leu residue at position 68 is expected to play an important role in phospholipid targeting (Takeda *et al.*, 2003). Among the three major phospholipids, non-bilayer phospholipid PE is important for the LolB-dependent incorporation of lipoproteins (Tsukahara *et al.*, 2009). However, it remains largely unknown how mLolB discharges a cargo on the lipid surface.

How Lol proteins interact with each other

In order to elucidate how lipoproteins are transferred through Lol proteins, a photo-cross-linking technique developed by Schultz and his collaborators was applied to the Lol system (Okuda and Tokuda, 2009). This technique enables the introduction of an unnatural photoreactive amino acid, *p*-benzoyl-phenylalanine (pBPA), into an amber (TAG) codon *in vivo* (Chin *et al.*, 2002; Ryu and Schultz, 2006; Wang *et al.*, 2006). A number of LolA or LolB mutants having an amber codon at desired positions were constructed. *E. coli* cells expressing such a mutant were irradiated with UV light for *in vivo* photo-cross-linking. Analyses of the cross-linked products revealed that LolA and LolB interact with each other at the entrances of their hydrophobic cavities. Moreover, the inside of the LolA cavity was found to interact with the outside of the LolB cavity. Lipoproteins were found to interact exclusively with the insides of the LolA and LolB cavities. NMR analyses revealed essentially the same mode of interaction between LolA and LolB (Nakada *et al.*, 2009). Taking these results together, it was proposed that lipoproteins are transferred from the cavity of LolA to that of LolB in a mouth-to-mouth fashion. In this model, the hydrophobic cavities of LolA and LolB are very close to each other, and therefore the transfer of lipoproteins occur smoothly in the direction of higher affinity, namely from LolA to LolB.

In vivo photo-cross-linking revealed that LolA interacts with LolC, but not with LolE, at the entrance of its cavity (Okuda and Tokuda, 2009). When pBPA was incorporated into lipoproteins, cross-linked products were obtained with LolE, but not LolC (unpublished data). From these results, it is speculated that LolC functions as a scaffold for LolA whereas LolE recognizes outer membrane-specific lipoproteins. LolC and LolE have similar topologies i.e. four membrane-spanning regions with a large loop exposed to the periplasm (Narita *et al.*, 2002; Yasuda *et al.*, 2009). Both proteins are essential for the growth of *E. coli*. These results indicate that LolC and LolE are functionally different, despite their structural and sequence similarities (26% identical). The periplasmic loops of LolC and LolE exhibit sequence similarity to LolA and/or LolB to some extent (~18% identical). It seems possible that these regions also have hydrophobic cavities. The transport of lipoproteins might be performed from LolCDE to LolA to LolB through the entrances, or mouths, of their hydrophobic cavities. The gene for LolE is only conserved in γ-proteobacteria i.e. not in other subdivisions (Narita, 2011). In these bacteria, the ABC transporters responsible for lipoprotein release are likely to comprise a LolCD homodimer, in which LolC can bind lipoproteins and interact with LolA.

The LolA–LolC interaction increases upon the binding of lipoproteins to LolCDE (Okuda and Tokuda, 2009). Interaction between the LolA–lipoprotein complex and LolB is not inhibited by an excess amount of free LolA (Watanabe *et al.*, 2007). The affinity of LolD for ATP increases upon lipoprotein binding (Ito *et al.*, 2006). Taking all these results together, the Lol system seems to have been designed to achieve the efficient one-way transfer of lipoproteins from the cytoplasmic to the outer membrane.

Lipoproteins involved in outer membrane biogenesis

Four major outer membrane components, phospholipids, LPS, β-barrel proteins, and lipoproteins, are hydrophobic and must be transported from the cytoplasmic membrane to the outer membrane via the hydrophilic periplasm. Then, the transported components must be correctly localized to the outer membrane. Since there is no energy source, such as ATP, in the periplasm, most of the transport reactions must take place in energy-independent manners. In addition to the Lol system described above, the machineries responsible for the transport and assembly of LPS and β-barrel proteins have been identified (Voulhoux *et al.*, 2003; Bos *et al.*, 2004; Wu *et al.*, 2005, 2006; Sklar *et al.*, 2007; Sperandeo *et al.*, 2007; Ruiz *et al.*, 2008). Various lipoproteins were found to play important roles in these transport-assembly reactions. On the other hand, little is known about how phospholipids are transported to the outer membrane although a mechanism underlying the transport of phospholipids from the outer to the cytoplasmic membrane was reported (Malinverni and Silhavy, 2009).

LPS is flipped from the cytoplasmic side to the periplasmic side of the inner membrane by an ABC transporter, MsbA, and then transported to the outer leaflet of the outer membrane by the Lpt (Lipopolysaccharide transport) system. This system is composed of the cytoplasmic membrane ABC transporter LptBFG complex, bitopic membrane protein LptC, which forms a complex with LptBFG, periplasmic protein LptA, and the complex comprising β-barrel protein LptD and lipoprotein LptE (Sperandeo *et al.*, 2009). All factors comprising the Lpt system and MsbA are essential for the growth of *E. coli*. Although LPS is essential for most Gram-negative bacteria, *Neisseria meningitidis* cells grow in the absence of LPS biogenesis, and therefore mutants lacking any one of above factors are viable. Moreover, the deletion of *lptE* does not inhibit the transport of LPS to the cell surface, indicating that the function of LptE is indirect in this Gram-negative bacterium (Bos and Tommassen, 2011). On the other hand, *E. coli* LptE was reported to bind to LPS strongly (Chng *et al.*, 2010). LptE has a chaperone-like function in the LptD–LptE complexes of both *E. coli* and *N. meningitidis*. The mechanism of the Lpt system might not be the same among Gram-negative diderms.

The Bam (β-barrel assembly machinery) complex is required for the outer membrane assembly of β-barrel proteins in *E. coli* (Hagan *et al.*, 2011). In addition, the periplasmic chaperone DegP, SurA and Skp are also involved in the targeting and assembly of β-barrel proteins. The Bam complex consists of one β-barrel protein, BamA, and four outer membrane-specific lipoproteins, BamB/C/D/E. A similar complex has been identified in *N. meningitidis*, although an outer membrane protein, RmpM, constitutes the complex instead of BamB (Volokhina *et al.*, 2009). Many components of the Bam complex are conserved in various bacteria (Gatsos *et al.*, 2008; Anwari *et al.*, 2010). BamA, a central component of the complex, is conserved in all Gram-negative bacteria. Among the four lipoproteins, only BamD is essential for the growth of *E. coli*, suggesting that BamD plays an important and direct role in the β-barrel protein assembly, while the functions of the respective Bam proteins remain to be clarified. The reconstitution of the Bam complex in proteoliposomes will help to solve these issues (Hagan *et al.*, 2010; Hagan and Kahne, 2011). The BamA orthologue is essential for the assembly of β-barrel proteins in the outer membrane of mitochondria (Gentle *et al.*, 2004; Voulhoux and Tommassen, 2004). Although some accessory components of the complexes differ between bacteria and mitochondria, which have no lipoproteins, the system for β-barrel protein assembly seems to have been evolutionary conserved.

As already mentioned, the Lol system mediates the sorting of lipoproteins to the outer membrane of *E. coli*, in which more than 90 species of lipoproteins are expressed (Tokuda, 2009). Most lipoproteins are predicted to be localized on the periplasmic side of the outer membrane. Although many lipoproteins have unknown functions, three essential lipoproteins, LolB, LptE and BamD, are involved in outer membrane biogenesis. Therefore, chemicals that inhibit the Lol system are expected to block the outer membrane biogenesis. Indeed, LolA has been shown to be a promising target of drugs (Pathania *et al.*, 2009).

Structural importance of lipoproteins

Bacterial lipoproteins are involved in a wide variety of cellular functions, such as the biogenesis and maintenance of cell surface structures, the transport of substrates, and drug efflux (Bernadac *et al.*, 1998; Clavel *et al.*, 1998; Ehrmann *et al.*, 1998; Nikaido, 1998). As discussed in later sections, some lipoproteins cause an inflammatory response in human host cells, and their lipid moieties account for the bioactivity (Scragg *et al.*, 2000; Ramesh *et al.*, 2003). Some lipoproteins have been shown to be functional without a lipid anchor (Tsukahara *et al.*, 2009; Chng *et al.*, 2010), although the activity of LolB decreases with a lack of the lipid anchor because it prevents the mislocalization of lipoproteins. The cytoplasmic membrane proteins have hydrophobic α-helical stretches, which cause the retention of the proteins in the cytoplasmic membrane. Proteins spanning the outer membrane have amphipathic β-strands, which do not cause the retention of the proteins in the cytoplasmic membrane. Posttranslational modification of proteins with a lipid anchor might have been evolved to generate proteins that can reach and associate with the outer membrane. The β-barrel proteins frequently form pores in the outer membrane, while many enzymatic functions are most likely due to lipoproteins and peripheral proteins.

It is known that perturbation of outer membrane biogenesis causes stress to *E. coli* cells and induces stress response systems. A LolA derivative, I93C/F140C, has two Cys residues in place of Ile at position 93 and Phe at position 140. Expression of this mutant in the absence of a reducing agent is lethal to *E. coli* because the hydrophobic cavity is closed by a disulfide bond, which activates the Cpx two-component system (Tao *et al.*, 2010). The Cpx system monitors the biogenesis of cell surface structures and plays an important role in the detection of misassembly of β-barrel proteins (Raivio and Silhavy, 2001; Gerken *et al.*, 2010). Since the Cpx pathway is generally induced by aberrant disulfide formation in the periplasm, the expression of LolA(I93C/F140C) might directly activate the pathway. Alternatively, inhibition of the localization of essential lipoproteins by dominant negative LolA(I93C/F140C) causes activation of the Cpx system. The overexpression of YafY, a cytoplasmic membrane-specific lipoprotein, whose function is unknown, induces DegP expression through activation of the Cpx pathway, as the outer membrane lipoprotein NlpE does (Miyadai *et al.*, 2004). Since delipidated NlpE does not induce the expression of DegP, overproduction of NlpE may cause its mislocalization to the cytoplasmic membrane, and then activation of the Cpx system. Rcs phosphorelay is one of the envelope stress response systems in *E. coli*. The outer membrane-specific lipoprotein RcsF, which transmits the signal from the cell surface to the cytoplasmic membrane histidine kinase RcsC, activates the Rcs phosphorelay system when overexpressed. The observation that the overexpression of LolA activates the Rcs pathway (Chen *et al.*, 2001) also suggests that the Rcs system might monitor correct lipoprotein sorting. In any event, involvement in the envelope stress response systems is one of the functions of lipoproteins.

Lipoproteins in Gram-positive diderm bacteria

Bacteria that exhibit Gram-positive staining belong to one of two divisions, *Actinobacteria* and *Firmicutes*. Formerly, these divisions were differentiated according to their G+C contents, i.e. high and low G+C Gram-positive bacteria, respectively. *Staphylococcus aureus*, *Micrococcus luteus* and *Bacillus anthracis* are typical members of the low G+C order *Firmicutes*. The cell envelope of these bacteria is characterized by a cytoplasmic membrane with a thick peptidoglycan layer and secondary glycopolymers, such as wall teichoic acid and lipoteichoic acid. Owing to the presence of a single membrane, these bacteria are classified as monoderm bacteria (Fig. 4.1c).

Within the division *Actinobacteria* and therein the order *Actinomycetales*, the genera *Mycobacterium*, *Corynebacterium*, *Rhodococcus* and *Nocardia* are grouped into the suborder *Corynebacterineae*. These genera share a common peptidoglycan structure and, in addition, are characterized by the presence of an outer membrane-like impermeable layer. Owing to the presence of a second membrane, these bacteria are considered to be diderm bacteria

(Fig. 4.1b). Thus, from a general point of view, the cell envelopes of *Corynebacterineae* and *Proteobacteria* (Gram-negative bacteria) are similar. However, their structures are analogous rather than homologous. Major differences exist with respect to the chemical composition and, as a result of this with respect to permeability. For example, the lipid content of the cell envelope of mycobacteria may represent up to 40% of the cellular dry mass, compared to only 10% in Gram-negative bacteria (Goren and Brennan, 1979). The outer membrane of the Gram-negative bacterium *Pseudomonas aeruginosa* is notably impermeable. However, the mycobacterial outer membrane exhibits 10- to 100-fold lower permeability than that of *P. aeruginosa* (Jarlier and Nikaido, 1990).

Mycobacterium tuberculosis – cell envelope composition and drug therapy

Within the subfamily *Corynebacterineae,* lipoprotein function and biosynthesis have mainly been investigated in *Mycobacterium tuberculosis* and the model organism *Mycobacterium smegmatis,* respectively. *M. tuberculosis,* the causative agent of human tuberculosis annually causes 1.7 million deaths and about one-third of the world's population is latently infected (http://www.who.int/mediacentre/factsheets/fs104/en/index.html). *Mycobacterium bovis* bacille Calmette–Guerin (BCG), an attenuated derivative of *Mycobacterium bovis* (the causative agent of bovine tuberculosis), is applied as a tuberculosis live vaccine in high incidence countries. It efficiently protects against disseminated tuberculosis in childhood but only shows limited protection against adult lung tuberculosis (Kaufmann *et al.*, 2010). *M. tuberculosis* is an intracellular pathogen that survives and multiplies in macrophages (Russell, 2011). The biology of *M. tuberculosis* is intimately linked to its complex cell envelope. The structure, function and synthesis of the mycobacterial cell envelope have been extensively discussed recently, and readers interested in further details are referred to two excellent books (Cole *et al.*, 2005; Daffe, 2008). The standard course therapy for tuberculosis is a four-drug regimen comprising isoniazid (INH), rifampicin (RIF), pyrazinamide (PZA), and ethambutol (EMB) administered for 2 months, followed by INH and RIF alone for an additional 4 months. A dramatic increase in drug-resistant tuberculosis has been observed in recent years. In the past decade, novel antituberculosis drug candidates have been developed intensively (Cole and Riccardi, 2011). However, the tuberculosis drug pipeline is insufficiently filled. Therapy for drug-resistant tuberculosis requires the substitution of first line drugs by second line drugs and extended medication. Second-line drugs include fluoroquinolones [ofloxacin (OFL) and moxifloxacin (MOX)], aminoglycosides [amikacin (AMK) and kanamycin (KAN)], peptide antibiotics [viomycin (VIO) and capreomycin (CAP)] and thioamides [e.g. ethionamide (ETH)], cycloserine (CYC) and *para*-aminosalicylic acid (PAS)]. Two of the first-line drugs (INH and EMB), several second-line drugs (ETH and CYC) as well as drugs under development (Makarov *et al.*, 2009) inhibit the cell envelope biogenesis. INH and ETH are pro-drugs that are activated within bacterial cells by catalase/peroxidase KatG and monooxygenase EthA, respectively (Da Silva and Palomino, 2011). Both INH and ETH form an adduct with nicotine adenine dinucleotide (NAD) and thereby inhibit the enoyl-acyl carrier protein (ACP) reductase component of a dissociable fatty acid synthase (FAS-II), which is involved in the synthesis of mycolic acid precursors. PA824 and OPC67683, two nitroimidazoles currently in phase II clinical trials, also interfere with mycolic acid biosynthesis (Cole and Riccardi, 2011). EMB interferes with arabinan biosynthesis. Most EMB-resistance associated mutations are found in the arabinosyltransferase EmbB. Benzothiazinone BTZ043 and dinitrobenzamide DNB are two promising antibiotic compounds under pre-clinical development. Both compounds inhibit the decaprenylphosphoryl-β-D-ribose 2′ epimerase, encoded by the *dprE1* and *dprE2* genes involved in D-arabinose synthesis. In particular, the enzymes transform decaprenylphosphoryl-D-ribose into decaprenylphosphoryl-D-arabinose, a precursor for the synthesis of mycobacterial cell-envelope polysaccharides arabinogalactan and lipoarabinomannan (LAM) (Cole and Riccardi, 2011). The number of antibiotics targeting cell envelope biogenesis emphasizes the importance of cell envelope biogenesis for *M. tuberculosis* viability and virulence.

The cell envelope structure and biogenesis of mycobacteria have been intensively studied by means of chemical analyses, microscopic techniques and genetic approaches during the past few decades. The structures surrounding the cytoplasm may be subdivided into several compartments, the cytoplasmic membrane, outer membrane-like structure (consisting of the cell wall core and extractable lipids) and the capsule.

The cytoplasmic membrane and peptidoglycan layer

The cytoplasmic membrane is mainly formed by phospholipids and proteins. The polar lipids are composed of hydrophilic head groups and fatty acid chains that usually consist of fatty acid residues of less than 20 carbon atoms. Palmitic acid ($C_{16:0}$), octadecenoic acid ($C_{18:1}$), and 10-methyloctadecanoic acid (tuberculostearic acid) are the major fatty acid constituents. The main phospholipids of the cytoplasmic membrane are phosphatidylinositol mannosides (PIM), PG, PE and CL (Daffe, 2008). The cytoplasmic membrane contains peripheral and integral membrane proteins as well as lipoproteins. It has a thickness of about seven nanometres, and it is supposed that it is structurally and functionally very similar to other bacterial cytoplasmic membranes (Daffe, 2008).

The cell wall core also called mycolyl-arabinogalactan-peptidoglycan (mAGP) complex is composed of peptidoglycan and molecules covalently linked to it such as arabinan and mycolylesters. This sacculus is an essential structure for mycobacteria and provides the bacteria with a formidable protective barrier against xenobiotics.

The mycobacterial peptidoglycan has some features that distinguish it from *E. coli* peptidoglycan. Its glycan chains are composed of alternating units of (β1→4)-linked *N*-acetylglucosamine (GlcNAc) and *N*-glycolylmuramic acid (MurNGlyc), whereas most other bacteria have *N*-acetylmuramic acid. A tetrapeptide (L-alanyl-D-isoglutaminyl-meso-diaminopimelyl-D-alanine) side chain substitutes for the carboxylic acid function of each muramic acid residue. The peptide chains are heavily cross-linked (70–80%) as compared to in *E. coli* (degree of cross-linking: 20–25%) (Mahapatra *et al.*, 2005). Particularly in non-replicating *M. tuberculosis*, the non-classical 3→3 linkages between two diaminopimelyl residues (rather than the classical 4→3 linkages between the terminal alanyl- and diaminopimelyl residues) predominate in the transpeptide network (Gupta *et al.*, 2010). C-6 of some muramic acid residues form phosphodiester bonds with C-1 of α-D-GlcNAc, which in turn is (1→3) linked to an α-L-rhamno-pyranose (Rha*p*) residue. Rha*p* provides the 'linker unit' between peptidoglycan and the galactan of arabinogalactan (Mahapatra *et al.*, 2005). The galactan consists of a linear chain of about 30 units of alternating 1→5 and 1→6 linked β-D-galactofuranose (gal*f*) residues. Two to three of the 1→6 linked galactan residues of the galactan chain are glycosylated at their C5 with an α-D-arabinofuranose (ara*f*) chain via an α 1→5 linkage. A tree-like arabinose structure is formed through the introduction of 1→2 or 1→3 branch points to the linear 1→5 linked stem. Mycolic acids are esterified to two-thirds of the terminal ara*f* (Mahapatra *et al.*, 2005). Peptidoglycan and arabinogalactan are hydrophilic, while mycolic acids are hydrophobic and form part of the mycobacterial outer membrane-like structure.

Outer membrane-like structure

Mycolic acids are long chain fatty acids, both α-branched and β-hydroxylated. In *M. tuberculosis*, they contain 70–90 carbon atoms with various modifications of the acyl chains. According to a recent model, which is a modification of that originally proposed by Minnikin (Minnikin, 1982), mycolic acids form the inner leaflet of an outer membrane-like structure, along with phospholipids (Zuber *et al.*, 2008). The acyl chain of mycolic acids is not straight but folds back to fit the thickness of the outer membrane-like structure of 7–8 nm, as observed on cryo-electron microscopy of vitreous sections (CEMOVIS) (Hoffmann *et al.*, 2008; Zuber *et al.*, 2008).

Besides forming the major part of the cell wall core component, mycolic acids also occur in unbound forms as esters of trehalose or glycerol and therefore are extractable with organic solvents. A huge variety of other complex lipids, such as phenolic glycolipids (trehalose monomycolate and trehalose dimycolate), phthiocerols and sulfolipids, interact with the covalently attached mycolic acids to build the outer leaflet of the outer membrane-like structure. The *M. tuberculosis* envelope lipids also contain phosphatidyl-*myo*-inositol mannosides (PIM) and their multiglycosylated counterparts, lipomannans (LM) and mannosylated lipoarabinomannans (ManLAM). In other Mycobacteria, LAM is modified differently. While PIM are found in the cytoplasmic membrane and also in the outer most layer, the capsule, the locations of LM and LAM remain controversial (Gilleron *et al.*, 2008). Recent fractionation experiments on *M. smegmatis* indicated that LAM and LM are mainly co-localized with mycolic acids (Dhiman *et al.*, 2011). More than 400 lipid structures representing 15 classes of lipids have been identified in whole lipid extracts of *M. tuberculosis* (Sartain *et al.*, 2011). The extraordinary ability of mycobacteria to degrade and synthesize lipids is reflected by the presence of 250 genes in the *M. tuberculosis* genome devoted to lipid metabolism, compared with 50 in *E. coli* (Cole *et al.*, 1998).

The presence of a highly impermeable outer layer explains, at least partially, the natural resistance of mycobacteria to many hydrophilic antibiotics. However, the existence of such a hydrophobic barrier also poses a serious problem with respect to nutrient uptake. Nutrient uptake by mycobacteria and Gram-negative bacteria is facilitated by water-filled pore proteins (porins) present in the outer membrane (Niederweis *et al.*, 2010). Other proteins of the outer membrane include proteins of the PE and PPE families, two highly polymorphic sets of proteins characterized by the presence of Pro-Glu (PE) and Pro-Pro-Glu (PPE) motifs near the N-terminus (Mukhopadhyay and Balaji, 2011), and lipoproteins (Mawuenyega *et al.*, 2005).

The capsule mainly contains polysaccharides composed of a glycogen-like saccharides and D-arabino-D-mannan heteropolysaccharides, little lipids and proteins also being found in culture filtrates (Mahapatra *et al.*, 2005).

Lipoprotein biosynthesis in mycobacteria

The genome of *M. tuberculosis* is of medium size (approximately 4 million basepairs) and encodes roughly 4000 genes. Approximately 2.5% of the open reading frames encode lipoproteins (Sutcliffe and Harrington, 2004). Lipoprotein biosynthesis in mycobacteria follows the general lipoprotein biosynthesis pathway, as described above for *E. coli* (Sutcliffe and Harrington, 2004; Rezwan *et al.*, 2007). Most preprolipoproteins are secreted via the general secretory pathway (*sec*) in a *secA1*- or *secA2*-dependent manner (Gibbons *et al.*, 2007); however, a significant number of preprolipoproteins are predicted to be translocated across the cytoplasmic membrane via the twin arginine transport pathway (*tat*) (McDonough *et al.*, 2008). After export, a conserved motif, the lipobox motif [LVI]-[ASTVI]-[GAS]-C (Babu *et al.*, 2006), is recognized by the membrane-bound preprolipoprotein diacylglycerol transferase (Lgt; Rv1614). A thioether linkage between Cys of the lipobox and diacylglycerol is formed. Attachment of the diacylglycerol anchor triggers lipoprotein signal peptidase (LspA; Rv1539)-dependent cleavage of the immature lipoprotein. This reaction generates a free amino group in the Cys carrying the thioether-linked diacylglycerol. It has long been assumed that mycobacterial lipoproteins are N-acylated. However, direct evidence of *N*-acylation of mycobacterial lipoproteins was only provided recently (Tschumi *et al.*, 2009).

Mycobacterial Lnt

A homologue of *E. coli* apo-lipoprotein *N*-acyltransferase (Lnt) is present in the genome of *M. tuberculosis* and in a wide variety of other GC-rich Gram-positive bacteria, but is absent from the genomes of low GC Gram-positive monoderms (Fig. 4.5). In *M. tuberculosis,* open reading frame Rv2051c is annotated as *ppm1,* for a polyprenol monophosphomannose (Ppm) synthase, which transfers mannose from GDP-mannose to endogenous polyprenol phosphate. Polyprenol phosphate is a metabolic intermediate in the synthesis of mycobacterial cell envelope constituents LM and LAM. The Ppm synthase domain is only encoded by the 3′ part of open reading frame 2051c. The 5′ part of this open reading frame encodes an amino terminal domain that exhibits considerable similarity to that of *E. coli* Lnt. *M. smegmatis,* a fast growing non-pathogenic *Mycobacterium,* which is tractable to genetic manipulation, is often used as a model organism for elucidating mycobacterial metabolic pathways. Heterologous expression of *M. tuberculosis* lipoprotein LppX in *M. smegmatis* and subsequent mass spectrometry analyses provided direct proof of N-acylation of mycobacterial lipoproteins (Tschumi *et al.*, 2009).

In *M. smegmatis,* orthologues of the two domains of *M. tuberculosis* Rv2051c are encoded by two distinct open reading frames, Msppm1 and Msppm2. Of these, Msppm2 (Msmeg3863) corresponds to Lnt. Targeted gene inactivation of Msppm2 results in a mutant that is unable to modify lipoprotein LppX with an *N*-acyl residue. Complementation of the mutant strain with either *M. smegmatis lnt* (Msmeg3863) or *M. tuberculosis* Rv2051c restores the capability of *N*-acylation of LppX (Tschumi *et al.*, 2009). Subsequently, expression and analyses of recombinant *M. tuberculosis* lipoprotein LprF in wild-type *M. smegmatis* and its *Δlnt* strains indicated that *lnt*-dependent *N*-acylation is not restricted to LppX but represents a general modification pathway for mycobacterial lipoproteins (Brülle *et al.*, 2010).

The structures of the membrane anchor of the two recombinant lipoproteins, LppX and LprF, have been resolved at the molecular level. In both cases, the thioether-linked diacylglycerol carries one esterified palmitic acid and one esterified tuberculostearic acid. Whether the tuberculostearic acid is at the *sn*-1 or *sn*-2 position has not been determined yet. The *N*-acyl residue is mainly derived from palmitic acid. Thus, the distribution of fatty

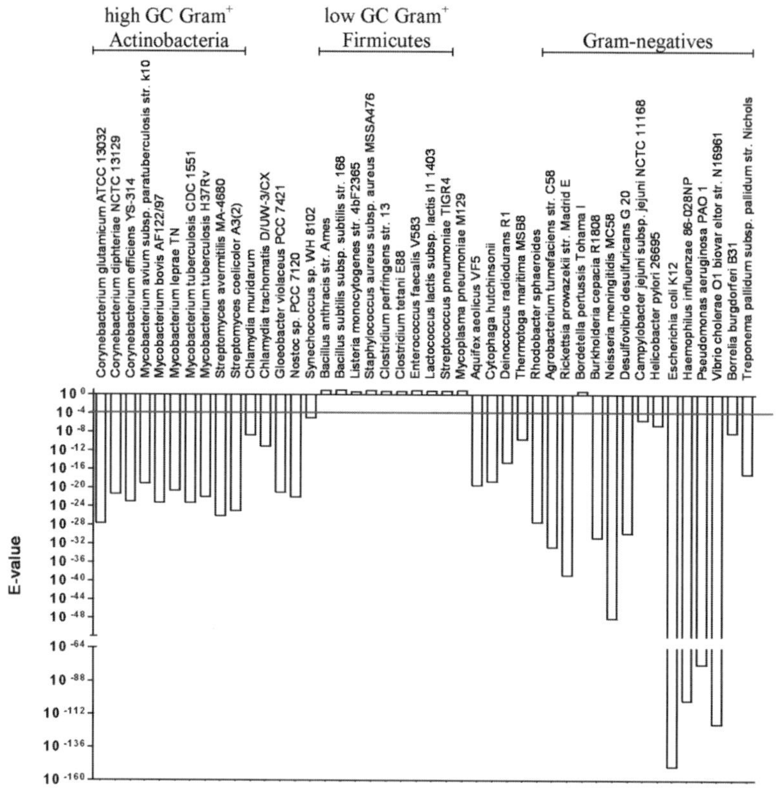

Figure 4.5 Distribution of Lnt in bacteria. *E. coli* Lnt was used as a query to identify homologues on the National Centre of Biotechnology Information BLASTp server (http://www.ncbi.nlm.nih.gov/sutils/genom_table.cgi). The sequence filtering option was switched off, and the expected value was set at 10 and the cut-off value at 10^{-4}. (Figure adapted from Tschumi *et al.*, J. Biol. Chem., 2009.)

acids in mycobacterial lipoproteins reflects the composition of fatty acids in membrane phospholipids.

In the genome of *M. bovis*, a second open reading frame, MB2285c, encoding a protein of 502 amino acids exhibits considerable homology to that of *E. coli* Lnt. Two of the amino acids of the catalytic triad (E267 and K335; *E. coli* numbering) are conserved, while the third residue (C387) is Ser (Vidal-Ingigliardi *et al.*, 2007). Most interestingly, the corresponding open reading frame in *M. tuberculosis* is split into two and the putative catalytic triad lies on two adjacent open reading frames. Of these, Rv2262c encodes a protein of 360 amino acids and Rv2261c encodes a protein of 140 amino acids. It remains to be elucidated whether or not proteins MB2285c and Rv2262c in combination with Rv2261c function as apolipoprotein N-acyl transferases, and whether or not they have the same specificity as Ppm1 (Rv2051c).

Lipoprotein biosynthesis and virulence

Lipoproteins have been shown to play a direct role in the interaction of bacterial pathogens with their hosts (Kovacs-Simon *et al.*, 2011). The virulence-related functions of lipoproteins include colonization, invasion, evasion of host defence, and immunomodulation. Additionally, lipoproteins have essential functions in general bacterial physiology, e.g. in nutrient acquisition or cell envelope biogenesis, and thereby indirectly affect host–pathogen interactions. The lipoprotein biosynthesis pathway is an important virulence factor of *M. tuberculosis*.

An *M. tuberculosis lspA* knock-out mutant exhibited a 3–4 log reduced number of colony forming units in an animal model of tuberculosis. Mice infected with the *lspA* mutant strain hardly show any lung pathology (Fig. 4.6) and the bacteria do not spread to secondary organs (Sander *et al.*, 2004; Rampini *et al.*, 2008). *M. tuberculosis* mutants deficient in *lgt*, the gating enzyme of the lipoprotein biosynthesis pathway, have not been described yet and high density transposon site hybridization mutagenesis (TRASH) and targeted mutagenesis support the view that this gene is essential in *M. tuberculosis* (Sassetti *et al.*, 2003b; Tschumi *et al.*, 2012). Although *lspA* had also been predicted to be essential, such a mutant could be generated (Sander *et al.*, 2004; Banaiee *et al.*, 2006). An *M. tuberculosis ppm1/lnt* mutant was isolated on screening for mutants unable to grow in an acidified medium. Detailed characterization of the mutant remains to be reported (Vandal *et al.*, 2008).

Mycobacteria and Gram-negative bacteria are diderms. Despite their similarities, these bacteria differ with respect to the essentiality of lipoprotein synthesis genes; all genes involved in lipoprotein synthesis and transport are essential for *E. coli* (Okuda and Tokuda, 2011). One reason for the viability of mycobacterial mutants defective in lipoprotein biosynthesis (and other monoderm Gram-positive bacteria) could be that some lipoprotein precursors retain functionality (Hutchings *et al.*, 2009). Globomycin, a cyclic peptide antibiotic produced by *Streptomyces*, inhibits LspA-dependent processing of lipoproteins in both Gram-negative and Gram-positive bacteria (Inukai *et al.*, 1978). A bactericidal effect of globomycin on Gram-negative bacteria but not on Gram-positive bacteria is consistent with the essentiality of LspA in the former bacteria, and a non-essential role in the latter ones. Globomycin exhibits a growth inhibitory effect on *M. tuberculosis*; however, this effect is not due to LspA inhibition since processing of lipoprotein Mpt83 is not affected, and an *M. tuberculosis lspA* mutant is as sensitive to globomycin as the parental strain (Banaiee *et al.*, 2007). Antibacterial activity of globomycin derivatives in monoderm Gram-positive bacteria, in which *lspA* is not essential, supports the view that globomycin may have additional targets (Kiho *et al.*, 2003, 2004).

Glycosylation of mycobacterial lipoproteins

Mycobacterial lipoproteins often contain additional post-translational modifications, besides the membrane anchor. Several *M. tuberculosis* lipoproteins (LpqH, SodC, LppX and LprF) have been shown to be modified with one or more glycosyl residues at their *N*-termini (Herrmann *et al.*, 1996; Sartain and Belisle, 2009; Tschumi *et al.*, 2009; Brülle *et al.*, 2010). O-Glycosylation occurs at Thr and Ser residues, respectively (Sartain and Belisle, 2009). Glycosylation increases protease resistance, while the exact molecular nature of the glycosylation, as well as its function, remain largely undefined.

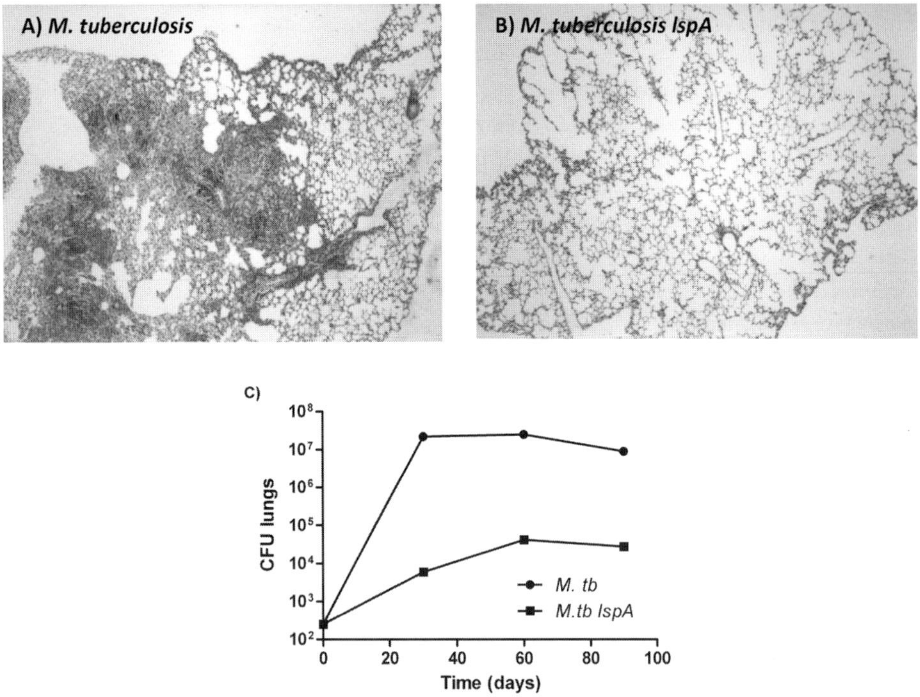

Figure 4.6 Virulence attenuation of *M. tuberculosis* lipoprotein biosynthesis mutant *lspA*. The lung pathology of BALB/c mice infected with *M. tuberculosis* (A) and *M. tuberculosis lspA* mutant (B) at 42 days after aerosol infection is shown, representative haematoxylin/eosin-stained views of lung cross sections (total magnification x 40) being shown. Note the almost complete absence of inflammatory infiltration in the lungs infected with the mutant strain lacking functional LspA. (C) Growth of *M. tuberculosis* and *M. tuberculosis lspA* in the lungs of CBA/J mice over a 90-day period after intranasal infection (figures from Sander *et al.*, 2004).

Transport and localization of lipoproteins in mycobacteria

Functional and structural investigations have suggested the localization of at least some *M. tuberculosis* lipoproteins in the mycobacterial outer membrane: *M. tuberculosis* lipoproteins have been described as adhesins, immune electron microscopy techniques have demonstrated their surface localization, and fractionation experiments have indicated co-localization with the fraction containing mycolic acids. Analyses of cytoplasmic membrane and outer membrane fractions allowed the identification of 21 lipoproteins (Mawuenyega *et al.*, 2005). More recently, 24 lipoproteins were identified in the detergent fraction of cell extracts (Wolfe *et al.*, 2010). The outer membrane transport of lipoproteins in Gram-negative bacteria depends on the localization of the lipoprotein Lol transport system and it has been suggested that sorting pathways for lipoproteins are critically important in all diderm bacteria (Okuda and Tokuda, 2011). The presence of lipoproteins in the outer membrane of mycobacteria suggests the existence of a lipoprotein transport system. However, its components remain to be identified. Mycobacteria release membrane vesicles *in vitro* and during infection. Interestingly, the membrane vesicles of pathogenic mycobacteria are enriched in lipoproteins and contribute to virulence (Prados-Rosales *et al.*, 2011).

Mycobacterial lipoproteins in health and disease

Manipulation of the immune system

Lipoproteins affect both innate and adaptive immunity, and have been identified as major antigens of *M. tuberculosis*. Mycobacterial lipoproteins trigger the activation of humoral and cellular immune responses to mycobacteria. Some lipoproteins (LpqH, PstS1) induce a protective immune response, while others are deleterious as to protection (LprG) (Hovav *et al.*, 2003). Successful immune evasion of *M. tuberculosis* has partly been attributed to Toll-like receptor 2 (TLR2)-dependent inhibition of antigen processing and presentation (Drage *et al.*, 2009; Harding and Boom, 2010). Lipoproteins (from different bacteria including mycobacteria) are potent agonists of TLR2. Although TLR signalling enhances both innate and adaptive immune responses, it can also down regulate some immune functions. TLR2, in particular, has been implicated in the down regulation or deviation of the immune response through the induction of interleukin 10 and T helper 2 cell or regulatory T cell responses. Prolonged TLR signalling might constitute homeostatic feedback regulation that limits the extent of the induced responses. TLR2 agonist activity has been demonstrated for several *M. tuberculosis* lipoproteins including LpqH, LprA, LprG and PstS1, and also for mycobacterial glycolipids such as PIM, LM and LAM. Recognition of these molecules may depend on different TLR2-coreceptors and accessory receptors (Drage *et al.*, 2009). *De novo* MHC class II antigen processing and presentation are inhibited by prolonged signalling with agonists of TLR2. Down-regulation of antigen presentation is not specific to *M. tuberculosis*, but it could be especially pronounced during infection with *M. tuberculosis*. *M. tuberculosis* survives and multiplies in an early endosomal compartment by arresting phagosome maturation. Thus, *M. tuberculosis* is persistently co-localized with TLRs. The prolonged residence of *M. tuberculosis* in phagosomes and the abundance of cell envelope ligands for TLR2 that are released from viable mycobacteria, and that can transfer out of the phagosomes and out of the infected cells into the neighbour cells, provide ample opportunity for TLR2 signalling in the phagosomes or on the surface of antigen-presenting cells. Thus, TLR2-dependent suppression of MHC class II expression and processing could be particularly relevant in *M. tuberculosis* (Harding and Boom, 2010).

Mycobacterial lipoproteins

The genome of *M. tuberculosis* encodes approximately 100 lipoproteins. Some of these proteins have annotated functions, which have been reviewed previously (Sutcliffe and Harrington, 2004; Rezwan *et al.*, 2007). Despite some progress, many of the open reading frames corresponding to lipoproteins still do not have annotated functions. Transposon site hybridization mutagenesis of *in vitro* grown *M. tuberculosis* mutants has suggested that several lipoprotein genes, i.e. *lpqW*, *lprB*, *lpqF* and *dppA*, are essential (Sassetti *et al.*, 2003). Additional lipoprotein genes, i.e. *lpqY*, *lpqZ*, *lprK*, *lpqT*, *lprG*, *lppX* and *lprN*, were shown to be required for virulence on corresponding *in vivo* screening (Sassetti and Rubin, 2003). Besides these genome-wide screenings, a variety of individually generated *M. tuberculosis* mutants deficient in certain lipoprotein genes were attenuated and several lipoproteins were characterized in more detail. SodC, LpqH, PstS2, PstS3, LppX, LprG, LpqN, LprC, LppK and LpqG were found to be abundantly expressed (Malen *et al.*, 2010). Among a large number of *M. tuberculosis* lipoproteins, some lipoproteins including ones most recently found are discussed here. The *M. tuberculosis* database TubercuList (http://tuberculist.epfl.

ch) hosted by the Institute of Global Health, École Polytechnique Fédérale de Lausanne, Switzerland, provides an excellent update for each *M. tuberculosis* open reading frame and is linked to other mycobacterial databases.

LprG (Rv1411c)

LprG has been identified in mycobacterial cell envelope fractions by means of various techniques. *lprG* is a non-essential gene of *M. tuberculosis,* but is required for the survival of *M. tuberculosis* in murine macrophages and for full virulence in mice. The gene is co-transcribed with Rv1410c, a protein of the major facilitator superfamily of small molecules. Inactivation of the homologous gene in *M. smegmatis* revealed a role of each of these genes in ethidium bromide resistance. Moreover, both mutants were deficient in sliding motility and exhibited an altered colony morphology. These observations suggest a role of the *lprG*-operon in the cell envelope function (Farrow and Rubin, 2008). *M. tuberculosis* LprG as well as other lipoproteins are potent TLR agonists. Surprisingly, a non-acylated LprG still exhibited TLR2 stimulation, despite the absence of a membrane anchor. The remaining TLR2-stimulating activity of LprG was attributed to glycolipid binding. The crystal structure of LprG revealed a binding pocket that could accommodate lipids with three acyl chains. Based on structural data, a role of LprG in outer membrane transport of mycobacterial lipoproteins has been considered. LAM, LM and PIM were co-purified with recombinant LprG. In contrast, mycobacterial lipoproteins can not be co-purified with LprG. These observations suggest a primary role of LprG in lipoglycan transport rather than in lipoprotein transport (Drage *et al.,* 2010). As mentioned in 'Lipoproteins in Gram-negative diderm bacteria' above, the crystal structures of LprG and LppX are very similar to those of LolA/LolB.

LppX (Rv2945c)

M. tuberculosis LppX, similar to LprG, is also involved in the transport of complex lipids. This protein has a hydrophobic cavity that binds phthiocerol dimycocerosates (DIM) and delivers them to the outer membrane (Sulzenbacher *et al.,* 2006). LppX orthologues are restricted to mycobacteria capable of synthesizing DIM, suggesting that their transport function is selective. An *M. tuberculosis* mutant deficient in *lppX* is attenuated in a mouse model of tuberculosis (Camacho *et al.,* 1999).

RpfB (Rv1009)

The genome of *M. tuberculosis* encodes five resuscitation-promoting factor (Rpf) proteins, one of them, RpfB, being a lipoprotein. The Rpf proteins have a c-type lysozyme-like fold, and are predicted to cleave the glycosidic bond between *N*-acetyl glucosamine and *N*-acetyl muramic acid in peptidoglycan. Characterization of single and multiple *rpf* mutants produced conflicting results with respect to the importance of individual genes in resuscitation, which was due to the use of different resuscitation and virulence models, respectively (Chao and Rubin, 2010).

LppA (Rv2543)

LppA is a lipoprotein confined to mycobacterial pathogens. One copy of its gene is present in the genome of *M. tuberculosis* H37Rv, while two copies (LppA and LprR) are present in the genome of *M. tuberculosis* CDC1551. LppA is highly homologous to its downstream gene, LppB (Rv2544), with which it shares 90% amino acid identity. The structure of LppA has been solved at 2 Å resolution. However, the lack of sequence and structural homologues of LppA hinders any functional assignments (Grana *et al.*, 2009).

LpqB (Rv3224)

LpqB (Rv3244c) is one of the 233 conserved signature proteins of the *Actinobacteria* (Gao *et al.*, 2006). The *lpqB* gene is located downstream of the essential signal transduction system, MtrAB. LpqB is part of a three-component system that co-ordinates cytokinetic and cell envelope homeostatic processes in mycobacteria, and directly interacts with the extracellular domain of MtrB. An *M. smegmatis lpqB* mutant was isolated during screening for multiple antibiotic resistance-defective phenotypes. The mutant exhibited increased cell–cell aggregation, and severe defects in surface motility and biofilm growth (Nguyen *et al.*, 2010).

LpqM (Rv0419)

Screening of *M. smegmatis* mutants deficient in conjugal DNA transfer revealed *lpqM*. LpqM is a putative lipoprotein with a metalloprotease signature sequence (HExxH). Both *M. smegmatis* and *M. tuberculosis lpqM* (Rv0419) restore DNA transfer. Secretion and lipidation of LpqM are mandatory for complementation of the phenotype. Expression of functional LpqM in the donor strain is sufficient for proficient transfer (Nguyen *et al.*, 2009). So far, a substrate of LpqM has not been identified.

LpqY(Rv1235)-SugA-SugB-SugC operon

Five putative carbon uptake permeases are encoded in the genome of *M. tuberculosis*. Of these, LpqY-SugA-SugB-SugC encodes an ABC transporter highly conserved in mycobacteria. It comprises periplasmic sugar-binding lipoprotein LpqY, ATP-binding protein SugC, and transmembrane proteins SugA and SugB. This ATP-binding cassette transporter is highly specific for the uptake of disaccharide trehalose. Disaccharide trehalose is not present in mammals. However, trehalose is released as a byproduct of the biosynthesis of the mycolic acids by the mycolyltransferase antigen 85 complex. The antigen 85 complex transfers the lipid moiety of the glycolipid trehalose monomycolate (TMM) to arabinogalactan or another molecule of TMM, yielding trehalose dimycolate. These reactions also lead to the concomitant extracellular release of the trehalose moiety of TMM. Disruption of retrograde trehalose transport and thus impairment of trehalose recycling attenuates *M. tuberculosis* (Kalscheuer *et al.*, 2010).

SodC (Rv0432)

SodC encodes a Cu,Zn-dependent superoxide dismutase and is annotated as a putative lipoprotein based on the presence of a lipoprotein lipid attachment site and radioactive labelling upon heterologous expression in *E. coli*. It detoxifies reactive oxygen intermediates and thereby contributes to the survival of *M. tuberculosis*, particularly in activated macrophages. The crystal structure of SodC has been solved by X-ray crystallography. The enzyme is glycosylated at its N-terminus, and is localized to the cytoplasmic and outer membrane fractions. However, its lipidation has recently been questioned due to the results of fractionation and mutagenesis experiments combined with 2D gel electrophoresis and mass spectrometry (Sartain and Belisle, 2009).

LpqH (Rv3763)

The 19-kDa antigen (LpqH, Rv3763) of *M. tuberculosis* has been recognized as an immunodominant lipoprotein and often has been used as a mycobacterial model lipoprotein, particularly as a TLR2 agonist. LpqH is glycosylated and functions as an adhesin by binding to the mannose receptor of monocytic cells (Diaz-Silvestre *et al.*, 2005). An *M. tuberculosis lpqH* knock-out mutant is attenuated in IFN-γ activated monocyte-derived macrophages and in mice. The *lpqH* mutant nearly does not multiply in C57BL/6 mice or even in IFN-γ deficient mice. When applied as a live vaccine, the *M. tuberculosis lpqH* mutant exhibits similar protective efficacy to BCG against an aerosol challenge with *M. tuberculosis* (Henao-Tamayo *et al.*, 2007).

Lipoprotein biosynthesis in *Streptomyces*

Streptomyces also has a GC-rich genome, exhibits Gram-positive staining and is a member of the order *Actinomycetales*. *Streptomyces* species are characterized by a complex secondary metabolism. They produce a wide variety of antibiotics, e.g. streptomycin and kanamycin, and antifungal and anti-parasitic drugs, and also immunosuppressants. Some of the 550 described species are pathogens of humans or plants. Unlike their relatives from the subfamily *Corynebacterineae*, they lack an outer membrane-like structure. *Streptomyces* export lipoproteins via the *sec* and *tat* pathways (Shruthi *et al.*, 2010; Thompson *et al.*, 2010). Interestingly, *Streptomyces coelicolor* encodes two functional Lgts (*lgt1* and *lgt2*). Both *lgt* genes could be disrupted separately, but a double deletion mutant was not viable. Likewise an *lspA* knock-out mutant of this species could only survive upon acquisition of suppressor mutation (Thompson *et al.*, 2010). In contrast, *lgt* and *lspA* deletion mutants of the plant pathogen *Streptomyces scabies* are viable. These mutants are defective in vegetative growth and spore development but only moderately attenuated in virulence (Widdick *et al.*, 2011). *lgt*-deficient mutants release some (non-lipidated) lipoproteins into the culture supernatant through shedding or shaving, while other lipoproteins are either not expressed or are degraded (Widdick *et al.*, 2011). The model lipoprotein SCO3484 was shown to be modified with three fatty acids: two ester-bound and one amide-bound fatty acid are attached to the N-terminal Cys. The length of the fatty acids varies between 15 and 18 carbon atoms. All *Streptomyces* species encode two Lnt-homologues. The *N*-acylation of SCO3484 is completely abolished in an *S. scabies lnt1* deletion mutant and diminished in an *lnt2* mutant. These data indicate that Lnt1 is a functional lipoprotein *N*-acyltransferase, while the role of Lnt2 remains elusive (Widdick *et al.*, 2011).

Lipoproteins in Gram-positive monoderm bacteria

Low GC-content Gram-positive bacterial species such as *Bacillus, Clostridium, Enterococcus, Listeria, Staphylococcus, Streptococcus* and cell wall-less mycoplasma have a single cellular membrane and are classified as monoderm bacteria. An *E. coli* Lnt homologue was not found in their genomes and therefore they had been expected to have only diacylated lipoproteins (Sutcliffe and Russell, 1995; Hutchings *et al.*, 2009; Kovacs-Simon *et al.*, 2011). However, earlier biochemical studies suggested that the lipoproteins of certain *Bacillus* and *Staphylococcus* species could be *N*-acylated (Nielsen and Lampen, 1983; Navarre *et al.*, 1996). It is important to determine whether or not the low-GC-content monoderm bacteria have an Lnt-like enzyme, and to clarify the exact structures subjected to lipid modification in these bacteria (Schmaler *et al.*, 2010; Kovacs-Simon *et al.*, 2011). To address these issues, Lee and his collaborators recently provided biochemical evidence that staphylococcal lipoproteins are *N*-acylated triacyl forms (Kurokawa *et al.*, 2009; Asanuma *et al.*, 2011). Serebryakova *et al.* (2011) also reported that lipoproteins of *Acholeplasma laidlawii*, a mycoplasma strain, are triacylated. These studies indicate that staphylococcal and some mycoplasma species must have another type of Lnt whose structure is distinct from that in *E. coli*. Furthermore, novel lipopeptide structures have been identified in low GC content Gram-positive bacteria (Kurokawa *et al.*, 2011). Their detailed structures, putative biosynthetic pathways, and biological functions are discussed in this section.

N-Terminal structures of *Staphylococcus aureus* lipoproteins

Hantke and Braun (1973) reported the structure of the major outer membrane lipoprotein, Lpp, of *E. coli* in 1973. In this classical study, the Lpp structure was clarified by a combination of several methods including chemical degradation, radioisotope incorporation and chemical synthesis. However, detailed structural analyses by means of these methods are not always easy because large amounts of purified lipoproteins are required. In contrast, mass spectrometric (MS) analysis, which was developed recently (Murphy and Gaskell, 2011), enables the direct analysis of the acylated states of bacterial lipoproteins. Furthermore, this method quantitatively reveals the lipid compositions of bacterial lipoproteins. Indeed, MS-based analysis has been used for the determination of lipoprotein/lipopeptide structures and their lipid compositions (Muhlradt *et al.*, 1997; Beermann *et al.*, 2000).

The MS-based strategy was applied to determine the molecular lipid moieties attached to the N-terminal Cys residue of *S. aureus* lipoproteins (Kurokawa *et al.*, 2009). *S. aureus* is a major human pathogen causing evasion of the host's protective immune responses and is known to affect human immunity (Foster, 2005). Furthermore, antibiotic-resistant strains such as methicillin- and vancomycin-resistant *S. aureus* ones are emerging worldwide (Lowy, 1998). Importantly, *S. aureus* lipoproteins/lipopeptides are recognized by a host pattern recognition receptor, Toll-like receptor 2 (TLR2), and activate the TLR signalling pathway, leading to the production of pro-inflammatory cytokines and the establishment of adaptive immunity (Iwasaki and Medzhitov, 2010; Takeuchi and Akira, 2010). Therefore, it is important to determine the exact structures of lipoproteins as immunostimulatory ligands in order to further understand host–pathogen interactions.

A lipoprotein-enriched fraction was prepared from *S. aureus* RN4220 cell lysates by means of Triton X-114 phase partitioning and then separated by SDS-PAGE (Kurokawa *et al.*, 2009). A 33-kDa protein band was identified as lipoprotein SitC on LC (liquid chromatography)-MS/MS analysis. The N-terminal peptides of SitC were prepared by in-gel digestion with trypsin and then analysed by matrix-assisted laser desorption/ionization time-of-flight mass spectrometry (MALDI-TOF MS). Surprisingly, a series of eight mass peaks exhibiting 14 mass differences corresponding to the theoretical *m/z* of *N*-acyl-*S*-(diacyl-glyceryl)-cysteinyl peptides were detected (peaks 1–8, Fig. 4.7a). These peaks consisted of those of three fatty acids and increasing numbers of methylene (-CH_2-) groups. Consistent with this, Edman degradation failed to reveal the N-terminal amino acid sequence. These results indicate that *S. aureus* SitC is a triacylated lipoprotein (Kurokawa *et al.*, 2009). In contrast, Tawaratsumida *et al.* (2009) reported that another lipoprotein, SA2202, of *S. aureus* SA113 had a diacylated (dipalmitoylated) *N*-terminus, based on the results of MS/MS analyses. Thus, the *N*-acylation of *S. aureus* lipoproteins is controversial.

a

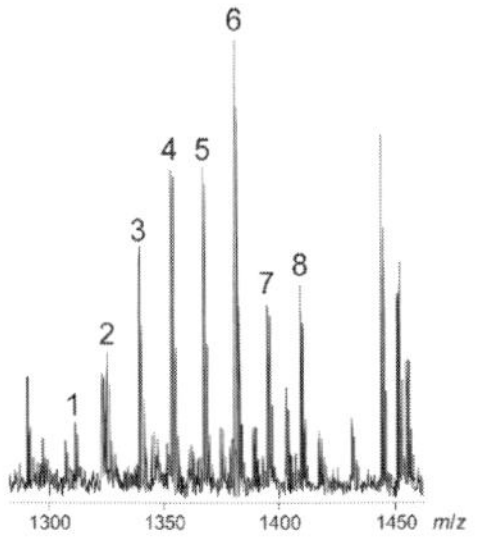

Pos.	*m/z* obtained	Corresponding structure & its *m/z*	
1	1310.9798	Pam3-CGTGGK,	1310.9604
2	1324.9886	(Pam3+CH2)-CGTGGK,	1324.9760
3	1338.9963	[Pam3+(CH2)2]-CGTGGK,	1338.9917
4	1353.0136	[Pam3+(CH2)3]-CGTGGK,	1353.0073
5	1367.0318	[Pam3+(CH2)4]-CGTGGK,	1367.0230
6	1381.0498	[Pam3+(CH2)5]-CGTGGK,	1381.0386
7	1395.0704	[Pam3+(CH2)6]-CGTGGK,	1395.0543
8	1409.0792	[Pam3+(CH2)7]-CGTGGK,	1409.0699

b

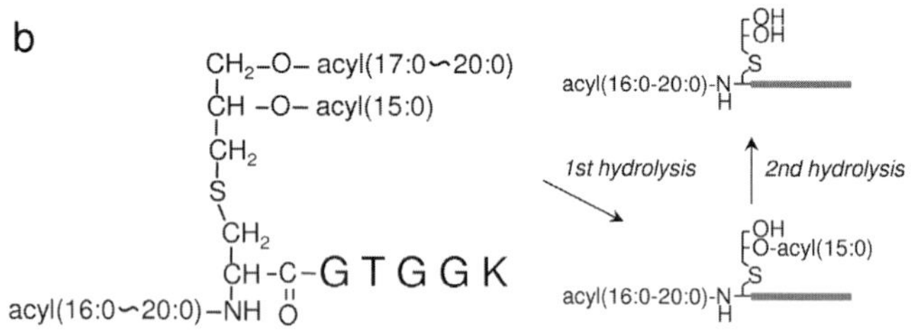

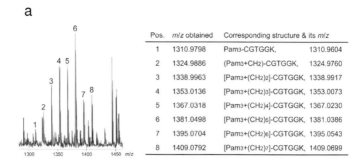

Figure 4.7 N-acylated triacyl structure of *S. aureus* lipoprotein SitC. (a) MALDI-TOF MS analysis pattern of SitC fragments, which were obtained by lysylendopeptidase digestion in an SDS-polyacrylamide gel slice. A series of mass peaks exhibiting 14 mass differences are numbered 1–8 in the left panel and listed in the right panel. The obtained *m/z* of peak 1 corresponded to the theoretical *m/z* of tripalmitic acid (Pam_3)-modified *N*-acyl-*S*-(diacyl-propyl)-cysteinyl-peptide, and those of peaks 2–8 corresponded to those of Pam_3-*N*-terminal peptides harbouring increasing numbers of methylene (CH_2) groups in their fatty acids. (b) Treatment of triacylated SitC with lipoprotein lipase (LPL). LPL specifically hydrolyses ester-bonded fatty acids; triacylated N-terminal SitC lipopeptides in the left panel were hydrolysed into diacylated and then monoacylated forms, as indicated on the right. The triacylated N-terminal SitC structure was determined by combined MS/MS analysis after treatment with lipoprotein lipase. The exact positions and lengths of ester-linked fatty acids could not be determined, but were predicted from additional information on the lipoprotein structures of *Bacillus* species (Kurokawa *et al.*, 2011) and the phospholipid structures of *S. aureus* (Fischer, 1994).

Asanuma *et al.* (2011) then examined whether the α-amino group of the *S*-diacylglyceryl Cys residue of *S. aureus* lipoproteins is acylated or not. Highly purified lipopeptides generated from SitC were analysed by MALDI-TOF MS and revealed to be triacylated (Fig. 4.7b, left). In this study, the N-terminal SitC lipopeptides were treated with lipoprotein lipase (LPL) and then analysed. LPL specifically degrades the ester-linked fatty acids of bacterial lipoproteins. This treatment generated a new series of 14-Da interval peaks corresponding to the diacyl-glyceryl CGTGGK SitC lipopeptide (Fig. 4.7b, lower right). After long-term treatment with LPL, peaks corresponding to monoacylglyceryl CGTGGK SitC lipopeptides (Fig. 4.7b, upper right) were also generated because of the release of two O-esterified fatty acids. Taken together, these results indicate that SitC is modified by two O-esterified fatty acids and one LPL-resistant fatty acid.

To confirm that the α-amino group of the *S*-diacyl-glyceryl-Cys residue is acylated, a peak corresponding to the octadecnoyl-glyceryl lipopeptide of SitC generated after long term treatment with LPL was further analysed by MALDI-ion trap MS/MS. The results strongly supported that the fatty acid modification site is the N-terminal Cys residue and that the octadecanoyl group is linked to the α-amino group of Cys via an amide bond. These results indicate that the N-terminal Cys of SitC from *S. aureus* RN4220 cells is N-acylated with a saturated C_{16} to C_{20} fatty acid and modified with a diacylglyceryl group containing two saturated fatty acids (Fig. 4.7b, left). Triacylation of SitC was confirmed with *S. aureus* SA113, clinically isolated *S. aureus* MW2, MSSA476, and *S. epidermidis* cells (Asanuma *et al.*, 2011). Furthermore, other *S. aureus* lipoproteins such as SA0739, SA0771, SA2074 (ModA), SA2158 and SA2202 were found to be triacylated. Taken together, these results indicate that staphylococcal species, typical monoderm bacteria, synthesize N-acylated *S*-diacylglyceryl lipoproteins even though an *E. coli* Lnt homologue is absent.

N-terminal lipopeptide structures of other monoderms

The structures of N-terminal lipopeptides generated from the lipoproteins of other low GC monoderms were analysed. Novel structures found in these bacteria (Fig. 4.8a) were

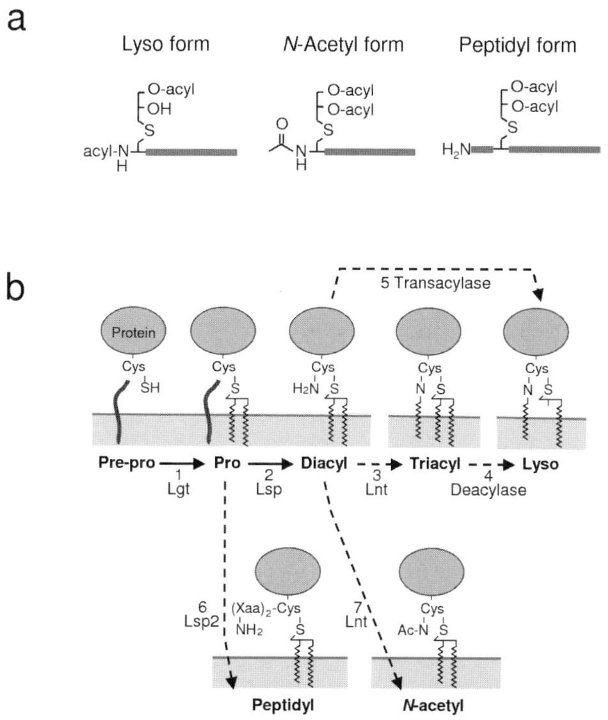

Figure 4.8 Novel N-terminal structures of lipoproteins and their possible biosynthetic pathways. (a) The structures of the lyso-form, *N*-acetyl form, and peptidyl form are illustrated. (b) Putative lipoprotein biogenesis pathways in low GC content monoderms are shown. The dotted lines indicate uncharacterized steps.

N-acyl-*S*-monoacyl-glyceryl-Cys (lyso form) in *Enterococcus faecalis*, *N*-acetyl-*S*-diacyl-glyceryl-Cys (*N*-acetyl form) in *Bacillus licheniformis* and dipeptidyl-*S*-diacyl-glyceryl-Cys (peptidyl form) in *Mycoplasma fermentans* (Kurokawa *et al.*, 2011).

The unexpected lyso-form structure reflects the loss of one of two esterified fatty acids linked to the glyceryl group (Fig. 4.8a). Three lipoproteins of *E. faecalis*, an intestinal bacterium, were found to be of the lyso-form (Kurokawa *et al.*, 2011) (Table 4.1). In addition, lyso-form lipoproteins were found in *Bacillus cereus*, a component of the microbiota in the human gut and also a food poisoning bacterium, *Lactobacillus delbrueckii* subsp. *bulgaricus*, a probiotic strain originating from Bulgarian yogurt, and *Streptococcus sanguinis*, a member of the human indigenous oral microflora. These results suggest that lyso-form lipoproteins are one of the common structures in low GC content Gram-positive bacteria. It might be possible that lyso-form lipoproteins function as immune regulatory molecules in the gut or play a role in the host gut mucosal immunity.

In addition to *Bacillus licheniformis*, a soil bacterium causing food poisoning, an *N*-acetyl form was found in *Bacillus subtilis* and strains living in extreme environments such as alkalophilic *Bacillus halodurans*, alkalophilic and extremely halotolerant *Oceanobacillus iheyensis*, and thermophilic *Geobacillus kaustophilus* (Kurokawa *et al.*, 2011). Therefore, both *N*-acetyl- and lyso-forms are found in the *Bacillaceae* family (Table 4.1).

The MBIO_0661 lipoprotein of *Mycoplasma fermentans* was found to be a dipeptidyl form comprising Ala-Gly (Kurokawa *et al.*, 2011). In contrast, the dipeptide of the MBIO_0319 lipoprotein was Ala-Ser, while the MBIO_0763 and MBIO_0869 lipoproteins were the conventional diacyl form (Table 4.1). These extension sequences coincided with the C-terminal sequences of the respective signal peptides, suggesting that *M. fermentans* Lsp has unusual substrate specificity. Lipoproteins purified from *M. genitalium* and *M. pneumoniae* were found to be triacylated (Kurokawa *et al.*, 2011), which was consistent with previous results obtained in immunological studies (Shimizu *et al.*, 2007; Shimizu *et al.*, 2008). These results indicate that lipoprotein structures are diverse in mycoplasma species and vary depending on even lipoprotein species in *M. fermentans*.

It should be noted that three lipoproteins of *Listeria monocytogenes* were found to be of the conventional diacyl form (Kurokawa *et al.*, 2011).

Putative lipoprotein biogenesis pathways

Based on the results of recent studies (Kurokawa *et al.*, 2009, 2012a; Asanuma *et al.*, 2011; Serebryakova *et al.*, 2011), the N-terminal structures of lipoproteins in monoderm bacteria and mycoplasma species are classified into three major classes (A to C) (Table 4.1). Possible biosynthetic pathways are shown in Fig. 4.8b. Class C lipoproteins having *S*-diacyl-glyceryl-Cys are generated by two enzymes, Lgt and Lsp (steps 1 and 2 or 6). A novel type of Lsp (Lsp 2) is postulated, which cleaves inside the signal peptide, probably after a serine residue, and leaves a peptide preceding the lipidated Cys (step 6). Class B lipoproteins having *N*-acyl-*S*-diacyl-glyceryl-Cys can be generated through sequential modification reactions by Lgt, Lsp and an unidentified Lnt (steps 1, 2, and 3 or 7). Since an *E. coli*-type Lnt homologue is not found in monoderm bacteria, not only the structure but also the properties of this putative Lnt seem to be significantly different from those of *E. coli*-type Lnt. Class A lipoproteins having *N*-acyl-*S*-monoacyl-glyceryl-Cys can be generated by either a putative lipoprotein O-deacylase from the triacyl form (step 4) or a putative lipoprotein transacylase from diacyl

Table 4.1 The biochemical characterization of recently identified bacterial lipoproteins

			Modified groups to the lipidated cysteine				
Bacterial species	Major habitat	Protein name	Total[1]	R1[2]	R2[2]	R3[2]	Reference
Class A. N-acyl-S-monoacylglyceryl-cysteine structures							
A-a. Lyso form							
B. cereus	Intestine	BC0200	32:0	17:0	H	15:0	Kurokawa *et al.* (2012a)
	Foods[3]	OppA	32:0	17:0, 18:0	H	14:0, 15:0	Kurokawa *et al.* (2012a)
		PrsA	32:0	17:0	H	15:0	Kurokawa *et al.* (2012a)
E. faecalis	Intestine	EF2256	34:1	18:1	H	16:0	Kurokawa *et al.* (2012a)
		EF3256	34:1	18:1	H	16:0	Kurokawa *et al.* (2012a)
		PnrA	34:1	18:0, 18:1	H	16:0, 16:1	Kurokawa *et al.* (2012a)
L. bulgaricus	Intestine	Ldb0202	36:2	18:1	H	18:1	Kurokawa *et al.* (2012a)
		Ldb2183	36:2	18:1	H	18:1	Kurokawa *et al.* (2012a)
S. sanguinis	Oral cavity	SSA_0375	34:1	18:1	H	16:0	Kurokawa *et al.* (2012a)
		SSA_1038	34:1	18:0, 18:1	H	16:0, 16:1	Kurokawa *et al.* (2012a)
Class B. N-acyl-S-diacylglyceryl-cysteine structures							
B-a. N-*acetyl form*							
B. halodurans	Soil	BH3460	32:0	15:0	15:0	2:0	Kurokawa *et al.* (2012a)
		MalE	32:0	15:0	15:0	2:0	Kurokawa *et al.* (2012a)
B. licheniformis	Soil	MntA	34:0	17:0	15:0	2:0	Kurokawa *et al.* (2012a)
		OppA	34:0	17:0	15:0	2:0	Kurokawa *et al.* (2012a)
B. subtilis	Soil	BSU01630	34:0	17:0	15:0	2:0	Kurokawa *et al.* (2012a)
		PrsA	34:0	17:0	15:0	2:0	Kurokawa *et al.* (2012a)
G. kaustophilus	Deep sea	GK0969	34:0	32:04		2:0	Kurokawa *et al.* (2012a)
		GK1283	34:0	17:0	15:0	2:0	Kurokawa *et al.* (2012a)
O. iheyensis	Deep sea	CtaC	32:0	15:0	15:0	2:0	Kurokawa *et al.* (2012a)

Table 4.1 (continued)

Bacterial species	Major habitat	Protein name	Modified groups to the lipidated cysteine				Reference
			Total[1]	R1[2]	R2[2]	R3[2]	
B-b. Conventional triacyl form							
A. laidlawii	Waste water	ACL_1223	48:0, 50:0[5]	16:0	16:0	16:0, 18:0	Serebryakova *et al.* (2011)
			50:0	(18:0)[6]	(16:0)[6]	(16:0)[6]	Nakayama *et al.* (2012)
		ACL_1410	48:0, 50:0[5]	16:0	16:0	16:0, 18:0	Serebryakova *et al.* (2011)
			50:0	(18:0)[6]	(16:0)[6]	(16:0)[6]	Nakayama *et al.* (2012)
M. genitalium	Genital	MG_040	50:1	18:1	16:0	16:0	Kurokawa *et al.* (2012a)
M. pneumoniae	Respiratory tracts	MPN052	50:1	18:1	16:0	16:0	Kurokawa *et al.* (2012a)
		MPN415	50:1	18:1	16:0	16:0	Kurokawa *et al.* (2012a)
S. aureus	Nares, skin	SA0739	52:0	32:0–37:0[4]		15:0–20:0	Asanuma *et al.* (2011)
		SA0771	52:0	32:0–36:0[4]		16:0–20:0	Asanuma *et al.* (2011)
		SA2074	51:0	31:0–36:0[4]		15:0–20:0	Asanuma *et al.* (2011)
		SA2202	52:0	32:0–36:0[4]		16:0–20:0	Asanuma *et al.* (2011)
		SitC	53:0	33:0–37:0[4]		16:0–20:0	Asanuma *et al.* (2011)
S. epidermidis	Skin	SitC	53:0	32:0–35:0[4]		17:0–20:0	Asanuma *et al.* (2011)
Class C. S-diacylglyceryl-cysteine structures							
C-a. Conventional diacyl form							
L. monocytogenes	Foods[3]	Lmo0135	32:0	17:0	15:0	H	Kurokawa *et al.* (2012a)
		Lmo2196	32:0	17:0	15:0	H	Kurokawa *et al.* (2012a)
		Lmo2219	32:0	17:0	15:0	H	Kurokawa *et al.* (2012a)
M. fermentans	Throat	MBIO_0763	34:0	34:0[4]		H	Kurokawa *et al.* (2012a)
		MBIO_0869	34:0	34:0[4]		H	Kurokawa *et al.* (2012a)
S. aureus	Nares, skin	SA1659	33:0	18:0	15:0	H	Kurokawa *et al.* (2012b)
		SitC	32:0–35:0	17:0–20:0	15:0	H	Kurokawa *et al.* (2012b)

Table 4.1 (continued)

Bacterial species	Major habitat	Protein name	Modified groups to the lipidated cysteine Total[1]	R1[2]	R2[2]	R3[2]	Reference
C-b. Peptidyl form							
M. fermentans	Throat	MBIO_0319	34:0	18:0	16:0	Ala-Ser-	Kurokawa *et al.* (2012a)
		MBIO_0661	34:0	18:0	16:0	Ala-Gly-	Kurokawa *et al.* (2012a)

[1]The total carbon number of modified acyl groups on the lipidated cysteine is calculated from the MS/MS-analyzed lipopeptide peak. The most abundant N-terminal lipopeptide ion in MALDI-TOF MS was usually analyzed by MS/MS.
[2]R1 and R2 denote a hydrogen or an acyl group attached to the *sn*-1 and *sn*-2 positions of the S-glyceryl group of the lipidated cysteine, respectively. R3 denotes a hydrogen, an acyl group, or a dipeptide attached to the α-amino group of the lipidated cysteine.
[3]Foods probably contained human, animal, and/or soil contaminants.
[4]The total carbon number of modified acyl groups on R1 and R2 is shown.
[5]Descending order in abundance.
[6]Our interpretation is shown in parenthesis.

form (step 5). In *E. coli, N*-acylation is essential for correct outer membrane localization of lipoproteins by the Lol system (Fukuda *et al.*, 2002; Robichon *et al.*, 2005). To understand the biological significance of lipoproteins in low GC content monoderms, it is important to characterize the enzymes involved in lipoprotein modification.

It seems interesting to consider the fatty acid preference of the putative Lnt enzymes involved in the formation of the triacyl form in monoderms (Kurokawa *et al.*, 2011). In *S. aureus*, 48% of fatty acids bound to the α-amino group of Cys (R3 position) are 18:0 (18 carbons with no double bond), whereas 18:0 fatty acid constitutes only 3.2% of total membrane lipids and 13.4% of PG. In contrast, 15:0 and 20:0 fatty acids occupy 40% and 22% of total membrane fatty acids, respectively. These results suggest that Lnt of *S. aureus* seems to prefer 18:0 fatty acid although the source of the fatty acid remains to be determined. On the other hand, the R3 position of the *M. pneumoniae* MPN052 lipoprotein carried only 16:0 fatty acid, while *M. pneumoniae* total membrane contained 18:0 and 18:1 fatty acids, each at about 20%, in addition to 16:0. Thus, the putative Lnt of *M. pneumonia* is specific to 16:0 fatty acid. Moreover, lyso form lipoproteins such as PrsA of *B. cereus*, PnrA of *E. faecalis*, and Ldb2183 of *L. bulgaricus* also exhibit strong fatty acid specificity at the R3 position (Kurokawa *et al.*, 2011). Lnt of *E. coli* was recently found to exhibit specificities for both fatty acid species and lipid head groups (Hillmann *et al.*, 2011). Kurokawa *et al.* (2012b) discovered conditions that alter between triacyl and diacyl forms of lipoproteins in *S. aureus*. Under the combination of low pH and post-log growth phase, *S. aureus* cells accumulated N-terminal free diacyl forms of lipoproteins. High temperatures and/or high salt concentrations additively increased the accumulation of diacyl forms. Interestingly, pH shift assays revealed that protein synthesis is required to the structural alterations, suggesting that the expression level or activity of unidentified Lnt should be regulated via protein synthesis.

Biological functions of monoderm-derived lipoproteins

Mammalian TLRs play an important role in recognizing microorganisms to activate host innate immune responses (Takeuchi and Akira, 2010). Eleven human TLRs and 13 mouse TLRs have been identified, and each TLR appears to recognize a pathogen-associated molecular pattern molecule derived from various microorganisms, including bacteria, viruses, protozoa, and fungi (Takeuchi and Akira, 2010). Among them, TLR4, TLR5 and TLR9 recognize a single class of pattern molecule, such as lipopolysaccharides (LPS), bacterial flagellin, and bacterial DNA, respectively (Ramos *et al.*, 2004; Ewald and Barton, 2011). Outer membrane LPS play a critical role in the activation of TLR4 in diderm Gram-negative bacteria (Hoshino *et al.*, 1999). TLR2 plays a major role in recognition of Gram-positive bacteria (Takeuchi *et al.*, 1999). TLR2 had been reported to recognize several molecules, including lipoproteins, synthetic lipopeptides, peptidoglycan, lipoteichoic acids (LTAs), lipomannans and lipoarabinomannans (Zähringer *et al.*, 2008). Since these molecules are structurally diverse, it appeared unlikely that TLR2 has the ability to react with all agonists to the same degree. To address this issue, two mutant strains of *S. aureus*, Δ*ltaS* (Gründling and Schneewind, 2007; Oku *et al.*, 2009) and Δ*lgt* (Stoll *et al.*, 2005), were used. It was found that bacterial lipoproteins, but neither LTA nor peptidoglycan, act as native TLR2 ligands (Kurokawa *et al.*, 2009). The *lgt*-deficient mutants of *S. aureus*, *L. monocytogenes*, and Group B *Streptococcus* also indicated that bacterial lipoproteins function as major ligands for TLR2 (Bubeck Wardenburg *et al.*, 2006; Hashimoto *et al.*, 2006; Henneke *et al.*, 2008; Machata *et al.*, 2008). Some lipoproteins tightly associate with peptidoglycan. When peptidoglycan-associated lipoproteins were enzymatically removed, TLR2 activation by peptidoglycan was completely abolished, while lipoproteins extracted from peptidoglycan remained active as to TLR2 activation (Kurokawa *et al.*, 2009). Taken together, these results indicate that true TLR2 ligands are lipoproteins.

Because *S. aureus* lipoproteins anchored to the cytoplasmic membrane are covered by a thick layer of peptidoglycan (Fig. 4.1), bacterial cells must be engulfed and delivered to acidic phagosomes for the efficient activation of TLR2. Enzymes in phagosomes digest the internalized bacteria, causing the release of bacterial lipoproteins. Thus, phagocytosis of *S. aureus* cells and following degradation of the cell wall are critical for an efficient lipoprotein–TLR2 interaction (Ip *et al.*, 2010; Shimada *et al.*, 2010; Kang *et al.*, 2011).

TLR2 was thought to function as a heterodimer with TLR1 or TLR6. It was then examined as to which heterodimer is involved in the cytokine production by triacylated SitC. Synthetic lipoprotein analogues, such as *N*-palmitoyl-*S*-dipalmitoylglyceryl (Pam_3)-Cys-Ser-Lys-Lys-Lys-Lys (Pam_3CSK_4) and *S*-dipalmitoylated macrophage-activating lipopeptide-2 kDa (MALP-2) (Mühlradt *et al.*, 1997), were used to examine the pro-inflammatory cytokine release. These examinations suggested that the TLR2/TLR1 heterodimer generally recognizes triacylated lipopeptides whereas the TLR2/TLR6 heterodimer responds to diacylated lipopeptides (Takeuchi *et al.*, 2001; Takeuchi *et al.*, 2002; Buwitt-Beckmann *et al.*, 2005, 2006; Omueti *et al.*, 2005). The crystal structures of the two TLR2 heterodimers complexed with synthetic lipopeptide support this model (Jin *et al.*, 2007; Kang *et al.*, 2009).

Induction of pro-inflammatory cytokines by purified triacylated SitC was examined using mouse thioglycolate-elicited peritoneal macrophages. The native SitC protein induced the production of both tumour necrosis factor-α (TNF-α) and interleukin 6 (IL-6) in wild-type mouse macrophages (Fig. 4.9) (Kurokawa *et al.*, 2009). In contrast, $TLR2^{-/-}$ mouse macrophages released neither TNF-α nor IL-6 upon the addition of SitC. Since both

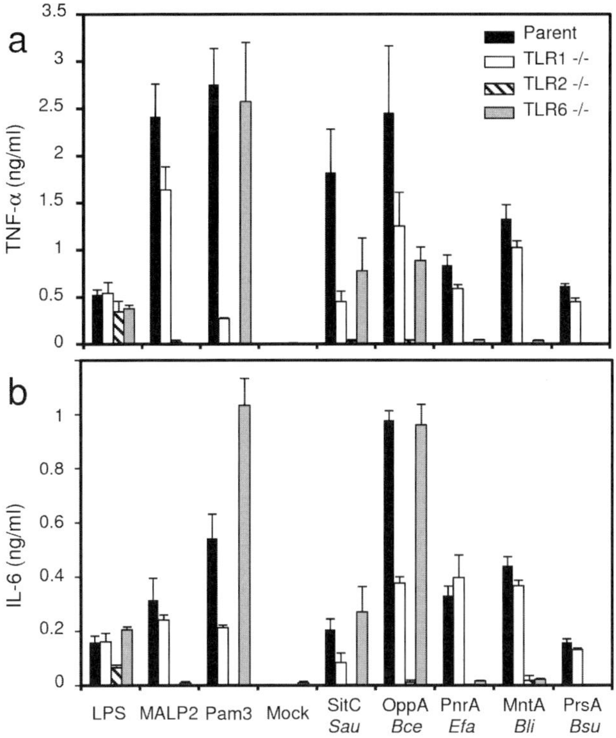

Figure 4.9 TLR2-dependent cytokine induction by purified native lipoproteins. Peritoneal macrophages (1 × 10^5 cells) were prepared from the parental C57BL/6 mice (filled bars). TLR1$^{-/-}$ (open bars), TLR2$^{-/-}$ (hatched bars), or TLR6$^{-/-}$ (grey bars) mice were stimulated for 24h without (Mock), or with 1 μg/ml lipopolysaccharide (LPS), 100 ng/ml MALP-2, 10 ng/ml synthetic Pam_3CSK_4 (Pam3), or 1 μg/ml each of the indicated purified lipoproteins/peptides. The lipoproteins/lipopeptides tested were triacyl SitC from *S. aureus*, lyso form OppA from *B. cereus*, lyso form PnrA from *E. faecalis*, N-acetyl form MntA from *B. licheniformis*, and the synthetic N-acetyl form lipopeptide derived from *B. subtilis* PrsA. The concentrations of tumour necrosis factor (TNF)-α (*a*) and interleukin 6 (IL–6) (*b*) in the culture supernatants were measured by ELISA. The data are shown as means + SD (*n*=4, duplicate or triplicate measurements).

cytokines were induced in macrophages from TLR1$^{-/-}$ and TLR6$^{-/-}$ mice, the two TLRs are not essential for SitC-mediated cytokine production while TLR2 is essential, suggesting that triacylated SitC stimulates immune cells through both the TLR2/TLR1 and TLR2/TLR6 heterodimers.

Lyso form lipoprotein OppA of *B. cereus* induced both TNF-α and IL-6 in a TLR2-dependent manner (Fig. 4.9) (Kurokawa *et al.*, 2011). Neither TLR1 nor TLR6 is required for this. In contrast, cytokine secretion induced by lyso form lipoprotein PnrA of *E. faecalis* is dependent on both TLR2 and TLR6 but independent of TLR1 (Fig. 4.9). *N*-acetyl lipoprotein/lipopeptide induced the release of TNF-α and IL-6 in TLR2- and TLR6-dependent and TLR1-independent manners (Fig. 4.9). Taken together, these results indicate that the lyso and *N*-acetyl form lipoproteins in monoderms activate TLR2 by forming a heterodimer with either TLR1 or TLR6. The choice of co-receptor seems to depend on the species of bacterial lipoproteins.

Problems to be answered

As discussed in this chapter, much new information has been obtained in relation to the N-terminal structures of lipoproteins, the enzymes involved in lipoprotein maturation, the physiological functions of lipoproteins, and the sorting of lipoproteins. However, there are many unanswered questions, as listed below.

Gram-negative diderms

- The Lol avoidance mechanism is probably general for the cytoplasmic membrane retention of lipoproteins. However, how do different residues function as Lol avoidance signals when the origin of LolCDE is different?
- How are lipoproteins transported to the cell surface of some bacteria? Does the Lol system contribute to this transport?

Gram-positive diderms

- How are lipoproteins transported to the mycobacterial outer membrane?
- Which lipoprotein motifs direct retention in the cytoplasmic membrane or transport to the mycobacterial outer membrane?
- What is the function of lipoprotein glycosylation in mycobacteria?
- Is the glycosylation of lipoproteins dependent on lipidation?

Low GC monoderms

- Why is the lipid modification of N-terminal Cys so diverse in low-GC monoderms?
- Are modification enzymes also diverse?
- Is there any physiological advantage of synthesizing *N*-acylated triacyl lipoproteins even though bacteria do not possess an outer membrane?
- Is structural alteration between lipid modifications unique in *S. aureus*, or common in other bacteria?

Acknowledgements

HT was supported by Grants-in-Aid from the Ministry of Education, Culture, Sports, Science and Technology of Japan. PS was supported by the Swiss National Science Foundation (SNF: 31003A_135705). LBL was supported by a Korean Research Foundation Grant (KRF: 2010-0020665).

References

Asanuma, M., Kurokawa, K., Ichikawa, R., Ryu, K.H., Chae, J.H., Dohmae, N., Lee, B.L., and Nakayama, H. (2011). Structural evidence of α-aminoacylated lipoproteins of *Staphylococcus aureus*. FEBS J. *278*, 716–728.

Anwari, K., Poggio, S., Perry, A., Gatsos, X., Ramarathinam, S.H., Williamson, N.A., Noinaj, N., Buchanan, S., Gabriel, K., Purcell, A.W., *et al.* (2010). A modular BAM complex in the outer membrane of the alpha-proteobacterium *Caulobacter crescentus*. PLoS One *5*, e8619.

Babu, M.M., Priya, M.L., Selvan, A.T., Madera, M., Gough, J., Aravind, L., and Sankaran, K. (2006). A database of bacterial lipoproteins (DOLOP) with functional assignments to predicted lipoproteins. J. Bacteriol. *188*, 2761–2773.

Banaiee, N., Kincaid, E.Z., Buchwald, U., Jacobs, W.R., and Ernst, J.D. (2006). Potent inhibition of macrophage responses to IFN-g by live virulent *Mycobacterium tuberculosis* is independent of mature mycobacterial lipoproteins but dependent on TLR2. J. Immunol. *176*, 3109–3027.

Banaiee, N., Jacobs, W.R., annd Ernst, J.D. (2007). LspA-independent action of globomycin on *Mycobacterium tuberculosis*. J. Antimicrob. Chemother. *60*, 414–416.

Bernadac, A., Gavioli, M., Lazzaroni, J.C., Raina, S., and Lloubes, R. (1998). *Escherichia coli* tol-pal mutants form outer membrane vesicles. J. Bacteriol. *180*, 4872–4878.

Bos, M.P., and Tommassen, J. (2011). The LptD chaperone LptE is not directly involved in lipopolysaccharide transport in *Neisseria meningitidis*. J. Biol. Chem. *286*, 28688–28696.

Bos, M.P., Tefsen, B., Geurtsen, J., and Tommassen, J. (2004). Identification of an outer membrane protein required for the transport of lipopolysaccharide to the bacterial cell surface. Proc. Natl. Acad. Sci. U.S.A. *101*, 9417–9422.

Brülle, J.K., Grau, T., Tschumi, A., Auchli, Y., Burri, R., Polsfuss, S., Keller, P.M., Hunziker, P., and Sander, P. (2010). Cloning, expression and characterization of *Mycobacterium tuberculosis* lipoprotein LprF. Biochem. Biophys. Res. Comm. *139*, 679–684.

Bubeck Wardenburg, J., Williams, W.A., and Missiakas, D. (2006). Host defenses against *Staphylococcus aureus* infection require recognition of bacterial lipoproteins. Proc. Natl. Acad. Sci. U.S.A. *103*, 13831–13836.

Buddelmeijer, N., and Young, R. (2010). The essential *Escherichia coli* apolipoprotein N-acyltransferase (Lnt) exists as an extracytoplasmic thioester acyl-enzyme intermediate. Biochemistry *49*, 341–346.

Buwitt-Beckmann, U., Heine, H., Wiesmüller, K.H., Jung, G., Brock, R., Akira, S., and Ulmer, A.J. (2005). Toll-like receptor 6-independent signaling by diacylated lipopeptides. Eur. J. Immunol. *35*, 282–289.

Buwitt-Beckmann, U., Heine, H., Wiesmüller, K.H., Jung, G., Brock, R., Akira, S., and Ulmer, A.J. (2006). TLR1- and TLR6-independent recognition of bacterial lipopeptides. J. Biol. Chem. *281*, 9049–9057.

Camacho, L.R., Ensergueix, D., Perez, E., Gicquel, B., and Guilhot, C. (1999). Identification of a virulence gene cluster of *Mycobacterium tuberculosis* by signature-tagged transposon mutagenesis. Mol. Microbiol. *34*, 257–267.

Chao, M.C., and Rubin, E.J. (2010). Letting sleeping dos lie: does dormancy play a role in tuberculosis? Annu. Rev. Microbiol. *64*, 293–311.

Chen, M.H., Takeda, S., Yamada, H., Ishii, Y., Yamashino, T., and Mizuno, T. (2001). Characterization of the RcsC→YojN→RcsB phosphorelay signaling pathway involved in capsular synthesis in *Escherichia coli*. Biosci. Biotechnol. Biochem. *65*, 2364–2367.

Chin, J.W., Martin, A.B., King, D.S., Wang, L., and Schultz, P.G. (2002). Addition of a photocrosslinking amino acid to the genetic code of Escherichiacoli. Proc. Natl. Acad. Sci. U.S.A. *99*, 11020–11024.

Chng, S.S., Ruiz, N., Chimalakonda, G., Silhavy, T.J., and Kahne, D. (2010). Characterization of the two-protein complex in *Escherichia coli* responsible for lipopolysaccharide assembly at the outer membrane. Proc. Natl. Acad. Sci. U.S.A. *107*, 5363–5368.

Clavel, T., Germon, P., Vianney, A., Portalier, R., and Lazzaroni, J.C. (1998). TolB protein of *Escherichia coli* K-12 interacts with the outer membrane peptidoglycan-associated proteins Pal, Lpp and OmpA. Mol. Microbiol. *29*, 359–367.

Cole, S.T., and Riccardi, G. (2011). New tuberculosis drugs on the horizon. Curr. Opinion Microbiol. *14*, 1–7.

Cole, S.T., Brosch, R., Parkhill, J., Garnier, T., Churcher, C., Harris, D., Gordon, S.V., Eigelmeier, K., Gas, S., Barry, C.E., *et al.* (1998). Deciphering the biology of *Mycobacterium tuberculosis* from the complete genome sequence. Nature *393*, 537–544.

Cole, S.T., Eisenach, K.D., McMurray, D.N., and Jacobs, W.R., ed. (2005). Tuberculosis and the Tubercle Bacillus (ASM Press, Washington, D.C.).

Da Silva, P.E.A., and Palomino, J.C. (2011). Molecular basis and mechanisms of drug resistance in *Mycobacterium tuberculosis*: classical and new drugs. J. Antimicrob. Chemother. *66*, 1417–1430.

Daffe, M. (2008). The global architecture of the mycobacterial cell envelope. In The Mycobacterial Cell Envelope, Daffe, M., and Reyrat, J.-M., eds. (ASM Press, Washington, D.C.), pp. 3–11.

Davidson, A.L., and Chen, J. (2004). ATP-binding cassette transporters in bacteria. Annu. Rev. Biochem. *73*, 241–268.

Dawson, R.J., and Locher, K.P. (2006). Structure of a bacterial multidrug ABC transporter. Nature *443*, 180–185.

Dawson, R.J., and Locher, K.P. (2007). Structure of the multidrug ABC transporter Sav1866 from *Staphylococcus aureus* in complex with AMP-PNP. FEBS Lett. *581*, 935–938.

Dawson, R.J., Hollenstein, K., and Locher, K.P. (2007). Uptake or extrusion: crystal structures of full ABC transporters suggest a common mechanism. Mol. Microbiol. *65*, 250–257.

DeChavigny, A., Heacock, P.N., and Dowhan, W. (1991). Sequence and inactivation of the pss gene of *Escherichia coli*. Phosphatidylethanolamine may not be essential for cell viability. J. Biol. Chem. *266*, 10710.

Desvaux, M., Hebraud, M., Talon, R., and Henderson, I.R. (2009). Secretion and subcellular localizations of bacterial proteins: a semantic awareness issue. Trends Microbiol. *17*, 139–145.

Dhiman, R.K., Dinadayala, P., Ryan, G.J., Lenaerts, A.J., Schenkel, A.R., and Crick, D.C. (2011). Lipoarabinomannan localization and abundance during *Mycobacterium smegmatis*. J. Bacteriol. *193*, 5802–5809.

Diaz-Silvestre, H., Espinosa-Cueto, P., Sanchez-Gonzalez, A., Esparza-Ceron, M.A., Pereira-Suarez, A.L., Bernai-Fernandez, G., Espitia, C., and Mancilla, R. (2005). The 19-kDa antigen of *Mycobacterium tuberculosis* is a major adhesin that binds the mannose receptor of THP-1 monocytic cells and promotes phagocytosis of mycobacteria. Microb. Pathog. *39*, 97–107.

Drage, M.G., Pecora, N.D., Hise, A.G., Febbraio, M., Silverstein, R.L., Golenbock, D.T., Boom, W.H., and Harding, C.V. (2009). TLR2 and its co-receptors determine responses of macrophages and dendritic cells to lipoproteins of *Mycobacterium tuberculosis*. Cell Immunol. *258*, 29–37.

Drage, M.G., Tsai, H.C., Pecora, N.D., Cheng, T.Y., Arida, A.R., Shukla, S., Rojas, R.E., Seshadri, C., Moody, D.B., Boom, W.H., *et al.* (2010). *Mycobacterium tuberculosis* lipoprotein LprG (Rv1411c) binds triacylated glycolipid agonists of Toll-like receptor 2. Nat. Struct. Mol. Biol. *17*, 1088–1095.

Ehrmann, M., Ehrle, R., Hofmann, E., Boos, W., and Schlosser, A. (1998). The ABC maltose transporter. Mol. Microbiol. *29*, 685–694.

Ewald, S.E., and Barton, G.M. (2011). Nucleic acid sensing Toll-like receptors in autoimmunity. Curr. Opin. Immunol. *23*, 3–9.

Farrow, M.F., and Rubin, E.J. (2008). Function of a mycobacterial major facilitator superfamily pump requires a membrane-associated lipoprotein. J. Bacteriol. *190*, 1783–1791.

Fischer, W. (1994). Lipoteichoic acid and lipids in the membrane of *Staphylococcus aureus*. Med. Microbiol. Immunol. *183*, 61–76.

Foster, T.J. (2005). Immune evasion by staphylococci. Nat. Rev. Microbiol. *3*, 948–958.

Fraser, C.M., Casjens, S., Huang, W.M., Sutton, G.G., Clayton, R., Lathigra, R., White, O., Ketchum, K.A., Dodson, R., Hickey, E.K., *et al.* (1997). Genomic sequence of a Lyme disease spirochaete, *Borrelia burgdorferi*. Nature *390*, 580–586.

Fukuda, A., Matsuyama, S., Hara, T., Nakayama, J., Nagasawa, H., and Tokuda, H. (2002). Aminoacylation of the N-terminal cysteine is essential for Lol-dependent release of lipoproteins from membranes but does not depend on lipoprotein sorting signals. J. Biol. Chem. *277*, 43512–43518.

Gao, B., Pramanathan, R., and Gupta, R.S. (2006). Signature proteins that are distinctive characteristics of Actinobacteria and their subgroups. Antonie Van Leeuwenhoek *90*, 69–91.

Gatsos, X., Perry, A.J., Anwari, K., Dolezal, P., Wolynec, P.P., Likic, V.A., Purcell, A.W., Buchanan, S.K., and Lithgow, T. (2008). Protein secretion and outer membrane assembly in Alphaproteobacteria. FEMS Microbiol. Rev. *32*, 995–1009.

Gennity, J.M., and Inouye, M. (1991). The protein sequence responsible for lipoprotein membrane localization in *Escherichia coli* exhibits remarkable specificity. J. Biol. Chem. *266*, 16458–16464.

Gentle, I., Gabriel, K., Beech, P., Waller, R., and Lithgow, T. (2004). The Omp85 family of proteins is essential for outer membrane biogenesis in mitochondria and bacteria. J. Cell Biol. *164*, 19–24.

Gerken, H., Leiser, O.P., Bennion, D., and Misra, R. (2010). Involvement and necessity of the Cpx regulon in the event of aberrant β-barrel outer membrane protein assembly. Mol. Microbiol. *75*, 1033–1046.

Gibbons, H.S., Wolschendorf, F., Abshire, M., Niederweis, M., and Braunstein, M. (2007). Identification of two *Mycobacterium smegmatis* lipoproteins exported by a SecA2-dependent pathway. J. Bacteriol. *189*, 5090–5100.

Gilleron, M., Jackson, M., Nigou, J., and Puzo, G. (2008). Structure, biosynthesis, and activities of the phosphatidyl-myo-inositol-based lipoglycans. In The Mycobacterial Cell Envelope, Daffe, M., and Reyrat, J.-M., eds. (ASM Press, Washington, D.C.), pp. 75–105.

Goren, M.B., and Brennan, P.J. (1979). Mycobacterial lipids: chemistry and biologic activities. In Tuberculosis, Yomans, G.P., ed. (W.B. Saunders Company, Philadelphia), pp. 63–193.

Grana, M., Bellinzoni, M., Bellalou, J., Haouz, A., Miras, I., Buschiazzo, A., Winter, N., and Alzari, P.M. (2009). Crystal structure of *Mycobacterium tuberculosis* LppA, a lipoprotein confined to pathogenic mycobacteria. Proteins *78*, 769–772.

Gründling, A., and Schneewind, O. (2007). Synthesis of glycerol phosphate lipoteichoic acid in *Staphylococcus aureus*. Proc. Natl. Acad. Sci. U.S.A. *104*, 8478–8483.
Gupta, R.S. (1998). What are archaebacteria: life's third domain or monoderm prokaryotes related to gram-positive bacteria? A new proposal for the classification of prokaryotic organisms. Mol. Microbiol. *29*, 695–707.
Gupta, R., Lavollay, M., Mainardi, J.-L., Arthur, M., Bishai, W.R., and Lamichhane, G. (2010). The *Mycobacterium tuberculosis* protein LdtMt2 is a nonclassical transpeptidase required for virulence and resistance to amoxicillin. Nat. Med. *16*, 466–469.
Haake, D.A. (2000). Spirochaetal lipoproteins and pathogenesis. Microbiology *146*, 1491–1504.
Hagan, C.L., and Kahne, D. (2011). The reconstituted *Escherichia coli* Bam complex catalyzes multiple rounds of β-Barrel assembly. Biochemistry *50*, 7444–7446.
Hagan, C.L., Kim, S., and Kahne, D. (2010). Reconstitution of outer membrane protein assembly from purified components. Science *328*, 890–892.
Hagan, C.L., Silhavy, T.J., and Kahne, D. (2011). β-Barrel membrane protein assembly by the Bam Complex. Annu. Rev. Biochem. *80*, 189–210.
Hantke, K., and Braun, V. (1973). Covalent binding of lipid to protein. Diglyceride and amide-linked fatty acid at the N-terminal end of the murein-lipoprotein of the *Escherichia coli* outer membrane. Eur. J. Biochem. *34*, 284–296.
Hara, T., Matsuyama, S., and Tokuda, H. (2003). Mechanism underlying the inner membrane retention of *Escherichia coli* lipoproteins caused by Lol avoidance signals. J. Biol. Chem. *278*, 40408–40414.
Harding, C.V., and Boom, W.H. (2010). Regulation of antigen presentation by *Mycobacterium tuberculosis*: a role for Toll-like receptors. Nat. Rev. Microbiol. *8*, 296–307.
Hashimoto, M., Asai, Y., and Ogawa, T. (2004). Separation and structural analysis of lipoprotein in a lipopolysaccharide preparation from *Porphyromonas gingivalis*. Int. Immunol. *16*, 1431–1437.
Hashimoto, M., Tawaratsumida, K., Kariya, H., Kiyohara, A., Suda, Y., Krikae, F., Kirikae, T., and Götz, F. (2006). Not lipoteichoic acid but lipoproteins appear to be the dominant immunobiologically active compounds in *Staphylococcus aureus*. J. Immunol. *177*, 3162–3169.
Henao-Tamayo, M., Junqueira-Kipnis, A.P., Ordway, D., Gonzalez-Juarrero, M., Stewart, G.R., Young, D.B., Wilkinson, R.J., Baaraba, R.J., and Orme, I.M. (2007). A mutant of *Mycobacterium tuberculosis* lacking the 19-kDa lipoprotein Rv3763 is highly attenuated *in vivo* but retains potent vaccinogenic properties. Vaccine *25*, 7153–7139.
Henneke, P., Dramsi, S., Mancuso, G., Chraibi, K., Pellegrini, E., Theilacker, C., Hübner, J., Santos-Sierra, S., Teti, G., Golenbock, D.T., *et al.* (2008). Lipoproteins are critical TLR2 activating toxins in group B streptococcal sepsis. J. Immunol. *180*, 6149–6158.
Herrmann, J.L., O'Gaora, P., Gallagher, A., Thole, J.E., and Young, D.B. (1996). Bacterial glycoproteins: a link between glycosylation and proteolytic cleavage of a 19 kDa antigen from *Mycobacterium tuberculosis*. EMBO J. *15*, 3547–3554.
Hillmann, F., Argentini, M., and Buddelmeijer, N. (2011). Kinetics and phospholipid specificity of apolipoprotein *N*-acyltransferase. J. Biol. Chem. *286*, 27936–27946.
Hirano, S., Asamizu, S., Onaka, H., Shiro, Y., and Nagano, S. (2008). Crystal structure of VioE, a key player in the construction of the molecular skeleton of violacein. J. Biol. Chem. *283*, 6459–6466.
Hoffmann, C., Leis, A., Niederweis, M., Plitzko, J.M., and Engelhardt, H. (2008). Disclosure of the mycobacterial outer membrane: cryo-electron tomography and vitreous sections reveal the lipid bilayer Structure Proc. Natl. Acad. Sci. U.S.A. *105*, 3963–3967.
Hoshino, K., Takeuchi, O., Kawai, T., Sanjo, H., Ogawa, T., Takeda, Y., Takeda, K., and Akira, S. (1999). Cutting edge: Toll-like receptor 4 (TLR4)-deficient mice are hyporesponsive to lipopolysaccharide: evidence for TLR4 as the *Lps* gene product. J. Immunol. *162*, 3749–3752.
Hovav, A.H., Mullerad, J., Davidvitch, L., Fishman, Y., Bigi, F., Cataldi, A., and Bercovier, H. (2003). The *Mycobacterium tuberculosis* recombinant 27-kilodalton lipoprotein induces a strong Th1-type immune response deleterious to protection. Infect. Immun. *71*, 3146–3154.
Hutchings, M.I., Palmer, T., Harrington, D.J., and Sutcliffe, I.C. (2009). Lipoprotein biogenesis in Gram-positive bacteria: knowing when to hold 'em, knowing when to fold 'em. Trends Microbiol. *17*, 13–21.
Inukai, M., Nakajima, M., Osawa, M., Haneishi, T., and Arai, M. (1978). Globomycin, a new peptide antibiotic with spheroplast-forming activity. J. Antibiot. *31*, 421–425.
Ip, W.K., Sokolovska, A., Charriere, G.M., Boyer, L., Dejardin, S., Cappillino, M.P., Yantosca, L.M., Takahashi, K., Moore, K.J., Lacy-Hulbert, A., *et al.* (2010). Phagocytosis and phagosome acidification are required for pathogen processing and MyD88-dependent responses to *Staphylococcus aureus*. J. Immunol. *184*, 7071–7081.

Ito, Y., Kanamaru, K., Taniguchi, N., Miyamoto, S., and Tokuda, H. (2006). A novel ligand bound ABC transporter, LolCDE, provides insights into the molecular mechanisms underlying membrane detachment of bacterial lipoproteins. Mol. Microbiol. *62*, 1064–1075.

Iwasaki, A., and Medzhitov, R. (2010). Regulation of adaptive immunity by the innate immune system. Science *327*, 291–295.

Jarlier, V., and Nikaido, H. (1990). Permeability barrier to hydrophilic solutes In Mycobacterium chelonei. J. Bacteriol. *172*, 1418–1423.

Jin, M.S., Kim, S.E., Heo, J.Y., Lee, M.E., Kim, H.M., Paik, S.G., Lee, H., and Lee, J.O. (2007). Crystal structure of the TLR1–TLR2 heterodimer induced by binding of a tri-acylated lipopeptide. Cell *130*, 1071–1082.

Juncker, A.S., Willenbrock, H., Von Heijne, G., Brunak, S., Nielsen, H., and Krogh, A. (2003). Prediction of lipoprotein signal peptides in Gram-negative bacteria. Protein Sci. *12*, 1652–1662.

Kalscheuer, R., Weinrick, B., Veeraraghavan, U., Besra, G.S., and Jacobs, W.R. (2010). Trehalose-recycling ABC transporter LpqY-SugA-SugB-SugC is essential for virulence of *Mycobacterium tuberculosis*. Proc. Natl. Acad. Sci. U.S.A. *107*, 21761–21766.

Kang, H.J., Ha, J.M., Kim, H.S., Lee, H., Kurokawa, K., and Lee, B.L. (2011). The role of phagocytosis in IL-8 production by human monocytes in response to lipoproteins on *Staphylococcus aureus*. Biochem. Biophys. Res. Commun. *406*, 449–453.

Kang, J.Y., Nan, X., Jin, M.S., Youn, S.J., Ryu, Y.H., Mah, S., Han, S.H., Lee, H., Paik, S.G., and Lee, J.O. (2009). Recognition of lipopeptide patterns by Toll-like receptor 2-Toll-like receptor 6 heterodimer. Immunity *31*, 873–884.

Kaufmann, S.H.E., Hussey, G., and Lambert, P.-H. (2010). New vaccines for tuberculosis. Lancet *375*, 2110–2119.

Kiho, T., Nakayama, M., Yasuda, K., Miyakoshi, S., Inukai, M., and Kogen, H. (2003). Synthesis and antimicrobial activity of novel globomycin analogues. Bioorg. Med. Chem. *13*, 2315–2318.

Kiho, T., Nakayama, M., Yasuda, K., Miyakoshi, S., Inukai, M., and Kogen, H. (2004). Structure–activity relationships of globomycin analogues as antibiotics. Bioorg. Med. Chem. *12*, 337–361.

Kim, D.Y., Jin, K.S., Kwon, E., Ree, M., and Kim, K.K. (2007). Crystal structure of RseB and a model of its binding mode to RseA. Proc. Natl. Acad. Sci. U.S.A. *104*, 8779–8784.

Klein, J.R., Henrich, B., and Plapp, R. (1991). Molecular analysis and nucleotide sequence of the envCD operon of *Escherichia coli*. Mol. Gen. Genet. *230*, 230–240.

Kobayashi, N., Nishino, K., Hirata, T., and Yamaguchi, A. (2003). Membrane topology of ABC-type macrolide antibiotic exporter MacB in *Escherichia coli*. FEBS Lett. *546*, 241–246.

Kovacs-Simon, A., Titball, R.W., and Michell, S.L. (2011). Lipoproteins of bacterial pathogens. Infect. Immun. *79*, 548–561.

Kurokawa, K., Lee, H., Roh, K.B., Asanuma, M., Kim, Y.S., Nakayama, H., Shiratsuchi, A., Choi, Y., Takeuchi, O., Kang, H.J., *et al.* (2009). The triacylated ATP binding cluster transporter substrate-binding lipoprotein of *Staphylococcus aureus* functions as a native ligand for Toll-like receptor 2. J. Biol. Chem. *284*, 8406–8411.

Kurokawa, K., Ryu, K.H., Ichikawa, R., Masuda, A., Kim, M.S., Lee, H., Chae, J.H., Shimizu, T., Saitoh, T., Kuwano, K., *et al.* (2012a). Novel bacterial lipoprotein structures conserved in low-GC content Gram-positive bacteria are recognized by Toll-like receptor 2. J. Biol. Chem. *287*, 13170–13181.

Kurokawa, K., Kim, M.S., Ichikawa, R., Ryu, K.H., Dohmae, N., Nakayama, H., and Lee, B.L. (2012b). Environment-mediated accumulation of diacyl lipoproteins over their triacyl counterparts in *Staphylococcus aureus*. J. Bacteriol. *194*, 3299–3306.

Lewenza, S., Vidal-Ingigliardi, D., and Pugsley, A.P. (2006). Direct visualization of red fluorescent lipoproteins indicates conservation of the membrane sorting rules in the family Enterobacteriaceae. J. Bacteriol. *188*, 3516–3524.

Lowy, F.D. (1998). *Staphylococcus aureus* infections. N. Engl. J. Med. *339*, 520–532.

Machata, S., Tchatalbachev, S., Mohamed, W., Jänsch, L., Hain, T., and Chakraborty, T. (2008). Lipoproteins of *Listeria monocytogenes* are critical for virulence and TLR2-mediated immune activation. J. Immunol. *181*, 2028–2035.

Mahapatra, S., Basu, J., Brennan, P.J., and Crick, D.C. (2005). Structure, biosynthesis, and genetics of the mycolic acid–arabinogalactan-peptidoglycan complex. In Tuberculosis and the Tubercle Bacillus, Cole, S.T., Eisenach, K.D., McMurray, D.N., and Jacobs, W.R., eds. (ASM Press, Washington, D.C.), pp. 275–285.

Makarov, V., Manina, G., Mikusova, K., Möllmann, U., Ryabova, O., Saint-Joanis, B., Dhar, N., Pasca, M.R., Buroni, S., Lucarelli, A.P., *et al.* (2009). Benzothiazinones kill *Mycobacterium tuberculosis* by blocking arabinan synthesis. Science *324*, 801–804.
Malen, H., Pathak, S., Softeland, T., de Souza, G.A., and Wiker, H.G. (2010). Definition of novel cell envelope associated proteins in Triton X-114 extracts of *Mycobacterium tuberculosis* H37Rv. BMC Microbiol. *10*, 132.
Malinverni, J.C., and Silhavy, T.J. (2009). An ABC transport system that maintains lipid asymmetry in the gram-negative outer membrane. Proc. Natl. Acad. Sci. U.S.A. *106*, 8009–8014.
Masuda, K., Matsuyama, S., and Tokuda, H. (2002). Elucidation of the function of lipoprotein-sorting signals that determine membrane localization. Proc. Natl. Acad. Sci. U.S.A. *99*, 7390–7395.
Matsuyama, S., Tajima, T., and Tokuda, H. (1995). A novel periplasmic carrier protein involved in the sorting and transport of *Escherichia coli* lipoproteins destined for the outer membrane. EMBO J. *14*, 3365–3372.
Matsuyama, S., Yokota, N., and Tokuda, H. (1997). A novel outer membrane lipoprotein, LolB (HemM), involved in the LolA (p20)-dependent localization of lipoproteins to the outer membrane of *Escherichia coli*. EMBO J. *16*, 6947–6955.
Mawuenyega, K.G., Forst, C.V., Dobos, K.M., Belisle, J.T., Chen, J., Bradbury, E.M., Bradbury, A.R.M., and Chen, X. (2005). *Mycobacterium tuberculosis* functional network analysis by global subcellular protein profiling. Mol. Biol. Cell *16*, 396–404.
McDonough, J.A., McCann, J.R., TekippeE., Silverman, J.S., Rigel, N.W., and Braunstein, M. (2008). Identification of functional Tat signal sequences in *Mycobacterium tuberculosis* proteins. J. Bacteriol. *190*, 6428–6438.
Minnikin, D.E. (1982). Lipids: complex lipids, their chemistry, biosynthesis and roles. In The biology of the Mycobacteria, Ratledge, C., and Stanford, J., eds. (Academic Press. Inc., New York), pp. 95–184.
Miyadai, H., Tanaka-Masuda, K., Matsuyama, S., and Tokuda, H. (2004). Effects of lipoprotein overproduction on the induction of DegP (HtrA) involved in quality control in the *Escherichia coli* periplasm. J. Biol. Chem. *279*, 39807–39813.
Miyamoto, S., and Tokuda, H. (2007). Diverse effects of phospholipids on lipoprotein sorting and ATP hydrolysis by the ABC transporter LolCDE complex. Biochim. Biophys. Acta *1768*, 1848–1854.
Miyamoto, A., Matsuyama, S., and Tokuda, H. (2001). Mutant of LolA, a lipoprotein-specific molecular chaperone of *Escherichia coli*, defective in the transfer of lipoproteins to LolB. Biochem. Biophys. Res. Commun. *287*, 1125–1128.
Mühlradt, P.F., Kieß, M., Meyer, H., Süßmuth, R., and Jung, G. (1997). Isolation, structure elucidation, and synthesis of a macrophage stimulatory lipopeptide from *Mycoplasma fermentans* acting at picomolar concentration. J. Exp. Med. *185*, 1951–1958.
Mukhopadhyay, S., and Balaji, K.N. (2011). The PE and PPE proteins of *Mycobacterium tuberculosis*. Tuberculosis *91*, 441–447.
Murphy, R.C., and Gaskell, S.J. (2011). New applications of mass spectrometry in lipid analysis. J. Biol. Chem. *286*, 25427–25433.
Navarre, W.W., Daefler, S., and Schneewind, O. (1996). Cell wall sorting of lipoproteins in *Staphylococcus aureus*. J. Bacteriol. *178*, 441–446.
Nakada, S., Sakakura, M., Takahashi, H., Okuda, S., Tokuda, H., and Shimada, I. (2009). Structural investigation of the interaction between LolA and LolB using NMR. J. Biol. Chem. *284*, 24634–24643.
Nakayama, H., Kurokawa, K., and Lee, B.L. (2012). Lipoproteins in bacteria: structures and biosynthetic pathways. FEBS J. *279*, 4247–4268.
Narita, S. (2011). ABC transporters involved in the biogenesis of the outer membrane in gram-negative bacteria. Biosci. Biotechnol. Biochem. *75*, 1044–1054.
Narita, S., and Tokuda, H. (2007). Amino acids at positions 3 and 4 determine the membrane specificity of *Pseudomonas aeruginosa* lipoproteins. J. Biol. Chem. *282*, 13372–13378.
Narita, S., and Tokuda, H. (2011). Overexpression of LolCDE allows deletion of the *Escherichia coli* gene encoding apolipoprotein N-acyltransferase. J. Bacteriol. *193*, 4832–4840.
Narita, S., Tanaka, K., Matsuyama, S., and Tokuda, H. (2002). Disruption of lolCDE, encoding an ATP-binding cassette transporter, is lethal for *Escherichia coli* and prevents release of lipoproteins from the inner membrane. J. Bacteriol. *184*, 1417–1422.
Narita, S., Kanamaru, K., Matsuyama, S., and Tokuda, H. (2003). A mutation in the membrane subunit of an ABC transporter LolCDE complex causing outer membrane localization of lipoproteins against their inner membrane-specific signals. Mol. Microbiol. *49*, 167–177.

Nguyen, H.T., Wolff, K.A., Cartabuke, R.H., Ogwang, S., and Nguyen, L. (2010). A lipoprotein modulates activity of the MtrAB two-component system to provide intrinsic multidrug resistance, cytokinetic control and cell wall homeostasis in Mycobacterium. Mol. Microbiol. *76*, 348–364.
Nguyen, K.T., Piastro, K., and Derbyhsire, K.M. (2009). LpqM, a mycobacterial lipoprotein-metalloproteinase, is required for conjugal DNA transfer in *Mycobacterium smegmatis*. J. Bacteriol. *191*, 2721–2727.
Niederweis, M., Danilchanka, O., Huff, J., Hoffmann, C., and Engelhardt, H. (2010). Mycobacterial outer membranes: in search of proteins. Trends Microbiol. *18*, 109–116.
Nielsen, J.B., and Lampen, J.O. (1983). Beta-lactamase III of *Bacillus cereus* 569: membrane lipoprotein and secreted protein. Biochemistry *22*, 4652–4656.
Nikaido, H. (1998). Multiple antibiotic resistance and efflux. Curr. Opin. Microbiol. *1*, 516–523.
Oguchi, Y., Takeda, K., Watanabe, S., Yokota, N., Miki, K., and Tokuda, H. (2008). Opening and closing of the hydrophobic cavity of LolA coupled to lipoprotein binding and release. J. Biol. Chem. *283*, 25414–25420.
Oku, Y., Kurokawa, K., Matsuo, M., Yamada, S., Lee, B.L., and Sekimizu, K. (2009). Pleiotropic roles of polyglycerolphosphate synthase of lipoteichoic acid in growth of *Staphylococcus aureus* cells. J. Bacteriol. *191*, 141–151.
Okuda, S., and Tokuda, H. (2009). Model of mouth-to-mouth transfer of bacterial lipoproteins through inner membrane LolC, periplasmic LolA, and outer membrane LolB. Proc. Natl. Acad. Sci. U.S.A. *106*, 5877–5882.
Okuda, S., and Tokuda, H. (2011). Lipoprotein Sorting in Bacteria. Annu. Rev. Microbiol. *65*, 239–259.
Okuda, S., Watanabe, S., and Tokuda, H. (2008). A short helix in the C-terminal region of LolA is important for the specific membrane localization of lipoproteins. FEBS Lett. *582*, 2247–2251.
Omueti, K.O., Beyer, J.M., Johnson, C.M., Lyle, E.A., and Tapping, R.I. (2005). Domain exchange between human Toll-like receptors 1 and 6 reveals a region required for lipopeptide discrimination. J. Biol. Chem. *280*, 36616–36625.
Pathania, R., Zlitni, S., Barker, C., Das, R., Gerritsma, D.A., Lebert, J., Awuah, E., Melacini, G., Capretta, F.A., and Brown, E.D. (2009). Chemical genomics in *Escherichia coli* identifies an inhibitor of bacterial lipoprotein targeting. Nat. Chem. Biol. *5*, 849–856.
Prados-Rosales, R., Baena, A., Martinez, L.R., Luque-Garcia, J., Kalscheuer, R., Veerarghavan, U., Camara, C., Nosanchuk, J.D., Besra, G.S., Chen, B., *et al.* (2011). Mycobacteria release active membrane vesciles that modulate immune responses in a TLR2-dependent manner in mice. J. Clin. Invest. *121*, 1471–1483.
Raivio, T.L., and Silhavy, T.J. (2001). Periplasmic stress and ECF sigma factors. Annu. Rev. Microbiol. *55*, 591–624.
Ramesh, G., Alvarez, A.L., Roberts, E.D., Dennis, V.A., Lasater, B.L., Alvarez, X., and Philipp, M.T. (2003). Pathogenesis of Lyme neuroborreliosis: *Borrelia burgdorferi* lipoproteins induce both proliferation and apoptosis in rhesus monkey astrocytes. Eur. J. Immunol. 33, 2539–2550.
Ramos, H.C., Rumbo, M., and Sirard, J.C. (2004). Bacterial flagellins: mediators of pathogenicity and host immune responses in mucosa. Trends Microbiol. *12*, 509–517.
Rampini, S.K., Selchow, P., Keller, C., Ehlers, S., Böttger, E.C., and Sander, P. (2008). LspA-inactivation in *Mycobacterium tuberculosis* results in attenuation without affecting phagosome maturation arrest. Microbiology *154*, 2991–3001.
Remans, K., Pauwels, K., van Ulsen, P., Buts, L., Cornelis, P., Tommassen, J., Savvides, S.N., Decanniere, K., and Van Gelder, P. (2010). Hydrophobic surface patches on LolA of *Pseudomonas aeruginosa* are essential for lipoprotein binding. J. Mol. Biol. *401*, 921–930.
Rezwan, M., Grau, T., Tschumi, A., and Sander, P. (2007). Lipoprotein synthesis in mycobacteria. Microbiology *153*, 652–658.
Rietveld, A.G., Koorengevel, M.C., and de Kruijff, B. (1995). Non-bilayer lipids are required for efficient protein transport across the plasma membrane of *Escherichia coli*. EMBO J. *14*, 5506–5513.
Robichon, C., Vidal-Ingigliardi, D., and Pugsley, A.P. (2005). Depletion of apolipoprotein N-acyltransferase causes mislocalization of outer membrane lipoproteins in *Escherichia coli*. J. Biol. Chem. *280*, 974–983.
Ruiz, N., Gronenberg, L.S., Kahne, D., and Silhavy, T.J. (2008). Identification of two inner-membrane proteins required for the transport of lipopolysaccharide to the outer membrane of *Escherichia coli*. Proc. Natl. Acad. Sci. U.S.A. *105*, 5537–5542.
Russell, D.G. (2011). *Mycobacterium tuberculosis* and the intimate discourse of a chronic infection. Immunol. Rev. *240*, 252–268.
Ryan, K.S., Balibar, C.J., Turo, K.E., Walsh, C.T., and Drennan, C.L. (2008). The violacein biosynthetic enzyme VioE shares a fold with lipoprotein transporter proteins. J. Biol. Chem. *283*, 6467–6475.

Ryu, Y., and Schultz, P.G. (2006). Efficient incorporation of unnatural amino acids into proteins in *Escherichia coli*. Nat. Methods *3*, 263–265.

Sakamoto, C., Satou, R., Tokuda, H., and Narita, S. (2010). Novel mutations of the LolCDE complex causing outer membrane localization of lipoproteins despite their inner membrane-retention signals. Biochem. Biophys. Res. Commun. *401*, 586–591.

Sander, P., Rezwan, M., Walker, B., Rampini, S.K., Kroppenstedt, R.M., Ehlers, S., Keller, C., Keeble, J.R., Hagemeier, M., Colston, M.J., *et al.* (2004). Lipoprotein processing is required for virulence of *Mycobacterium tuberculosis*. Mol. Microbiol. *52*, 1543–1552.

Sankaran, K., and Wu, H.C. (1994). Lipid modification of bacterial prolipoprotein. Transfer of diacylglyceryl moiety from phosphatidylglycerol. J. Biol. Chem. *269*, 19701–19706.

Sartain, M.J., and Belisle, J.T. (2009). N–terminal clustering of the O-glycosylation sites in the *Mycobacterium tuberculosis* lipoprotein SodC. Glycobiology *19*, 38–51.

Sartain, M.J., Dick, D.L., Rithner, C.D., Crick, D.C., and Belisle, J.T. (2011). Lipidomic analyses of *Mycobacterium tuberculosis* based on accurate mass measurements and the novel 'Mtb LipidDB'. J. Lipid Res. *52*, 861–872.

Sassetti, C.M., and Rubin, E.J. (2003). Genetic requirements for mycobacterial survival during infection. Proc. Natl. Acad. Sci. U.S.A. *100*, 12989–12994.

Sassetti, C.M., Boyd, D.H., and Rubin, E.J. (2003). Genes required for mycobacterial growth defined by high density mutagenesis. Mol. Microbiol. *48*, 77–84.

Schmaler, M., Jann, N.J., Götz, F., and Landmann, R. (2010). Staphylococcal lipoproteins and their role in bacterial survival in mice. Int. J. Med. Microbiol. *300*, 155–160.

Schulze, R.J., Chen, S., Kumru, O.S., and Zuckert, W.R. (2010). Translocation of *Borrelia burgdorferi* surface lipoprotein OspA through the outer membrane requires an unfolded conformation and can initiate at the C-terminus. Mol. Microbiol. *76*, 1266–1278.

Scragg, I.G., Kwiatkowski, D., Vidal, V., Reason, A., Paxton, T., Panico, M., Dell, A., and Morris, H. (2000). Structural characterization of the inflammatory moiety of a variable major lipoprotein of *Borrelia recurrentis*. J. Biol. Chem. *275*, 937–941.

Seiffer, D., Klein, J.R., and Plapp, R. (1993). EnvC, a new lipoprotein of the cytoplasmic membrane of *Escherichia coli*. FEMS Microbiol. Lett. *107*, 175–178.

Serebryakova, M.V., Demina, I.A., Galyamina, M.A., Kondratov, I.G., Ladygina, V.G., and Govorun, V.M. (2011). The acylation state of surface lipoproteins of mollicute *Acholeplasma laidlawii*. J. Biol. Chem. *286*, 22769–22776.

Shimada, T., Park, B.G., Wolf, A.J., Brikos, C., Goodridge, H.S., Becker, C.A., Reyes, C.N., Miao, E.A., Aderem, A., Götz, F., *et al.* (2010). *Staphylococcus aureus* evades lysozyme-based peptidoglycan digestion that links phagocytosis, inflammasome activation, and IL-1β secretion. Cell Host Microbe 7, 38–49.

Shimizu, T., Kida, Y., and Kuwano, K. (2007). Triacylated lipoproteins derived from *Mycoplasma pneumoniae* activate nuclear factor-κB through Toll-like receptors 1 and 2. Immunology *121*, 473–483.

Shimizu, T., Kida, Y., and Kuwano, K. (2008). A triacylated lipoprotein from *Mycoplasma genitalium* activates NF-κB through Toll-like receptor 1 (TLR1) and TLR2. Infect. Immun. *76*, 3672–3678.

Shruthi, H., Babu, M.M., and Sankaran, K. (2010). TAT-pathway-dependent lipoproteins as a niche-based adaptation in prokaryotes. J. Mol. Evol. *70*, 359–370.

Sklar, J.G., Wu, T., Gronenberg, L.S., Malinverni, J.C., Kahne, D., and Silhavy, T.J. (2007). Lipoprotein SmpA is a component of the YaeT complex that assembles outer membrane proteins in *Escherichia coli*. Proc. Natl. Acad. Sci. U.S.A. *104*, 6400–6405.

Smith, P.C., Karpowich, N., Millen, L., Moody, J.E., Rosen, J., Thomas, P.J., and Hunt, J.F. (2002). ATP binding to the motor domain from an ABC transporter drives formation of a nucleotide sandwich dimer. Mol. Cell *10*, 139–149.

Sperandeo, P., Cescutti, R., Villa, R., Di Benedetto, C., Candia, D., Deho, G., and Polissi, A. (2007). Characterization of lptA and lptB, two essential genes implicated in lipopolysaccharide transport to the outer membrane of *Escherichia coli*. J. Bacteriol. *189*, 244–253.

Sperandeo, P., Deho, G., and Polissi, A. (2009). The lipopolysaccharide transport system of Gram-negative bacteria. Biochim. Biophys. Acta *1791*, 594–602.

Stoll, H., Dengjel, J., Nerz, C., and Götz, F. (2005). *Staphylococcus aureus* deficient in lipidation of prelipoproteins is attenuated in growth and immune activation. Infect. Immun. *73*, 2411–2423.

Sulzenbacher, G., Canaan, S., Bordat, Y., Neyrolles, O., Stadthagen, G., Roig-Zamboni, V., Rauzier, J., Maurin, D., Laval, F., Daffe, M., *et al.* (2006). LppX is a lipoprotein required for the translocation of phthiocerol dimycocerosates to the surface of *Mycobacterium tuberculosis*. EMBO J. *25*, 1436–1444.

Sutcliffe, I.C., and Harrington, D.J. (2004). Lipoproteins of *Mycobacterium tuberculosis*: an abundant and functionally diverse class of cell envelope components. FEMS Microbiol. Rev. *28*, 645–659.

Sutcliffe, I.C., and Russell, R.R. (1995). Lipoproteins of Gram-positive bacteria. J. Bacteriol. *177*, 1123–1128.

Tajima, T., Yokota, N., Matsuyama, S., and Tokuda, H. (1998). Genetic analyses of the *in vivo* function of LolA, a periplasmic chaperone involved in the outer membrane localization of *Escherichia coli* lipoproteins. FEBS Lett. *439*, 51–54.

Takeda, K., Miyatake, H., Yokota, N., Matsuyama, S., Tokuda, H., and Miki, K. (2003). Crystal structures of bacterial lipoprotein localization factors, LolA and LolB. EMBO J. *22*, 3199–3209.

Takeuchi, O., and Akira, S. (2010). Pattern recognition receptors and inflammation. Cell *140*, 805–820.

Takeuchi, O., Hoshino, K., Kawai, T., Sanjo, H., Takada, H., Ogawa, T., Takeda, K., and Akira, S. (1999). Differential roles of TLR2 and TLR4 in recognition of Gram-negative and Gram-positive bacterial cell wall components. Immunity *11*, 443–451.

Takeuchi, O., Kawai, T., Mühlradt, P.F., Morr, M., Radolf, J.D., Zychlinsky, A., Takeda, K., and Akira, S. (2001). Discrimination of bacterial lipoproteins by Toll-like receptor 6. Int. Immunol. *13*, 933–940.

Takeuchi, O., Sato, S., Horiuchi, T., Hoshino, K., Takeda, K., Dong, Z., Modlin, R.L., and Akira, S. (2002). Cutting edge: role of Toll-like receptor 1 in mediating immune response to microbial lipoproteins. J. Immunol. *169*, 10–14.

Tanaka, K., Matsuyama, S.I., and Tokuda, H. (2001). Deletion of lolB, encoding an outer membrane lipoprotein, is lethal for *Escherichia coli* and causes accumulation of lipoprotein localization intermediates in the periplasm. J. Bacteriol. *183*, 6538–6542.

Tanaka, S.Y., Narita, S., and Tokuda, H. (2007). Characterization of the *Pseudomonas aeruginosa* Lol system as a lipoprotein sorting mechanism. J. Biol. Chem. *282*, 13379–13384.

Taniguchi, N., Matsuyama, S., and Tokuda, H. (2005). Mechanisms underlying energy-independent transfer of lipoproteins from LolA to LolB, which have similar unclosed β-barrel structures. J. Biol. Chem. *280*, 34481–34488.

Tao, K., Watanabe, S., Narita, S., and Tokuda, H. (2010). A periplasmic LolA derivative with a lethal disulfide bond activates the Cpx stress response system. J. Bacteriol. *192*, 5657–5662.

Tawaratsumida, K., Furuyashiki, M., Katsumoto, M., Fujimoto, Y., Fukase, K., Suda, Y., and Hashimoto, M. (2009). Characterization of N-terminal structure of TLR2-activating lipoprotein in *Staphylococcus aureus*. J. Biol. Chem. *284*, 9147–9152.

Terada, M., Kuroda, T., Matsuyama, S.I., and Tokuda, H. (2001). Lipoprotein sorting signals evaluated as the LolA-dependent release of lipoproteins from the cytoplasmic membrane of *Escherichia coli*. J. Biol. Chem. *276*, 47690–47694.

Thompson, B.J., Widdick, D.J., Hicks, M.G., Chandra, G., Sutcliffe, I.C., Palmer, T., and Hutchings, M.I. (2010). Investigating lipoprotein biogenesis and function in the model Gram-positive bacterium *Streptomyces coelicolor*. Mol. Microbiol. *77*, 943–957.

Tokuda, H. (2009). Biogenesis of outer membranes in Gram-negative bacteria. Biosci. Biotechnol. Biochem. *73*, 465–473.

Tokuda, H., and Matsuyama, S. (2004). Sorting of lipoproteins to the outer membrane in *E. coli*. Biochim. Biophys. Acta *1693*, 5–13.

Tschumi, A., Nai, C., Auchli, Y., Hunziker, P., Gehrig, P., Keller, P., Grau, T., and Sander, P. (2009). Identification of apolipoprotein N-acyltransferase (Lnt) in mycobacteria. J. Biol. Chem. *284*, 27146–27156.

Tschumi, A., Grau, T., Albrecht, D., Rezwan, M., Antelmann, H., and Sander, P. (2012). Functional analyses of mycobacterial lipoprotein diacylglyceryl transferase and comparative secretome analysis of a mycobacterial lgt mutant. J. Bacteriol. *194*, 3938–3949.

Tsukahara, J., Mukaiyama, K., Okuda, S., Narita, S., and Tokuda, H. (2009). Dissection of LolB function – lipoprotein binding, membrane targeting and incorporation of lipoproteins into lipid bilayers. FEBS J. *276*, 4496–4504.

Vandal, O.H., Pierini, L.M., Schnappinger, D., Nathan, C.F., and Ehrt, S. (2008). A membrane protein preserves intrabacterial pH in intraphagosomal *Mycobacterium tuberculosis*. Nat. Med. *14*, 849–854.

Vidal-Ingigliardi, D., Lewenza, S., and Buddelmeijer, N. (2007). Identification of essential residues in apolipoprotein N-acyl transferase, a member of the CN hydrolase family. J. Bacteriol. *189*, 4456–4464.

Volokhina, E.B., Beckers, F., Tommassen, J., and Bos, M.P. (2009). The β-barrel outer membrane protein assembly complex of *Neisseria meningitidis*. J. Bacteriol. *191*, 7074–7085.

Voulhoux, R., and Tommassen, J. (2004). Omp85, an evolutionarily conserved bacterial protein involved in outer-membrane-protein assembly. Res. Microbiol. *155*, 129–135.

Voulhoux, R., Bos, M.P., Geurtsen, J., Mols, M., and Tommassen, J. (2003). Role of a highly conserved bacterial protein in outer membrane protein assembly. Science *299*, 262–265.

Wang, L., Xie, J., and Schultz, P.G. (2006). Expanding the genetic code. Annu. Rev. Biophys. Biomol. Struct. *35*, 225–249.

Watanabe, S., Oguchi, Y., Yokota, N., and Tokuda, H. (2007). Large-scale preparation of the homogeneous LolA lipoprotein complex and efficient *in vitro* transfer of lipoproteins to the outer membrane in a LolB-dependent manner. Protein Sci. *16*, 2741–2749.

Watanabe, S., Oguchi, Y., Takeda, K., Miki, K., and Tokuda, H. (2008). Introduction of a lethal redox switch that controls the opening and closing of the hydrophobic cavity in LolA. J. Biol. Chem. *283*, 25421–25427.

Widdick, D.A., Hicks, M.G., Thompson, B.J., Tschumi, A., Chandra, G., Sutcliffe, I.C., Brülle, J.K., Sander, P., Palmer, T., and Hutchings, M.I. (2011). Dissecting the complete lipoprotein biogenesis pathway in Streptomyces scabies. Mol. Microbiol. *80*, 1395–1412.

Wolfe, L.M., Mahaffey, S.B., Kruh, N.A., and Dobos, K.M. (2010). Proteomic definition of the cell wall of *Mycobacterium tuberculosis*. J. Proteome Res. *9*, 5816–5826.

Wu, C.H. (1966). Biosynthesis of lipoproteins. In *Escherichia coli* and Salmonella, Cell Mol. Biol., Neidhardt, F.C., Curtiss, R. III, Ingraham, J.L., Lin, E.C.C., Low, K.B., Magasanik, B., Reznikoff, W.S., Riley, M., Schaechter, M., and Umbarger, H.E., eds. (ASM Press, Washington DC), 2nd ed, pp. 1005–1014.

Wu, T., Malinverni, J., Ruiz, N., Kim, S., Silhavy, T.J., and Kahne, D. (2005). Identification of a multicomponent complex required for outer membrane biogenesis in *Escherichia coli*. Cell *121*, 235–245.

Wu, T., McCandlish, A.C., Gronenberg, L.S., Chng, S.S., Silhavy, T.J., and Kahne, D. (2006). Identification of a protein complex that assembles lipopolysaccharide in the outer membrane of *Escherichia coli*. Proc. Natl. Acad. Sci. U.S.A. *103*, 11754–11759.

Yakushi, T., Yokota, N., Matsuyama, S., and Tokuda, H. (1998). LolA-dependent release of a lipid-modified protein from the inner membrane of *Escherichia coli* requires nucleoside triphosphate. J. Biol. Chem. *273*, 32576–32581.

Yakushi, T., Masuda, K., Narita, S., Matsuyama, S., and Tokuda, H. (2000). A new ABC transporter mediating the detachment of lipid-modified proteins from membranes. Nat. Cell Biol. *2*, 212–218.

Yamaguchi, K., Yu, F., and Inouye, M. (1988). A single amino acid determinant of the membrane localization of lipoproteins in *E. coli*. Cell *53*, 423–432.

Yamanaka, H., Kobayashi, H., Takahashi, E., and Okamoto, K. (2008). MacAB is involved in the secretion of *Escherichia coli* heat-stable enterotoxin II. J. Bacteriol. *190*, 7693–7698.

Yasuda, M., Iguchi-Yokoyama, A., Matsuyama, S., Tokuda, H., and Narita, S. (2009). Membrane topology and functional importance of the periplasmic region of ABC transporter LolCDE. Biosci. Biotechnol. Biochem. *73*, 2310–2316.

Yuan, Y.R., Blecker, S., Martsinkevich, O., Millen, L., Thomas, P.J., and Hunt, J.F. (2001). The crystal structure of the MJ0796 ATP-binding cassette. Implications for the structural consequences of ATP hydrolysis in the active site of an ABC transporter. J. Biol. Chem. *276*, 32313–32321.

Zähringer, U., Lindner, B., Inamura, S., Heine, H., and Alexander, C. (2008). TLR2 – promiscuous or specific? A critical re-evaluation of a receptor expressing apparent broad specificity. Immunobiology *213*, 205–224.

Zuber, B., Chami, M., Houssin, C., Dubochet, J., Griffiths, G., and Daffe, M. (2008). Direct visualization of the outer membrane of mycobacteria and corynebacteria in their native state. J. Bacteriol. *190*, 5672–5680.

The Fascinating Coat Surrounding Mycobacteria

5

Mamadou Daffé and Benoît Zuber

Abstract

The mycobacterial cell envelope is fascinating in several ways. First, its composition is unique by the exceptional lipid content, which consists of very long-chain (up to C90) fatty acids, the so-called mycolic acids, and a variety of exotic compounds. Second, these lipids are atypically organized into a Gram-negative-like outer membrane (mycomembrane) in these Gram-positive bacteria, as recently revealed by CEMOVIS, and this mycomembrane also contains pore-forming proteins. Third, the mycolic acids esterified a holistic heteropolysaccharide (arabinogalactan), which in turn is linked to the peptidoglycan to form the cell wall skeleton (CWS). In slow-growing pathogenic mycobacterial species, this giant structure is surrounded by a capsular layer composed mainly of polysaccharides, primarily a glycogen-like glucan. The CWS is separated from the plasma membrane by a periplasmic space. A challenging research avenue for the next decade comprises the identification of the components of the uptake and secretion machineries and the isolation and biochemical characterization of the mycomembrane.

Introduction

The bacterial cell envelope of mycobacteria, defined as the structure that surrounds the cytosol, is crucial for the bacterial physiology since it controls the transfer of materials into and out of the bacterium and protects the micro-organism from its environment. The genus *Mycobacterium* belongs to the order of *Corynebacteriales,* which also includes corynebacteria, nocardia, rhodococci and other related micro-organisms. Mycobacteria are responsible of important human diseases such as tuberculosis and leprosy; the etiologic agent of the former disease infects 2 billion people worldwide and kills 2 million people annually. Most of the pathogenic mycobacterial species grow very slowly (e.g. generation times of 24 h and 13 days for the tubercle and leprosy bacilli, respectively) whereas saprophytes are relatively rapid growers (e.g. doubling time of 6 h for the widely used genetically tractable *Mycobacterium smegmatis*).

Based on phylogenetic criteria, mycobacteria and related genera are classified as Gram-positive bacteria. Importantly, however, the chemical nature of their envelope is different from those of both Gram-positive and Gram-negative bacteria; for instance, the lipid content of the cell envelope of mycobacteria may represent up to 40% of the cell dry mass, compared to less than 5% in other Gram-positive bacteria and only 10% in Gram-negative bacteria (Goren and Brennan, 1979). Such a lipid-rich coat could explain both the tendency of mycobacteria to grow in clumps and their distinctive property of acid-fastness. Furthermore, mycobacteria exhibit a very low permeability to nutrients and antibacterial drugs, 10–100-fold lower than that of the notably impermeable bacillus *Pseudomonas aeruginosa* (Jarlier and Nikaido, 1990), a property attributed to the singularity of their cell envelope. Consistently, members of the *Corynebacteriales* order contain pore-forming proteins (porins) in their cell walls (Neiderweis *et al.*, 2010). It is only recently that the occurrence of a typical outer membrane in mycobacteria and corynebacteria consistent with the unique properties of these Gram-positive bacteria has been firmly established, thanks to the use of cryo-electron microscopy of vitreous section (CEMOVIS) (Hoffman *et al.*, 2008; Zuber *et al.*, 2008; Sani *et al.*, 2010). In addition, the chemical structure of many envelope major components has been shown to be chemically very different from what is found elsewhere. A schematic representation of the mycobacterial envelope based on chemical data and recent CEMOVIS data is shown in Fig. 5.2.

Architecture of the mycobacterial cell envelope

The structure of the mycobacterial cell envelope has been studied by transmission electron microscopy (TEM) of thin sections (Fig. 5.1). It always appears as a multi-layered structure but the aspect and the arrangement of the layers dramatically varies depending on the specimen preparation method (Bleck *et al.*, 2010). Obviously, this has hindered an accurate depiction of the molecular organization of the cell envelope. In conventional preparations for TEM, the specimen is first fixed, either by chemical cross-linking or by freezing. Water is then replaced by organic solvents prior to embedding in a hydrophobic resin. Finally, the resin is hardened by polymerization, ultrathin sections (50 nm) are produced and stained with heavy-metal salts. Lipids that compose the mycobacterial cell envelope are inefficiently

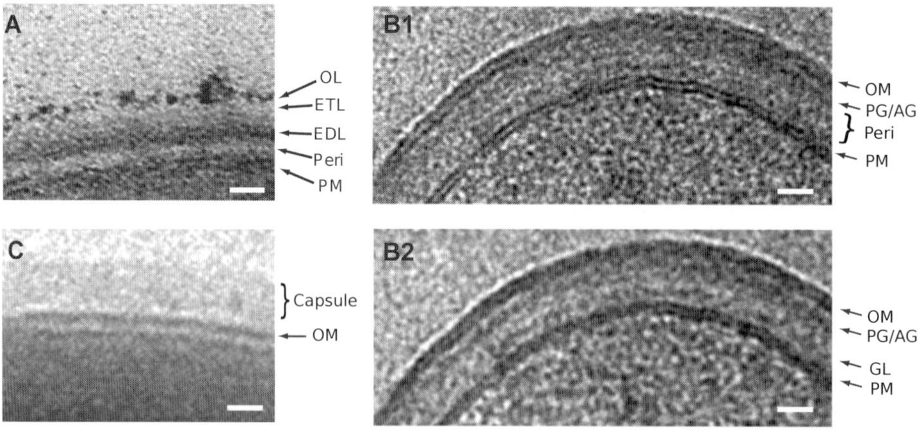

Figure 5.1 TEM micrographs of mycobacteria. (A) Conventional preparation. (B) CEMOVIS. B1 was acquired with a lower defocus than B2. The bilayer aspect of the plasma membrane and the outer membrane is clearly visible in B1, whereas the granular layer and the peptidoglycan/arabinogalactan layers are more obvious in B2. (C) Whole cell cryoEM showing the native capsule. EDL, electron-dense layer; ETL, electron-transparent layer; GL, granular layer; OL, outer layer; Peri, periplasm; PM, plasma membrane; OM, outer membrane; PG/AG, peptidoglycan/arabinogalactan. Bars, 20nm. A and B are adapted from Zuber (2008) and C is adapted from Sani (2010).

fixed by chemical cross-linkers and they can easily be extracted by solvents during the dehydration step. Furthermore, during that step, water-rich structures may collapse (Dubochet and Blanc, 2001). The use of different fixation protocols or different solvents can lead to the extraction and shrinkage of different components. Last but not least, the affinity of the various stains for the cell envelope components varies and is not well defined.

Recently, CEMOVIS of mycobacteria and cryo-electron microscopy (cryoEM) of whole-mount mycobacteria has revealed the native structure of the mycobacterial cell envelope (Hoffman *et al.*, 2008; Zuber *et al.*, 2008; Sani *et al.*, 2010) (Fig. 5.1). In these techniques, the specimens are observed at liquid nitrogen temperature, which allows them to be imaged in their fully hydrated and unstained state. Thus, the specimens do not suffer from extraction or shrinkage of their components. Moreover the contrast seen in the micrographs is directly related to the mass density distribution in the specimen. Images of the native mycobacterial cell envelope look significantly different from conventional images and have unequivocally revealed for the first time the presence of a long-predicted mycobacterial outer membrane (Fig. 5.1). In the next sections we will describe the molecular structure of the mycobacterial cell envelope in the light of these recent results and of the wealth of biochemical data.

From the combination of chemical and microscopic data, it is well established that the mycobacterial envelope consists of a plasma membrane, which is homologous to plasma membranes of other bacteria, surrounded by a complex wall of carbohydrates and lipids organized in an outer membrane bilayer, which is in turn surrounded by an outermost layer, called 'capsule' in the case of pathogenic species (Daffé and Draper, 1998). The occurrence of the latter outermost structure has been recently and elegantly demonstrated by CEMOVIS in both pathogenic and saprophytic mycobacterial species (Sani *et al.*, 2010). A compartment analogous to the periplasmic space in Gram-negative bacteria also exists in mycobacteria (Zuber *et al.*, 2008).

The plasma membrane

The basic structure of the plasma membrane of the mycobacterial cell envelope does not seems to differ from that of the other biological plasma membranes, as judged from their appearance in ultra-thin sections (Zuber *et al.*, 2008), the characterization of classical metabolic functions, and their known chemical composition. Isolated plasma membranes are typically obtained by breaking the cells by mechanical stress, e.g. sonication or shearing in the French pressure cell, followed by fractionation using differential centrifugation or density gradients (Rezwan *et al.*, 2007). Chemical and biochemical analyses of the fractions are essential to ensure their purity. As far as lipid composition of the plasma membranes is concerned, no obvious difference was found between those of rapid- and slow-growing *Mycobacterium* species examined. Polar lipids, mainly phospholipids, assemble, in association with proteins, into a lipid bilayer (see Minnikin, 1982).

The polar lipids of the mycobacterial plasma membrane are composed of hydrophilic head groups and fatty acid chains that usually consist of mixtures of straight-chain, unsaturated and mono-methyl branched fatty acid residues having less than 20 carbons. Palmitic ($C_{16:0}$), octadecenoic ($C_{18:1}$) and 10-methyloctadecanoic (or tuberculostearic, C_{19r}) are the major fatty acid constituents of the isolated plasma membranes. The main phospholipids of the plasma membrane are phosphatidylinositol mannosides (PIM), phosphatidylglycerol, cardiolipin and phosphatidylethanolamine; phosphatidylinositol occurs in small amounts.

Incidentally, PIM and phosphatidylethanolamine have also been identified among the lipids extracted from the cell surface layers by means of gentle mechanical treatment of cells with glass beads (Ortalo-Magné *et al.*, 1996) but the function of the phospholipids in those layers is unknown.

Besides the phospholipids, other lipids, such as menaquinones, may be present in the plasma membrane (see Minnikin, 1982). The mycobacterial lipopolysaccharide lipoarabinomannan (LAM; Fig. 5.2) may be partly located in the plasma membrane, presumably consisting of biosynthetic precursors and/or material that has yet to be transported to the outer membrane, the primarily location of the molecule in *M. smegmatis* (Dhiman *et al.*, 2011). In sharp contrast, LAM and its putative precursor lipomannan (LM) were most exclusively found in the purified plasma membrane fraction of *Corynebacterium glutamicum* (Marchand *et al.*, 2012). Such a localization is consistent with the chemical structure of

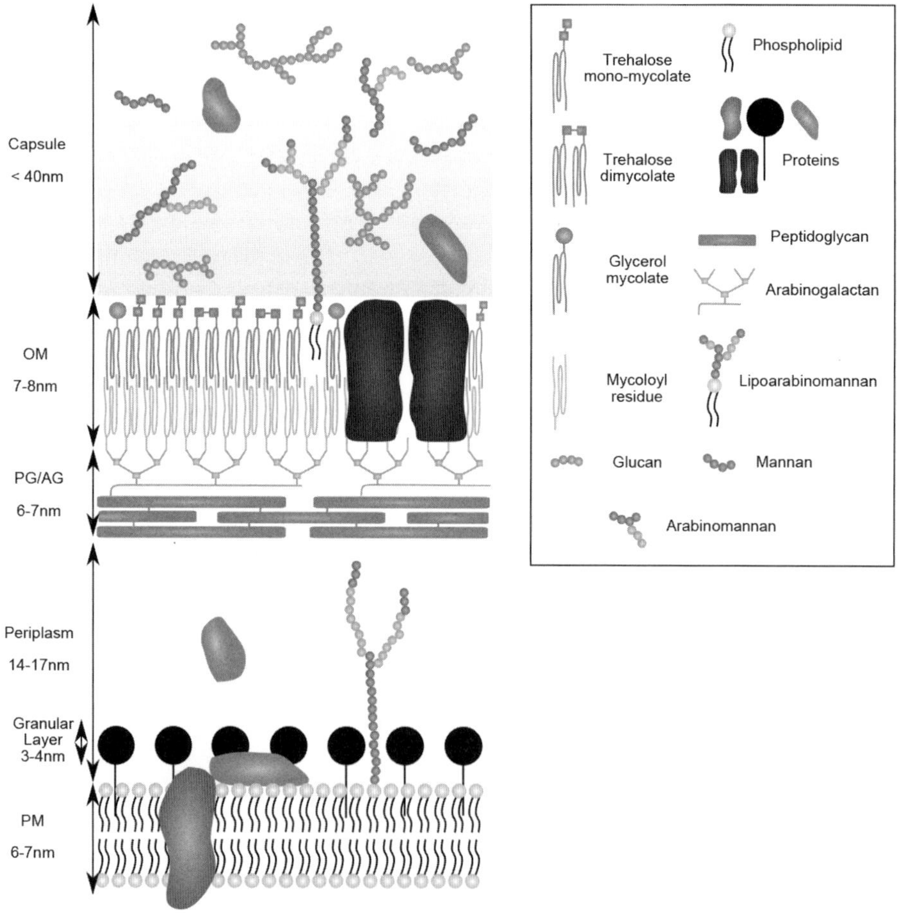

Figure 5.2 Model of the cell envelope and the capsule of mycobacteria. The cell envelope consists of (i) the plasma membrane, (ii) the periplasmic space, which contains the granular layer, (iii) the cell wall skeleton, which is made of peptidoglycan, arabinogalactan and mycoloyl residues, and (iv) the outer membrane. A colour version of this figure is located in the plate section at the back of the book.

LAM, which is composed of a phosphatidylinositol group (Hunter and Brennan, 1990) covalently linked to arabinomannan.

The mycobacterial plasma membrane appears symmetrical in CEMOVIS pictures but asymmetrical in images of some conventionally prepared cells, with the accumulation of material in its outer leaflet (Silva and Macedo, 1983a, 1984; Daffé *et al.*, 1989). This appearance, which depends on conditions of fixation (Silva and Macedo, 1983b), has been attributed to the presence of excess glycoconjugates in the thicker outer leaflet. Also, CEMOVIS shows a granular layer a few nm beyond the outer leaflet of the plasma membrane that might represent the globular head of membrane-bound molecules, such as membrane proteins (Fig. 5.1). It is also possible that the apparent thicker outer leaflet in conventional preparations is due to the collapse of the granular layer against the plasma membrane (Zuber *et al.*, 2008).

The periplasm

A low-density periplasmic space has long been described in Gram-negative bacteria. It is defined as the region located between the plasma membrane and the peptidoglycan cell wall and it contains numerous enzymes and other proteins. Recently the optimal sample preservation provided by CEMOVIS has revealed that Gram-positive bacteria also possess a periplasmic space between their plasma membrane and peptidoglycan (Matias and Beveridge, 2005, 2006) and that it contains a granular layer (Zuber *et al.*, 2006). Mycobacteria possess an analogous periplasmic space between the plasma membrane and the peptidoglycan-arabinogalactan cell wall (Zuber *et al.*, 2008). The periplasmic spaces of mycobacteria and of the other Gram-positive bacteria have a very similar aspect: they are approximately 20-nm thick and they contain a granular layer (potentially representing membrane proteins as explained above).

We proposed that the maintenance of the low-density periplasmic space of Gram-positive bacteria is ensured by the presence of large flexible polymers in it: they could generate the same osmolarity as the cytoplasmic osmolarity, thus preventing the cell from swelling until its plasma membrane is crushed against the cell wall; and because of their large size, these molecules could not diffuse away from the periplasm through the cell wall (Zuber *et al.*, 2006). Later, Matias and Beveridge demonstrated by CEMOVIS that the main constituent of the periplasmic space of *B. subtilis*, a Gram-positive bacterium, is plasma membrane-anchored lipoteichoic acids (Matias and Beveridge, 2008). Similarly, the analogous LAMs found in mycobacteria and related bacterial genera are probably localized in the periplasmic space. Both lipoteichoic acids and LAMs are long polymers, and therefore have the properties mentioned above and required for periplasmic space maintenance.

Cell wall

The cell wall skeleton

The wall of mycobacteria consists of a covalently linked 'cell wall skeleton' (CWS), an abundant variety of wall-associated lipids and a few polypeptides. The CWS is defined as the material remaining after removal of all non-covalently bound wall-associated substances: soluble proteins, lipids and glycans (Kotani *et al.*, 1959). It is a giant macromolecule surrounding the entire cell and chemically composed of three covalently linked constituents:

peptidoglycan, arabinogalactan and mycolates. Because of its rigidity, the CWS defines the shape of the mycobacterial cell. Mycobacterial walls are readily isolated and purified from fragments of plasma membrane and cytosolic material by breaking the cells with mechanical stress, followed by purification using differential centrifugation or density gradients to remove unbroken cells (see Rezwan *et al.*, 2007). It can be dissected into its constituent parts by relatively gentle methods, so that each part may be studied separately (see Daffé *et al.*, 1990; McNeil *et al.*, 1990, 1991).

The peptidoglycan is composed of repeating units of *N*-acetylglucosamine and *N*-acetyl/glycolylmuramic acid cross-linked by short peptides. The arabinogalactan is a complex, branched heteropolysaccharide that contains a galactan chain composed of alternating 5- and 6-linked D-galactofuranosyl residues (Fig. 5.2); three D-arabinan chains substituted the D-galactan chain in *M. tuberculosis*. In this species, two thirds of the non-reducing termini of the pentaarabinosyl motifs are esterified by mycolic acids, whereas half of these are occupied by mycoloyl residues in other mycobacterial species, including *M. leprae* and *M. bovis* BCG (McNeil *et al.*, 1991). Detailed structural and quantitative analysis of these mycoloyl residues has recently shown more subtle differences between *M. tuberculosis* and *M. leprae* (Bhamidi *et al.*, 2011). The cell wall of the *in vivo*-grown leprosy bacillus (an as yet *in vitro* non-cultivatable bacterium) was found to contain significantly more arabinogalactan-mycolic acids attached to peptidoglycan than that of *in vitro*-grown *M. tuberculosis*. Mycolic acids are very long-chain (up to C_{90}) α-branched and β-hydroxylated fatty acids (Daffé and Draper, 1998); they esterified the four hydroxyl groups at position 5 of both terminal and 2-linked D-arabinofuranosyl of the pentaarabinosyl motifs of arabinogalactan (McNeil *et al.*, 1991).

The wall appears in conventional transmission electron microscopy as an 'electron-dense' layer (EDL) surrounded more externally by an 'electron-transparent' layer (ETL) (Fig. 5.1). The electron density of the inner layer makes it likely to contain the peptidoglycan; for physical reasons, part of the covalently linked arabinogalactan must be located within the EDL as well (Daffé and Draper, 1998). The interpretation of the nature of the EDL has been elusive: the low density of the outer layer of the wall (the ETL) has usually been interpreted as an evidence that it is composed of mycolates; however, the occurrence of this layer in conventional sections of *Corynebacterineae* devoid of mycolic acids, e.g. *Corynebacterium amycolatum* (Puech *et al.*, 2001), has questioned the former interpretation.

The outer membrane and the arrangement of mycolates within it

Minnikin was the first to propose a model of a mycobacterial asymmetric outer membrane. In this model, mycolic acids are the main constituents of the inner leaflet, which is formed by the CWS mycolic acids covalently bound to the peptidoglycan-arabinogalactan. The outer leaflet is constituted of free polar mycolate trehaloses (Minnikin, 1982). In addition, he suggested that, because the mycolic acid chains are found with variable lengths, there would be gaps to accommodate some of the cell envelope polar and apolar lipids. This original model of an outer membrane analogous to the one found in Gram-negative bacteria, implied the existence of a second hydrophobic layer in addition to the plasma membrane and could explain the very low permeability to polar molecules. It could also provide a good explanation for the additional and main fracture plane observed in freeze fractures of mycobacteria (Barksdale and Kim, 1977; Benedetti *et al.*, 1984; Rulong *et al.*, 1991): the fracture would occur between the two leaflets of the outer membrane. This model was later supported by experimental data (see Draper, 1998), notably the work of Nikaido and his colleagues, who showed that an aqueous suspension of purified walls of *M. chelonae* produced an X-ray diffraction pattern with reflections characteristic of close-packed hydrocarbon chains in a semicrystalline monolayer (Nikaido *et al.*, 1993). Several variants on the basic Minnikin model have been proposed (McNeil and Brennan, 1991; Rastogi, 1991; Liu *et al.*, 1995) and discussed elsewhere (Draper, 1998). However, the complete lack of observation of this outer membrane bilayer in transmission electron micrographs of sectioned mycobacterial cells seriously questioned the validity of this model.

The issue has recently been resolved when the existence of a true outer membrane was established independently by three research groups (Hoffman *et al.*, 2008; Zuber *et al.*, 2008; Sani *et al.*, 2010) Using CEMOVIS, the authors have clearly visualized a symmetrical 7- to 8 nm-thick bilayer whose thickness was, surprisingly in view of the Minnikin's model, comparable to that of the plasma membrane in the analysed mycobacterial species (Fig. 5.1). The thickness of the outer membrane was even less in corynebacteria, 4–5 nm (Hoffman *et al.*, 2008; Zuber *et al.*, 2008). That the outer membrane contains mycolates has been proved by the absence of this structure in a mycolate-less corynebacterium mutant. The 7–8 nm thickness of the outer membrane layer raises the question of the arrangement of mycolates. In the original Minnikin's model, the long alkyl chains of mycolates organize parallel to one another; as mycolates have two alkyl chains of very unequal length (C_{40-60} versus C_{22-24}), this would create spaces in which alkyl chains of associated lipids would fit, resulting in a > 40 nm-thick asymmetrical bilayer (Minnikin, 1982). Consequently, alkyl chains have to be arranged in an intercalate or folded manner to be consistent with the observed micrographs. Consistent with the published chemical structures and fold of mycobacterial mycolates (Villeneuve *et al.*, 2005, 2007), we have proposed a model (Fig. 5.2) in which mycoloyl chains are intercalate in a zipper-like manner and where the long mycoloyl chains are folded in an ω-shape (Zuber *et al.*, 2008). Recently, LAM and LM have been shown to be primarily located in the outer membrane of *M. smegmatis*, presumably associated with its outer leaflet in analogy to the location of LPS in Gram-negative organisms (Dhiman *et al.*, 2011). This is not the case in *C. glutamicum* where only a tiny amount of LM, but no LAM, and phospholipids was found associated with the outer membrane.

The porins and putative outer membrane proteins

The existence of an outer membrane, as proposed in the various Minnikin-based models, poses the problem of getting small polar molecules across it, notably those needed for nutrition. Gram-negative bacteria solve this problem by producing specialized proteins called porins, which form hydrophilic pores through the structure. Pore-forming proteins have been found in the walls of both slow- and rapid-growing mycobacterial species such as *M. chelonae* and *M. smegmatis*, two rapidly growing species (Trias *et al.*, 1992; Trias and Benz, 1994; Niederweis *et al.*, 2010). The mycobacterial porins assemble to form pores spontaneously in artificial bilayer membranes, and have other properties characteristic of the Gram-negative porins (Niederweis *et al.*, 2010). The X-ray structure for the *M. smegmatis* major outer membrane protein MspA reveals a β-barrel trans-membrane channel, as is found for most outer membrane proteins in gram-negative bacteria (Faller *et al.*, 2004) (Fig. 5.3). MspA forms an octameric composite β-barrel where each monomer comprises two β-strands (Fig. 5.3). Based on amino acid distribution, the ~4 nm high β-barrel is believed to span the hydrophobic core of the mycobacterial OM bilayer, whilst the vestibule on top of

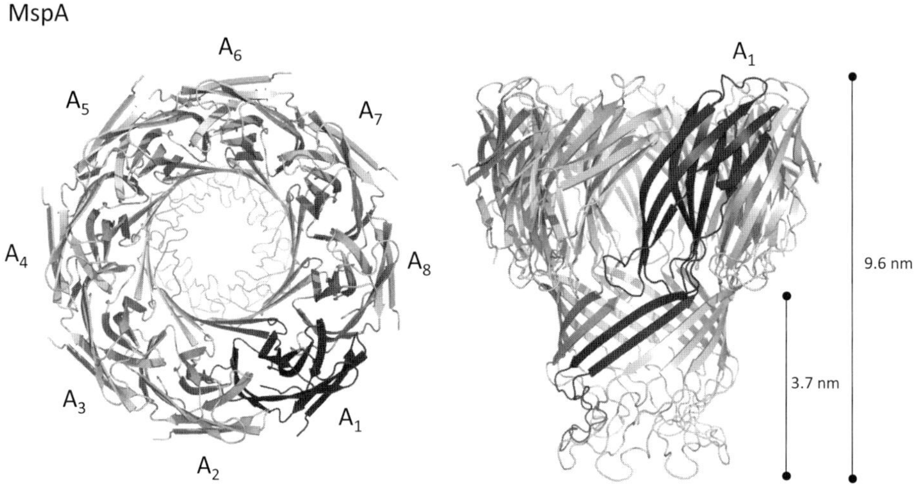

Figure 5.3 Structure of the *M. smegmatis* major outer membrane protein MspA (PDB: 1UUN). Ribbon diagram of top (left) and side (right) view of the octameric channel (monomers indicated $A_1 - A_8$). One monomer is represented in dark grey for clarity.

the channel is hydrophilic and is likely partly embedded in the mycolate hydrophilic heads and partly sticking out of the bilayer.

In Gram-negative bacteria, such as *Escherichia coli*, there are about 60 outer membrane proteins (OMPs). Nearly all integral OMPs adopt a β-barrel conformation and most of them function as open (porins) or gated channels that allow the uptake or the secretion of hydrophilic molecules. From *in silico* analysis two independent studies have proposed a list of putative OMPs of *M. tuberculosis* based essentially on β-barrel computational predictions (Niederweis *et al.*, 2010). Out of other 100 putative mycobacterial outer membrane proteins, only a few have been identified and characterized. These include mycoloyltransferases (11), some glycosylated proteins (Rpf proteins) (Hartmann *et al.*, 2004) or some virulence factors such as mycobacterial lipoprotein or Erp protein (Kocincova *et al.*, 2004). Among the numerous putative outer membrane proteins predicted in *C. glutamicum*, only the pore-forming proteins (PorA, PorB) and five mycoloyltransferases (cMytA, cMytB, cMytC, cMytD, cMytF) have been clearly identified (Marchand *et al.*, 2012).

The capsule

Microscopical aspects

Chapman was the first to describe the existence of a space between the wall of *M. lepraemurium* infecting eukaryotic cells and enclosed in phagosomes and the phagosomal membrane, which he termed the 'capsular space' (Chapman *et al.*, 1959). Soon after, Hanks tried to correlate the presence of an unstainable 'halo' observed around some pathogenic mycobacteria using light microscopic techniques with their permeability to dyes (Hanks, 1961b,c; Hanks *et al.*, 1961). He also has speculated that this corresponds to the 'electron-transparent zone' (ETZ) (Hanks, 1961a) that is seen in several published electron micrographs surrounding pathogenic mycobacteria living within host cells. A capsule was always seen around intracellular mycobacterial pathogens but not around non-pathogenic mycobacteria or pathogenic mycobacteria grown *in vitro* in standard conditions (Ryter *et al.*, 1984; Fréhel *et al.*, 1986). The knowledge that pathogenic mycobacterial species secrete abundant amounts of polysaccharides and proteins (Ortalo-Magné *et al.*, 1995; Lemassu *et al.*, 1996) prompted us to postulate that this material forms the ETZ, reasoning that what is shed into the culture medium of *in-vitro*-grown mycobacteria is retained by the phagosomal membrane (Daffé and Draper, 1998). The above hypothesis is supported by the appearance of a capsule around *in vitro*-grown bacteria that were first coated with either specific antibodies or pre-embedded in gelatine prior the different drastic treatments involved in conventional electron microscopy studies (Fréhel *et al.*, 1988; Daffé and Etienne, 1999). Under these conditions, a capsule-like structure, resembling the capsule observed *in vivo*, was observed around the pathogenic mycobacterial species examined but not around the non-pathogenic *M. smegmatis* and *M. aurum* (Fréhel *et al.*, 1988). When the same bacteria are instead processed by the conventional method commonly used to observe mycobacteria by transmission electron microscopy, the outer layer probably collapses by dehydration to give a thin dark layer (see Zuber *et al.*, 2008). Surprisingly, however, this layer was not observed in CEMOVIS pictures where a cryo-protectant must be used (Hoffman *et al.*, 2008; Zuber *et al.*, 2008). We have postulated that this lack of observation was due (1) to the density of the capsule probably matching the density of the cryo-protectant-containing freezing solution used and (2) to the fact that, under standard culture conditions used to prevent mycobacterial cell clumping, the capsular material would be shed in the medium. Consistently, a capsular layer was seen around all mycobacterial species examined when whole-mount plunge-frozen mycobacteria were imaged by cryoEM, which does not require the use of a cryo-protectant, and when they were grown without detergent and shaking (Sani *et al.*, 2010). Furthermore, when the outermost capsular compounds were extracted by using Tween 80 for short periods of time, the outermost layer disappeared (Sani *et al.*, 2010).

Note that plunge-frozen cells cannot be cryo-sectioned like in CEMOVIS and must be observed as whole cells. Consequently their thickness only allows their very outer part to be well resolved; the bilayer aspect of the plasma membrane, the granular layer, and the peptidoglycan–arabinogalactan layer are not visible; furthermore, the bilayer aspect of the outer membrane can be seen only in a few cases.

Chemical composition

The chemical identity of the capsular layer has been available for two host-grown mycobacterial species, namely *M. leprae* and *M. lepraemurium*. The 'fibrillar' substance that surrounds bacteria inside the phagocytic vacuoles of the host cell was found to be a glycopeptidolipid (GPL) in the case of *M. lepraemurium* (Draper, 1971) and a phenolic glycolipid (PGL) in the case of *M. leprae* (Boddingius and Dijkman, 1990). That the capsule is not always composed of lipids is deduced from the observation that both GPL-positive and GPL-negative strains of *M. avium-intracellulare* elaborate capsules (Rastogi *et al.*, 1989). Moreover, most strains of *M. tuberculosis* are devoid of PGL (Daffé *et al.*, 1987).

Mycobacterial species secrete abundant amounts of polysaccharides and proteins in their environment (Ortalo-Magné *et al.*, 1995; Lemassu *et al.*, 1996) and more extracellular material was recovered from the culture filtrates of the pathogenic species, e.g. *M. tuberculosis* and *M. kansasii*, than those of saprophytic and non-pathogenic strains such as *M. smegmatis* and

M. aurum (Lemassu and Daffé, 1994; Lemassu *et al.*, 1996). Shaking gently the bacterial pellicles with glass beads and the use of Tween 80 for short period of times (Ortalo-Magné *et al.*, 1995, 1996; Raynaud *et al.*, 1998) extract the outermost capsular compounds. Importantly, these treatments do not affect the viability of cells that are declumped, as judged by electron microscopy (Ortalo-Magné *et al.*, 1995, 1996). The main components of the outermost capsular layer of slow-growing mycobacterial species, most of which are pathogens (e.g. *M. tuberculosis* and *M. kansasii*), are polysaccharides whereas the major components of the outer layer of rapid growers (e.g. *M. phlei* and *M. smegmatis*) are proteins (Ortalo-Magné *et al.*, 1995; Lemassu *et al.*, 1996).

The major extracellular and capsular component of slow-growing mycobacterial species is a glucan (Fig. 5.2) that is composed of repeating units of five or six α(1->4) linked D-glucosyl residues substituted at position 6 with mono- or oligoglucosyl residues (Lemassu and Daffé, 1994; Ortalo-Magné *et al.*, 1995; Dinadayala *et al.*, 2003). The surface-exposed and extracellular materials also contain a D-arabino-D-mannan (Fig. 5.2), a heteropolysaccharide that exhibits an apparent molecular mass of 13 kDa, and a mannan chain composed of a α(1->6)-D-mannosyl core, with some units substituted with α-D-mannose at position 2.

The outermost part of the mycobacterial cell envelope contains only a tiny amount of lipid (2–3% of the surface-exposed material), and progressive removal of the capsular material shows that most of the lipids are in the inner rather than the outer part of the capsule (Ortalo-Magné *et al.*, 1996). Some of the species- and type-specific lipids and glycolipids, e.g. phthiocerol dimycocerosates, PGL and GPL, can be found on the surface of the capsule, in agreement with serological and ultrastructural findings. Most of the other lipids occur in the capsule in deeper compartments since they are extracted by a 1-hour treatment with Tween 80 (Ortalo-Magné *et al.*, 1996).

The outermost capsular proteins of *in vitro*-grown *M. tuberculosis* are a complex mixture of polypeptides (Ortalo-Magné *et al.*, 1995). Some proteins seem to correspond to secreted polypeptides found in short-term culture filtrates whereas other appear as cell wall-associated (Daffé and Etienne, 1999). In fact, most of the hundreds of proteins identified in the short-term culture filtrate of *M. tuberculosis* by 2-D SDS-PAGE (Sonnenberg and Belisle, 1997) are also found in significant amounts in the surface-exposed material extracted by mild mechanical treatment of cells. This observation supports the concept that what is found in the culture filtrate of *in vitro*-grown cells is probably shed from the surface of the bacilli and, in an *in vivo* context, would be confined around the cells in the capsular layer (Daffé and Brennan, 1998). Components of the ESAT (transport) system have been recently identified by immuno-electron microscopy of cryosectioned mycobacteria (Sani *et al.*, 2010).

Concluding remarks

The chemical structure of most of the core components of the mycobacterial cell envelope has been determined a long time ago, establishing the singularity of the lipid coat. However, it is only recently the combination of CEMOVIS and whole-cell cryoEM, together with very gentle extraction conditions, has revealed the true structure of the mycobacterial cell envelope, suggesting the organization of the core. Although data that are consistent with the existence of a second lipid barrier in the cell envelope had been published, no sign of an outer lipid layer had ever been reported in electron micrographs of sectioned mycobacterial cells before their study by CEMOVIS. Its application to the study of various species of mycobacteria and wide-type and isogenic mycolate-less mutant of the related *Corynebacterium glutamicum* has proved the existence of (i) a Gram-negative-like outer membrane in these Gram-positive bacteria and (ii) a capsule-like outermost layer of protein and carbohydrate. Nearly 30 years after its introduction by Minnikin, the elegant 'chemical' model of an outer membrane could be validated and modified.

Characterization of the mycomembrane is essential for understanding the physiology and the pathogenicity of these bacteria. Identification of the components of the uptake and secretion machineries is a challenging research avenue for the next decade. Despite tremendous efforts on deciphering the mycobacterial cell envelope, the biochemical characterization of the mycomembrane is still far from being completed and only a few proteins has been identified and characterized so far. In mycobacteria, MspA is the archetype of mycomembrane protein and represents the major diffusion pathway for small hydrophilic solutes in *M. smegmatis*. Additional proteins are putatively anchored, or associated to the mycomembrane but their identification and precise localization have not been always carefully examined and awaits future studies. Recent evidence has shown that mycobacteria have developed novel and specialized secretion systems for the transport of extracellular proteins, notably ESX systems, across their hydrophobic and highly impermeable cell envelope. This implies the existence of a dedicated secretion machinery to cross the OM that remains to be discovered (Abdallah *et al.*, 2007).

The mycobacterial cell envelope is important for the bacterial physiology, as proved by the essentiality of the synthesis of several of its components, e.g. mycolic acids and arabinogalactan, which are known to be produced inside the infected host cells (Asselineau *et al.*, 1981; Hunter *et al.*, 1986; Daffé *et al.*, 1993). As the structures of these compounds are exotic, compared to mammalian products, enzymes involved in the building of the cell envelope remains an attractive source for new targets in the context of the development of inhibitors that may serve as lead compounds for the development of anti-tuberculosis drugs.

References

Abdallah, A.M., van Pittius, N.C.G., Champion, P.A.D., Cox, J., Luirink, J., Vandenbroucke-Grauls, C.M.J.E., Appelmelk, B.J., and Bitter, W. (2007). Type VII secretion – mycobacteria show the way. Nat. Rev. *5*, 883–891.

Asselineau, C., Clavel, S., Clément, F., Daffé, M., David, H.L., Lanéelle, M.-A., and Promé, J.-C. (1981). Constituants lipidiques de *Mycobacterium leprae* isolé de tatou infecté expérimentalement. Ann. Microbiol. (Inst. Pasteur) *132A*, 19–30.

Bhamidi, S., Scherman, M.S., Jones, V., Crick, D.C., Belisle, J.T., Brennan, P.J., and McNeil, M. (2011). Detailed structural and quantitative analysis reveals the spacial organization of the cell walls of *in vivo* grown *Mycobacterium leprae* and *in vitro* grown *Mycobacterium tuberculosis.* J. Biol. Chem. *286*, 23168–23177.

Barksdale, L., and Kim, K.-S. (1977). Mycobacterium. Bacteriol. Rev. *41*, 217–372.

Benedetti, E.L., Dunia, L., Ludosky, M.A., Man, N.V., Trach, D.D., Rastogi, N., and David, H.L. (1984). Freeze-etching and freeze-fracture structural features of cell envelopes in mycobacteria and leprosy derived corynebacteria. Acta Leprologica *95*, 237–248.

Bleck, C.K.E., Merz, A., Gutierrez, M.G., Walther, P., Dubochet, J., Zuber, B., and Griffiths, G. (2010). Comparison of different methods for thin section EM analysis of *Mycobacterium smegmatis*. J. Microsc. *237*, 23–38.

Boddingius, J., and Dijkman, H. (1990). Subcellular localization of *Mycobacterium leprae*-specific phenolic glycolipid (PGL-I) antigen in human leprosy lesions and in *M.leprae* isolated from armadillo liver. J. Gen. Microbiol. *136*, 2001–2012.

Chapman, G.B., Hanks, J.H., and Wallace, J.H. (1959). An electron microscope study of the disposition and fine structure of *Mycobacterium lepraemurium* in mouse spleen. J. Bacteriol. *77*, 205–211.

Daffé, M., and Draper, P. (1998). The envelope layers of mycobacteria with reference to their pathogenicity. Adv. Microbial Phys. *39*, 131–203.

Daffé, M., and Etienne, G. (1999). The capsule of *Mycobacterium tuberculosis* and its implications for pathogenicity. Tuberc. Lung Dis. *79*, 153–169.

Daffé, M.,. Lacave, C., Lanéelle, M.-A., and Lanéelle, G. (1987). Structure of the major triglycosyl phenolphthiocerol of *Mycobacterium tuberculosis* (strain Canetti). Eur. J. Biochem. *167*, 155–160.

Daffé, M., Dupont, M.-A., and Gas, N. (1989). The cell envelope of *Mycobacterium smegmatis*: cytochemistry and architectural implications. FEMS Microbiol. Lett. *61*, 89–94.

Daffé, M., Brennan., P.J., and McNeil, M. (1990). Predominant structural features of the cell wall arabinogalactan of *Mycobacterium tuberculosis* as revealed through characterization of oligoglycosyl alditol fragments by gas chromatography/mass spectrometry and by ^{1}H- and ^{13}C-NMR analyses. J. Biol. Chem. *265*, 6734–6743.

Daffé, M., Brennan, P.J., and McNeil, M. (1993). Major structural features of the cell wall arabinogalactans of *Mycobacterium, Rhodococcus,* and *Nocardia* spp. Carbohydr. Res. *249*, 383–398.

Dhiman, R.K., Dinadayala, P., Ryan, G.J., Lenaerts, A.J., Schenkel, A.R., and Crick, D.C. (2011). Lipoarabinomannan localization and abundance during growth of *Mycobacterium smegmatis.* J. Bacteriol. *193*, 5802–5809.

Dinadayala, P., Laval, F., Raynaud, C., Lemassu, A., Lanéelle, M.-A., Lanéelle, G., and Daffé, M. (2003). Tracking the putative precursors of oxygenated mycolates of *Mycobacterium tuberculosis.* Structural analysis of fatty acids of a mutant devoid of methoxy- and ketomycolates. J. Biol Chem. *278*, 7310–7319.

Draper, P. (1971). The walls of *Mycobacterium lepraemurium*: chemistry and ultraStructure J. Gen. Microbiol. *69*, 313–332.

Draper, P. (1998). The outer parts of the mycobacterial envelope as permeability barriers. Front. Biosci. *3*, d1253–1261.

Dubochet, J., and Sartori Blanc, N. (2001). The cell in absence of aggregation artifacts. Micron. *32*, 91–99.

Faller, M., Niederweis, M., and Schulz, G.E. (2004). The structure of a mycobacterial outer-membrane channel. Science *303*, 1189–1192.

Fréhel, C., Ryter, A., Rastogi, N., and David, H.L. (1986). The electron-transparent zone in phagocytised *Mycobacterium avium* and other mycobacteria: formation, persistence and role in bacterial survival. Ann. Microbiol. (Inst. Pasteur) *137B*, 239–257.

Fréhel, C., Rastogi, N., Bénichou, J.-C., and Ryter, A. (1988). Do test tube-grown pathogenic mycobacteria possess a protective capsule? FEMS Microbiol. Lett. *56*, 225–230.

Gagliardi, M.C., Lemassu, L., Teloni, R., Mariotti, S., Sargentini, V., Pardini, M., Daffé, M., and Nisini, R. (2007). Cell-wall associated alpha-glucan is instrumental for *Mycobacterium tuberculosis* to block CD1 molecule expression and disable the function of dendritic cell derived from infected monocyte. Cell. Microbiol. *9*, 2081–2092.

Goren, M.B., and Brennan, P.J. (1979). Mycobacterial lipids: chemistry and biologic activities. In Tuberculosis, Youmans, G.P., ed. (W.B. Saunders Company, Philadelphia), pp. 63–193.

Hanks, J.H. (1961a). Capsules in electron micrographs of *Mycobacterium leprae.* Int. J. Lepr. *29*, 84–87.

Hanks, J.H. (1961b). The problem of preserving internal structures in pathogenic mycobacteria by conventional methods of fixation. Int. J. Lepr. *29*, 175–178.

Hanks, J.H. (1961c). Demonstration of capsules on *M.leprae* during carbol-fuchsin staining mechanism of the Ziehl-Neelsen stain. Int. J. Lepr. *26*, 179–182.

Hanks, J.H., Moore, J.T., and Michaels, J.E. (1961). Significance of capsular components of *Mycobacterium leprae* and other mycobacteria. Int. J. Lepr. *26*, 74–83.

Hartmann, M., Barsch, A., Niehaus, K., Puhler, A., Tauch, A., and Kalinowski, J. (2004). The glycosylated cell surface protein Rpf2, containing a resuscitation-promoting factor motif, is involved in intercellular communication of *Corynebacterium glutamicum*. Arch. Microbiol. *182*, 299–312.

Hoffmann, C., Leis, A., Niederweis, M., Plitzko, J.M., and Engelhardt, H. (2008). Disclosure of the mycobacterial outer membrane: cryo-electron tomography and vitreous sections reveal the lipid bilayer structure. Proc. Natl. Acad. Sci. U.S.A. *105*, 3963–3967.

Hunter, S.W., and Brennan, P.J. (1990). Evidence for the presence of a phosphatidylinositol anchor on the lipoarabinomannan and lipomannan of *Mycobacterium tuberculosis*. J. Biol. Chem. *265*, 9272–9279.

Hunter, S.W., Gaylord, H., and Brennan, P.J. (1986). Structure and antigenicity of the phosphorylated lipopolysaccharides from the leprosy and tubercle bacilli. J. Biol. Chem. *261*, 12345–12351.

Jarlier, V., and Nikaido, H. (1990). Permeability barrier to hydrophilic solutes in *Mycobacterium chelonei* (sic). J. Bacteriol. *172*, 1418–1423.

Kocincova, D., Sonden, B., Bordat, Y., Pivert, E., de Mendonca-Lima, L., Gicquel, B., and Reyrat, J.M. (2004). The hydrophobic domain of the Mycobacterial Erp protein is not essential for the virulence of *Mycobacterium tuberculosis*. Infect. Immun. *72*, 2379–2382.

Kotani, S., Kitaura, T., Hirano, T., and Tanaka, A. (1959). Isolation and chemical composition of the cells walls of BCG. Biken's J. *2*, 129–141.

Lemassu, A., and Daffé, M. (1994). Structural features of the exocellular polysaccharides of *Mycobacterium tuberculosis*. Biochem. J. *297*, 351–357.

Lemassu, A., Ortalo-Magné, A., Bardou, F., Silve, G., Lanéelle, M.-A., and Daffé, M. (1996). Extracellular and surface-exposed polysaccharides of non-tuberculous mycobacteria. Microbiology *142*, 1513–1520.

Liu, J., Rosenberg, E.Y., and Nikaido, H. (1995). Fluidity of the lipid domain of cell wall from *Mycobacterium chelonae*. Proc. Natl. Acad. Sci. U.S.A. *92*, 11254–11258.

Marchand, C.H., Salmeron, C., Bou Raad, R., Méniche, X., Chami, M., Masi, M., Blanot, D., Daffé, M., Tropis, M., Huc, E., *et al.* (2012). Biochemical disclosure of the mycolate outer membrane of *Corynebacterium glutamicum*. J. Bacteriol. *194*, 587–597.

Matias, V.R., and Beveridge, T.J. (2005). Cryo-electron microscopy reveals native polymeric cell wall structure in *Bacillus subtilis* 168 and the existence of a periplasmic space. Mol. Microbiol. *64*, 195–206.

Matias, V.R., and Beveridge, T.J. (2006). Native cell wall organization shown by cryo-electron microscopy confirms the existence of a periplasmic space in *Staphylococcus aureus*. J. Bacteriol. *188*, 1011–1021.

Matias, V.R., and Beveridge, T.J. (2008). Lipoteichoic acid is a major component of the *Bacillus subtilis* periplasm. J. Bacteriol. *190*, 7414–7418.

McNeil, M.R., and Brennan, P.J. (1991). Structure, function and biogenesis of the cell envelope of mycobacterial in relation to bacterial physiology, pathogenesis and drug resistance; some thoughts and possibilities arising from recent structural information. Res. Microbiol. *142*, 451–463.

McNeil, M., Daffé, M., and Brennan, P.J. (1990). Evidence for the nature of the link between the arabinogalactan and peptidoglycan of mycobacterial cell walls. J. Biol. Chem. *265*, 18200–18206.

McNeil, M., Daffé, M., and Brennan, P.J. (1991). Location of the mycoloyl ester substituents in the cell walls of mycobacteria. J. Biol. Chem. *266*, 13217–13223.

Minnikin, D.E. (1982). Lipids: complex lipids, their chemistry, biosynthesis and roles. In The Biology of the Mycobacteria, Vol. 1. Physiology, Identification and Classification, Ratledge, C., and Stanford, J., eds. (Academic Press, London, UK), pp. 95–184.

Niederweis, M., Danilchanka, O., Huff, J., Hoffmann, C., and Engelhardt, H. (2010). Mycobacterial outer membranes: in search of proteins. Trends Microbiol. *18*, 109–116.

Nikaido, H., Kim, S.-H., and Rosenberg, E.Y. (1993). Physical organization of lipids in the cell wall of *Mycobacterium chelonae*. Mol. Microbiol. *8*, 1025–1030.

Ortalo-Magné, A., Dupont, M.-A., Lemassu, A., Andersen, Å.B., Gounon, P., and Daffé, M. (1995). Molecular composition of the outermost capsular material of the tubercle bacillus. Microbiology *141*, 1609–1620.

Ortalo-Magné, A., Lemassu, A., Lanéelle, M.A., Bardou, F., Silve, G., Gounon, P., Marchal, G., and Daffé, M. (1996). Identification of the surface-exposed lipids on the cell envelope of *Mycobacterium tuberculosis* and other mycobacterial species. J. Bacteriol. *178*, 456–461.

Puech, V., Chami, M., Lemassu, A., Lanéelle, M.A., Schiffler, B., Gounon, P., Bayan, N., Benz, R., and Daffé, M. (2001). Structure of the cell envelope of corynebacteria : importance of the non-covalently bound lipids in the formation of the cell wall permeability barrier and fracture plane. Microbiology *147*, 1365–1382.

Rastogi, N. (1991). Recent observations concerning structure and function relationships in the mycobacterial cell envelope: elaboration of a model in terms of mycobacterial pathogenicity, virulence and drug-resistance. Res. Microbiol. *142*, 464–476.

Rastogi, N., Lévy-Frebault, V., Blom-Potar, M.-C., and David, H.L. (1989). Ability of smooth and rough variants of *Mycobacterium avium* and *M. intracellulare* to multiply and survive intracellularly: role of C-mycosides. Zbl. Bakt. Hyg. *270*, 345–360.

Raynaud, C., Etienne, G., Peyron, P., Lanéelle, M.-A., and Daffé, M. (1998). Extracellular enzyme activities potentially involved in the pathogenicity of *Mycobacterium tuberculosis*. Microbiology *144*, 577–587.

Rezwan, M., Lanéelle, M.-A., Sander, P., and Daffé, M. (2007). Breaking down the wall: fractionation of mycobacteria. J. Microbiol. Methods *68*, 32–39.

Rulong, S., Aguas, A.P., Da Silva, P.P., and Silva, T.S. (1991). Intramacrophagic *Mycobacterium avium* bacilli are coated by a multiple lamellar structure: freeze fracture anlysis of infected mouse liver. Infect. Immun. *59*, 3895–3902.

Ryter, A., Fréhel, C., Rastogi, N., and David, H.L. (1984). Macrophage interaction with mycobacteria including *M. leprae*. Acta Leprologica 95, 211–235.

Sani, M., Houben, E.N.G., Geurtsen, J., Pierson, J., de Punder, K., van Zon, M., Wever, B., Piersma, S.R., Jiménez, C.R., Daffé, M., *et al.* (2010). Direct visualization by cryo-EM of the mycobacterial capsular layer: a labile structure containing ESX-1-secreted proteins. PLoS Pathog. *6*, e1000794.

Silva, M.T., and Macedo, P.M. (1983a). A comparative ultrastructural study of the membranes of *Mycobacterium leprae* and of cultivable *Mycobacteria* (sic). Biol. Cell *47*, 383–386.

Silva, M.T., and Macedo, P.M. (1983b). The interpretation of the ultrastructure of mycobacterial cells in transmission electron microscopy in ultrathin sections. Int. J. Lepr. *51*, 225–234.

Silva, M.T., and Macedo, P.M. (1984). Ultrastructural characterization of normal and damaged membranes of *Mycobacterium leprae* and cultivable mycobacteria. J. Gen. Microbiol. *130*, 369–380.

Sonnenberg, M.G., and Belisle, J.T. (1997). Definition of *Mycobacterium tuberculosis* culture filtrate proteins by two-dimentional polyacrylamide gel electrophoresis, N-terminal amino-acid sequencing and electrospray mass spectrometry. Infect. Immun. *65*, 4515–4524.

Trias, J., and Benz, R. (1993). Characterization of the channel formed by the mycobacterial porin in lipid bilayer membranes. Demonstration of voltage gating and of negative point charges at the channel mouth. J. Biol. Chem. *268*, 6234–6240.

Trias, J., Jarlier, V., and Benz, R. (1992). Porins in the cell wall of mycobacteria. Science *258*, 1479–1481.

Villeneuve, M., Kawai, M., Kanashima, H., Watanabe, M., Minnikin, D.E., and Nakahara, H. (2005). Temperature dependence of the Langmuir monolayer packing of mycolic acids from *Mycobacterium tuberculosis*. Biochim. Biophys. Acta *1715*, 71–80.

Villeneuve, M., Kawai, M., Watanabe, M., Aoyagi, Y., Hitotsuyanagi, Y., Takeya, K., Gouda, H., Hirono, S., Minnikin, D.E., and Nakahara, H. (2007). Conformational behavior of oxygenated mycobacterial mycolic acids from *Mycobacterium bovis* BCG. Biochim. Biophys. Acta *1768*, 1717–1726.

Zuber, B., Haenni, M., Ribeiro, T., Minnig, K., Lopes, F., Moreillon, P., and Dubochet, J. (2006). Granular layer in the periplasmic space of Gram-positive bacteria and fine structures of Enterococcus gallinarum and *Streptococcus gordonii* septa revealed by cryo-electron microscopy of vitreous sections. J. Bacteriol. *188*, 6652–6660.

Zuber, B., Chami, M., Houssin, C., Dubochet, J., Griffiths, G., and Daffé, M. (2008). Direct visualization of the outer membrane of native mycobacteria and corynebacteria. J. Bacteriol. *190*, 5672–5680.

Part II

Protein–Lipid Interactions

The Role of Lipid Composition on Bacterial Membrane Protein Conformation and Function

6

Vinciane Grimard, Marc Lensink, Fabien Debailleul, Jean-Marie Ruysschaert and Cédric Govaerts

Abstract

Cellular, biochemical and biophysical studies have shown over the last years that membrane proteins interact intimately with the lipid bilayer and that the structure and the activity of these proteins can be modulated by the type of lipids that surround them. Studies have demonstrated that physico-chemical properties of the membrane affect protein function. In fact some proteins directly sense the membrane properties, such as fluidity, tension, curvature stress or hydrophobic mismatch, for example as a signal for temperature or osmotic stress. In other cases, specific interactions between defined protein motifs and given lipids have been evidenced. These interactions often fulfil structural and functional role and are intimately linked to the biological function of membrane proteins. This chapter aims at summarizing the elaborate interplay existing between proteins and lipids and the influence of both specific interactions and bulk properties of the membrane on the stability, the structure and the activity of bacterial membrane proteins.

Introduction

The structure and function of proteins are intimately linked to their direct environment. This is long established for soluble proteins where water is a driving force for folding and maintaining structural stability. In addition, solvent molecules take part in numerous interactions with polar moieties of the polypeptide that are crucially required for function, such as composing enzymatic sites, protein–protein interfaces, channel protons etc. This implies that direct interaction with the environment has been included during the evolution of biological functions of proteins.

It thus seems sensible that membrane proteins must also have adapted specifically to their direct environment in order to achieve their required biological function. A dramatic difference is that instead of a simple molecule like H_20, the membrane is composed of a wide diversity of lipid molecules, with different structures and physico-chemical properties, making for a highly complex landscape. Not only a single biological membrane is composed of many different species, but heterogeneity in composition can occur within a single bilayer, including in prokaryotes, giving rise to the formation of microdomains of defined lipid composition (Mileykovskaya *et al.*, 2009). In addition, when considering a biological membrane, one should bear in mind that the proteins constitute about 50% of the whole mass, and as such will modify the physico-chemical properties of the lipid bilayer.

Composition and structure of bacterial membranes

Bacterial membrane composition can vary strongly from one bacterial species to another. Gram-negative bacteria are surrounded by two membranes. *Escherichia coli* has been used as a model to study the lipid composition of these bacteria. Its inner membrane is mostly constituted of phosphatidylethanolamine (PE, typically around 70%) as well as phosphatidylglycerol (PG, around 20%) and cardiolipin (CL, around 10%), while the outer leaflet of the outer membrane contains a high percentage of lipopolysaccharides. Using mass spectrometry lipidomic approaches, the various species present in *E. coli* were analysed (Smith *et al.*, 1995; Oursel *et al.*, 2007; Gidden *et al.*, 2009). They could show that in addition to more common fatty acid such as palmitoyl ($C_{16:0}$) and palmitoleyl acid ($C_{16:1}$), another prominent fatty acid in *E. coli* is the methylene-hexadecanoic acid ($cyC_{17:0}$), which contains a cyclopropane ring. Although a lot of Gram-negative bacteria have a lipid composition similar to *E. coli*, there may be significant variations. For example, phosphatidylcholine (PC) is found as a minor component in *Pseudomonas aeruginosa* membranes, but can represent up to 73% of *Acetinobacter aceti* inner membrane (Sohlenkamp *et al.*, 2003). It is also found in several photosynthetic bacteria, such as in *Rhodobacter spheroides* (Zhang *et al.*, 2011). Sphingolipids have also been found in some bacteria. For example, *Sphingomonas capsulata* is devoid of lipopolysaccharides in its outer membrane, which are replaced by glycosphingolipids (Geiger *et al.*, 2010).

Gram-positive bacteria are surrounded by only one bilayer, itself surrounded by a cell wall. Although several Gram-positive bacterial membranes also contain PE, PG and CL (Gbaguidi *et al.*, 2007; Gidden *et al.*, 2009), they are often enriched with other lipids such as glycolipids or lipoamino acids, derivatives of PG where the glycerol moiety is esterified to an amino acid such as ornithine, alanine or lysine. Lysyl-PG is even a major lipid in a lot of Gram-positive bacteria, such as *Staphylococcus aureus*, *Bacillus subtilis*, *Listeria monocytogenes* or *Lactococcus plantarum* (Gidden *et al.*, 2009; Geiger *et al.*, 2010), but also in a few Gram-negative bacteria such as *P. aeruginosa*. A wide variety of acyl chains, saturated, mono- or polyunsaturated or even branched can be coupled to these lipids. The branching and unsaturation of the acyl chains will play a major role in the membrane fluidity, by increasing the flexibility of the chain.

Bacteria in particular have adapted their membrane composition according to the environment. Extremophiles are remarkable examples, where, in order to maintain given physico-chemical properties of the cellular membrane (fluidity, permeability, etc.), specific lipids have been selected. Most extremophiles membranes are formed with diether lipids, where the ester linkage present in traditional phospholipids has been replaced by a more stable ether linkage. Furthermore, to withstand high temperature, thermophiles decrease the fluidity of their membranes either by cyclization of their acyl chains, or by formation of tetraether lipids resulting probably from the formation of covalent bonds at the extremities of the acyl chains of lipids from both leaflets, thereby creating a lipid spanning both leaflets of the bilayer (Sprott, 1992). Furthermore, in one given organism, dynamic adaptation of the lipid composition to the change in the outside conditions can also happen (Nicholas, 1984). For instance, upon lowering of the growth temperature, bacteria will typically shift their lipid tails from unsaturated to monounsaturated, shorten chain length and modify branching or cyclization so as to lower the temperature of phase transition from gel to liquid crystalline state, a process called *homeoviscous adaptation*. It is intriguing that membrane proteins must have evolved their sequences to achieve proper interplay with the bilayer but at the same time must maintain integrity and most likely (some) biological function upon changes in the composition of the bilayer.

From bulk to detail

Our view of the biological membranes has significantly evolved during the last decades, as its complex organization and its ability to adapt are continuously being uncovered. Forty years ago, Singer and Nicolson profoundly changed the view on biological membranes by integrating seemingly diverging experimental data into one comprehensive model, called the mosaic fluid model that constituted the basis of our current conception of the membrane bilayer. In the model, the biological membrane is presented as a 'two-dimensional solutions of oriented globular proteins and lipids' (Singer *et al.*, 1972), where proteins move freely to perform their physiological duties and are physically regulated by the laws of fluid dynamics, albeit in two dimensions only. However, in the same paper, Nicolson and Singer also highlighted some of the key ideas that have found experimental demonstration over the last years. In particular, the authors hypothesized that 'With any one membrane protein, the tightly coupled lipid might be specific; that is, the interaction might require that the phospholipid contain specific fatty acid chains or particular polar head groups'. This has now been demonstrated for a number of cases where interactions between a membrane protein and given lipidic species are required for proper structure and function.

Singer and Nicolson have therefore put forward both the idea that the fluid properties (e.g. bulk or material properties) and specific protein–lipid interactions are required for proper physiological behaviour of the biological membrane. These two views, coined by Anthony Lee as, respectively, 'lipid-based' and 'protein-based' approaches (Lee, 2011) have both been used over the years to describe how the composition of the membrane may modulate its biological function. Interestingly, many authors tend to oppose these descriptions, either considering a material/energetic view of the membrane and 'dispense with the notion of specific lipid–protein interactions' (Andersen *et al.*, 2007) or conversely focusing solely on an atomic description of lipid binding to proteins.

The purpose of the present review is to illustrate that the two views are complementary and should both be taken into account when considering structure and function of membrane proteins in their native environment.

Sensing the bulk: mismatch, frustration and heat

It had been known for years that changing the physico-chemical properties of the membrane modulate activity of membrane properties; in fact, some proteins achieve their function specifically by sensing changes in the bulk properties of the bilayer.

A striking example comes from the family of mechanosensitive channels (Msc) which, upon variations in membrane tension, open and conduct ions. In bacteria, mechanosensitive channels such as MscL (Large) and MscS (Small) protect the cell from turgor pressure and thus from osmotic cell lysis (Poolman *et al.*, 2004; Nomura *et al.*, 2006). Low osmotic pressure in the external medium give rise to an influx of water and cell expansion, inducing an increase in the surface tension of the membrane, leading to the opening of the MscS channel in case of a small osmotic stress and the opening of McsL in case of a stronger stress.

How does a change in membrane tension lead to the opening of the channel? McsL is a homopentameric protein, with each subunit containing two transmembrane (TM) helices, the TM1 forming an inner bundle and the TM2 an outer bundle interacting with the lipids

(Chang *et al.*, 1998). Applying negative pressure in a patch-clamp pipette is sufficient to induce the opening of the channel (Lee, 2004). With a fixed lipid composition, the increase in membrane area will result in a decrease of the membrane thickness. This would induce a tilt of the TM1 that would give rise to the opening of the channel (Perozo *et al.*, 2002). Indeed, decreasing the membrane thickness correlates with a decrease of the activation energy of the channel (Perozo *et al.*, 2002). Similarly, by introducing tryptophan residues at the end of the helices, it was shown by fluorescence that the environment of these tryptophans was not changing in bilayer of various chain length (from C_{12} to C_{24}), suggesting hydrophobic matching via an adaptation of the tilt of the helices in these various environments (Powl *et al.*, 2005). Thus it appears that McsL is effectively sensing changes in the thickness of the bilayer that induce conformational adaptation of the protein resulting in channel opening or closing.

Obviously, correlation between the thickness of the membrane and the span of the protein transmembrane domains is essential as, if they are of different length a *hydrophobic mismatch* would occur, resulting in either lipid acyl chain or hydrophobic amino-acid residues exposure to the polar environment of the surrounding solvent. The thickness of the membrane will mostly depend on the length of the acyl chains of the lipids as well as their level of unsaturation. Analysis of crystal structures of membrane proteins demonstrate that the hydrophobic domain embedded in the bilayer (as assessed from the positions of Arg, Lys and Trp sidechains) varies greatly in thickness, ranging from about 20 Å to over 35 Å (Lee, 2003). Thus, considering the observed thickness ranges of both lipid bilayers and membrane proteins, hydrophobic divergence is bound to happen.

Thus how will the system adapt to prevent hydrophobic mismatch? Adaptation of both the lipids and the protein are theoretically possible, but the compressibility of a protein is typically 2 orders of magnitude lower than that of the bilayer. Stretching and bending of the lipids should therefore be favoured and lipid chain extension or compression has been observed by nuclear magnetic resonance (NMR), upon insertion of a model transmembrane peptide in PC bilayers of various acyl chain lengths (de Planque *et al.*, 1998). Bending around the protein will also alleviate locally the mismatch. Obviously lipids with spontaneous curvature, such as PE, would be particularly appropriate to correct the curvature imposed by the membrane protein (Lee, 2004), indicating that the energetic cost of adaptation of the bilayer to the protein hydrophobic span will depend on the lipidic composition. Indeed, bilayer deformation comes at a significant cost (Andersen *et al.*, 2007) and while proteins will not directly compress, limited conformational changes, such as helix tilting or rotation, may significantly contribute to the adaptation of the mismatch at a reasonable energetic cost. As mentioned above, this could be at the basis of the molecular mechanism of MscL opening upon change in membrane tension. However, such adaptation of the system to a hydrophobic mismatch, whether by the lipids or the protein, will be limited and excessive differences will not be compensated. Indeed, by reconstituting membrane proteins in liposomes made of phospholipids of defined chain length it was shown in a number of cases that the enzymatic activity of the protein depends on the membrane thickness. Usually, optimal activity is observed with acyl chain length around 16–20, which is typical of biological membranes, and will dramatically decrease upon shortening or lengthening of the hydrophobic segments indicating that the system has reached the limits of its adaptation capabilities (In't Veld *et al.*, 1991; Andersen *et al.*, 2007). An extreme case comes from the atypical gramicidin A channels, where rigorous hydrophobic match must be obtained to observe channel formation (Mobashery *et al.*, 1997) and explains why gramicidin activity rely on the presence of non-lamellar forming phospholipids (see below). In contrast, the outer membrane protein OmpF shows a maximal binding constant to lipids of shorter chain length (C_{14}), which is expected considering that the outer membrane is thought to be thinner (O'Keeffe *et al.*, 2000).

The adaptation to hydrophobic mismatch may also come from release in the curvature frustration that results from the intrinsic propensity of a lipid bilayer to form non-lamellar structures (see Fig. 6.1). A lipid of cylindrical shape (i.e. whose projected surface of the headgroup matches that of the acyl chain) like phosphatidylcholine will naturally tend to form planar (lamellar) bilayer (zero curvature). In contrast, if the shape is conical, the lipid layer will depart from planarity, with negative curvature when the headgroup projected area is smaller than the acyl chain projection (such as for phosphatidylethanolamine) and positive curvature when the headgroup is larger (such as phosphatidylserine). While the natural arrangement of such monolayer is non planar (e.g. hexagonal H_{II} phase for PE), a bilayer can be formed into planar structures with both bilayer balancing each other but at a significant energetic cost, stored into the membrane and traditionally denoted as *curvature frustration*. Bearing in mind that many bacterial membranes contain large proportion of PE (i.e. about 70% in *E. coli*) such contribution is likely to play a significant role in the thermodynamics of the membrane. The introduction of a membrane protein in a 'frustrated' bilayer can release (some of) the stored energy, especially if the hydrophobic span of the protein mismatches that of the bilayer: in order to resolve the difference, the bilayer will curve or splay around the protein, thus releasing locally the curvature stress. Such adaptation has been proposed to be crucial for gramicidin channel formation in membranes of thickness greater than that of the channels. In phosphatidylserine bilayers, the curvature stress can be modulated by changing the ionic force; increasing cation concentration will decrease electrostatic headgroup repulsion and thus decrease positive curvature without affecting the bilayer thickness. Lundbæk *et al.* have shown that such manoeuver strongly decreases

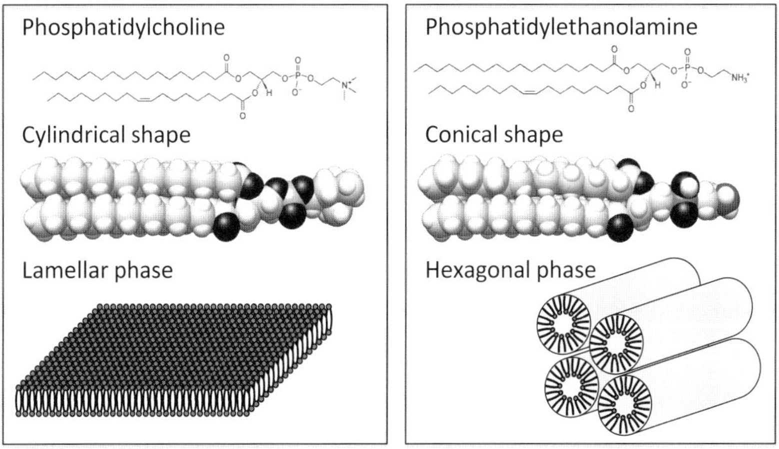

Figure 6.1 The shape of a lipid dictates its assembly behaviour. Cylindrical lipids like phosphatidylcholine will spontaneously form lamellar structures, while for conical lipids such as phosphatidylethanolamine, local curvature is induced, leading to hexagonal structures.

channel activity, indicating that local curvature is required to allow for gramicidin channel formation (Lundbaek *et al.*, 1997). Thus, clearly, modulating the local membrane curvature is an efficient strategy to resolve hydrophobic mismatch and, as a corollary, is potentially an important means to modulate transmembrane protein function.

Indeed, it is remarkable to note that bacteria regulate efficiently the amount of non-lamellar lipids present in their membrane over a wide range of growth temperatures (Cronan, 2003). This suggests that the curvature frustration and its interconnection with membrane protein incorporation are of important functional relevance.

Another remarkable example of protein adapted to sense changes in the membrane properties comes from the temperature sensor DesK of *B. subtilis* (Mansilla *et al.*, 2004).

DesK is a membrane protein involved in the biological response of the bacteria upon changes in temperature. This protein acts as a kinase at cold temperatures (~25°C), autophosphorylating itself in its cytoplasmic kinase domain. This phosphorylation activates a cascade reaction leading to the transcription of the *des* gene coding for a lipid desaturase. Desaturated lipids generated this way allow maintenance of the overall membrane fluidity at low temperatures.

DesK is constituted of five transmembrane domains and a c-terminal catalytic part. Its first TM segment (TM1) is unusually composed of both polar and hydrophobic residues constituting more an amphipathic stretch than a regular TM-spanning helix. Furthermore, deletion of this TM1 leads to a kinase-on state. Taken together, these observations point to a regulating role for TM1. By successive deletions and chimera constructions, the laboratory of Mendoza was able to reduce the thermosensor to only a chimeric TM1/TM5 transmembrane domain of 31 residues (14 coming from the N-terminus of TM1 and 17 coming from the C-terminus of TM5) fused with the catalytic domain (Cybulski *et al.*, 2010). This minimal sensor was expressed *in vitro* and inserted in liposomes of *E. coli* phospholipids. These lipids undergo a change from a fluid to a more ordered state when the temperature is lowered from 37°C to 25°C. Under these conditions, the minimal sensor coupled to the kinase domain was able to react to the temperature change and phosphorylates itself.

In order to decipher the molecular mechanism of cold sensing, the research group focused on the amphipathic nature of TM1. Upon temperature lowering and fluidity switch, the membrane thickness increases and the three hydrophilic residues of TM1 are buried deeper in the hydrophobic bilayer. The hydration of these amino acids is more costly on a thermodynamic point of view and destabilizes the protein leading to its autophosphorylation. This was shown by direct mutation of the hydrophilic motif by hydrophobic residues, which impairs the *in vivo* activity of the protein. But also by burying deeper one of the three residues in the membrane, which leads to a kinase-dominant state. Finally, by reconstituting the protein in a series of phosphatidylcholines of increasing fatty acyl chain length (14, 16 and 20 carbons), it was established that the protein acts as a thickness-measuring device and this is used as cold-sensing (Cybulski *et al.*, 2010). These data illustrate how physical properties of the lipid bilayer can be detected by proteins and how it can be used by the cell to apprehend its environment.

The extend of modulation of protein functions by the bulk characteristics of the membrane is illustrated by proteins that have specifically evolved to sense variations in these properties, such a mechanosensitive channels, gramicidin channel or temperature sensors. The material properties changes often result locally in variations of the bilayer thickness and trigger conformational rearrangements of both the protein and the lipid bilayer to surmount the thermodynamical cost of a mismatch. Beyond these extreme cases, a body of evidence shows that structure and function of bacterial membrane proteins are modulated by the physico-chemical properties of the bilayer and begs the question as to why mismatch and frustration are so common. The fact that bacteria maintain a high proportion of non-lamellar-lipids and that membrane proteins often show some apparent mismatch with the expected hydrophobic thickness of the bilayer is indicative that these seemingly non-ideal conditions may in fact be required for efficient function. By storing energy through mismatch adaptation or curvature frustration the system may provide the appropriate thermodynamic conditions for efficient and rapid rearrangement of proteins and lipids as required physiologically.

The first shell

Spectroscopic techniques such as electron paramagnetic resonance (EPR) and NMR have demonstrated that, when it comes to the interaction with membrane proteins, all lipids are not equal. Using nitroxide spin-labelled lipids, Marsh and colleagues have shown that the presence of embedded proteins into a lipid bilayer triggers the emergence of a population of slow moving lipids (Marsh, 2008). These lipids form the first shell of lipids around the protein and are sometimes called *annular lipids*, to evoke the image of a ring of lipids surrounding the proteins. They constitute the physical interface between the surface of the protein and the rest of the bilayer, and thus transmit the material properties of the bulk lipids to the protein, and vice versa, as the presence of embedded protein may also modulate the physical properties of the membrane. NMR and EPR have shown that the acyl chains of annular lipids show a drastic slowing of their rotational motion compared to that of the bulk lipids but they do not show preferential orientation however, suggesting that they distort in various directions when adapting to the surface of the protein. Nevertheless, these lipids still exchange relatively fast with the bulk lipids, with an off-rate calculated by EPR of approximately 1–2 × 10^7/s, about five times slower than the exchange rate between lipids of the bulk phase. It was also possible to calculate the number of lipids present in the annular shell of various proteins by EPR (recapitulated in Marsh, 2008), which was in good agreement with the expected surface of the transmembrane domain of the protein.

Using EPR, Marsh and colleagues have also been able to determine the association constant of several lipids to a number of integral membrane proteins and to establish significant differences in the relative affinities (Marsh, 2008). Lee and colleagues also observed differences in relative lipid association to the KcsA channel using Trp fluorescence (Williamson *et al.*, 2002; Marius *et al.*, 2005). Interestingly, these results indicate that the affinity of first-shell lipids to the protein depends both on the headgroup and the acyl chains. In particular, headgroup charge can induce changes in local electrostatic properties and pH, thus anionic lipids will modify the local proton concentration, thereby increasing the protonation of acidic residues. Conversely, modifying properties of the aqueous phase (pH, ionic strength), either *in vitro*, or locally *in vivo* upon activation of proton or ion channels, can modulate the headgroup effect and thus relative association of annular lipids. Dependence of the association constant to the acyl chain lengths underlines the role of the first lipid shell in modulation of hydrophobic mismatch. Indeed, EPR data confirmed that lipids with acyl chain length matching the hydrophobic core of the protein show higher affinities than those with significantly longer or shorter tails.

Annular lipids are thus regarded as weakly interacting with the protein surface during short residency time, in contrast to 'non-annular' lipids that will, as described below, bind specifically to the protein with direct interactions involving headgroup moieties and with a much longer residency time (albeit not clearly established). Nevertheless, it has been proposed that annular lipids have been identified in crystal structures of membrane proteins such as bacteriorhodopsin. This was supported by the lack of densities of the headgroups with visible interactions solely at the level of the acyl chain. In the bacteriorhodopsin trimer of *Halobacterium salinarum*, fatty acyl chains are highly distorted to fit the grooves on the surface of the protein, thereby ensuring continuity of the permeability barrier at the points of insertion of the membrane protein (Fig. 6.2; Belrhali *et al.*, 1999). The case for crystallized annular lipids is even stronger for mammalian lens specific aquaporin where a 1.9 Å resolution electron crystallography structure shows a well-defined layer of lipid molecules surrounding each functional unit (tetramer) (Gonen *et al.*, 2005). 2D crystals were grown by providing dimyristoyl phosphatidylcholine (DMPC) lipids to the purified protein samples but as crystals can be obtained with various types of lipid, specific interactions are unlikely. Rather, the annular lipids appear to mediate tetramer–tetramer interactions, which appear to tightly constrain the conformation of the lipid molecules and thus allow for high resolution of their individual structures. This was further confirmed by a high resolution structure using *E. coli* polar lipids as additives, where the acyl chains adopted similar conformations as in the DMPC structure with the headgroups stabilized in various ways, depending on the tail length and biochemical moiety type (Hite *et al.*, 2010).

A particular case of crystallized annular lipids was observed in the case of the motor ring of the Na^+-ATPase from *Enterococcus hirae* where a shell of well-defined lipids is bound on the inner surface (Murata *et al.*, 2005). The protein surface forming the inside of the pore-like structure is covered with 10 molecules of dipalmitoylphosphatidylglycerol and 10 molecules of dipalmitoylglycerol, in both leaflets of the bilayer. A combination of electrostatic interactions with lysine residues and hydrophobic interactions with leucine and phenylalanine residues stabilize this inner lipid ring. External phospholipids however were lost, probably due to faster exchange with detergent molecules.

Although annular lipids are generally considered as non-specific, early studies suggested that the coat protein of the R12 bacteriophage specifically sequesters cardiolipin in its

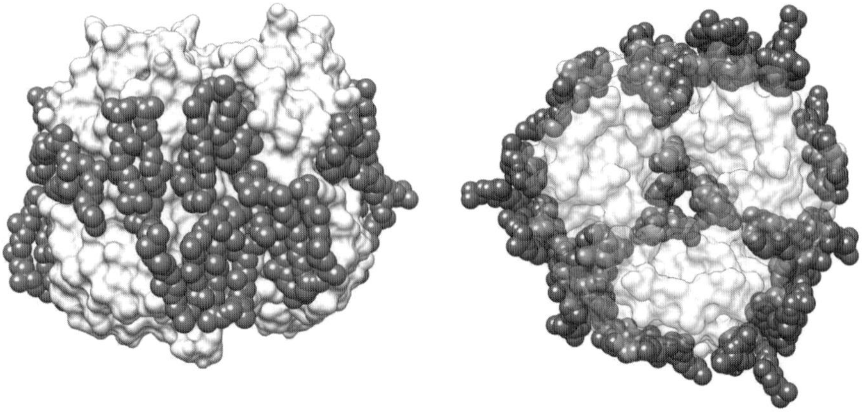

Figure 6.2 Annular lipids bound to the trimer of bacteriorhodopsin (PDB 1QHJ), viewed from the side (left) and top (right). The protein surface is represented as solid white, with the lipids represented as grey space-filling spheres.

annular ring once inserted in the *E. coli* membrane, inducing an increase in CL synthesis and a decrease in CL turnover to account for the sequestered molecules (Chamberlain *et al.*, 1976). However, although this specificity is observed *in vivo, in vitro* studies showed that the coat protein was still able to insert in pure PC liposomes, suggesting that the specificity of this interaction is not absolute (Nozaki *et al.*, 1976).

Co-crystallized lipids: intimate and functional

Over the last years, significant developments in membrane protein crystallography have increasingly evidenced at an atomic level that given lipidic species may bind to specific sites, underlining the need for a molecular details in the description of protein–membrane interactions. Indeed, in most cases, the lipids visible in the electron densities were dragged along from the initial membrane, thus resisting solubilization by detergent and removal during the various purification steps. Furthermore, as these lipids contribute to the diffraction patterns they must be structured in an identical manner in all units of the crystal lattice, implying a strong and specific binding mode. Are these interactions specifically evolved? What are their structural and functional relevance?

Recent examples on mammalian protein structure have identified specific binding sites with functional relevance for a number of lipids such a cardiolipin or cholesterol (Pebay-Peyroula *et al.*, 2003; Nury *et al.*, 2005; Hanson *et al.*, 2008) and have been reviewed extensively (Hunte *et al.*, 2008; Adamian *et al.*, 2011; Lee, 2011).

However, a number of cases have been observed in bacterial proteins as well, demonstrating that structural and functional adaptation of membrane proteins to the lipidic composition of their cognate membrane is an ubiquitous process in all taxonomic domains.

Phosphatidylglycerol

An early but remarkable example came from KcsA, a 160-amino-acid bacterial potassium channel that is activated by low pH, or in the presence of a strong outward K^+ gradient (Roux *et al.*, 2000). It exists as a homotetramer, each subunit containing two transmembrane helices. A large central pore is formed by the four subunits, with a narrow opening on the extracellular side, serving as a selectivity filter with just the right diameter to allow the translocation of a potassium ion (Valiyaveetil *et al.*, 2002). The structure of KcsA was originally obtained by McKinnon and colleagues (Zhou *et al.*, 2001) and showed the presence of an electron dense component, shaped as a lipid molecule. Biochemical analysis of the protein crystals, using phosphate assay and thin-layer chromatography confirmed the association of one PG molecule to each KcsA monomer (Valiyaveetil *et al.*, 2002). Considering that PE is at least three times more abundant in *E. coli* membrane, the sole presence of PG in the structure suggests a specific interaction. Interestingly, various lipid species supported proper refolding of the protein, but channel activity was dependent on the exact nature of the lipid. Indeed, while PG and PC allowed for tetramer formation in refolding assays, the presence of PE was essential for ion uptake in proteoliposome of KcsA (Valiyaveetil *et al.*, 2002). The specificity of the lipid binding site was further confirmed using tryptophan fluorescence quenching by brominated lipids. Although both brominated PC and PG were able to quench the fluorescence of a tryptophan residue located in an annular binding site, only brominated anionic lipids could quench a tryptophan residue located in the non-annular binding site (Alvis *et al.*, 2003). It was even suggested recently that a minimum occupancy of three out of four non-annular binding sites was necessary for opening of the channel, a result in agreement with the low open probability of the channel in *E. coli* (Marius *et al.*, 2008). It is proposed that the anionic lipid would reduce the electrostatic repulsion between the arginine residues of the adjacent monomer.

The crystal structure shows (Fig. 6.3A, B) that the lipid molecule is located at the interface between two monomers, with one of its acyl chain located between helices of adjacent monomers, thus taking part of the hydrophobic helix–helix interface. The headgroup of the PG molecules, although not fully resolved, is located near the extracellular surface, and appears to be coordinated by Arg89 and maybe (as the headgroup is not fully resolved) Arg64 (Valiyaveetil *et al.*, 2002). Interestingly, sequence analysis of various bacterial species shows that Arg89 is extremely conserved among KcsA channels. Indeed sequence conservation of putative lipid binding site is indicative of a functional role. Using statistical analysis, Adamian *et al.* (2011 analysed evolutionary selection pressure at lipid-binding sites in various proteins. In the case of KcsA, the authors showed strong evolutionary conservation for nine residues participating in binding of the PG molecule, including Arg89. Furthermore, a similar lipid binding site has been observed in the crystal structure of the mammalian

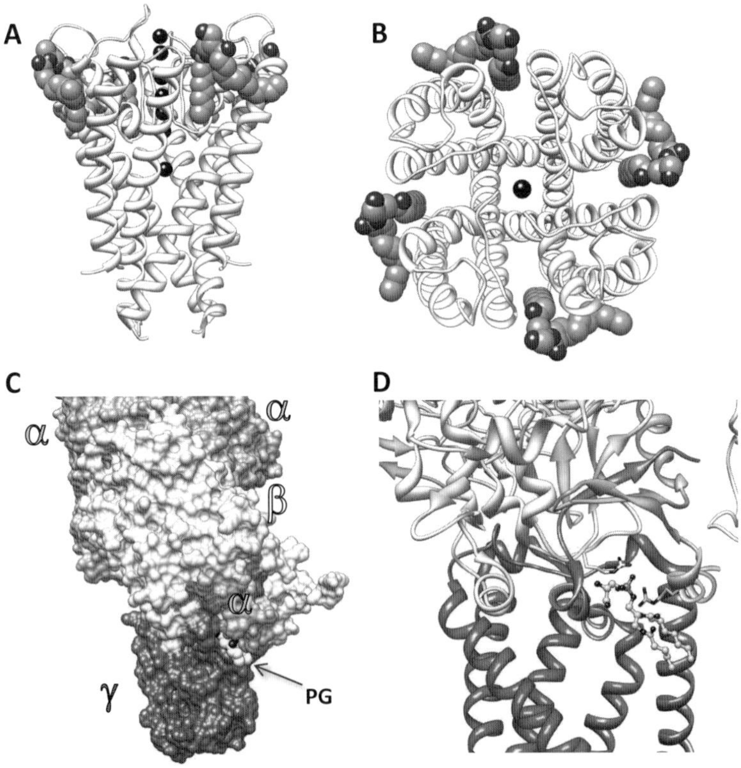

Figure 6.3 Phosphatidylglycerol bound to protein. (A and B) Side and top views of the KcsA homotetramer with PG bound (PDB 1K4C). The coordinated ions are shown as black sphere in the centre. (C) Surface representation of Nitrate Reductase (NAR) (PDB 1Q16) bound to PG shown as grey space-filling spheres. (D) Close-up of the interaction between NAR and PG, highlighting the two arginines.

voltage-gated Kv1.2 channel (Long *et al.*, 2007), suggesting that this lipid binding site was conserved throughout evolution, thus confirming the importance of lipid–protein interactions in the mechanism of potassium transport.

PG is also seen in the crystal structure of nitrate reductase A from *E. coli* (Bertero *et al.*, 2003). While the functional importance of PG has not been clearly established for this enzyme, the crystal structure (Fig. 6.3C, D) indicates a structural role, as the lipid appears to stabilize the heterotrimer by forming, on one hand, hydrophobic contacts between the acyl chain and the TM domain of the γ-subunit and, on the other hand, polar interactions between Arg6 of the α-subunit and the phosphate and Arg218 of the β-subunit and the glycerol moiety.

Three molecules of PG, as well as a monogalactosyldiacylglycerol (Jordan *et al.*, 2001) are also seen in the structure of the cyanobacterial photosystem I. The presence of these different lipids in symmetrical position in the complex has been suggested to be responsible for modulation of the rate of electron transfer that is different for both halves of the molecules. However, they could also fulfil a more structural role as interfacial lipids between the various subunits of the complex (Jones, 2007).

Cardiolipin

Functional dependence on the anionic lipid cardiolipin has been described for a number of eukaryotic proteins such as cytochrome bc1 (Lange *et al.*, 2001) or mitochondrial carriers (Pebay-Peyroula *et al.*, 2003). While a relatively minor component of the bacterial membrane, cardiolipin has been observed in several structures of prokaryotic membrane proteins.

The purple bacterial reaction centre uses light energy to power the synthesis of ATP, as well as other energy-requiring processes. It is composed of three protein subunits. The subunits L and M have five transmembrane helices and are organized in a pseudo-twofold symmetry. The third subunit, the subunit H is located on the cytoplasmic face of the two other subunits and is anchored in the membrane via a single transmembrane helix (Jones *et al.*, 2002). The complex interacts with 10 cofactors, also organized in a twofold symmetry, but only half of them are functionally active, and implicated in the electron transfer process (Camara-Artigas *et al.*, 2002). Cardiolipin has been observed in the crystal structure of the complex isolated from *Rb. spheroides* (Fig. 6.4A, B), in association with the α-helix of the H subunit (McAuley *et al.*, 1999). Interestingly, amino acids that bind the cardiolipin headgroup are highly conserved in purple photosynthetic bacteria (Wakeham *et al.*, 2001). Mutation of the arginine residue interacting with cardiolipin did not affect the activity of the reaction centre, but rather its thermal stability (Fyfe *et al.*, 2004). The importance of cardiolipin for thermal stability was also observed in native membranes enriched in CL upon osmotic shock (De Leo *et al.*, 2009). It was suggested that cardiolipin could play a role in stabilizing interaction of adjacent membrane helices that do not directly interact through protein-protein contacts (Jones *et al.*, 2002).

Formate dehydrogenase-N (Fdh-N) is a bacterial heterotrimeric enzyme required for the redox loops of the oxygen respiratory chain involved in generating the proton-motive force (Jormakka *et al.*, 2003). The α subunit is a soluble domain located in the periplasm, the β subunit is a membrane protein with a periplasmic extension while the γ subunit is an integral membrane domain. The crystal structure of Fdh-N shows a trimeric quaternary structure, a trimer of heterotrimers, in a mushroom-like shape (Jormakka *et al.*, 2002).

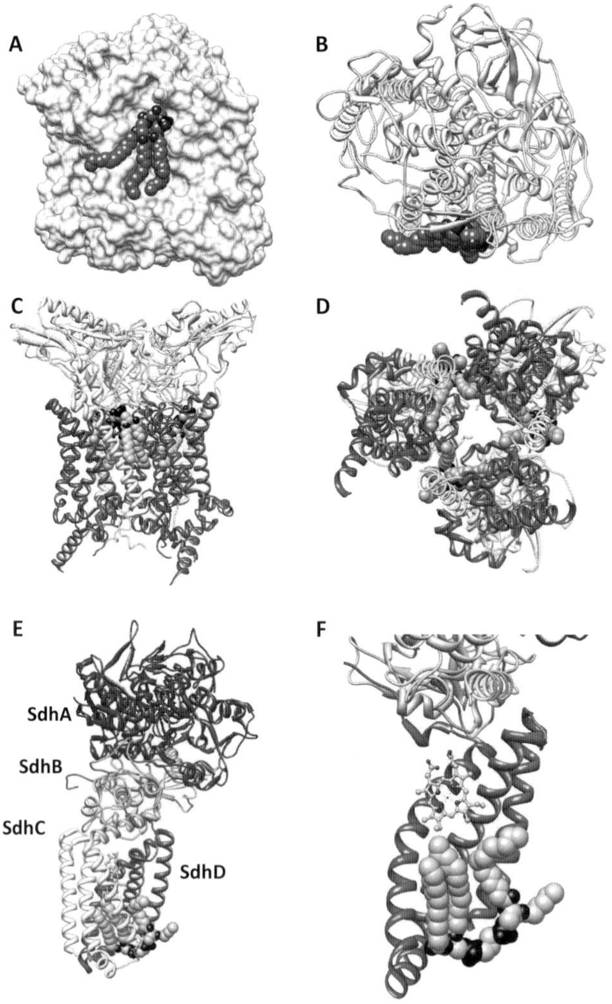

Figure 6.4 Cardiolipin binding to proteins. (A and B) Cardiolipin bound to photoreaction centre of Rb. Spheroides in surface representation (side view, A) and ribbons (top views B). CL is shown as grey space-filling spheres. (C and D) side and top view of Formate dehydrogenase-N homotrimer showing the lipid at the interface. (E) Ribbon representation of succinate dehydrogenase showing CL embedded between subunits *C* and *D*, just below the haem. (F) Close-up shown the proximity and steric constrains between the haem and CL, subunit *C* was removed for clarity.

Three cardiolipin molecules are observed, one at each interface between protomers. Within each protomer, the β subunit makes polar contacts with the headgroup of CL and hydrophobic contacts with one of the four acyl chains while another acyl tail extends into the groove separating two α-helices of the γ-subunit, thus participating actively in the β–γ interface. But more importantly, the remaining two acyl chains of CL make extensive contacts with the other transmembrane helices of the γ-subunit from the neighbouring protomer (Fig. 6.4C, D), thus actively stabilizing the homotrimer.

Interestingly, it has been shown that the other enzyme involved in the bacterial anaerobic redox loop, nitrate reductase (Nar) depends on the presence of cardiolipin for its enzymatic activity (Arias-Cartin *et al.*, 2011). Cardiolipin is also specifically enriched during the purification of this enzyme, accounting for 50% of the co-purified lipids. The authors have reinterpreted the crystallographic data of Nar structures and suggest that a cardiolipin molecule is tightly bound to the enzyme, in a conserved hydrophobic motif, close to the haem group and propose that it is important in maintaining the heterotrimer.

A cardiolipin molecule is also found deeply buried in the transmembrane domain of another bacterial enzyme involved in the aerobic respiratory chain of *E. coli*, succinate dehydrogenase (SQR) (Yankovskaya *et al.*, 2003). The lipid occupies an extensive pocket between subunits C and D, just below the haem group (see Fig. 6.4E, F). The tail of CL is actually sterically constraining the haem molecule and thus could potentially be involved in some functional regulation. Interestingly, in the homologous fumarate reductase from *Wolinella succinogenes*, the CL pocket is occupied by a second haem group. While it appears that CL takes an active part in the structural organization of SQR, its exact requirement for enzymatic activity remains to be established.

Therefore, functional and structural importance of cardiolipin in respiratory complexes appears to be conserved both in eukaryotes and prokaryotes.

Phosphatidylethanolamine

Being the most abundant lipid in bacterial membranes, it is not surprising that PE has been found in a number of crystal structures. Most bacterial structures containing annular lipids will typically show bound PE molecules, as for instance observed for Aquaporin Z (Jiang *et al.*, 2006) and sensory rhodopsin (Vogeley *et al.*, 2004). On the other hand, a number of crystal structures also show non-annular PE molecules as defined by apparent specific coordination motifs. The structure of the reaction centre of *Thermochromatium tepidum* shows a well defined PE molecule contacting subunits H, M and L through hydrophobic contacts with the acyl chains (Fig. 6.5A, B) (Nogi *et al.*, 2000). The polar headgroup is coordinated by interaction between the phosphate and the side chains of Arg31, Lys35 and Tyr39 from the H subunit while the amine interacts with the backbone carbonyl of Gly256 on subunit M. In order to allow for this interaction, the backbone of Gly256 adopts a peculiar conformation with dihedral angles $(\Phi,\Psi) = (84°,-3)$ that are mostly restricted to Glycine and rarely found for other residues. As a Gly256 is highly conserved in subunit M of reaction centres from purple bacterium, it is likely that binding of PE has been evolutionary favoured.

Cytochrome C oxidase, the terminal enzyme in the respiratory chain in mitochondria and bacteria, has been found to functionally depend on the presence of cardiolipin in several eukaryotic cases (Robinson, 1993). Interestingly, the structure of cytochrome C oxidase from *Rb. spheroides* (Fig. 6.5C, D) shows a series of lipid molecules that have been identified as PE (Svensson-Ek *et al.*, 2002). The molecules are deeply buried inside the quaternary assembly, providing extensive interface between subunits I, III and IV. In fact, subunit IV, a single transmembrane helix, interacts with subunit I and II solely via PE-mediated hydrophobic contacts. No direct residue–residue interactions are observed. This very strongly argues for a major structural role of bound lipids in maintaining the quaternary assembly of cytochrome C oxidase. Oddly, the crystal structure does not show coordination of the amine by the protein and thus the headgroup specificity may not be highly stringent.

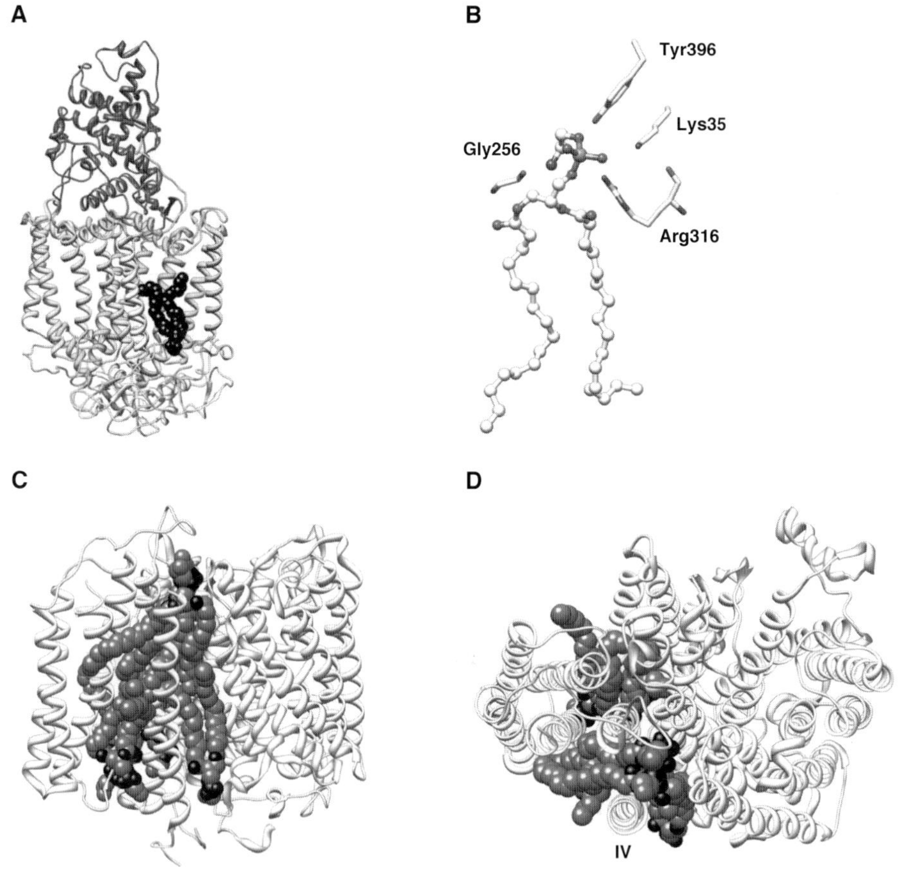

Figure 6.5 (A) Phosphatidylethanolamine bound to the reaction centre of *T. tepidum* (PDB 1EYS). (B) Close-up of the coordination of PE by Gly256, Arg31, Lys35 and Tyr39. (C) Cytochrome C oxidase from Rb. Spheroides with two PE molecules bound (PDB 1M56) (D) Top view of the protein–lipid complex structure, highlighting that subunit IV contacts the rest of the protein solely via the lipid.

The recent crystal structure of $ASBT_{NM}$, the bacterial homologue of bile acid sodium transporter shows two PE molecules bound on the surface of the helical bundle (Hu *et al.*, 2011). While one of the PE shows an annular-like conformation with no apparent specific interaction, the other is nicely coordinated by polar interactions between Y304 and K308 and the phosphate, while the amine, although resolved, does not show specific interactions with the protein, and thus the lipid specificity at this site remains open.

Although it was not observed in the crystal structure, specific enrichment of PE was observed during the isolation of the peripheral antenna complex LH2 of *Rb. spheroides*. Sequence alignments confirmed the presence of a putative PE-binding site, suggesting that enrichment could be due to a non-annular binding site (Kwa *et al.*, 2008).

The role of PE as modulator of membrane protein function is discussed below in the case of the secondary transporters of the major facilitator superfamily (MFS).

Lipopolysaccharide

Specific adaptation to the peculiar lipid composition of the outer membrane of bacteria has been observed in the case of ferric hydroxamate uptake receptor (FhuA), the siderophore ferrichrome receptor that adopts a typical porin β-barrel fold. The crystal structure of the protein shows a single lipopolysaccharide (LPS) molecule bound to the side of the barrel (Fig. 6.6) (Ferguson *et al.*, 2000). LPS, the main lipidic component of the outer membrane of *E. coli*, is a complex lipoglycan typically composed of two linked phosphorylated glucosamines and six fatty acid chains and is responsible for innate anti-bacterial immune response in higher organisms. The crystal structure shows multiple polar interactions between the sugar groups and several charged and polar side chains sticking out of the barrel (Fig. 6.6). The acyl chains achieve typical hydrophobic contacts with the aliphatic and aromatic residues located in the middle of the barrel. Adamian *et al.* (2011) have identified that 11 of the 24 contacting residues are under strong selection pressure, suggesting a functional role for the binding site. Furthermore, the authors indicate that the strands 7–9 of the porin, which are involved in contacting the LPS, form an unstable domain of the β-barrel, due to the presence of numerous neighbouring Arg and Lys residues that would be unfavoured in most cases. However, these side chains interact tightly with the charged headgroup of LPS, which therefore stabilizes this region of the protein. In addition, stoichiometric amounts of LPS are required for crystallization of FhuA further suggesting that binding of the polysaccharide thermodynamically stabilizes the protein. While the exact functional role of LPS in iron transport is not clear, it is more than likely that this unusual domain of FhuA helps to anchor it in the outer membrane while, as a corollary, binding the LPS molecule stabilizes the β-barrel.

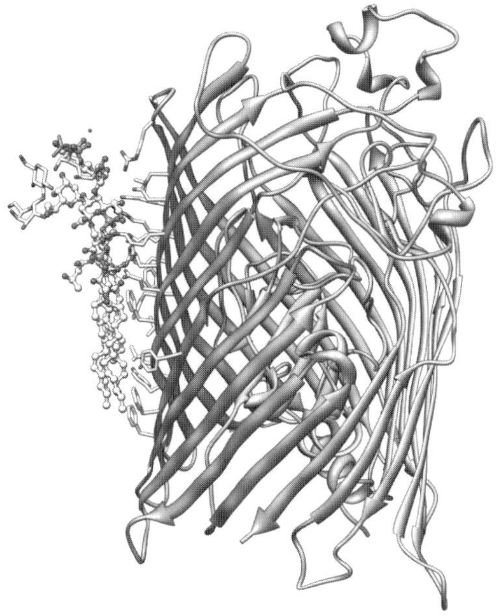

Figure 6.6 Structure of the FhuA–LPS complex (PDB 2FCP). The protein is shown as ribbon representation with side chains interacting with the lipid being explicitly shown as sticks. LPS is shown as ball-and-sticks.

Other lipids

E. coli being by far the most used model for protein expression and crystallization, the majority of lipids found in crystal structures will be, as illustrated above, either PE, PG, CL or LPS. However, a few structures of bacterial membrane proteins coming from other species and/or expression systems show the presence of other lipids. For instance the structure of the bacterial reaction centre from *Rb. spheroides* reveals the presence of a phosphatidylcholine and a glucosylgalactosyl diacylglycerol (Camara-Artigas *et al.*, 2002). These lipids interact predominantly with hydrophobic residues, at the interface between the various subunits and in close contact with some of the cofactor. It was suggested that the glycolipids could participate to the hydrophobic environment of the cofactor on one side of the protein, thereby favouring the electron transfer, while PC would participate in the inhibition of the electron transfer on the other side (Camara-Artigas *et al.*, 2002). This was further supported by studies of the energetic of the electron transfer (Nagy *et al.*, 2004).

The crystal structure of the cyanobacterial photosystem II is particular rich in lipids, with 14 lipids found per monomer, namely six monogalactosyldiacylglycerol, four digalactosyldiacylglycerol, three sulfoquinovosyldiacylglycerol and one PG. Eleven of these lipids are sandwiched between the core of the reaction centre and the surrounding protein subunits and cofactors, serving as a molecular glue between the different elements of the complex (Loll *et al.*, 2005).

MFS transporters: the role of PE

As phosphatidylethanolamine represents by far the major constituent of the bacterial inner membrane, it is not surprising that a number of bacterial proteins require the presence of PE in the bilayer for proper structure and activity. For instance, when the lactose permease LacY was reconstituted in proteoliposomes composed of *E. coli* lipids, uphill and downhill transport of sugar was observed, but when liposomes lacked PE and contained only PG, CL, and/or PC only kinetically downhill transport was seen (Chen *et al.*, 1984). A similar functional dependence was seen *in vivo* for other proteins such as gamma-aminobutyric acid permease GabP (Zhang *et al.*, 2005), the phenylalanine permease PheP (Zhang *et al.*, 2003), both from *E. coli* and *in vitro* for the multidrug transporter from *Lactococcus lactis*, LmrP (Gbaguidi *et al.*, 2007). PE has also been shown to increase the transport activity of the leucine transport system from *L. lactis* (Driessen *et al.*, 1988).

Are these proteins requiring given bulk properties of PE membranes to function or are some specific protein–PE interactions required? The different studies offer different explanations.

LmrP is a secondary multidrug transporter of *L. lactis* that confers resistance to an array of structurally diverse cytotoxic compounds and belongs to the major facilitator superfamily. The use of fluorescent substrates allows to follow the transport activity of the protein reconstituted in different lipidic environments. When reconstituted in a synthetic lipid mixture mimicking the bacterial membrane (i.e. 70% PE, 20% PG, 10% CL), LmrP shows significant extrusion of substrate upon imposition of a proton gradient, while replacement of PE by PC leads to complete loss of transport (Gbaguidi *et al.*, 2007). Although the secondary structure of the protein is identical in both lipid compositions, hydrogen/deuterium exchange measurements showed increased solvent exposure in PC-containing proteoliposomes suggesting a more dynamic conformational state. The use of PE analogues that differ solely in the methylation state of the ethanolamine headgroup offers a way to differentiate specific interactions from effects related to the physicochemical state of the membrane, as many of such physicochemical properties have been determined for the analogues as well. Reconstituting LmrP in proteoliposomes where PE bears one or two methylations still permit wild type function and structure, in sharp contrast with the PC situation (where all three hydrogens are methylated) (Hakizimana *et al.*, 2008). On the other hand, physicochemical properties like the transition temperatures of the two methylation intermediates are evenly distributed between PE and PC indicating that the structural and functional impairments observed when going from PE to PC liposomes are due to a specific PE–LmrP interaction. It was suggested that the functional differences observed are likely the consequence of changes in the hydrogen bonding capability in the lipid headgroup although a combination of specific and more global effects cannot be excluded.

How would hydrogen donor ability of PE regulate structure and function of LmrP? We have proposed that aspartate 68 in the first intracellular loop (very close to the membrane according to the available structures of homologues) that is highly conserved in the MFS family, would interact with the hydrophilic moiety of PE (Hakizimana *et al.*, 2008). This interaction would thus allow conformational changes of this functionally important domain, probably following protonation-deprotonation cycles of the aspartate side chain. This was supported by EPR measurements as the spin label at position 67 shows increased mobility in PC environment compared with PE (Masureel *et al.*, unpublished observation). This hypothesis was further supported by molecular dynamics simulations using LacY structure that identified the formation of a strong hydrogen bond between (non-, mono-, and dimethylated) palmitoyl oleoyl phosphatidylethanolamine (POPE) amine groups and Asp68 that is significantly weaker in the case of PC. The bond is formed to a free hydrogen of the amine group with a speed of formation being inversely related to the degree of methylation. Asp68-bound PE is initially bound to Lys69 through its phosphate entity before being recruited by Asp68 (Lensink *et al.*, 2010). Mutation of this conserved aspartate in other MFS transporters consistently lead to functional defect, suggesting a conserved mechanism (Yamaguchi *et al.*, 1992; Jessen-Marshall *et al.*, 1995; Pazdernik *et al.*, 2000).

This model is in contrast to that proposed by Dowhan and colleagues to explain the functional and structural behaviour displayed by LacY in the absence of PE, as they proposed that PE is required as folding chaperone for LacY, as illustrated in the next chapter.

Lipids and protein folding

Considering the highly hydrophobic nature of transmembrane domains of proteins, it is not surprising that lipids play a major role in the proper folding of membrane proteins. A body of data is shedding light on the phospholipid species involved at various folding intermediates. Notably, β-barrel and α-helical membrane proteins have evolved different folding pathways, as it could be expected in light of their profound topological and biochemical differences.

The lack of ATP as an energy source in the periplasm explains the spontaneous insertion of barrels into bacterial outer membranes after secretion into the periplasmic space whereas helical membrane proteins require the intervention of the translocon machinery.

Lipids and folding of helical proteins

The Sec-translocon drives the translocation and insertion of membrane proteins in bacterial membranes. When the first transmembrane segment emerges from the ribosome, it is recognized by the signal recognition particle (SRP). The complex will interact with the membrane through the SRP receptor, the FtsY GTPase. SecYEG is an integral membrane protein complex, which constitutes the core channel for translocation across the membrane, while the ATPase SecA mediates the preprotein translocation (van Dalen *et al.*, 2004).

The influence of the lipid environment on the translocation of membrane proteins has been investigated *in vivo* using *E. coli* mutants defective in the synthesis of specific phospholipids, as well as *in vitro*. Although protein translocation was fully functional in the strain depleted in PE, protein translocation was very poor in isolated PE-deficient inner membranes (Rietveld *et al.*, 1995). However, reintroducing the curvature stress induced by PE, either using other non-bilayer forming lipids, or divalent cations, was sufficient to restore translocation (Rietveld *et al.*, 1995). Non-bilayer lipids could therefore influence the functional assembly of the SecYEG channel in a conformation required for translocation. In addition, non-bilayer lipids happen to stimulate ATPase activity of SecA (Ahn *et al.*, 1998), which could also influence protein translocation.

Protein translocation is also strongly impaired in *E. coli* strains depleted in PG (2% PG compared to wild-type levels) (de Vrije *et al.*, 1988), but supplementation with other anionic lipid was sufficient to restore translocation (Kusters *et al.*, 1991). Anionic phospholipids seem to mostly affect SecA, and in particular its membrane insertion and binding to SecY (Hendrick *et al.*, 1991; Breukink *et al.*, 1992). Notably, the equilibrium between dimeric and monomeric SecY is shifted in the presence of anionic lipids (Or *et al.*, 2002) and this equilibrium could affect SecA functional state. Moreover, anionic lipids could also be important for binding of the nascent peptide chain positive N-terminus (Phoenix *et al.*, 1993).

In addition to their role in the regulation of the insertion machinery, lipids have also been directly involved in the final topology of several secondary transporters. This could be characterized *in vivo* thanks to the availability of *E. coli* strains where membrane phospholipid composition can be controlled and varied systematically over a broad range while maintaining cell viability. Evidence demonstrating the role for lipids as determinants of membrane protein topology comes from null mutants of *E. coli* that result in the complete lack of PE (and almost exclusively includes PG and CL) (DeChavigny *et al.*, 1991). Coupling strains of *E. coli* in which PE content can be controlled with methods, designed to determine the orientation of TMs with respect to the plane of the membrane bilayer such as substituted cysteine accessibility method, enabled to analyse LacY organization in the membrane as a function of lipid composition (Bogdanov *et al.*, 2002; Dowhan *et al.*, 2004; Bogdanov *et al.*, 2005). LacY assembled in PE-containing cells showed a topological organization in agreement with X-ray crystallography (Abramson *et al.*, 2003). LacY expressed in PE-lacking cells is misassembled with the N-terminal six-TM helical bundle (TM I–VI) and adjacent extramembrane domains completely inverted with respect to the C-terminal five-TM helical bundle (Bogdanov *et al.*, 2002).

LacY reconstituted into *E. coli* liposomes adopts a wild type orientation whatever the source of LacY, either from PE-containing or PE-lacking cells, suggesting that protein–lipid interactions are an important determinant of membrane protein organization but only after protein domains have been extruded from the translocon (Wang *et al.*, 2002). Topological organization was incorrect when reconstitution was carried out from phospholipids extracted from PE-lacking cells showing only downhill transport. The fact that addition of dioleoyl-PC to the latter liposomes corrects the topological organization in the central domain of LacY provides some evidence of the potential for large topological changes in membrane protein post-assembly due to changes in lipid environment underwent during intracellular trafficking.

It is interesting to mention that *E. coli* tolerates foreign lipids like monoglucosyl diacylglycerol (GlcDAG) and diglucosyl diacylglycerol (GlcGlcDAG). PE-lacking cells expressing the gene encoding the GlcDAG synthase contain 30–40 mol% GlcDAG with a proportional reduction of PG and CL levels. Remarkably, the overall topology of LacY is normal in these cells (Xie *et al.*, 2006). PE and PC and the glycolipids GlcDAG and GlcGlcDAG are uncharged and it has been proposed that it is the dilution of the negative charge character of the membrane surface that is the topological determinant rather than the phase-forming properties of the lipids. The balance between anionic and neutral lipids could govern topological orientation of polytopic membrane proteins. This contradicts however recent modelling and dynamics data. Indeed, docking studies showed that the glucosyl moiety GlcDAG interact with both Asp68 and Lys69 in a PE-like fashion. In the absence of the phosphate group, Lys69 interacts with the oxygen O4 of the sugar ring, whereas Asp68 interacts with O3, leading to a motif conformation very similar to the one observed in the POPE simulation. In addition, O2 is found to be hydrogen bonded to the backbone NH of Lys69. Note that in GlcGlcDAG, the O2 atom is involved in the connection between the two sugar rings and would therefore not be available for binding, and binding of the second sugar ring to the motif would be prohibited by the bulkiness of GlcGlcDAG as a whole (Lensink *et al.*, 2010).

Lipids and barrel folding

Extensive studies of β-barrel folding have been carried out on OmpA, a protein of the outer membrane of *E. coli*. This porin is made of a N-terminal β-barrel membrane-anchoring domain and a globular periplasmic domain that interacts with the peptidoglycan. The transmembrane domain of OmpA has been solved by X-ray crystallography (Pautsch *et al.*, 1998) and NMR spectroscopy (Arora *et al.*, 2000) and used successfully as a model to study the *in vitro* folding of β-barrels in different lipid model membranes. OmpA insertion into bilayers was monitored by time-resolved Trp fluorescence quenching (Kleinschmidt *et al.*, 1999). Phospholipids were selectively labelled with bromines or nitroxide spin labels at defined positions of the acyl chain and used as depth-specific fluorescence quenchers (Kleinschmidt *et al.*, 1999; Ramakrishnan *et al.*, 2004). The data reveal a mechanism of insertion in several steps.

1. A binding phase to the lipid bilayer independent of temperature.
2. An insertion step into the lipid bilayer which is temperature-dependent with time constants in the 15 min to 3 h range. In this intermediate state, the protein adopts a 'molten globule conformation in which the correct tertiary fold has still not been achieved'.
3. A final phase leads to completion of the β-barrel structure.

Urea-induced unfolding curves were recorded in lipid bilayers whose composition was modified in terms of lipid properties like polar headgroup, acyl chains and physical parameters like bilayer thickness or fluidity. The effect of the bilayer thickness was investigated by measuring the rates of folding and insertion of OmpA into PC bilayers with different acyl chain lengths (Kleinschmidt *et al.*, 2002). Insertion rates increase significantly as the bilayer thickness decreases. Interestingly, outer bacterial membranes are thought to be thinner than the internal membrane, and this could be important for the spontaneous insertion of the proteins in the bilayer. On the other hand, the thermodynamic stability of OmpA has been shown to increase in mono-unsaturated PC bilayer of increasing chain length (to C_{18}) (Hong *et al.*, 2004). However, the increase observed is only 20% of the value expected to account for the hydrophobic effect. This is probably due to the counteraction of hydrophobic mismatch and the energetic cost of the induced lipid deformation (Tamm *et al.*, 2004). Modification of the tilt of the barrel could be followed by infrared spectroscopy in bilayer of increasing chain length, which represents another mechanism to account for the hydrophobic mismatch generated by increased bilayer thickness (Ramakrishnan *et al.*, 2005). Similar results were obtained for larger β-barrel proteins, such as FhuA, OmpG or FomA (Ramakrishnan *et al.*, 2005; Anbazhagan *et al.*, 2008a,b). Another physico-chemical parameter analysed is the effect of curvature stress and lateral pressure. PC species with cis-unsaturation on both fatty acyl-chains present a conical shape, an effect that is stronger for shorter chain length. In this case, stability increased with decreased chain length, evidencing the stabilizing effect of lateral pressure on OmpA (Hong *et al.*, 2004). This may also explain why OmpA is able to insert in small unilamellar vesicles of various bilayer thicknesses, but only in large unilamellar vesicles made of PC species with short chain lengths (Kleinschmidt *et al.*, 2002).

Typical bacterial membranes, however, do not contain PC, but are mostly composed of PE. Including increasing amounts of POPE in a bilayer composed of POPC and 7.5% POPG increased OmpA stability when POPE was included up to 40 mol% (Hong *et al.*, 2004). When the physiological POPE content was reached, OmpA became so stable that the reversible folding experiment could no longer be carried out. Similarly, including PG or CL up to 30 and 15 mol%, increased the cooperativity of OmpA folding (Tamm *et al.*, 2004). It is likely that the favourable effect of POPE on the thermodynamic stability of OmpA is due to its conical shape that induces a curvature stress and an internal lateral pressure on embedded membrane proteins, similarly to what was observed with poly-unsaturated PC.

In Gram-negative bacteria, LPS, present in the outer leaflet of the outer membrane might contribute to the insertion and folding of barrels. However, to study the influence of this lipid *in vitro*, it would require making vesicles having an asymmetric lipid distribution with LPS only present in the inner leaflet, in order to mimic the situation prevailing *in vivo*. This raises technical problems that have not been solved so far, limiting our understanding of the role of LPS in β-barrel folding.

Simulations: towards a molecular understanding

While crystallography provides atomic description in a static fashion, dynamic understanding of protein–lipid interaction at the molecular detail can be obtained using computational approaches. Molecular dynamics (MD) simulations solve Newton's equations of motion to provide a dynamical picture of the movement of individual atoms and molecules. The interaction parameters are described in a force field that is a combination of spectroscopic measurements, quantum mechanical calculations and empirical calibration. The integration of the equations of motion results in a trajectory of atomic positions, providing a correlated and time-resolved ensemble of interaction snapshots. MD simulations of proteins originated in the seventies, but the incorporation of biological membranes is a more recent endeavour (Tieleman *et al.*, 1997). In most simulation studies, especially in the early days, the lipid bilayer is no more than a hydrophobic host for the (membrane) protein. Only a single and preferentially non-charged lipid type was used, with for current standards poorly developed interaction parameters. Yet for all practical purposes this provided a sufficient description to study the behaviour and properties of the embedded membrane protein or peptide. Subsequent years have seen an increased focus on the development of interaction parameters for lipid bilayer systems and their application in the simulation of biomolecular systems (Scott, 2002). This includes the simulation of polyunsaturated fatty acids (Feller *et al.*, 2002), of cholesterol (Hofsass *et al.*, 2003) and sphingolipids (Pandit *et al.*, 2006), or the effect of salt on a lipid bilayer structure (Bockmann *et al.*, 2003; Gurtovenko, 2005). In addition, the increase in computing power has allowed for the study of larger and more complicated systems. The simulation and analysis of such systems has now become a routine matter (Kandt *et al.*, 2007; Biggin *et al.*, 2008; Lensink, 2008; Wolf *et al.*, 2010).

The effect of the membrane on protein structure has been studied in the simulation of the bacterial outer membrane protein OmpA by comparing with the dynamics of OmpA in a detergent micelle (Bond *et al.*, 2003). The short 10 ns simulations showed only subtle differences between the nature of OmpA-detergent and OmpA–lipid interactions, but larger protein flexibility was observed in the micellar environment, which was hypothesized as to explain an experimentally observed channel formation. The simulations illustrate how a different hydrophobic environment may have an effect on structure, and therefore activity, of integral membrane proteins.

The effect of varying lipid composition has been considered in a series of MD simulations of MscL (Elmore *et al.*, 2003), varying both headgroup (PC or PE) and/or lipid tail length (18:16 to 10:10). In all cases, the system was found to adapt quickly to the new lipid composition, showing a significant hydrophobic matching for the shorter lipid molecules. Intriguingly though, a significant correlation was found between observed protein–lipid interactions and experimental mutagenesis data, implying an increased weight of protein–lipid interactions over intersubunit interactions.

A study of the *E. coli* outer membrane protease OmpT focused on the environment of active site residues and in particular their accessibility to the intra- or extracellular water medium (Baaden *et al.*, 2004). It was found that water from the intracellular medium could penetrate into the β-barrel and participate in the reaction mechanism, whereas the outer membrane pore remained closed. The outer membrane of bacteria is a complex lipid environment consisting mainly of LPS. In absence of a force field description for LPS, outer bacterial membrane proteins are typically embedded in POPC bilayers. However, an inspection of protein–lipid interactions in the OmpT simulations did appear to reveal specific lipid interaction sites on the outer surface of the barrel.

These findings testify to the ability of MD to identify relevant protein–lipid interactions. Important interaction may be found, even when the 'wrong' lipid species is used, due to the chemical similarity of functional groups in different types of lipids. For instance, hydrophobic patches on the surface of the protein will show affinity for the lipid tails whatever the nature of the lipid headgroup and positively charged side chains will show affinity for the phosphate moiety of phospholipids. In a MD study of pentapeptides in three different membrane-mimetic systems the authors mainly focused on the influence of the membrane environment onto the structure of the peptide, but this study also noted an interaction between the tryptophan side chain and the lipid choline group (Aliste *et al.*, 2003).

Tryptophan is generally attributed a preference for the membrane, due to its π-orbitals that favour a position in the electrostatically complex interface between membrane and solvent (Yau *et al.*, 1998). More detail as to such interactions is provided in a thorough study of the association of the cationic translocating peptide penetratin to both neutral and negatively charged lipid bilayers (Lensink *et al.*, 2005). The close to 0.25-μs-long MD simulations shed a first light onto the penetratin translation mechanism, with the aromatic residues forming a cation/π cluster with arginine, effectively shielding them not only from the aqueous phase, but also from the strong bidentate hydrogen bonds they can make with the lipid phosphate groups. Such bonds are responsible for anchoring a membrane protein in its environment and their annulment can certainly contribute to the translocating properties of cell-penetrating peptides. The study also attributed a significant membrane-interacting role to the lysine residues of the peptide, which were found to be responsible for initial binding to the lipid phosphate groups. The flexible lysine side chains were subsequently able to displace the lipid molecule headgroups to allow the cationic peptide to interact with the hydrophobic core of the membrane. Many other MD studies also evidence the anchoring role of arginine and lysine at the molecular level (Deol *et al.*, 2004; Patargias *et al.*, 2005).

The anchoring role of positively charged side chains has been used to optimally position LacY in a lipid bilayer (Lensink *et al.*, 2010). The crystal structure of LacY is of medium-quality resolution (2.95 Å) and shows no lipid associated (Guan *et al.*, 2007), but, as discussed above, there is strong evidence of a PE-dependence for LacY and other MFS transporters (Dowhan *et al.*, 2009), with PE being responsible for both proper functioning as well as topological organization. We have performed a systematic characterization and analysis of protein–lipid interactions in lipid bilayers of varying composition. In particular, we have simulated the protein in POPE, POPC and POPG bilayers, but also in bilayers of POPE with increasing degrees of methylation on the ethanolamine group. Using the notion of persistent protein–lipid interactions we have introduced the lipid-mediated salt bridge, where a zwitterionic lipid (PC or PE) binds both a positively as well as negatively charged residue through its phosphate and ethanolamine or choline moieties, respectively. A classification of such interactions using residence time and strength of interaction resulted in the identification of a single and specific protein–lipid interaction, involving a PE or PC, but not PG lipid. The interaction with PE was found to be significantly stronger than with PC, whereas no such interaction with PG could be identified. On the protein side, the interaction was with Asp68 and Lys69, two residues that are strongly conserved throughout the family and even superfamily of secondary transporters (Jessen-Marshall *et al.*, 1995).

As detailed above, experiments on reconstituted MFS transporters where the degree of methylation of the PE headgroup was modified, demonstrated that at least one resident hydrogen atom moiety must be present on the ethanolamine to observe transport. We have subsequently performed additional simulations of LacY in lipid bilayers composed of mono- and of dimethylated phosphatidylethanolamine in an effort to better understand the experimentally found preference of PE over PC and its implications. Interestingly, in both additional simulations the same observation could be made, namely that a PE lipid participated in a lipid-mediated salt bridge between Asp68 and Lys69. However, the speed of formation was found to be inversely proportional to the degree of methylation, with the last step in the salt bridge formation being a rotation of the amine group to point the remaining resident hydrogen (or one of them) towards Asp-68. In the case of PC, where no hydrogen is left on the amine atom, the methyl groups provides such steric hindrance as to push the amine group a full Angstrom farther away, making the rather significant distinction between a strong and a weak hydrogen bond. In a parallel MD study of the intramembrane rhomboid protease GlpG, the different hydrogen-bonding capabilities of the ethanolamine and choline groups have been found to be a determinant of lipid-dependent dynamics of the supposedly regulatory L1 loop (Bondar *et al.*, 2009).

Our LacY simulations constitute the first computational study ever to identify a lipid species-specific protein–lipid interaction that can be correlated with functional dependence and residue conservation (Lensink *et al.*, 2010). It is also an excellent example of how molecular dynamics simulations and relatively simple analysis techniques can be used to provide insight at the molecular level of the types of interactions between individual lipids of the membrane and individual residues of the protein that govern both membrane protein function as well as activity. A fine example of such is the recently published study of the molecular recognition of a single sphingolipid species by the transmembrane domain of the COPI machinery protein p24 (Contreras *et al.*, 2012). The 1 μs MD simulations show spontaneous diffusion of sphingomyelins to the transmembrane domain, with a highest affinity residence lifetime of approximately 250 ns for SM 18; the signature sequence of binding is proposed to represent a conserved sphingolipid-binding cavity in mammalian membrane proteins.

Lipid–protein interactions and crystallization of membrane proteins

While crystal structures have told us a lot about how lipids bind and stabilize membrane proteins, it is now clear that lipids are often very valuable to obtain crystal structures of such proteins. Indeed, by stabilizing thermodynamically the protein, given lipids may favour crystallization, and in a number of cases, the presence of lipids is absolutely required to obtain diffracting crystals. In the crystal structures of bacteriorhodopsin and aquaporin 0, Hite and colleagues could show that, binding to lipids rather than detergent molecules decreased the B-factor (crystallographic stability factor) of the interacting amino-acids, confirming the stabilizing effect of lipids on the protein structure (Hite *et al.*, 2008).

For instance, by changing the solubilization conditions, Kaback and colleagues have manipulated the amount of lipids that copurified with the lactose permease (Guan *et al.*, 2007). They have shown that the actual amount of bound lipids modifies the quality of the crystals. Delipidated proteins produce hexagonal crystals of poor diffraction quality, between 18 and 25 lipids per protein gave orthorhombic crystal diffracting at 3 Å while keeping between 9 and 16 lipids per protein gave the best results, with tetragonal crystals diffracting at 2.6 Å. Modulation of the lipid content could also be performed by supplementing delipidated lactose permease with lipid preparation (typically *E. coli* lipid extracts).

A similar effect was observed for the crystallization of the cytochrome b_6f complex from cyanobacteria. DOPC was essential to obtain crystals diffracting at 3 Å while crystals diffracted only at 12 Å in the absence of supplemented lipids (Kurisu *et al.*, 2003; Jones, 2007).

For another MFS transporter, GlpT, crystals were not observed after IMAC purification, when the amount of bound lipid was estimated around 42/protein, but only after a subsequent size-exclusion (SEC) step where the lipid content was reduced by half (Lemieux *et al.*, 2003). Adding another SEC step led to precipitation of the protein, most likely due to excessive delipidation.

Maintaining a significant amount of bound lipids or targeting given lipid–detergent ratio is now a common approach in membrane protein crystallography (Wenz *et al.*, 2009; Stroud, 2011). Detergent and lipid mixes have now been standardized as either an adjuvant during the purification or as additive in the crystallization drops (Long *et al.*, 2007).

One of the technical difficulties is to evaluate the amount of bound lipids to a given sample. Classical approaches will require to extract the lipids from the sample and quantify the amount of lipids by either thin-layer-chromatography (Lemieux *et al.*, 2003) or phosphorus assay (Guan *et al.*, 2007). We have used infrared spectroscopy as a fast measurement technique, allowing to estimate the amount of co-purified lipids by using moderate amounts of proteins. Concentrated protein samples are deposited on a diamond reflection element and Attenuated-Total-Reflection FT-IR measurements are performed. The relative intensities of both the protein and the lipid signals will directly depend on the relative amounts of the molecular species in the sample. Interestingly the absorption bands of the lipid C=O is spectrally distinct from the protein signal, allowing for easy estimation of lipid–protein ratio (A. Troupiotis, unpublished observation). We can show that the infrared signal correlates both with the phosphorus assay and the SEC profile, confirming that this fast method allows for comparable measurement.

Conclusions

The adaptation of membrane proteins to the complexity of their environment is remarkable. What may at first sight appear as evolutionary contradictions, such as hydrophobic mismatches, curvature stress, presence of non-lamellar lipids, or seemingly destabilizing sequence motifs are, in fact, required for proper biological function. Dynamical adaptation in changes of membrane thickness can lead to appropriate channel opening while binding to specific lipids can provide important stabilizing forces or contribute to activation mechanisms.

Biological membranes are composed of hundreds of different lipid molecules that, on one hand, bear individual physico-chemical properties, and on the other hand contribute to complex ensemble properties. In our opinion, this dual character is reflected by how the membrane modulates structure and function of embedded proteins, both through material properties of the bulk and by specific interactions with given lipidic species. As such the lipid-based and protein-based approaches are complementary and must be definitely both taken into account when considering the molecular basis of membrane protein function. Structural, biochemical and theoretical characterizations of these interactions have long been performed independently from each other, but the recent trend to combine these approaches will pave the way for a better understanding of the complex interplay between proteins and lipids in biological pathways.

Acknowledgements

This work was supported by the Fonds National de la Recherche Scientifique (MIS F.4523.12).

References

Abramson, J., Smirnova, I., Kasho, V., Verner, G., Kaback, H.R., and Iwata, S. (2003). Structure and mechanism of the lactose permease of *Escherichia coli*. Science *301*, 610–615.

Adamian, L., Naveed, H., and Liang, J. (2011). Lipid-binding surfaces of membrane proteins: evidence from evolutionary and structural analysis. Biochim. Biophys. Acta *1808*, 1092–1102.

Ahn, T., and Kim, H. (1998). Effects of nonlamellar-prone lipids on the ATPase activity of SecA bound to model membranes. J. Biol. Chem. *273*, 21692–21698.

Aliste, M.P., MacCallum, J.L., and Tieleman, D.P. (2003). Molecular dynamics simulations of pentapeptides at interfaces: salt bridge and cation–pi interactions. Biochemistry *42*, 8976–8987.

Alvis, S.J., Williamson, I.M., East, J.M., and Lee, A.G. (2003). Interactions of anionic phospholipids and phosphatidylethanolamine with the potassium channel KcsA. Biophys. J. *85*, 3828–3838.

Anbazhagan, V., Qu, J., Kleinschmidt, J.H., and Marsh, D. (2008a). Incorporation of outer membrane protein OmpG in lipid membranes: protein–lipid interactions and beta-barrel orientation. Biochemistry *47*, 6189–6198.

Anbazhagan, V., Vijay, N., Kleinschmidt, J.H., and Marsh, D. (2008b). Protein–lipid interactions with *Fusobacterium nucleatum* major outer membrane protein FomA: spin-label EPR and polarized infrared spectroscopy. Biochemistry *47*, 8414–8423.

Andersen, O.S., and Koeppe, R.E. (2007). Bilayer thickness and membrane protein function: an energetic perspective. Annu. Rev. Biophys. Biomol. Struct. *36*, 107–130.

Arias-Cartin, R., Grimaldi, S., Pommier, J., Lanciano, P., Schaefer, C., Arnoux, P., Giordano, G., Guigliarelli, B., and Magalon, A. (2011). Cardiolipin-based respiratory complex activation in bacteria. Proc. Natl. Acad. Sci. U.S.A. *108*, 7781–7786.

Arora, A., Rinehart, D., Szabo, G., and Tamm, L.K. (2000). Refolded outer membrane protein A of *Escherichia coli* forms ion channels with two conductance states in planar lipid bilayers. J. Biol. Chem. *275*, 1594–1600.

Baaden, M., and Sansom, M.S. (2004). OmpT: molecular dynamics simulations of an outer membrane enzyme. Biophys. J. *87*, 2942–2953.

Belrhali, H., Nollert, P., Royant, A., Menzel, C., Rosenbusch, J.P., Landau, E.M., and Pebay-Peyroula, E. (1999). Protein, lipid and water organization in bacteriorhodopsin crystals: a molecular view of the purple membrane at 1.9 A resolution. Structure *7*, 909–917.

Bertero, M.G., Rothery, R.A., Palak, M., Hou, C., Lim, D., Blasco, F., Weiner, J.H., and Strynadka, N.C. (2003). Insights into the respiratory electron transfer pathway from the structure of nitrate reductase A. Nat. Struct. Biol. *10*, 681–687.

Biggin, P.C., and Bond, P.J. (2008). Molecular dynamics simulations of membrane proteins. Methods Mol. Biol. *443*, 147–160.

Bockmann, R.A., Hac, A., Heimburg, T., and Grubmuller, H. (2003). Effect of sodium chloride on a lipid bilayer. Biophys. J. *85*, 1647–1655.

Bogdanov, M., Heacock, P.N., and Dowhan, W. (2002). A polytopic membrane protein displays a reversible topology dependent on membrane lipid composition. EMBO J. *21*, 2107–2116.

Bogdanov, M., Zhang, W., Xie, J., and Dowhan, W. (2005). Transmembrane protein topology mapping by the substituted cysteine accessibility method (SCAM(TM)): application to lipid-specific membrane protein topogenesis. Methods *36*, 148–171.

Bond, P.J., and Sansom, M.S. (2003). Membrane protein dynamics versus environment: simulations of OmpA in a micelle and in a bilayer. J. Mol. Biol. *329*, 1035–1053.

Bondar, A.N., del Val, C., and White, S.H. (2009). Rhomboid protease dynamics and lipid interactions. Structure *17*, 395–405.

Breukink, E., Demel, R.A., de Korte-Kool, G., and de Kruijff, B. (1992). SecA insertion into phospholipids is stimulated by negatively charged lipids and inhibited by ATP: a monolayer study. Biochemistry *31*, 1119–1124.

Camara-Artigas, A., Brune, D., and Allen, J.P. (2002). Interactions between lipids and bacterial reaction centers determined by protein crystallography. Proc. Natl. Acad. Sci. U.S.A. *99*, 11055–11060.

Chamberlain, B.K., and Webster, R.E. (1976). Lipid–protein interactions in *Escherichia coli*. Membrane-associated f1 bacteriophage coat protein and phospholipid metabolism. J. Biol. Chem. *251*, 7739–7745.

Chang, G., Spencer, R.H., Lee, A.T., Barclay, M.T., and Rees, D.C. (1998). Structure of the MscL homolog from *Mycobacterium tuberculosis*: a gated mechanosensitive ion channel. Science *282*, 2220–2226.

Chen, C.C., and Wilson, T.H. (1984). The phospholipid requirement for activity of the lactose carrier of *Escherichia coli*. J. Biol. Chem. *259*, 10150–10158.

Contreras, F.X., Ernst, A.M., Haberkant, P., Bjorkholm, P., Lindahl, E., Gonen, B., Tischer, C., Elofsson, A., von Heijne, G., Thiele, C., *et al.* (2012). Molecular recognition of a single sphingolipid species by a protein's transmembrane domain. Nature *481*, 525–529.

Cronan, J.E. (2003). Bacterial membrane lipids: where do we stand? Annu. Rev. Microbiol. *57*, 203–224.

Cybulski, L.E., Martin, M., Mansilla, M.C., Fernandez, A., and de Mendoza, D. (2010). Membrane thickness cue for cold sensing in a bacterium. Curr. Biol. *20*, 1539–1544.

De Leo, V., Catucci, L., Ventrella, A., Milano, F., Agostiano, A., and Corcelli, A. (2009). Cardiolipin increases in chromatophores isolated from Rhodobacter spheroides after osmotic stress: structural and functional roles. J. Lipid Res. *50*, 256–264.

de Planque, M.R., Greathouse, D.V., Koeppe, R.E., Schafer, H., Marsh, D., and Killian, J.A. (1998). Influence of lipid/peptide hydrophobic mismatch on the thickness of diacylphosphatidylcholine bilayers. A 2H NMR and ESR study using designed transmembrane alpha-helical peptides and gramicidin A. Biochemistry *37*, 9333–9345.

de Vrije, T., de Swart, R.L., Dowhan, W., Tommassen, J., and de Kruijff, B. (1988). Phosphatidylglycerol is involved in protein translocation across *Escherichia coli* inner membranes. Nature *334*, 173–175.

DeChavigny, A., Heacock, P.N., and Dowhan, W. (1991). Sequence and inactivation of the pss gene of *Escherichia coli*. Phosphatidylethanolamine may not be essential for cell viability. J. Biol. Chem. *266*, 10710.

Deol, S.S., Bond, P.J., Domene, C., and Sansom, M.S. (2004). Lipid–protein interactions of integral membrane proteins: a comparative simulation study. Biophys. J. *87*, 3737–3749.

Dowhan, W., and Bogdanov, M. (2009). Lipid-dependent membrane protein topogenesis. Annu. Rev. Biochem. *78*, 515–540.

Dowhan, W., Mileykovskaya, E., and Bogdanov, M. (2004). Diversity and versatility of lipid–protein interactions revealed by molecular genetic approaches. Biochim. Biophys. Acta *1666*, 19–39.

Driessen, A.J., Zheng, T., In't Veld, G., Op den Kamp, J.A., and Konings, W.N. (1988). Lipid requirement of the branched-chain amino acid transport system of Streptococcus cremoris. Biochemistry *27*, 865–872.

Elmore, D.E., and Dougherty, D.A. (2003). Investigating lipid composition effects on the mechanosensitive channel of large conductance (MscL) using molecular dynamics simulations. Biophys. J. *85*, 1512–1524.

Feller, S.E., Gawrisch, K., and MacKerell, A.D. Jr. (2002). Polyunsaturated fatty acids in lipid bilayers: intrinsic and environmental contributions to their unique physical properties. J. Am. Chem. Soc. *124*, 318–326.

Ferguson, A.D., Welte, W., Hofmann, E., Lindner, B., Holst, O., Coulton, J.W., and Diederichs, K. (2000). A conserved structural motif for lipopolysaccharide recognition by procaryotic and eucaryotic proteins. Structure *8*, 585–592.

Fyfe, P.K., Isaacs, N.W., Cogdell, R.J., and Jones, M.R. (2004). Disruption of a specific molecular interaction with a bound lipid affects the thermal stability of the purple bacterial reaction centre. Biochim. Biophys. Acta *1608*, 11–22.

Gbaguidi, B., Hakizimana, P., Vandenbussche, G., and Ruysschaert, J.M. (2007). Conformational changes in a bacterial multidrug transporter are phosphatidylethanolamine-dependent. Cell Mol. Life Sci. *64*, 1571–1582.

Geiger, O., Gonzalez-Silva, N., Lopez-Lara, I.M., and Sohlenkamp, C. (2010). Amino acid-containing membrane lipids in bacteria. Prog. Lipid Res. *49*, 46–60.

Gidden, J., Denson, J., Liyanage, R., Ivey, D.M., and Lay, J.O. (2009). Lipid compositions in *Escherichia coli* and *Bacillus subtilis* during growth as determined by MALDI-TOF and TOF/TOF mass spectrometry. Int. J. Mass Spectrom. *283*, 178–184.

Gonen, T., Cheng, Y., Sliz, P., Hiroaki, Y., Fujiyoshi, Y., Harrison, S.C., and Walz, T. (2005). Lipid–protein interactions in double-layered two-dimensional AQP0 crystals. Nature *438*, 633–638.

Guan, L., Mirza, O., Verner, G., Iwata, S., and Kaback, H.R. (2007). Structural determination of wild-type lactose permease. Proc. Natl. Acad. Sci. U.S.A. *104*, 15294–15298.

Gurtovenko, A.A. (2005). Asymmetry of lipid bilayers induced by monovalent salt: atomistic molecular-dynamics study. J. Chem. Phys. *122*, 244902.

Hakizimana, P., Masureel, M., Gbaguidi, B., Ruysschaert, J.M., and Govaerts, C. (2008). Interactions between phosphatidylethanolamine headgroup and LmrP, a multidrug transporter: a conserved mechanism for proton gradient sensing? J. Biol. Chem. *283*, 9369–9376.

Hanson, M.A., Cherezov, V., Griffith, M.T., Roth, C.B., Jaakola, V.P., Chien, E.Y., Velasquez, J., Kuhn, P., and Stevens, R.C. (2008). A specific cholesterol binding site is established by the 2.8 A structure of the human beta2-adrenergic receptor. Structure *16*, 897–905.

Hendrick, J.P., and Wickner, W. (1991). SecA protein needs both acidic phospholipids and SecY/E protein for functional high-affinity binding to the *Escherichia coli* plasma membrane. J. Biol. Chem. *266*, 24596–24600.

Hite, R.K., Gonen, T., Harrison, S.C., and Walz, T. (2008). Interactions of lipids with aquaporin-0 and other membrane proteins. Pflugers Arch. *456*, 651–661.

Hite, R.K., Li, Z., and Walz, T. (2010). Principles of membrane protein interactions with annular lipids deduced from aquaporin-0 2D crystals. EMBO J. *29*, 1652–1658.

Hofsass, C., Lindahl, E., and Edholm, O. (2003). Molecular dynamics simulations of phospholipid bilayers with cholesterol. Biophys. J. *84*, 2192–2206.

Hong, H., and Tamm, L.K. (2004). Elastic coupling of integral membrane protein stability to lipid bilayer forces. Proc. Natl. Acad. Sci. U.S.A. *101*, 4065–4070.

Hu, N.J., Iwata, S., Cameron, A.D., and Drew, D. (2011). Crystal structure of a bacterial homologue of the bile acid sodium symporter ASBT. Nature *478*, 408–411.

Hunte, C., and Richers, S. (2008). Lipids and membrane protein structures. Curr. Opin. Struct. Biol. *18*, 406–411.

In't Veld, G., Driessen, A.J., Op den Kamp, J.A., and Konings, W.N. (1991). Hydrophobic membrane thickness and lipid–protein interactions of the leucine transport system of *Lactococcus lactis*. Biochim. Biophys. Acta *1065*, 203–212.

Jessen-Marshall, A.E., Paul, N.J., and Brooker, R.J. (1995). The conserved motif, GXXX(D/E)(R/K) XG[X](R/K)(R/K), in hydrophilic loop 2/3 of the lactose permease. J. Biol. Chem. *270*, 16251–16257.

Jiang, J., Daniels, B.V., and Fu, D. (2006). Crystal structure of AqpZ tetramer reveals two distinct Arg-189 conformations associated with water permeation through the narrowest constriction of the water-conducting channel. J. Biol. Chem. *281*, 454–460.

Jones, M.R. (2007). Lipids in photosynthetic reaction centres: structural roles and functional holes. Prog. Lipid Res. *46*, 56–87.

Jones, M.R., Fyfe, P.K., Roszak, A.W., Isaacs, N.W., and Cogdell, R.J. (2002). Protein–lipid interactions in the purple bacterial reaction centre. Biochim. Biophys. Acta *1565*, 206–214.

Jordan, P., Fromme, P., Witt, H.T., Klukas, O., Saenger, W., and Krauss, N. (2001). Three-dimensional structure of cyanobacterial photosystem I at 2.5 A resolution. Nature *411*, 909–917.

Jormakka, M., Byrne, B., and Iwata, S. (2003). Protonmotive force generation by a redox loop mechanism. FEBS Lett. *545*, 25–30.

Jormakka, M., Tornroth, S., Byrne, B., and Iwata, S. (2002). Molecular basis of proton motive force generation: structure of formate dehydrogenase-N. Science *295*, 1863–1868.

Kandt, C., Ash, W.L., and Tieleman, D.P. (2007). Setting up and running molecular dynamics simulations of membrane proteins. Methods *41*, 475–488.

Kleinschmidt, J.H., and Tamm, L.K. (1999). Time-resolved distance determination by tryptophan fluorescence quenching: probing intermediates in membrane protein folding. Biochemistry *38*, 4996–5005.

Kleinschmidt, J.H., and Tamm, L.K. (2002). Secondary and tertiary structure formation of the beta-barrel membrane protein OmpA is synchronized and depends on membrane thickness. J. Mol. Biol. *324*, 319–330.

Kurisu, G., Zhang, H., Smith, J.L., and Cramer, W.A. (2003). Structure of the cytochrome b6f complex of oxygenic photosynthesis: tuning the cavity. Science *302*, 1009–1014.

Kusters, R., Dowhan, W., and de Kruijff, B. (1991). Negatively charged phospholipids restore prePhoE translocation across phosphatidylglycerol-depleted *Escherichia coli* inner membranes. J. Biol. Chem. *266*, 8659–8662.

Kwa, L.G., Wegmann, D., Brugger, B., Wieland, F.T., Wanner, G., and Braun, P. (2008). Mutation of a single residue, beta-glutamate-20, alters protein–lipid interactions of light harvesting complex II. Mol. Microbiol. *67*, 63–77.

Lange, C., Nett, J.H., Trumpower, B.L., and Hunte, C. (2001). Specific roles of protein–phospholipid interactions in the yeast cytochrome bc1 complex Structure EMBO J. *20*, 6591–6600.

Lee, A.G. (2003). Lipid–protein interactions in biological membranes: a structural perspective. Biochim. Biophys. Acta *1612*, 1–40.

Lee, A.G. (2004). How lipids affect the activities of integral membrane proteins. Biochim. Biophys. Acta *1666*, 62–87.

Lee, A.G. (2011). Lipid–protein interactions. Biochem. Soc. Trans. *39*, 761–766.

Lemieux, M.J., Song, J., Kim, M.J., Huang, Y., Villa, A., Auer, M., Li, X.D., and Wang, D.N. (2003). Three-dimensional crystallization of the *Escherichia coli* glycerol-3-phosphate transporter: a member of the major facilitator superfamily. Protein Sci. 12, 2748–2756.

Lensink, M.F. (2008). Membrane-associated proteins and peptides. Methods Mol. Biol. *443*, 161–179.

Lensink, M.F., Christiaens, B., Vandekerckhove, J., Prochiantz, A., and Rosseneu, M. (2005). Penetratin–membrane association: W48/R52/W56 shield the peptide from the aqueous phase. Biophys. J. *88*, 939–952.

Lensink, M.F., Govaerts, C., and Ruysschaert, J.M. (2010). Identification of specific lipid-binding sites in integral membrane proteins. J. Biol. Chem. *285*, 10519–10526.

Loll, B., Kern, J., Saenger, W., Zouni, A., and Biesiadka, J. (2005). Towards complete cofactor arrangement in the 3.0 A resolution structure of photosystem II. Nature *438*, 1040–1044.

Long, S.B., Tao, X., Campbell, E.B., and MacKinnon, R. (2007). Atomic structure of a voltage-dependent K+ channel in a lipid membrane-like environment. Nature *450*, 376–382.

Lundbaek, J.A., Maer, A.M., and Andersen, O.S. (1997). Lipid bilayer electrostatic energy, curvature stress, and assembly of gramicidin channels. Biochemistry *36*, 5695–5701.

Mansilla, M.C., Cybulski, L.E., Albanesi, D., and de Mendoza, D. (2004). Control of membrane lipid fluidity by molecular thermosensors. J. Bacteriol. *186*, 6681–6688.

Marius, P., Alvis, S.J., East, J.M., and Lee, A.G. (2005). The interfacial lipid binding site on the potassium channel KcsA is specific for anionic phospholipids. Biophys. J. *89*, 4081–4089.

Marius, P., Zagnoni, M., Sandison, M.E., East, J.M., Morgan, H., and Lee, A.G. (2008). Binding of anionic lipids to at least three nonannular sites on the potassium channel KcsA is required for channel opening. Biophys. J. *94*, 1689–1698.

Marsh, D. (2008). Protein modulation of lipids, and vice-versa, in membranes. Biochim. Biophys. Acta *1778*, 1545–1575.

McAuley, K.E., Fyfe, P.K., Ridge, J.P., Isaacs, N.W., Cogdell, R.J., and Jones, M.R. (1999). Structural details of an interaction between cardiolipin and an integral membrane protein. Proc. Natl. Acad. Sci. U.S.A. *96*, 14706–14711.

Mileykovskaya, E., and Dowhan, W. (2009). Cardiolipin membrane domains in prokaryotes and eukaryotes. Biochim. Biophys. Acta *1788*, 2084–2091.

Mobashery, N., Nielsen, C., and Andersen, O.S. (1997). The conformational preference of gramicidin channels is a function of lipid bilayer thickness. FEBS Lett. *412*, 15–20.

Murata, T., Yamato, I., Kakinuma, Y., Leslie, A.G., and Walker, J.E. (2005). Structure of the rotor of the V-Type Na+-ATPase from Enterococcus hirae. Science *308*, 654–659.

Nagy, L., Milano, F., Dorogi, M., Agostiano, A., Laczko, G., Szebenyi, K., Varo, G., Trotta, M., and Maroti, P. (2004). Protein/lipid interaction in the bacterial photosynthetic reaction center: phosphatidylcholine and phosphatidylglycerol modify the free energy levels of the quinones. Biochemistry *43*, 12913–12923.

Nicholas, J. (1984). Mechanisms of thermal adaptation in bacteria: blueprints for survival. Trends Biochem. Sci. *9*, 108–112.

Nogi, T., Fathir, I., Kobayashi, M., Nozawa, T., and Miki, K. (2000). Crystal structures of photosynthetic reaction center and high-potential iron–sulfur protein from *Thermochromatium tepidum*: thermostability and electron transfer. Proc. Natl. Acad. Sci. U.S.A. *97*, 13561–13566.

Nomura, T., Sokabe, M., and Yoshimura, K. (2006). Lipid–protein interaction of the MscS mechanosensitive channel examined by scanning mutagenesis. Biophys. J. *91*, 2874–2881.

Nozaki, Y., Chamberlain, B.K., Webster, R.E., and Tanford, C. (1976). Evidence for a major conformational change of coat protein in assembly of fl bacteriophage. Nature *259*, 335–337.

Nury, H., Dahout-Gonzalez, C., Trezeguet, V., Lauquin, G., Brandolin, G., and Pebay-Peyroula, E. (2005). Structural basis for lipid-mediated interactions between mitochondrial ADP/ATP carrier monomers. FEBS Lett. *579*, 6031–6036.

O'Keeffe, A.H., East, J.M., and Lee, A.G. (2000). Selectivity in lipid binding to the bacterial outer membrane protein OmpF. Biophys. J. *79*, 2066–2074.

Or, E., Navon, A., and Rapoport, T. (2002). Dissociation of the dimeric SecA ATPase during protein translocation across the bacterial membrane. EMBO J. *21*, 4470–4479.

Oursel, D., Loutelier-Bourhis, C., Orange, N., Chevalier, S., Norris, V., and Lange, C.M. (2007). Lipid composition of membranes of *Escherichia coli* by liquid chromatography/tandem mass spectrometry using negative electrospray ionization. Rapid Commun. Mass Spectrom. *21*, 1721–1728.

Pandit, S.A., and Scott, H.L. (2006). Molecular-dynamics simulation of a ceramide bilayer. J. Chem. Phys. *124*, 14708.

Patargias, G., Bond, P.J., Deol, S.S., and Sansom, M.S. (2005). Molecular dynamics simulations of GlpF in a micelle vs in a bilayer: conformational dynamics of a membrane protein as a function of environment. J. Phys. Chem. B *109*, 575–582.

Pautsch, A., and Schulz, G.E. (1998). Structure of the outer membrane protein A transmembrane domain. Nat. Struct. Biol. *5*, 1013–1017.

Pazdernik, N.J., Matzke, E.A., Jessen-Marshall, A.E., and Brooker, R.J. (2000). Roles of charged residues in the conserved motif, G-X-X-X-D/E-R/K-X-G-[X]-R/K-R/K, of the lactose permease of *Escherichia coli*. J. Membr. Biol. *174*, 31–40.

Pebay-Peyroula, E., Dahout-Gonzalez, C., Kahn, R., Trezeguet, V., Lauquin, G.J., and Brandolin, G. (2003). Structure of mitochondrial ADP/ATP carrier in complex with carboxyatractyloside. Nature *426*, 39–44.

Perozo, E., Kloda, A., Cortes, D.M., and Martinac, B. (2002). Physical principles underlying the transduction of bilayer deformation forces during mechanosensitive channel gating. Nat. Struct. Biol. *9*, 696–703.

Phoenix, D.A., Kusters, R., Hikita, C., Mizushima, S., and de Kruijff, B. (1993). OmpF-Lpp signal sequence mutants with varying charge hydrophobicity ratios provide evidence for a phosphatidylglycerol-signal sequence interaction during protein translocation across the *Escherichia coli* inner membrane. J. Biol. Chem. *268*, 17069–17073.

Poolman, B., Spitzer, J.J., and Wood, J.M. (2004). Bacterial osmosensing: roles of membrane structure and electrostatics in lipid-protein and protein–protein interactions. Biochim. Biophys. Acta *1666*, 88–104.

Powl, A.M., Carney, J., Marius, P., East, J.M., and Lee, A.G. (2005). Lipid interactions with bacterial channels: fluorescence studies. Biochem. Soc. Trans. *33*, 905–909.

Ramakrishnan, M., Pocanschi, C.L., Kleinschmidt, J.H., and Marsh, D. (2004). Association of spin-labeled lipids with beta-barrel proteins from the outer membrane of *Escherichia coli*. Biochemistry *43*, 11630–11636.

Ramakrishnan, M., Qu, J., Pocanschi, C.L., Kleinschmidt, J.H., and Marsh, D. (2005). Orientation of beta-barrel proteins OmpA and FhuA in lipid membranes. Chain length dependence from infrared dichroism. Biochemistry *44*, 3515–3523.

Rietveld, A.G., Koorengevel, M.C., and de Kruijff, B. (1995). Non-bilayer lipids are required for efficient protein transport across the plasma membrane of *Escherichia coli*. EMBO J. *14*, 5506–5513.

Robinson, N.C. (1993). Functional binding of cardiolipin to cytochrome c oxidase. J. Bioenerg. Biomembr. *25*, 153–163.

Roux, B., Berneche, S., and Im, W. (2000). Ion channels, permeation, and electrostatics: insight into the function of KcsA. Biochemistry *39*, 13295–13306.

Scott, H.L. (2002). Modeling the lipid component of membranes. Curr. Opin. Struct. Biol. *12*, 495–502.

Singer, S.J., and Nicolson, G.L. (1972). The fluid mosaic model of the structure of cell membranes. Science *175*, 720–731.

Smith, P.B., Snyder, A.P., and Harden, C.S. (1995). Characterization of bacterial phospholipids by electrospray ionization tandem mass spectrometry. Anal. Chem. *67*, 1824–1830.

Sohlenkamp, C., Lopez-Lara, I.M., and Geiger, O. (2003). Biosynthesis of phosphatidylcholine in bacteria. Prog. Lipid Res. *42*, 115–162.

Sprott, G.D. (1992). Structures of archaebacterial membrane lipids. J. Bioenerg. Biomembr. *24*, 555–566.

Stroud, R.M. (2011). New tools in membrane protein determination. F1000 Biol. Rep. *3*, 8.

Svensson-Ek, M., Abramson, J., Larsson, G., Tornroth, S., Brzezinski, P., and Iwata, S. (2002). The X-ray crystal structures of wild-type and EQ(I-286) mutant cytochrome c oxidases from Rhodobacter spheroides. J. Mol. Biol. *321*, 329–339.

Tamm, L.K., Hong, H., and Liang, B. (2004). Folding and assembly of beta-barrel membrane proteins. Biochim. Biophys. Acta *1666*, 250–263.

Tieleman, D.P., Marrink, S.J., and Berendsen, H.J. (1997). A computer perspective of membranes: molecular dynamics studies of lipid bilayer systems. Biochim. Biophys. Acta *1331*, 235–270.

Valiyaveetil, F.I., Zhou, Y., and MacKinnon, R. (2002). Lipids in the structure, folding, and function of the KcsA K+ channel. Biochemistry *41*, 10771–10777.

van Dalen, A., and de Kruijff, B. (2004). The role of lipids in membrane insertion and translocation of bacterial proteins. Biochim. Biophys. Acta *1694*, 97–109.

Vogeley, L., Sineshchekov, O.A., Trivedi, V.D., Sasaki, J., Spudich, J.L., and Luecke, H. (2004). Anabaena sensory rhodopsin: a photochromic color sensor at 2.0 A. Science *306*, 1390–1393.

Wakeham, M.C., Sessions, R.B., Jones, M.R., and Fyfe, P.K. (2001). Is there a conserved interaction between cardiolipin and the type II bacterial reaction center? Biophys. J. *80*, 1395–1405.

Wang, X., Bogdanov, M., and Dowhan, W. (2002). Topology of polytopic membrane protein subdomains is dictated by membrane phospholipid composition. EMBO J. *21*, 5673–5681.

Wenz, T., Hielscher, R., Hellwig, P., Schagger, H., Richers, S., and Hunte, C. (2009). Role of phospholipids in respiratory cytochrome bc(1) complex catalysis and supercomplex formation. Biochim. Biophys. Acta *1787*, 609–616.

Williamson, I.M., Alvis, S.J., East, J.M., and Lee, A.G. (2002). Interactions of phospholipids with the potassium channel KcsA. Biophys. J. *83*, 2026–2038.

Wolf, M.G., Hoefling, M., Aponte-Santamaria, C., Grubmuller, H., and Groenhof, G. (2010). g_membed: efficient insertion of a membrane protein into an equilibrated lipid bilayer with minimal perturbation. J. Comput. Chem. *31*, 2169–2174.

Xie, J., Bogdanov, M., Heacock, P., and Dowhan, W. (2006). Phosphatidylethanolamine and monoglucosyldiacylglycerol are interchangeable in supporting topogenesis and function of the polytopic membrane protein lactose permease. J. Biol. Chem. *281*, 19172–19178.

Yamaguchi, A., Nakatani, M., and Sawai, T. (1992). Aspartic acid-66 is the only essential negatively charged residue in the putative hydrophilic loop region of the metal-tetracycline/H+ antiporter encoded by transposon Tn10 of *Escherichia coli*. Biochemistry *31*, 8344–8348.

Yankovskaya, V., Horsefield, R., Tornroth, S., Luna-Chavez, C., Miyoshi, H., Leger, C., Byrne, B., Cecchini, G., and Iwata, S. (2003). Architecture of succinate dehydrogenase and reactive oxygen species generation. Science *299*, 700–704.

Yau, W.M., Wimley, W.C., Gawrisch, K., and White, S.H. (1998). The preference of tryptophan for membrane interfaces. Biochemistry *37*, 14713–14718.

Zhang, W., Bogdanov, M., Pi, J., Pittard, A.J., and Dowhan, W. (2003). Reversible topological organization within a polytopic membrane protein is governed by a change in membrane phospholipid composition. J. Biol. Chem. *278*, 50128–50135.

Zhang, W., Campbell, H.A., King, S.C., and Dowhan, W. (2005). Phospholipids as determinants of membrane protein topology. Phosphatidylethanolamine is required for the proper topological organization of the gamma-aminobutyric acid permease (GabP) of *Escherichia coli*. J. Biol. Chem. *280*, 26032–26038.

Zhang, X., Tamot, B., Hiser, C., Reid, G.E., Benning, C., and Ferguson-Miller, S. (2011). Cardiolipin deficiency in Rhodobacter spheroides alters the lipid profile of membranes and of crystallized cytochrome oxidase, but structure and function are maintained. Biochemistry *50*, 3879–3890.

Zhou, Y., Morais-Cabral, J.H., Kaufman, A., and MacKinnon, R. (2001). Chemistry of ion coordination and hydration revealed by a K+ channel–Fab complex at 2.0 A resolution. Nature *414*, 43–48.

Part III

Transport Across Bacterial Membranes

Bacterial ABC Transporters: Structure and Function

7

Anthony M. George and Peter M. Jones

Abstract

ATP-binding cassette (ABC) membrane transporters belong to one of the largest and most ancient gene families, occurring in bacteria, archaea, and eukaryota. In addition to nutrient uptake, ABC transporters are involved in other diverse processes such as the export of toxins, peptides, proteins, antibiotics, polysaccharides and lipids, and in cell division, bacterial immunity and nodulation in plants. While prokaryotic ABC transporters encompass both importers and exporters, eukaryotes harbour only exporters. Bacterial ABC transporters are intricately involved either directly or indirectly in all aspects of cellular physiology, metabolism, homeostasis, drug resistance, secretion, and cellular division. Whilst several complete ABC transporter structures have been solved over the past decade, their functional mechanism of transport is still somewhat controversial and this aspect is discussed in detail.

Introduction

ABC transporters constitute a large and diverse group of integral membrane proteins that translocate solutes across cellular or intracellular biological membranes (Higgins, 1992). Included among the many export functions and substrates are drug resistance, nodulation, cellular immunity, and the extrusion of noxious compounds, toxins, bacteriocins, peptides, proteins, proteases, lipases, polysaccharides, and lipids; and in bacteria only, the uptake of essential nutrients and ions such as amino acids, mono- and oligosaccharides, vitamins, metals, organic and inorganic ions, opines, iron-siderophores, and polyamine cations (Davidson *et al.*, 2008). ABC transporters are also involved in cell signalling, viability, pathogenicity, osmotic homeostasis, immunity, and cell division; and in many phenomena of biomedical importance including cystic fibrosis, Stargardt disease, adrenoleukodystrophy, multidrug resistance to cytotoxic drugs, and multidrug resistance in bacteria, yeasts and parasites. ABC transporters are transmembrane ATPase primary active transporters, along with P-type ATPase ion pumps, F-type ATP synthases, and V-type vacuolar ATPases.

A great number of scholarly reviews have been written on ABC transporters (Ames *et al.*, 1990, 1992; Higgins, 1992; Doige and Ames, 1993; Fath and Kolter, 1993; Dean and Allikmets, 1995; Bolhuis *et al.*, 1997; Mourez *et al.*, 1997; Croop, 1998; Hrycyna and Gottesman, 1998; Schneider and Hunke, 1998; van Veen and Konings, 1998; Holland and Blight, 1999; Jones and George, 1999, 2004; Saurin *et al.*, 1999; Dawson *et al.*, 2007; Hollenstein *et al.*, 2007a; Linton and Higgins, 2007; Lubelski *et al.*, 2007; Davidson *et al.*, 2008; Oldham *et al.*, 2008; Jones *et al.*, 2009; Kos and Ford, 2009; Nikaido and Takatsuka, 2009; Rees *et al.*,

2009; Seeger and van Veen, 2009; Kerr *et al.*, 2010). For the purposes of this chapter, we will confine the descriptions and discussion to bacterial ABC transporters, diverging only out of necessity to draw structural comparisons and parallels with some of their eukaryotic cousins. Nevertheless, many eukaryotes would have acquired their ABC systems from symbiotic bacteria that were the putative ancestors of cellular organelles such as peroxisomes, mitochondria, chloroplasts, and the endoplasmic reticulum; or from independent duplication-fusion events (Davidson *et al.*, 2008). Typical members of bacterial ABC importer and exporter families are listed in Tables 7.1 and 7.2, respectively.

Table 7.1 Prokaryotic ABC transporter uptake families[1]

Uptake families[2]	Host	Transporter
Carbohydrates		
Maltooligosaccharide	Escherichia coli	MalEFGK
Glycerol-phosphate	Escherichia coli	UgpABCE
Sucrose/maltose/trehalose	Sinorhizobium meliloti	AglEFGK
Glucose/mannose/galactose	Sulfolobus solfataricus	GlcSTUV
Amino acids		
Glutamate/glutamine/aspartate/asparagine	Rhodobacter capsulatus	BztABCD
Histidine; arginine/lysine/ornithine	Salmonella typhimurium	HisJ-ArgT-HisMPQ
Leucine/proline/alanine/serine/glycine	*Synechocystis* sp.	NatABCDE
Peptides		
Oligopeptides	Salmonella typhimurium	OppABCDF
Dipeptides	Bacillus subtilis	ppABCDE
Sulfate/tungstate		
Vanadate	Anabaena variabilis	VupABC
Sulfate	Mycobacterium tuberculosis	SubI-CysAWT
Phosphate		
	Escherichia coli	PhoS-PstABC
Molybdate		
Molybdate	Escherichia coli	ModABC
Polyamine/opine/phosphonate		
Putrescine/spermidine	Escherichia coli	PotABCD
Chrysopine	Agrobacterium tumefaciens	ChtGHIJK
Quaternary amines		
Glycine betaine	Bacillus subtilis	OpuAABC
Choline, glycine betaine, proline betaine	Bacillus subtilis	OpuCABCD
Siderophore-Fe^{3+}		
Fe^{3+}-carboxymycobactin	Mycobacterium tuberculosis	IrtAB

[1]Adapted from the TCDB: http://www.tcdb.org/search/result.php?tc=3.A.1.
[2]Other uptake families and substrates not listed include: phosphonate; iron, iron chelates/haem; manganese/zinc/iron chelates; nitrate/nitrite/cyanate; taurine; cobalt; vitamin B_{12}; thiamine; nickel; nickel/cobalt; methionine; biotin; cholesterol; steroids.

Table 7.2 Prokaryotic ABC transporter export families[1]

Export families[2]	Host	Transporter
Polysaccharides		
Capsular polysaccharide	Escherichia coli	KpsMT
Lipooligosaccharide	Rhizobium galegae	NodIJ
Lipopolysaccharide	Klebsiella pneumoniae	RfbAB
Drugs		
Daunorubicin/doxorubicin	Streptomyces peucetius	DrrAB
Oleandomycin	Streptomyces antibioticus	OleC4-OleC5
Multidrug	Lactococcus lactis	LmrA
Multidrug	Lactococcus lactis	LmrCD
Lipids (Lipid E)		
Phospholipid, LPS, lipid A, drugs	Escherichia coli	MsbA
Doxorubicin, verapamil, ethidium, TPP+, vinblastine, Hoechst 33342	Staphylococcus aureus	Sav1866
Peptide antimicrobials, nisin and polymyxin	Brevibacterium longum	ABC 1/2
Norfloxacin, DAPI, Hoechst 33342, tetracycline	Serratia marcescens	SmdAB
Proteins		
α-Haemolysin	Escherichia coli	HlyB
Colicin V	Enteric bacteria	CvaB
Glycanases, adhering proteins	Rhizobium leguminosarum	PrsD/PrsE
Peptides		
Subtilin	Bacillus subtilis	SpaB
Nisin	Lactococcus lactis	NisT
Pyoverdin	Pseudomonas aeruginosa	PvdE
Pep5 lantibiotic	Staphylococcus epidermidis	PepT
Aureocin A70	Staphylococcus aureus	AurT

[1]Adapted from the TCDB: http://www.tcdb.org/search/result.php?tc=3.A.1.
[2]Other export families and substrates not listed include teichoic acid; heteropolysaccharide O-antigen; β-glucan; haem; cytochrome c; adhesin; biofilm induction; S-layer proteins; lipases; bacteriocin immunity protein; competence factor; bacteriocins; hop resistance protein; macrolide antibiotics; peptides; lipoproteins; cysteine/glutathione; haemolysin; bacitracin; ethyl (methyl, benzyl) viologen; glycolipids.

Origins, phylogeny and diversity

ATP-binding cassette (ABC) transporters belong to one of the largest and most ancient gene families, occurring in bacteria, archaea, and eukaryota (Higgins and Gottesman, 1992; Fath and Kolter, 1993; Holland and Blight, 1999; Lee *et al.*, 2007; Davidson *et al.*, 2008). The acronym 'ABC' derives from ATP-binding cassette (Hyde *et al.*, 1990). ABC transporters were first discovered and characterized in prokaryotes from the 1970s, with the histidine (Ames and Lever, 1970) and maltose (Ferenci *et al.*, 1977) permease systems (Gilson *et al.*, 1982) in *Salmonella typhimurium* and *Escherichia coli*, respectively. From this beginning, many bacterial proteins were discovered and described that displayed related sequences and, in addition to nutrient uptake, were involved in other diverse processes such as the export of toxins, peptides, proteins, antibiotics, polysaccharides and lipids, and in cell division, bacterial immunity and nodulation in plants, and DNA repair (Higgins *et al.*, 1986).

ABC transporters make up one of the four major gene superfamilies in humans (Tatusov *et al.*, 1997). Examples of the numerical diversity of ABC transporters include over 80 ABC proteins in *Escherichia coli*, 46 in the thermophile *Thermus thermophilus*, 48 in humans, 51 in the mouse, 31 in yeast, 56 in the fly, and 126 in the plant *Arabidopsis*; indicating that diversity and numbers of ABC transporters do not follow any phlyogenetic order of ascendency in numbers. There are several websites that have lists and descriptions of ABC transporter genes and proteins. Prokaryotic ABC transporters are expansively listed at: http://www.tcdb.org/search/result.php?tc=3.A.1, with links and cross-references to GenBank, UniProtKB/Swiss-Prot, KEGG, and NCBI-Entrez; while human ABC proteins are best represented at: http://nutrigene.4t.com/humanabc.htm. Phylogenetic studies of the ABC superfamily have organized over 600 ABC-ATPases into 34 clusters (Saurin *et al.*, 1999; Dassa and Bouige, 2001). The separation of eukaryotic from prokaryotic systems does not occur at the root of the clusters, with homologous systems from the three kingdoms present at the tips of the branches of the cluster tree, suggesting that ABC systems began to specialize very early, probably before the separation of the three kingdoms of living organisms. Bacteria harbour both importers and exporters, whereas eukaryotes express only exporters (Fig. 7.1). The *Escherichia coli K12*, for example, contains 65 experimentally verified ABC transporters of which 50 are importers and 15 exporters (Moussatova *et al.*, 2008). Bacterial importers are also known as Traffic ATPases or periplasmic permeases (Ames *et al.*, 1990).

The main difference between prokaryotic ABC importers and exporters is that the former have extracellular domains or substrate-binding proteins that are essential for capturing nutrients to present to the membrane-embedded components of ABC importers. Gram-negative bacteria deploy periplasmic-binding proteins for solute capture (Neu and Heppel, 1965), whilst Gram-positive bacteria and Archaea use surface-anchored lipoproteins or cell surface-associated proteins, bound to the external membrane via electrostatic interactions (Bouige *et al.*, 2002). An unusual adaptation of these extracellular solute capture domains is found in two families of ABC importers in which one or two periplasmic receptors are fused covalently to the translocating membrane-embedded domains (van der Heide and Poolman, 2002). Phylogenic and bioinformatics analyses of prokaryotic ABC transporters have established that differences in substrate specificity do not correlate with evolutionary relatedness, suggesting that a large number of events and adaptations have occurred during the diversification of this superfamily.

Structure of bacterial ABC transporters

Despite a plenitude of solute types and processes with which they are involved, and two major divisions of importers and exporters, ABC transporters have the conserved core architecture of four domains, being two transmembrane domains (TMDs) and two cytosolic ATP-binding cassette domains, also commonly known as nucleotide-binding domains (NBDs) (Higgins, 1992; Schneider and Hunke, 1998; Jones and George, 2004; Hollenstein *et al.*, 2007a; Jones *et al.*, 2009; Kos and Ford, 2009). The four domains may comprise one, two or four polypeptide chains, encoded by the same or different genes, enabling the transporters to assemble into single polypeptide monomers, half-transporter homo- or

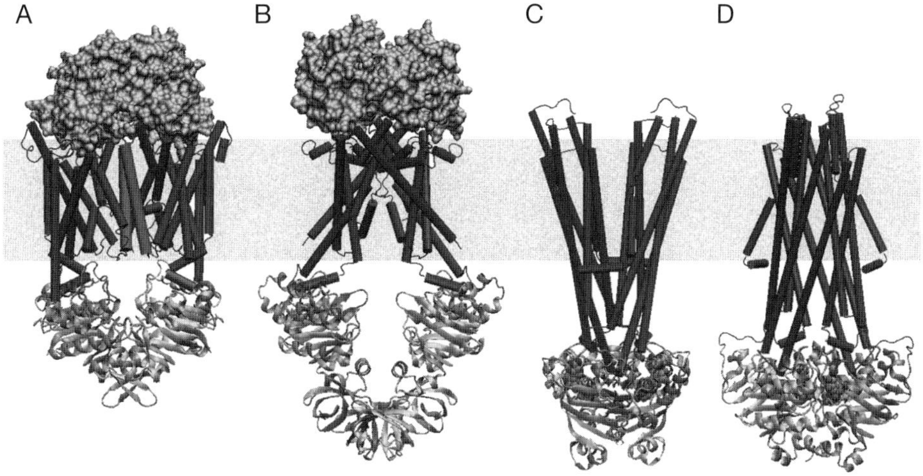

Figure 7.1 Prokaryotic ABC transporter architectures. The TMDs (blue and purple) of the binding-protein dependent ABC importers have two general conformation types: (A) that of the densely packed 10-TM helix HI1470/1 (2NQ2.pdb), with only subtle tilt changes in the central pore lining helices (red) throughout the transport cycle; and (B) the inverted tepee conformation of the *M. acetivorans* ModBC (with BP docked) (2ONK.pdb), in which the TMDs undergo large-scale structural rearrangements induced by NBD (orange and gold) dimerization and separation during the transport cycle. The two ABC exporters crystallized, Sav1866 and murine MDR1A, depict TMD helical domain swapping and the same NBD:TMD interface, but differ in the conformation of the TMDs, with the nucleotide-free MDR1A adopting an inverted tepee TMD configuration (not shown). (C) Side view of Sav1866 (2HYD.pdb) depicting domain swapping in the TMDs, which form divergent open wings in the nucleotide-bound state. (D) Front view of Sav1866 depicting the two coupling helices (short, almost horizontal purple and blue cylinders just above the NBDs). (Reprinted with permission from Jones *et al.*, 2009.) A colour version of this figure is located in the plate section at the back of the book.

heterodimers, or tetramers of four separately folded polypeptides. The NBDs are mostly C-terminal to the TMDs, but some exceptions deploy the opposite polarity (NBD-TMD). Prokaryotic importers also express extracellular accessory receptors or binding proteins, which are separately folded polypeptides, but in some cases may be covalently linked to the TMDs. The TMDs, which contain the putative substrate-binding sites, are comprised of multiple hydrophobic α-helical segments that form the transmembrane pore, and these often extend well into the cytosol, where they fold into mostly helical intra-cytoplasmic loop domains (ICLs) that form the physical interface with the NBDs. The consensus configuration for TMDs is of two sets of six α-helices, although there are notable exceptions among importers, with some transporters having ten or more whereas others have fewer than six (Fig. 7.1). The TMDs of canonical and atypical ABC transporters are thus not well conserved in length or sequence, probably reflecting their roles in binding diverse substrates. At any point in time, the TMDs are gated so that they are closed to one side of the membrane, preventing passive diffusion of substrates and ensuring unidirectional movement. The body of work – biochemical, biophysical, structural – that has ensued over more than a decade has struggled to answer key questions about ABC transporters: Where are the substrate sites and how many are there per transporter? How are substrates shuttled through the transporter? What is the coordinated allosteric mechanism of coupling of ATP binding and hydrolysis that drives the process and is it the same for importers and exporters? These questions will be addressed in the last section of this chapter on functional mechanisms.

The NBDs are molecular motors that coordinate ATP binding/hydrolysis with substrate binding and translocation across the TMDs, in either direction, depending on the import or export function of the specific transporter. NBDs also communicate with the extracellular accessory domains to signal substrate capture and release (Dawson *et al.*, 2007, 2008). The association of NBDs with the other the major domains of ABC transporters is depicted in Fig. 7.1, but a more detailed view of an NBD dimer and the motifs that define its highly conserved sequence and structure are depicted in Fig. 7.2. The geometry of the highly conserved NBDs is common to all ABC transporters. Each NBD is roughly an L-shape, with two lobes. The larger lobe I is a RecA and F_1-ATPase-like ATP-binding core subdomain (Vetter and Wittinghofer, 1999), containing the Walker A (GxxGxGKS/T) and B (ϕϕϕϕDE) motifs, where 'x' is any residue and 'ϕ' is any aliphatic residue (Walker *et al.*, 1982). The glycine-rich Walker A motif is also known as the phosphate-binding loop or P-loop that is common to all ATPases and binds the triphosphate moiety of the nucleotide via a number of electrostatic interactions that include the invariant lysine. The A-loop, located N-terminal to the Walker A sequence, contains a conserved aromatic (hence 'A' loop) residue (usually tyrosine) that stacks against the adenine ring of the nucleotide, helping to position it in the catalytic site. The Walker B conserved aspartate hydrogen bonds to coordinating ligands of the catalytic Mg^{2+} ion, assisting in establishing and maintaining the geometry of the active site. Lobe I also contains a β-subdomain that is peculiar to ABC-ATPases and that plays a predominantly structural role (Karpowich *et al.*, 2001). Lobe II is the α-helical subdomain, which is attached flexibly to lobe I and contains the 'LSGGQ' signature sequence that is unique to ABC-ATPase and defines the family. Unlike the A and B motifs that bind nucleotide in each monomer catalytic site, the signature sequence is located remotely within the same monomer (Fig. 7.2). However, in all ABC proteins, the two NBDs associate as a head-to-tail dimer, with two ATPs sandwiched between the A and B motifs of one monomer and the signature sequence of the other monomer (Jones and George, 1999; Smith *et al.*, 2002). The LSGGQ sequence makes several main-chain and side-chain hydrogen-bonded contacts to the nucleotide and is involved in catalysis of the ATP hydrolysis reaction, possibly in a role similar to that of the arginine finger observed in other P-loop ATPases, which extends from one domain into the active site of the opposite domain (Ye *et al.*, 2004). In addition to the three major motifs, a number of additional conserved motifs and residues are involved in the ATP binding and hydrolysis process, namely the Q-, D-, and H-loops that lie in the vicinity of the γ-phosphate and help to produce an 'induced fit' effect in each catalytic site between the sandwiched NBDs.

The first and most pointed statement to be made about the TMDs of ABC transporters is that they are not conserved in sequence in the same way as the NBDs, but generally speaking, the majority deploy the canonical 6+6 TMD configuration, with notable exceptions (Fig. 7.1). The various caveats and exceptions to the consensus arrangement are discussed in detail in several reviews (Dawson *et al.*, 2007; Hollenstein *et al.*, 2007a; Jones *et al.*, 2009; Kos and Ford, 2009; Rees *et al.*, 2009; Seeger and van Veen, 2009; Kerr *et al.*, 2010). In summary, among the resolved bacterial ABC structures to date, there are three architectural types of TMD configurations (Fig. 7.1; Jones and George, 2009). The first comprises: the *E. coli* BtuCD vitamin B_{12} importer (Locher *et al.*, 2002) and the *Haemophilus influenzae* HI1470/71 metal chelate importer (Pinkett *et al.*, 2007) that have two block-like 10-helix

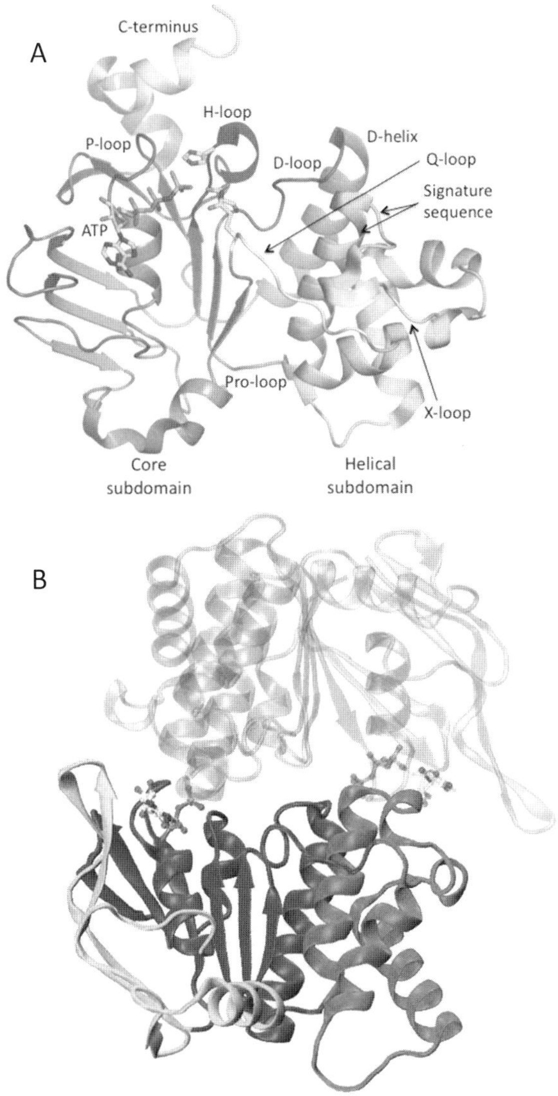

Figure 7.2 NBD monomer and dimer showing the major conserved motifs and residues. (A) NBD monomer in ribbon representation with ATP in stick format. The core subdomain (CSD) is coloured grey and the helical subdomain (HSD) is green. The Walker A (P-loop) motif is coloured magenta. The catalytic glutamate (in stick form) is depicted at the start of the D-loop (dark orange) followed immediately by the D-helix (light orange). The Q-loop is in yellow with the glutamine side chain in stick form. The H-loop (histidine side chain in stick form) is in dark blue; and the Pro-loop is in light blue at the bottom of the figure connecting the two subdomains and leading into the signature sequence (vertical green helix and loop adjacent to the orange D-helix). The C-terminus is coloured cyan. The short orange helix and loop in the centre of the HSD is the X-loop. (B) The ATP-binding cassette dimer. Ribbon diagram of the MJ0796 ABC ATPase dimer (1L2T.pdb). One monomer is coloured with the ATP-binding core subdomain blue and the antiparallel β subdomain green, together comprising lobe I. The α-helical subdomain or lobe II is coloured red. ATP is shown in ball-and-stick form with carbon in yellow, nitrogen in blue, oxygen in red and phosphorus in tan. The opposite monomer and ATP are shown in ghost representation. (Reprinted with permission from Jones *et al.*, 2009.). A colour version of this figure is located in the plate section at the back of the book.

bundles forming a narrow chamber within the membrane (Fig. 7.1A). The second group comprises the *Archaeoglobus fulgidus* and *Methanosarcina acetivorans* ModBC molybdate/tungstate importers (Hollenstein *et al.*, 2007b; Gerber *et al.*, 2008), the *E. coli* MetNI methionine importer (Kadaba *et al.*, 2008) and the *E. coli* MBP-MalFGK$_2$ maltose permease (Oldham *et al.*, 2007; Khare *et al.*, 2009). This group has five to eight curved TMD helices that form a relatively wide tepee-shaped pore (Fig. 7.1B). The third group comprises the *Staphylococcus aureus* Sav1866 multidrug exporter (Dawson and Locher, 2006) and the *E. coli*, *S. typhimurium*, and *Vibrio cholerae* MsbA lipid flippase exporters (Ward *et al.*, 2007). These exporters deploy the canonical 6+6 TMD helix arrangement that traces two large wing-like arcs (Fig. 7.1C). Sav1866, MsbA, and the closely related *Lactococcus lactis* ABC multidrug transporter LmrA (van Veen *et al.*, 1996), whose structure has not been solved, are the best studied. All are homodimer half-transporters.

There are six resolved structures for Sav1866 (two) and MsbA (four) at resolutions from 3.0 to 5.5 Å. Each depicts six TMD helices in each protomer, five of which extend well beyond the membrane into the cytoplasm and provide three main contact points to the NBDs (Fig. 7.3), two of which constitute the antiparallel ICL1 and ICL2 'coupling helices', the third being the extension to TM6, which has a direct covalent linkage to the NBD (Dawson and Locher, 2006). In these multidrug-resistant (MDR) bacterial ABC transporters, the second coupling helix in each half (at the base of TM helices 4 and 5) exclusively contacts the NBD of the other protomer in the regions C-terminal to the Walker A motif and N-terminal to the signature sequence, the latter being the conserved 'X-loop' motif (Dawson and Locher, 2006) found exclusively in ABC exporters and that may function to enable the mechanical domain swapping of the ICL coupling helices at the TMD:NBD transmission interface (Kerr *et al.*, 2010). This domain-swapping feature is continued in the TMDs in which TM helices 4 and 5 are splayed away from their own protomer TMD and make the majority of their interhelical contacts with helices in the opposite protomer (Fig. 7.3). In the middle of the membrane, TM1/TM2 and TM3/TM6 diverge into discrete 'wings' that point away from one another towards the cell exterior in an outward-facing conformation (Dawson and Locher, 2006) that has been observed in all ABC MDR exporter structures to date. Bacterial ABC importers do not show this domain swapping interaction in either the TMDs or the ICL:NBD interface. Importers appear to mainly interact via ICL2 only in each TMD containing the 'EAA' motif (Mourez *et al.*, 1997), with structural evidence for this interaction provided in the crystal structures of the importers, BtuCD (Locher *et al.*, 2002), ModABC (Hollenstein *et al.*, 2007b), HI1470/71 (Pinkett *et al.*, 2007), MalFGK$_2$ (Oldham *et al.*, 2007), and MetNI (Kadaba *et al.*, 2008).

The unified model of translocation across membranes

An allosteric model for membrane pumps was first proposed as long ago as 1966 (Jardetzky, 1966). To function as a pump, a membrane protein need only meet three structural conditions, which are (i) contain a cavity in the interior large enough to admit the solute; (ii) be able to assume inward- and outward-facing configurations so that the cavity is alternately open to one side of the membrane; and (iii) contain a binding site within the cavity for the solute for which the affinity is different in the two configurations. A number of recent structural studies on secondary active proton pump membrane transporters lend some support to this alternating access model (Boudker and Verdon, 2010). NorM is a cation-bound MDR

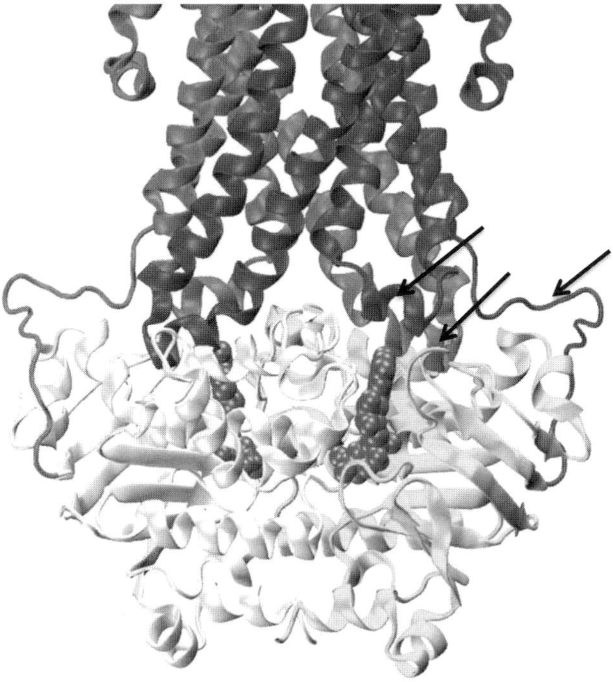

Figure 7.3 The cross-protomer interaction of Sav1866. The intracellular portion of a Sav1866 homodimer is represented in cartoon fashion; the two Sav1866 molecules are coloured yellow and red, and blue and green. Bound nucleotide is rendered in grey space-filling representation. The cross-protomer ('domain swapping') interaction is illustrated by the intracellular loops of one TMD (blue) interacting primarily with the NBD of the opposite protomer (yellow). Five helices in each protomer of Sav1866 provide three main contact points to the NBD. The extension to TM6 has a direct covalent linkage into the NBD itself, whereas the linker regions (ICL1 and ICL2) between TM helices 2 and 3 and between TM helices 4 and 5 provide non-covalent interactions with sites on the NBD. These three regions are arrowed in the figure. (Reprinted with permission from Kerr *et al.*, 2010.) A colour version of this figure is located in the plate section at the back of the book.

and toxin extrusion transporter belonging to the 'multidrug and toxic compound extrusion' (MATE) family and its solution structure from *V. cholerae* depicts an outward-facing conformation (He *et al.*, 2010). CusCBA is a tripartite efflux pump of the 'resistance–nodulation–division' (RND) family that extrudes biocidal copper and silver ions in Gram-negative bacteria. The solution structure of the inner membrane CusA pump component from *E. coli* (Long *et al.*, 2010) identifies a three-methionine putative metal–binding cluster within a cleft region of the periplasmic domain. The cleft is closed in the apo-CusA form, but open in the CusA-Cu(I) and CusA-Ag(I) structures, suggesting that metal binding triggers substantial conformational changes in the periplasmic and transmembrane domains. The LeuT family Mhp1 sodium-benzylhydantoin symporter from *Microbacterium liquefaciens* has been solved in outward-facing and occluded (Weyand *et al.*, 2008) and inward-facing forms (Shimamura *et al.*, 2010). Switching from outward- to inward-facing configurations involves rigid-body movements of 4 of the 12 TMD helices. The 'major facilitator superfamily' (MFS) members LacY (Guan *et al.*, 2007) and FucP (Dang *et al.*, 2010), both from *E.*

coli, are both sugar/H^+ symporters, and have been solved in inward- and outward-facing configurations, respectively. The FucP working hypothesis proposes that fucose import is mediated by a proton relay that alternately neutralizes accessible aspartate and glutamate residues on one side on the TMD cavity. Aspartate must firstly be neutralized to lower the energy barrier for entry of fucose; then a glutamate, which is closer to the internal side of the membrane, is protonated, triggering outward- to inward-facing switching to allow solute release to the interior.

The next two examples of secondary active membrane transporters are intriguingly atypical in that rather than conforming to the alternating inward–outward pump model, they more closely resemble channel families such as aquaporins and bacterial outer membrane porins. The 'chloride channel' CLC family deploy aqueous channels for anion diffusion and ion-coupling chambers that coordinate Cl^-/H^+ antiport in transporters, with the ion movement in these proteins being contained wholly within each subunit of the homodimer (Robertson *et al.*, 2010). This has been illustrated by depicting ion movement within a CLC transporter that was straitjacketed by cross-linking across the dimer interface. The inference from this result was that the cyclic mechanism of ion movement resides within each protomer and does not require rigid body rearrangements between the TMDs of each subunit. The CLC-ec1 Cl^-/H^+ exchanger from *E. coli* is one such homodimer, in which though the dimer interface was destabilized by mutations, the structure and independent transport function of each subunit was preserved, proving that the CLC subunit protomer alone was the basic functional unit for transport and that cross-subunit dimer interactions were not required for Cl^-/H^+ exchange (Robertson *et al.*, 2010). This result prompts the still unanswered question of why these proteins need to assemble as dimers, while drawing parallels with other families that contain individual solute pathways in each protomer of oligomeric assemblies such as trimeric bacterial porins and urea channels, tetrameric aquaporins, and pentameric formate channels. Indeed, among fourteen such families, only the trimeric Na^+/asparate-coupled transporter of the EAAT family has been shown to require a multisubunit architecture as essential for subunit transport (Reyes *et al.*, 2009). The second atypical example is the *E. coli* AdiC proton pump (Fang *et al.*, 2009) that belongs to the 'amino acid-polyamine-organocation' (APC) superfamily. AdiC is an antiporter that enables *E. coli* to survive in extremely acidic environments by expelling protons in the form of agmatine (Agm^{2+}) for extracellular arginine (Arg^+). A substrate-bound resolved structure of AdiC (Gao *et al.*, 2010) depicts an outward-facing homodimer with 12 helices in each TMD and an arginine molecule bound within each protomer in an acidic chamber that binds the Arg^+ head group, with its aliphatic side chain stacked against highly conserved hydrophobic AdiC residues. This binding induces pronounced rearrangement of TM helix 6 and minor changes in helices 2 and 10, resulting in an occluded conformation. There are three potential 'gates' involving aromatic residues and a glutamate that might work to regulate the uptake and release of Arg^+ and Agm^{2+}. These descriptions of secondary active proton pumps may deploy inward- and outward-facing conformations that satisfy the Jardetsky allosteric model though to date only the Mhp1 sodium-benzylhydantoin symporter has been obtained in both conformations. The two atypical examples of CLC and AdiC pose the puzzling situation of suggesting compliance with the model but substrate pathways within rather than between the TMD protomers.

The Jardetsky model has been adapted into a unified model for primary active ABC importers and exporters, based on a comparative analysis of several full-length ABC structures (Hollenstein *et al.*, 2007a). The essentials of this ABC transporter model, which parallels that of the secondary active transporters just described, are depicted in Fig. 7.4. The main problem with ABC transporters is a dearth of alternate states for any given transporter, with only the maltose permease crystallized in two different conformations. An alternating access mechanism is suggested that involves rigid-body rotations of the TMDs, coupled to the closure and opening of the NBD interface (Khare *et al.*, 2009). Substrate transport involves structural reshuffling of the TMDs, physically closing them to the substrate delivery side and opening them on the opposite side of the membrane. ATP-driven closure of the NBD dimer interface decreases the distance between the intracellular sides of the TMDs, mediated through their tethered coupling helices that interact directly with the NBDs. The positional changes of the coupling helices triggers flipping of the TMDs from an inward- to an outward-facing conformation in a clothes peg-like motion. ABC importers can now accept substrates from their cognate periplasmic binding proteins, whereas ABC exporters can extrude drugs or other solutes to the extracellular side of the membrane. ATP hydrolysis drives the NBDs apart, flipping the TMDs from the outward- to inward-facing conformation to reset the cycle. For ABC importers, the transporter is deployed in the outward-facing conformation during transfer of substrate from the docked binding protein to the vestibule between the TMDs, with ATPs sandwiched within the closed NBDs and primed for hydrolysis (Davidson *et al.*, 2008); though the actual distances moved and degree of separation, if any, of the NBDs are currently in dispute (Jones *et al.*, 2009).

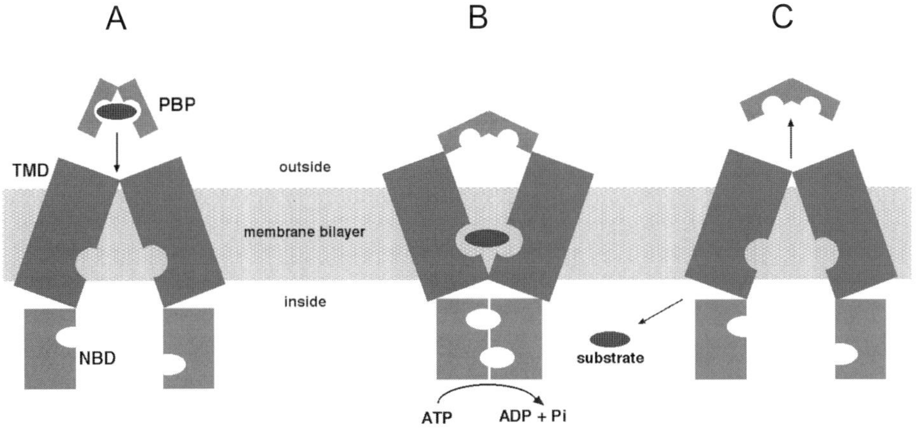

Figure 7.4 Cartoon of the mechanism for ABC importers, and the unified model of membrane translocation. Simple two-state scheme for ABC importers. (A) In the absence of substrate, the transporter is in a conformation in which the NBD dimer interface is open and the translocation pathway is exposed only to the cytoplasm. (B) Interaction of substrate-bound BP with the closed extracellular side of the TMDs in the presence of ATP triggers a global conformational change in which the NBDs close to promote ATP hydrolysis, substrate-bound BP becomes tightly bound to the TMDs, and both BP and TMDs open at the periplasmic surface of the membrane to facilitate substrate transfer from the BP to a binding site in the membrane. (C) Following ATP hydrolysis, which destabilizes the NBD dimer, the transporter returns to the inward facing state and the substrate completes its translocation across the membrane. (Reprinted with permission from Jones *et al.*, 2009.)

Despite the obvious appeal of this model, there remain unanswered questions regarding substrate translocation in ABC exporters. For MDR exporters, vacuum cleaner/flippase models (discussed above) predict that hydrophobic drugs are partitioned directly from the lipid bilayer into the inward-facing TMDs of the transporter. Presumably, cellular substrates, many of which would be hydrophilic, would enter the TMD vestibule directly from the cytoplasm, but not via a route requiring substantive separation of the NBDs. Why? Because the models would need substantial modifications to encompass the different substrate routes; and there is the complication of the transporter 'managing' the shuttling of exporter substrates, ATP, and its hydrolysis products within the restricted NBD-ICL regions. However, if the NBDs moved apart significantly, as depicted in 'snapshots' of some nucleotide-free crystallized ABC transporters, and supported by the switch and processive clamp models, then the substrate and ATP/ADP/Pi would access the wide (~20–50 Å) cavity alternately, and in a manner completely divorced from the mechanism/model that proposes substrates to be accessed from the bilayer. How does the interior cavity differ for different substrates? What are the repulsive forces that drive the TMDs and/or NBDs apart and how is the extent of domain separation controlled? What are the attractive forces that bring the domains back into contact – is electrostatic attraction across a solvent filled gap sufficient to enable NBD re-association? The Sav1866 structure is constrained by the intertwined TMD 'wings' and domain-swapped ICL–NBDs, prompting the authors to suggest that the protomers are unlikely to move independently and their maximum separation during the transport cycle is limited (Dawson and Locher, 2006). A detailed mechanistic description of substrate translocation through the TMDs of MDR-type ABC exporters and its allosteric linkage to ATP binding and hydrolysis within the NBDs will require their structural characterization in multiple states, including bound nucleotide and drug substrate.

Bacterial ABC importers: binding proteins as 'hunter-gatherers'

Receptor binding proteins (BPs) bind their substrates with high affinities that reflect the efficiency of the importers at low substrate concentrations with cells being capable of capturing and concentrating nutrients to >100-fold when they are present at submicromolar concentrations in the external milieu (Dippel and Boos, 2005). Most ABC importers are specific for single substrates, or for several related substrates, but some importers are remarkably versatile in being able to capture structurally unrelated substrates (Table 7.1). This versatility is achieved by the binding receptor protein having diverse specificity (Russell *et al.*, 1992); or by the docking of multiple BPs, with individual substrate specificities, to a single ABC transporter (Higgins and Ames, 1981). In either iteration of the BP type, they are monomeric, are open in the absence and closed in the presence of substrate, and bind substrates into single sites largely through H-bonding (Quiocho and Ledvina, 1996; Davidson *et al.*, 2008). The majority of BPs are comprised of two domains connected by a flexible hinge (Quiocho and Ledvina, 1996) that enables the BP to adopt an open unliganded state with high affinity that upon substrate binding, undergoes a conformational change to a closed state with low affinity for substrate (van der Heide and Poolman, 2002). In the closed, substrate-bound state, the BP interacts with its cognate membrane translocator, triggering changes in the TMDs and NBDs that enable the BP to open, release substrate and disengage from the crown of the membrane-embedded TMDs. The most studied example of this process is that of the maltose permease (Davidson *et al.*, 2008).

In Gram-negative bacteria, BPs are present in the periplasm in large excess over the membrane-embedded translocator proteins (Ames *et al.*, 1996). In bacteria that lack a periplasmic space, the BP is a lipoprotein that uses its lipid moiety as an anchor to the membrane in Gram positives (Gilson *et al.*, 1988; Sutcliffe and Russell, 1995), or uses a transmembrane segment as the anchor in Archaea (Albers *et al.*, 1999). Most BPs are separate subunits of ABC transporter permease complexes, but there is a subset of ABC permeases for which there is one or two substrate capture domains fused to one or both TMDs (Fig. 7.5). Illustrative examples of fused BPs are the OpuA (Biemans-Oldehinkel and Poolman, 2003) and GlnPQ (Schuurman-Wolters and Poolman, 2005) systems of *L. lactis*. ABC importers such as these have either two or four substrate-binding domains tethered to the transporter TMDs. Such BP fusions offer positive cooperativity that is manifested as enhanced proximity, orientation, binding, and transfer of substrate to the TMDs (van der Heide and Poolman, 2002). ABC permeases with chimeric TMD-fused BPs are found in the PAAT and QAT families (http://www.tcdb.org) and are widespread with members within the Gram-positive families *Clostridiaceae*, *Streptococcaceae*, *Listeriaceae*, *Staphylococcaceae*, Actinobacteria, and among Gram-negative *Helicobacteraceae*.

The best-described system is that of the maltose permease for which crystal structures are available that show the full transporter in different configurations with the BP attached and substrate maltose within the TMDs (Oldham *et al.*, 2007; Ward *et al.*, 2007; Khare *et al.*, 2009; Oldham and Chen, 2011). The structure reveals heterodimeric TMDs with a core of 6 helices in each half, but with one TMD (MalF) having two additional helices to its MalG partner. The conserved core is related by a pseudo-twofold symmetry that is similar to the ModB and MetI importer TMDs. The maltose permease has been captured in a nucleotide-bound, outward-facing configuration, with the open BP docked at the periplasmic entrance to the TMDs and a maltose molecule in a solvent-filled cavity about halfway into the lipid bilayer (Oldham *et al.*, 2007). It has also been resolved in the open nucleotide-free state with the substrate-binding pocket only accessible from the cytoplasm (Khare *et al.*, 2009). Taken together, these conformations support the alternating access model of membrane transport

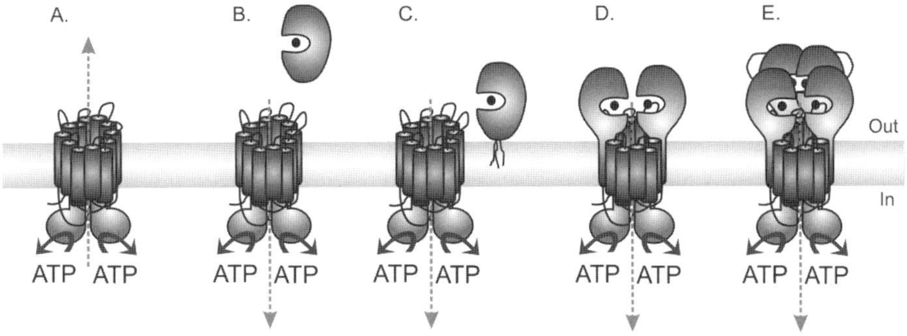

Figure 7.5 Receptor binding protein types, including fused BPs. Schematic representation of the domain organization of ABC transporters. An SBP-independent efflux system is shown in (A), and a conventional SBP-dependent uptake system is shown in (B), with the anchoring of the BP depicted in (C). The newly identified chimeric substrate-binding/translocator systems with two and four substrate-binding sites per functional complex are shown in (D) and (E), respectively. ATP-binding domains (ABC cassettes) are shown in orange.

(Jardetzky, 1966), discussed above (Fig. 7.4). In the outward-facing conformation, maltose is bound to ten residues within MalF, but makes no contact with its partner TMD, MalG. Genetic studies have shown that six of these ten MalF residues severely decrease or eliminate maltose transport, thus lending weight to this region as forming at least part of the binding site (Oldham *et al.*, 2007). Nevertheless, an interesting contrast is made from observations that bacterial ABC importers that translocate a number of different substrates may lack substrate-binding sites within the TMDs since specificity is governed only by the specificity of the BP that docks to the crown of the TMDs (Moussatova *et al.*, 2008; Goetz *et al.*, 2009). Despite this seeming exception, maltose translocation is proffered as the current model for ABC importers (Austermuhle *et al.*, 2004; Orelle *et al.*, 2008). The process begins with ATP binding to the NBDs that triggers BP high-affinity binding to the TMDs with closure of the NBD dimer interface, forming a tepee-shaped configuration that is now primed to receive maltose from the BP. Movement of the TMDs to the outward-facing configuration releases a hydrophobic gate that prises open the BP. This is followed by the concerted actions of ATP hydrolysis, release of the BP, translocation of maltose to the centre of the TMD cavity, rigid-body rotations of the TMDs to the inward-facing configuration, opening of the NBD dimer interface, and release of substrate to the cytoplasm (shown schematically in Fig. 7.4).

In contrast to the maltose permease, ModBC, and MetNI, the metal-chelate importers BtuCD (Locher *et al.*, 2002) and HI1470/71 (Pinkett *et al.*, 2007) require more significant rearrangements of their TMDs and NBDs to effect substrate translocation, chiefly because they have two 10-helix bundles that form a narrow vestibule at the membrane entrance. The translocation pathways for their vitamin B_{12} and metal chelate substrates present special problems that are still puzzling investigators (Hvorup *et al.*, 2007; Sonne *et al.*, 2007; Weng *et al.*, 2008; Goetz *et al.*, 2009; Kandt and Tieleman, 2010; Dibartolo and Booth, 2011). Consider, for instance, the BtuC TMDs that are considerably longer (326 residues) than those of MetI (208 residues). Curiously, the 118 extra residues are used in adding more helices to the densely packed TMD bundles rather than in expanding the membrane chamber, posing the question of what attributes these 'blockaded' TMDs have that make them suited to translocating their substrates? What is clear is that BtuCD, HI1470/71, and homologues are atypical importers in terms of structure, if not function. All importers whose structures have been resolved seem to present narrower extracellular inward-facing entrances to their substrates, whereas in the nucleotide-free open structures, the water-filled outward-facing TMDs have much wider vestibules, which holds true for both importers and exporters, despite the fact that the importers do not require such a separation from the inward out, but rather from the outward in. It is true that the maltose permease has been resolved with a maltose molecule in the translocation pore, albeit bound only to one of the two TMD partners. Nevertheless, if this is truly one of several intermediate structures in the pathway, then what is lacking presently is a structure that 'captures' its substrate just released from its cognate BP and positions it in the membrane vestibule created by well-separated TMDs – a feature that has not been observed to date.

Bacterial ABC exporters: multidrug resistance

ABC exporters have none of the accommodation problems of importers, or so it seems, as the nucleotide-free crystal structures of the bacterial exporters Sav1866 (Dawson and Locher, 2006) and MsbA (Ward *et al.*, 2007), and the MDR murine homologue ABCB1 (Aller *et al.*, 2009) all depict wide inward-facing 'tepee-shaped' conformations (Fig. 7.3A).

Interestingly, the domain swapped docking of two ICL coupling helices in exporters, versus only one in importers, might suggest some 'crowding' around and between the NBDs, ICLs, and cytoplasmic entrance to the TMDs, but the gap of variously 20–50 Å created by the separation of the nucleotide-free NBDs in the resolved structures should allow simple entrance and egress of substrates to be exported. However there are three caveats against this simple explanation. Firstly, the popular 'hydrophobic vacuum cleaner' or 'flippase model' for drug efflux (Higgins and Gottesman, 1992) proposes that since many of the drugs exported by ABC transporters are lipophilic, binding most likely takes place within the membrane from whence they are either translocated from the inner to the outer leaflet (in the manner of a lipid flippase) or extruded from the membrane into the extracellular space (hydrophobic vacuum cleaner). This model presents the rather intuitive paradox that lipid-soluble drugs are extruded from the bilayer into a large water-filled cavity between the splayed TMDs, which requires an unfavourable displacement of bilayer lipids (Higgins and Linton, 2001). Two recent computational molecular dynamics studies have provided contradictory evidence for the entry point of drugs into the TMDs in the exporters MbsA (Weng *et al.*, 2010) and Sav1866 (Oliveira *et al.*, 2011). In the former, a cytoplasmic 'gate' is located in the ICL region and its formation is a consequence of dissociation of the dimer interface as the first step in the transition from inward- to outward-facing TMDs. In the second study, a similar cytoplasmic gate was observed, but in contrast to the first study, the gate entrance was visible in the ICL region leading to the exposure of the TMDs internal cavity to the cytoplasm, which was large enough to accommodate drug substrates such as doxorubicin. Importantly, the gate was observed in the transporter only when the NBD dimer was in semi-open or closed configurations. Secondly, if drugs are indeed largely 'flipped' across the bilayer, why is there a requirement for the NBDs to separate to such a large distance post-hydrolysis? If the gap is not required to accommodate incoming or outgoing substrate, could it be necessary to allow for entry and egress of nucleotide? This does not seem to be the case either, since all studies that have broached this topic have concluded that such a large separation was unnecessary as it is far in excess of the gap required for nucleotides to move into and out of their catalytic sites. Further to this is the point that the NBDs function in a cooperative allosteric mode that would seem to be somewhat dissipated by their physical separation; unless their remote displacement is 'rescued' by being tethered to the TMDs and ICLs, in a tweezers-like fashion (Chen *et al.*, 2003).

Thirdly, do the NBDs actually separate? Recent studies support a constant-contact model for the allosteric operation of NBD dimers (Jones and George, 2007, 2009). Prior to the first crystal structures revealing the nucleotide-free NBDs at the extremities of tepee-like inward-facing TMDs, there was popular regard for an alternating sites model (Senior *et al.*, 1995), in which the ATPs were hydrolysed alternately and the active sites were coupled closely throughout the catalytic cycle, consistent with an asymmetrical mechanism. Thus, in contrast to the variously named 'switch', 'processive clamp', or tweezers-like' model (Chen *et al.*, 2003; Higgins and Linton, 2004; van der Does and Tampe, 2004), the alternating sites or constant contact model envisages the two ATP sites operating 180 degrees out of phase with ATP hydrolysis alternating between the opposite sites. The NBDs remain in contact throughout the catalytic cycle, with opening and closing of the active sites occurring by way of intrasubunit conformational changes within the NBD monomers, specifically by rotation of the core domain of the NBD in which ATP has been hydrolysed down and away from the ICLs while the helical subdomain remains approximately stationary. The opposite NBD site remains bound and charged with ATP. Thus, in this model, which is supported by a number of recent studies (Sauna *et al.*, 2007; Gottesman *et al.*, 2009; Becker *et al.*, 2010; Loo *et al.*, 2010; Siarheyeva *et al.*, 2010; Schultz *et al.*, 2011; Verhalen and Wilkens, 2011), the NBDs only separate at one ATP catalytic site alternately as it changes from low- to high-affinity for ATP during the hydrolytic cycle (Jones *et al.*, 2009). We are at a crossroads of sorts where, on the one hand, evidence is mounting that the NBD dimer does not dissociate during the transport cycle, but this is incompatible with the open structures of ABC transporters with the NBDs well separated that indicate unambiguously that the NBDs must move apart in a rigid body fashion to enable opening of the TMDs to the cytoplasm, and this scenario is supported by a number of studies (Aittoniemi *et al.*, 2010; Doshi *et al.*, 2010; Weng *et al.*, 2010). One question that remains to be answered is whether the open structures are physiological or simply crystallographic snapshots of the transporters purified and crystallized in detergent states. One very recent study poses the intriguing question that events within the cell can induce a stable, closed conformation of the MsbA homodimer that does not reopen *even in the absence* of nucleotide (Schultz *et al.*, 2011).

Bacterial ABC exporters: secretion systems

Gram-negative bacteria possess several multicomponent secretion pathways to transport proteins across the cell envelope to the extracellular medium (Economou *et al.*, 2006). Of six systems identified, the Type 1 secretion system (T1SS) is common among Gram-negative bacteria. T1SS directs the secretion of proteins of different sizes and activities including pore-forming haemolysins, adenylate cyclases, lipases, proteases, surface layers, and haemophores. HasA haemophores are small extracellular proteins that scavenge extracellular haem and deliver it to outer membrane receptors. One of the best described and recent models for secretion is the HasA T1SS system of *Serratia marcescens* (Masi and Wandersman, 2010). The HasA haemophore of *S. marcescens* is secreted by a T1SS tripartite complex, comprising an inner membrane ABC protein (HasD), a periplasmic adaptor (HasE), and an outer membrane channel-forming protein of the TolC family (HasF). The T1SS assembly is transient and is triggered by binding of substrate (HasA) to the ABC component (HasD) of the complex. This binding requires multiple interaction sites between HasA and HasD, but in addition, the C-terminus of HasA triggers HasD-driven ATP hydrolysis that also signals disassembly of the tripartite complex. A secretion model has been proposed, beginning with an interaction of HasA with SecB then of HasA with HasD through multiple recognition sites, recruitment of the TolC-like HasF channel, and finally the C-terminal release/dissociation signal (Fig. 7.6). SecB is only needed for its antifolding activity, suggesting tight coupling between HasA and HasD synthesis and HasA secretion.

In contrast to the general protein secretion system (Sec system), very little is known of bacterial ABC secretion mechanisms for molecules such as toxins, bacteriocins, proteins, and lipids. Indeed, many such systems have been, and continue to be, studied not for their physiologic properties, but for their recruitment as multidrug transporters, with the examples of Sav1866, MsbA, and LmrA being the most prominently profiled, particularly as there are now several resolved structures of the first two, discussed in detail above. The third member of this group, LmrA from *L. lactis*, is the subject of some controversy and bears brief comment here, being described as an ABC MDR exporter and a close homologue of the human MDR transporter, P-glycoprotein (van Veen *et al.*, 1996, 2000; Konings and Poelarends, 2002). A later study performed *in silico* analysis of the *L. lactis* genome to

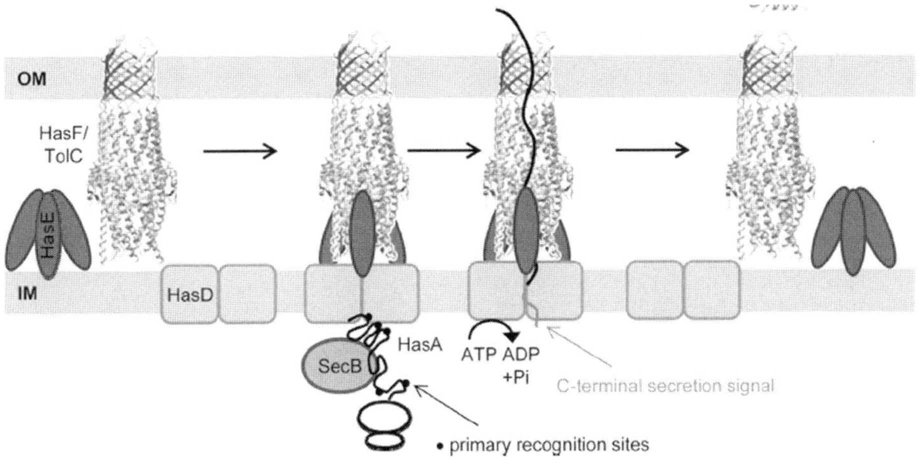

Figure 7.6 HasA secretion system of *S. marcescens*. Steps of the Type I secretion pathway. The model focuses on the functions of different regions of HasA in the assembly dynamics of the tripartite TolC-HasD-HasE system. Step 1: Outer membrane (OM) TolC is recruited to HasDE only when HasA interacts with the ABC protein HasD (IM, inner membrane). These primary interactions between HasD and newly synthesized HasA are driven by linear sites, which are sequentially exposed on the unfolded molecules during or soon after protein translation. Step 2: The secondary interaction between HasD and HasA via C-terminal signalling induces ATP hydrolysis and separation of TolC to its pre-engagement state.

identify 40 putative MDR transporters in this organism, of which only three, namely the ABC transporters LmrA, LmrCD and the major facilitator LmrP, have been associated experimentally with MDR (Lubelski *et al.*, 2006). Concomitant with this result was a global transcriptome analysis of four independently isolated drug-resistant strains with wild-type, which showed a significant up-regulation of *lmrC* and *lmrD* genes in all four strains, while mRNA levels of other putative MDR transporters were unaltered. Deletion of *lmrCD* rendered *L. lactis* sensitive to several toxic compounds. Also up-regulated in all mutant strains was LmrR (YdaF), the transcriptional repressor of *lmrCD* that binds to the *lmrCD* promoter region. These results strongly suggested that the heterodimeric ABC transporter LmrCD is the major determinant of both acquired and intrinsic drug resistance in *L. lactis* (Lubelski *et al.*, 2006).

References

Aittoniemi, J., de Wet, H., Ashcroft, F.M., and Sansom, M.S. (2010). Asymmetric switching in a homodimeric ABC transporter: a simulation study. PLoS., Comput. Biol. *6*, e1000762.

Albers, S.V., Elferink, M.G., Charlebois, R.L., Sensen, C.W., Driessen, A.J., and Konings, W.N. (1999). Glucose transport in the extremely thermoacidophilic *Sulfolobus solfataricus* involves a high-affinity membrane-integrated binding protein. J. Bacteriol. *181*, 4285–4291.

Aller, S.G., Yu, J., Ward, A., Weng, Y., Chittaboina, S., Zhuo, R., Harrell, P.M., Trinh, Y.T., Zhang, Q., Urbatsch, I.L., *et al.* (2009). Structure of P-glycoprotein reveals a molecular basis for poly-specific drug binding. Science *323*, 1718–1722.

Ames, G.F., and Lever, J. (1970). Components of histidine transport: histidine-binding proteins and hisP protein. Proc. Natl. Acad. Sci. U.S.A. *66*, 1096–1103.

Ames, G.F., Mimura, C.S., and Shyamala, V. (1990). Bacterial periplasmic permeases belong to a family of transport proteins operating from *Escherichia coli* to human: traffic ATPases. FEMS Microbiol. Rev. *6*, 429–446.

Ames, G.F., Mimura, C.S., Holbrook, S.R., and Shyamala, V. (1992). Traffic ATPases: a superfamily of transport proteins operating from *Escherichia coli* to humans. Adv. Enzymol. Relat. Areas Mol. Biol. *65*, 1–47.

Ames, G.F., Liu, C.E., Joshi, A.K., and Nikaido, K. (1996). Liganded and unliganded receptors interact with equal affinity with the membrane complex of periplasmic permeases, a subfamily of traffic ATPases. J. Biol. Chem. *271*, 14264–14270.

Austermuhle, M.I., Hall, J.A., Klug, C.S., and Davidson, A.L. (2004). Maltose-binding protein is open in the catalytic transition state for ATP hydrolysis during maltose transport. J. Biol. Chem. *279*, 28243–28250.

Becker, J.P., Van Bambeke, F., Tulkens, P.M., and Prevost, M. (2010). Dynamics and structural changes induced by ATP binding in SAV1866, a bacterial ABC exporter. J. Phys. Chem. B *114*, 15948–15957.

Bieker, K.L., Phillips, G.J., and Silhavy, T.J. (1990). The sec and prl genes of *Escherichia coli*. J. Bioenerg. Biomembr. *22*, 291–310.

Biemans-Oldehinkel, E., and Poolman, B. (2003). On the role of the two extracytoplasmic substrate-binding domains in the ABC transporter OpuA. EMBO J. *22*, 5983–5993.

Bolhuis, H., van Veen, H.W., Poolman, B., Driessen, A.J., and Konings, W.N. (1997). Mechanisms of multidrug transporters. FEMS Microbiol. Rev. *21*, 55–84.

Boudker, O., and Verdon, G. (2010). Structural perspectives on secondary active transporters. Trends Pharmacol. Sci. *31*, 418–426.

Bouige, P., Laurent, D., Piloyan, L., and Dassa, E. (2002). Phylogenetic and functional classification of ATP-binding cassette (ABC) systems. Curr. Protein Pept. Sci. *3*, 541–559.

Chen, J., Lu, G., Lin, J., Davidson, A.L., and Quiocho, F.A. (2003). A tweezers-like motion of the ATP-binding cassette dimer in an ABC transport cycle. Mol. Cell *12*, 651–661.

Croop, J.M. (1998). Evolutionary relationships among ABC transporters. Methods Enzymol. *292*, 101–116.

Dang, S., Sun, L., Huang, Y., Lu, F., Liu, Y., Gong, H., Wang, J., and Yan, N. (2010). Structure of a fucose transporter in an outward-open conformation. Nature *467*, 734–738.

Dassa, E., and Bouige, P. (2001). The ABC of ABCS: a phylogenetic and functional classification of ABC systems in living organisms. Res. Microbiol. *152*, 211–229.

Davidson, A.L., Dassa, E., Orelle, C., and Chen, J. (2008). Structure, function, and evolution of bacterial ATP-binding cassette systems. Microbiol. Mol. Biol. Rev. *72*, 317–364.

Dawson, R.J., and Locher, K.P. (2006). Structure of a bacterial multidrug ABC transporter. Nature *443*, 180–185.

Dawson, R.J., Hollenstein, K., and Locher, K.P. (2007). Uptake or extrusion: crystal structures of full ABC transporters suggest a common mechanism. Mol. Microbiol. *65*, 250–257.

Dean, M., and Allikmets, R. (1995). Evolution of ATP-binding cassette transporter genes. Curr. Opin. Genet. Dev. *5*, 779–785.

Dibartolo, N., and Booth, P.J. (2011). Unravelling the folding and stability of an ABC (ATP-binding cassette) transporter. Biochem. Soc. Trans. *39*, 751–760.

Dippel, R., and Boos, W. (2005). The maltodextrin system of *Escherichia coli*: metabolism and transport. J. Bacteriol. *187*, 8322–8331.

Doige, C.A., and Ames, G.F. (1993). ATP-dependent transport systems in bacteria and humans: relevance to cystic fibrosis and multidrug resistance. Annu. Rev. Microbiol. *47*, 291–319.

Doshi, R., Woebking, B., and van Veen, H.W. (2010). Dissection of the conformational cycle of the multidrug/lipidA ABC exporter MsbA. Proteins *78*, 2867–2872.

Economou, A., Christie, P.J., Fernandez, R.C., Palmer, T., Plano, G.V., and Pugsley, A.P. (2006). Secretion by numbers: protein traffic in prokaryotes. Mol. Microbiol. *62*, 308–319.

Fang, Y., Jayaram, H., Shane, T., Kolmakova-Partensky, L., Wu, F., Williams, C., Xiong, Y., and Miller, C. (2009). Structure of a prokaryotic virtual proton pump at 3.2 A resolution. Nature *460*, 1040–1043.

Fath, M.J., and Kolter, R. (1993). ABC transporters: bacterial exporters. Microbiol. Rev. *57*, 995–1017.

Ferenci, T., Boos, W., Schwartz, M., and Szmelcman, S. (1977). Energy-coupling of the transport system of *Escherichia coli* dependent on maltose-binding protein. Eur. J. Biochem. *75*, 187–193.

Gao, X., Zhou, L., Jiao, X., Lu, F., Yan, C., Zeng, X., Wang, J., and Shi, Y. (2010). Mechanism of substrate recognition and transport by an amino acid antiporter. Nature *463*, 828–832.

Gerber, S., Comellas-Bigler, M., Goetz, B.A., and Locher, K.P. (2008). Structural basis of trans-inhibition in a molybdate/tungstate ABC transporter. Science *321*, 246–250.

Gilson, E., Higgins, C.F., Hofnung, M., Ferro-Luzzi Ames, G., and Nikaido, H. (1982). Extensive homology between membrane-associated components of histidine and maltose transport systems of *Salmonella typhimurium* and *Escherichia coli*. J. Biol. Chem. *257*, 9915–9918.
Gilson, E., Alloing, G., Schmidt, T., Claverys, J.P., Dudler, R., and Hofnung, M. (1988). Evidence for high affinity binding-protein dependent transport systems in gram-positive bacteria and in Mycoplasma. EMBO J. *7*, 3971–3974.
Goetz, B.A., Perozo, E., and Locher, K.P. (2009). Distinct gate conformations of the ABC transporter BtuCD revealed by electron spin resonance spectroscopy and chemical cross-linking. FEBS Lett. *583*, 266–270.
Gottesman, M.M., Ambudkar, S.V., and Xia, D. (2009). Structure of a multidrug transporter. Nat. Biotechnol. *27*, 546–547.
Guan, L., Mirza, O., Verner, G., Iwata, S., and Kaback, H.R. (2007). Structural determination of wild-type lactose permease. Proc. Natl. Acad. Sci. U.S.A. *104*, 15294–15298.
He, X., Szewczyk, P., Karyakin, A., Evin, M., Hong, W.X., Zhang, Q., and Chang, G. (2010). Structure of a cation-bound multidrug and toxic compound extrusion transporter. Nature *467*, 991–994.
Higgins, C.F. (1992). ABC transporters; from microorganisms to man. Annu. Rev. Cell Biol. *8*, 67–113.
Higgins, C.F., and Ames, G.F. (1981). Two periplasmic transport proteins which interact with a common membrane receptor show extensive homology: complete nucleotide sequences. Proc. Natl. Acad. Sci. U.S.A. *78*, 6038–6042.
Higgins, C.F., and Gottesman, M.M. (1992). Is the multidrug transporter a flippase? Trends Biochem. Sci. *17*, 18–21.
Higgins, C.F., and Linton, K.J. (2001). Structural biology. The xyz of ABC transporters. Science *293*, 1782–1784.
Higgins, C.F., and Linton, K.J. (2004). The ATP switch model for ABC transporters. Nat. Struct. Mol. Biol. *11*, 918–926.
Higgins, C.F., Hiles, I.D., Salmond, G.P., Gill, D.R., Downie, J.A., Evans, I.J., Holland, I.B., Gray, L., Buckel, S.D., Bell, A.W., *et al.* (1986). A family of related ATP-binding subunits coupled to many distinct biological processes in bacteria. Nature *323*, 448–450.
Holland, I.B., and Blight, M.A. (1999). ABC-ATPases, adaptable energy generators fuelling transmembrane movement of a variety of molecules in organisms from bacteria to humans. J. Mol. Biol. *293*, 381–399.
Hollenstein, K., Dawson, R.J., and Locher, K.P. (2007a). Structure and mechanism of ABC transporter proteins. Curr. Opin. Struct. Biol. *17*, 412–418.
Hollenstein, K., Frei, D.C., and Locher, K.P. (2007b). Structure of an ABC transporter in complex with its binding protein. Nature *446*, 213–216.
Hrycyna, C.A., and Gottesman, M.M. (1998). Multidrug ABC transporters from bacteria to man: an emerging hypothesis for the universality of molecular mechanism and function. Drug Resist. Updat. *1*, 81–83.
Hvorup, R.N., Goetz, B.A., Niederer, M., Hollenstein, K., Perozo, E., and Locher, K.P. (2007). Asymmetry in the structure of the ABC transporter-binding protein complex BtuCD-BtuF. Science *317*, 1387–1390.
Hyde, S.C., Emsley, P., Hartshorn, M.J., Mimmack, M.M., Gileadi, U., Pearce, S.R., Gallagher, M.P., Gill, D.R., Hubbard, R.E., and Higgins, C.F. (1990). Structural model of ATP-binding proteins associated with cystic fibrosis, multidrug resistance and bacterial transport. Nature *346*, 362–365.
Jardetzky, O. (1966). Simple allosteric model for membrane pumps. Nature *211*, 969–970.
Jones, P.M., and George, A.M. (1999). Subunit interactions in ABC transporters: towards a functional architecture. FEMS Microbiol. Lett. *179*, 187–202.
Jones, P.M., and George, A.M. (2004). The ABC transporter structure and mechanism: perspectives on recent research. Cell Mol. Life Sci. *61*, 1–18.
Jones, P.M., and George, A.M. (2007). Nucleotide-dependent allostery within the ABC transporter ATP-binding cassette: a computational study of the MJ0796 dimer. J. Biol. Chem. *282*, 22793–22803.
Jones, P.M., and George, A.M. (2009). Opening of the ADP-bound active site in the ABC transporter ATPase dimer: evidence for a constant contact, alternating sites model for the catalytic cycle. Proteins *75*, 387–396.
Jones, P.M., O'Mara M.L., and George, A.M. (2009). ABC transporters: a riddle wrapped in a mystery inside an enigma. Trends Biochem. Sci. *34*, 520–531.
Kadaba, N.S., Kaiser, J.T., Johnson, E., Lee, A., and Rees, D.C. (2008). The high-affinity *E. coli* methionine ABC transporter: structure and allosteric regulation. Science *321*, 250–253.
Kandt, C., and Tieleman, D.P. (2010). Holo-BtuF stabilizes the open conformation of the vitamin B12 ABC transporter BtuCD. Proteins *78*, 738–753.

Karpowich, N., Martsinkevich, O., Millen, L., Yuan, Y.R., Dai, P.L., MacVey, K., Thomas, P.J., and Hunt, J.F. (2001). Crystal structures of the MJ1267 ATP binding cassette reveal an induced- fit effect at the ATPase active site of an ABC transporter. Structure *9*, 571–586.
Kerr, I.D., Jones, P.M., and George, A.M. (2010). Multidrug efflux pumps: the structures of prokaryotic ATP-binding cassette transporter efflux pumps and implications for our understanding of eukaryotic P-glycoproteins and homologues. FEBS J. *277*, 550–563.
Khare, D., Oldham, M.L., Orelle, C., Davidson, A.L., and Chen, J. (2009). Alternating access in maltose transporter mediated by rigid-body rotations. Mol. Cell *33*, 528–536.
Konings, W.N., and Poelarends, G.J. (2002). Bacterial multidrug resistance mediated by a homologue of the human multidrug transporter P-glycoprotein. IUBMB Life *53*, 213–218.
Kos, V., and Ford, R.C. (2009). The ATP-binding cassette family: a structural perspective. Cell Mol. Life Sci. *66*, 3111–3126.
Linton, K.J., and Higgins, C.F. (2007). Structure and function of ABC transporters: the ATP switch provides flexible control. Pflugers Arch *453*, 555–567.
Locher, K.P., Lee, A.T., and Rees, D.C. (2002). The *E. coli* BtuCD structure: a framework for ABC transporter architecture and mechanism. Science *296*, 1091–1098.
Long, F., Su, C.C., Zimmermann, M.T., Boyken, S.E., Rajashankar, K.R., Jernigan, R.L., and Yu, E.W. (2010). Crystal structures of the CusA efflux pump suggest methionine-mediated metal transport. Nature *467*, 484–488.
Loo, T.W., Bartlett, M.C., and Clarke, D.M. (2010). Human P-glycoprotein is active when the two halves are clamped together in the closed conformation. Biochem. Biophys. Res. Commun. *395*, 436–440.
Lubelski, J., de Jong, A., van Merkerk, R., Agustiandari, H., Kuipers, O.P., Kok, J., and Driessen, A.J. (2006). LmrCD is a major multidrug resistance transporter in *Lactococcus lactis*. Mol. Microbiol. *61*, 771–781.
Lubelski, J., Konings, W.N., and Driessen, A.J. (2007). Distribution and physiology of ABC-type transporters contributing to multidrug resistance in bacteria. Microbiol. Mol. Biol. Rev. *71*, 463–476.
Masi, M., and Wandersman, C. (2010). Multiple signals direct the assembly and function of a type 1 secretion system. J. Bacteriol. *192*, 3861–3869.
Mourez, M., Hofnung, M., and Dassa, E. (1997). Subunit interactions in ABC transporters: a conserved sequence in hydrophobic membrane proteins of periplasmic permeases defines an important site of interaction with the ATPase subunits. EMBO J. *16*, 3066–3077.
Moussatova, A., Kandt, C., O'Mara, M.L., and Tieleman, D.P. (2008). ATP-binding cassette transporters in *Escherichia coli*. Biochim. Biophys. Acta *1778*, 1757–1771.
Neu, H.C., and Heppel, L.A. (1965). The release of enzymes from *Escherichia coli* by osmotic shock and during the formation of spheroplasts. J. Biol. Chem. *240*, 3685–3692.
Nikaido, H., and Takatsuka, Y. (2009). Mechanisms of RND multidrug efflux pumps. Biochim. Biophys. Acta *1794*, 769–781.
Oldham, M.L., and Chen, J. (2011). Crystal structure of the maltose transporter in a pretranslocation intermediate state. Science *332*, 1202–1205.
Oldham, M.L., Khare, D., Quiocho, F.A., Davidson, A.L., and Chen, J. (2007). Crystal structure of a catalytic intermediate of the maltose transporter. Nature *450*, 515–521.
Oldham, M.L., Davidson, A.L., and Chen, J. (2008). Structural insights into ABC transporter mechanism. Curr. Opin. Struct. Biol. *18*, 726–733.
Oliveira, A.S., Baptista, A.M., and Soares, C.M. (2011). Conformational changes induced by ATP-hydrolysis in an ABC transporter: a molecular dynamics study of the Sav1866 exporter. Proteins *79*, 1977–1990.
Orelle, C., Ayvaz, T., Everly, R.M., Klug, C.S., and Davidson, A.L. (2008). Both maltose-binding protein and ATP are required for nucleotide-binding domain closure in the intact maltose ABC transporter. Proc. Natl. Acad. Sci. U.S.A. *105*, 12837–12842.
Park, E., and Rapoport, T.A. (2011). Preserving the membrane barrier for small molecules during bacterial protein translocation. Nature *473*, 239–242.
Pinkett, H.W., Lee, A.T., Lum, P., Locher, K.P., and Rees, D.C. (2007). An inward-facing conformation of a putative metal-chelate-type ABC transporter. Science *315*, 373–377.
Quiocho, F.A., and Ledvina, P.S. (1996). Atomic structure and specificity of bacterial periplasmic receptors for active transport and chemotaxis: variation of common themes. Mol. Microbiol. *20*, 17–25.
Rapoport, T.A. (2007). Protein translocation across the eukaryotic endoplasmic reticulum and bacterial plasma membranes. Nature *450*, 663–669.
Rees, D.C., Johnson, E., and Lewinson, O. (2009). ABC transporters: the power to change. Nat. Rev. Mol. Cell Biol. *10*, 218–227.

Reyes, N., Ginter, C., and Boudker, O. (2009). Transport mechanism of a bacterial homologue of glutamate transporters. Nature *462*, 880–885.

Robertson, J.L., Kolmakova-Partensky, L., and Miller, C. (2010). Design, function and structure of a monomeric ClC transporter. Nature *468*, 844–847.

Russell, R.R., Aduse-Opoku, J., Sutcliffe, I.C., Tao, L., and Ferretti, J.J. (1992). A binding protein-dependent transport system in Streptococcus mutans responsible for multiple sugar metabolism. J. Biol. Chem. *267*, 4631–4637.

Sauna, Z.E., Kim, I.W., Nandigama, K., Kopp, S., Chiba, P., and Ambudkar, S.V. (2007). Catalytic cycle of ATP hydrolysis by P-glycoprotein: evidence for formation of the E.S reaction intermediate with ATP-gamma-S, a nonhydrolyzable analogue of ATP. Biochemistry *46*, 13787–13799.

Saurin, W., Hofnung, M., and Dassa, E. (1999). Getting in or out: early segregation between importers and exporters in the evolution of ATP-binding cassette (ABC) transporters. J. Mol. Evol. *48*, 22–41.

Schneider, E., and Hunke, S. (1998). ATP-binding-cassette (ABC) transport systems: functional and structural aspects of the ATP-hydrolyzing subunits/domains. FEMS Microbiol. Rev. *22*, 1–20.

Schultz, K.M., Merten, J.A., and Klug, C.S. (2011). Characterization of the E506Q and H537A Dysfunctional Mutants in the *E. coli* ABC Transporter MsbA. Biochemistry

Schuurman-Wolters, G.K., and Poolman, B. (2005). Substrate specificity and ionic regulation of GlnPQ from *Lactococcus lactis*. An ATP-binding cassette transporter with four extracytoplasmic substrate-binding domains. J. Biol. Chem. *280*, 23785–23790.

Seeger, M.A., and van Veen, H.W. (2009). Molecular basis of multidrug transport by ABC transporters. Biochim. Biophys. Acta *1794*, 725–737.

Shimamura, T., Weyand, S., Beckstein, O., Rutherford, N.G., Hadden, J.M., Sharples, D., Sansom, M.S., Iwata, S., Henderson, P.J., and Cameron, A.D. (2010). Molecular basis of alternating access membrane transport by the sodium-hydantoin transporter Mhp1. Science *328*, 470–473.

Siarheyeva, A., Liu, R., and Sharom, F.J. (2010). Characterization of an asymmetric occluded state of P-glycoprotein with two bound nucleotides: implications for catalysis. J. Biol. Chem. *285*, 7575–7586.

Smith, P.C., Karpowich, N., Millen, L., Moody, J.E., Rosen, J., Thomas, P.J., and Hunt, J.F. (2002). ATP binding to the motor domain from an ABC transporter drives formation of a nucleotide sandwich dimer. Mol. Cell *10*, 139–149.

Sonne, J., Kandt, C., Peters, G.H., Hansen, F.Y., Jensen, M.O., and Tieleman, D.P. (2007). Simulation of the coupling between nucleotide binding and transmembrane domains in the ATP binding cassette transporter BtuCD. Biophys. J. *92*, 2727–2734.

Sutcliffe, I.C., and Russell, R.R. (1995). Lipoproteins of gram-positive bacteria. J. Bacteriol. *177*, 1123–1128.

Tatusov, R.L., Koonin, E.V., and Lipman, D.J. (1997). A genomic perspective on protein families. Science *278*, 631–637.

van der Does, C., and Tampe, R. (2004). How do ABC transporters drive transport? Biol. Chem. *385*, 927–933.

van der Heide, T., and Poolman, B. (2002). ABC transporters: one, two or four extracytoplasmic substrate-binding sites? EMBO Rep. *3*, 938–943.

van Veen, H.W., and Konings, W.N. (1998). The ABC family of multidrug transporters in microorganisms. Biochim. Biophys. Acta *1365*, 31–36.

van Veen, H.W., Venema, K., Bolhuis, H., Oussenko, I., Kok, J., Poolman, B., Driessen, A.J., and Konings, W.N. (1996). Multidrug resistance mediated by a bacterial homolog of the human multidrug transporter MDR1. Proc. Natl. Acad. Sci. U.S.A. *93*, 10668–10672.

van Veen, H.W., Margolles, A., Muller, M., Higgins, C.F., and Konings, W.N. (2000). The homodimeric ATP-binding cassette transporter LmrA mediates multidrug transport by an alternating (two-cylinder engine) mechanism. EMBO J. *19*, 2503–2514.

Verhalen, B., and Wilkens, S. (2011). P-glycoprotein retains drug-stimulated ATPase activity upon covalent linkage of the two nucleotide binding domains at their C-terminal ends. J. Biol. Chem. *286*, 10476–10482.

Vetter, I.R., and Wittinghofer, A. (1999). Nucleoside triphosphate-binding proteins: different scaffolds to achieve phosphoryl transfer. Q. Rev. Biophys. *32*, 1–56.

Walker, J.E., Saraste, M., Runswick, M.J., and Gay, N.J. (1982). Distantly related sequences in the alpha- and beta-subunits of ATP synthase, myosin, kinases and other ATP-requiring enzymes and a common nucleotide binding fold. EMBO J. *1*, 945–951.

Ward, A., Reyes, C.L., Yu, J., Roth, C.B., and Chang, G. (2007). Flexibility in the ABC transporter MsbA: alternating access with a twist. Proc. Natl. Acad. Sci. U.S.A. *104*, 19005–19010.

Weng, J., Ma, J., Fan, K., and Wang, W. (2008). The conformational coupling and translocation mechanism of vitamin B12 ATP-binding cassette transporter BtuCD. Biophys. J. *94*, 612–621.

Weng, J.W., Fan, K.N., and Wang, W.N. (2010). The conformational transition pathway of ATP binding cassette transporter MsbA revealed by atomistic simulations. J. Biol. Chem. *285*, 3053–3063.

Weyand, S., Shimamura, T., Yajima, S., Suzuki, S., Mirza, O., Krusong, K., Carpenter, E.P., Rutherford, N.G., Hadden, J.M., O'Reilly, J., *et al.* (2008). Structure and molecular mechanism of a nucleobase-cation-symport-1 family transporter. Science *322*, 709–713.

Ye, J., Osborne, A.R., Groll, M., and Rapoport, T.A. (2004). RecA-like motor ATPases: lessons from structures. Biochim. Biophys. Acta *1659*, 1–18.

Energy-coupled Transport Across the Outer Membrane of Gram-negative Bacteria

8

Volkmar Braun

Abstract

Active, energy-coupled transport across the outer membrane (OM) of Gram-negative bacteria is intriguing. Because there is no energy source in the periplasm, the energy required for transport is instead provided by the cytoplasmic membrane proton-motive force, to which a particular class of transport proteins responds. Such an energy-coupled transporter in the OM forms a β-barrel that is tightly closed on the periplasmic side by the N-terminal domain plug. This plug domain distinguishes active transporters from porins – other β-barrel proteins that form permanently open pores, through which substrates passively diffuse. The energy from the cytoplasmic membrane is transferred to the OM transporters by the TonB–ExbB–ExbD protein complex (Ton system). Crystal structures of transporters and of transporters with bound TonB fragments provide clues as to how OM transport is energized, but experimental evidence of how the Ton system responds to the proton motive force, how it transfers energy from the cytoplasmic membrane to the OM, and how the transporters react to energy input is largely lacking. Substrates actively taken up across the OM include haem, Fe^{3+} incorporated into siderophores, transferrin, lactoferrin, haptoglobin, haemopexin, lipocalin, vitamin B_{12}, Ni^{2+}, Zn^{2+}, various oligosaccharides, and aromatic compounds. OM transporters also regulate transcription of genes from the cell surface and serve as receptors for toxic peptides and protein toxins. The transporters are particularly abundant in environmental and gut bacteria (up to 140 per strain), which use a large variety of substrates, but are absent in intracellular pathogens and obligate parasites.

Introduction

The outer membrane (OM) of Gram-negative bacteria forms a permeability barrier to noxious compounds and is involved in multiple interactions between bacteria and their environments. Substrates pass through the OM by passive diffusion, facilitated diffusion, or active transport. Permanently open porin proteins mediate diffusion, whereas energy-coupled transport is catalysed by transport proteins composed of a β-barrel with a pore that is tightly closed by a globular domain, called the plug.

Substrates in low abundance or that are too large to diffuse through the pores of the porins (or both) are actively transported. They include haem, Fe^{3+} bound to siderophores or various host proteins, vitamin B_{12}, Ni^{2+}, Zn^{2+}, oligosaccharides, sialic acid, and aromatic compounds. Many more are expected to be actively transported. Such substrates are extracted from the environment through tight binding to the transporters, from where they are taken up into the periplasm. Subsequent transport across the cytoplasmic membrane occurs independently of transport across the OM.

In the resting state (when no substrates are transported), the pore of the transport protein is closed. This is crucial for maintaining the permeability barrier of the OM. When substrates are transported, energy input is required to release the tightly bound substrates from their primary binding sites at the OM transport proteins and to open the pores of the transporters so that the substrates can enter the periplasm. The required energy cannot be generated in the OM because the permanently open porins prevent the formation of a transmembrane potential and no energy source such as ATP is known to exist outside the cytoplasm. Instead, the energy is derived from the cytoplasmic membrane proton-motive force. The energy is transferred from the cytoplasmic membrane to the OM by three proteins – TonB, ExbB, and ExbD –, which form a complex of unknown structure (Ton system) (Fig. 8.1). TonB directly contacts OM transport proteins. TonB and ExbD each have a single N-terminal hydrophobic segment that anchors them to the cytoplasmic membrane. ExbB has three transmembrane segments inserted in the cytoplasmic membrane, the N-terminus in the periplasm, and the C-terminus in the cytoplasm. Most of ExbB is exposed to the cytoplasm (Fig. 8.1).

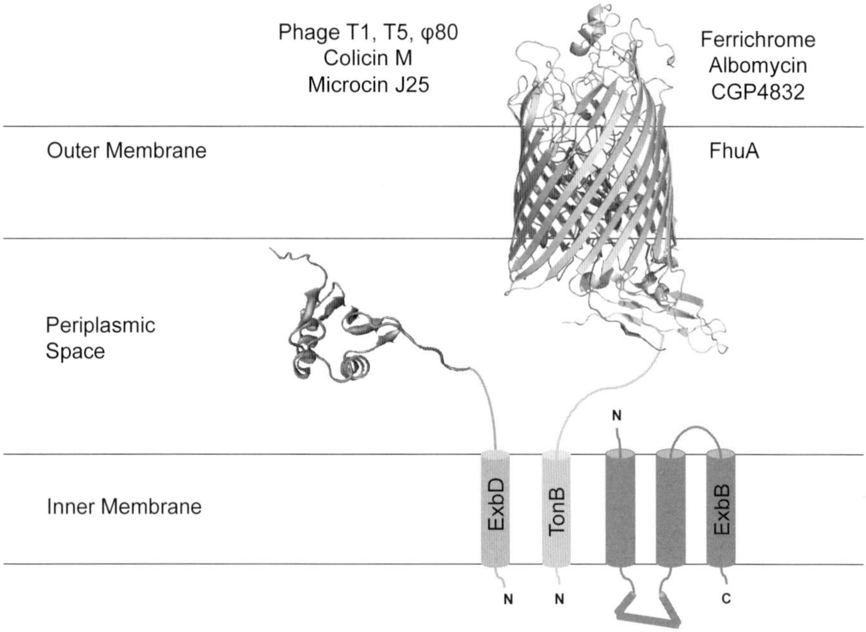

Figure 8.1 Proteins involved in ferrichrome transport in *Escherichia coli.* Shown are the crystal structure of the FhuA protein (Ferguson *et al.*, 1998; Locher *et al.*, 1998) with the ligands that use FhuA to enter cells; the crystal structure of a C-terminal portion of the TonB protein bound to the TonB box of FhuA (Pawelek *et al.*, 2006); the NMR structure of a C-terminal portion of ExbD (residues 64–133; Garcia-Herrero *et al.*, 2007); and the transmembrane topology of TonB, ExbB, and ExbD. The TonB, ExbB, and ExbD proteins form a complex of at least six ExbB proteins and one ExbD protein (Pramanik *et al.*, 2010, 2011; see also Higgs *et al.*, 2002). The structure of ExbB is not known. A colour version of this figure is located in the plate section at the back of the book.

The OM transport proteins have additional functions other than substrate importers. They also serve as signal input devices of nutrients that regulate gene transcription from the cell surface, and they function as receptors for phages and colicins. Although they are involved in phage infection and the uptake of microcins (Braun *et al.*, 2002) and colicins (Cascales *et al.*, 2007), acting as the primary binding sites, it has not been shown that these compounds pass through the pores of the transporters to enter the periplasm. It is known that the binding of phages and the uptake of microcins and colicins are energy-coupled processes because they require the Ton system.

Besides the Ton system, another system couples the energy between the cytoplasmic membrane and the OM – the Tol system, consisting of the proteins TolA, TolB, TolQ, TolR, and Pal. TolA and TonB are functionally equivalent, as are TolQ and ExbB, and TolR and ExbD. TolQ and TolR can functionally replace to some extent ExbB and ExbD, respectively. The Tol system is required for infection by single-stranded filamentous DNA phages and colicin uptake and plays a role in the stability of the OM and in cell division. However, no nutrient uptake system that requires energy input via the Tol system has been identified.

Many OM transporters have been predicted by genome sequencing, but only a few have been studied in some detail. This chapter will discuss what is known about energy-coupled OM transport, illustrated by representative examples, and is confined to the import of low-molecular-weight substrates and of biopolymers that pass the OM at the expense of energy. Certain aspects of the topics discussed herein are covered in recent reviews in more detail than is possible here (Ferguson and Deisenhofer, 2004; Koebnik, 2005; Wiener, 2005; Braun and Mahren, 2007; Cascales *et al.*, 2007; Schauer *et al.*, 2007; Brooks and Buchanan, 2008; Braun, 2010; Braun and Hantke, 2010; Cornelis and Andrews, 2010; Kleanthous, 2010; Mirus *et al.*, 2010; Noinaj *et al.*, 2010; Cornelissen and Hollander, 2011).

Structure and function of the TonB–ExbB–ExbD protein complex

The β-barrel of the energy-coupled OM transporter consists of 22 anti-parallel β-strands. The pore in the barrel is tightly closed on the periplasmic side by a N-terminal globular domain. Well above the cell surface, substrates strongly bind to residues of the transporters. The key questions of OM transport are how the substrates are released from their binding sites and how they are translocated across the OM. Both of these steps require structural alterations that require energy not contained in the OM. The energy is derived from the electrochemical potential of the cytoplasmic membrane. This poses the questions of the energy transfer from the cytoplasmic membrane to the OM. What kind of energy is transferred, and how does energy transfer affect the OM transporters such that they are converted from receptors to active transporters?

Three proteins of the Ton system couple the energy usage at the cytoplasmic membrane and the transport process at the OM. Mutations in any of the TonB, ExbB, and ExbD proteins inactivate energy-coupled OM transport and most receptor functions of the OM transport proteins. The subcellular location of the Ton system (Fig. 8.1) supports a role in transducing energy from the cytoplasmic membrane to the OM. The three proteins interact with each other, as shown by stabilization of TonB and ExbD by ExbB, and by formation of TonB homodimers, TonB–ExbD heterodimers, ExbD homodimers, and ExbB–ExbD heterodimers in cells treated with formaldehyde. Cysteine cross-linking of TonB and ExbB depends on an unimpaired response to the proton-motive force, which indicates a functionally relevant dimerization (summarized in Postle and Larsen, 2007). Larger cross-linked complexes have not been identified, possibly because the low yields of cross-linked products prevent their observation. Dimerization of TonB has also been deduced from an *in vivo* assay in which TonB is fused to ToxR, which in its dimeric form activates the *ctx* cholera toxin promoter (Sauter *et al.*, 2003). The ToxR–TonB construct was designed such that full-length TonB dimerizes ToxR. Functional interactions of the three proteins are indicated by mutations in one protein that can be suppressed by mutations in the other proteins (Swayne and Postle, 2011).

The ExbB, ExbD, and TonB proteins are found in cells at a ratio of 7:2:1 (Higgs *et al.*, 2002). It is unclear whether these numbers reflect the stoichiometry of the proteins in the complex or if they also include proteins not yet assembled. Recently, ExbB has been isolated in an oligomeric form. Size-exclusion chromatography, native gel electrophoresis, transmission electron microscopy, small-angle X-ray scattering, and laser-induced liquid bead ion desorption mass spectrometry (LILBID) indicate a hexamer (Pramanik *et al.*, 2010). Co-purification of ExbB with ExbD results in an $ExbB_6$–$ExbD_1$ complex (Pramanik *et al.*, 2011). Less than stoichiometric amounts of TonB are co-purified with ExbB and ExbD. The results indicate a rather large complex of the three proteins, with $ExbB_6$ forming the platform on which the final complex is built. Since ExbD dimers are obtained by formaldehyde cross-linking, additional ExbD may be bound to ExbB, but less firmly because it dissociates from ExbB during solubilization in undecyl maltoside. Alternatively, the final complex could be even larger, consisting of $ExbB_{12}$–$ExbD_2$–$TonB_2$.

His20, the only hydrophilic amino acid in the transmembrane region of TonB, has been considered to be the amino-acid through which TonB reacts to the proton-motive force and changes conformation in response to protonation/deprotonation. In support of this notion, replacement of His20 by some other amino-acids inactivates TonB (Traub *et al.*, 1993; Swayne and Postle, 2011). However, substitution of His20 by non-protonatable asparagine results in an active TonB, which excludes His20 serving as an essential amino acid for proton conductance through TonB.

ExbD contains one polar amino acid, Asp25, which is found in the single transmembrane region. Replacement of Asp25 by Asn abolishes Ton-related activities, which suggests that protonation/deprotonation of Asp25 may be associated with energization/de-energization of the complex in response to the proton-motive force (Braun *et al.*, 1996). In light of the results obtained with His20 of TonB, a systematic replacement of Asp25 by site-directed mutagenesis is required to validate its essential role in ExbD function. Plasmid-encoded ExbD(D25N) negatively complements a strain with low amounts of chromosomally encoded wild-type ExbD, which indicates that mutant ExbD displaces wild-type ExbD in the TonB–ExbB–ExbD complex (Braun *et al.*, 1996).

The transmembrane helix 3 of ExbB contains a Glu residue at position 176. When this residue is mutated to Ala, all Ton-related activities are abolished. This residue is strictly conserved in ExbB and TolQ proteins and may play an essential role in the response of the TonB–ExbB–ExbD complex to the proton-motive force, as is predicted for Asp25 of ExbD (Braun and Herrmann, 2004a).

The fundamental question regarding the mechanism of action of the energy-coupled OM proteins is whether and how the plug moves inside the β-barrel to open the pore or whether the entire plug is released into the periplasm. Steered molecular dynamic simulations of membrane-embedded TonB in complex with BtuB, an OM transport protein of vitamin B_{12}, reveal that mechanical force applied to TonB is transmitted to BtuB without disruption of the connection between the two proteins, which supports a mechanical mode of coupling. Unfolding of the plug is energetically favoured over pulling the intact plug out of the β-barrel (Gumbart *et al.*, 2007). An alternative model considers TonB as an allosteric regulatory protein that assumes an 'energized' conformation in response to the proton-motive force. In the energized state, TonB interacts with the OM transport proteins such that it induces conformational changes in the transport proteins, with the consequence that the binding sites of the ligands are weakened and the pore is opened. After transfer of conformational energy from TonB to the transport proteins, TonB folds back into the resting state and can again be energized. Any model of how the Ton system functions must take into account that there are about 10^5 transport proteins in the induced/derepressed state in the OM, but only 700 TonB molecules per cell. This means that TonB can only interact with a minor fraction of the transport proteins. Why are there then so many copies of transport proteins? The transport proteins bind substrates with very high affinity and make them available for uptake into the cells. A large number of substrate-loaded transport proteins increase the probability of uptake, and indeed an increase in their synthesis increases the transport rates.

Well-studied examples of energy-coupled substrate transport across the OM

Crystal structures of OM transporters with and without ligands bound and two structures in complex with a C-terminal TonB fragment have provided information that can be used in functional studies, but, unfortunately, they do not reveal the transport mechanism. The ligands of the co-crystals are ferric siderophores, haem bound to haemophore, vitamin B_{12}, and colicin receptor domains through which colicins bind to OM proteins. A few examples of structural changes that occur upon ligand binding and that are representative for most transporters will be presented.

Transport of siderophores

Bacteria and fungi produce and release compounds of low molecular weight that tightly bind Fe^{3+} and are therefore designated siderophores. The concentration of Fe^{3+} in culture media of neutral pH is extremely low. Fe^{3+} is bound by siderophores and the Fe^{3+}-siderophores are then bound and transported into cells by a dedicated OM transport protein, resulting in enough Fe^{3+} to fulfil the iron requirements of the cells. The extremely high affinity of siderophores to Fe^{3+} (free Fe^{3+} concentration 10^{-24} M) and the extremely high affinity of Fe^{3+}-siderophores to the transport proteins ($K_d = 0.1$ nM) allows growth on very low concentrations of available iron.

FepA transports Fe^{3+}-enterobactin

The most definitive quantitative results on Fe^{3+}-siderophore transport was recently published (Newton *et al.*, 2010). These authors devised a method to measure transport of Fe^{3+}-enterobactin across the OM of *Escherichia coli* independently of subsequent transport across the cytoplasmic membrane. A single Fe^{3+}-enterobactin molecule passes FepA (the Fe^{3+}-enterobactin transport protein) in 10–15 s. If bacteria are saturated with $^{56}Fe^{3+}$-enterobactin at 0°C, at which they do not transport Fe^{3+}-enterobactin, and then rapidly shifted to 37°C for 1 min, FepA transports Fe^{3+}-enterobactin and vacant FepA can be loaded with $^{59}Fe^{3+}$-enterobactin. The extent of $^{59}Fe^{3+}$-enterobactin loading reveals the number of active transporter molecules. The technique also allows the measurement of initial transport rates in wild-type and mutant cells. When chilled cells are rapidly transferred to 37°C, the initial rate of transport of Fe^{3+}-enterobactin through FepA increases from 83 pmol/min per 10^9 cells to 172 pmol min^{-1} per10^9 cells, and to 208 pmol/min per 10^9 cells in exponentially growing cells; a turnover number of 2 for the 1-min assay period is derived. Uptake of radioactive Fe^{3+}-enterobactin over 90 min shows a 30-s initial phase at maximum rate (V_{max} = 150 pmol/min per10^9 cells), followed by a 10-min secondary phase (V_{max} = 78 pmol/min per 10^9 cells) and a final steady-state phase (V_{max} = 37 pmol/min per 10^9 cells). The phases may reflect loading of FepA with Fe^{3+}-enterobactin, subsequent interaction of loaded FepA with TonB, and transfer of Fe^{3+}-enterobactin to the FepB binding protein in the periplasm and from there to the FepCDG ABC transporter in the cytoplasmic membrane.

All 35,000 FepA proteins per cell transport Fe^{3+}-enterobactin, which translates to a low turnover number of 5/min. Since in the experiments the ratio of FepA to TonB was 35:1, each TonB protein energizes within 80 s of transport approximately 30 FepA proteins, during which FepA transports all initially bound Fe^{3+}-enterobactin molecules. The rate at which TonB interacts with FepA probably determines the low turnover number. Co-transport of Fe^{3+}-enterobactin with ferrichrome (saturating concentrations), the latter through FhuA, either does not change the K_m and V_{max} of Fe^{3+}-enterobactin transport or decreases the V_{max} by 20%, depending on the assay used. In contrast, Fe^{3+}-enterobactin at saturating concentrations decreases ferrichrome transport to about 50%. These experiments show that TonB can rapidly engage various receptors and energizes various transport systems simultaneously. TonB is a limiting factor in energy-coupled OM transport, but to a much lower degree than might be expected from its low abundance compared with the number of the transport proteins.

For transport of Fe^{3+}-enterobactin, rather high activation energy of 33–36 kcal/mol is calculated. This value equates to hydrolysis of about 4 ATP molecules per Fe^{3+}-enterobactin molecule transported through FepA. Cells contain an intracellular iron concentration of approximately 1 mM (10^6 ions per cell), which when taken up during one generation time would require 4×10^6 ATP molecules to transport Fe^{3+}-enterobactin through the OM. Additional ATP is required for transport across the cytoplasmic membrane by the ABC transporter FepCDG.

No OM transport is observed in energy-depleted cells, in *tonB* mutants, or at 0°C. Unexpectedly, also no OM transport is found in a *fepB* mutant that lacks the periplasmic Fe^{3+}-enterobactin binding protein, which delivers Fe^{3+}-enterobactin to the FepCDG transporter. Without binding to FepB, Fe^{3+}-enterobactin is either not transported across the OM or not retained in the periplasm. In an assay in which labelling of a surface residue of FepA by a fluorescent molecule is inhibited by Fe^{3+}-enterobactin, *fepB* mutant cells become fluorescent under transport conditions, which shows that Fe^{3+}-enterobactin is taken up by the *fepB* mutant as by wild-type *fepB*. In a *tonB* mutant, fluorescence labelling does not occur since Fe^{3+}-enterobactin is not released from FepA after import. The *fepB* mutant does not retain Fe^{3+}-enterobactin in the periplasm. When not bound to FepB, Fe^{3+}-enterobactin escapes into the medium through TolC, an OM export protein with broad substrate specificity. In a mutant lacking TolC, Fe^{3+}-enterobactin remains in the periplasm. Since FepB is required to maintain Fe^{3+}-enterobactin in the periplasm, it either binds free Fe^{3+}-enterobactin in the periplasm or takes over Fe^{3+}-enterobactin from FepA by physical interaction with FepA. When FepA is precipitated with an anti-FepA antibody, FepB is not co-precipitated regardless of whether Fe^{3+}-enterobactin is in the assay or not. This excludes direct transfer of Fe^{3+}-enterobactin from FepA to FepB and indicates that FepB collects soluble Fe^{3+}-enterobactin in the periplasm.

FhuA of *E. coli* transports ferrichrome and serves as multifunctional receptor

Typically, OM transporters serve multiple functions as transporters of substrates and receptors for phages and colicins. FhuA is such an example and has been studied since the dawn of molecular biology (Braun, 2009). FhuA is an attractive study protein since it not only transports a variety of Fe^{3+}-siderophores of the ferrichrome type, including the antibiotic albomycin, a structurally unrelated rifamycin derivative (CGP 4832), and microcin J25, but also serves as receptor for the phages T1, T5, φ80, and UC-1 and for colicin M (Fig. 8.1).

FhuA was the first energy-coupled OM transporter for which the crystal structure was determined (Ferguson *et al.*, 1998; Locher *et al.*, 1998). It consists of a β-barrel composed of 22 antiparallel β-strands with long loops at the cell surface and short turns in the periplasm. The β-barrel forms a continuously open pore across the membrane. The pore in the β-barrel is tightly closed by the N-terminal globular domain, designated plug, cork, or hatch, which enters the β-barrel from the periplasm. The FhuA structure is typical for this class of transporters.

Ferrichrome, a Fe^{3+}-siderophore transported by FhuA, binds to residues on several surface loops and on loops of the plug that extend into an extracellular pocket accessible from the outside. Binding of ferrichrome causes mostly small changes in the plug and β-barrel but does not open a pore. One large movement of 17 Å occurs, that of residue 19, which is exposed to the periplasm and is the first residue seen in the crystal structure. Residues 1–18 are flexible and give no defined diffraction pattern. It is thought that this movement facilitates interaction of FhuA with the TonB protein in the periplasm. The segment of residues 6–10 is called the TonB box since point mutations in this segment inactivate FhuA and the mutations can be suppressed by mutations at residue 160 in TonB, restoring FhuA transport and receptor functions (Schöffler and Braun, 1989; Endriß *et al.*, 2003). Replacement of residues in the TonB box and of residue 160 in TonB by cysteine cross-links the two proteins. The two regions form an interface in the co-crystal of FhuA and fragment 158–235 of TonB (Pawelek *et al.*, 2006). In the co-crystal, the FhuA N-terminus is organized in a β-strand that together with three β-strands of TonB forms a β-sheet. The TonB box bound to TonB is positioned close to the β-sheet of the plug and could impose a pulling or shearing force to the plug that locally unfolds the plug or partially disrupts interaction between the plug and the β-barrel, thus opening a pore in the β-barrel (Gumbart *et al.*, 2007). In the crystal structure, the plug is still inside the β-barrel and occludes the pore. Binding of TonB to FhuA is not sufficient to move the plug. When the plug *in vivo* is fixed to the β-barrel by a disulfide bridge between introduced cysteine residues in the plug (T27C) and the β-barrel (P533C), with both residues exposed to the periplasm, most of the TonB-dependent activities of FhuA are abolished (ferrichrome transport, sensitivity of cells to albomycin, phage T1, colicin M, microcin J25); only sensitivity to the TonB-independent phage T5 is fully retained. Opening of the disulfide bond by reduction restores transport and sensitivity to all FhuA ligands (Endriß *et al.*, 2003; Braun and Endriß, 2007). These data indicate that the TonB-coupled activities of FhuA require a flexible plug in contrast to the TonB-independent activity. Since the plug is highly solvated and rather loosely packed, minimal structural changes could result in an open pore (Faraldo-Gomez *et al.*, 2003; Chimento *et al.*, 2005).

TonB not only alters the structure and position of the plug but also alters FhuA at the outer surface located well above the OM boundary. The first evidence for an altered structure is derived from irreversible binding of the phages T1 and φ80 to the surface of FhuA; binding depends on TonB and the proton-motive force (Hancock and Braun, 1976). The phages bind reversibly to unenergized cells or to *tonB* mutants, which results in no infection. Phage host-range mutants that infect cells in the absence of TonB and that spontaneously release DNA have been isolated. This suggests that the phage DNA transfer into cells does not require energy and TonB. The binding to an energized conformation of FhuA triggers DNA release from the phage head. In another approach, the *in vitro* kinetics of binding of monoclonal antibodies to surface loops of FhuA and extrinsic fluorescence emission of a fluorophore attached to an introduced cysteine in loop 4 was altered in the presence of TonB (James *et al.*, 2008). Upon binding of ferrichrome no TonB-dependent fluorescence change was observed *in vivo* with another fluorophore at the same residue (Bös *et al.*, 1998).

Similar structural transitions observed in the loop region of FhuA and in the TonB box upon substrate binding have been observed in all examined OM transporters (Chimento *et al.*, 2005; Noinaj *et al.*, 2010). Substrate binding transduces a signal across the OM that increases the mobility (unfolding) of the TonB box, indicating that the transporter is ligand-loaded and ready for transport.

The antibiotic albomycin consists of a thioribosyl pyrimidine linked to a siderophore carrier of the hydroxamate type. It inhibits seryl-tRNA synthetase (Stefanska *et al.*, 2000). Active transport of albomycin across the OM and the ferrichrome ABC transporter spanning the cytoplasmic membrane reduce the minimal inhibitory concentration of the antibiotic (*E. coli*, IC_{50} 8 ng/ml) 30,000-fold compared with the antibiotic moiety without siderophore carrier (IC_{50} 256 μg/ml). Co-crystals of albomycin bound to FhuA reveal two conformations of albomycin – an extended and a compact form (Ferguson *et al.*, 2000). Albomycin occupies the same binding sites as ferricrocin, a hydroxamate-type siderophore similar to ferrichrome, with some additional binding sites for the antibiotic moiety.

Another antibiotic that is taken up by FhuA is a synthetic rifamycin derivative that is 200-fold more active than rifamycin because it is active transported across the OM. Although it has no structural similarity to the other FhuA ligands, it shares 9 of its 17 FhuA binding sites with the siderophore ferricrocin (Ferguson *et al.*, 2001).

BtuB of *E. coli* transports vitamin B_{12}

BtuB is more than 100 residues smaller than the other energy-coupled OM transporters of which crystal structures have been determined. Nevertheless, the BtuB crystal structure reveals a typical OM β-barrel that is closed from the periplasmic side by a plug (Chimento *et al.*, 2003a). Its special feature is two calcium ions that order three extracellular loops. High-affinity vitamin B_{12} binding (K_d ~0.3 nM) and calcium coordination are coupled (Chimento *et al.*, 2003b). Bound vitamin B_{12} is not excluded from the extracellular space as is observed upon binding of diferric dicitrate to FecA (Yue *et al.*, 2003). In FecA, diferric dicitrate is sequestered within the barrel as the rather large extracellular loops fold over and shield the small substrate from its extracellular milieu. Vitamin B_{12} is a rather large substrate, and the loops in BtuB are small. Electron paramagnetic resonance (EPR) spectroscopy of BtuB reveals highly mobile loops in the nanosecond timescale that become strongly restricted upon substrate binding. Solutes such as polyethylene glycol (e.g. PEG 3350) used for protein crystallization result in a more compact and ordered state of the external loops of BtuB (Kim *et al.*, 2008).

The structure of the TonB box in the Ca^{2+}-B_{12}-BtuB structure differs from that in the Ca^{2+}-BtuB structure. For example, residues 6 and 7 rotate ~180°, but this shift does not further extend the TonB box (DTLVVTA, residues 6–12) into the periplasmic space. Cysteine mutagenesis has been used to determine disulfide cross-linking with cysteine-labelled TonB (Cadieux and Kadner, 1999), to label with biotin maleimide (Cadieux *et al.*, 2003), and for EPR spectroscopy of spin-probe-labelled cysteine mutants (Merianos *et al.*, 2000). These experiments reveal the structural dynamics of the TonB box upon binding of vitamin B_{12}. Binding of substrate causes changes in the cross-linking pattern and strongly increases the degree of labelling. The EPR line widths and collision parameters are consistent with a structured TonB box bound to the barrel. Addition of substrate converts the TonB box into an extended, highly dynamic, and disordered structure. The biochemical experiments with living cells and the biophysical experiments with isolated OMs reveal changes in the TonB box stronger than those observed in the crystal structures, in which crystal forces may limit structural transitions. The structural changes indicate that the TonB box cycles between sequestered and accessible states in a substrate-dependent manner.

HasR/HasA of *Serratia marcescens* transport haem

The haem uptake system of *S. marcescens* involves the OM transport protein HasR, the haemophore HasA, the sigma factor HasI, the anti-sigma factor HasS, TonB, ExbB, ExbD, and the TonB homologue HasB. The special feature of the Has system is the haemophore HasA, a small protein of 19 kDa that is released from *S. marcescens* into the medium by a haemophore-specific type I secretion system. The crystal structure reveals the residues that bind haem (Krieg *et al.*, 2009). HasA also accepts haem from haemoglobin, haemopexin, myoglobin, and other haem-binding proteins; their haem is passively transferred to HasA because of its extremely high haem affinity.

HasA in the haem-loaded and unloaded states binds to the HasR OM transport protein with equal high affinity. Whereas 1 μM haem and haemoglobin concentrations are sufficient for transport via HasR, transport via HasA occurs at 0.01 μM concentration. More energy and larger amounts of TonB, ExbB, and ExbD are required for haemophore-mediated haem uptake than for free haem uptake through HasR. Although HasR has a much lower affinity for haem ($K_d = 2 \times 10^{-1}$ μM) than HasA ($K_d = 10^{-5}$ μM), haem is transferred from HasA to HasR. This is made possible by an interaction of the two proteins. Upon docking to HasR (K_d of a few nanomolar), one of the two axial haem coordinates of HasA is disrupted, followed by breaking of the second haem coordination site of HasA through steric displacement of haem by Ile671 of HasR (Krieg *et al.*, 2009). After transfer of haem from HasA to HasR, HasA is released from HasR into the medium; this requires energy provided by the electrochemical potential of the cytoplasmic membrane through the action of the Ton system. In the haemophore-mediated haem uptake, the Ton complex functions twice, once for transport of haem across the OM through HasR and once for the release of unloaded HasA. Interestingly, haem must be present to achieve energy-dependent HasA release (Cescau *et al.*, 2007).

In addition to *S. marcescens*, in which haemophore-mediated haem uptake and regulation has been studied in the most detail, haemophore systems of the HasA type have been identified in *Pseudomonas aeruginosa, Pseudomonas fluorescens, Yersinia pestis* and *Yersinia enterocolitica*.

HuxC/HuxA of *Haemophilus influenzae* transports haem

H. influenzae expresses a different type of haemophore than *S. marcescens*. HxuA is a 100-kDa extracellular protein that binds haem-loaded haemopexin (Cope *et al.*, 1998). The complex of haem, haemopexin, and HuxA binds to the OM transport protein HuxC. Haem is transported by HuxC in a TonB-dependent fashion. HuxC also transports haem provided by haemoglobin (Cope *et al.*, 2010). *H. influenzae* depends on haem uptake since it lacks haem biosynthesis.

Most bacteria contain multiple haem transport systems that use haem, haemoglobin, or haem bound to haemopexin or haptoglobin as haem and iron sources (Wandersman and Stojiljkovic, 2000).

TbpA/B and LbpA/B of *Neisseria gonorrhoeae* transport transferrin/lactoferrin iron

N. gonorrhoeae does not produce siderophores and instead relies on iron sources of the human host: human transferrin, human lactoferrin, haemoglobin and haem. It can also use bacterial siderophores, such as aerobactin, enterobactin, and other catecholates, e.g. dimers and trimers of dihydroxybenzoylserine and salmochelin S2 (Cornelissen and Hollander, 2011).

The peculiar feature of iron uptake by *N. gonorrhoeae* is the use of two OM proteins – a transporter with a β-barrel and a plug domain, and a lipoprotein largely exposed at the cell surface to which it is anchored by a lipid moiety. Iron is released from transferrin and lactoferrin, and haem is mobilized from haemoglobin at the cell surface. Transferrin tightly binds to the TbpA transporter ($K_d \sim 10$ nM). TbpA is sufficient for the uptake of iron provided by transferrin. The TbpB lipoprotein enhances the efficiency of transport. The crystal structure of TbpB of *Actinobacillus pleuropneumoniae* consists of two lobes (an N lobe and a C lobe with similar folds), two small β-barrels, and two β-strand-rich structures, which differ in morphologies and electrostatic properties. The N lobe is mainly responsible for binding of TbpB to TbpA (Moraes *et al.*, 2009). TbpB of *N. gonorrhoeae* preferentially binds to iron-loaded transferrin. Although TbpA and TbpB remove iron from human transferrin, only TbpA transports iron into the periplasm. Both proteins are essential for colonizing animals and for virulence.

The crystal structure of the TbpA-human transferring (hTF) complex, small-angle X-ray scattering of the TbpB–hTF complex, and electron microscopy of the TbpA–TbpB-hTF complex reveal detailed insights on how TbpA promotes iron release specifically from human hTF and how TbpB facilitates this process (Noinaj *et al.*, 2012). Unlike other OM transporters in which the plug domain is contained within the β-barrel, the plug domain of TbpA of *Neisseria meningitidis* protrudes ~25 Å above the cell surface. The TbpA plug domain interacts with the C1 domain of hTF. Additional surface areas contribute to the large interface (~2800 Å^2 of buried surface) between TbpA and hTF, such as a TbpA α-helix of loop 3 which is inserted into the cleft between the C1 and C2 subdomains of hTF. These interactions induce a partial opening of the cleft of the hTF C lobe which destabilizes the iron coordination site and facilitates the release of iron from the C-lobe to TbpA. The C1 and C2 subdomains contain residues unique to hTF which explains the specificity of TbpA for hTF. Like TbpA, TbpB binds to the C-lobe of hTF but at sites different from TbpA. The triple complex of TbpA, TbpB, and hTF forms an enclosed chamber that is located above the plug domain of TbpA. This arrangement suggests that the released iron does not diffuse away from hTF but is guided towards the β-barrel of TbpA for uptake into the periplasm.

Transport of iron delivered by human lactoferrin requires LbpA and LbpB. LbpA is required for iron transport, but LbpB is dispensable. LbpB does not bind lactoferrin independent of LbpA, in contrast to TbpA/TbpB, which bind human transferrin. The Lbp proteins are predicted to be structurally similar to the Tbp proteins.

Energy-coupled sugar transport across the OM

It can be predicted that substrates other than Fe^{3+}, Fe^{3+}-siderophores, haem and vitamin B_{12} are taken up by energy-coupled transport across the OM, provided they are present in very low amounts in the growth medium and are too large to freely diffuse through the porins. Indeed, this has been first shown for transport of oligosaccharides and other substrates by various bacterial species.

SusC/SusD of *Bacteroides thetaiotaomicron* transport oligosaccharides

B. thetaiotaomicron is a human intestinal symbiont that uses as nutrients a large variety of host-derived and dietary polysaccharides. Genome analysis predicts 107 TonB-dependent OM proteins (SusC paralogues) and 102 lipoproteins (SusD paralogues). SusC and SusD interact with each other (Cho and Salyers, 2001). The dual nature of the OM transporters is reminiscent of the transferrin and lactoferrin transporters. The bacterium also contains an unusually large set of 51 extracytoplasmic function (ECF) σ factors and 26 anti-sigma factors that sense carbohydrates in the environment and coordinately regulate genes involved in their metabolism. SusC-type proteins contain signalling domains, which indicates that they regulate transcription of carbohydrate transport and utilization genes, similarly as FecA, FecR, and FecI of the ferric citrate transport system. The genes encoding the SusC and SusD paralogues are typically positioned adjacent to each other and are often part of larger gene clusters encoding regulatory proteins and enzymes of carbohydrate metabolism (Cho and Salyers, 2001; Bjursell *et al.*, 2006). Transport and energization have not been studied.

MalA and NagA of *Caulobacter crescentus* transport maltodextrins and chitin oligosaccharides, respectively

The genome of *C. crescentus* predicts 67 OM transporters, which reflect its habitat in freshwater lakes, streams, and ponds with low nutrient content. *C. crescentus* takes up a large variety of nutrients by energy-coupled OM transport, as first shown by the identification of a transport system with a specific OM transport protein for maltooligosaccharides (maltodextrins), MalA. Growing cells in a maltose medium induces synthesis of MalA. Deletion of the *malA* gene reduces maltose transport to 1% of the wild-type rate. *malA* mutants can grow on maltose and maltodextrins such as maltotetraose, but no transport of larger maltodextrins through MalA is observed.

Transport of maltodextrins requires the electrochemical potential of the cytoplasmic membrane and the proteins TonB, ExbB, and ExbD (Neugebauer *et al.*, 2005; Lohmiller *et al.*, 2008). Transport displays biphasic kinetics with an initial K_d of 0.2 μM and a second K_d of 5 μM. This is typical for transport through OM transporters, whereby the lower K_d

reflects binding to the transporter and the larger K_d reflects subsequent transport. The K_d values are 1000-fold smaller than the K_d value of the facilitated diffusion of maltose through the LamB porin of *E. coli*.

Deletion of the *tonB* gene encoded adjacent to the *exbB exbD* genes and deletion of the *exbB exbD* genes abolishes MalA-dependent maltose transport. Replacement of the conserved Val15 in the TonB box of MalA by proline eliminates maltose transport. The data prove that MalA is a Ton-dependent transporter of maltodextrins.

C. crescentus grows on *N*-acetyl-β-D-glucosamine (GlcNAc) and larger chitin-derived oligosaccharides. Growth on GlcNAc induces synthesis of an OM protein, designated NagA, with a typical β-barrel/plug structure of OM transporters and a TonB box (Eisenbeis *et al.*, 2008). It is encoded adjacent to genes that determine a PTS sugar transport system, a GlcNAc deacetylase, a GlcNAc deaminase, and two enzymes that convert GlcNAc-6-P to fructose-6-P, which is further metabolized by glycolysis.

In contrast to maltose transport by MalA, transport of GlcNAc by NagA does not seem to depend on TonB, ExbB, and ExbD. *C. crescentus* encodes two TonB proteins and one distantly related TonB-like protein. Single and double *tonB* insertion mutants and an *exbB exbD* insertion mutant still transport GlcNAc. Whether TolQ TolR functionally replace ExbB ExbD cannot be studied since TolQ and TolR seem to be essential as no mutants can be isolated. However, insertion mutants in ExbB and ExbD grow much slower on GlcNAc trimers and pentamers, which shows an effect of energy uncoupling. These data are consistent with a pore activity of NagA for GlcNAc and a transporter function for the larger chitin oligosaccharides. It is unclear how transport across the OM is energized.

Other TonB-dependent substrate transporters

SuxA of *Xanthomonas campestris* pv. *campestris* is one of the 72 predicted energy-coupled OM transporters that transports sucrose (Banvillain *et al.*, 2007), and TanO/NanU of *Tannerella forsythia* transport sialic acid (Roy eta l.; 2010). FrB4 of *Helicobacter pylori* transports Ni^{2+} (Schauer *et al.*, 2007), and ZnuD of *Neisseria meningitidis* transports Zn^{2+}. Among the 140 predicted OM transporters *Sphingomonas wittichii* most likely contains transporters for aromatic hydrocarbons (Miller *et al.*, 2010).

Energy-coupled OM proteins as signalling molecules: transcription initiation from the cell envelope (transenvelope signalling)

Transenvelope signalling in *E. coli* K-12

A novel signal-receiving and signal-transmitting OM protein, FecA, was discovered in studies on the transcription regulation of the ferric citrate transport genes. Ferric citrate serves as an iron source for *E. coli* and a number of other gram-negative bacteria. Ferric citrate – in fact diferric dicitrate, as revealed by the crystal structure of FecA with bound ferric citrate (Ferguson *et al.*, 2002) – induces transcription of the *fec* genes encoding the OM transport protein FecA, the periplasmic protein FecB, and the ABC transporter FecCDE in the cytoplasmic membrane (Fig. 8.2). Binding of ferric citrate elicits strong movements of loop 7 by 11 Å and loop 8 by 15 Å, which occlude the entry site of ferric citrate. Both loops are essential for transport and induction by ferric citrate (Sauter and Braun, 2004). Binding of iron-free citrate to a region overlapping the ferric citrate binding site does not cause this movement (Yue *et al.*, 2003).

Ferric citrate does not have to be taken up into the cytoplasm for induction to occur, as shown by *fecBCDE* mutants that do not synthesize a functional ABC transporter but whose *fec* genes are still induced by ferric citrate. Ferric citrate also does not have to be transported from the cell surface to the periplasm since *fecA* point mutants constitutively express *fec* transport genes in the absence of ferric citrate and one mutant does not require TonB for induction (Härle *et al.*, 1995). Wild-type FecA requires the Ton system for induction.

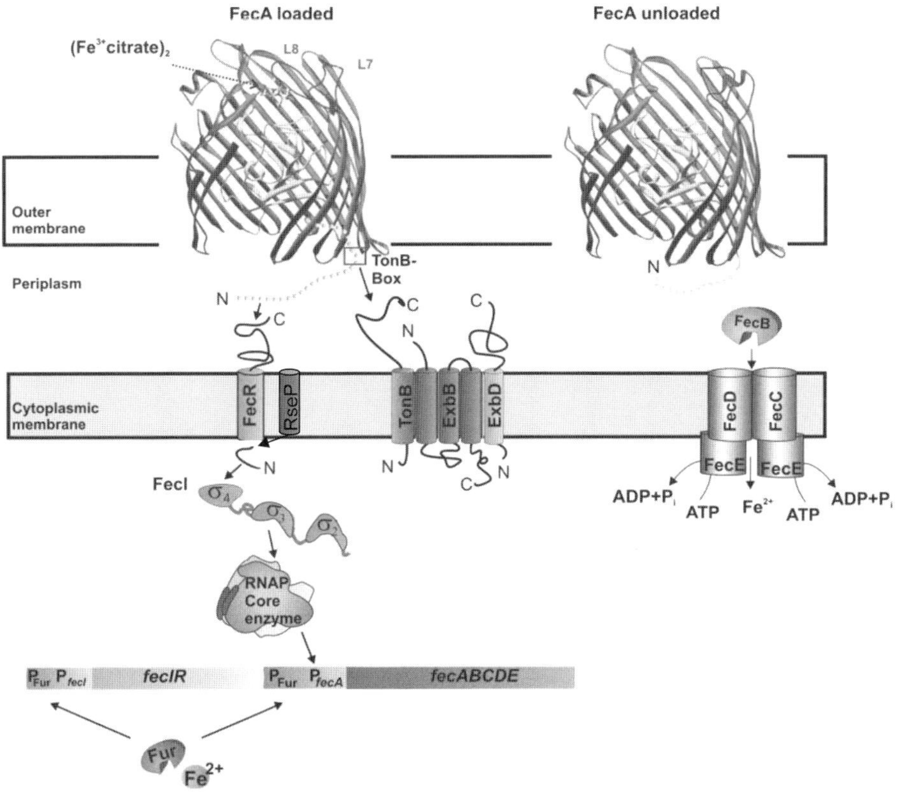

Figure 8.2 Transenvelope regulation of the *fec* genes (left side) and transport of ferric citrate (right side). The N-terminal region of FecA (yellow dashed line) is not seen in the crystal structure but has been determined by NMR. TonB–ExbB–ExbD are required for both regulation and transport. A signal elicited by binding of diferric dicitrate to FecA is transmitted across the OM to FecR, which transmits the signal across the cytoplasmic membrane to the ECF sigma factor FecI. It is predicted that the signal changes the structure of FecR such that it is cleaved by the RseP protease and the resulting cytoplasmic FecR fragment binds to FecI which then binds to the *fecABCDE* promoter. Transcription of the *fecIRA* regulatory genes and the *fecABCDE* transport genes is repressed by the Fur repressor protein loaded with Fe^{2+}. Under iron-limiting growth conditions, Fe^{2+} dissociates from Fur and transcription of *fecIR* and in turn *fecABCDE* is initiated. After transport of diferric dicitrate into the periplasm, iron is transported across the cytoplasmic membrane by the ABC transporter FecBCDE. A colour version of this figure is located in the plate section at the back of the book.

Signalling and transport are distinct FecA functions. In contrast to energy-coupled OM transport proteins that display no signalling activity, FecA contains an N-terminal extension located in the periplasm. This signalling domain comprises residues 1–79 and is essential for signalling but dispensable for transport (Kim *et al.*, 1997). The presence of the signalling domain shifts the location of the TonB box to residues 81–86, in contrast to its N-terminal location in transport proteins that contain no signalling domain. The crystal structures of FecA, FecA loaded with citrate, and FecA loaded with ferric citrate do not show the signalling domain because the link to the FecA plug domain is flexible, thereby not allowing a defined position of the signalling domain in the FecA crystal. The NMR structure of the isolated signalling domain shows a globular form consisting of two α-helices and two β-sheets (Garcia-Herrero and Vogel, 2005). Random mutagenesis of the signalling domain results in mutants with a reduced transcription initiation capacity. The mutations are clustered mostly along one side of the signalling domain, which probably forms the interface to FecR (Breidenstein *et al.*, 2006), the transmembrane regulatory protein in the cytoplasmic membrane. This conclusion is supported by two point mutations in the predicted interface of FecA that are suppressed by point mutations in FecR (Enz *et al.*, 2003). Interaction of FecA with FecR has been further demonstrated by co-elution of FecA and His_{10}-FecR from a Ni-NTA agarose column, by use of a bacterial two-hybrid system that defined the periplasmic regions that interact ($FecA_{1-79}$ and $FecR_{101-317}$), and by randomly generated point mutants of FecR that are impaired in interaction with FecA and ferric–citrate-dependent induction. In the absence of FecR, transcription is not initiated. N-terminal fragments of FecR located in the cytoplasm induce *fecABCDE* transcription in the absence of ferric citrate. Point mutations in the fragments abolish induction. Without doubt, FecR is required for induction. Since induction requires FecR, the term sigma regulator is more appropriate than anti-sigma factor. Usually, anti-sigma factors must be inactive to initiate transcription by the related sigma factors.

Regulation by FecR occurs similar to σ^E regulation. In response to envelope stress, σ^E is activated. In the absence of stress, σ^E is held in an inactive state by interaction with the cytoplasmic domain of RseA, which is anchored to the cytoplasmic membrane. Binding of RseA to σ^E occludes the two major binding sites of σ^E at the RNA polymerase. Under stress conditions, RseA is cleaved by DegS in the periplasm and subsequently in the cytoplasmic membrane by the RseP protease, which abolishes inactivation of σ^E by RseA. *degS* and *rseA* mutants have an unaltered level of ferric-citrate-induced FecA synthesis, whereas an *rseP* mutant has a strongly reduced synthesis (Braun *et al.*, 2005). The cytoplasmic cleavage product of FecR activates FecI, as is observed with genetically constructed N-terminal FecR fragments. An N-terminal fragment of FecR, $FecR_{1-85}$, forms a complex with FecI and β'_{1-317} of the RNA polymerase. This finding suggests that the FecR N-terminus remains attached to FecI while FecI binds to the *fec* promoter in front of the *fecABCDE* operon. Based on these results, the following signalling pathway from the cell surface to the cytoplasm is envisaged. Binding of ferric citrate to FecA changes the structure of FecA to enable binding of FecA to FecR, thereby causing a structural change in FecR. The altered FecR structure is no longer resistant to cleavage by RseP. Cleavage of an N-terminal fragment of FecR releases FecI from the cytoplasmic membrane to which it is bound by complete FecR. FecI with the bound FecR fragment binds to the RNA polymerase, which is directed to the *fec* promoter, where transcription of the *fec* genes is initiated (Fig. 8.2).

The *E. coli* K-12 *fec* operon is flanked upstream by an IS911 element that is disrupted by an IS30 element and a truncated IS2 insertion, which suggests that it has been acquired by horizontal gene transfer. This conclusion is supported by an only occasional finding of the *fec* operon in other *E. coli* strains. The *fec* genes do not belong to the house-keeping genes. *Shigella flexneri* has a *fec* transport system encoded on a chromosomal pathogenicity island inserted in a tRNA gene; such an insertion is frequently observed for horizontally acquired pathogenicity islands (Luck *et al.*, 2001). Incomplete and truncated *fecIRA* genes have been identified on the chromosome of *Enterobacter aerogenes* and on a plasmid of *Klebsiella pneumoniae*. *E. coli* B, three *K. pneumoniae* strains, and *Photorhabdus luminescens* contain complete *fecIRA* genes with high sequence similarity to those of *E. coli* K-12 (Mahren *et al.*, 2005).

The ferric citrate transport system is regulated by two signals: the lack of iron and the presence of ferric citrate. Iron starvation is not sufficient to transcribe the *fec* transport genes at a sufficient level for citrate-mediated iron supply under low-iron conditions. The system functions economically as it synthesizes the transport system only when the substrate is in the medium. Iron starvation unloads the Fur protein, which acts as repressor of the *fecIR* and *fecABCDE* operons in the iron-loaded state. This causes transcription of the *fecIR* regulatory genes and a low-level transcription of the *fecABCDE* genes. This leads to sufficient FecA in the OM so that transcription initiation via FecI is started. As more FecAIR are formed, transcription of the *fec* transport genes increases. When sufficient iron has entered the cells, Fur is loaded with iron and shuts off transcription of *fecIR* and *fecABCDE*. The Fec regulatory system is described in greater detail in Braun *et al.* (2003, 2005), Braun and Mahren (2005) and Brooks and Buchanan (2008).

This mechanism, with some modifications, is realized in many iron-regulated ECF transcription systems, as will be discussed with a few well-studied examples in the following. OM transport proteins that regulate transcription can be recognized by their N-terminal signalling domain and the location of their structural gene adjacent to genes for ECF sigma factors and sigma regulatory genes. The frequency of occurrence of OM transducers among OM transporters has been analysed. Among 110 genomes analysed in 2005, 84 encode transporters and 26 of these are predicted to function also as transducers (Koebnik, 2005). The relative frequency of the Fec type regulatory mechanism may remain when more genomes are analysed.

Transenvelope signalling in *Pseudomonas aeruginosa*

In contrast to *E. coli* K-12, which encodes only one *fec* operon, the genome of *P. aeruginosa* predicts two *fecI*, four *fecIR* and eight *fecIRA* homologues (Braun *et al.*, 2005). One of these regulatory systems has been particularly well studied. The siderophore Fe^{3+}-pyoverdine binds to the FpvA OM transport protein and induces transcription of the structural genes of FpvA, pyoverdine biosynthesis, and synthesis of exotoxin A and the protease PrpL (Visca *et al.*, 2006). FpvA interacts with the anti-sigma factor FpvR, which spans the cytoplasmic membrane, and FpvR interacts with two sigma factors, PvdS and FpvI. FpvI mediates transcription of the *fpvA* gene. PvdS mediates transcription of the pyoverdine biosynthesis genes and the genes that encode exotoxin A and the protease PrpL. *P. aeruginosa* in the exponential growth phase in iron-poor medium contains 582 copies of PvdS, 30% of which are bound to the cytoplasmic membrane. Signalling by Fe^{3+}-pyoverdine facilitates release of PvdS from the membrane, thereby increasing the cytosolic concentration from 35% to 70% in the wild-type and to 83% in the *fpvR* mutant (Tiburzi *et al.*, 2008).

The crystal structure of FpvA without bound Fe^{3+}-pyoverdine reveals the structure of the signalling domain (Brillet *et al.*, 2007), which, as already mentioned, cannot be observed in *E. coli* FecA. Two molecules of FpvA form an asymmetric unit. The protein molecules differ in the position of the signalling domain. Molecule B lacks the electron density connecting the signalling domain to the plug domain. In molecule A, the mixed-stranded β-sheet of the signalling domain forms a four-stranded β-sheet with a folded β-strand of the FpvA TonB box. This structure is reminiscent of a C-terminal TonB fragment bound to the TonB box of FhuA and BtuB. In this conformation, there is not enough space around the TonB box to bind to TonB. In FpvA loaded with Fe^{3+}-pyoverdine, the signalling domain moves to a position that allows the TonB box to interact with TonB. In molecule B, the position of the signalling domain allows a substantial flexibility of the TonB box. This flexibility allows FpvA to interact with FpvR to induce gene transcription and to interact with TonB for energizing transport. Signalling also requires TonB activity, which suggests a sequential binding of the TonB box to the signalling domain for transcription initiation and to TonB for energizing transport. The functional model of FpvA probably applies to other OM transport proteins that regulate transcription of genes by transmembrane signalling. However, there must be also mechanistic differences because FpvR functions as an anti-sigma factor, whereas FecR and many other transmembrane sigma regulators activate ECF sigma factors.

Mutations in FpvA created by linker insertion mutagenesis prevent transport but not signalling. Conversely, mutations that affect signalling do not affect transport. The properties of these mutants demonstrate that signalling and transport are independent events and that transport of Fe^{3+}-pyoverdine into the periplasm is not required for transcription initiation (James *et al.*, 2005).

A system that regulates virulence of *P. aeruginosa* involves the OM protein VreA (*vre* from virulence regulator involving ECF sigma factor), the ECF sigma factor VreI, and the sigma regulator VreR (Llamas *et al.*, 2009). VreR inhibits VreI but is also required for VreI activity. A *vreR* mutant has a much higher amount of VreI. VreI associates with the cytoplasmic membrane through VreR. VreR functions as an anti-sigma factor under non-inducing conditions but is required for induction under inducing conditions. The OM receptor protein is much smaller (23 kDa) than energy-dependent OM transport proteins (75–85 kDa). VreA does not contain a β-barrel. Sequence similarity searches predict a TonB-box-like sequence in VreA, 88-DALTR-92, in α-helix 3 of the signalling domain. The C-terminal domain resembles that of the TolA/TonB protein superfamily. Microarray analysis has revealed 27 genes that are controlled by the Vre system, most of which are located downstream of *vreAIR*. Substrates that elicit *vre* genes transcription have not been identified. VreA synthesis is induced upon interaction of *P. aeruginosa* with human airway epithelial cells. Infection of zebra fish embryos with a *P. aeruginosa* strain that overexpresses VreI increases zebra fish mortality. Overexpression of VreI in the absence of VreR does not increase mortality, which shows that the sigma regulator is essential for VreI sigma factor activity in the virulence model. The structural properties of VreA suggest that it regulates transcription from the cell surface but does not transport a substrate.

Transenvelope signalling in *Pseudomonas putida*

The genome of *P. putida* predicts 19 ECF sigma factors, 13 of which display sequence similarities to FecI (Martinez-Bueno *et al.*, 2002). Eleven of the sigma factor genes are flanked by OM transport genes and anti-sigma factor genes. The first operon to be studied was *pupIRB;* the operon is regulated by the cognate ferric siderophore (Koster *et al.*, 1994). Pseudobactin BN8 induces transcription of the *pupB* gene, which encodes the OM transport protein for pseudobactin BN8. PupB contains an N-terminal extension in the periplasm that is required for signal transduction. The structure of the signalling domain was determined by NMR and is very similar to that of FecA and FpvA (Ferguson *et al.*, 2007). Deletion of the *pupR* gene results in constitutive expression of *pupB* in the absence of pseudobactin BN8, which suggests that PupR functions as an anti-sigma factor. However, transcription of *pupB* by pseudobactin in a *pupI pupR* wild-type is two- to three fold higher than in the *pupR* mutant without pseudobactin BN8, which indicates that PupR not only acts as an anti-sigma factor but also enhances PupI activity. PupR thus functions as a positive and negative regulator of PupI and lies between the pure anti-sigma factors FpvR and HasS (see below) and the positive sigma regulators FecR, FiuR, FoxR, BhuR, RhuR, and PrhR.

P. putida expresses another OM protein, PupA, that transports pseudobactin 358. Fusion of the PupB signalling domain to the plug domain of PupA results in a hybrid protein, PupBA, that takes up pseudobactin 358, as can be expected since the ferric siderophore binding sites are in the PupA β-barrel and plug domain. Pseudobactin 358 induces *pupB* transcription via PupBA, which indicates that signal transfer from the cell surface into the periplasm and proper interaction with PupR occurs in the hybrid protein.

Transenvelope signalling in *Serratia marcescens*

Induction of *has* haem transport genes transcription in *S. marcescens* requires very low iron concentrations, around 25 nM, and haem bound to the haemophore HasA (Rossi *et al.*, 2003). Although free haem and haemoglobin bind to the OM transport protein HasR, they do not induce *has* gene transcription. Sigma factor HasI is required for induction. The activity of HasI is negatively controlled by HasS, the anti-sigma factor. Insertion mutants in *hasS* transcribe *hasR* in the absence of haem-loaded haemophore. Iron deprivation results in iron-free Fur repressor protein, which does not bind to the Fur box of the promoters upstream of *hasI* and the adjacent *hasRADEB* operon. HasI and HasRADEB are synthesized. Transcription of the anti-sigma gene *hasS* depends on HasI (Biville *et al.*, 2004); such a dependency has not been observed in other iron-related ECF systems. Binding of haem-loaded haemophore to the OM transport protein elicits a signal that is transmitted by HasR across the OM to HasS, which transmits the signal across the cytoplasmic membrane. In the cytoplasm, repression of HasI by HasS is relieved, and HasI binds to the RNA polymerase, which then transcribes the *has* transport genes at a high level.

The *has* operon encodes a TonB homologue, HasB. Disruption of *hasB* strongly decreases HasR expression. HasB is important for induction of the *has* operon and cannot be replaced by TonB. Only HasB induces conformational changes in HasR to initiate the signalling cascade that induces *has* operon transcription by HasI (Benevides-Matos and Biville, 2010).

Transenvelope signalling in other bacteria

Transenvelope signalling has also been detected in *Ralstonia solanacearum* (Brito *et al.*, 2002), *Bacteroides thetaiotaomicron* (Martens *et al.*, 2009), and in *Bordetella* strains (Vanderpool and Armstrong, 2003; Kirby *et al.*, 2004).

In vivo reconstitution of active transport proteins from plugs and β-barrels

Reconstitution of the FhuA protein

If plugs occlude the pores of OM transport proteins, it would be expected that deletion of the plugs converts the β-barrels into open pores. This prediction was first examined with the FhuA protein (Braun *et al.*, 2003). The plug was genetically deleted, and the remaining β-barrel was equipped with a signal sequence for secretion into the periplasm. The cells incorporate this β-barrel into the OM, as shown by growth of the cells expressing this mutant FhuA with a Δ*tonB* background on ferrichrome as sole iron source. However, the amount of ferrichrome required by this mutant for growth (0.3 mM) is tenfold higher than the amount needed to support growth of a FhuA TonB wild-type strain. Active transport is more efficient than diffusion. Evidence that the β-barrel without the plug actually forms an open pore includes the increased sensitivity of these cells to large antibiotics that poorly diffuse through the porins (erythromycin 734 kDa, rifamycin 823 kDa) or not at all (bacitracin 1421 kDa); the channel of the β-barrel is wider (~25x35 $Å^2$) than the channel of porins (7×11 $Å^2$). However, the isolated β-barrel does not form stable pores in artificial lipid bilayer membranes, as shown by determination of the K^+ conductance (Braun *et al.*, 2002). Cells that express the β-barrel without the plug do not respond to any of the FhuA ligands (albomycin, microcin J25, phages T1 and T5, colicin M). The isolated β-barrel does not bind phage T5, which does not require TonB for infecting cells.

Co-expression of the plug with the β-barrel, both equipped with a signal peptide for secretion into the periplasm, results in an active FhuA that TonB-dependently transports ferrichrome with nearly half the rate of wild-type FhuA. This reconstituted FhuA confers full or almost full sensitivity to all FhuA ligands except microcin J25, to which cells remain insensitive. Microcin J25 sensitivity requires rather high concentrations of FhuA, which are not met by the reconstituted FhuA. A *tonB* mutant that separately expresses the plug and the β-barrel is fully sensitive to phage T5 but resistant to the TonB-dependent ligands. Plugs with mutations in the TonB box that inactivate FhuA do not complement the β-barrel to a functional FhuA. The proper construction of functional FhuA is further supported by the lack of ferrichrome transport by reconstituted FhuA with disulfide cross-links between the periplasmically exposed residues T27C and P533C and restoration of transport after reduction of the disulfide bond. The reconstituted FhuA displays the properties of wild-type FhuA. The cross-linked FhuA was then used to confirm the size of the reconstituted molecule by SDS polyacrylamide gel electrophoresis in the absence of β-mercaptoethanol. The plug and the β-barrel are separately secreted across the cytoplasmic membrane and inserted into the OM in an active form. The β-barrel inserts into the OM independently of the plug.

Two observations, however, are puzzling. Not only the plug (residues 1–160) complements the β-barrel to an active transporter and receptor but also a much larger fragment (residues 1–357). Whether the larger fragment is proteolytically truncated to the size of the plug is not known but is feasible since the FhuA plug is prone to proteolysis. Even more astonishing is the restoration of transport activity with one of two inactive FhuA β-barrel mutants (E522R, E571R) and a wild-type β-barrel. This means that the wild-type plug of the β-barrel mutants inserts into the wild-type β-barrel, possibly after release of a portion of the plug from the β-barrel mutants by proteolysis. The wild-type plug does not restore activity of FhuA derivatives carrying mutations in the plug (V11D, Δ5–17) as the wild-type plug cannot displace the mutant plug incorporated into the β-barrel. These results suggest that the β-barrel can insert independently of the plug into the OM. The plug and the β-barrel form independent domains. The plug may assume its final structure when it incorporates into the β-barrel. This may occur in the periplasm or while the β-barrel is incorporated into the OM.

Reconstitution of the FpvA protein

As already discussed, FpvA is an energy-coupled OM transporter of *P. aeruginosa* that takes up ferric pyoverdines. Pyoverdines are siderophores composed of a fluorescent dihydroxyquinoline chromophore linked to peptides of various lengths. They serve as iron-complexing molecules that are taken up by the energy-coupled OM transporter FpvA into the periplasm, where iron is released and transported into the cytoplasm by an unknown system. Pyoverdine is released into the medium by the PvdRT-OmQ efflux pump and reused (Hannauer *et al.*, 2010). Removal of the FpvA plug results in a β-barrel that incorporates into the OM but does not bind and transport ferric pyoverdine (Nader *et al.*, 2011). Co-expression of the plug and the β-barrel results in a reconstituted FpvA in the OM that binds ferric pyoverdine with the same affinity as wild-type FpvA but transports ferric pyoverdine at only 25% of the rate of wild-type FpvA. Transport is energy dependent, as shown by inhibition by the protonophore carbonyl cyanide *m*-chlorophenylhydrazone (CCCP). The kinetics of ferric pyoverdine transport into the periplasm can be determined by fluorescence spectroscopy. Iron-loaded pyoverdine does not fluoresce, but pyoverdine without iron does. When FpvA is reconstituted, fluorescence increases owing to the release of iron from pyoverdine in the periplasm. In contrast, the fluorescence of the isolated β-barrel does not increase, which indicates that no ferric pyoverdine diffuses into the periplasm. Nader *et al.* (2011) take their results as an indication that the plug does not move out of the β barrel to form an open pore in the β-barrel. These results are similar to those obtained with FhuA.

Reconstitution of the HasR protein

As discussed earlier, HasR is the energy-coupled OM transport protein of *S. marcescens* that takes up haem or haem from a haem-binding protein (haemophore) into the periplasm. Genetic removal of the HasR plug results in a β-barrel that incorporates into the OM and binds the haemophore but does not actively transport haem. In contrast to FhuA and FpvA, the β-barrel forms a diffusion pore that is specific for its ligand, haem. The pore formed by the β-barrel does not increase sensitivity of cells to SDS and vancomycin, for which the OM forms a permeability barrier (Létoffé *et al.*, 2005). That haem diffusion is specific is further supported by its dependence on His603, which is essential for haem transport by wild-type HasR. Replacement of His603 by Ala abolishes haem transport and haem diffusion.

In vitro reconstitution of energy-coupled transporters in planar lipid bilayer membranes

Whether FhuA, Cir, and BtuB form channels in planar lipid bilayer membranes has been tested. Cir transports ferric catecholates, and BtuB transports vitamin B_{12}. Cir also serves as receptor of colicin I. Wild-type FhuA (Killmann *et al.*, 1993), Cir, and BtuB (Udho *et al.*, 2009) do not increase the conductance across planar lipid bilayer membranes, as can be expected from the crystal structures of FhuA (Ferguson *et al.*, 1998; Locher *et al.*, 1998), Cir (Buchanan *et al.*, 2007), and BtuB (Chimento *et al.*, 2003), in which the plug completely occludes the pore in the β-barrel. Addition of 4 M urea to the *cis* side of the planar lipid bilayer membranes, the compartment to which the protein is added, increases conductance by the three proteins continuously with some single steps (Udho *et al.*, 2009). Removal of urea reduces conductance through FhuA and Cir by 60–90%. This increase and decrease cycle of conductance can be repeated several times by adding and removing 4 M urea. In contrast, removal of urea stops the increase in conductance by BtuB, which then declines only slightly. When 4 M urea or 3 M glycerol is added to the *trans* compartment, along with addition of 4 M urea to the *cis* compartment, conductance by FhuA is not increased. Removal of urea or glycerol from the *trans* compartment results in an increase in conductance. These results suggest that 4 M urea unfolds the plug and establishes an osmotic pressure gradient across the black lipid membrane.

Phage T5 binds to FhuA, and it has been reported that it opens a channel in planar phospholipid bilayers endowed with FhuA (Bonhivers *et al.*, 1996). However, the obvious conclusion that the phage expels the plug of FhuA to get its DNA across the OM does not hold true. Cryo-electron tomography reveals that FhuA inserted into liposomes serves only as a docking site for phage T5. The tip of the phage tail acts like an injection needle, creating a passageway at the periphery of FhuA, through which the phage DNA crosses the membrane (Böhm *et al.*, 2001). FhuA added to the *cis* compartment increases conductance when T5 is added either to the *cis* or the *trans* compartment, provided 3 M glycerol has been added to the *cis* compartment. The osmotic gradient created by 3 M glycerol orientates FhuA such that phage binding sites are exposed at both the *cis* and the *trans* side. The native state of FhuA after treatment with 4 M urea is derived from binding of ferrichrome to FhuA and of colicin Ia to Cir. Ferrichrome added to the *trans* side stops the increase of the FhuA conductance and somewhat reverses it. Colicin Ia added to the *trans* side slows or stops conductance increase through Cir elicited by 4 M urea. Both proteins bind their cognate ligands, which indicates that an appreciable proportion of the protein binding sites are not denatured by 4 M urea. Most likely, 4 M urea changes the structure of the proteins so that the plugs move within or out of the β-barrels and a channel is formed through which the ions in the *cis* and *trans* solutions (100 mM KCl, 5 mM $CaCl_2$) move across the lipid bilayer membrane. It is tempting to propose that the action of 4 M urea on the plug somehow mimics the function of TonB.

OM transporters as colicin receptors

Structural and functional characteristics of colicins

Colicins belong to the class of bacteriocins – bacterial protein toxins that kill bacteria. Only cells closely related to the bacteriocin-producing cells are killed. A comprehensive review on colicins has been recently published (Cascales *et al.*, 2007). Colicins are formed by nearly half of *E. coli* natural isolates, and they predominantly kill *E. coli* cells. Colicins are composed of three functional domains that can be recognized in the crystal structures. The rather short N-terminal domain is required for translocation of colicins across the OM, the large central domain binds to the OM receptor proteins, and the C-terminal domain forms the active centre. Colicin-producing cells are not killed because they co-synthesize immunity proteins that bind to and inactivate the cognate colicins. Colicins only kill cells when they are imported from the medium; colicins that rest inside cells do not kill cells. Colicins degrade DNA or RNA in the cytoplasm, cleave a murein precursor in the periplasm, or dissipate the transmembrane potential of the cytoplasmic membrane by forming small pores. The uptake systems are specified by OM proteins to which colicins bind for subsequent import.

Import of colicins

Colicins are the only proteins that are taken up by *E. coli* cells. This import is fascinating since proteins as large as 70 kDa are specifically translocated across the OM to their target sites in the periplasm, the cytoplasmic membrane, or the cytoplasm. Uptake requires a set of proteins that use intricate mechanisms, including energy transfer from the cytoplasmic membrane into the OM. Crystal structures of colicins reveal elongated structures with compact portions or only compact structures (listed in Cascales *et al.*, 2007; Zeth *et al.*, 2008; Arnold *et al.*, 2009) (Fig. 8.3). Colicins must unfold to be transported across the OM and the cytoplasmic membrane. Evidence for colicin unfolding has been obtained from cross-linking between introduced cysteine residues, which when cross-linked lose *in vivo* activity but retain *in vitro* activity. Upon reduction of the cysteine bridges, the *in vivo* activity is recovered. Colicins bound to receptors are sensitive to added proteases. For example, colicin M is resistant to trypsin but is degraded by trypsin when bound to cells (Hullmann *et al.*, 2008). Colicin M modelled on the FhuA receptor demonstrates the need for unfolding for translocation through the pore of FhuA or along FhuA into the periplasm. Colicin M is modelled with its α1 helix (Helbig and Braun, 2011) in the surface cavity of FhuA, where also ferrichrome binds (Fig. 8.3b).

Colicins are imported by specific uptake systems. They bind to OM transporters or porins. Their initial characterization was based on receptor specificities and uptake requirements. For example, the E colicins, E1 to E9, were catalogued according to their common receptor protein BtuB and the requirement for the Tol proteins, which transduce energy from the cytoplasmic membrane into the OM, similar to the TonB ExbB ExbD proteins (Cascales *et al.*, 2000, 2001). The transmembrane topologies of TolA, TolQ, and TolR resemble the transmembrane topologies of TonB, ExbB, and ExbD, respectively (Kampfenkel and Braun, 1993). The TolQR proteins show sequence similarities to the ExbBD proteins and can functionally replace to some extent the ExbB and ExbD proteins and vice versa (Braun and Herrmann, 2004b). Only the N-terminal transmembrane region of TolA is sufficiently similar to that of TonB to replace it, resulting in a functional TonB (Karlsson *et al.*, 1993). Despite the absence of sequence similarities, the crystal structures of the TolA and TonB C-termini display a rather high degree of similarity (Witty *et al.*, 2002). The Tol system is more complex than the Ton system as it consists of the proteins TolA, TolB, TolQ, TolR, and Pal. For colicin uptake, Pal is not required (Sharma *et al.*, 2009).

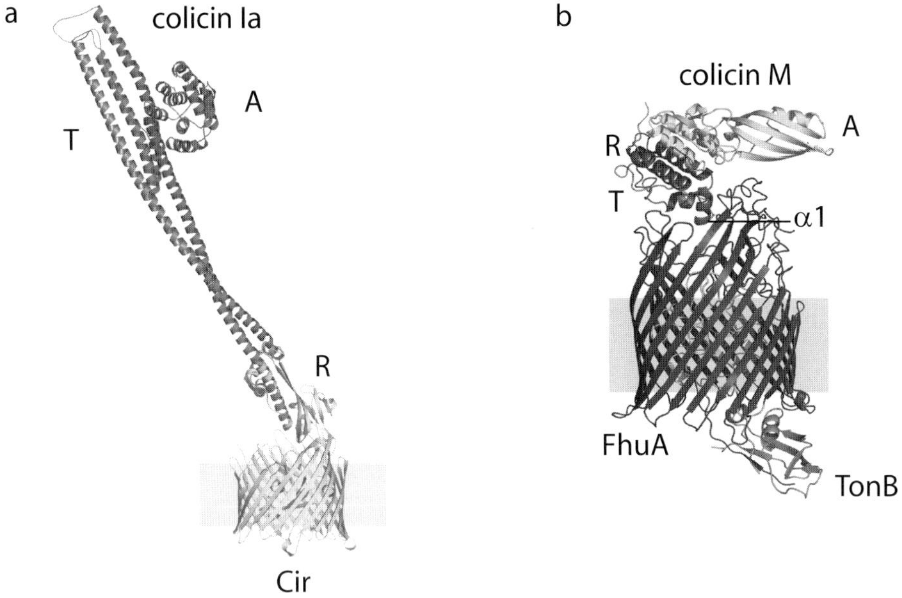

Figure 8.3 (a) Crystal structure of the Cir OM transporter with bound colicin Ia. The crystal structure of Cir with the colicin Ia receptor binding domain (designated R) has been determined (Buchanan *et al.*, 2007). The translocation domain (T) and the activity domain (A) are shown. The proteins are drawn to scale. Cir has a height of 40 Å, and colicin Ia is 210 Å long. (Courtesy of Susan Buchanan, NIH, Bethesda, Maryland, USA). (b) Crystal structure of colicin M modelled on the crystal structure of the FhuA OM transporter such that the α1 helix of colicin M contacts FhuA in its surface pocket. Colour code: green, colicin M activity domain (A); magenta, receptor binding domain (R); orange, translocation domain (T); blue, FhuA; yellow, bound C-terminal TonB fragment. For the sake of clarity the drawing scale of (b) is larger than that of (a). A colour version of this figure is located in the plate section at the back of the book.

According to their uptake requirements, colicins are divided into two groups: group A and group B. Uptake of group A colicins is energized by the Tol system; uptake of the group B colicins is energized by the Ton system.

Group A colicins

Five of the nine group A colicins bind to BtuB, but TonB is not required for their uptake. BtuB only serves as a receptor. These colicins require for uptake the OmpF porin as a second OM protein, with the exception of colicin E1, which uses TolC for import.

The uptake mechanism across the OM is not known for any colicin, but the following scenario can be envisaged for the uptake of the group A colicins. After tight binding to BtuB (K_d of colicin E9 ~1–2 nM), OmpF is recruited by the colicins with their natively unstructured translocation domain; this domain is often disordered in crystal structures and therefore not observed. The colicins remain bound to BtuB through their receptor binding domains. The translocation domain then enters the periplasm through the OmpF pore or along the surface of OmpF (Baboolay *et al.*, 2008) and interacts with TolB. Colicins mediate contact between TolA and TolB, which triggers the proton-motive-force-linked import into cells, as shown with colicin E9 (Bonsor *et al.*, 2009). The flexible translocation domain contains two regions that bind to OmpF with dissociation constants of 2 and 24 μM. An intrinsically unstructured 83 residues of the 37 kDa translocation domain delivers a 16-residue peptide in the centre of the translocation domain into the periplasm, where it binds to TolB. One of the translocation regions that bind to OmpF, residues 2–18, is found inside the OmpF lumen of an OmpF-peptide co-crystal structure. The unstructured nature of the translocation domain and its narrow cross-sectional area allows the peptide (6 kDa) to be taken up through the lumen of OmpF, which forms a cut-off filter of 0.6 kDa (Housden *et al.*, 2010). This mechanism also applies for colicin E3; an 83-residue N-terminal peptide from the translocation domain of colicin E3 has been found inside an OmpF crystal structure (Yamashita *et al.*, 2008). These results suggest that at least the translocation domain of the E colicins enters the periplasm through the pore of OmpF. It remains to be determined how the cytotoxic domain finds its way through the OM and the cytoplasmic membrane to get access to its target in the cytoplasm. For colicin E2, it has been shown that its receptor binding and translocation domain remain bound to the receptor BtuB and the Tol complex during killing of cells. Competition experiments with colicin A, which uses the same BtuB and OmpF OM proteins as E2 to enter cells, show that E2 prevents rapid K^+ efflux caused by colicin A as long as 30 min after E2 addition (Duché, 2007).

Colicin N uses OmpF twice – once as a primary binding site and once as a translocator across the OM. In addition, and unique for colicin N, the inner core region of lipopolysaccharide is required (Sharma *et al.*, 2009). Electron microscopy evidence and biochemical data suggest that the colicin N translocator domain is deposited along the OmpF trimer (Baboolay *et al.*, 2008).

Colicin S4 contains two receptor-binding regions, as revealed by duplication of a 32-residue sequence in the receptor binding domain (Pilsl *et al.*, 1999) and the crystal structure (Arnold *et al.*, 2009). Both receptor binding domains bind to the OM OmpW protein, but one is sufficient to kill cells.

Group B colicins

Colcins B, D, Ia, Ib, M, 5, and 10 belong to the group B colicins. They require only a single OM receptor protein belonging to the class of the energy-coupled transporters: FepA for colicins B and D, Cir for colicins Ia and Ib, and FhuA for colicin M.

The receptors and the colicins contain TonB boxes, both of which must interact with TonB for colicin uptake (Mende and Braun, 1990; Pilsl *et al.*, 1993; Mora *et al.*, 2005; Buchanan *et al.*, 2007). Interesting exceptions are colicins 5 and 10, which bind to the Tsx protein, a nucleoside-specific porin that serves as receptor for phage T6 and colicin K (Hantke, 1976). Tsx is not coupled to TonB, but TonB is required for sensitivity of cells to colicins 5 and 10, which contain a TonB box (Pilsl and Braun, 1995). Interaction of these colicins with TonB alone is sufficient for uptake. Tsx is the initial binding site, and TolC is probably required for translocation. These are the only two TonB-dependent colicins that require an additional porin for uptake, as is typical for group A colicins. These two colicins therefore stand somehow between the group A and group B colicins. This notion is supported by conversion of colicin 10 to a group A colicin by exchange of the TonB box through a conserved sequence of colicin E1 (Tol box), a group A colicin. After mutual replacement of region 1–67 of colicin 10 by region 1–58 of E1, the hybrid protein Col10-E1 is carried into cells by the Ton system and BtuB, the receptor of E1, and E1-Col10 is imported via the Tol system and Tsx (Pilsl and Braun, 1995). A Tol-dependent colicin is partially artificially changed to a Ton-dependent colicin by replacing the DGTGW motif in the translocation domain of the Tol-dependent colicin U by the DTMVV TonB box of colicin B. The resulting hybrid protein is still taken up via TolA and TolQ but no longer requires TolB. Instead, uptake requires TonB and ExbB (Pilsl and Braun, 1998). These examples further support the evolutionary connection between the Ton and the Tol systems.

Import of colicin D requires an RNase that cleaves $tRNA^{Arg}$ isoforms, the N-terminal TonB box, and a second site of interaction with TonB through a 45-residue region within the central domain of colicin D (Mora *et al.*, 2008).

A new concept for the uptake of group B colicins arose from studies of a hybrid protein between colicin E3 and colicin Ia (Jakes and Finkelstein, 2009). The receptor binding domain of colicin Ia was replaced by the receptor binding domain of E3. The hybrid protein requires the BtuB receptor for E3 to kill cells but retains some killing activity in a *btuB* mutant background, which is completely resistant to E3. Unexpectedly, a *cir* mutant lacking the Cir protein required for sensitivity to Ia is resistant to the hybrid protein. Sensitivity of cells to the hybrid protein depends on the Ton system and not on the Tol system. Deletion of the receptor binding domain of Ia results in a protein in which the translocation domain is fused to the activity domain. This hybrid protein has the same specific activity as the E3–Ia fusion protein in the *btuB* mutant background. It still requires Cir and TonB to kill cells. This implies that independent of the receptor domain, the translocation domain binds to Cir. This conclusion is supported by the inhibition of the Ia activity and the activity of the hybrid proteins by the isolated translocation domain. However, binding of the translocation domain to Cir is weak, as a 3×10^6 fold excess of the translocation domain over colicin Ia is required to obtain 60% survival. These results lead to the conclusion that Cir plays a dual role in colicin Ia entry, with one Cir serving as the primary binding site and a second serving in the actual translocation into the cells. The requirement for two OM proteins resembles the uptake of group A colicins. However, it is not certain whether the findings with colicin Ia/Cir apply to the other group B colicins. For example, high concentrations of colicin Ia that saturate all Cir receptors inhibit killing. When all receptors are occupied with colicin Ia, none are free for colicin Ia translocation. We have not observed inhibition of killing by high concentrations of colicin M (C. Römer and V. Braun, unpublished). More experiments are required to resolve this discrepancy between colicins M and Ia.

Colicin M has the particular property that it specifically requires a periplasmic chaperone, FkpA, to kill cells (Hullmann *et al.*, 2008). FkpA is a prolyl *cis–trans* isomerase that can be deleted without affecting *E. coli* growth. Its unknown function can probably be substituted by other chaperones in the periplasm. It is not known whether after translocation across the OM refolding of colicin M to an active toxin requires FkpA and whether refolding provides the energy for translocation. In the latter case, FkpA would be part of the uptake machinery, which consists of FhuA, TonB, ExbB, and ExbD. Point mutants in the isomerase domain of FkpA confer resistance to colicin M and a proline residue in colicin M has been identified that might be *cis–trans* isomerized (Helbig *et al.*, 2011). However, definitive proof of FkpA catalysed *cis–trans* isomerization of colicin M isolated from the periplasm is lacking.

Crystal structures of colicins bound to OM receptors

Co-crystal of colicin I bound to the Cir OM transporter

Comparison of the crystal structure of the colicin I receptor Cir with the crystal of Cir and the bound receptor binding domain (R-domain) of colicin Ia reveals large conformational changes upon binding of the R-domain (Fig. 8.3a) (Buchanan *et al.*, 2007). Cir is exposed at the extracellular side and open, with the R-domain positioned directly above it. The tip of the R-domain sits in the binding pocket where iron-siderophores bind. Only two arginine residues of loops L7 and L8 of Cir mainly serve for binding of the R-domain. Loops 7 and 8 move together to open outwards by 37° relative to uncomplexed Cir. An additional translation and rotation displaces residue 437 of loop 7 by 17 Å and residue 511 of loop 8 by 21 Å; these are the largest movements observed in OM transporters upon ligand binding. Binding of the R-domain opens Cir to the extracellular milieu. In contrast, binding of (ferric citrate)$_2$ to FecA induces movement of loops 7 and 8 by 11 and 14 Å, respectively, towards diferric dicitrate, which closes the binding pocket so that the substrate can no longer escape to the medium (Ferguson *et al.*, 2002; Yue *et al.*, 2003). The conformational changes induced by R-domain binding do not expel the plug from the β-barrel, nor do they open a pore in the plug or at the interface between the plug and the β-barrel. The plug remains fixed to the barrel by two salt bridges and 39 hydrogen bonds. The plug-barrel interface is highly solvated with 23 bridging water molecules and 70 non-bridging water molecules, posing no large energy barrier to the movement of the plug within or out of the barrel. However, this does not occur when the R-domain binds to Cir. This is in agreement with all transporter structures with bound ligands in which the plug does not move to an extent that a pore is formed. The usually observed greater exposure of the TonB box, or a higher flexibility of the TonB box upon ligand binding to transporters is not observed in Cir, where the TonB box (residues 6–10 in the mature form) adopts an extended conformation in the liganded and unliganded structure.

The R-domain retains the conformation it has in the complete colicin Ia (Wiener *et al.*, 1997). Modelling of the complete colicin Ia to Cir positions the colicin translocation and activity domains 150 Å above the OM (Fig. 8.3a). Translocation of the activity domain into the cytoplasmic membrane, where it kills cells by forming pores, involves a huge distance. It is assumed that colicin Ia bends over the cell surface to contact the translocation domain of a second Cir molecule, where it triggers the TonB-dependent uptake of the activity domain into the periplasm, from where it spontaneously inserts into the cytoplasmic membrane to form a pore. Uptake of the translocation domain and finally of the activity domain is proposed to occur through the pore of the Cir β-barrel (Jakes and Finkelstein, 2009).

Co-crystals of colicins E2 and E3 bound to the BtuB OM transporter

The crystal structure of the coiled-coil receptor binding domain of colicin E3 (R135) bound the BtuB transporter reveals 27 residues near the tip of the coiled-coil involved in binding to 29 residues at the BtuB extracellular surface (Kurisu *et al.*, 2003). The interface of R135 with BtuB is thus much larger than the interface between colicin Ia and Cir. The N- and C-terminal ends of R135 are unfolded, which may be important for unfolding the translocation domain of E3. BtuB serves as the primary binding site, but uptake of E3 is mediated by a second OM protein, porin OmpF. The channel of OmpF incorporated into planar lipid bilayer membranes is occluded by E3. Occlusion requires a *cis*-negative potential of

~50 mV, which unlikely exists across the OM. The model of E3 uptake proposes bending of E3 bound to BtuB over OmpF, through which it is translocated by binding of the translocation domain, followed by translocation of the unfolded activity domain released from the translocation domain by proteolysis (Shi *et al.*, 2005).

The co-crystal of the colicin E2 receptor binding domain bound to BtuB is similar to the co-crystal of the colicin E3 binding domain bound to BtuB; this similarity can be expected from the very similar structures of E2 (DNase) and E3 (RNase). No displacement of the plug domain of BtuB is observed. OmpF channels in planar lipid bilayer membranes are occluded by E2 (Sharma *et al.*, 2007), similar to E3 occlusion.

Do colicins expel the plug of transporters?

For the group A colicins, it seems to be clear that those that bind to the BtuB transporter do not expel the BtuB plug. BtuB serves only as the primary binding site, and translocation across the OM is mediated by the porins OmpF, OmpC, or TolC. Occlusion of the porin channels by colicins E2 and E3 and the finding of engineered fragments of the translocation domains within the OmpF pore argue in favour of a translocation through the pore of the porins. Binding of the colicin Ia receptor domain to Cir also does not open a pore in the energy-coupled transporters. Whether uptake through the proposed second Cir occurs through the lumen of Cir or along the surface of Cir is an open question. Identification of the binding site(s) of the Ia translocation domain on Cir would provide clues as to which route the translocation domain and subsequently the activity domain takes. However, very weak binding of the translocation domain to Cir makes it difficult to determine the Cir binding site.

Two labelling approaches have been used to attempt to answer the question whether the plug leaves the barrel of FepA when colicin B is bound. Cysteines introduced into the plug domain are used to measure their reactions towards the biotin maleimide derivative BMCC (Devanathan and Postle, 2007). Addition of colicin B increases labelling in a TonB-dependent manner. The TonB box of FepA but not the TonB box of colicin B is required for the increase in labelling. However, the cysteine residues are not uniformly labelled as might be expected when the entire plug moves out of the barrel. Only 5 of the 11 cysteines are labelled 25-fold, the rest only 2-fold. Upon colicin B binding, a portion of the FepA molecules are degraded at the N-terminus by protease K; degradation depends on TonB. It has been postulated that an unknown bound protein inhibits labelling of those residues that are only weakly labelled or that under the experimental conditions used, the plug is only partially extruded from the β-barrel.

A second approach has used fluorescein maleimide to label cysteines introduced into FepA (Smallwood *et al.*, 2009). In the presence of colicin B, no increase in fluorescein maleimide modification is observed at 35 cysteine sites, including 15 in the plug domain, regardless whether TonB is present or not. Three sites at which colicin B strongly increases labelling by BMCC are the same as those studied with fluorescein maleimide. If the plug moves out of the β-barrel, different experimental designs should yield similar results.

We introduced cysteine residues into the plug domain of FhuA and determined FhuA labelling in metabolizing cells with BMCC in the presence and absence of colicin M (Endriß and Braun, unpublished). Residues I9C in the TonB box, T27C at the periplasmic surface, and R81C in the cavity at the cell surface, where ferrichrome binds, were labelled with BMCC, regardless whether colicin M had been added prior to labelling or not. S51C, S65C,

Table 8.1 Discussed energy-coupled OM transporters and their ligands

Transporter	Ligands
FepA	Fe^{3+}-enterobactin, colicin, B, D
FhuA	Ferrichrome, albomycin, rifamycin CGP4832, colicin M, microcin J25, phages T1, T5, Φ80, UC-1
BtuB	Vitamin B_{12}, colicins E2, E3, E9
FecA	Diferric dicitrate
OmpW	Colicin S4
PupB	Pseudobactin BN8
FpvA	Fe^{3+}-pyoverdine
HasR	Haem, haemoglobin, Has A haemophore
HuxC	Haem, haemopexin, HuxA
TbpA/TbpB	Transferrin
LbpA/LbpB	Lactoferrin
SusC/SusD	Sugars
MalA	$(\text{Maltodextrins})_{1-4}$
NagA	$(\text{GlcNAc})_{1-5}$
SuxA	Sucrose
TanO/TanU	Sialic acid
FrpB4	Ni^{2+}
ZnuD	Zn^{2+}

S77C, R93C, and S137C located within the β-barrel were not labelled under any conditions. Residues from within the β-barrel do not become exposed upon binding of colicin M to FhuA, which suggests no displacement of the plug out of the β-barrel by colicin M. The lack of plug displacement in Cir with the bound colicin Ia receptor fragment does not solve the issue as the crystal structure is uncoupled from TonB.

Outlook

Although much insight has been gained into the structure and function of OM transporters, fundamental questions remain to be answered. How is the proton-motive force across the cytoplasmic membrane harvested by the TonB–ExbB–ExbD protein complex? How do these proteins structurally react to the proton-motive force? In which protein (most likely TonB) is the energy stored and what role do ExbB and ExbD play? What distinguishes the energized form of the proteins from the unenergized form? How does energization affect interaction of TonB with the OM transporters? What happens structurally in the interacting proteins when energy is transferred from TonB to the OM transport proteins? Does the plug move inside the β-barrel to open a pore, or does it move out of the β-barrel? Do substrates once released from the binding site at the cell surface freely diffuse through the pores, or do they sequentially bind to amino acid side chains as they pass through the pores (diffusion or facilitated diffusion)? How large is the variety of substrates transported? What kind of structural transitions are induced in transport proteins by substrates that trigger gene transcription in the cytoplasm while they stay bound at the cell surface? How is the signal transmitted across the OM, to the regulatory protein in the cytoplasmic membrane, and finally to the ECF sigma factor? Do the transporters only function as primary binding sites for phages and colicins or do they also translocate phage DNA and the activity domains of colicins through their pores, along their surface, or through additional proteins? Which domains of colicins and to what extent are they translocated across the OM and how much do they unfold? The answering of these questions requires a multitude of different experimental approaches, in particular *in vitro* reconstituted systems. Beyond understanding of energy-coupled OM transport, insights of general relevance will be gained on how energy is transferred between adjacent membranes and how proteins are translocated across membranes.

Acknowledgements

The author thanks the Max Planck Society and the German Science Foundation for continuous support and Andrei Lupas for his generous hospitality. Karen A. Brune and Klaus Hantke critically read the paper, and Stephanie Helbig drew figures.

References

Arnold, T., Zeth, K., and Linke, D. (2009). Structure and function of colicin S4, a colicin with a duplicated receptor binding domain. J. Biol. Chem. *284*, 6403–6413.

Babbolai, T.G., Conroy, M.J., Gill, K., Ridley, H., Visudtiphole, V., Bullough, P.A., and Lakey, J.H. (2008). Colicin N binds to the periphery of its receptor and translocator OM protein F. Structure (Camb) *16*, 371–379.

Benevides-Matos, N., and Biville, F. (2010). The heme and Has heme uptake systems in *Serratia marcescens*. Microbiology *156*, 1749–1757.

Biville, F., Cwermann, H., Létoffé, S., Rossi, M.-S., Drouet, V., Ghigo, J.M., and Wandersman, C. (2004). Haemophore-mediated signaling in *Serratia marcescens*: a new mode of an extracytoplasmic function (ECF) sigma factor involved in heme acquisition. Mol. Microbiol. *53*, 1267–1277.

Bjursell, M.K., Martens, E.C., and Gordon, J.I. (2006). Functional genomic and metabolic studies of the adaptation of a prominent adult human gut syMBio.nt, *Bacteroides thataiotaomicron*, to the suckling period. J. Biol. Chem. *281*, 36269–36279.

Blanvillain, S., Meyer, D., Boulanger, A., Lautier, M., Guynet, C., Denacé, N., Lauber, E., and Arlat, M. (2007). Plant carbohydrate scavenging through *tonB*-dependent receptors: a feature shared by phytopathogenic and aquatic bacteria. PLoS One *2*, e224.

Böhm, J., Lambert, O., Letellier, L., Baumeister, W., and Rigaud, J.L. (2001). FhuA mediated phage genome transfer into liposomes: a cryo-electron tomography study. Curr. Biol. *7*, 1168–1175.

Bonhivers, M., Ghazi, A., Boulanger, P., and Letellier, L. (1996). FhuA, a transporter of the *Escherichia coli* OM, is converted into a channel upon binding of bacteriophage T5. EMBO J. *15*, 1850–1856.

Bonsor, D.A., Hecht, O., Vankemmelbeke, M., Sharma, A., Krachler, A.M., Housden, N.G., Lilliy, K.J., James, R., Moore, R.G., and Kleanthous, C. (2009). Allosteric beta–propeller signaling in TolB and its manipulation by colicins. EMBO J. *28*, 2846–2857.

Bös, C., Lorenzen, D., and Braun, V. (1998). Specific *in vivo* labeling of cell surface-exposed loops: reactive cysteines in the predicted gating loop mark a ferrichrome binding site and a ligand-induced conformational change of the *Escherichia coli* FhuA protein. J. Bacteriol. *180*, 605–613.

Braun, M., Killmann, H., Maier, E., Benz, R., and Braun, V. (2002). Diffusion through channel derivatives of the *Escherichia coli* FhuA transport protein. Eur. J. Biochem. *269*, 4948–4859.

Braun, M., Endriss, F., Killmann, H., and Braun, V. (2003). *In vivo* reconstitution of the FhuA transport protein of *Escherichia coli* K-12. J. Bacteriol. *185*, 5508–5518.

Braun, V. (2009). FhuA (TonA), the career of a protein. J. Bacteriol. *191*, 3431–3436.

Braun, V. (2010). OM signaling in gram-negative bacteria. In Bacterial Signaling, Krämer, R., and Jung, K., eds. (Wiley-VCH Verlag, Weinheim), pp. 117–133.

Braun, V., and Herrmann, C. (2004a). Point mutations in transmembrane helices 2 and 3 of ExbB and TolQ affect their activities in *Escherichia coli* K-12. J. Bacteriol. *186*, 4402–4406.

Braun, V., and Herrmann, C. (2004b). Evolutionary relationship of uptake systems for biopolymers in *Escherichia coli*: cross-complementation between the TonB-ExbB-ExbD and the TolA-TolQ-TolR proteins. Mol. Microbiol. *8*, 261–268.

Braun, V., and Mahren, S. (2005). Transmembrane transcriptional control (surface signaling) of the *Escherichia coli* Fec type. FEMS Microbiol. Rev. *29*, 673–684.

Braun, V., and Mahren, S. (2007). Transfer of energy and information across the periplasm in iron transport and regulation. In The Periplasm, Ehrmann, M., ed. (ASM Press, Washington D.C.), pp. 276–286.

Braun, V., Patzer, S.I., and Hantke. K. (2002). TonB dependent colicins and microcins: modular design and evolution. Biochim. *84*, 365–380.

Braun, V., Mahren, S., and Ogierman, M. (2003). Regulation of the FecI type ECF sigma factor by transmembrane signaling. Curr. Opin. Microbiol. *6*, 173–180.

Braun, V., Mahren, S., and Sauter, A. (2005). Gene regulation by transmembrane signaling. Biometals *18*, 507–517.

Brillet, K., Journet, L., Celia, H., Paulus, L., Stahl, A., Pattus, F., and Cobessi, D. (2007). A β strand lock exchange for signal transduction in TonB-dependent transducers on the basis of a common structural motif. Structure *15*, 1383–1391.

Brito, B., Aldon, D., Barberis, P., Boucher, C., and Genin, S. (2002). A signal transfer system through three compartments transduces the plant contact-dependent signal controlling *Ralstonia solanacearum hrp* genes. Mol. Plant Microbiol. *15*, 109–119.

Brooks, B.E., and Buchanan, S.K. (2008). Signaling mechanisms for activation of extracytoplasmic function (ECF) sigma factors. Biochim. Biophys. Acta *1778*, 1930–1945.

Breidenstein, E., Mahren, S., and Braun, V. (2006). Residues involved in FecR binding are localized on one side of the FecA signaling domain of *Escherichia coli*. J. Bacteriol. *188*, 6440–6442.

Buchanan, S.K., Lukacik, P., Grizot, S., Ghirlano, R., Jakes, K.S., Klienker, P.K., and Esser, L. (2007). Structure of colicin I receptor bound to the R-domain of colicin Ia: implications for protein import. EMBO J. *26*, 2594–2604.

Cadieux, N., and Kadner, R.J. (1999). Site-directed disulfide bonding reveals an interaction site between energy-coupling protein TonB and BtuB, the OM cobalamin transporter. Proc. Natl. Acad. Sci, USA *96*, 10673–10678.

Cadieux, N., Phan, P.G., Cafiso, D.S., and Kadner, R.I. (2003). Different substrate-induced signaling through the TonB-dependent transporter BtuB. Proc. Natl. Acad. Sci. U.S.A. *100*, 10688–10693.

Cascales, E., Gavioli, M., Sturgis, J.N., and Lloubès, R. (2000). Proton motive force drives the interaction of the inner membrane TolA protein and OM Pal proteins in *Escherichia coli*. Mol. Microbiol. *38*, 904–915.

Cascales, E., Lloubès, R., and Sturgis, J.N. (2001). The TolQ-TolR proteins energize TolA and share homologies with the flagellar motor proteins MotA-MotB. Mol. Microbiol. *42*, 795–807.

Cascales, E., Buchanan, S.K., Duché, D., Kleanthous, C., Lloubès, R., Postle, K., Riley, M., Slatin, S., and Cavard, D. (2007). Colicin biology. Microbiol. Molec. Biol. Rev. *71*, 158–229.

Cescau, S., Cwerman, H., Létoffé, S., Delepelaire, P., Wandersman, C., and Biville, F. (2007). Heme acquisition by hemophores. Biometals *20*, 603–613.

Chimento, D.P., Mohanty, A.K., Kadner, R.J., and Wiener, M.C. (2003a). Substrate-induced transmembrane signaling in the cobalamin transporter BtuB. Nat. Struct. Biol. *10*, 394–401.

Chimento, D.P., Kadner, R.J., and Wiener, M.C. (2003b). The *Escherichia coli* OM cobalamin transporter BtuB: structural analysis of calcium and substrate binding, and identification of orthologous transporters by sequence/structure conservation. J. Mol. Biol. *332*, 999–1014.

Chimento, D.P., Kadner, R.J., and Wiener, M.C. (2005). Comparative structural analysis of TonB-dependent OM transporters: implications for the transport cycle. Proteins *59*, 240–251.

Cho, K.H., and Salyers, A.A. (2001). Biochemical analysis of interactions between OM proteins that contribute to starch utilization by *Bacteroides thetaiotaomicron*. J. Bacteriol. *183*, 7224–7230.

Cope, L.D., Love, R.P., Guinn, S.E., Gilep, A., Usanov, S., Estabrook, R.W., Hrkal, Z., and Hansen, E.J. (2001). Involvement of the HxuC OM protein in utilization of hemoglobin by *Haemophilus influenzae*. Infect. Immun. *69*, 2353–2363.

Cope, L.D., Thomas, S.D., Hrkal, Z., and Hansen, E.J. (1998). Binding of heme–hemopexin complexes by soluble HxuA protein allows utilization of this complexed heme by *Haemophilus influenzae*. Infect. Immun. *66*, 4511–4516.

Cornelis, P. (2010). Iron uptake and metabolism in Pseudomonads. Appl. Microbiol. Biotechnol. *86*, 1637–1645.

Cornelis, P., and Andrews, S.C. (2010). Iron uptake and homeostasis in microorganisms (Caister Academic Press, Norfolk, UK).

Cornelissen, C.N., and Hollander, A. (2011). TonB-dependent transporters expressed in *Neisseria gonorrhoeae*. Front. Microbiol. *2*, 1–13.

Devanathan, S., and Postle, K. (2007). Studies on colicin B translocation: FepA is gated by TonB. Mol. Microbiol. *65*, 441–453.

Duché, D. (2007). Colicin E2 is still in contact with its receptor and import machinery when its nuclease domain enters the cytoplasm. J. Bacteriol. *189*, 4217–4222.

Eisenbeis, S., Lohmiller, S., Valdebenito, M., Leicht, S., and Braun, V. (2008). NagA-dependent uptake of N-acetyl-glucosamine and N-acetyl-chitin oligosaccharides across the OM of *Caulobacter crescentus*. J. Bacteriol. *190*, 5230–5238.

Endriß, F., Braun, M., Killmann, H., and Braun, V. (2003). Mutant analysis of the *Escherichia coli* FhuA protein reveals sites of FhuA activity. J. Bacteriol. *185*, 4683–4692.

Enz, S., Brand, H., Oreland, C., Mahren, S., and Braun, V. (2003). Sites of interaction between FecA and FecR signal transduction proteins of ferric citrate transport in *Escherichia coli* K-12. J. Bacteriol. *185*, 3745–3752.

Ferguson, A.D., Hofmann, E., E., Colton, J.E.W., Dietrich's, K., and Welted, W. (1998). Siderophore-mediated iron transport: crystal structure of FhuA with bound lipopolysaccharide. Science *282*, 2215–2220.

Ferguson, A.D., Braun, V., Fiedler, H.-P., Coulton, J.W., Diederichs, K., and Welte, W. (2000). Crystal structure of the antibiotic albomycin in complex with the OM transporter FhuA. Prot. Sci. *9*, 956–963.

Ferguson, A.D., Ködding, J., Walker, G., Bös, C., Coulton, J.W., Diederichs, K., Braun, V., and Welte, W. (2001). Active transport of an antibiotic rifamycin derivative by the outer-membrane protein FhuA. Structure *9*, 707–716.

Ferguson, A.D., Chakraborty, R., Smith, B.S., Esser, L., van der Helm, D., and Deisenhofer, J. (2002). Structural basis of gating by the OM transporter FecA. Science *295*, 1715–1719.

Ferguson, A.D., Amicus, C.A., Alibi, N.M., Celia, Y., Rosen, M.R., Rang Nathan, R., and Deisenhofer, J. (2007). Signal transduction pathway of TonB-dependent transporters. Proc. Natl. Acad. Sci. U.S.A. *104*, 513–518.

Garcia-Herrero, A., and Vogel, H.J. (2005). Nuclear magnetic resonance structure of the periplasmic signaling domain of the TonB-dependent OM transporter FecA. Mol. Microbiol. *58*, 1226–1237.

Garcia-Herrero, A., Peacock, R.S., Howard, S.P., and Vogel, H.J. (2007). The solution structure of the periplasmic domain of theTonB system ExbD protein reveals an unexpected structural homology with siderophore binding proteins. Mol. Microbiol. *66*, 872–889.

Gumbart, J., Wiener, M.C., and Tajkhorshid, E. (2007). Mechanism of force propagation in TonB-dependent OM transport. Biophys. J. *93*, 496–504.

Härle, C., InSook, K., Angerer, A., and Braun, V. (1995). Signal transfer through three compartments: transcription initiation of the *Escherichia coli* ferric citrate transport system from the cell surface. EMBO J. *14*, 1430–1438.

Hannauer, M., Yeterian, E., Martin, L.W., Lamont, I.L., and Schalk, I., J. (2010). An efflux pump is involved in secretion of newly synthesized siderophore by *Pseudomonas aeruginosa*. FEBS Lett. *584*, 4451–4455.

Hantke, K. (1976). Phage T6-colicin K receptor and nucleoside transport in *Escherichia coli*. FEBS Lett. *70*, 109–112.

Helbig, S., Patzer, I.A., Schiene-Fischer, C., Zeth, K., and Braun, V. (2011). Activation of colicin M by the FkpA prolyl *cis–trans* isomerase/chaperone. J. Biol. Chem. *286*, 6280–6290.

Higgs, P.I., Larsen, R.A., and Postle, K. (2002). Quantification of known components of the *Escherichia coli* TonB energy transducing system: TonB, ExbB, ExbD, and FepA. Mol. Microbiol. *44*, 271–281.

Housden, N.G., Wojdyla, J.A., Korczynska, J., Grishkovskaya, I., Kirkpatrick, N., Brzozowski, A.M., and Kleanthous, C. (2010). Directed epitope delivery across the *Escherichia coli* OM through the porin OmpF. Proc. Natl. Acad. Sci. U.S.A. *107*, 21412–21417.

Hullmann, J., Patzer, S.I., Römer, C., Hantke, K., and Braun, V. (2008). Periplasmic chaperone FkpA is essential for imported colicin M toxicity. Mol. Microbiol. *69*, 926–937.

James, H.E., Beare, P.A., Martin, L.W., and Lamont, I.L. (2005). Mutational analysis of bifunctional ferrisiderophore receptor and signal-transducing protein from *Pseudomonas aeruginosa*. J. Bacteriol. *187*, 4514–4520.

James, K.J., Hancock, M.A., Moreau, V., Molina, F., and Coulton, J.W. (2008). TonB induces conformational changes in surface-exposed loops of FhuA, OM receptor of *Escherichia coli*. Protein Sci. *17*, 1679–1688.

Kampfenkel, K., and Braun, V. (1993). Membrane topologies of the TolQ and TolR proteins of *Escherichia coli*: inactivation of TolQ by a missense mutation in the proposed first transmembrane segment. J. Bacteriol. *175*, 4485–4491.

Karlsson, M., Hannavy, K., and Higgins, C.F. (1993). A sequence-specific function of the N-terminal signal-like sequence of theTonB protein. Mol. Microbiol. *8*, 370–388.

Kim, I., Stiefel, A., Plantör, S., Angerer, A., and Braun, V. (1997). Transcription induction of the ferric citrate transport genes via the N-terminus of the FecA OM protein, the Ton system, and the electrochemical potential of the cytoplasmic membrane. Mol. Microbiol. *23*, 333–344.

Kim, M., Xu, Q., Murray, D., and Cafiso, D.S. (2008). Solutes alter conformation of ligand binding loops in OM transporters. Biochemistry *47*, 670–679.

Kirby, A.E., King, N.D., and Connell, T.D. (2004). RhuR, an extracytoplasmic sigma factor activator, is essential for heme.dependent expression of the OM heme and hemoprotein receptor of *Bortetella avium*. Infect. Immun. *72*, 896–907.

Kleanthous, C. (2010). Swimming against the tide: progress and challenges in our understanding of colicin translocation. Nat. Rev. Microbiol. *8*, 843–848.

Koebnik, R. (2005). TonB-dependent trans-envelope signaling: the exception of the rule? Trends Microbiol. *13*, 343–347.

Koster, M., van Klombenburg, W., Bitter, W., Leong, J., and Weisbeek, P. (1994). Role of the OM ferric siderophore receptor PupB in signal transduction across the bacterial cell envelope. EMBO J. *13*, 2805–2813.

Krieg, S., Huché, F., Diederichs, K., Izadi-Pruneyre, N., Lecroisey, A., Wandersman, C., Delepelaire, P., and Welte, W. (2009). Heme uptake across the OM as revealed by crystal structures of the receptor–heme complex. Proc. Natl. Acad. Sci. U.S.A. *106*, 1045–1050.

Kuriso, G., Zakharov, S.D., Zhalnina, M.V., Bano, S., Eroukova, V.Y., Rokitskaya, T.I., Antonenko, Y.N., Wiener, M.C., and Cramer, W.A. (2003). The structure of BtuB with bound colicin E3 R-domain implies a translocon. Nat. Struct. Biol. *10*, 948–954.

Létoffé, S., Wecker, K., Delepelaire, M., Delepelaire, P., and Wandersman, C. (2005). Activities of the *Serratia marcescens* heme receptor HasR and isolated plug and β-barrel domains: the β-barrel forms a heme-specific channel. J. Bacteriol. *187*, 4637–4645.

Llamas, M.A., Mooij, M.J., Sparrius, M., Vandenbroucke-Grauls, C.M., Ratledge, C., and Bitter, W. (2008). Characterization of five novel *Pseudomonas aeruginosa* cells surface signalling systems. Mol. Microbiol. *67*, 458–472.

Llamas, M.A., van der Sar, A., Chu, B.C.H., Sparrius, M., Vogel, H.J., and Bitter, W. (2009). A novel extracytoplasmic function (ECF) sigma factor regulates virulence in *Pseudomonas aeruginosa*. PLoS Pathog. *5*, 1–15.

Locher, K.P., Rees, B., Koebnik, R., Mitschler, A., Moulinier, L., Rosenbusch, J.P., and Moras, D. (1998). Transmembrane signaling across the ligand-gated FhuA receptor: crystal structures of free and ferrichrome-bound states reveal allosteric changes. Cell *95*, 771–778.

Luck, S.N., Turner, S.A., Rayakumar, K., Sakellaris, H., and Adler, B. (2001). Ferric citrate transport system (Fec) of *Shigella flexneri* 2a YSH6000 is encoded on a novel pathogenicity island carrying multiple antibiotic resistance genes. J. Bacteriol. *69*, 6012–6021.

Mahren, S., Schnell, H., and Braun, V. (2005). Occurrence and regulation of the ferric citrate transport system in *Escherichia coli* B, *Klebsiella pneumoniae, Enterobacter aerogenes,* and *Photorhabdus luminescens.* Arch. Microbiol. *184*, 175–186.

Martens, E.C., Roth, R., Heuser, J.E., and Gordon, J.I. (2009). Coordinated regulation of glycan degradation and polysaccharide capsule biosynthesis by a prominent human gut syMBio.nt. J. Biol. Chem. *284*, 18445–18457.

Martinez-Bueno, M.A., Tobes, R., Rey, M., and Ramos, J.L. (2002). Detection of multiple extracytoplasmic function (ECF) sigma factors in the genome of *Pseudomonas putida* KT2440 and their counterparts in *Pseudomonas aeruginosa* PA01. Environ. Microbiol. *4*, 842–855.

Mende, J., and Braun, V. (1990). Import-defective colicin B derivatives mutated in the TonB box. Mol. Microbiol. *4*, 1523–1539.

Merianos, H.J., Cadieux, N., Lindy, C.H., Kadner, R.I., and Cafiso, D.S. (2000). Substrate-induced exposure of an energy-coupling motif of a membrane transporter. Nat. Struct. Biol. *7*, 205–209.

Mettrick, K.A., and Lamont, I.L. (2009). Different roles for anti-sigma factors in siderophore signaling pathways of *Pseudomonas aeruginosa*. Mol. Microbiol. *74*, 1257–1271.

Miller, T.R., Delcher, A.L., Salzberg, S.L., Saunders, E., Detter, J.C., and Halden, R. (2010). Genomic sequence of the dioxin-mineralizing bacterium *Sphingomonas wittichii* RW1. J. Bacteriol. *192*, 6101–6102.

Mora, L., Diaz, N., Buckingham, R.H., and de Zamaroczy, M. (2005). Import of the transfer RNase colicin D requires site-specific interaction with the energy-transducing protein TonB. J. Bacteriol. *187*, 2693–2697.

Mora, L., Klepsch, M., Buckingham, R.H., Heurgué-Hamard, V., Kervestin, S., and de Zamaroczy, M. (2008). Dual roles of the central domain of colicin D tRNase in TonB-mediated import and in immunity. J. Biol. Chem. *283*, 4993–5003.

Moraes, T.F., Yu, R.H., Strynadka, N.C., and Schryvers, A.B. (2009). Insights into the bacterial transferrin receptor: the structure of transferrin-binding protein B from *Actinobacillus pleuropneumoniae*. Mol. Cell *35*, 523–533.

Nader, M., Journet, L., Meksem, A., Guillon, L., and Schalk, I.J. (2011). Mechanism of ferripyoverdine uptake by *Pseudomonas aeruginosa* OM transporter FpvA: no diffusion channel formed at any time during ferrisiderophore uptake. Biochemistry *50*, 2530–2540.

Newton, S.M., Trinh, V., Pi, H., and Klebba, P.E. (2010). Direct measurements of the OM stage of ferric enterobactin transport. Postuptake binding. J. Biol. Chem. *285*, 17488–17497.

Noinaj, N., Guillier, M., Barnard, T.J., and Buchanan, S.K. (2010). TonB dependent transporters: regulation, structure and function. Annu. Rev. Microbiol. *64*, 43–60.

Noinaj. N., Easley, N.C., Oke, M., Mizuno, N., Gumbart, J., Boura, E., Steere, A.N., Zak, O., Aisen, P., Tajkhorshid, E., *et al.* (2012). Structural basis for iron piracy by pathogenic *Neisseria*. Nature *483*, 53–58.

Pawelek, P.D., Croteau, N., Ng-Tow-Hing, C., Khursigara, C., Moiseeva, N., Allaire, M., and Coulton, J.W. (2006). Structure of TonB in complex with FhuA, E. coli OM receptor. Science *312*, 1399–1402.

Pilsl, H., and Braun, V. (1995). Novel colicin 10: assignment of four domains to TonB- and TolC-dependent uptake via the Tsx receptor and to pore formation. Mol. Microbiol. *16*, 57–67.

Pilsl, H., Glaser, C., Groß, P., Killmann, H., Ölschläger, T., and Braun, V. (1993). Domains of colicin M involved in uptake and activity. Mol. Gen. Genet. *240*, 103–112.

Pilsl, H., Šmajs, D., and Braun, V. (1999). Characterization of colicin S4 and its receptor, OmpW, a minor protein of the *Escherichia coli* OM. J. Bacteriol. *181*, 3578–3581.

Postle, K., and Larsen, R.A. (2007). TonB-dependent energy transduction between outer and cytoplasmic membranes. Biometals *20*, 453–465.

Pramanik, A., Zhang, F., Schwarz, H., Schreiber, F., and Braun, V. (2010). ExbB protein in the cytoplasmic membrane of *Escherichia coli* forms a stable oligomer. Biochemistry *49*, 8721–8728.

Pramanik, A., Hauf, W., Hoffmann, J., Cernescu, M., Brutschy, B., and Braun, V. (2011). Oligomeric structure of ExbB and ExbB-ExbD isolated from *Escherichia coli* as revealed by LILBID mass spectrometry. Biochemistry *50*, 8950–8956.

Rossi, M.-S., Paquelin, A., Gigo, J.M., and Wandersman, C. (2003). Haemophore-mediated signal transduction across the bacterial cell envelope in *Serratia marcescens*: the inducer and the transported substrate are different molecules. Mol. Microbiol. *48*, 1467–1480.

Roy, S., Douglas, C.W., and Stafford, G.P. (2010). A novel sialic acid utilization and uptake system in the periodontal pathogen *Tannerella forsythia*. J. Bacteriol. *192*, 2285–2293.

Sauter, A., and Braun, V. (2004). Defined inactive FecA derivatives mutated in functional domains of the OM signaling and transport protein of *Escherichia coli* K-12. J. Bacteriol. *186*, 5303–5310.

Sauter, A., Howard, S.P., and Braun, V. (2003). *In vivo* evidence for TonB dimerization. J. Bacteriol. *185*, 5747–5754.

Schöffler, H., and Braun, V. (1989). Transport across the OM of *Escherichia coli* K12 via the FhuA receptor is regulated by the TonB protein of the cytoplasmic membrane. Mol. Gen. Genet. *217*, 378–383.

Schauer, K., Gouget, B., Carrière, M., Labigne, A., and de Reuse, H. (2007). Novel nickel transport mechanism across the bacterial OM energized by the TonB/ExbB/ExbD machinery. Mol. Microbiol. *63*, 1054–1068.

Schauer, K., Rodionov, D.A., and de Reuse, H. (2008). New substrates for TonB-dependent transport: do we only see the 'tip of the iceberg'? Trends Biochem. Sci. *33*, 330–338.

Sharma, O., Yamashita, E., Zhalnina, M.V., Zakharov, S.D., Datsenko, K.A., Warner, B.L., and Cramer, W.A. (2007). Structure of the complex of the colicin E2 R-domain and its BtuB receptor. The OM colicin translocon. J. Biol. Chem. *282*, 23163–23170.

Sharma, O., Datsenko, K.A., Ess, S.C., Zhalnina, M.V., Wanner, B.L., and Cramer, W.A. (2009). Genome-wide screens: novel mechanisms of colicin import and cytotoxicity. Mol. Microbiol. *73*, 571–585.

Shi, Z., Chak, K.F., and Yuan, H.S. (2005). Identification of an essential cleavage site in ColE7 required for import and killing of cells. J. Biol. Chem. *280*, 24663–24668.

Smallwood, C.R., Marco, A.G., Xiao, Q., Trinh, V., Newton, S.M.C., and Klebba, P.E. (2009). Fluoresceination of FepA during colicin B killing: effects of temperature, toxin and TonB. Mol. Microbiol. 72, 1171–1180.

Stefanska, A.L., Fulston, M., Houge-Frydrych, C.S.V., Jones, J.J., and Warr, S.P. (2000). A potent seryl tRNA synthetase inhibitor SB-217452 isolated from a *Streptomyces* species. J. Antibiot. *53*, 1346–1353.

Stork, M., Bos, M.P., Jongerius, I., de Kok, N., Schilders, I., Weynants, V.E., Poolman, J.T., and Tommassen, J. (2010). An OM receptor of *Neisseria meningitidis* involved in zink acquisition with vaccine potential. PLoS Pathog. *7*, e100969.

Swayne, C., and Postle, K. (2011). Taking the *Escherichia coli* TonB transmembrane domain 'offline'? Nonprotonatable Asn substitutes fully for TonB His20. J. Bacteriol. *193*, 6393–6701.

Tiburzi, F., Imperi, F., and Visca, P. (2008). Intracellular levels and activity of PvdS, the major iron starvation sigma factor of *Pseudomonas aeruginosa*. Mol. Microbiol. *67*, 213–227.

Traub, I., Gaisser, S., and Braun, V. (1993). Activity domains of the TonB protein. Mol. Microbiol. *8*, 409–423.

Udho, E., Jakes, K.S., Buchanan, S.K., James, K.J., Jiang, X., Klebba, P.E., and Finkelstein, A. (2009). Reconstitution of bacterial OM TonB-dependent transporters in planar lipid bilayer membranes. Proc. Natl. Acad. Sci. U.S.A. *106*, 21990–21995.

Vanderpool, C.K., and Armstrong, S.K. (2003). Heme-responsive transcriptional activation of *Bordetella bhU* genes. J. Bacteriol. *185*, 909–917.

Visca, P., Imperi, F., and Lamont, I.L. (2006). Pyoverdine siderophores, from biogenesis to biosignificance. Trends Microbiol. *15*, 22–30.

Wandersman, C., and Stojiljkovic, I. (2000). Bacterial heme sources: the role of heme, heme protein receptors and hemophores. Curr. Opin. Microbiol. *3*, 215–220.

Wiener, M.C. (2005). TonB-dependent OM transport; going for Baroque. Curr. Opin. Struct. Biol. *15*, 394–400.

Wiener, M., Freymann, D., Ghosh, P., and Stroud, R.M. (1997). Crystal structure of colicin Ia. Nature *385*, 461–464.

Witty, M., Sanz, C., Shah, A., Grossmann, J.G., Perham, R.N., and Luisi, B. (2002). Structure of the periplasmic domain of *Pseudomonas aeruginosa* TolA: evidence for an evolutionary relationship with the TonB transporter protein. EMBO J. *21*, 4207–4218.

Yamashita, E., Zhalnina, M.V., Zakharov, D., Sharma, O., and Cramer, W. (2008). Crystal structures of the OmpF porin: function in a colicin translocon. EMBO J. *27*, 2171–2180.

Yue, W.W., Grizot, S., and Buchanan, S.K. (2003). Structural evidence for iron-free citrate and ferric citrate binding to the TonB-dependent OM transporter FecA. J. Mol. Biol. *332*, 353–368.

Zeth, K., Römer, C., Patzer, S.I., and Braun, V. (2008). Crystal structure of colicin M, a novel phosphatase specifically imported by *Escherichia coli*. J. Biol. Chem. *283*, 25324–25331.

The Permeability Barrier: Passive and Active Drug Passage Across Membranes 9

Kozhinjampara R. Mahendran, Robert Schulz, Helge Weingart, Ulrich Kleinekathöfer and Mathias Winterhalter

Abstract

Under antibiotic stress a reduced permeability for drugs to enter and an enhanced active extrusion of undesired compounds was observed in Gram-negative bacteria. With respect to the influx of drugs the first line of defence is the outer membrane containing a number of channel-forming proteins called porins allowing passive penetration of water-soluble compounds into the periplasmic space. Mutations in the constriction zone or a reduction of the number of porins reduce the flux. In addition porins coupled to efflux pumps allow the cell to eject antibiotics actively into the extracellular space. In order to understand the function of the involved proteins, a quantification of the individual transport elements is necessary. Here we describe experimental and computational biophysical methods to characterize molecular transport of antibiotics and small compounds across bacterial membranes.

Introduction

The cell envelope of Gram-negative bacteria consists of two membranes and typical pathways of antibiotic molecules through the bacterial cell wall are schematically shown in Fig. 9.1. First an antibiotic molecule must overcome the outer membrane (pathways *1* and *2*) consisting of an outer layer containing lipopolysaccharides (LPS) and an inner layer containing phospholipids (Simonet *et al.*, 2000; Nikaido, 2003; Pages *et al.*, 2008; Weingart *et al.*, 2008). Hydrophobic molecules (*2*) may dissolve in the lipid layer, accumulate there (*3*), and cross the membrane (*4*). Bacteria may minimize the permeation of hydrophobic agents by tight packing of the outer membrane. In contrast, water soluble drugs may benefit from hydrophilic pathways across porin channels (*1*) to effectively reach the target site. The influx of small water-soluble substances through porins can be lowered by reducing the number of the hydrophilic channels located in the outer membrane, by mutational changes reducing the size of the porin channels or in the selectivity of the constriction zone inside the channel. Once the drug reached the periplasmic space, enzymes may degrade the active compound (*5*). Depending on the target the drug eventually needs to translocate the inner membrane (*6*–*8*). At the same time, efflux pumps may harvest hydrophobic molecules and eject them actively against the gradient (*10*–*11*).

To date, the efficiencies of antibiotics are mostly characterized by their minimal inhibitory concentration (MIC). The MIC is defined as the lowest concentration of an antimicrobial

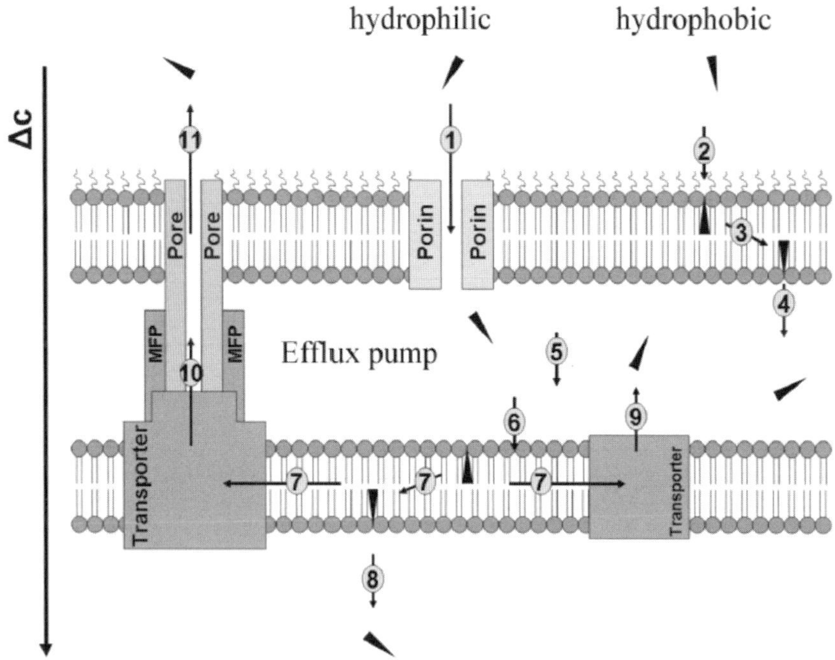

Figure 9.1 Schematic view of the different barriers along the pathway of an antibiotic from the outside to the cellular target. The drug follows the concentration gradient Δc. *1*, Entry of hydrophilic drugs through a channel; *2*, partitioning of hydrophobic drugs into the outer membrane; *3*, diffusion or flip-flop across the hydrophobic layer; *4*, partitioning into the periplasmic space; *5*, diffusion through the periplasmic space; *6*, partitioning into the inner membrane, *7*, lateral diffusion, flip-flop across the inner membrane; *8*, partitioning into the cytoplasm; *9*, active efflux into the periplasmic space, *1*, active efflux bypassing the periplasmic space, *11*, diffusion into the extracellular space. All processes are either in equilibrium or approaching the equilibrium. Only steps *9* and *10* require energy supply using the proton motive force. Adapted from Weingart *et al.* (2008).

compound sufficient to prevent growth of a microorganism. This value reflects the integral success of an antibiotic to cross all barriers and to fulfil its activity in a given type of bacteria.

Several resistance mechanisms may contribute to the MIC. Moreover, as antibiotic action is rather a kinetic effect, synergies between the 'players' may be expected. The molecular mechanism and potential cooperative effects cannot be deduced without detailed measurements of the individual components. Inhibition of one resistance mechanism by an antibiotic may increase the susceptibility of the bacterium to a specific antibacterial agent; however, it may also restore or enhance the activity of a second class of antibiotics. In order to optimize the drug efficiency it is crucial to quantify the individual contribution of each resistance mechanisms to the overall resistance of an organism. Below we describe the current state in direct evaluation of transport processes.

Lipid–drug interactions

To distinguish possible translocation pathways, the partitioning of the drug into a lipid can be measured spectroscopically or by using calorimetry (Rodrigues *et al.*, 2003; Howe *et al.*,

2007; Su and Yu, 2007). It should be noted that particular hydrophobic drugs rather partition and accumulate in the lipid membrane than showing a specific binding site with lipids. Moreover, the affinity may change in the presence of already bound molecules as in case of charged molecules. Thus, a binding constant does often not exist and instead a binding constant, one needs to determine a binding isotherm. Usually, such an analysis is not included in the standard software packages but it is straightforward (Schwarz *et al.*, 2007). Moreover, reaching the inner membrane, antibiotics may bind to efflux pumps. For example, the binding of hydrophobic compounds like ethidium to the multidrug efflux transporter EmrE has been studied by Isothermal Titration Calorimetry ITC (Sikora and Turner, 2005). In contrast to other binding assays under favourable conditions ITC allows the direct quantification of the entropic and enthalpic contributions to the binding. Another possibility for such a separation is to quantify the binding constant at different temperatures. Applying the Van't Hoff equation $\ln K = -\Delta G^{\circ}/RT$, relating the temperature-dependent binding constant K with the free standard enthalpy $-\Delta G^{\circ}$, allows one to separate entropic from enthalpic contributions. However, equilibrium measurements do not allow direct conclusions on the kinetics of transport.

Translocation across liposomes

Liposomes are used as model system for membrane transport using radioactive compounds for uptake assays. However, the time resolution is limited by the separation method necessary to distinguish free molecules from bound molecules and by the availability of radioactive-labelled substrates (Hong and Kaback, 1972; Michea-Hamzehpour *et al.*, 1991). Recently, a study has been presented which revealed the slow kinetics of an H,K-ATPase in oocytes. Atomic absorption spectroscopy allowed quantifying the uptake of Rb^{+} with a precision of 0.1 pmol (Dürr *et al.*, 2008).

Permeation of β-lactams across the lipid bilayer has been analysed by liposome swelling assays (Luckey and Nikaido, 1980; Nikaido and Rosenberg, 1983; Nikaido, 2003). This method detects a change in the optical density caused by osmotic swelling. For example, ampicillin does not permeate across lipid membranes. As β-lactams are water soluble, they may translocate more rapidly through membrane channels. Subsequently, iso-osmotic addition of millimolar concentrations of β-lactams to a liposome preparation does not change the internal osmotic pressure and the optical density will not change. It is expected that the permeation is even less for tighter LPS-lipid membranes but to date this has not been characterized. In contrast, hydrophobic antibiotics like some of the quinolones (e.g. nalidixic acid) permeate significantly across lipid bilayers. Permeation rates of a large variety of antibiotics have been observed by measuring the swelling of liposomes containing reconstituted porins (see Table 9.1). It should be noted that the success of this type of experiment depends crucially on the sample preparation and small variation will result in arbitrary values. Moreover, the swelling assay provides only relative numbers to assess how the translocation varies from one compound to the other for one preparation. In principle, quantification is possible but rather tedious. In addition, the underlying effect is still under debate. It is expected that an increase in osmotic pressure should stretch the liposome giving rise to an enhanced scattering or enhanced optical density. However, all experiments revealed a reduction in optical density. From mechanical stretching experiment it is known that liposome will leak or rupture after stretching by 2–4% (Needham and Nunn, 1990). We performed a number of

Table 9.1 Permeation of antibiotics through bacterial porins. Various antibiotics have been tested for possible interactions with the general diffusion porins OmpF and OmpC from *E. coli* using ion current-fluctuation analysis. In brackets, the values for the corresponding liposome swelling assays are indicated (data from Weingart *et al.*, 2008).

Antibiotics	
Penicillins	OmpF
Ampicillin	Yes (46)
Amoxicillin	Yes
Carbenicillin	No translocation (5)
Azlocillin	No translocation
Piperacillin	No translocation (<5)
Penicillin G	Translocation could not be resolved
Fluoroquinolones	
Moxifloxacin	Yes
Nalidixic acid	No translocation
Enrofloxacin	Yes

tests to probe for the possible underlying mechanism. It seems that the increasing osmotic pressure cause leakage and thus a decreasing optical density. Thus, the influx is creating a stress causing rupture accompanied by a reduction in optical density. In this case, we expect a large scattering as the interlayer distance will vary on buffer composition and multilamellar liposome preparation.

Characterization of active energy-requiring transport is also possible. An elegant method is to create a transmembrane electric potential to drive an efflux pump. For example, transport proteins are reconstituted in a potassium buffer into liposomes and dialysed with a potassium free buffer. Addition of valinomycin will increase the permeability of the lipid membrane for potassium and thus create a gradient able to drive the transporter (Nikaido *et al.*, 1998; Ward *et al.*, 2001). Similar systems have been used to characterize a variety of uptake systems including efflux pumps (Benz *et al.*, 1978; Nishino *et al.*, 2003; Gbaguidi *et al.*, 2007).

Permeation revealed by electrophysiology

The method of choice to characterize channel-forming proteins is electrophysiology. As a lipid membrane is a very good insulator the extremely low conductance of a pure lipid membrane compared to a membrane in presence of a channel provides an outstanding signal-to-noise ratio allowing detection of Angstrom-sized defects. Unfortunately, the small size of bacteria and the presence of a thick hydrophilic LPS layer does not allow direct recording of porin channels through patching a natural membrane leaflet (Patch-Clamp). However, isolation and reconstitution of membrane channels into planar lipid membranes has been successfully in practice since almost 35 years (Benz *et al.*, 1978). Briefly, an orifice ranging from 50–100 μm is pretreated with an organic solvent rendering the pore area lipophilic (Fig. 9.2). In one technique, known as Mueller–Rudin technique, the membrane lipid is dissolved in hydrocarbon (decane or similar solvent) and brushed across the hole

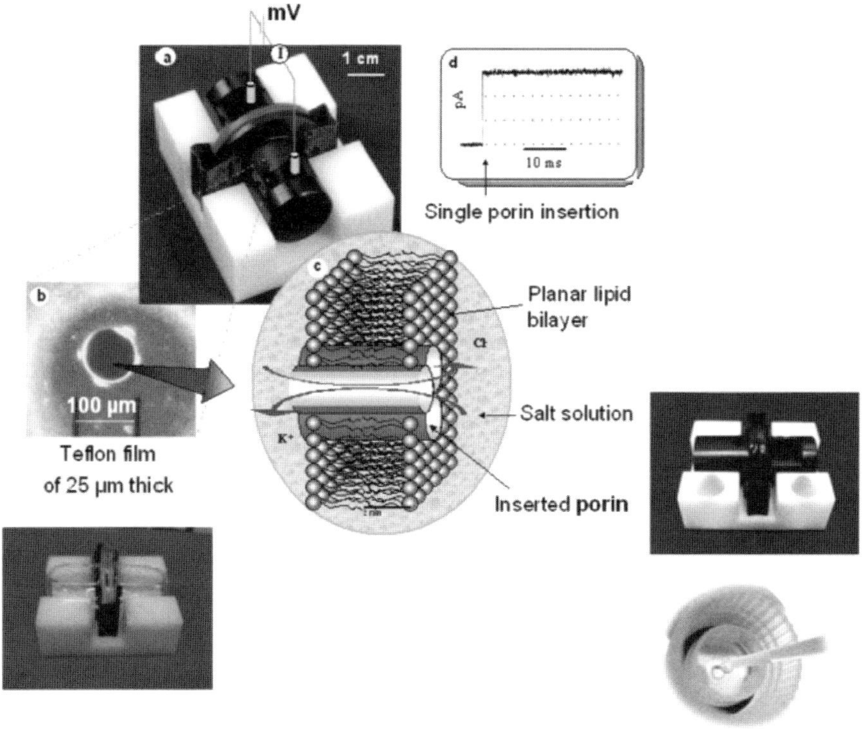

Figure 9.2 Set-up of the planar lipid bilayer. A planar bilayer is formed across an aperture separating two aqueous solutions. A single porin spontaneously inserts, allowing a constant ionic current. (a) Measurement cell. The volume of each cell is about 250 µl. Further miniaturization is possible; (b) View of the hole under the microscope; (c) Scheme of the bilayer with an incorporated porin; (d) Typical current recording of a single porin insertion (adapted from Danelon *et al.*, 2003, and Schwarz *et al.*, 2003). Below typical cuvettes are shown: left side, 3-ml cuvette with a Teflon septum in the middle; in the middle, two different half cells made of delrine (0.2ml and 1.5ml); right side, a glass chip (Nanion) requiring only a few microlitres on both sides.

(Mueller and Rudin, 1963). These so-called painted membranes are considered to be more flexible and allow easy reconstitution of channels. Later this technique was adopted by Montal and Mueller (Montal and Mueller, 1972). Using this technique the lipids are spread with organic solvent across the air water interface. The membrane is formed by lowering and raising the lipid monolayer across the hole. During the past fifty years both basic approaches have been adapted to particular needs coming either from the design of the set-up or from requirements for the protein. Typically, the smaller the membrane, the larger is its stability which allows longer recordings but requires more patience during protein reconstitution. During most set-ups, the capacity of the system is limited by the capacity of the lipid membrane. Subsequently, better time resolutions can be achieved using smaller membranes. In addition, an acoustic screening of the set-up, in combination with a small membrane area and an anti-vibration table enables ion conductance measurements with time resolution better than milliseconds (Nestorovich *et al.*, 2002; Danelon *et al.*, 2006). With the upcoming nanotechnology, micrometre-sized holes have been drilled reproducibly into glass slides

(Kreir *et al.*, 2008). Here the lipid membrane is attached via liposome adsorption to the glass support. Another approach is based on patch clamping. Here the micropipettes are pulled down to 100 nm diameter. The latter methods may reduce the membrane capacity down to picofarads and subsequently, the time resolution is only limited by the amplifier (Gornall *et al.*, 2011).

Channel reconstitution is usually done by adding very small amounts of proteins stock solutions containing usually a detergent at a concentration above the critical micellar concentration (CMC). Addition of proteins into the aqueous phase causes a rapid dilution below the CMC causing protein aggregation and precipitation. Only a few porins will spontaneously insert into the lipid bilayer. Depending on the concentration of the protein stock, the first insertion will occur after a few minutes. To avoid multiple insertions the cuvette is quickly flushed with buffer. Insertion of a porin is visible by a sudden jump in the ionic current. More experience is required for reconstitution of planar lipid bilayers from giant proteoliposomes. Here an unknown number of channels will be present in the lipid bilayer which forms across the hole. The particular difficulty of this approach is to distinguish the conductance of an unknown channel from the leak current.

Application of a transmembrane voltage induces an ion current flowing through the porin channel. In a first approximation this current is proportional to channel size. In a simplified manner the channel conductance G is given by

$$G = \kappa A / l$$

with κ as the bulk ion conductance, A the pore area and l the pore length. However, in particular for narrow channels a substantial deviation occurs. The conductance of large channels is mainly governed by the bulk conductance, whereas the pore conductance of smaller channels is predominantly influenced by the surface pattern of the channel wall. Depending on the surface charge pattern, selectivity for cations or anions is observed. Numerous theoretical models have been suggested allowing an interpretation of the experimental data. For OmpF and OmpC high resolution crystal structures are available. Together with a new generation of powerful computer and new software, the modelling of the transport pathway became possible (Aksimentiev and Schulten, 2005; Sotomayor *et al.*, 2007; Chimerel *et al.*, 2008; Pezeshki *et al.*, 2009; Biro *et al.*, 2010; Mahendran *et al.*, 2010). Molecular dynamics simulations allow the investigation of a number of processes at atomic level by integrating Newton's equation of motion. However, the current available computation time limits the system to about 100 000 atoms and a few tens of nanoseconds. Application of comparable high voltages will allow enough ions to pass the channel within the simulation time to achieve reasonable statistics. Nevertheless, the modelling of ion fluxes is in good agreement with the experimentally measured conductance (Fig. 9.3a). Moreover, even ion selectivity can be approximately calculated (Fig. 9.3b). Not only the conduction of simple salts like KCl can be modelled but also the transport of more complicated ions like the ionic liquid 1-butyl-3-methyl-imidazolium chloride (BMIM-Cl) in aqueous solution (Modi *et al.*, 2011). Again good agreement with experimental findings is obtained explaining at atomic detail features like reorientation of the BMIM ions before passing the constriction zone of OmpF which are inaccessible with current experimental techniques.

The typical size of a porin channel is large enough to cause an ion current of several hundreds of picoamperes. Thus, it is possible to detect the presence of single porins in a

a

b

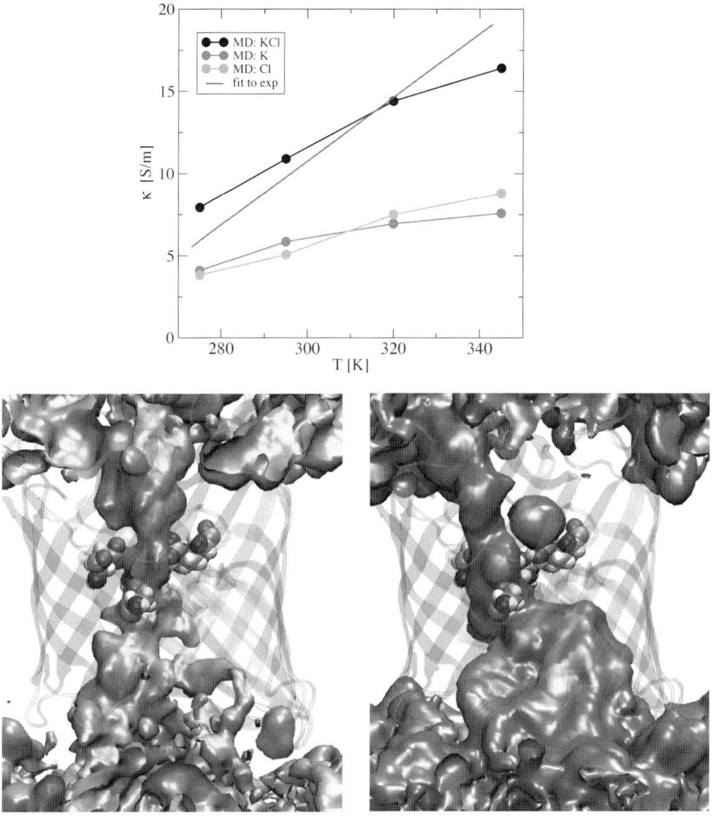

Figure 9.3 (a) Temperature-dependent conductivity of KCl in bulk water. Shown are the experimental findings compared to MD simulations. In the simulations one also gets the individual currents of the two ion types. For details see Pezeshki *et al.* (2009). (b) Iso-density surfaces of the Cl– (left) and K+ ion densities (right) in an OmpF channel averaged over the full trajectory. The important residues in the constriction zone are shown as well. For details see Pezeshki *et al.* (2009). A colour version of this figure is located in the plate section at the back of the book.

lipid bilayer by standard conductance measurements. Carbohydrates or antibiotics are often neutral and their permeation through a membrane does not provide a direct electric signal. However, under the particular condition that the penetration of uncharged molecules into the porin channel causes a detectable reduction in conductance, hereby induced fluctuation in the ion current can be used as a sensor for the binding. Almost 20 years ago the group of Roland Benz introduced such an analysis of the ion current fluctuation as a measure of binding (Benz *et al.*, 1985; Nekolla *et al.*, 1994). The penetration of sugars into the maltose-specific channel maltoporin blocks the channel for several milliseconds, thereby interrupting the ion current in a concentration-dependent manner. In a multichannel experiment the power density spectra can be fitted to a Lorentzian with the corner frequency as fitting parameter. Chemical relaxation theory relates the concentration-dependent corner frequency to the on- and off-rate. Such analysis can also be applied to a single trimer of the porin (Danelon *et al.*, 2003; Schwarz *et al.*, 2003). Here, the ion current of a single channel is interrupted and the on- and off-rate can be obtained by a simple statistical analysis. Both approaches are equivalent. A single molecule analysis allows revealing deviations in the behaviour. Reconstitution of many channels in large membranes is often accompanied by certain randomness in the orientation of reconstituted proteins. Obviously single-channel recordings account for asymmetries within a protein or variations in the substrate affinity from protein to protein, for example due to partial denaturation or misfolding of the proteins, to be easily discriminated (Kullman *et al.*, 2006). Blocking of the ion current after addition of a compound to a porin-containing bilayer reveals binding sites present in the porin channel; however, the reverse conclusion is not possible as binding of the compound to the porin does not have to reduce the ion current if the channel is not noticeable blocked for ions, for example, if the binding site is at the entry of the channel.

The above described method was adopted to analyse the interaction of antibiotics with outer membrane proteins on a single-molecular level. Time-dependent ion current fluctuations can be analysed to identify binding events. If the channel has a similar size as the diffusing particle, the ion flow will be interrupted for a short period. Fig. 9.4a shows the

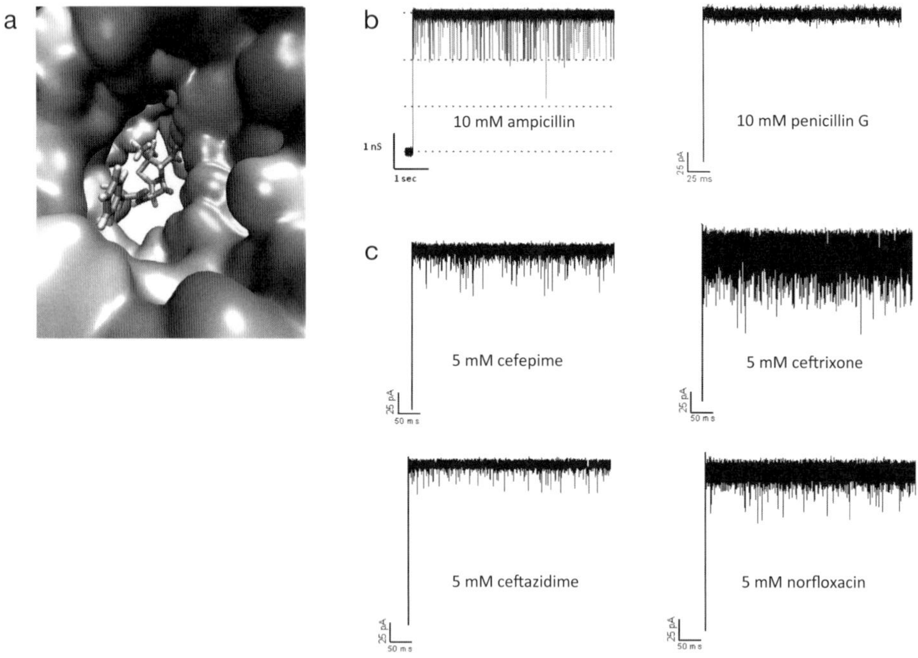

Figure 9.4 (a) Schematic representation of antibiotic translocation through the constriction zone of a single OmpF channel. (b) Typical tracks of ion currents through single trimeric OmpF channels reconstituted into planar lipid membranes in the presence of zwitterionic and anionic penicillins. Membrane bathing solutions contained 1 M NaCl (pH 5.0) and 5.7mM of the indicated antibiotic, and the applied voltage was 100mV (adopted from Danelon *et al.*, 2006). (c) Typical tracks of ion conductance through single trimeric OmpF channels reconstituted into miniaturized free standing bilayer setup in the presence of cephalosporins and fluoroquinolones. The membrane was formed from DPhPC; membrane bathing solutions contained 1 M KCl, pH6 and few millimolar antibiotics. Applied voltage was –50mV (adapted from Mahendran *et al.*, 2010). A colour version of this figure is located in the plate section at the back of the book.

constriction zone of OmpF harbouring an ampicillin molecule. The binding of the antibiotic to the constriction zone inside the porin occludes the channel for ions which is detectable by fluctuation in the ion current, reflecting the molecular interactions with the channel wall. Typical beta-lactam antibiotics are shown in Fig. 9.4b giving raise to short interruption in the ion current during their interaction with the *E. coli* porin OmpF. Increasing the concentration of the antibiotic increases the number of blocking events but not the mean residence time of the antibiotic molecules inside the pore. More recently a new method was introduced for the formation of lipid bilayers on a glass substrate with an approximately 1 μm size aperture. Stable lipid bilayers are formed by bursting a giant unilamellar vesicle (GUV) on the glass surface forming a free-standing portion above the hole (Mahendran *et al.*, 2010). OmpF can be directly reconstituted into GUVs before the formation of the lipid bilayer, resulting in patching and subsequent rupture of proteo-GUVs. Single channel recordings in the presence of the antibiotics are possible by perfusing the chip with the relevant antibiotic solution (Fig. 9.4c). There are several advantages of using patch clamp chips containing small micrometre-sized holes compared with the classical Montal–Mueller membrane of larger size. The chip system allows successful high resolution single channel measurements, rapid perfusion of antibiotics and continuous measurements over a long time. In addition, lipid bilayers formed on this system were stable up to 500 mV. With the help of the perfusion system, we could rapidly screen different antibiotics in the same single trimeric channel. Developing this higher throughput screening possibility for antibiotics could place planar lipid bilayers as a first step in screening new antibiotics on a larger scale.

Translocation through a channel: limiting factors

OmpF and OmpC, the major outer membrane channels found in *E. coli*, are attractive candidates for the study of the influence of the channel surface on the permeation of antibiotics. For both a high resolution crystal structure is available which reveals a sixteen β-strands trimeric channel with a pore size of about 1 nm, as expected from earlier conductance studies (Basle *et al.*, 2006). The fundamental question is how to design antibiotic molecules in order to optimize their influx through the porin channels. In a macroscopic approach the flux of molecules J (substrate molecules per unit time) through a cylinder is given by

$$J = -(D\pi a^2/l)\Delta c \tag{9.1}$$

with D as the diffusion constant of the particle, usually in the order of 10^{-9} to $10^{-10}\,m^2/s$ in aqueous solution depending on the viscosity and size of the drug, a is the pore radius of about 0.5 nm, l is the approximate channel length of about 4 nm and Δc is the concentration gradient. Inserting the above-mentioned approximate values and presuming a concentration gradient of $\Delta c = 1\,\mu M$ results in a flux of about 10–100 molecules per second. This corresponds to the fastest possible permeation through a cylinder of this size and is limited only by diffusion towards the channel. However, at distances below 1 nm molecular interaction with the channel becomes dominant. This is reflected by the molecular specificity: evidence suggests that it is the chemical structure, rather than the size, that exerts the dominant influence on permeation pathways (Berezhkovskii and Bezrukov, 2005). Indirect measurements on maltose translocation through maltoporin revealed two orders of magnitudes slower translocation compared to free diffusion (Danelon *et al.*, 2003).

Recently the affinity of molecules with the porin channel has been measured (Nestorovich *et al.*, 2002; Danelon *et al.*, 2006; Mahendran *et al.*, 2010). An analysis of the equilibrium fluctuation readily provides the chemical rates for entering and leaving the binding site inside the channel. Determining the on- and off-rates allows the net permeation to be obtained. In a similar study, the analysis of maltose translocation through maltoporin leads to the development of a symmetric one-binding-site model (Nekolla *et al.*, 1994; Schwarz *et al.*, 2003). The specific experimental conditions of one trimer surrounded by a few millilitre of buffer solution allow quasi-equilibrium measurements to be performed. To simplify, we assume a symmetric barrier and a one side drug addition giving rise to a concentration gradient. The measured rates can be included in equation 9.1 in which the geometrical factors are replaced by kinetic rates. The flux is given by

$$J = [K \times k_{off} / (2 + K\,\Delta c)]\ \Delta c \approx [k_{on} / (2 + K\,\Delta c)]\ \Delta c$$

where K is the binding constant and k_{off}, and k_{on} are the off- and on-rate, respectively (Danelon *et al.*, 2003; Mahendran *et al.*, 2010). The net flux J, in molecules per second, is at low concentration (non-saturated condition) proportional to the concentration gradient Δc. It should be noted that for a strong binding the permeation is limited by the off-rate and that for a large gradient Δc the flux is independent on the concentration gradient. More sophisticated models have been suggested and allow the inclusion of detailed binding parameters (Berezhkovskii and Bezrukov, 2005; Bauer and Nadler, 2006; Bezrukov *et al.*, 2007). However, within the limited accuracy of the measurements and the limited knowledge of molecular parameters the above simplified model may satisfy most of the current needs. It is also interesting to discuss the relevance of substrate binding for translocation. Inspection of the experimental values for the on- and off-rate in the biological relevant concentration gradient range (μM) yields that $k_{off} >> k_{on}\ \Delta c$. This simplifies the flux to $J = k_{on}\ \Delta c/2$. In other words, increasing the on-rate will directly result in an increase in translocation and subsequently, the drug efficiency should increase. The translocation of several β-lactams (carbapenems, cephalosporins) through Omp36 (a homologue of OmpC from *Enterobacter aerogenes*) has been quantified (James *et al.*, 2009). Inserting experimental values for $k_{on} = 10^6$/M s and $K = 1$ M s yields about 3 molecules/second at 1 μM concentration gradient per monomer. This value is about a factor of ten lower than the above derived diffusion limited flux. In order to discuss the uptake efficiency of a given drug, one has to understand more about the molecular interactions involved in diffusion at the nanoscale level, for which molecular details are lacking.

Antibiotic transport across membranes

Compared to ion conductance, which is in the pA to nA range corresponding to more than 10^6 events per second, antibiotic translocation occurs less than hundred times per second. Owing to these statistics, antibiotics translocation is not yet accessible by standard MD simulations and would require a less detailed approach like a coarse-grained description. An alternative solution to catch such rare-events is called metadynamics and is based on an algorithm that limits backward motions. In a first investigation, Ceccarelli and coworkers applied a new algorithm to elucidate possible pathways of the antibiotic through the channel. In this way, the pathway of five different β-lactams (ampicillin, amoxicillin, piperacillin, azlocillin and carbenicillin) through OmpF has been evaluated (Ceccarelli *et al.*, 2004; Danelon *et al.*, 2006). It is interesting to note that the mid-channel constriction loop L3 of OmpF hardly moves at all in the presence of antibiotics. Azlocillin and piperacillin did not permeate through the channel, likely because they are too bulky. In contrast, ampicillin and amoxicillin permeate nicely and two binding pockets have been elucidated. The energy gain related to the binding pocket has to be related to the kinetics of permeation. This is approximated following the transition state theory through

$$k = (1/\tau)\exp(-\Delta G/k_B T)$$

with k being the corresponding rate constant and $1/\tau$ as the pre-exponential. In the exponent ΔG denotes the energy difference calculated from the molecular modelling and $k_B T$ the thermal energy. Originally, this theory was derived for reactions in the gas phase and in this scenario the pre-exponential factor corresponds to the number of hits. In a condensed phase environment this values is to be replaced by a more complex relation. Another approach to model antibiotic translocation through porins is to apply a force to the antibiotic molecule and drag it through the channel. In both cases the pathway of the antibiotic molecule through the channel and possible interactions with the amino acid residues of the channel wall can be obtained. However, conclusions from simulations must be taken with a certain caution because the permeation values obtained by these approaches are in the nanosecond range and have to be extrapolated into the experimentally available millisecond range. Nevertheless, current efforts to model the flux of ions through OmpF suggest that this approach is able to predict macroscopic observables like ion currents.

Contribution of efflux pumps

Efflux pumps belonging to the resistance-nodulation-cell division (RND) family play a key role in multidrug resistance in Gram-negative bacteria, which have a cell envelope made up of two membranes. In contrast to specific transporters which mediate the extrusion of a given drug or class of drugs, RND-type pumps are so-called multidrug efflux transporters able to transport a wide range of structurally unrelated compounds. Most of the transporters belonging to the RND family interact with a membrane fusion protein (MFP) and an outer membrane protein (OMP) to allow drug transport across both the inner and the outer membranes of Gram-negative bacteria (Tseng *et al.*, 1999; Zgurskaya and Nikaido, 2000). In *Escherichia coli*, the most widely studied prokaryotic model organism, the RND-type pump AcrAB-TolC has been recognized as one of the major multidrug efflux systems (Okusu *et al.*, 1996). The inner membrane component, AcrB, is a proton-dependent drug antiporter that determines the substrate specificity and actively exports the substrates. AcrB cooperates with an outer membrane channel, TolC and a membrane fusion protein, AcrA (Tikhonova and Zgurskaya, 2004). Crystallographic studies of AcrB crystals described an asymmetric trimer in which each monomer was suggested to correspond to a distinct functional state of a proposed three-step transport cycle reminiscent of a peristaltic pump (Murakami *et al.*, 2006; Seeger *et al.*, 2006). The AcrB monomer in the 'access' or 'loose' state forms a tunnel which is open to the periplasm. The tunnel starts just above the outer leaflet of the cytoplasmic membrane and extends halfway towards the centre of the trimer allowing potential substrates to access the transporter. In the 'binding' or 'tight' state, the tunnel extends diagonally upwards towards a voluminous hydrophobic binding pocket. The conversion from the 'binding' to the 'extrusion' or 'open' state leads to a complete closure of the entry tunnel and creates an exit guiding the substrate towards TolC. These conformational changes are driven by a proton translocation across the membrane. The crystal structure for TolC revealed a homotrimer forming a hollow tapered cylinder called a channel-tunnel (Koronakis *et al.*, 2004). TolC is anchored in the outer membrane by a β-barrel domain spanning the outer membrane. The tunnel domain is formed by four α-helical strands per monomer and projects into the periplasmic space. A third domain, a mixed α/β-structure forms a strap around the equator of the tunnel. The assembled channel forms a single pore with a diameter of 35 Å.

The membrane fusion protein AcrA establishes a functional connection between AcrB and TolC in the periplasm. AcrA has an elongated structure comprising a β-barrel, a lipoyl domain, and a long α-helical hairpin that projects into the periplasm (Mikolosko *et al.*, 2006) In particular, the α-hairpin domain of AcrA is responsible for binding to TolC, whereas the other domains bind to AcrB (Husain *et al.*, 2004; Touze *et al.*, 2004). Cross-linking studies indicate that the α-helical coiled-coil hairpin of AcrA fits into intraprotomer grooves of the open state of TolC, whereas its β-lipoyl domains interact with the periplasmic domain of AcrB at the interface between adjacent subunits (Fig. 9.5) (Lobedanz *et al.*, 2007; Symmons *et al.*, 2009). In contrast, recent studies using a chimeric AcrA in which the α-helical domain was substituted with that of MacA (MacA is the membrane fusion protein of the macrolide-specific efflux pump MacAB-TolC) indicate that the AcrA forms a hexameric funnel-like channel that covers the periplasmic region of AcrB. In these studies, a tip-to-tip binding of the AcrA α-hairpin domain and the periplasmic tip of TolC was suggested (Xu *et al.*, 2011).

Characterization of efflux

The search for inhibitors of multidrug efflux pumps is pharmacologically important; compounds able to block such transporter may restore the susceptibility of resistant clinical strains to antibiotics. The efficiencies of such inhibitors are mainly determined by their effect on the minimal inhibitory concentration (MIC) values. As discussed above, the MIC value reflects the integral success of an antibiotic to cross all barriers and to fulfil its activity on a given target inside a bacterial cell. In order to quantify the contribution of efflux pumps to the resistance of an organism, accumulation assays using fluorescent dyes are frequently used. Comparisons between the fluorescence of the wild-type and AcrAB-TolC deletion strains in the presence of dyes added to the medium demonstrate the activity of the efflux pump. An increased cellular fluorescence after addition of putative efflux pump inhibitors to the assay mixture indicates inhibitory activity against the efflux pumps. Another approach is to preload energy-depleted cells with dyes in the presence of the proton conductor carbonyl cyanide *m*-chlorophenylhydrazone (CCCP). After removal of the CCCP, efflux is triggered by energization of the cell with glucose leading to a time-dependent decrease in fluorescence intensity. Efflux pump inhibitors and antimicrobials can be tested for the ability to slow down efflux of the dye. Lipophilic membrane-partitioning dyes, e.g. DASPEI (Murakami *et al.*, 2004) or Nile Red (Bohnert *et al.*, 2010) were shown to be well suited for this type of efflux assay. However, accumulation assays can give only indirect evidence of AcrAB-TolC activity and often additional efflux systems are present in a given cell. These problems could be circumvented by performing efflux assays with purified proteins reconstituted into proteoliposomes. Zgurskaya and Nikaido (1999) reconstituted AcrB into proteoliposomes. In the presence of a pH gradient as energy source, AcrB catalysed the

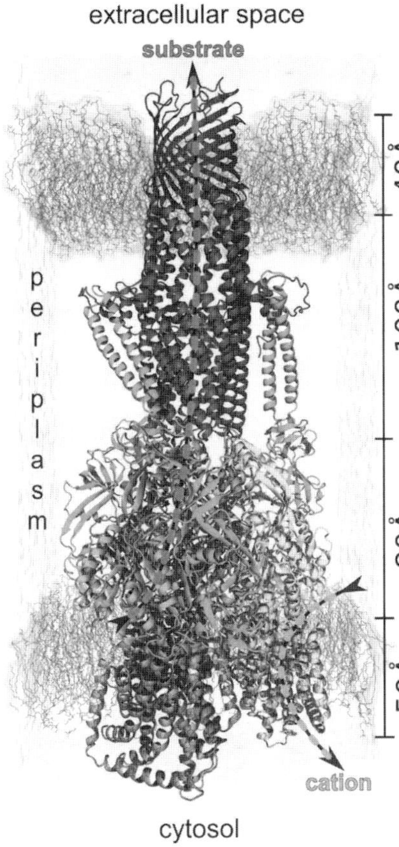

Figure 9.5 The tripartite efflux system ArcAB-TolC spanning two membranes according to the model of Symmons *et al.* (2009). A colour version of this figure is located in the plate section at the back of the book.

extrusion of fluorescent substrates from the liposomes. However, attempts to demonstrate directly the efflux of lipophilic dyes from within the liposome were mostly unsuccessful, presumably because they re-entered spontaneously into the bilayer. Therefore, the authors incorporated fluorescent derivatives of phosphatidylethanolamine into the AcrB-containing vesicles at high concentrations so that the fluorescence was quenched. Active extrusion of these fluorescent phospholipids by AcrB decreased their concentrations in the bilayer. This leads to a decreased quenching and the liposomes increased fluorescence. The re-entry of expelled fluorescent phospholipids into the bilayer of the original vesicles was prevented by addition of protein-free 'trap' vesicles in excess that contained no fluorescent lipids. Known substrates of AcrAB, including lipophilic antibiotics and bile salts inhibited the extrusion of fluorescent phospholipids by AcrB.

However, in order to quantify the contribution of multidrug efflux to drug resistance of an organism, knowledge of its kinetic behaviour is essential. Su *et al.* (Su and Yu, 2007) used a steady-state fluorescence polarization assay to determine the affinity of four fluorescent ligands to purified AcrB in a detergent environment. The principle underpinning this assay is that the polarization of fluorescence of a ligand bound to the AcrB protein increases due to a decrease of its rotational motion. The obtained results indicate that the four ligands bind to AcrB with K_D values of 5.5 μM (rhodamine 6G), 8.7 μM (ethidium), 14.5 μM (proflavin), and 74.1 μM (ciprofloxacin), respectively.

However, it is unclear whether assays using the isolated AcrB protein reflect the behaviour of the protein in this tripartite protein complex which the transporter forms with an outer membrane channel and a periplasmic adaptor protein spanning both the cytoplasmic and the outer membrane. Recently, two studies described an approach to estimate kinetic constants of the AcrB pump using intact cells, in which the tripartite structure of the efflux system is maintained (Nagano and Nikaido, 2009; Lim and Nikaido, 2010).The difficulty with assays using whole cells is to determine the concentration of the drug within the periplasm of the bacterial cells, where the drugs enter the transporter. The authors solved this problem by assessing the periplasmic concentration of the drug from the rate of hydrolysis by a periplasmic β-lactamase and the rate of efflux as the difference between the influx rate and the hydrolysis rate. When a β-lactam is added to the external medium at a given concentration C_o, it diffuses spontaneously across the outer membrane into the periplasm. A certain number of molecules will be expelled by the AcrAB-TolC system. Other antibiotic molecules will be hydrolysed by the periplasmic β-lactamases. The authors determined the periplasmic hydrolysis (V_h) of the β-lactams spectrophotometrically, and from the known Michaelis–Menten constants of the enzyme, they could calculate the periplasmic concentration of the drug (C_p). The influx of the drugs across the outer membrane (V_{in}) was calculated by using the permeability coefficient (P). The permeability coefficient was determined by using cells in which the AcrB pump was inactivated by de-energization with the proton uncoupler carbonyl cyanide *m*-chlorophenylhydrazone (CCCP). According to Fick's first law of diffusion, the influx can be calculated using the following equation

$$V_{in} = P \times A \times (C_o - C_p)$$

where A denotes the surface area of the cells. Finally, the rate of efflux, V_e, is the difference between the (calculated) V_{in} and the (measured) V_h. However, the authors had to use an optimized *E. coli* strain to carry out the assay with the required precision. To ensure efficient influx of the drugs, the *E. coli* strain was modified to produce an outer membrane porin with an expanded channel. The gene encoding the β-lactamase was introduced on a plasmid. The K_m of the β-lactamase must be reasonably high, so that V_h becomes a sensitive indicator of C_p of the drug. The strain was further improved by increasing the expression of the AcrAB pump by the inactivation of the *acrR* repressor gene.

Using this approach, the kinetic behaviour of several β-lactams in efflux mediated by the AcrAB-TolC system of *E. coli* was analysed. The penicillins tested had a much stronger affinity to the efflux pump and higher efflux rates than the cephalosporins. The K_m values of the penicillins were about 1 μM, whereas those of the cephalosporins were between 5 to 300 μM. The results are in line with the hypothesis that the more lipophilic penicillins are better substrates of the AcrAB pump. A comparison of various β-lactam antibiotics showed a quantitative relationship between the β-lactam side chain lipophilicity and MIC values (Nikaido *et al.*, 1998).

The usual criterion to evaluate the impact of efflux on antibiotic resistance is to compare the MIC of the wild-type strain with that of the pump-deficient mutant. Some drugs appear to be good substrates of an efflux pump, because deletion of the pump significantly decreased their MIC. However, the studies of Nagano and Nikaido (2009) and Lim and Nikaido (2010) on the kinetic behaviour of AcrAB revealed that some compounds are good substrates of this system although the deletion of the *acrB* gene had no effect on their MIC. For example, the efflux kinetics of cloxacillin whose MIC decreased 512-fold in an *acrB*-deficient *E. coli* strain, were quite similar to those of ampicillin, whose MIC decreased only 2-fold. Analysis of this phenotype showed that the decrease in the MIC for cloxacillin is primarily due to the low permeation of the drug. These results lead to the conclusion that the synergy with the outer membrane permeability barrier makes RND-type multidrug efflux pumps so effective.

Molecular modelling of efflux pumps

Within the last years, computational scientists have started to evaluate different approaches to complement the experimental results on efflux pumps on a more detailed, atomistic level. Therefore, the major tools are all-atom MD simulations as well as so-called coarse-grained models of the proteins. While the first approach keeps the atomistic structure information of the X-ray crystal structures, the latter one simplifies these data by merging several atoms into one pseudo-particle, thereby decreasing the computational demand and increasing the feasible simulation time scale. Both of them can help to understand the crucial conformational changes that lead to the ejection of noxious compounds, such as antibiotics.

Owing to the complexity of such a protein complex, each single protein has been investigated separately to understand its structure–function relationship without its partners in most of the publications. Since the membrane-fusion protein is structurally rather simple and is supposed to stabilize the complex, it is more important to investigate it in complex with at least one of the partners. Nevertheless, an initial study of its structural elements and their flexibility was performed by Vaccaro *et al.* (2006). The outer membrane channel was mainly studied focusing on the opening of the periplasmic aperture as well as the flexibility of the extracellular loops. While the periplasmic tip is closed without the presence of the protein partners, the outer loops seem to be able to disturb the influx of molecules by blocking the entrance, which is established by hydrogen bonding between polar residues on each of the three loops (Vaccaro *et al.*, 2008). In one study, the wild-type structure was compared with mutated structures at the aperture (Schulz and Kleinekathöfer, 2009). These mutated structures opened during a few tenths of nanoseconds due to the application of an electric field. Interestingly, the ions of the solution were able to restabilize the hydrogen bond network at the periplasmic tip which was initially disturbed by the mutation. The applied electric field pushed these ions away from these regions allowing the aperture to open. The dynamics of the transporter of such an efflux pump are more difficult to understand. These transporters are usually driven by the proton gradient across the inner membrane. To induce the large transitions for the pumping, the local conformational changes by the de-/protonation in the transmembrane domain need to be transmitted towards the pumping domain. This is one of the issues which are not yet fully understood. Based on a functional rotation hypothesis, which was deduced from an asymmetric structure obtained by X-ray crystallography (Sennhauser *et al.*, 2007), the individual monomers undergo transitions between three states of a cycle. These transitions are driven by the previously mentioned proton gradient. Using these pieces of information, different approaches have been used to enhance the understanding of this transporter.

In two of the studies, the cycle transitions mentioned above have been induced using so-called targeted MD simulations to observe their effects on an unsteered substrate molecule initially in the binding pocket. As proposed by crystallographers, a peristaltic pumping was observed that squeezed the drug out of the binding pocket (Schulz *et al.*, 2010). Moreover, a water stream was found from the entrance towards the binding pocket and then towards the exit gate at the domain docked to the outer membrane channel (Schulz *et al.*, 2011). Such an analysis of water flow was also recently performed on the transmembrane domain by Fischer *et al.* (Fischer and Kandt, 2011). Therein, three distinct influx pathways from the periplasm and one towards the cytoplasm have been found in the three monomers of an asymmetric structure using unbiased MD simulations.

In addition to these computational studies, there are two studies which combine docking approaches with either experimental methods (Takatsuka *et al.*, 2010) or extended MD methods (Vargiu *et al.*, 2011) to understand the relation between structure and substrate binding in more detail. While the first study described the different binding spots for various substrates, the latter indicated specific interactions of the amino acids in the binding pocket with a substrate. This study yielded strong hints why the F610A mutation changes the MIC values for certain substrates so strongly in experiments. Another, less detailed approach was used by two groups applying different methods of coarse graining to study this transporter (Yao *et al.*, 2010; Wang *et al.*, 2011). While Yao *et al.* focused on AcrB and a probabilistic description of the transition between the states of the cycle, Wang *et al.* described the conformational couplings in the tripartite complex. All of these studies highlighted the importance of associating the computational findings with experimental ones and vice versa.

Conclusions

During recent years we gained a first molecular view on transport processes across protein channels. High resolution electrophysiology readily provides information on possible interaction with the channel interior. Miniaturization enhances the time resolution and allows even screening of a larger number of molecules. However, interaction does not allow direct conclusions on translocation. In case of charged molecules some conclusions on translocation can be made whereas for uncharged molecules no direct distinction can be made. Moreover, the binding site can be out of the constriction zone and does not modulate the ion current. In order to understand electrophysiological data at a molecular level, computational modelling is necessary. Unfortunately, this unique approach is restricted to membrane proteins with known high-resolution X-ray structure. Computational studies can also help to understand the detailed structure–function relationship of efflux pumps. With this knowledge a (more) rational design of new antibiotics or molecules blocking the antibiotics extrusion should be possible.

Acknowledgements

The investigations were supported by the Deutsche Forschungsgemeinschaft (grant Wi 2278/18-1).

References

Aksimentiev, A., and Schulten, K. (2005). Imaging alpha-hemolysin with molecular dynamics: ionic conductance, osmotic permeability, and the electrostatic potential map. Biophys. J. *88*, 3745–3761.

Basle, A., Rummel, G., Storici, P., Rosenbusch, J.P., and Schirmer, T. (2006). Crystal structure of osmoporin OmpC from *E. coli* at 2.0 A. J. Mol. Biol. *362*, 933–942.
Bauer, W.R., and Nadler, W. (2006). Molecular transport through channels and pores: effects of in-channel interactions and blocking. Proc. Natl. Acad. Sci. U.S.A. *103*, 11446–11451.
Benz, R., Janko, K., Boos, W., and Läuger, P. (1978). Formation of large, ion-permeable membrane channels by the matrix protein (porin) of *Escherichia coli*. Biochim. Biophys. Acta *511*, 305–319.
Benz, R., Schmid, A., and Hancock, R.E. (1985). Ion selectivity of gram-negative bacterial porins. J. Bacteriol. *162*, 722–727.
Berezhkovskii, A.M., and Bezrukov, S.M. (2005). Optimizing transport of metabolites through large channels: molecular sieves with and without binding. Biophys. J. *88*, L17–19.
Bezrukov, S.M., Berezhkovskii, A.M., and Szabo, A. (2007). Diffusion model of solute dynamics in a membrane channel: mapping onto the two-site model and optimizing the flux. J. Chem. Phys. *127*, 115101.
Biro, I., Pezeshki, S., Weingart, H., Winterhalter, M., and Kleinekathöfer, U. (2010). Comparing the temperature-dependent conductance of the two structurally similar *E. coli* porins OmpC and OmpF. Biophys. J. *98*, 1830–1839.
Bohnert, J.A., Karamian, B., and Nikaido, H. (2010). Optimized Nile Red efflux assay of AcrAB-TolC multidrug efflux system shows competition between substrates. Antimicrob. Agents Chemother. *54*, 3770–3775.
Ceccarelli, M., Danelon, C., Laio, A., and Parrinello, M. (2004). Microscopic mechanism of antibiotics translocation through a porin. Biophys. J. *87*, 58–64.
Chimerel, C., Movileanu, L., Pezeshki, S., Winterhalter, M., and Kleinekathöfer, U. (2008). Transport at the nanoscale: temperature dependence of ion conductance. Eur. Biophys. J. *38*, 121–125.
Danelon, C., Brando, T., and Winterhalter, M. (2003). Probing the orientation of reconstituted maltoporin channels at the single-protein level. J. Biol. Chem. *278*, 35542–35551.
Danelon, C., Nestorovich, E.M., Winterhalter, M., Ceccarelli, M., and Bezrukov, S.M. (2006). Interaction of zwitterionic penicillins with the OmpF channel facilitates their translocation. Biophys. J. *90*, 1617–1627.
Dürr, K.L., Tavraz, N.N., Zimmermann, D., Bamberg, E., and Friedrich, T. (2008). Characterization of Na,K-ATPase and H,K-ATPase enzymes with glycosylation-deficient beta-subunit variants by voltage-clamp fluorometry in *Xenopus* oocytes. Biochemistry *47*, 4288–4297.
Fischer, N., and Kandt, C. (2011). Three ways in, one way out: water dynamics in the trans-membrane domains of the inner membrane translocase AcrB. Proteins *79*, 2871–2885.
Gbaguidi, B., Hakizimana, P., Vandenbussche, G., and Ruysschaert, J.M. (2007). Conformational changes in a bacterial multidrug transporter are phosphatidylethanolamine-dependent. Cell Mol. Life Sci. *64*, 1571–1582.
Gornall, J.L., Mahendran, K.R., Pambos, O.J., Steinbock, L.J., Otto, O., Chimerel, C., Winterhalter, M., and Keyser, U.F. (2011). Simple reconstitution of protein pores in nano lipid bilayers. Nano Lett. *11*, 3334–3340.
Hong, J.S., and Kaback, H.R. (1972). Mutants of Salmonella typhimurium and *Escherichia coli* pleiotropically defective in active transport. Proc. Natl. Acad. Sci. U.S.A. *69*, 3336–3340.
Howe, J., Andra, J., Conde, R., Iriarte, M., Garidel, P., Koch, M.H., Gutsmann, T., Moriyon, I., and Brandenburg, K. (2007). Thermodynamic analysis of the lipopolysaccharide-dependent resistance of gram-negative bacteria against polymyxin B. Biophys. J. *92*, 2796–2805.
Husain, F., Humbard, M., and Misra, R. (2004). Interaction between the TolC and AcrA proteins of a multidrug efflux system of *Escherichia coli*. J. Bacteriol. *186*, 8533–8536.
James, C.E., Mahendran, K.R., Molitor, A., Bolla, J.M., Bessonov, A.N., Winterhalter, M., and Pages, J.M. (2009). How beta-lactam antibiotics enter bacteria: a dialogue with the porins. PLoS One *4*, e5453.
Koronakis, V., Eswaran, J., and Hughes, C. (2004). Structure and function of TolC: the bacterial exit duct for proteins and drugs. Annu. Rev. Biochem. *73*, 467–489.
Kreir, M., Farre, C., Beckler, M., George, M., and Fertig, N. (2008). Rapid screening of membrane protein activity: electrophysiological analysis of OmpF reconstituted in proteoliposomes. Lab Chip *8*, 587–595.
Kullman, L., Gurnev, P.A., Winterhalter, M., and Bezrukov, S.M. (2006). Functional subconformations in protein folding: evidence from single-channel experiments. Phys. Rev. Lett. *96*, 038101.
Lim, S.P., and Nikaido, H. (2010). Kinetic parameters of efflux of penicillins by the multidrug efflux transporter AcrAB-TolC of *Escherichia coli*. Antimicrob. Agents Chemother. *54*, 1800–1806.
Lobedanz, S., Bokma, E., Symmons, M.F., Koronakis, E., Hughes, C., and Koronakis, V. (2007). A periplasmic coiled–coil interface underlying TolC recruitment and the assembly of bacterial drug efflux pumps. Proc. Natl. Acad. Sci. U.S.A. *104*, 4612–4617.

Luckey, M., and Nikaido, H. (1980). Diffusion of solutes through channels produced by phage lambda receptor protein of *Escherichia coli*: inhibition by higher oligosaccharides of maltose series. Biochem. Biophys. Res. Commun. *93*, 166–171.

Mahendran, K.R., Kreir, M., Weingart, H., Fertig, N., and Winterhalter, M. (2010). Permeation of antibiotics through *Escherichia coli* OmpF and OmpC porins: screening for influx on a single-molecule level. J. Biomol. Screen. *15*, 302–307.

Michea-Hamzehpour, M., Furet, Y.X., and Pechere, J.C. (1991). Role of protein D2 and lipopolysaccharide in diffusion of quinolones through the outer membrane of *Pseudomonas aeruginosa*. Antimicrob. Agents Chemother. *35*, 2091–2097.

Mikolosko, J., Bobyk, K., Zgurskaya, H.I., and Ghosh, P. (2006). Conformational flexibility in the multidrug efflux system protein AcrA. Structure *14*, 577–587.

Modi, N., Singh, P.R., Mahendran, K.R., Schulz, R., Winterhalter, M., and Kleinekathöfer, U. (2011). Probing the transport of ionic liquids in aqueous solution through nanopores. J. Phys. Chem. Lett. *2*, 2331–2336.

Montal, M., and Mueller, P. (1972). Formation of bimolecular membranes from lipid monolayers and a study of their electrical properties. Proc. Natl. Acad. Sci. U.S.A. *69*, 3561–3566.

Mueller, P., and Rudin, D.O. (1963). Induced excitability in reconstituted cell membrane structure. J. Theor. Biol. *4*, 268–280.

Murakami, S., Tamura, N., Saito, A., Hirata, T., and Yamaguchi, A. (2004). Extramembrane central pore of multidrug exporter AcrB in *Escherichia coli* plays an important role in drug transport. J. Biol. Chem. *279*, 3743–3748.

Murakami, S., Nakashima, R., Yamashita, E., Matsumoto, T., and Yamaguchi, A. (2006). Crystal structures of a multidrug transporter reveal a functionally rotating mechanism. Nature *443*, 173–179.

Nagano, K., and Nikaido, H. (2009). Kinetic behavior of the major multidrug efflux pump AcrB of *Escherichia coli*. Proc. Natl. Acad. Sci. U.S.A. *106*, 5854–5858.

Needham, D., and Nunn, R.S. (1990). Elastic deformation and failure of lipid bilayer membranes containing cholesterol. Biophys. J. *58*, 997–1009.

Nekolla, S., Andersen, C., and Benz, R. (1994). Noise analysis of ion current through the open and the sugar-induced closed state of the LamB channel of *Escherichia coli* outer membrane: evaluation of the sugar binding kinetics to the channel interior. Biophys. J. *66*, 1388–1397.

Nestorovich, E.M., Danelon, C., Winterhalter, M., and Bezrukov, S.M. (2002). Designed to penetrate: time-resolved interaction of single antibiotic molecules with bacterial pores. Proc. Natl. Acad. Sci. U.S.A. *99*, 9789–9794.

Nikaido, H. (2003). Molecular basis of bacterial outer membrane permeability revisited. Microbiol. Mol. Biol. Rev. *67*, 593–656.

Nikaido, H., and Rosenberg, E.Y. (1983). Porin channels in *Escherichia coli*: studies with liposomes reconstituted from purified proteins. J. Bacteriol. *153*, 241–252.

Nikaido, H., Basina, M., Nguyen, V., and Rosenberg, E.Y. (1998). Multidrug efflux pump AcrAB of Salmonella typhimurium excretes only those beta-lactam antibiotics containing lipophilic side chains. J. Bacteriol. *180*, 4686–4692.

Nishino, K., Yamada, J., Hirakawa, H., Hirata, T., and Yamaguchi, A. (2003). Roles of TolC-dependent multidrug transporters of *Escherichia coli* in resistance to beta-lactams. Antimicrob. Agents Chemother. *47*, 3030–3033.

Okusu, H., Ma, D., and Nikaido, H. (1996). AcrAB efflux pump plays a major role in the antibiotic resistance phenotype of *Escherichia coli* multiple-antibiotic-resistance (Mar) mutants. J. Bacteriol. *178*, 306–308.

Pages, J.M., James, C.E., and Winterhalter, M. (2008). The porin and the permeating antibiotic: a selective diffusion barrier in Gram-negative bacteria. Nat. Rev. Microbiol. *6*, 893–903.

Pezeshki, S., Chimerel, C., Bessonov, A.N., Winterhalter, M., and Kleinekathöfer, U. (2009). Understanding ion conductance on a molecular level: an all-atom modeling of the bacterial porin OmpF. Biophys. J. *97*, 1898–1906.

Rodrigues, C., Gameiro, P., Prieto, M., and de Castro, B. (2003). Interaction of rifampicin and isoniazid with large unilamellar liposomes: spectroscopic location studies. Biochim. Biophys. Acta *1620*, 151–159.

Schulz, R., and Kleinekathöfer, U. (2009). Transitions between closed and open conformations of TolC: the effects of ions in simulations. Biophys. J. *96*, 3116–3125.

Schulz, R., Vargiu, A.V., Collu, F., Kleinekathöfer, U., and Ruggerone, P. (2010). Functional rotation of the transporter AcrB: insights into drug extrusion from simulations. PLoS Comput. Biol. *6*, e1000806.

Schulz, R., Vargiu, A.V., Ruggerone, P., and Kleinekathöfer, U. (2011). Role of water during the extrusion of substrates by the efflux transporter AcrB. J. Phys. Chem. B *115*, 8278–8287.

Schwarz, G., Danelon, C., and Winterhalter, M. (2003). On translocation through a membrane channel via an internal binding site: kinetics and voltage dependence. Biophys. J. *84*, 2990–2998.

Schwarz, G., Damian, L., and Winterhalter, M. (2007). Model-free analysis of binding at lipid membranes employing micro-calorimetric measurements. Eur. Biophys. J. *36*, 571–579.

Seeger, M.A., Schiefner, A., Eicher, T., Verrey, F., Diederichs, K., and Pos, K.M. (2006). Structural asymmetry of AcrB trimer suggests a peristaltic pump mechanism. Science *313*, 1295–1298.

Sennhauser, G., Amstutz, P., Briand, C., Storchenegger, O., and Grutter, M.G. (2007). Drug export pathway of multidrug exporter AcrB revealed by DARPin inhibitors. PLoS Biol *5*, e7.

Sikora, C.W., and Turner, R.J. (2005). Investigation of ligand binding to the multidrug resistance protein EmrE by isothermal titration calorimetry. Biophys. J. 88, 475–482.

Simonet, V., Mallea, M., and Pages, J.M. (2000). Substitutions in the eyelet region disrupt cefepime diffusion through the *Escherichia coli* OmpF channel. Antimicrob. Agents Chemother. *44*, 311–315.

Sotomayor, M., Vasquez, V., Perozo, E., and Schulten, K. (2007). Ion conduction through MscS as determined by electrophysiology and simulation. Biophys. J. *92*, 886–902.

Su, C.C., and Yu, E.W. (2007). Ligand–transporter interaction in the AcrB multidrug efflux pump determined by fluorescence polarization assay. FEBS Lett. *581*, 4972–4976.

Symmons, M.F., Bokma, E., Koronakis, E., Hughes, C., and Koronakis, V. (2009). The assembled structure of a complete tripartite bacterial multidrug efflux pump. Proc. Natl. Acad. Sci. U.S.A. *106*, 7173–7178.

Takatsuka, Y., Chen, C., and Nikaido, H. (2010). Mechanism of recognition of compounds of diverse structures by the multidrug efflux pump AcrB of *Escherichia coli*. Proc. Natl. Acad. Sci. U.S.A. *107*, 6559–6565.

Tikhonova, E.B., and Zgurskaya, H.I. (2004). AcrA, AcrB, and TolC of *Escherichia coli* form a stable intermembrane multidrug efflux complex. J. Biol. Chem. *279*, 32116–32124.

Touze, T., Eswaran, J., Bokma, E., Koronakis, E., Hughes, C., and Koronakis, V. (2004). Interactions underlying assembly of the *Escherichia coli* AcrAB-TolC multidrug efflux system. Mol. Microbiol. *53*, 697–706.

Tseng, T.T., Gratwick, K.S., Kollman, J., Park, D., Nies, D.H., Goffeau, A., and Saier, M.H. Jr. (1999). The RND permease superfamily: an ancient, ubiquitous and diverse family that includes human disease and development proteins. J. Mol. Microbiol. Biotechnol. *1*, 107–125.

Vaccaro, L., Koronakis, V., and Sansom, M.S. (2006). Flexibility in a drug transport accessory protein: molecular dynamics simulations of MexA. Biophys. J. *91*, 558–564.

Vaccaro, L., Scott, K.A., and Sansom, M.S. (2008). Gating at both ends and breathing in the middle: conformational dynamics of TolC. Biophys. J. *95*, 5681–5691.

Vargiu, A.V., Collu, F., Schulz, R., Pos, K.M., Zacharias, M., Kleinekathöfer, U., and Ruggerone, P. (2011). Effect of the F610A mutation on substrate extrusion in the AcrB transporter: explanation and rationale by molecular dynamics simulations. J. Am. Chem. Soc. *133*, 10704–10707.

Wang, B., Weng, J., Fan, K., and Wang, W. (2011). Elastic network model-based normal mode analysis reveals the conformational couplings in the tripartite AcrAB-TolC multidrug efflux complex. Proteins *79*, 2936–2945.

Ward, A., Hoyle, C., Palmer, S., O'Reilly, J., Griffith, J., Pos, M., Morrison, S., Poolman, B., Gwynne, M., and Henderson, P. (2001). Prokaryote multidrug efflux proteins of the major facilitator superfamily: amplified expression, purification and characterisation. J. Mol. Microbiol. Biotechnol. *3*, 193–200.

Weingart, H., Petrescu, M., and Winterhalter, M. (2008). Biophysical characterization of in- and efflux in Gram-negative bacteria. Curr. Drug Targets *9*, 789–796.

Xu, Y., Lee, M., Moeller, A., Song, S., Yoon, B.Y., Kim, H.M., Jun, S.Y., Lee, K., and Ha, N.C. (2011). Funnel-like hexameric assembly of the periplasmic adapter protein in the tripartite multidrug efflux pump in gram-negative bacteria. J. Biol. Chem. *286*, 17910–17920.

Yao, X.Q., Kenzaki, H., Murakami, S., and Takada, S. (2010). Drug export and allosteric coupling in a multidrug transporter revealed by molecular simulations. Nat. Commun. *1*, 117.

Zgurskaya, H.I., and Nikaido, H. (1999). Bypassing the periplasm: reconstitution of the AcrAB multidrug efflux pump of *Escherichia coli*. Proc. Natl. Acad. Sci. U.S.A. *96*, 7190–7195.

Zgurskaya, H.I., and Nikaido, H. (2000). Multidrug resistance mechanisms: drug efflux across two membranes. Mol. Microbiol. *37*, 219–225.

Targeting and Integration of Bacterial Membrane Proteins

10

Patrick Kuhn, Renuka Kudva, Thomas Welte, Lukas Sturm and Hans-Georg Koch

Abstract

Membrane proteins execute a plethora of essential functions in bacterial cells and therefore bacteria utilize efficient strategies to ensure that these proteins are properly targeted and inserted into the membrane. Most bacterial inner membrane proteins are recognized early during their synthesis, i.e. co-translationally by the bacterial signal recognition particle (SRP), which delivers the ribosome-nascent chain (RNC) via its interaction with the membrane-bound SRP receptor to the SecYEG translocon, a highly dynamic and evolutionarily conserved protein conducting channel. Membrane protein insertion via SecYEG is coupled to on going polypeptide chain elongation at the ribosome and the emerging transmembrane helices exit the SecYEG channel laterally into the lipid phase. Lateral release and folding of transmembrane helices is most likely facilitated by YidC, which transiently associates with the SecYEG translocon. YidC has also been shown to facilitate insertion of inner membrane proteins independently of the SecYEG translocon. The targeting of outer membrane proteins to SecYEG occurs predominantly post-translationally by the SecA/SecB pathway and thus follows the same route as periplasmic proteins. In this chapter, we summarize the current knowledge on membrane protein targeting and transport/integration by either SecYEG or YidC.

Introduction

Although bacteria lack the subcellular organization of eukaryotic cells, they are challenged with the fact that protein synthesis takes place exclusively in the cytosol while approximately 20–30% of all synthesized proteins operate outside the cytosol. Thus, efficient transport machineries need to be engaged to deliver proteins to their correct destination (Papanikou *et al.*, 2007). These machineries operate either in a co-translational or post-translational mode, both of which are employed for the transport of bacterial membrane proteins (Fig. 10.1). Bacteria contain two types of membrane proteins:

1. those that contain α-helical transmembrane domains (TMs) that are targeted co-translationally as ribosome-nascent chains (RNCs) by the signal recognition particle (SRP);
2. those containing multiple β-strands, typical for β-barrel proteins of the outer membrane of Gram-negative bacteria, which follow a post-translational route to their destination via the SecA/SecB pathway after they are completely synthesized. (du Plessis *et al.*, 2011).

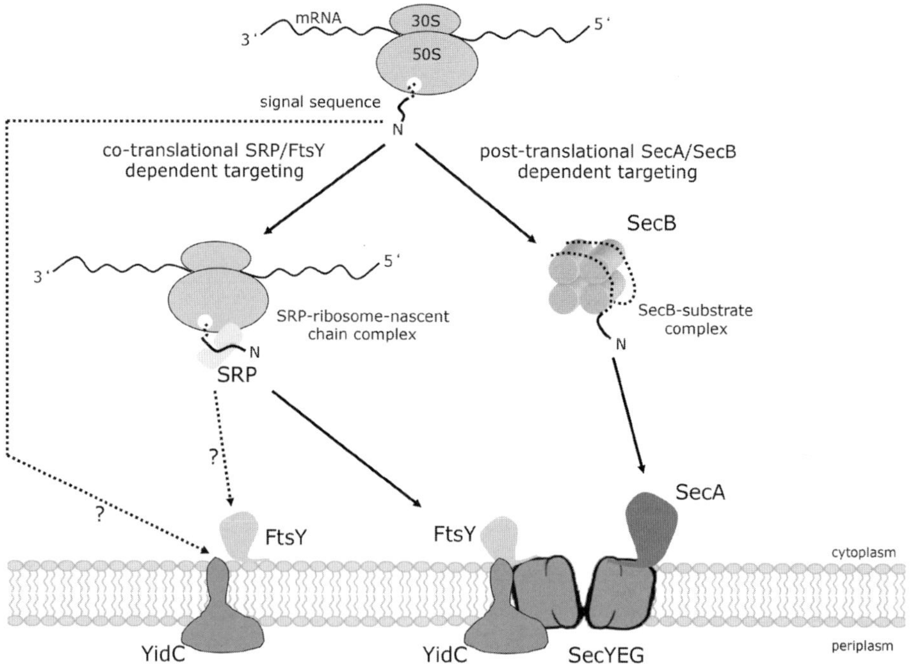

Figure 10.1 Co- and post-translational protein targeting pathways in bacteria.

Since protein synthesis is much slower than protein transport, the post-translational mode probably allows a higher transport rate but is unsuitable for substrates that rapidly lose their transport competence due to aggregation.

The SRP and SecA/SecB targeting pathways converge at the SecYEG translocon that directs substrate proteins either into the cytoplasmic membrane or into the periplasmic space for further targeting to the outer membrane (Fig. 10.1). Although SecYEG probably constitutes the major integration site for membrane proteins, the insertion of some membrane proteins is facilitated by YidC, which together with the mitochondrial Oxa1 and the chloroplast Alb3 constitutes a conserved family of proteins (Funes *et al.*, 2011). The integration function of the bacterial YidC was first demonstrated for phage coat proteins (Samuelson *et al.*, 2000), which were previously shown to insert independently of SecYEG into the membrane (Geller and Wickner, 1985) and therefore thought to insert spontaneously. In later studies, it was shown that YidC in *E. coli* also functions as an insertase for small membrane proteins of the respiratory chain (Dalbey *et al.*, 2011). Recent studies show that multi-spanning membrane proteins can also be inserted via YidC if they lack large periplasmic loops (Welte *et al.*, 2012). Targeting of these proteins to either SecYEG or YidC is SRP-dependent, but the determinants which route closely spaced membrane proteins to either SecYEG or YidC are currently unknown. An additional function of YidC is in association with the SecYEG translocon where it is thought to assist in the folding and integration of SecYEG-dependent membrane proteins.

How are substrates selected for the SRP- or SecA-pathway?

The selective recognition of inner and outer membrane proteins occurs at the ribosomal tunnel exit and the ribosome-bound chaperone trigger factor was considered to be the first protein contacting an emerging nascent chain (Deuerling *et al.*, 1999). Trigger factor is the only ribosome-bound chaperone in *E. coli* and due to its abundance is probably bound to most of the 50,000–70,000 *E. coli* ribosomes (Kramer *et al.*, 2009). Trigger factor binds to the L23 protein of the large ribosomal subunit and forms a cage-like structure below the ribosomal exit tunnel (Fig. 10.2) (Ferbitz *et al.*, 2004). Predominantly hydrophilic interactions of trigger factor with its substrates are suggested to support folding of cytosolic substrates within the cage before they acquire their native structure with the help of the DnaK and GroEL chaperone systems (Lakshmipathy *et al.*, 2007). In contrast, hydrophobic interactions of trigger factor with signal sequence containing proteins are suggested to postpone folding of nascent chains allowing their subsequent interaction with SRP or SecA/SecB. In contrast to these *in vitro* studies that suggested that trigger factor is pre-bound to ribosomes waiting for polypeptides to emerge from the ribosomal exit tunnel, a recent *in vivo* ribosome profiling study indicated that trigger factor engages ribosomes only after approximately 100 amino acids are translated (Oh *et al.*, 2011). This *in vivo* study also indicated that trigger factor has a high preference for β-barrel outer membrane proteins.

Like trigger factor, SRP binds to the L23 subunit of the ribosome and efficiently recognizes its substrates co-translationally (Fig. 10.2). Binding of SRP and trigger factor to L23 appears to be non-exclusive (Buskiewicz *et al.*, 2004; Raine *et al.*, 2004); however, owing to the low copy number of SRP, only about 1% of all *E. coli* ribosomes have SRP bound. Recent data indicate that SecA, the motor protein of the post-translational protein transport machinery, can also bind directly to the L23 subunit (Huber *et al.*, 2011). Whether SecA competes with trigger factor or SRP for access to L23 is currently unknown, but it is possible that post-translational protein transport by SecA/SecB is initiated by co-translational substrate recognition. Thus, the ribosomal proteins at or close to the exit tunnel provide a platform for other proteins that are involved in co-translational folding (trigger factor) and targeting (SRP and SecA) of newly synthesized proteins.

SecA-dependent bacterial outer membrane proteins are generally synthesized as pre-proteins with a tripartite signal sequence which contains a positively charged N-terminal region of up to five amino acids that is followed by a hydrophobic core of 10 to 15 amino acids, referred to as the H-domain. The C-terminal C-domain of 3–7 amino acids usually contains a signal peptidase cleavage site and is more polar (von Heijne, 1990; Hegde and Bernstein, 2006). The N-terminal part of the signal sequence adopts a stable α-helical structure in a membrane environment while the C-terminal part appears to be more flexible (Clérico *et al.*, 2008). The positive charges at the N-terminus interact with negatively charged phospholipids and orient the signal sequence in a hairpin-like structure in the SecYEG translocon. This orientation is also facilitated by the presence of helix-breaking amino acids like glycine or proline present at the centre of the hydrophobic H-domain. Signal sequences are often interchangeable even between species, facilitating the use of eukaryotic signal sequences in bacterial expression systems (Talmadge *et al.*, 1980). The signal sequence of outer membrane proteins is cleaved by signal peptidase I (SPase I) during transport across the SecYEG translocon. SPase I typically recognizes an Ala-X-Ala motif at the −1 to −3 positions relative to the cleavage site of the signal sequence and cleaves after the (−1)-Ala residue. The catalytic domain of SPase I is located in the periplasm and utilizes a Lys-Ser catalytic dyad for signal

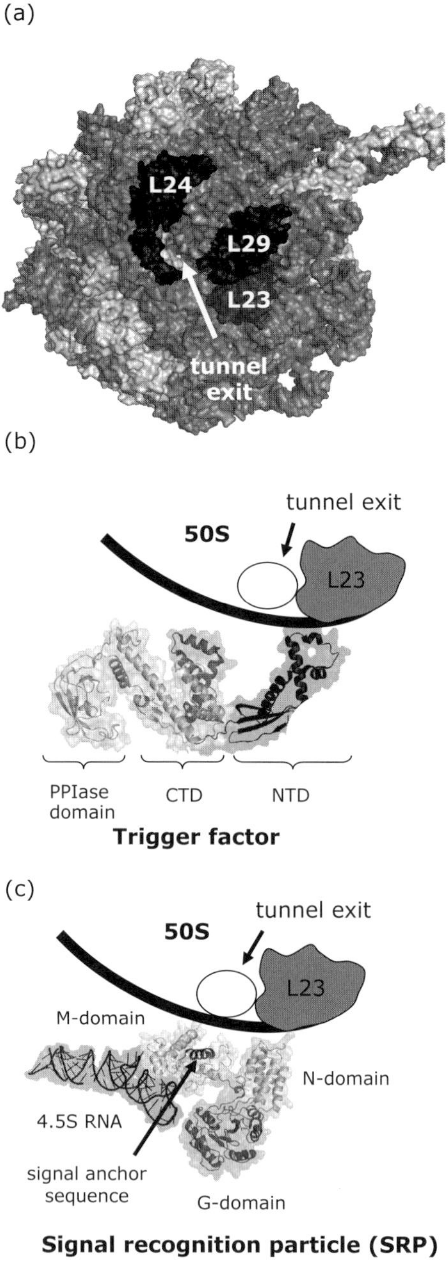

Figure 10.2 Substrate selection at the ribosomal tunnel exit. (a) View from the bottom onto the tunnel exit of the 70S ribosome. Ribosomal RNA is displayed in dark grey and ribosomal proteins in light grey. Indicated in black are the ribosomal proteins L23, L24 and L29. The model was generated using PyMol and the PDB coordinates 3OAT. (b) The ribosome-bound chaperone trigger factor (PDB: 1W26) binds via L23 to the 50S ribosomal subunit and is probably the first protein contacting an emerging nascent chain. Note that the tertiary structure does not represent the linear domain order. (c) *E. coli* SRP binds to L23 and recognizes mainly hydrophobic signal anchor sequences of inner membrane proteins. (PDB: 2J28).

sequence cleavage (Paetzel *et al.*, 2002). A second signal peptidase SPaseII is specifically involved in the cleavage of lipoproteins and depends on diacyl-glycerol modification of a cysteine residue at position (+1) of the signal sequence (Paetzel *et al.*, 2002). After the initial cleavage by signal peptidase, signal peptides are degraded by peptidases, like oligopeptidase IV and oligopeptidase A or RseP (Saito *et al.*, 2011). A unique signal sequence is found in some members of the autotransporter family of outer membrane proteins, which contains an N-terminal extension of 30 to 40 amino acids (Dautin and Bernstein, 2007). These proteins are also post-translationally targeted by SecA (Chevalier *et al.*, 2004) and replacing the extended signal sequence of the *E. coli* autotransporter protein EspP with a classical signal sequence does not influence its transport across the inner membrane (Szabady *et al.*, 2005). However prolonged contact of the extended signal sequence to the SecYEG translocon is suggested to prevent the formation of a transport-incompetent conformation of the protein in the periplasm. Finally, some proteins with cleavable signal sequences can also use the SRP-pathway for targeting to the SecYEG translocon (Huber *et al.*, 2005). These proteins still require SecA for translocation across the inner membrane and it is likely that proteins which tend to fold rapidly in the cytosol require the coordinated activity of both SRP and SecA. However the number of SRP-dependent secretory proteins is probably low considering the low numbers of SRP/Ffh molecules in *E. coli* cells (approximately 100 per cell).

Inner membrane proteins usually do not contain cleavable signal sequences; instead their α-helical transmembrane domains (TM) serve as a recognition signal for SRP. The TMs are significantly more hydrophobic than cleavable signal sequences and so it is assumed that cells discriminate inner and outer membrane proteins based on their hydrophobicity (Beha *et al.*, 2003; Luirink *et al.*, 2005). This is consistent with results showing that SRP preferentially binds to hydrophobic stretches of at least seven consecutive hydrophobic amino acids (Valent *et al.*, 1997; Lee and Bernstein, 2001). The presence of helix-breaking amino acids within this region prevents the interaction with SRP, which also explains why hydrophobic cleavable signal sequences, like that of β-lactamase are not recognized by SRP (Beha *et al.*, 2003). The first TM is not necessarily located at the N-terminus of inner membrane proteins and SRP also recognizes hydrophobic regions that do not function as TMs. This has been shown for the potassium sensor KdpD, in which the first TM is preceded by a 400 amino acid long cytosolic domain that contains a hydrophobic recognition sequence for SRP (Maier *et al.*, 2008).

Co-translational targeting of inner membrane proteins by the SRP-system

SRP-dependent protein targeting to the SecYEG translocon is the only protein transport pathway that is universally conserved in all living organisms (Koch *et al.*, 2003). In Gram-negative bacteria, the SRP system consists of three components: the GTPase Ffh (*fifty-four homologue*; homologous to the SRP54 subunit of eukaryotic SRP) which together with the 4.5S RNA constitutes the bacterial SRP, while the GTPase FtsY functions as a membrane-bound SRP receptor. These three components represent the minimal SRP system and are essential for cell viability in Gram-negative bacteria. The SRP-system in many Gram-positive bacteria contains additional proteins like the histone-like protein HBsu in *B. subtilis* (Nakamura *et al.*, 1999) or the Ffh-related GTPase FlhF (Bange *et al.*, 2007, 2011). Despite its presence in all Gram-positive bacteria, it is not entirely clear whether the SRP-pathway is essential for cell viability (Zanen *et al.*, 2006). For example, the oral pathogen *Streptococcus mutans* survives SRP-deletion (Hasona *et al.*, 2005) by down-regulating protein-synthesis (Hasona *et al.*, 2007) and by employing an insertion pathway via YidC2. YidC2 contains in contrast to YidC an extended C-terminal ribosome binding site, which probably provides sufficient affinity for nascent chains, therefore diminishing the need for SRP-dependent targeting.

A structural view of the bacterial signal recognition particle and its receptor

Ffh and FtsY belong to the SIMIBI-family (after signal recognition particle (SRP), MinD and BioD) of nucleotide-hydrolysing enzymes and are similar in their modular organization. They are composed of three domains and their respective N- and G-domains exhibit high sequence conservation (Fig. 10.3). The N-domain forms a four-helix bundle, while the G-domain harbours the Ras-like GTPase-domain. The G-domain also contains a unique insertion (I-box), which facilitates nucleotide exchange and stabilizes the nucleotide-free protein (Moser *et al.*, 1997; Rapiejko *et al.*, 1997). As a result, Ffh and FtsY do not depend on external guanine nucleotide exchange factors (GEFs). Complex formation between SRP and FtsY proceeds via their NG-domains (Fig. 10.3) and leads to the formation of an active site that promotes GTP hydrolysis (Egea *et al.*, 2004; Focia *et al.*, 2004). Thus, FtsY and Ffh belong to a growing number of GTPases that are activated by nucleotide-dependent dimerization (GADs, Gasper *et al.*, 2009). The third domains of Ffh and FtsY show no similarity to each other, but serve as function-related modules that are fused to the conserved NG-core.

The C-terminal M-domain of Ffh is responsible for substrate binding and is characterized by an unusually high number of methionine residues in most mesophilic bacteria. This led to the *methionine-bristle hypothesis* (Bernstein *et al.*, 1989), in which the methionine side-chains provide a flexible hydrophobic surface for interaction with diverse signal sequences. An early X-ray structure of the M-domain in the thermophilic bacterium *Thermus* (*T.*) *aquaticus* revealed a putative signal sequence binding groove, which is lined with mainly hydrophobic residues (Keenan *et al.*, 1998). Cryo-EM studies on SRP-RNCs have identified an electron dense region in the signal sequence binding groove, which was attributed to the signal sequence (Halic *et al.*, 2004; Schaffitzel *et al.*, 2006). Two recent crystal structures of a signal sequence in complex with SRP have confirmed the presence of an α-helical signal sequence in the hydrophobic groove and have also revealed significant conformational changes of SRP upon signal sequence recognition (Janda *et al.*, 2010; Hainzl *et al.*, 2011). In addition to its role in substrate binding, the M-domain also interacts with the 4.5S RNA via a helix–turn–helix motif, which is located opposite the hydrophobic substrate binding groove (Zheng *et al.*, 1997; Batey *et al.*, 2000) (Fig. 10.3).

The function-related module in FtsY is its N-terminal A-domain. Determining its exact function has been difficult because it varies in length and amino acid sequence even between closely related species (Maeda *et al.*, 2003; Weiche *et al.*, 2008). Furthermore, it has been shown that the A-domain is not essential for co-translational targeting in *E. coli* (Eitan and Bibi, 2004). Suggestions that the A-domain is involved in membrane binding (de Leeuw *et al.*, 1997) have been confirmed by the identification of two polybasic lipid-binding helices in *E. coli* FtsY (MTS, membrane-targeting sequence; Fig. 10.3) (Parlitz *et al.*, 2007; Weiche *et al.*, 2008; Braig *et al.*, 2009). One is located at the N-terminus of FtsY and is highly conserved among enterobacteria (Weiche *et al.*, 2008), the other is located at

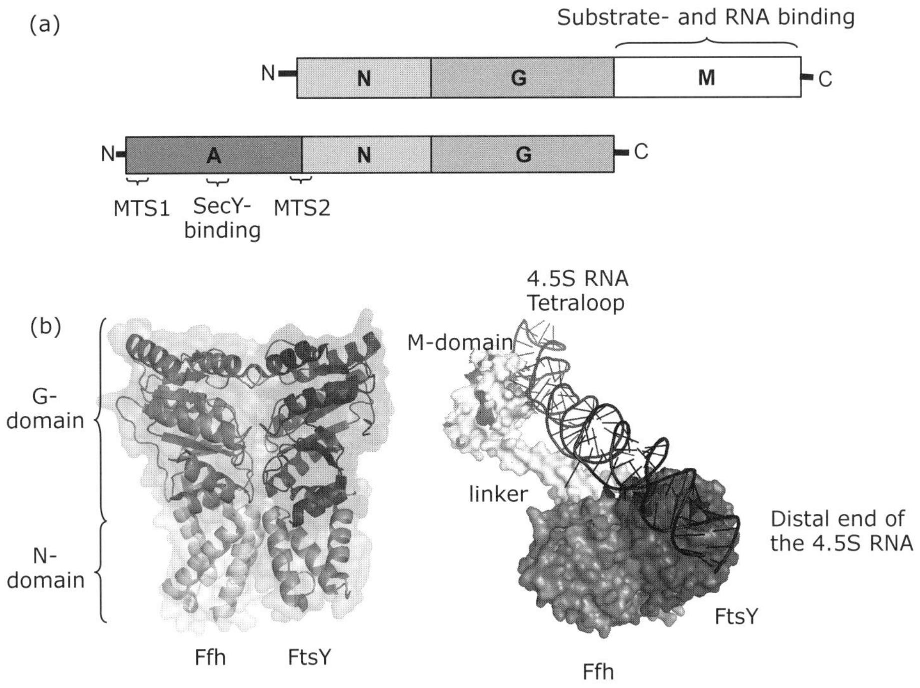

Figure 10.3 The signal recognition particle (SRP) and the SRP receptor FtsY are responsible for co-translational targeting. (a) Domain structure of Ffh, the protein component of the bacterial SRP and FtsY. MTS 1 and 2 indicate the membrane targeting sequences. In *E. coli*, the A-domain is also involved in SecY binding. (b) The heterodimer of the NG-domains of Ffh and FtsY from *T. aquaticus* (left, PDB: 1Okk). The structure shows a large common surface area, which also includes a composite GTPase site. Right: The complete X-ray structure of the FtsY–SRP complex from *E. coli* (PDB 2XXA). The M-domain is connected via a flexible linker to the NG-domain of Ffh. The tetraloop of the 4.5S RNA is in contact with the M-domain but the distal part of the RNA is also in contact with FtsY.

the interface between the A-domain and the N-domain, appears to be universally conserved (Parlitz *et al.*, 2007) and influences the interaction of FtsY with SRP (Braig *et al.*, 2011). Both helices bind preferentially to negatively charged phospholipids (Parlitz *et al.*, 2007; Braig *et al.*, 2009), which primes FtsY for subsequent interaction with the SRP-RNCs (Lam *et al.*, 2010; Braig *et al.*, 2011; Stjepanovic *et al.*, 2011). Overexpressing negatively charged phospholipids has been shown to restore the function of inactive FtsY derivatives demonstrating their importance on FtsY function (Erez *et al.*, 2010). In *E. coli*, the helices are separated by about 180 amino acids with low sequence conservation. In solution, this region appears to be unstructured (Stjepanovic *et al.*, 2011) and attempts to determine its structure by X-ray crystallography or Cryo-EM reconstruction have been unsuccessful. Nevertheless, a recent *in vivo* cross-linking study has shown that this region interacts with SecY, the central component of the SecYEG translocon (Kuhn *et al.*, 2011). This confirms previous reports of a direct interaction between FtsY and the SecYEG translocon (Angelini *et al.*, 2005, 2006; Bahari *et al.*, 2007). Whether the SecY–FtsY interaction also involves the NG-domain of FtsY is currently unknown. Thus, two populations of FtsY exist; one bound to phospholipids and the second bound to the SecYEG translocon. Additionally, cell fractionation studies have indicated that a significant portion of FtsY is located in the cytosolic fraction of the cell (Luirink *et al.*, 1994; Koch *et al.*, 1999). This led to the idea that FtsY partitions between the membrane and the cytosol. However, fluorescence microscopy studies have shown that FtsY is exclusively membrane localized in both *E. coli* and *B. subtilis* cells (Mircheva *et al.*, 2009). How FtsY reaches the membrane is currently unknown; some reports indicate that FtsY is co-translationally tethered to the membrane in an SRP-independent process (Bibi, 2011).

The *E. coli* SRP contains the 114-nucleotide-long 4.5S RNA (Poritz *et al.*, 1990), which forms a hairpin-like structure comprising two universally conserved helices (Schmitz *et al.*, 1999). The crucial GGAA tetra-loop of helix 8 is located at the tip, while the distal helix 5 contains the conserved GUGCCG motif (Fig. 10.3) (Rosenblad *et al.*, 2009). Originally considered to be a stabilizing structural element, it is now evident that the 4.5S RNA is a crucial component of the SRP cycle. In the presence of a signal sequence, a transient interaction between FtsY and the 4.5S RNA (Shen and Shan, 2010) accelerates complex formation between SRP and FtsY (Jagath *et al.*, 2001; Neher *et al.*, 2008). In the FtsY–SRP complex, the NG-heterodimer contacts the distal part of the 4.5S RNA, which is suggested to stimulate GTP hydrolysis (Ataide *et al.*, 2011). In addition, the M-domain of Ffh binds close to the RNA-tetraloop and it has been proposed that the negatively charged RNA participates in tethering the signal anchor sequence to the hydrophobic groove of Ffh (Batey *et al.*, 2000).

Co-translational recognition and targeting of substrates

Free SRP has a compact, L-shaped structure considered to be the closed state with a shielded hydrophobic signal sequence binding groove. In the presence of the ribosome, the flexible linker between the NG- and M-domains adopts an extended helical structure (Halic *et al.*, 2004; Schaffitzel *et al.*, 2006). Cross-linking studies have shown that the ribosomal protein L23 contacts SRP (Pool *et al.*, 2002; Gu *et al.*, 2003; Ullers *et al.*, 2003) (Fig. 10.2). Subsequent cryo-EM studies have shown that three helices of the N-domain of Ffh are in close contact to L23 and to a lesser extent to L29 (Halic *et al.*, 2004; Schaffitzel *et al.*, 2006) both located at the ribosomal exit tunnel. The M-domain additionally contacts the 23S rRNA and the ribosomal protein L22 (Halic *et al.*, 2004). SRP has an intrinsic affinity for ribosomes with a K_d value of about 50 nM (Bornemann *et al.*, 2008), but the affinity of SRP for isolated signal sequences appears to be lower with K_d values of 1–2 μM for signal sequences with low hydrophobicity (Bradshaw *et al.*, 2009). The affinity of SRP for RNCs depends on the hydrophobicity of the exposed signal sequence and varies between 1 nM for the signal anchor sequence of leader peptidase (Lep) (Bornemann *et al.*, 2008) and 80–100 nM for the cleavable signal sequence of the periplasmic protein alkaline phosphatase (Zhang *et al.*, 2010). High-affinity SRP binding to RNCs therefore is mainly determined by hydrophobicity, which is in agreement with early cross-linking studies (Valent *et al.*, 1998; Neumann-Haefelin *et al.*, 2000). SRP binding to RNCs is stimulated while the signal anchor sequence is still inside the ribosomal exit tunnel (Woolhead *et al.*, 2004; Bornemann *et al.*, 2008; Berndt *et al.*, 2009). The 70S ribosomal exit tunnel is approximately 10 nm long (Ban *et al.*, 2000; Berisio *et al.*, 2003) and can accommodate 28–30 amino acids in a fully extended conformation. Cross-linking studies have shown that RNCs of 24 amino acids can contact L23 (Houben *et al.*, 2005), indicating that L23 extends into the ribosomal tunnel. Signal sequences as short at 37 amino acid residues cross-link to SRP i.e. when the sequence is only half-exposed outside the ribosomal tunnel (Houben *et al.*, 2005). Additionally, high-affinity binding of SRP to RNCs is already observed at a chain length of 27 amino acids (Bornemann *et al.*, 2008), i.e. before SRP can contact the signal sequence directly. This indicates that the ribosome is able to sense the physical properties of nascent chains and somehow transmits this information to the exit tunnel for early recruitment of SRP (Wilson and Beckmann, 2011) and for preparing the Sec translocon for the insertion process (Lin *et al.*, 2011). The X-ray structure of SRP bound to a signal sequence shows that SRP has an extended conformation where the GTPase domain is closer to the tetraloop region of the 4.5S RNA (Janda *et al.*, 2010; Hainzl *et al.*, 2011). In this conformation, the RNC-bound SRP is probably primed for subsequent interaction with FtsY (Zhang *et al.*, 2009; Ataide *et al.*, 2011).

One of the hallmarks of the co-translational recognition of substrates by the eukaryotic SRP is the induction of a transient elongation arrest, which is thought to open a time window during which SRP can deliver substrates to the correct membrane (Lakkaraju *et al.*, 2008). Translation arrest is probably achieved because the Alu-domain of eukaryotic SRP blocks access of elongation factors to the peptidyl-transferase centre of the ribosome (Halic *et al.*, 2004). The bacterial SRP does not contain the Alu-domain and whether bacteria utilize an Alu-domain independent translational arrest is unknown (Bürk *et al.*, 2009; Yosef *et al.*, 2010). A recent study indicated that cycloheximide-induced slower translation elongation rates facilitate the interaction of SRP with substrates bearing suboptimal signal sequences (Zhang and Shan, 2012). Although the physiological significance of this finding needs to be further explored, it indicates that substrate recognition in bacteria is influenced by the rate of protein synthesis.

After substrate recognition, SRP has to deliver RNCs to the membrane-bound FtsY, which in *E. coli* is present in 100-fold excess over SRP (10,000 FtsY/cell versus 100 Ffh/cell). The ultimate consequences of the interaction between FtsY and the SRP-RNCs are:

1 SRP is first delocalized on the ribosome, which exposes the SecY binding site of the ribosome.
2 The delocalization of SRP probably favours the release of the signal anchor sequence, which can then insert into the SecYEG translocon.
3 The GTPase activity of the FtsY–SRP complex increases due to the formation of an active composite GTPase site. GTP hydrolysis leads to FtsY-SRP dissociation and facilitates the start of a new SRP cycle.

Biochemical studies have indicated that FtsY can interact with SRP-RNCs even in the absence of membranes (Jagath *et al.*, 2001; Shan *et al.*, 2007) and so it was suggested that soluble FtsY interacts with SRP-RNCs in the cytosol and subsequently recruits them to the membrane. However, soluble i.e. non-membrane bound FtsY is almost undetectable in living bacterial cells and *in vitro* studies have shown that soluble FtsY is unable to target SRP-RNCs to the membrane (Mircheva *et al.*, 2009), which is also supported by cross-linking studies (Scotti *et al.*, 1999; Neumann-Haefelin *et al.*, 2000). The need for membrane-bound FtsY is explained by the recent observation that lipids trigger a conformational switch (Stjepanovic *et al.*, 2011) that accelerates the subsequent FtsY–SRP complex formation approximately 100-fold (Lam *et al.*, 2010; Braig *et al.*, 2011). After formation of the FtsY–SRP-RNC complex at the membrane, RNCs have to be transferred to the SecYEG translocon prior to GTP hydrolysis. Based on kinetic analyses, a *pausing step* in GTP hydrolysis has been proposed, in which correct substrates and negatively charged phospholipids delay GTP-dependent dissociation (Lam *et al.*, 2010; Zhang *et al.*, 2010). Thus, SRP–FtsY complex formation and its stability are stimulated by two external signals: the presence of a correct substrate that primes SRP for interaction with FtsY and contacts to the membrane primes FtsY for the interaction with SRP. Both stimuli appear to have a stronger effect on complex formation than on GTPase activity. These proof-reading steps are thought to promote the formation of a stable FtsY–SRP-RNC complex and prevent futile SRP–FtsY interactions.

The transfer of RNCs from the SRP–FtsY complex to the SecYEG translocon is probably the most important and the least understood step of co-translational targeting (Fig. 10.4). The transfer has to occur before the emerging RNC reaches a critical length (Flanagan *et al.*, 2003); otherwise transport would be significantly compromised. The 23S ribosomal RNA and the ribosomal proteins L23, L24 and L29 interact with the surface exposed cytoplasmic loops 4, 5 and 6 of SecY (Prinz *et al.*, 2000; Cheng *et al.*, 2005; Frauenfeld *et al.*, 2011; Kuhn *et al.*, 2011). Thus, in the SRP–RNC complex, the translocon binding site of the ribosome is occupied by SRP and exposed only upon conformational rearrangements which are induced by binding to FtsY (Halic *et al.*, 2006; Ataide *et al.*, 2011; Estrozi *et al.*, 2011). FtsY interacts with the cytoplasmic loops 4 and 5 of SecY which constitute the ribosome binding site of the translocon (Kuhn *et al.*, 2011). Here, FtsY is positioned to align the incoming SRP-RNCs with the protein conducting channel of SecY but would then need to dissociate from the cytoplasmic loops to allow for binding of the ribosome to SecY. This occurs after the interaction of FtsY with SRP which potentially results in a delocalization of SRP on the ribosome and a delocalization of FtsY on the SecYEG translocon (Fig. 10.4). This coordinated movement then allows the ribosome to dock onto the SecYEG translocon. The FtsY-SecY contact could also be a trigger for signal-sequence release from SRP and subsequent dissociation of the SRP–FtsY complex.

Post-translational targeting of outer membrane proteins by the SecA/SecB system

Secretory proteins such as periplasmic proteins and outer membrane proteins are targeted to the SecYEG translocon post-translationally. Post-translational transport in Gram-negative bacteria is promoted by two proteins, the essential ATPase SecA and the non-essential, transport-specific chaperone SecB. SecB is predominantly found in proteobacteria and appears to be completely absent in Gram-positive bacteria (van der Sluis and Driessen, 2006). SecB is proposed to keep fully synthesized substrates which are released from the ribosome in a translocation competent conformation until they are transferred to SecA. SecA then threads the substrate in ATP-dependent cycles through the protein conducting channel of the SecYEG translocon. However, there is some indication that SecA can also bind to substrates co-translationally (Eisner *et al.*, 2003; Karamyshev and Johnson, 2005; Huber *et al.*, 2011) but whether co-translational recognition of substrates by SecA is a general feature remains to be determined. SecB can also interact with its substrates while they are still attached to the ribosome (Randall *et al.*, 1997).

SecB is not essential for targeting of all secretory proteins in *E. coli* and other Gram-negative bacteria, *e.g.* β-lactamase and it was thought that these proteins require SRP for targeting. However, it has been shown that SecB-independent proteins are not necessarily

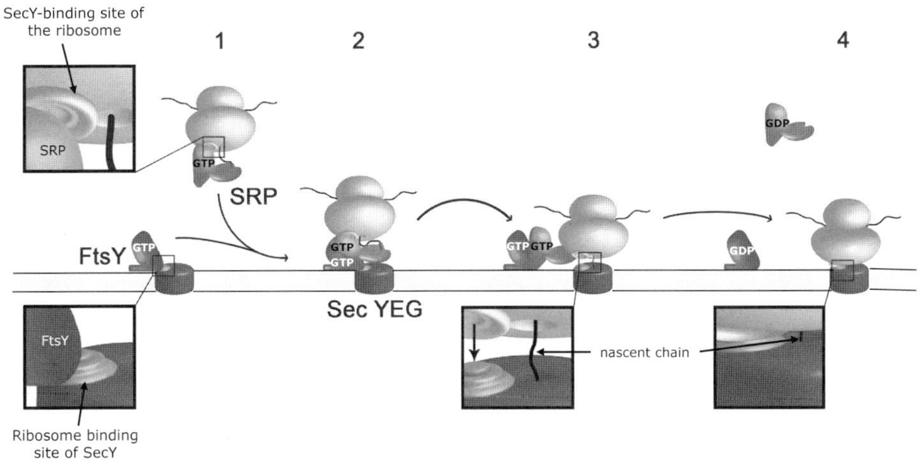

Figure 10.4 Model for the SRP-dependent co-translational targeting. (1) SRP is bound to the ribosome primarily via the L23 ribosomal subunit, which also serves as a major contact site for the cytoplasmic loops of SecY. SRP binds to an emerging signal anchor sequence via its M-domain and targets the ribosome-nascent chain complex to the membrane-bound FtsY. (2) FtsY probably exists in two membrane-bound pools: one bound to negatively charged lipids (not shown) and one bound directly to the cytoplasmic loops of SecY. (3) The interaction between FtsY and SRP induces a delocalization of SRP on the ribosome, which exposes the SecY binding site. Likewise, FtsY is delocalized on the SecYEG translocon, which exposes the cytosolic loops of SecY and allows stable docking of the translating ribosome onto the SecYEG translocon (4). The FtsY–SRP interaction also reciprocally stimulates their GTPase activity, which induces their subsequent dissociation.

routed into the SRP-pathway (Beha *et al.*, 2003). Instead it is possible that the interaction of cytosolic SecA with nascent chains diminishes the need for SecB, or that other chaperones take over its function.

SecA-dependent protein transport does not seem to occur in archaea. Instead, archaea appear to prefer the twin-arginine-translocase (TAT) for translocating secretory proteins (Müller and Klösgen, 2005).

A structural view of SecA and SecB

SecA structures from different bacteria (Sardis and Economou, 2010; du Plessis *et al.*, 2011) show a modular composition (Fig. 10.5). Six distinct domains have been identified in SecA; the N-terminal nucleotide-binding domain I (NBD1) which is followed by the peptide-binding-domain (PBD), the nucleotide-binding domain 2 (NDB2), the helical scaffold domain (HSD), the helical wing domain (HWD) and finally, the C-terminal domain (CTD). SecA belongs to superfamily 2 of the DExH/D proteins, which comprises nucleic acid-modifying enzymes and helicases (Sardis and Economou, 2010). The conserved DEAD helicase motor of SecA contains the nucleotide-binding sites NBD1 and NBD2 (also called IRA2, *intramolecular regulator of ATPase2*) and is the site of ATP-binding and hydrolysis. A single ATP molecule binds at the NBD1/NBD2 interface and its hydrolysis drives conformational changes in the motor domain and in the peptide-binding domain (also called *preprotein-cross-linking domain*, PPXD), which is thought, to link ATP hydrolysis

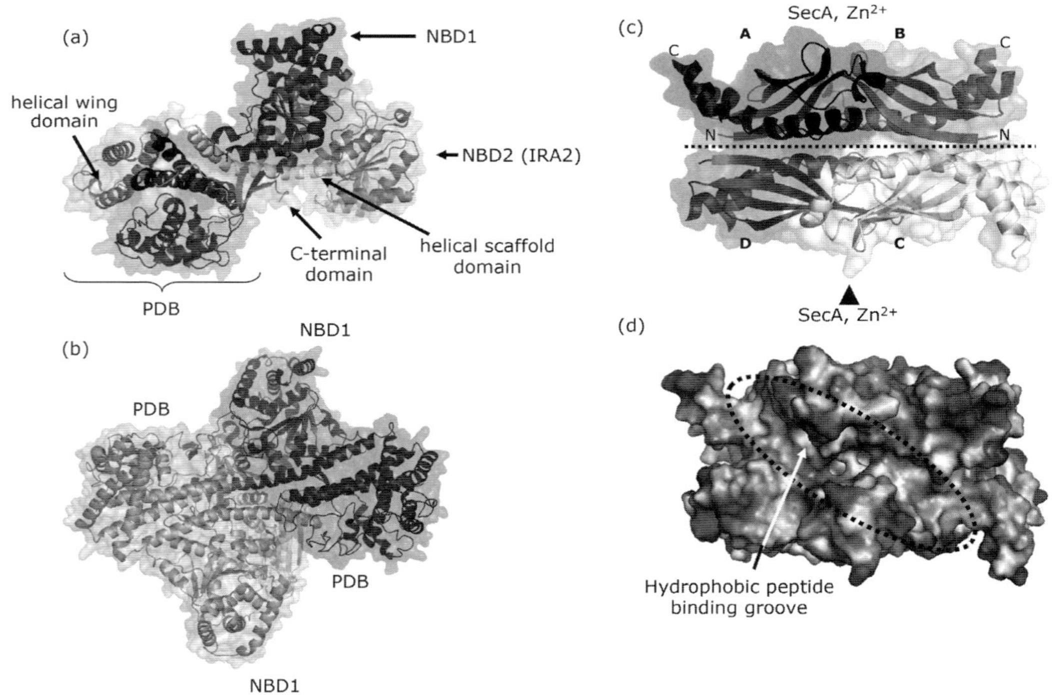

Figure 10.5 Structural insights into the transport-specific chaperone SecB and into SecA, the motor protein of the post-translational targeting. (a) Structure of the SecA monomer from *B. subtilis* (PDB: 1M6N). The SecA-domains are indicated: NDB1 and NDB2, nucleotide-binding domains 1 and 2, NDB2 is also called Ira2 (intra-molecular regulator of ATPase); PBD corresponds to the peptide binding domain. (b) Structure of the SecA-dimer from *B. subtilis* (PDB: 2IBM). The structure shows an anti-parallel orientation of the two protomers, which is also observed for most other dimeric SecA structures. However, the positioning of each protomer relative to one another appears to be unique in every SecA dimer (reviewed in Sardis and Economou, 2010). (c) and (d) Structure of the *E. coli* SecB tetramer (PDB: 1QYN), which is organized as a dimer of dimers. SecA binds its C-terminal domain to the indicated surface of SecB. Binding is mediated by zinc, which is complexed by three cysteine residues within the C-terminal domain of SecA. Indicated is also the hydrophobic groove that is suggested to bind to substrate proteins.

to translocation. The PBD consists of an anti-parallel β-strand and a globular region and is involved in substrate binding and in binding to the SecY translocon (Zimmer *et al.*, 2008). The SecA–SecY interaction involves multiple SecA residues, which map to all domains except the HWD (Mori and Ito, 2006; Zimmer *et al.*, 2008; Das and Oliver, 2011). Substrates are most likely trapped in a clamp that is formed by the PBD, NBD2 and the helical scaffold domain (HSD) (Cooper *et al.*, 2008; Zimmer and Rapoport, 2009). The two-helix finger motif of the HSD, reaches into the SecY channel and could provide the mechanical energy that pushes the substrate across the channel (Erlandson *et al.*, 2008; Zimmer *et al.*, 2008). The C-terminal domain (CTD) of SecA is involved in SecB binding (Breukink *et al.*, 1995) and in binding to phospholipids (Lill *et al.*, 1990). The CTD is not essential for catalysis, but has been suggested to inhibit ATP-hydrolysis in solution preventing futile ATP cycles in the absence of the SecYEG translocon (Keramisanou *et al.*, 2006). In most structures, SecA forms an antiparallel dimer, but whether the SecA-dimer is the active species in protein translocation is still controversially discussed (see below; Jilaveanu *et al.*, 2005; Or and Rapoport, 2007; Kusters *et al.*, 2011).

SecB exists as a homo-tetramer (a dimer of dimers) in the cytosol of many Gram-negative bacteria (Fig. 10.5) (Xu *et al.*, 2000). Each monomer has a molecular mass of 17 kDa and comprises an α/β-fold with four β-strands and two α-helices. The four monomers interact via their α-helices and expose an eight-stranded β-sheet on each side of the tetramer. SecB does not bind directly to signal sequences of secretory proteins instead it recognizes patches enriched in hydrophobic and basic residues (Topping and Randall, 1994; Knoblauch *et al.*, 1999). Since these patches are not only found in secretory proteins but also in many cytosolic proteins, the exact mechanism of substrate recognition is still not entirely clear (Hardy and Randall, 1991; Bechtluft *et al.*, 2010). However it is thought that SecB binds to these regions via a hydrophobic peptide binding groove, which contains many acidic residues at the periphery (Zhou and Xu, 2005; Crane *et al.*, 2006). SecB displays two important functions in *E. coli*: it prevents folding and aggregation of proteins (Bechtluft *et al.*, 2010) and facilitates protein targeting by binding directly to the C-terminus of the motor protein SecA (Hartl *et al.*, 1990). Structural studies using C-terminal SecA peptides indicate that one SecB tetramer binds to two SecA peptides (Zhou and Xu, 2005) leading to a SecB–preprotein-SecA complex of a 4:1:2 stoichiometry.

Protein targeting by the SecA/SecB system

In many studies, SecA is considered to be a subunit of the SecYEG translocon because the activities of both proteins are closely linked. SecA exhibits a dual function during translocation: it serves as the initial receptor for the SecB–preprotein complex and acts as the motor for pushing the substrate into the periplasm. With regards to its receptor function SecA and the bacterial SRP receptor FtsY show a remarkable similarity:

1 both function as nucleotide-dependent receptors for their cognate partner protein in complex with a substrate;
2 both have to be membrane-bound to function in targeting;
3 both bind to negatively charged phospholipids and to the surface exposed loops of SecY.

Nevertheless, there is no sequence or structural homology between SecA and FtsY and nucleotide hydrolysis appears to serve different functions. GTP-hydrolysis by the FtsY–SRP complex induces its dissociation and allows its recycling, while ATP hydrolysis by SecA is thought to provide the energy for translocation.

The fidelity of the SecA/SecB targeting system is maintained by kinetic proof-reading steps similar to the ones described for the SRP/FtsY system. Soluble SecA has a low affinity for SecB, with a K_d in the μM range (den Blaauwen *et al.*, 1997). However, the binding of SecA to SecYEG increases its affinity for SecB to approximately 30 nM and to 10 nM for a SecB–substrate complex (Fekkes *et al.*, 1997). In the ternary SecB–substrate-SecA complex, the substrate is probably no longer bound to SecB but is transferred to SecA, which has a high affinity for signal sequences (Bechtluft *et al.*, 2010). ATP-binding by SecA is believed to induce protein translocation via SecYEG and the release of SecB from SecA (Fekkes *et al.*, 1997). ATP hydrolysis by SecA provides the energy for protein translocation and SecA probably undergoes multiple ATP-hydrolysis cycles for transporting a secretory protein across the SecYEG translocon. Each ATP-hydrolysis is suggested to push about 30 amino acids across the SecYEG translocon and kinetic analyses have shown that the time required for translocation is directly proportional to the length of the substrate (Tomkiewicz *et al.*, 2008).

Conflicting results exist as to a possible dissociation of the SecA dimer into monomers during translocation/interaction with SecYEG (Alami *et al.*, 2007). Signal sequences bind to monomeric and dimeric SecA with the same affinity as indicated by FRET experiments (Auclair *et al.*, 2010). SecA can bind as a monomer or a dimer to SecYEG-dimers in detergent-solution when the SecYEG-dimer is antibody-stabilized (Tziatzios *et al.*, 2004). Tethering both SecA protomers permanently together, either by chemical cross-linking or by genetic fusion resulted in both active (de Keyzer *et al.*, 2005; Jilaveanu *et al.*, 2005, 2006; Wang *et al.*, 2008) as well as inactive dimeric SecA species (Or and Rapoport, 2007). Recent single-molecule studies favour dimeric SecA as the active species (Kusters *et al.*, 2011), but the X-ray structure shows a monomeric SecA bound to monomeric SecYEG, indicating that a SecA-dimer is not required for a stable SecY–SecA complex (Zimmer *et al.*, 2008). It therefore still remains an open question whether the active form of SecA during translocation is a monomer or dimer.

The SecYEG translocon constitutes the major protein conducting channel in bacteria

The SecYEG translocon is a universally conserved protein conducting channel for transporting proteins out of the cytosol (du Plessis *et al.*, 2011). It is located in the endoplasmic reticulum membrane of eukaryotes (Sec61αβγ complex) and in the cytoplasmic membrane of bacteria (SecYEG complex). The Sec61/SecYEG complex transports secretory proteins and inserts membrane proteins. Therefore it needs to switch between opening laterally for the release of transmembrane domains and opening vertically to facilitate transport of hydrophilic domains of membrane proteins and periplasmic and outer membrane proteins.

The core of the bacterial SecYEG translocon is composed of the three membrane proteins SecY, SecE and SecG. Only SecY and SecE are essential in bacteria and homologous to the Sec61α and Sec61γ subunits of the eukaryotic Sec61 complex (Fig. 10.6). SecY is a hydrophobic multiple spanning membrane protein of 48 kDa with 10 TM domains. SecE in *E. coli* is a 14-kDa protein containing three predicted transmembrane domains (TMs). Only the third TM is essential for function and some bacteria contain SecE with only one

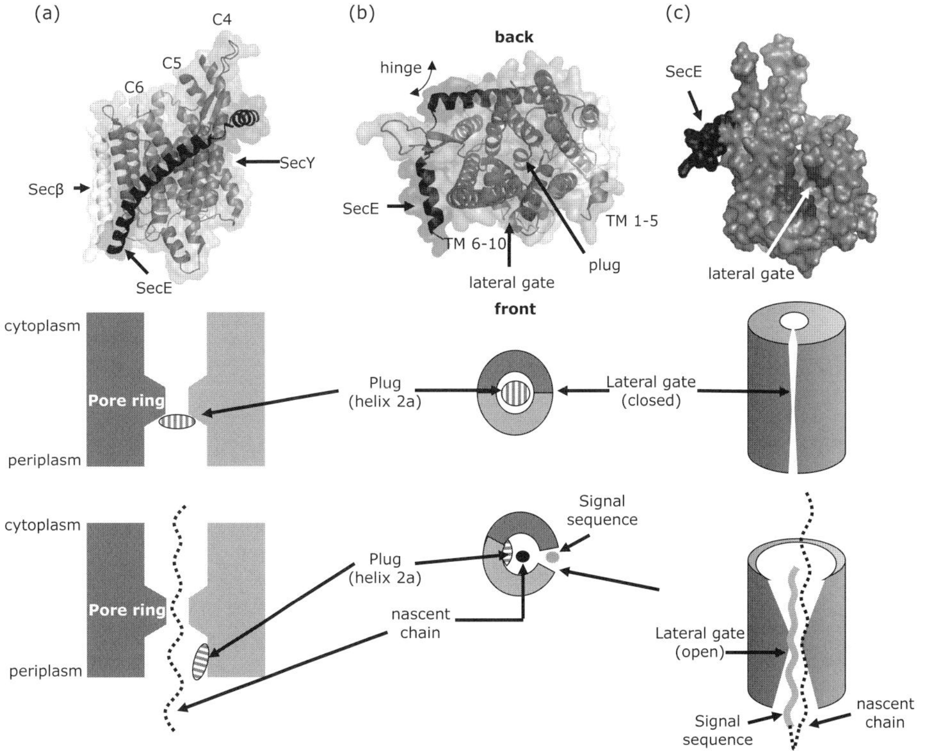

Figure 10.6 Structure of the SecYEG translocon. (a) Top: Side view of the SecYEβ complex of *M. jannaschii* (PDB: 1RH5), the first resolved X-ray structure of a Sec translocon. While SecY and SecE are universally conserved, eukaryotes and archaea contain Sec61β as third subunit, instead of SecG, which is found in bacteria. The relative positioning of SecG and Sec61β to the SecYE core appears to be similar. C4, C5 and C6 correspond to the cytosolically exposed loops of SecY, which provide the contact sites for FtsY, SecA and ribosomal subunits. Bottom: Schematic cross-sectional views of the SecYEG translocon in the closed and open states. The plug corresponds to the small helix 2a, which blocks the hydrophobic pore ring. (b) Top: Top view from the cytosol onto the *M. jannaschii* SecYEβ complex. The SecE subunit stabilizes the back of the two halves of SecY. One half corresponds to TMs 1–5 and the second to TM2 6–10. Bottom: Schematic top view of the SecYEG translocon in the closed and open states. (c) Top: Side view of the *P. furiosus* SecYEβ complex, showing the lateral gate (PDB: 3MP7). IMPs probably exit the SecY channel via the lateral gate into the membrane. Bottom: Schematic side views of the SecYEG translocon in the closed and open states.

TM. In the stoichiometric SecYE complex, SecY forms the protein conducting channel that is stabilized by the clamp-like structure of SecE (van den Berg *et al.*, 2004). In the absence of SecE, SecY is rapidly degraded by FtsH (Kihara *et al.*, 1995). The non-essential SecG subunit appears to be specifically linked to SecA-dependent translocation (Nishiyama *et al.*, 1996) particularly at low temperatures or in the absence of the proton motive force (PMF). SecG is proposed to support the dynamic SecA–SecY interaction by reversible topology changes (Nishiyama *et al.*, 1996), but in the available SecA-SecYEG structure the SecG topology appears to be unaltered (Zimmer *et al.*, 2008). SecG can also bind directly to SecY (van der Sluis *et al.*, 2002; Kuhn *et al.*, 2011), but is not required for the transport of

SecA-independent membrane proteins (Koch and Müller, 2000). SecG in eukaryotes and archaea is replaced by the Sec61β subunit, which shows no homology to SecG. In eukaryotes, Sec61β is involved in binding to the β-subunit of the SRP receptor, which is absent in prokaryotes (Fulga *et al.*, 2001; Jiang *et al.*, 2008).

In bacteria, the SecYEG complex associates at least transiently with another membrane protein complex, the SecDFYajC complex. In *E. coli*, SecD is a 67 kDa protein with six transmembrane domains and an extended periplasmic loop between TM1 and TM2. The topology of the 35 kDa SecF is very similar to SecD, while YajC is a 12 kDa single-spanning membrane protein with a large cytoplasmic domain (Fig. 10.7). There are about 20–50 copies of the SecDFYajC complex in *E. coli*, but the exact function has been a mystery. Cells lacking SecD and SecF are cold sensitive and impaired in protein translocation (Pogliano and Beckwith, 1994). Although YajC associates with SecDF, deleting it does not cause a growth phenotype. Recent data indicate that the SecDFYajC complex could connect the SecYEG complex to YidC with residues 215–265 of the periplasmic loop of *E. coli* YidC

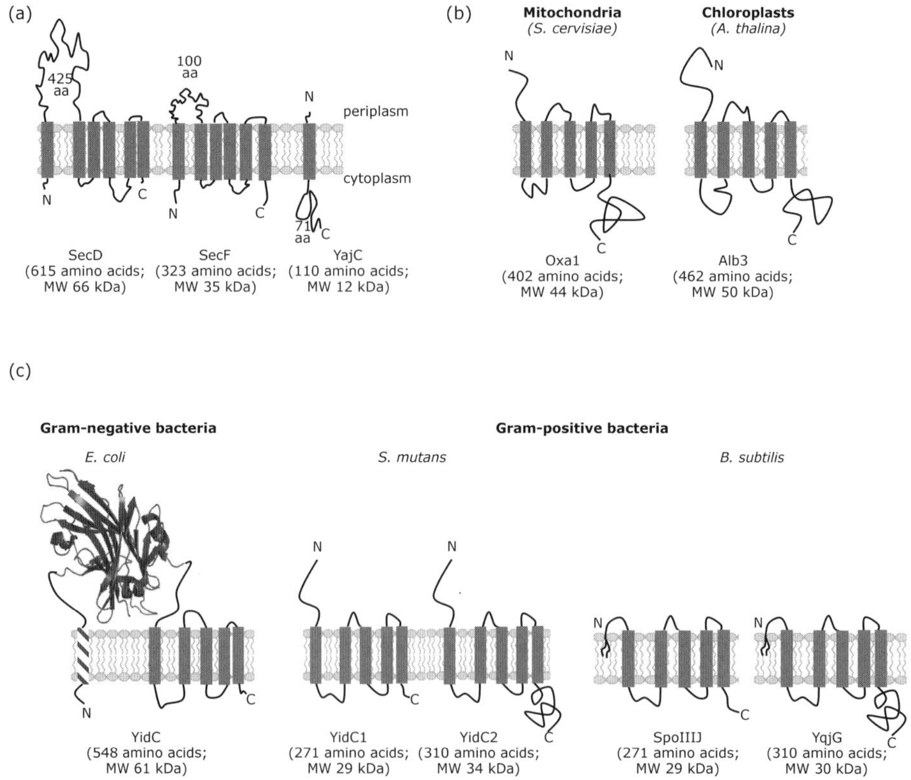

Figure 10.7 Membrane topology of the SecYEG associated proteins SecD, SecF, YajC and YidC. (a) Topology of the *E. coli* proteins. The X-ray structure of a fused SecDF protein from *T. thermophilus* has recently been resolved (Tsukazaki *et al.*, 2011). (b) Proposed topology of Oxa1/Alb3. (c) Proposed topology of YidC from different bacterial species. In *E. coli*, the crystal structure of the first periplasmic loop is shown (PDB: 3BS6). In *B. subtilis* the N-terminus is probably membrane-anchored via a lipid-modification.

shown to be a potential SecF binding region (Nouwen and Driessen, 2002; Xie *et al.*, 2006). However, YidC–SecYEG interaction is also observed in the absence of the SecDFYajC complex (Boy and Koch, 2009; Sachelaru *et al.*, unpublished data). An X-ray structure of the SecDF complex of *T. thermophilus* has recently been published and it was proposed that SecDF functions as a PMF-powered chaperone that completes translocation after the ATP-dependent SecA functions (Tsukazaki *et al.*, 2011).

YidC is an essential protein that transiently associates with SecYEG and has been suggested to facilitate insertion and folding of membrane proteins (Dalbey *et al.*, 2011). It has also been shown to co-purify with components of the Sec translocon indicating an associative function (Scotti *et al.*, 2000). However, *in vitro* data suggests that the association of YidC with SecYEG is not required for the insertion of membrane proteins via SecYEG (Braig *et al.*, 2011; Welte *et al.*, 2012). YidC is universally conserved and is found in the cytoplasmic membrane of bacteria and in the inner membranes of mitochondria (Oxa1) and chloroplasts (Alb3) (Yen *et al.*, 2001; Dalbey *et al.*, 2011) (Fig. 10.7). All YidC homologues share a conserved region of five predicted transmembrane helices (Funes *et al.*, 2011). In Gram-negative bacteria, an additional N-terminal transmembrane domain is fused via a 320 amino acid long periplasmic loop to the central core. The X-ray structure of the periplasmic loop has been elucidated (Oliver *et al.*, 2008; Ravaud *et al.*, 2008) and will be referred to later along with functions of *E. coli* YidC. Some Gram-positive bacteria, like *S. mutans* contain two YidC homologues and either one appears to be sufficient for cell viability under normal growth conditions, but YidC2 seems to be specifically required under stress conditions (Hasona *et al.*, 2005).

The structure and oligomeric state of the SecYEG translocon

First attempts to reveal the structure of the translocon by electron microscopy showed a quasi-pentagonal structure of purified Sec61 complexes with a diameter of about 8.5 nm and a predicted central pore of about 2 nm in diameter (Hanein *et al.*, 1996) consistent with EM studies on SecYE complexes (Meyer *et al.*, 1999). These studies and the first 3D cryo-EM reconstruction of the Sec61 complex (Beckmann *et al.*, 1997) indicated that the protein conducting channel is formed by an oligomer presumably containing 2–4 SecYEG copies. A SecYEG dimer was also predicted based on a 3D reconstruction of 2D *E. coli* SecYEG crystals (Breyton *et al.*, 2002). However, the first high-resolution structure obtained from the archaeon *Methanocaldococcus jannaschii* showed a monomeric SecYEβ translocon (van den Berg *et al.*, 2004) that superimposed onto the 3D reconstruction of the *E. coli* SecYEG complex (Bostina *et al.*, 2005). The 10 transmembrane domains of SecY are arranged like a clam-shell of which TMs 1–5 form one half and TMs 6–10 the second half (Fig. 10.6). The loop connecting TM5 with TM6 probably hinges the two halves together, which is further stabilized by SecE. This hinged region is called the 'back' of the SecYEβ translocon. A lateral gate is formed on the opposite site – the 'front' region – by helices 2b and 7 and is thought to allow partitioning of hydrophobic domains into the lipid phase. The SecY channel has an hourglass-like conformation with cytoplasmic and periplasmic vestibules demarcated by a constriction in the middle. The constriction (or 'pore ring') is lined with hydrophobic residues with their side chains directed towards the centre of the constriction. These residues could provide a seal for maintaining the permeability barrier of the membrane during protein translocation (Park and Rapoport, 2011). In the *M. jannaschii* structure, the diameter of the constriction is approximately 0.5–0.8 nm and thus too small for allowing α-helical polypeptides to pass through. On the periplasmic side of the membrane, the constriction is blocked by a small helix (TM helix 2a) predicted to function as a plug. The structure therefore likely represents the closed state of the Sec translocon. Two additional structures probably represent the pre-open state of the Sec translocon; the *Thermus thermophilus* SecYE structure was resolved with a Fab fragment bound to the cytosolic loops 4 and 5 of SecY (Tsukazaki *et al.*, 2008), which was suggested to mimic the SecA-bound state of SecY. In comparison to the *M. jannaschii* structure, helices 6, 8 and 9 were seen to be displaced. The structure of the *Thermotoga maritima* SecYEG was resolved in complex with SecA and shows an open lateral gate and dislocation of the plug domain towards the periplasmic side of the translocon. Nevertheless, the central pore is still closed and therefore the SecYEG is probably still in a pre-open state. SecG in the *T. maritima* SecYEG complex has two transmembrane domains, with both termini located in the periplasm. The C-terminal transmembrane domain has the same orientation and position as the β-subunit in *M. jannaschii*. In the SecYEG-SecA structure, the loop connecting both transmembrane domains is located between SecY and SecA, which probably explains its protease protection during translocation (Nishiyama *et al.*, 1996). The pore size of the SecY channel appears be to be flexible in the open state. Based on fluorescence quenching experiments a pore size of 4–6 nm has been proposed for the eukaryotic Sec61 complex (Hamman *et al.*, 1997), which was also supported by electrophysiological experiments (Wirth *et al.*, 2003). Molecular dynamics simulations of the *E. coli* SecYEG complex have suggested a maximal pore size of approx. 1.6 nm (Gumbart and Schulten, 2007). On the other hand, a pore size of at least 2.2–2.4 nm has been proposed for the *E. coli* SecYEG based on experiments using sizable spherical molecules attached to a pre-protein, (Bonardi *et al.*, 2011). These differences in pore size might reflect certain plasticity in pore geometry, which could be influenced by the substrate and lateral gate opening.

Although the oligomeric state of the SecYEG translocon remains a controversial issue it has been suggested that in a SecYEG dimer only one channel is used for translocation, while the second copy serves as a high-affinity docking site for SecA (Osborne and Rapoport, 2007; Deville *et al.*, 2011). A possible oligomeric organization of the SecYEG translocon is supported by several biochemical studies (e.g. Manting *et al.*, 2000; Osborne and Rapoport, 2007; Boy and Koch, 2009; Deville *et al.*, 2011), while X-ray crystallography and recent high-resolution cryo-EM studies (Becker *et al.*, 2009; Frauenfeld *et al.*, 2011) point to a SecYEG monomer as the active species. A recent single-molecule study also indicates that a single copy of SecYEG is sufficient for translocation (Kedrov *et al.*, 2011). However, this does not necessarily exclude that the active SecYEG translocon exists in a substrate-dependent dynamic equilibrium between monomers, dimers and probably even higher order oligomers.

The ribosome–SecY interaction and the mechanism of membrane protein integration

Cryo-EM reconstruction of the Sec61 translocon bound to empty or translating ribosomes have shown that the ribosomal exit tunnel aligns precisely with the protein conducting channel (Beckmann *et al.*, 1997; Becker *et al.*, 2009). The ribosome–SecY interaction involves the cytoplasmic loops C4, C5 and C6 of SecY and the ribosomal proteins L23, L24, and L29 (Cheng *et al.*, 2005; Menetret *et al.*, 2007; Becker *et al.*, 2009). The SecY loops C4 and C5 extend into the ribosomal exit tunnel and contact the 23S rRNA namely helices H50,

H59 and H7 (Fig. 10.8) Ribosome binding to SecY requires the delocalization of SRP on the ribosome to expose the L23 subunit (Halic *et al.*, 2006) and probably the delocalization of FtsY on the SecY translocon to expose the cytosolic loops C4 and C5 (Kuhn *et al.*, 2011).

The contact between the ribosome and SecYEG probably switches the translocon into the pre-open state, in which the lateral gate is partially opened but the central pore still occluded by the plug domain (Becker *et al.*, 2009). Simultaneously, the diameter of the pore ring seems to widen and the cytoplasmic vestibule increases (Egea and Stroud, 2010). The signal anchor sequence moves through the cytoplasmic vestibule and is sandwiched between helices 2b and 7 of the lateral gate (Fig. 10.6), as indicated by both structural (Becker *et al.*, 2009; Egea and Stroud, 2010) as well as cross-linking studies (Plath *et al.*, 1998). Blocking the opening of the lateral gate by disulfide bridges inhibits protein transport (du Plessis *et al.*, 2009). The interaction between the signal anchor sequence and the lateral gate probably further destabilizes the connection between the pore ring and the plug domain, causing it to move into the periplasmic vestibule. This is supported by cross-links between the transmembrane domain of SecE and the plug domain during transport (Harris and Silhavy, 1999). This open state of the translocon is probably fixed when the on going protein synthesis aligns the next polypeptide segment distal to the signal anchor sequence.

Transport of soluble proteins across the channel probably occurs in the same manner with two major exceptions:

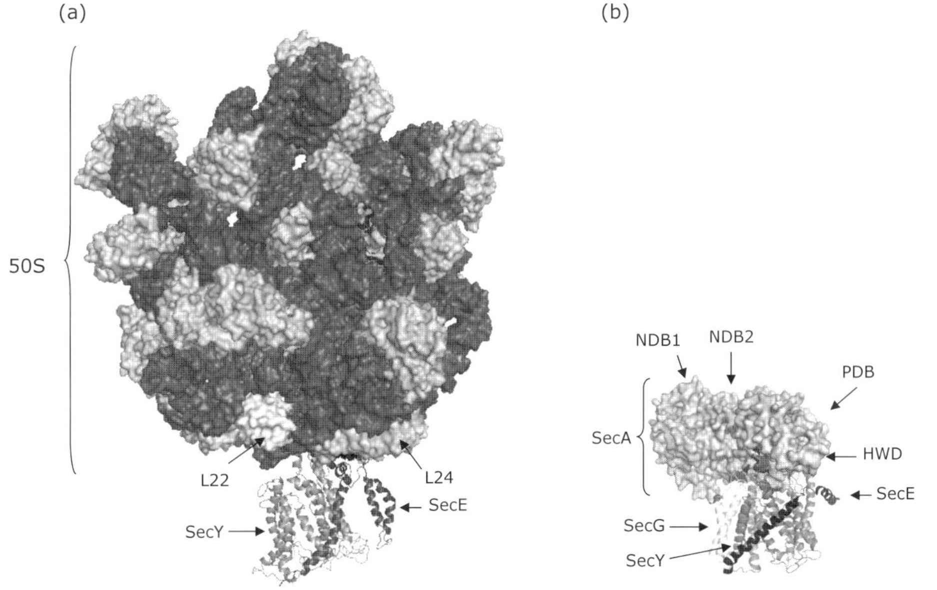

Figure 10.8 Interaction of the SecYEG complex with ribosomes and SecA (a) The ribosome–SecYEG complex of *E. coli*. The structure is based on a cryo-EM reconstruction (PDB: 3J01). The ribosomal RNA is displayed in dark grey and the ribosomal protein in light grey. The L22 and L24 subunits are indicated, the L23 and L29 subunits are not visible because they are located at the back. (b) The SecA-SecYEG structure from *T. maritima* (PDB: 3DIN). The labelling corresponds to the labelling in Fig. 10.5.

1. SecA induces the pre-open state of the translocon instead of the ribosome, and
2. Plug movement appears to be different during transport of soluble and membrane proteins (Nijeholt-Lycklama *et al.*, 2011).

This raises the question of how the translocon differentiates between protein segments that need to be laterally released into the lipids and those that need to be transported across. In the proposed pre-open state, signal-anchor sequences or cleavable signal sequences are in contact with lipids and thermodynamic analysis has indicated that if a protein segment is sufficiently hydrophobic, e.g. a signal anchor sequence or a TM, it will partition into the lipids while less hydrophobic segments will be threaded through the aqueous channel onto the periplasmic side (Hessa *et al.*, 2007). Thus, protein–lipid interactions assist in this decision. Owing to its width, the SecY channel can probably accommodate only one transmembrane domain, indicating that during the insertion of polytopic membrane proteins, transmembrane domains exit the channel in the order they are synthesized (Sadlish *et al.*, 2005). However, the insertion of moderately hydrophobic transmembrane domains depends on the preceding transmembrane domain and it has been suggested that multiple transmembrane domains assemble at or close to the SecYEG translocon before they are released *en bloc* into the lipid bilayer (Heinrich and Rapoport, 2003). Concerted insertion is probably favoured if the transmembrane domains are closely spaced, while transmembrane domains which are separated by large hydrophilic loops leave the translocon sequentially (Houben *et al.*, 2005).

YidC facilitates insertion, folding and assembly of SecYEG-dependent membrane proteins

Independently of a sequential or concerted release of TM domains from SecY, the inner membrane protein YidC probably associates with the SecYEG translocon (Scotti *et al.*, 2000; Boy and Koch, 2009). YidC either assists transmembrane domains during lipid insertion (Houben *et al.*, 2005) or assembles multiple transmembrane domains before their release into the lipid phase (Beck *et al.*, 2001). However, this SecYEG-associated function of YidC is probably not essential for membrane protein integration; the polytopic membrane protein mannitol permease (MtlA) (Beck *et al.*, 2001) and the single spanning membrane protein FtsQ (Houben *et al.*, 2005) have been cross-linked to YidC, but proteoliposome studies have shown that their integration does not require YidC (van der Laan *et al.*, 2001; Braig *et al.*, 2011; Welte *et al.*, 2012). On the other hand, the integration of subunit *a* of the F_1F_0 ATP synthase is dependent on both SecY and YidC (Kol *et al.*, 2008). YidC has also been shown to control the topology of SecYEG-dependent membrane proteins such as the *E. coli* Lactose/H^+-symporter LacY (Nagamori *et al.*, 2004).

YidC has been implicated in the assembly of membrane protein complexes based on the functions of its mitochondrial homologue Oxa1. Oxa1 is required for the functional assembly of respiratory chain complexes in mitochondria like cytochrome oxidase (Bonnefoy *et al.*, 1994; Hell *et al.*, 2001) and ATP synthase (Altamura *et al.*, 1996). Since mitochondria lack the SecYEG translocon, it was also proposed that Oxa1 is required for the integration of mitochondrially encoded proteins and their subsequent assembly into functional complexes (Bonnefoy *et al.*, 1994). This assembly function appears to be also universally conserved; in *E. coli*, some protein complexes, e.g. cytochrome bo_3 oxidase (van der Laan *et al.*, 2003), NADH dehydrogenase I (Price and Driessen, 2008), and ATP synthase (Wang *et al.*, 2010) are severely impaired under YidC depletion conditions, most likely because individual subunits of these complexes depend on YidC for integration. However, the assembly function of YidC is not limited to respiratory chain complexes, as the maltose transporter has been found to be severely reduced in YidC-depleted membranes (Wagner *et al.*, 2008).

YidC is also probably involved in protein quality control. In *E. coli*, misfolded membrane proteins are degraded by the AAA-type membrane protease FtsH (Sauer and Baker, 2011) and other membrane proteases like YaeL (RseP) (Akiyama *et al.*, 2004) and GlpG (Maegawa *et al.*, 2005). FtsH has been shown to co-purify and cross-link with YidC which might indicate that that FtsH and YidC function co-operatively in the quality control of membrane proteins (van Bloois *et al.*, 2008). Here, two models are conceivable:

1 YidC binds to incompletely folded membrane proteins and protects them against proteolysis. This model is supported by data that indicate that YidC keeps HbP in a transport-competent state and upon YidC depletion the protein gets degraded by DegP (Jong *et al.*, 2010). HbP belongs to the autotransporter family of outer membrane proteins (Dautin and Bernstein, 2007) and is so far the only YidC-dependent outer membrane protein.
2 YidC delivers misfolded proteins to membrane proteases like FtsH. This is supported by the observation that upon YidC depletion a subset of membrane proteins is increased in the inner membrane (Price *et al.*, 2010). A possible interpretation would be that under YidC-depletion conditions the degradation of non-functional proteins in decreased.

However, details about quality control mechanisms for bacterial membrane proteins need to be further explored.

The integration of inner membrane proteins with large periplasmic loops requires SecA

On going chain elongation at the ribosome is sufficient for transferring small periplasmic loops of inner membrane proteins across the SecYEG channel, while periplasmic loops longer than 30 amino acids seem to require the ATPase activity of SecA to be pushed across the translocon (Andersson and v. Heijne, 1993). This has been shown for the 320 amino acid long periplasmic loop of YidC (Koch *et al.*, 2002; Deitermann *et al.*, 2005) and for the periplasmic loop of the single spanning membrane protein FtsQ (van der Laan *et al.*, 2004b) Both proteins are co-translationally targeted to the SecYEG translocon by SRP and FtsY. Thus, during the integration of these membrane proteins, the binding of SecA and the ribosome to the SecYEG translocon have to be coordinated. The simultaneous binding of SecA and the ribosome to SecYEG (Zito and Oliver, 2003) seems unlikely if the SecYEG channel functions as a monomer since both SecA and the ribosome use overlapping binding sites (Mori and Ito, 2006, Zimmer *et al.*, 2008; Kuhn *et al.*, 2011; Wu *et al.*, 2012). For single spanning membrane proteins, protein synthesis is probably completed and the ribosome released before SecA translocates the hydrophilic loop (Deitermann *et al.*, 2005). The situation is more complex for polytopic membrane proteins when the large loop is sandwiched between two transmembrane domains; here SecA-dependent translocation occurs while the ribosome is still attached (Deitermann *et al.*, 2005). It is likely that the ribosome-SecY contact is weakened allowing SecA access to the hydrophilic loop, but this needs to be determined in future studies.

The passage of outer membrane proteins through the SecYEG translocon

After SecA/SecB dependent targeting to the SecYEG translocon, periplasmic and outer membrane proteins have to traverse the SecYEG translocon completely. By interacting with SecY, SecA probably switches the translocon from the closed to the pre-open state. SecA functions as a receptor for the SecB–preprotein complex and as a motor protein that hydrolyses ATP during translocation. In addition, the proton motive force (PMF) is involved in driving translocation either by orienting the signal sequence in the SecYEG translocon (van Dalen *et al.*, 1999) or by being involved in opening the SecYEG channel (Nouwen *et al.*, 1996). After initial ATP binding to SecA and pre-protein release from SecB, the signal sequence of the pre-protein acquires an α-helical conformation (Chou and Gierrasch, 2005) that is probably bound to the substrate binding clamp of SecA (Zimmer *et al.*, 2008). Simultaneous conformational changes cause SecA to penetrate deeper into the channel, where the signal sequence is intercalated by the lateral gate of SecY (Papanikou *et al.*, 2007) and remains till it is cleaved by signal peptidase and degraded. The opening of the lateral gate by the signal sequence is thought to destabilize the interaction between the plug and pore-ring of SecY and causes the channel to open. The open-state is then fixed by the hair-pin like orientation of N-terminus of the pre-protein in the channel. ATP is then hydrolysed, the substrate is released and the SecA–SecY interaction weakened. Whether SecA completely dissociates from SecY at this stage is unclear.

A stepwise translocation of secretory proteins through the SecYEG translocon has been observed in biochemical assays, which has led to the hypothesis that repeated ATP hydrolysis cycles of SecA would push pre-proteins in segments of about 30 amino acids across the membrane (Schiebel *et al.*, 1991; Uchida *et al.*, 1995; van der Wolk *et al.*, 1997). The recent crystal structure of the SecYEG–SecA complex provides a possible explanation for this power-stroke model (Zimmer *et al.*, 2008) (Fig. 10.8). The substrate binding clamp of SecA is partially formed by the essential two-helix-finger domain (Jarosik and Oliver, 1991; Karamanou *et al.*, 1999) that can be cross-linked to a polypeptide (Erlandson *et al.*, 2008). In the SecA-SecYEG structure, the two-helix-finger is inserted into the SecY channel (Zimmer *et al.*, 2008) and thought to move up and down inside the channel during the ATP-hydrolysis cycle. This mechanical force would push the pre-protein across the channel. Nevertheless, whether the two-helix-finger moves in response to ATP-binding and hydrolysis and how this relates to stepwise translocation remains to be determined. As an alternative to the power-stroke model, the Brownian ratchet model was proposed. Here SecA mainly functions in preventing the back-sliding of the substrate in an ATP-dependent manner and ATP hydrolysis is not directly coupled to precursor movement. Additional models have been proposed which have been reviewed in Kusters and Driessen (2011).

During post-translational targeting in eukaryotes, substrates that emerge from the Sec translocon interact with ER luminal chaperones like members of the Hsp70 family (Rapoport, 2007) while in bacteria, periplasmic chaperones like Skp or PpiD bind to substrate proteins when they are still in contact to the SecYEG translocon (Schäfer *et al.*, 2002; Antonoaea *et al.*, 2008). These chaperones might fulfil several functions; in addition to facilitating protein folding, they might also be involved in preventing back-sliding or in substrate release from the periplasmic side of the SecY channel. However, analysis of the exact function of periplasmic chaperones in protein translocation is difficult due to their redundancy.

YidC forms a SecYEG-independent integration site for some membrane proteins

While it is generally believed that the majority of inner membrane proteins are inserted via SecYEG, some inner membrane proteins seem to use YidC as an alternative insertion site (Dalbey *et al.*, 2011; Wang *et al.*, 2011; Welte *et al.*, 2012). This was first described for small phage proteins (Samuelson *et al.*, 2000) and later for native *E. coli* inner membrane proteins (Dalbey *et al.*, 2011). As described earlier, YidC appears to execute multiple functions during membrane protein integration in bacteria. Considering its multiple functions, it is not surprising that YidC is essential in bacteria and that YidC depletion in *E. coli* results in global changes in cell physiology (Price *et al.*, 2010; Wang *et al.*, 2010; Wickström *et al.*, 2011b).

The members of the YidC family of proteins have been described and characterized in chloroplast membranes (Alb3), mitochondria (Oxa1) and bacteria (Funes *et al.*, 2011). In archaea, hypothetical proteins with sequence homology to YidC have been identified (Yen *et al.*, 2001), but not yet characterized. Phylogenetic studies have classified the members of the YidC family into six subgroups, which probably evolved by independent gene duplication events (Funes *et al.*, 2011). Conservation of the YidC family of proteins across kingdoms allows complementation of YidC depletion mutants in *E. coli* with Oxa1 or Alb3 (Jiang *et al.*, 2002; van Bloois *et al.*, 2005). Complementation of a yeast Oxa1-deletion strain with *E. coli* YidC is also possible if the C-terminal ribosome-binding domain of Oxa1 is fused to the C-terminus of YidC (Preuss *et al.*, 2005). This indicates that during evolution, additional functions were appended to the five-transmembrane core of YidC. Most prominent is the addition of the C-terminal extension in Oxa1, which binds to ribosomes (Szyrach *et al.*, 2003) and the N-terminal extension in YidC of Gram-negative bacteria (Sääf *et al.*, 1998), for which a function still needs to be assigned. In the Gram-positive bacteria *B. subtilis* and *S. mutans*, two YidC-homologues have been identified, but either one is sufficient for cell viability (Tjalsma *et al.*, 2003; Hasona *et al.*, 2005). YidC2 has a C-terminal extension like Oxa1 that is likely to bind ribosomes as YidC2 can complement an Oxa1 mutant (Funes *et al.*, 2011). YidC2 appears to be required for growth under acid stress and at high salt concentrations, but *E. coli* YidC can substitute for *S. mutans* YidC2 even under stress conditions (Dong *et al.*, 2008).

YidC most likely exists as dimer in *E. coli* membranes (Nouwen *et al.*, 2002; Lotz *et al.*, 2008; Boy and Koch, 2009), but structural information is so far limited to the X-ray structure of the large periplasmic loop (Oliver and Paetzel, 2008; Ravaud *et al.*, 2008). The periplasmic loop crystallized as a dimer, where each monomer is characterized by a β-supersandwich folding motif and a C-terminal α-helical region. A potential substrate binding site is occupied by polyethylene glycol in the X-ray structure, which could indicate that peptides or acyl side chains interact with the periplasmic loop (Ravaud *et al.*, 2008). Tryptophan fluorescence measurements show conformational changes of the periplasmic loop upon binding of Pf3, a phage protein inserted via YidC (Imhof *et al.*, 2011). Nevertheless, all available data indicate that the periplasmic loop is not essential for YidC function in *E. coli* (Jiang *et al.*, 2003) and its exact function still needs to be defined.

A striking feature of YidC is its tolerance towards mutagenesis; mutations in TM4 and TM5 or even replacing them with unrelated TMs do not interfere with function (Jiang *et al.*, 2003). However, TM2, TM3 and TM6 seem to be important for activity (Yu *et al.*, 2008).

Cold-sensitive TM3 mutants have been described that can be suppressed by second-site mutations within TM2, which led to the assumption that TM2 and TM3 cooperate during membrane protein integration (Yuan *et al.*, 2007). A recent cysteine cross-linking study has shown that the Pf3 coat protein contacts TM1, TM3, TM4 and TM5 of YidC (Klenner and Kuhn, 2012). These data, together with the previously described cryo-EM structure of a ribosome-nascent chain bound to YidC (Kohler *et al.*, 2009), suggest that in the dimeric *E. coli* YidC, TM2 and 3 of each protomer are involved in building an insertion pore.

Substrates for the YidC insertase and their targeting

The common feature of known YidC-only substrates is that they are closely spaced inner membrane proteins, lacking large periplasmic or cytoplasmic loops. Initially, mainly small subunits of respiratory complexes were analysed as YidC substrates due to the established link between the mitochondrial Oxa1 and the respiratory chain (Bonnefoy *et al.*, 1994). So far the number of substrates, which clearly use the YidC-only pathway is still low (Dalbey *et al.*, 2011), but recent data demonstrate that YidC can also insert polytopic membrane proteins *in vitro* (Welte *et al.*, 2012), suggesting that the number of bona fide YidC substrates is probably significantly higher.

The first proteins identified as substrates for the YidC-only insertion pathway were the coat proteins of the M13 and Pf3 bacteriophages (Samuelson *et al.*, 2000, 2001). YidC was shown to interact directly with the transmembrane helices of Pf3- and M13-coat protein and proteoliposomes containing only YidC inserted these substrates (Serek *et al.*, 2004). It was assumed that these phage proteins hijacked a so far unknown membrane insertion pathway. Owing to the established role of Oxa1 in the insertion of mitochondrial respiratory chain complexes, the assembly of respiratory chain proteins was studied in YidC-depletion strains and in YidC proteoliposomes. The activity of the F_0F_1 ATP-Synthase and the cytochrome bo_3 oxidase was reduced upon YidC depletion and it was shown that the F_0c subunit of the F_0F_1-ATP Synthase and subunit II (CyoA) of the cytochrome bo_3 oxidase required YidC for insertion (van der Laan *et al.*, 2003, 2004a; Yi *et al.*, 2003). However, while the F_0c subunit required only YidC, CyoA required both YidC and SecYEG. CyoA insertion appears to follow a rather unusual pathway where the N-terminal region of CyoA was shown to insert via the YidC only pathway whereas the C-terminus is inserted into the membrane by the SecYEG complex (Celebi *et al.*, 2006). The insertion of the C-terminus by SecYEG requires previous insertion of the N-terminus by YidC (Celebi *et al.*, 2006; du Plessis *et al.*, 2006; van Bloois *et al.*, 2006). Why CyoA requires YidC and SecYEG as separate entities during integration and whether also other proteins follow this pathway is currently unknown. Studies using CyoA and F_0c mutants suggest that the charge distribution of their periplasmic loops plays an important role in YidC binding (Celebi *et al.*, 2008; Kol *et al.*, 2008).

Another putative YidC-only substrate is the mechano-sensitive channel of low conductance, MscL. MscL insertion has been shown to be YidC dependent and SecYEG independent (Facey *et al.*, 2007; Price *et al.*, 2010), however, there is some controversy whether YidC functions as insertase for MscL or whether it facilitates the assembly of MscL monomers into stable pentamers (Pop *et al.*, 2009). Recently, it has also been suggested that MscL might insert spontaneously, i.e. independently of any proteinaceous factor (Berrier *et al.*, 2011).

In recent studies, a more general role of YidC in membrane protein integration was demonstrated. *In vitro* YidC is sufficient to integrate the polytopic membrane proteins TatC,

MtlA, as well as a YidC-mutant lacking the large periplasmic loop (Welte *et al.*, 2012). All three polytopic membrane proteins can also be inserted via SecYEG, indicating that some polytopic membrane proteins can be inserted by either SecYEG or YidC. In contrast, YidC is unable to integrate polytopic membrane proteins with large periplasmic loops that are SecA-dependent. These data are supported by membrane proteome analyses of YidC- and SecYEG depleted cells, which show that YidC-depletion preferentially affects closely spaced membrane proteins, while SecYEG depletion mainly effects membrane proteins with large hydrophilic loops (Baars *et al.*, 2008; Wickström *et al.*, 2011b).

For the YidC substrates MtlA and TatC, a clear dependence on SRP for targeting has been demonstrated (Welte *et al.*, 2012). For other YidC substrates targeting is less clear. The phage proteins M13 and Pf3 are too short to allow co-translational SRP interaction and their targeting does not require SRP/FtsY. Nevertheless, the transmembrane helix of Pf3 can be cross-linked to SRP if it is fused to leader peptidase (Chen *et al.*, 2002). On the other hand, MscL insertion has been shown to be impaired in conditional SRP and FtsY mutants (Facey *et al.*, 2007). Controversial results were obtained for the SRP-dependence of the F_0c subunit (van Bloois *et al.*, 2004; van der Laan *et al.*, 2004a; Yi *et al.*, 2004) and therefore its targeting mode is currently uncertain. A direct interaction between components of the SRP-FtsY pathway and members of the YidC family has been observed in *E. coli* (Welte *et al.*, 2012) and for the chloroplast Alb3, which interacts with FtsY (Moore *et al.*, 2003) and with the chloroplast SRP subunit cpSRP43 (Falk *et al.*, 2010; Dünschede *et al.*, 2011). An involvement of the SRP pathway in targeting substrates to YidC indicates a co-translational process. However, YidC lacks the extended C-terminus responsible for ribosome binding and for co-translational insertion of substrates via Oxa1 (Szyrach *et al.*, 2003). A high-affinity ribosome binding motif might however be required for Oxa1, because mitochondria lack the SRP pathway for membrane targeting of ribosomes (Funes *et al.*, 2011). A recent cryo-EM reconstruction of F_0c-ribosome nascent chains bound to YidC or Oxa1 shows that the ribosome contacts a YidC dimer via the ribosomal subunits L23 and L29 and via helix 59 of the 23S rRNA (Kohler *et al.*, 2009). This indicates that the short C-terminus together with the cytoplasmic loops is sufficient for ribosome binding under the conditions tested. These data also suggest the same ribosomal contact sites are used for YidC and SecY-binding (Becker *et al.*, 2009).

Subcellular localization of the protein transport machinery

Although much is known about biochemical and structural aspects of protein targeting in bacteria, little is known about the localization of the protein transport machinery in living cells. This is partly related to the erroneous assumption that bacterial cells lack subcellular organization. However, recent advances in live cell imaging have disclosed that bacteria, like eukaryotic cells use compartmentalization as a basic principle in cell physiology (Gitai, 2006).

The importance of using live cell imaging for studying protein transport is outlined by studies on the localization of the bacterial SRP receptor FtsY. Based on cell fractionation studies, it had been suggested that FtsY is evenly distributed between the cytosol and the membrane (Luirink *et al.*, 1994; Koch *et al.*, 1999). As no specific function could be assigned to soluble FtsY, it was assumed that soluble FtsY would bind to SRP-RNCs in the cytosol. However, it has been demonstrated *in vivo* that an FtsY–GFP fusion protein predominantly localizes to the cytoplasmic membrane in *E. coli*, which has also been confirmed by immunofluorescence (Fig. 10.9) (Mircheva *et al.*, 2009). Thus, despite the lack of a transmembrane domain, the synergy between protein-protein and protein–lipid interactions allows FtsY to bind strongly to the membrane. In non-sporulating cells of the Gram-positive bacterium *B. subtilis*, FtsY is mainly localized to the membrane showing an accumulation in defined foci along the lateral membrane and occasionally at the cell poles (Mircheva *et al.*, 2009). FtsY was also shown to transiently localize at early sporulation septa in *B. subtilis* (Rubio *et al.*, 2005), which led to the idea that the SRP-pathway might deliver proteins to specific regions within the cell. SRP appears to be evenly distributed in the cytosol of *E. coli*, although a septal accumulation is also occasionally observed (Fig. 10.9; Sturm *et al.*, unpublished).

Membrane binding of SecA in *E. coli* has been shown to be almost identical to that of FtsY (Lill *et al.*, 1990; Zimmer *et al.*, 2008) and, as for FtsY, biochemical data have indicated that SecA is partly soluble and partly membrane associated (Cabelli *et al.*, 1991). There has been no microscopic analysis on SecA in *E. coli*, but it is likely that a significant portion is released from the membrane after cell fractionation and therefore the amount of soluble SecA *in vitro* appears to be higher than it might be *in vivo*. However, this needs to be further analysed. SecA in *B. subtilis* shows a very specific localization pattern: in growing cells SecA-GFP seems to localize at defined foci at the cell poles, the septum and at several positions in between. This paired dot appearance is typical for a helical orientation with at least two helices running longitudinally through the cell (Campo *et al.*, 2004). The localization of SecA-GFP in *B. subtilis* seems to be growth-phase dependent and appears to be dependent of transcription and translation (Campo *et al.*, 2004). Since SecA binds to negatively charged phospholipids, its localization pattern is significantly changed in phosphatidylglycerol deficient *B. subtilis* cells or in cardiolipin-deficient *B. subtilis* cells (Campo *et al.*, 2004). In *Streptococcus pyogenes*, SecA localizes to a single spot, which probably constitutes a secretion-specific microdomain called 'ExPortal' and which is assumed to contain also

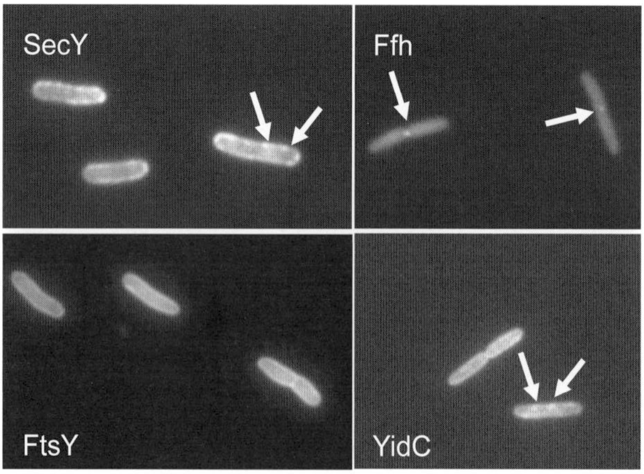

Figure 10.9 The localization of the protein transport machinery in *E. coli*. Functional fluorescence-tagged proteins were expressed from a plasmid and images were obtained using an Olympus BX-51 fluorescence microscope at a 100-fold magnification with a numerical aperture of 1.4. For further information see Mircheva *et al.* (2009). Arrows indicate the local accumulation of fluorescently labelled proteins.

SecY (Rosch and Caparon, 2004). In contrast to these data, recent studies on another gram-positive bacterium *Streptococcus pneumonia* using immune-fluorescence indicate that SecA and SecY do not localize to a single spot, but show a specific localization pattern dependent on the growth phase of the cells (Tsui *et al.*, 2011). According to this study SecA and SecY move together and localize at the equator of single predivisional cells, at the septa of early divisional cells and at hemispheres in late-divisional cells. Furthermore, it was shown that the localization of SecA and SecY is dependent on the anionic phospholipids PG and CL.

There are two contradicting views on the localization of the SecYEG translocon *in vivo*. Using C-terminally GFP-labelled versions of SecY and SecE, i.e. SecY-GFP and SecE-GFP respectively, an even distribution of the SecYEG complex within the cytoplasmic membrane of *E. coli* has been suggested (Brandon *et al.*, 2003). In contrast, another study employing fluorescence microscopy and 3D reconstruction using a SecE–GFP construct has indicated a helical organization of the SecYEG complex. (Shiomi *et al.*, 2006) which was also confirmed using immuno-fluorescence microscopy and by co-localization studies using fluorescently labelled SecYEG substrates (CFP-MalE and YFP-Tar) (Shiomi *et al.*, 2006). The mechanism of formation of this helical array remains to be discovered. These contradicting observations (even versus helical distribution) might be explained by different expression levels of the fluorescently labelled proteins and by growth phase-dependent variations. It is also possible that the presence of defined SecYEG foci represents oligomeric states of the SecYEG translocon. Furthermore the lipid composition, especially the content of negatively charged phospholipids like cardiolipin and phosphatidylglycerol in bacterial membranes appears to have an impact on SecYEG distribution (Gold *et al.*, 2010). In *B. subtilis*, GFP-SecY exhibits a similar distribution as SecA-GFP i.e. at defined foci at the cell poles, the septum and several positions in between. However, GFP-SecY seems to form curved structures instead of the complete helices found for SecA-GFP. A clustered organization of SecY in *B. subtilis* has also been confirmed by immuno-gold labelling (Campo *et al.*, 2004). On the other hand, SecD and SecF, which are thought to transiently associate with the SecYEG complex, have been found to be evenly distributed in *B. subtilis* membranes. (Rubio *et al.*, 2005).

Although it is tempting to speculate that a helical organization of the SecYEG translocon is achieved via interaction with the cytoskeleton, the available data indicate that the localization of SecYEG is not compromised in MreB or Mbl mutants of *E. coli* and *B. subtilis* and no co-localization of MreB/Mbl and SecY/SecE was observed (Campo *et al.*, 2004; Shiomi *et al.*, 2006).

Fluorescently labelled YidC was shown to localize predominantly at the cell poles in *E. coli* (Urbanus *et al.*, 2002). However this could be related to the expression level, as YidC seems to show a uniform membrane localization when its expression level is tightly regulated (Sturm *et al.*, unpublished). Analysing the localization of the *B. subtilis* YidC homologue SpoIIIJ has led to two different observations. On the one hand, there are data showing that SpoIIIJ-GFP is enriched at the septal area of the membrane (Murakami *et al.*, 2002), but a different study has indicated that SpoIIIJ-GFP is uniformly distributed in the membrane (Rubio *et al.*, 2005). Presumably, the initially observed enrichment of SpoIIIJ-GFP in the septal area is related to the fact that the forespore is surrounded by three membranes after engulfment, increasing the local SpoIIIJ-GFP concentration.

It is apparent that the available microscopic data do not provide an entirely comprehensive picture of the spatial organization of the bacterial transport machinery, but the available data indicate that its distribution is not entirely random. Nevertheless, as the resolution of light microscopy is increasing and many initial problems of live bacterial imaging have been solved, microscopic techniques will provide invaluable tools for understanding highly dynamic processes like protein transport.

Conclusions and future trends

Our understanding of how proteins are inserted into bacterial membranes has made tremendous progress during the last two decades. Originally considered to occur spontaneously, it is now obvious that the insertion of inner membrane proteins requires sophisticated machinery that monitors and possibly corrects each individual step during membrane protein assembly. With the exception of YidC, high resolution structures of the essential players have been obtained and the mechanistic details on how they work have begun to emerge. However, additional high-resolution structures are required, e.g. of a polypeptide inside of the SecYEG channel, of a holo-translocon comprising SecYEG-SecDFYajC-YidC, and of an FtsY–SecYEG complex. The coordination of the SecA/SecB and SRP targeting processes needs to be revealed, e.g. how binding of SecA, SRP and trigger factor to a single ribosomal subunit (L23) is coordinated and how SecA and FtsY binding to very same residues in cytosolic loops of SecY is regulated. The mechanistic details on how SecA translocates periplasmic loops of inner membrane proteins need to be resolved, i.e. how SecA gets access to a periplasmic loop that is emerging from a 1 MDa ribosome sitting on 75 kDa SecYEG complex. The observation that SRP can target multi-spanning membrane proteins to either SecYEG or YidC needs to be further explored in order to understand how integrations sites are selected and how mistargeting of substrates that can use either SecYEG or YidC is prevented. This might also answer the question of why bacteria maintain two seemingly independent integration sites for inner membrane proteins. The number of dedicated protein transport pathways in bacteria is increasing continuously and many appear to function simultaneously (Papanikou *et al.*, 2007). Whether this is related to the bacterial domain as an evolutionary playground or whether specific features in the cognate substrates necessitate dedicated transport machineries, requires further analysis.

Some rather general questions have still not been answered, such as the contribution of lipids to membrane protein activity, folding and stability and these questions need to be further explored. Lipids have been shown to control the topology of membrane proteins but it is still unclear whether this occurs independently of SecYEG (Dowhan and Bogdanov, 2009). With respect to protein transport, the proposed role of cardiolipin in activating SecA/SecY needs to be analysed (Gold *et al.*, 2010). Proteomic analyses of SRP, SecY, YidC or SecB depleted cells have revealed substantial physiological changes (Baars *et al.*, 2008; Wickström *et al.*, 2011b), but surprisingly the membrane proteome of SRP-depleted cells revealed no drastic changes (Wickström *et al.*, 2011a). Considering that so far only a few model substrates of the SRP pathway like MtlA or FtsQ have been analysed in detail, it will be important to extend our analyses to additional substrates. It is also important to emphasize that some observations are still difficult to relate to the general view on SRP-dependent targeting. As an alternate scenario, it has been proposed that FtsY targets ribosomes independently of SRP to the membrane and that SRP then facilitates SecYEG-dependent insertion and folding of membrane proteins (Bibi, 2011). Whether SRP can function downstream of FtsY, as suggested by this scenario, requires further *in vivo* and *in vitro* exploration. Post-insertion processes like membrane protein quality control or the assembly into higher order complexes are largely a molecular *terra incognita* and invite further studies (reviewed in Dalbey *et al.*, 2011). A rather unconventional model of protein translocation in *E. coli* has also been proposed in which SecA forms the protein conducting channel and SecYEG provides some proofreading for signal sequences (Hsieh *et al.*, 2011)

And then there are new (or forgotten) questions about membrane protein integration arising such as: how common translation-independent membrane targeting of mRNAs is in bacteria and how are these mRNAs recognized and targeted (Prilusky and Bibi, 2009; Nevo-Dinur *et al.*, 2011)? It is also not known how proteins that were synthesized from membrane-bound mRNAs reach the translocon. A possible explanation is offered by the recent observation of a signal sequence independent SRP–FtsY complex formation at the membrane (Braig *et al.*, 2011). These data indicate that substrate recognition by SRP does not necessarily need to take place in the cytosol but can also occur at the membrane by a pre-assembled FtsY–SRP complex. The spatial organization of the protein transport machinery in bacterial cells is largely unknown and the question of whether microdomains for integration exist and if so how they are assembled and maintained is unsolved. Finally, our knowledge on protein transport in bacteria is mainly based on studies using *E. coli* and a few other bacteria, e.g. *B. subtilis* or *S. mutans.* It is unlikely that these few model organisms will allow us to obtain an entirely comprehensive view on protein transport systems in bacteria.

Acknowledgements

This work was funded by the Deutsche Forschungsgemeinschaft (FOR 929, FOR 967, GRK 1478), the German-French University (UFA-04-07), the FF-Nord Foundation and the Excellence Initiative of the Federal and State Governments (GSC-4, Spemann-Graduate School of Biology and Medicine).

References

Akiyama, Y., Kanehara, K., and Ito, K. (2004). RseP (YaeL), an *Escherichia coli* RIP protease, cleaves transmembrane sequences. EMBO J. *23*, 4434–4442.

Alami, M., Dalal, K., Lelj-Garolla, B., Sligar, S.G., and Duong, F. (2007). Nanodiscs unravel the interaction between the SecYEG channel and its cytosolic partner SecA. EMBO J. *26*, 1995–2004.

Altamura, N., Capitanio, N., Bonnefoy, N., Papa, S., and Dujardin, G. (1996). The *Saccharomyces cerevisiae* OXA1 gene is required for the correct assembly of cytochrome c oxidase and oligomycin-sensitive ATP synthase. FEBS Lett. *382*, 111–115.

Andersson, H., and von Heijne, G. (1993). Sec dependent and sec independent assembly of *E. coli* inner membrane proteins: the topological rules depend on chain length. EMBO J. *12*, 683–691.

Angelini, S., Deitermann, S., and Koch, H.G. (2005). FtsY, the bacterial signal-recognition particle receptor, interacts functionally and physically with the SecYEG translocon. EMBO Rep. *6*, 476–481.

Angelini, S., Boy, D., Schiltz, E., and Koch, H.G. (2006). Membrane binding of the bacterial signal recognition particle receptor involves two distinct binding sites. J. Cell Biol. *174*, 715–724.

Antonoaea, R., Fürst, M., Nishiyama, K., and Müller, M. (2008). The periplasmic chaperone PpiD interacts with secretory proteins exiting from the SecYEG translocon. Biochemistry *47*, 5649–5656.

Ataide, S.F., Schmitz, N., Shen, K., Ke, A., Shan, S.O., Doudna, J.A., and Ban, N. (2011). The crystal structure of the signal recognition particle in complex with its receptor. Science *331*, 881–886.

Auclair, S.M., Moses, J.P., Musial-Siwek, M., Kendall, D.A., Oliver, D.B., and Mukerji, I. (2010). Mapping of the signal peptide-binding domain of *Escherichia coli* SecA using Förster resonance energy transfer. Biochemistry *49*, 782–792.

Baars, L., Wagner, S., Wickström, D., Klepsch, M., Ytterberg, A.J., van Wijk, K.J., and de Gier, J.W. (2008). Effects of SecE depletion on the inner and outer membrane proteomes of *Escherichia coli*. J. Bacteriol. *190*, 3505–3525.

Bahari, L., Parlitz, R., Eitan, A., Stjepanovic, G., Bochkareva, E.S., Sinning, I., and Bibi, E. (2007). Membrane targeting of ribosomes and their release require distinct and separable functions of FtsY. J. Biol. Chem. *282*, 32168–32175.

Ban, N., Nissen, P., Hansen, J., Moore, P.B., and Steitz, T.A. (2000). The complete atomic structure of the large ribosomal subunit at 2.4 A resolution. Science *289*, 905–920.

Bange, G., Petzold, G., Wild, K., Parlitz, R.O., and Sinning, I. (2007). The crystal structure of the third signal-recognition particle GTPase FlhF reveals a homodimer with bound GTP. Proc. Natl. Acad. Sci. U.S.A. *104*, 13621–13625.

Bange, G., Kümmerer, N., Grudnik, P., Lindner, R., Petzold, G., Kressler, D., Hurt, E., Wild, K., and Sinning, I. (2011). Structural basis for the molecular evolution of SRP-GTPase activation by protein. Nat. Struct. Mol. Biol. *18*, 1376–1380.

Batey, R.T., Rambo, R.P., Lucast, L., Rha, B., and Doudna, J.A. (2000). Crystal structure of the ribonucleoprotein core of the signal recognition particle. Science *287*,1232–1239.

Bechtluft, P., Kedrov, A., Slotboom, D.J., Nouwen, N., Tans, S.J., and Driessen, A.J. (2010). Tight hydrophobic contacts with the SecB chaperone prevent folding of substrate proteins. Biochemistry *49*, 2380–2388.

Beck, K., Eisner, G., Trescher, D., Dalbey, R.E., Brunner, J., and Müller, M. (2001). YidC, an assembly site for polytopic *Escherichia coli* membrane proteins located in immediate proximity to the SecYE translocon and lipids. EMBO Rep. *2*, 709–714.

Becker, T., Bhushan, S., Jarasch, A., Armache, J.P., Funes, S., Jossinet, F., Gumbart, J., Mielke, T., Berninghausen, O., Schulten, K., *et al.* (2009). Structure of monomeric yeast and mammalian Sec61 complexes interacting with the translating ribosome. Science *326*, 1369–1373.

Beckmann, R., Bubeck, D., Grassucci, R., Penczek, P., Verschoor, A., Blobel, G., and Frank, J. (1997). Alignment of conduits for the nascent polypeptide chain in the ribosome–Sec61 complex. Science *278*, 2123–2126.

Beha, D., Deitermann, S., Müller, M., and Koch, H.G. (2003). Export of beta-lactamase is independent of the signal recognition particle. J. Biol. Chem. *278*, 22161–22167.

Berisio, R., Schluenzen, F., Harms, J., Bashan, A., Auerbach, T., Baram, D., and Yonath, A. (2003). Structural insight into the role of the ribosomal tunnel in cellular regulation. Nat. Struct. Biol. *10*, 366–370.

Berndt, U., Oellerer, S., Zhang, Y., Johnson, A.E., and Rospert, S. (2009). A signal-anchor sequence stimulates signal recognition particle binding to ribosomes from inside the exit tunnel. Proc. Natl. Acad. Sci. U.S.A. *106*,1398–1403.

Bernstein, H.D., Poritz, M.A., Strub, K., Hoben, P.J., Brenner, S., and Walter, P. (1989). Model for signal sequence recognition from amino-acid sequence of 54K subunit of signal recognition particle. Nature *340*, 482–486.

Berrier, C., Guilvout, I., Bayan, N., Park, K.H., Mesneau, A., Chami, M., Pugsley, A.P., and Ghazi, A. (2011). Coupled cell-free synthesis and lipid vesicle insertion of a functional oligomeric channel MscL MscL does not need the insertase YidC for insertion *in vitro.* Biochim. Biophys. Acta. *1808*, 41–46.

Bibi, E. (2011). Early targeting events during membrane protein biogenesis in *Escherichia coli.* Biochim. Biophys. Acta. *1808*, 841–850.

Bonardi, F., Halza, E., Walko, M., Du Plessis, F., Nouwen, N., Feringa, B.L., and Driessen, A.J. (2011). Probing the SecYEG translocation pore size with preproteins conjugated with sizable rigid spherical molecules. Proc. Natl. Acad. Sci. U.S.A. *108*, 7775–7780.

Bonnefoy, N., Chalvet, F., Hamel, P., Slonimski, P.P., and Dujardin, G. (1994). OXA1, a *Saccharomyces cerevisiae* nuclear gene whose sequence is conserved from prokaryotes to eukaryotes controls cytochrome oxidase biogenesis. J. Mol. Biol. *239*, 201–212.

Bornemann, T., Jöckel, J., Rodnina, M.V., and Wintermeyer. W. (2008). Signal sequence-independent membrane targeting of ribosomes containing short nascent peptides within the exit tunnel. Nat. Struct. Mol. Biol. *15*, 494–499.

Bostina, M., Mohsin, B., Kühlbrandt, W., and Collinson, I. (2005). Atomic model of the *E. coli* membrane-bound protein translocation complex SecYEG. J. Mol. Biol. *352*, 1035–1043.

Boy, D., and Koch, H.G. (2009). Visualization of distinct entities of the SecYEG translocon during translocation and integration of bacterial proteins. Mol. Biol. Cell *20*, 1804–1815.

Bradshaw, N., Neher, S., Booth, D., and Walter, P. (2009). Signal sequences activate the catalytic switch of SRP RNA. Science *323*, 127–130.

Braig, D., Bär, C., Thumfart, J.O., and Koch, H.G. (2009). Two cooperating helices constitute the lipid-binding domain of the bacterial SRP receptor. J. Mol. Biol. *390*, 401–413.

Braig, D., Mircheva, M., Sachelaru, I., van der Sluis, E.O., Sturm, L., Beckmann, R., and Koch, H.G. (2011). Signal-sequence independent SRP–SR complex formation at the membrane suggests an alternative targeting pathway within the SRP cycle. Mol. Biol. Cell *22*, 2309–2323.

Brandon, L.D., Goehring, N., Janakiraman, A., Yan, A.W., Wu, T., Beckwith, J., and Goldberg. M.B. (2003). IcsA, a polarly localized autotransporter with an atypical signal peptide, uses the Sec apparatus for secretion, although the Sec apparatus is circumferentially distributed. Mol. Microbiol. *50*, 45–60.

Breukink, E., Nouwen, N., van Raalte, A., Mizushima, S., Tommassen, J., and de Kruijff, B. (1995). The C terminus of SecA is involved in both lipid binding and SecB binding. J. Biol. Chem. *270*, 7902–7907.

Breyton, C., Haase, W., Rapoport, T.A., Kühlbrandt, W., and Collinson, I. (2002). Three-dimensional structure of the bacterial protein–translocation complex SecYEG. Nature *418*, 662–665.

Bürk, J., Weiche, B., Wenk, M., Boy, D., Nestel, S., Heimrich, B., and Koch, H.G. (2009). Depletion of the signal recognition particle receptor inactivates ribosomes in *Escherichia coli*. J. Bacteriol. *191*, 7017–7026.

Buskiewicz, I., Deuerling, E., Gu, S.Q., Jöckel, J., Rodnina, M.V., Bukau, B., and Wintermeyer, W. (2004). Trigger factor binds to ribosome-signal-recognition particle (SRP) complexes and is excluded by binding of the SRP receptor. Proc. Natl. Acad. Sci. U.S.A. *101*, 7902–7906.

Cabelli, R.J., Dolan, K.M., Qian, L.P., and Oliver, D.B. (1991). Characterization of membrane-associated and soluble states of SecA protein from wild-type and SecA51(TS) mutant strains of *Escherichia coli*. J. Biol. Chem. *266*, 24420–24427.

Campo, N., Tjalsma, H., Buist, G., Stepniak, D., Meijer, M., Veenhuis, M., Westermann, M., Müller, J.P., Bron, S., Kok, J., *et al.* (2004). Subcellular sites for bacterial protein export. Mol. Microbiol. *53*, 1583–1599.

Celebi, N., Yi, L., Facey, S.J., Kuhn, A., and Dalbey, R.E. (2006). Membrane biogenesis of subunit II of cytochrome bo oxidase: contrasting requirements for insertion of N-terminal and C-terminal domains. J. Mol. Biol. *357*, 1428–1436.

Celebi, N., Dalbey, R.E., and Yuan, J. (2008). Mechanism and hydrophobic forces driving membrane protein insertion of subunit II of cytochrome bo 3 oxidase. J. Mol. Biol. *375*, 1282–1292.

Chen, M., Samuelson, J.C., Jiang, F., Muller, M., Kuhn, A., and Dalbey, R.E. (2002). Direct interaction of YidC with the Sec-independent Pf3 coat protein during its membrane protein insertion. J. Biol. Chem. *277*, 7670–7675.

Cheng, Z., Jiang, Y., Mandon, E.C., and Gilmore, R. (2005). Identification of cytoplasmic residues of Sec61p involved in ribosome binding and cotranslational translocation. J. Cell Biol. *168*, 67–77.

Chevalier, N., Moser, M., Koch, H.G., Schimz, K.L., Willery, E., Locht, C., Jacob-Dubuisson, F., and Müller, M. (2004). Membrane targeting of a bacterial virulence factor harbouring an extended signal peptide. J. Mol. Microbiol. Biotechnol. *8*, 7–18.

Chou, Y.T., and Gierasch, L.M. (2005). The conformation of a signal peptide bound by *Escherichia coli* preprotein translocase SecA. J. Biol. Chem. *280*, 32753–32760.

Clérico, E.M., Maki, J.L., and Gierasch, L.M. (2008). Use of synthetic signal sequences to explore the protein export machinery. Biopolymers *90*, 307–319.

Cooper, D.B., Smith, V.F., Crane, J.M., Roth, H.C., Lilly, A.A., and Randall, L.L. (2008). SecA, the motor of the secretion machine, binds diverse partners on one interactive surface. J. Mol. Biol. *382*, 74–87.

Crane, J.M., Suo, Y., Lilly, A.A., Mao, C., Hubbell, W.L., and Randall, L.L. (2006). Sites of interaction of a precursor polypeptide on the export chaperone SecB mapped by site-directed spin labeling. J. Mol. Biol. *363*, 63–74.

Dalbey, R.E., Wang, P., and Kuhn, A. (2011). Assembly of bacterial inner membrane proteins. Annu. Rev. Biochem. *80*, 167–187.

Das, S., and Oliver, D.B. (2011). Mapping of the SecA-SecY and SecA-SecG Interfaces by Site-directed *in vivo* Photocross-linking. J. Biol. Chem. *286*, 12371–12380.

Dautin, N., and Bernstein, H.D. (2007). Protein secretion in gram-negative bacteria via the autotransporter pathway. Annu. Rev. Microbiol. *61*, 89–112.

de Keyzer, J., van der Sluis, E.O., Spelbrink, R.E., Nijstad, N., de Kruijff, B., Nouwen, N., van der Does, C., and Driessen, A.J. (2005). Covalently dimerized SecA is functional in protein translocation. J. Biol. Chem. *280*, 35255–35260.

de Leeuw, E., Poland, D., Mol, O., Sinning, I., ten Hagen-Jongman, C.M., Oudega, B., and Luirink, J. (1997). Membrane association of FtsY, the *E. coli* SRP receptor. FEBS Lett. *416*, 225–229.

Deitermann, S., Sprie, G.S., and Koch, H.G. (2005). A dual function for SecA in the assembly of single spanning membrane proteins in *Escherichia coli*. J. Biol. Chem. *280*, 39077–39085.

den Blaauwen, T., de Wit, J.G., Gosker, H., van der Does, C., Breukink, E.J., de Leij, L., and Driessen, A.J. (1997). Inhibition of preprotein translocation and reversion of the membrane inserted state of SecA by a carboxyl terminus binding mAb. Biochemistry *36*, 9159–9168.

Deuerling, E., Schulze-Specking, A., Tomoyasu, T., Mogk, A., and Bukau, B. (1999). Trigger factor and DnaK cooperate in folding of newly synthesized proteins. Nature *400*, 693–696.

Deville, K., Gold, V.A., Robson, A., Whitehouse, S., Sessions, R.B., Baldwin, S.A., Radford, S.E., and Collinson, I. (2011). The oligomeric state and arrangement of the active bacterial translocon. J. Biol. Chem. *286*, 4659–4669.

Dong, Y., Palmer, S.R., Hasona, A., Nagamori, S., Kaback, H.R., Dalbey, R.E., and Brady, L.J. (2008). Functional overlap but lack of complete cross-complementation of Streptococcus mutans and *Escherichia coli* YidC orthologs. J. Bacteriol. *190*, 2458–2469.

Dowhan, W., and Bogdanov, M. (2009). Lipid-dependent membrane protein topogenesis. Annu. Rev. Biochem. *78*, 515–540.

Dünschede, B., Bals, T., Funke, S., and Schünemann, D. (2011). Interaction studies between the chloroplast signal recognition particle subunit cpSRP43 and the full-length translocase Alb3 reveal membrane-embedded binding region in Alb3 protein. J. Biol. Chem. *286*, 35187–35195.

du Plessis, D.J., Nouwen, N., and Driessen, A.J. (2006). Subunit a of cytochrome o oxidase requires both YidC and SecYEG for membrane insertion. J. Biol. Chem. *281*, 12248–12252.

du Plessis, D.J., Berrelkamp, G., Nouwen, N., and Driessen, A.J. (2009). The lateral gate of SecYEG opens during protein translocation. J. Biol. Chem. *284*, 15805–15814.

du Plessis, D.J., Nouwen, N., and Driessen, A.J. (2011). The Sec translocase. Biochim. Biophys. Acta. *1808*, 851–865.

Egea, P.F., and Stroud, R.M. (2010). Lateral opening of a translocon upon entry of protein suggests the mechanism of insertion into membranes. Proc. Natl. Acad. Sci. U.S.A. *107*, 17182–17187.

Egea, P.F., Shan, S.O., Napetschnig, J., Savage, D.F., Walter, P., and Stroud, R.M. (2004). Substrate twinning activates the signal recognition particle and its receptor. Nature *427*, 215–221.

Eisner, G., Koch, H.G., Beck, K., Brunner, J., and Muller, M. (2003). Ligand crowding at a nascent signal sequence. J. Cell Biol. *163*, 35–44.

Eitan, A., and Bibi, E. (2004). The core *Escherichia coli* signal recognition particle receptor contains only the N and G domains of FtsY. J. Bacteriol. *186*, 2492–2494.

Erez, E., Stjepanovic, G., Zelazny, A.M., Brugger, B., Sinning, I., and Bibi, E. (2010). Genetic evidence for functional interaction of the *Escherichia coli* signal recognition particle receptor with acidic lipids *in vivo*. J. Biol. Chem. *285*, 40508–40514.

Erlandson, K.J., Miller, S.B., Nam, Y., Osborne, A.R., Zimmer, J., and Rapoport, T.A. (2008). A role for the two-helix finger of the SecA ATPase in protein translocation Nature *455*, 984–987.

Estrozi, L.F., Boehringer, D., Shan, S.O., Ban, N., and Schaffitzel, C. (2011). Cryo-EM structure of the *E. coli* translating ribosome in complex with SRP and its receptor. Nat. Struct. Mol. Biol. *18*, 88–90.

Facey, S.J., Neugebauer, S.A., Krauss, S., and Kuhn, A. (2007). The mechanosensitive channel protein MscL is targeted by the SRP to the novel YidC membrane insertion pathway of *Escherichia coli*. J. Mol. Biol. *365*, 995–1004.

Fekkes, P., van der Does, C., and Driessen, A.J. (1997). The molecular chaperone SecB is released from the carboxy-terminus of SecA during initiation of precursor protein translocation. EMBO J. *16*, 6105–6113.

Ferbitz, L., Maier, T., Patzelt, H., Bukau, B., Deuerling, E., and Ban, N. (2004). Trigger factor in complex with the ribosome forms a molecular cradle for nascent proteins. Nature *431*, 590–596.

Flanagan, J.J., Chen, J.C., Miao, Y., Shao, Y., Lin, J., Bock, P.E., and Johnson, A.E. (2003). Signal recognition particle binds to ribosome-bound signal sequences with fluorescence-detected subnanomolar affinity that does not diminish as the nascent chain lengthens. J. Biol. Chem. *278*, 18628–18637.

Focia, P.J., Shepotinovskaya, I.V., Seidler, J.A., and Freymann, D.M. (2004). Heterodimeric GTPase core of the SRP targeting complex. Science *303*, 373–377.

Frauenfeld, J., Gumbart, J., Sluis, E.O., Funes, S., Gartmann, M., Beatrix, B., Mielke, T., Berninghausen, O., Becker, T., Schulten, K., *et al.* (2011). Cryo-EM structure of the ribosome–SecYE complex in the membrane environment. Nat. Struct. Mol. Biol. *18*, 614–621.

Fulga, T.A., Sinning, I., Dobberstein, B., and Pool, M.R. (2001). SRbeta coordinates signal sequence release from SRP with ribosome binding to the translocon. EMBO J. *20*, 2338–2347.

Funes, S., Kauff, F., van der Sluis, E.O., Ott, M., and Herrmann, J.M. (2011). Evolution of YidC/Oxa1/Alb3 insertases: three independent gene duplications followed by functional specialization in bacteria, mitochondria and chloroplasts. Biol. Chem. *392*, 13–19.

Gasper, R., Meyer, S., Gotthardt, K., Sirajuddin, M., and Wittinghofer, A. (2009). It takes two to tango: regulation of G proteins by dimerization. Nat. Rev. Mol. Cell Biol. *10*, 423–429.

Geller, B.L., and Wickner, W. (1985). M13 procoat inserts into liposomes in the absence of other membrane proteins. J. Biol. Chem. *260*,13281–13285.

Gitai, Z. (2006). The new bacterial cell biology: moving parts and subcellular architecture. Cell *120*, 577–586.

Gold, V.A., Robson, A., Bao, H., Romantsov, T., Duong, F., and Collinson, I. (2010). The action of cardiolipin on the bacterial translocon. Proc. Natl. Acad. Sci. U.S.A. *107*, 10044–10049.

Gu, S.Q., Peske, F., Wieden, H.J., Rodnina, M.V., and Wintermeyer, W. (2003). The signal recognition particle binds to protein L23 at the peptide exit of the *Escherichia coli* ribosome. RNA *9*, 566–573.

Gumbart, J., and Schulten, K. (2007). Structural determinants of lateral gate opening in the protein translocon. Biochemistry *46*, 11147–11157.

Hainzl, T., Huang, S., Meriläinen, G., Brännström, K., and Sauer-Eriksson, A.E. (2011). Structural basis of signal-sequence recognition by the signal recognition particle. Nat. Struct. Mol. Biol. *18*, 389–391.

Halic, M., Becker, T., Pool, M.R., Spahn, C.M., Grassucci, R.A., Frank, J., and Beckmann, R. (2004). Structure of the signal recognition particle interacting with the elongation-arrested ribosome. Nature *427*, 808–814.

Halic, M., Blau, M., Becker, T., Mielke, T., Pool, M.R., Wild, K., Sinning, I., and Beckmann, R. (2006). Following the signal sequence from ribosomal tunnel exit to signal recognition particle. Nature *444*, 507–511.

Hamman, B.D., Chen, J.C., Johnson, E.E., and Johnson, A.E. (1997). The aqueous pore through the translocon has a diameter of 40–60 A during cotranslational protein translocation at the ER membrane. Cell *89*, 535–544.

Hanein, D., Matlack, K.E., Jungnickel, B., Plath, K., Kalies, K.U., Miller, K.R., Rapoport, T.A., and Akey, C.W. (1996). Oligomeric rings of the Sec61p complex induced by ligands required for protein translocation. Cell *87*, 721–732.

Hardy, S.J., and Randall, L.L. (1991). A kinetic partitioning model of selective binding of nonnative proteins by the bacterial chaperone SecB Science *251*, 439–443.

Harris, C.R., and Silhavy, T.J. (1999). Mapping an interface of SecY (PrlA) and SecE (PrlG) by using synthetic phenotypes and *in vivo* cross-linking. J. Bacteriol. *181*, 3438–3444.

Hartl, F.U., Lecker, S., Schiebel, E., Hendrick, J.P., and Wickner, W. (1990). The binding cascade of SecB to SecA to SecY/E mediates preprotein targeting to the *E. coli* plasma membrane. Cell *63*, 269–279.

Hasona, A., Crowley, P.J., Levesque, C.M., Mair, R.W., Cvitkovitch, D.G., Bleiweis, A.S., and Brady, L.J. (2005). Streptococcal viability and diminished stress tolerance in mutants lacking the signal recognition particle pathway or YidC2. Proc. Natl. Acad. Sci. U.S.A. *102*, 17466–17471.

Hasona, A., Zuobi-Hasona, K., Crowley, P.J., Abranches, J., Ruelf, M.A., Bleiweis, A.S., and Brady, L.J. (2007). Membrane composition changes and physiological adaptation by *Streptococcus mutans* signal recognition particle pathway mutants. J. Bacteriol. *189*, 1219–1230.

Hegde, R.S., and Bernstein, H.D. (2006). The surprising complexity of signal sequences. Trends Biochem. Sci. *31*, 563–571.

Heinrich, S.U., and Rapoport, T.A. (2003). Cooperation of transmembrane segments during the integration of a double-spanning protein into the ER membrane. EMBO J. *22*, 3654–3663.

Hell, K., Neupert, W., and Stuart, R.A. (2001). Oxa1p acts as a general membrane insertion machinery for proteins encoded by mitochondrial DNA. EMBO J. *20*, 1281–1288.

Hessa, T., Meindl-Beinker, N.M., Bernsel, A., Kim, H., Sato, Y., Lerch-Bader, M., Nilsson, I., White, S.H., and von Heijne, G. (2007). Molecular code for transmembrane-helix recognition by the Sec61 translocon Nature *450*, 1026–1030.

Houben, E.N., Zarivach, R., Oudega, B., and Luirink, J. (2005). Early encounters of a nascent membrane protein: specificity and timing of contacts inside and outside the ribosome. J. Cell Biol. *170*, 27–35.

Hsieh, Y.H., Zhang, H., Lin, B.R., Cui, N., Yang, H., Jiang, C., Sui, S.F., and Tai, P.C. (2011). SecA alone can promote translocation and ion channels activity: SecYEG increases efficiency and signal peptide specificity. J. Biol. Chem. *286*, 44702–44709.

Huber, D., Boyd, D., Xia, Y., Olma, M.H., Gerstein, M., and Beckwith, J. (2005). Use of thioredoxin as a reporter to identify a subset of *Escherichia coli* signal sequences that promote signal recognition particle-dependent translocation. J. Bacteriol. *187*, 2983–2991.

Huber, D., Rajagopalan, N., Preissler, S., Rocco, M.A., Merz, F., Kramer, G., and Bukau, B. (2011). SecA interacts with ribosomes in order to facilitate posttranslational translocation in bacteria. Mol. Cell *41*, 343–353.

Imhof, N., Kuhn, A., and Gerken, U. (2011). Substrate-dependent conformational dynamics of the *Escherichia coli* membrane insertase YidC. Biochemistry *50*, 3229–3239.

Jagath, J.R., Matassova, N.B., de Leeuw, E., Warnecke, J.M., Lentzen, G., Rodnina, M.V., Luirink, J., and Wintermeyer, W. (2001). Important role of the tetraloop region of 4.5S RNA in SRP binding to its receptor FtsY. RNA *7*, 293–301.

Janda, C.Y., Li, J., Oubridge, C., Hernández, H., Robinson, C.V., and Nagai, K. (2010). Recognition of a signal peptide by the signal recognition particle. Nature *465*, 507–510.

Jarosik, G.P., and Oliver, D.B. (1991). Isolation and analysis of dominant secA mutations in *Escherichia coli.* J. Bacteriol. *173*, 860–868.

Jiang, F., Chen, M., Yi, L., de Gier, J.W., Kuhn, A., and Dalbey, R.E. (2003). Defining the regions of *Escherichia coli* YidC that contribute to activity. J. Biol. Chem. *278*, 48965–48972.

Jiang, F., Yi, L., Moore, M., Chen, M., Rohl, T., Van Wijk, K.J., De Gier, J.W., Henry, R., and Dalbey, R.E. (2002). Chloroplast YidC homolog Albino3 can functionally complement the bacterial YidC depletion strain and promote membrane insertion of both bacterial and chloroplast thylakoid proteins. J. Biol. Chem. *277*, 19281–19288.

Jiang, Y., Cheng, Z., Mandon, E.C., and Gilmore, R. (2008). An interaction between the SRP receptor and the translocon is critical during cotranslational protein translocation. J. Cell Biol. *180*, 1149–1161.

Jilaveanu, L.B., and Oliver, D. (2006). SecA dimer cross-linked at its subunit interface is functional for protein translocation. J. Bacteriol. *188*, 335–338.

Jilaveanu, L.B., Zito, C.R., and Oliver, D. (2005). Dimeric SecA is essential for protein translocation. Proc. Natl. Acad. Sci. U.S.A. *102*, 7511–7516.

Jong, W.S., ten Hagen-Jongman, C.M., Ruijter, E., Orru, R.V., Genevaux, P., and Luirink, J. (2010). YidC is involved in the biogenesis of the secreted autotransporter hemoglobin protease. J. Biol. Chem. *285*, 39682–39690.

Karamanou, S., Vrontou, E., Sianidis, G., Baud, C., Roos, T., Kuhn, A., Politou, A.S., and Economou, A. (1999). A molecular switch in SecA protein couples ATP hydrolysis to protein translocation. Mol. Microbiol. *34*, 1133–1145.

Karamyshev, A.L., and Johnson, A.E. (2005). Selective SecA association with signal sequences in ribosome-bound nascent chains: a potential role for SecA in ribosome targeting to the bacterial membrane. J. Biol. Chem. *280*, 37930–37940.

Kedrov, A., Kusters, I., Krasnikov, V.V., and Driessen, A.J. (2011). A single copy of SecYEG is sufficient for preprotein translocation. EMBO J. 30, 4387–4397.

Keenan, R.J., Freymann, D.M., Walter, P., and Stroud, R.M. (1998). Crystal structure of the signal sequence binding subunit of the signal recognition particle. Cell *94*,181–191.

Keramisanou, D., Biris, N., Gelis, I., Sianidis, G., Karamanou, S., Economou, A., and Kalodimos, C.G. (2006). Disorder-order folding transitions underlie catalysis in the helicase motor of SecA. Nat. Struct. Mol. Biol. *13*, 594–602.

Kihara, A., Akiyama, Y., and Ito, K. (1995). FtsH is required for proteolytic elimination of uncomplexed forms of SecY, an essential protein translocase subunit. Proc. Natl. Acad. Sci. U.S.A. 92, 4532–4536.

Klenner, C., and Kuhn, A. (2012). Dynamic disulfide-scanning of the membrane-inserting PF3 coat protein reveals multiple YidC-substrate contacts. J. Biol. Chem, *287*, 3769–3776.

Knoblauch, N.T., Rüdiger, S., Schönfeld, H.J., Driessen, A.J., Schneider-Mergener, J., and Bukau, B. (1999). Substrate specificity of the SecB chaperone. J. Biol. Chem. *274*, 34219–34225.

Koch, H.G., and Müller, M. (2000). Dissecting the translocase and integrase functions of the *Escherichia coli* SecYEG translocon. J. Cell. Biol. *150*, 689–694.

Koch, H.G., Hengelage, T., Neumann-Haefelin, C., MacFarlane, J., Hoffschulte, H.K., Schimz, K.L., Mechler, B., and Müller, M. (1999). *In vitro* studies with purified components reveal signal recognition particle (SRP) and SecA/SecB as constituents of two independent protein-targeting pathways of *Escherichia coli*. Mol. Biol. Cell *10*, 2163–2173.

Koch, H.G., Moser, M., Schimz, K.L., and Muller, M. (2002). The integration of YidC into the cytoplasmic membrane of *Escherichia coli* requires the signal recognition particle, SecA and SecYEG. J. Biol. Chem. *277*, 5715–5718.

Koch, H.G., Moser, M., and Müller, M. (2003). Signal recognition particle-dependent protein targeting, universal to all kingdoms of life. Rev. Physiol. Biochem. Pharmacol. *146*, 55–94.

Kohler, R., Boehringer, D., Greber, B., Bingel-Erlenmeyer, R., Collinson, I., Schaffitzel, C., and Ban, N. (2009). YidC and Oxa1 form dimeric insertion pores on the translating ribosome. Mol. Cell *34*, 344–353.

Kol, S., Nouwen, N., and Driessen, A.J. (2008). The charge distribution in the cytoplasmic loop of subunit C of the F1F0 ATPase is a determinant for YidC targeting. J. Biol. Chem. *283*, 9871–9877.

Kramer, G., Boehringer, D., Ban, N., and Bukau, B. (2009). The ribosome as a platform for co-translational processing, folding and targeting of newly synthesized proteins. Nat. Struct. Mol. Biol. *16*, 589–597.

Kuhn, P., Weiche, B., Sturm, L., Sommer, E., Drepper, F., Warscheid, B., Sourjik, V., and Koch, H.G. (2011). The bacterial SRP receptor, SecA and the ribosome use overlapping binding sites on the SecY translocon. Traffic *12*, 563–578.

Kusters, I., and Driessen, A.J. (2011). SecA, a remarkable nanomachine. Cell. Mol. Life Sci. *68*, 2053–2066.

Kusters, I., van den Bogaart, G., Kedrov, A., Krasnikov, V., Fulyani, F., Poolman, B., and Driessen, A.J. (2011). Quaternary structure of SecA in solution and bound to SecYEG probed at the single molecule level. Structure *19*, 430–439.

Lakkaraju, A.K., Mary, C., Scherrer, A., Johnson, A.E., and Strub, K. (2008). SRP keeps polypeptides translocation-competent by slowing translation to match limiting ER-targeting sites. Cell *133*, 440–451.

Lakshmipathy, S.K., Tomic, S., Kaiser, C.M., Chang, H.C., Genevaux, P., Georgopoulos, C., Barral, J.M., Johnson, A.E., Hartl, F.U., and Etchells, S.A. (2007). Identification of nascent chain interaction sites on trigger factor. J. Biol. Chem. *282*, 12186–12193.

Lam, V.Q., Akopian, D., Rome, M., Henningsen, D., and Shan, S.O. (2010). Lipid activation of the signal recognition particle receptor provides spatial coordination of protein targeting. J. Cell Biol. *190*, 623–635.

Lee, H.C., and Bernstein, H.D. (2001). The targeting pathway of *Escherichia coli* presecretory and integral membrane proteins is specified by the hydrophobicity of the targeting signal. Proc. Natl. Acad. Sci. U.S.A. *98*, 3471–3476.

Lill, R., Dowhan, W., and Wickner, W. (1990). The ATPase activity of SecA is regulated by acidic phospholipids, SecY, and the leader and mature domains of precursor proteins. Cell *60*, 271–280.

Lin, P.J., Jongsam, C.G., Pool, M.R., and Johnson, A.E. (2011). Polytopic membrane protein folding at L17 in the ribosome tunnel initiates cyclical changes at the translocon. J. Cell Biol. *195*, 55–70.

Lotz, M., Haase, W., Kühlbrandt, W., and Collinson, I. (2008). Projection structure of yidC: a conserved mediator of membrane protein assembly. J. Mol. Biol. *375*, 901–907.

Luirink, J., ten Hagen-Jongman, C.M., van der Weijden, C.C., Oudega, B., High, S., Dobberstein, B., and Kusters, R. (1994). An alternative protein targeting pathway in *Escherichia coli*: studies on the role of FtsY. EMBO J. *13*, 2289–2296.

Luirink, J., von Heijne, G., Houben, E., and de Gier, J.W. (2005). Biogenesis of inner membrane proteins in *Escherichia coli*. Annu. Rev. Microbiol. *59*, 329–355.

Maeda, I., Hirose, N., Yamashiro, H., Ichibori, A., Ohshima, S., Fujimoto, T., Kawase, M., and Yagi, K. (2003). Comparative study of the N-terminal hydrophilic region in Streptomyces lividans and *E. coli* FtsY. Curr. Microbiol. *47*, 22–25.

Maier, K.S., Hubich, S., Liebhart, H., Krauss, S., Kuhn, A., and Facey, S.J. (2008). An amphiphilic region in the cytoplasmic domain of KdpD is recognized by the signal recognition particle and targeted to the *Escherichia coli* membrane. Mol. Microbiol. *68*, 1471–1484.

Manting, E.H., van Der Does, C., Remigy, H., Engel, A., and Driessen, A.J. (2000). SecYEG assembles into a tetramer to form the active protein translocation channel. EMBO J. *19*, 852–861.

Maegawa, S., Ito, K., and Akiyama, Y. (2005). Proteolytic action of GlpG, a rhomboid protease in the *Escherichia coli* cytoplasmic membrane. Biochemistry 44, 13543–13552.

Ménétret, J.F., Schaletzky, J., Clemons, W.M. Jr., Osborne, A.R., Skånland, S.S., Denison, C., Gygi, S.P., Kirkpatrick, D.S., Park, E., Ludtke, S.J., *et al.* (2007). Ribosome binding of a single copy of the SecY complex: implications for protein translocation. Mol. Cell *28*, 1083–1092.

Meyer, T.H., Ménétret, J.F., Breitling, R., Miller, K.R., Akey, C.W., and Rapoport, T.A. (1999). The bacterial SecY/E., translocation complex forms channel-like structures similar to those of the eukaryotic Sec61p complex. J. Mol. Biol. *285*, 1789–1800.

Mircheva, M., Boy, D., Weiche, B., Hucke, F., Graumann, P., and Koch, H.G. (2009). Predominant membrane localization is an essential feature of the bacterial signal recognition particle receptor. BMC Biol. *7*, 76.

Moore, M., Goforth, R.L., Mori, H., and Henry, R. (2003). Functional interaction of chloroplast SRP/FtsY with the ALB3 translocase in thylakoids: substrate not required. J. Cell Biol. *162*, 1245–1254.

Mori, H., and Ito, K. (2006). Different modes of SecY–SecA interactions revealed by site-directed *in vivo* photo-cross-linking. Proc. Natl. Acad. Sci. U.S.A. *103*, 16159–16164.

Moser, C., Mol, O., Goody, R.S., and Sinning, I. (1997). The signal recognition particle receptor of *Escherichia coli* (FtsY) has a nucleotide exchange factor built into the GTPase domain. Proc. Natl. Acad. Sci. U.S.A. *94*, 11339–11344.

Müller, M., and Klösgen, R.B. (2005). The Tat pathway in bacteria and chloroplasts (review). Mol. Membr. Biol. *22*,113–121.

Nagamori, S., Smirnova, I.N., and Kaback, H.R. (2004). Role of YidC in folding of polytopic membrane proteins. J. Cell Biol. *165*, 53–62.

Nakamura, K., Yahagi, S., Yamazaki, T., and Yamane, K. (1999). *Bacillus subtilis* histone-like protein, HBsu, is an integral component of a SRP-like particle that can bind the Alu domain of small cytoplasmic RNA. J. Biol. Chem. *274*, 13569–13576.

Neher, S.B., Bradshaw, N., Floor, S.N., Gross, J.D., and Walter, P. (2008). SRP RNA controls a conformational switch regulating the SRP-SRP receptor interaction. Nat. Struct. Mol. Biol. *15*, 916–923.

Neumann-Haefelin, C., Schäfer, U., Müller, M., and Koch, H.G. (2000). SRP-dependent co-translational targeting and SecA-dependent translocation analyzed as individual steps in the export of a bacterial protein. EMBO J. *19*, 6419–6426.

Nevo-Dinur, K., Nussbaum-Shochat, A., Ben-Yehuda, S., and Amster-Choder, O. (2011). Translation-independent localization of mRNA in *E. coli*. Science *331*, 1081–1084.

Nijeholt-Lycklama, J.A., Wu, Z.C., and Driessen, A.J.M. (2011). Conformational dynamics of the plug domain of the SecYEG protein-conducting channel. J. Biol. Chem. *286*, 43881–43890.

Nishiyama, K., Suzuki, T., and Tokuda, H. (1996). Inversion of the membrane topology of SecG coupled with SecA-dependent preprotein translocation. Cell *85*, 71–81.

Nouwen, N., and Driessen, A.J. (2002). SecDFyajC forms a heterotetrameric complex with YidC. Mol. Microbiol. *44*, 1397–1405.

Nouwen, N., de Kruijff, B., and Tommassen, J. (1996). prlA suppressors in *Escherichia coli* relieve the proton electrochemical gradient dependency of translocation of wild-type precursors. Proc. Natl. Acad. Sci. U.S.A. *93*, 5953–5957.

Oh, E., Becker, A.H., Sandikci, A., Huber, D., Chaba, R., Gloge, F., Nichols, R.J., Typas, A., Gross, C.A., Kramer, G., *et al.* (2011). Selective ribosome profiling reveals the cotranslational chaperone action of trigger factor *in vivo*. Cell *147*, 1295–1308.

Oliver, D.C., and Paetzel, M. (2008). Crystal structure of the major periplasmic domain of the bacterial membrane protein assembly facilitator YidC. J. Biol. Chem. *283*, 5208–5216.

Or, E., and Rapoport, T. (2007). Cross-linked SecA dimers are not functional in protein translocation. FEBS Lett. *581*, 2616–2620.

Osborne, A.R., and Rapoport, T.A. (2007). Protein translocation is mediated by oligomers of the SecY complex with one SecY copy forming the channel. Cell *129*, 97–110.

Paetzel, M., Karla, A., Strynadka, N.C., and Dalbey, R.E. (2002). Signal peptidases. Chem. Rev. *102*, 4549–4580.

Papanikou, E., Karamanou, S., and Economou, A. (2007). Bacterial protein secretion through the translocase nanomachine. Nat. Rev. Microbiol. *5*, 839–851.

Park, E., and Rapoport, T.A. (2011). Preserving the membrane barrier for small molecules during bacterial protein translocation. Nature *473*, 239–242.

Parlitz, R., Eitan, A., Stjepanovic, G., Bahari, L., Bange, G., Bibi, E., and Sinning, I. (2007). *Escherichia coli* signal recognition particle receptor FtsY contains an essential and autonomous membrane-binding amphipathic helix. J. Biol. Chem. *282*, 32176–32184.

Plath, K., Mothes, W., Wilkinson, B.M., Stirling, C.J., and Rapoport, T.A. (1998). Signal sequence recognition in posttranslational protein transport across the yeast ER membrane. Cell *94*, 795–807.

Pogliano, J.A., and Beckwith, J. (1994). SecD and SecF facilitate protein export in *Escherichia coli*. EMBO J. *13*, 554–561.

Pool, M.R., Stumm, J., Fulga, T.A., Sinning, I., and Dobberstein, B. (2002). Distinct modes of signal recognition particle interaction with the ribosome. Science *297*, 1345–1348.

Pop, O.I., Soprova, Z., Koningstein, G., Scheffers, D.J., van Ulsen, P., Wickström, D., de Gier, J.W., and Luirink, J. (2009). YidC is required for the assembly of the MscL homopentameric pore. FEBS J. *276*, 4891–4899.

Poritz, M.A., Bernstein, H.D., Strub, K., Zopf, D., Wilhelm, H., and Walter, P. (1990). An *E. coli* ribonucleoprotein containing 4.5S RNA resembles mammalian signal recognition particle. Science *250*, 1111–1117.

Preuss, M., Ott, M., Funes, S., Luirink, J., and Herrmann, J.M. (2005). Evolution of mitochondrial oxa proteins from bacterial YidC. Inherited and acquired functions of a conserved protein insertion machinery. J. Biol. Chem. *280*, 13004–13011.

Price, C.E., and Driessen, A.J. (2008). YidC is involved in the biogenesis of anaerobic respiratory complexes in the inner membrane of *Escherichia coli*. J. Biol. Chem. *283*, 26921–26927.

Price, C.E., Otto, A., Fusetti, F., Becher, D., Hecker, M., and Driessen, A.J. (2010). Differential effect of YidC depletion on the membrane proteome of *Escherichia coli* under aerobic and anaerobic growth conditions. Proteomics *10*, 3235–3247.

Price, C.E., Kocer, A., Kol, S., van der Berg, J.P., and Driessen, A.J. (2011). *In vitro* synthesis and oligomerization of the mechanosensitive channel of large conductance, MscL, into a functional ion channel. FEBS Lett. *585*, 249–254.

Prilusky, J., and Bibi, E. (2009). Studying membrane proteins through the eyes of the genetic code revealed a strong uracil bias in their coding mRNAs. Proc. Natl. Acad. Sci. U.S.A. *106*, 6662–6666.

Prinz, A., Behrens, C., Rapoport, T.A., Hartmann, E., and Kalies, K.U. (2000). Evolutionarily conserved binding of ribosomes to the translocation channel via the large ribosomal RNA. EMBO J. *19*, 1900–1906.

Raine, A., Ivanova, N., Wikberg, J.E., and Ehrenberg, M. (2004). Simultaneous binding of trigger factor and signal recognition particle to the *E. coli* ribosome. Biochimie *86*, 495–500.

Randall, L.L., Topping, T.B., Hardy, S.J., Pavlov, M.Y., Freistroffer, D.V., and Ehrenberg, M. (1997). Binding of SecB to ribosome-bound polypeptides has the same characteristics as binding to full-length, denatured proteins. Proc. Natl. Acad. Sci. U.S.A. *94*, 802–807.

Rapiejko, P.J., and Gilmore, R. (1997). Empty site forms of the SRP54 and SR alpha GTPases mediate targeting of ribosome-nascent chain complexes to the endoplasmic reticulum. Cell *89*, 703–713.

Rapoport, T.A. (2007). Protein translocation across the eukaryotic endoplasmic reticulum and bacterial plasma membranes. Nature *450*, 663–669.

Ravaud, S., Wild, K., and Sinning, I. (2008). Purification, crystallization and preliminary structural characterization of the periplasmic domain P1 of the *Escherichia coli* membrane-protein insertase YidC. Acta Crystallogr. Sect. F Struct. Biol. Cryst. Commun. *64*, 144–148.

Rosch, J.W., and Caparon, M.G. (2005). The ExPortal: an organelle dedicated to the biogenesis of secreted proteins in Streptococcus pyogenes. Mol. Microbiol. *58*, 959–968.

Rosenblad, M.A., Larsen, N., Samuelsson, T., and Zwieb, C. (2009). Kinship in the SRP RNA family. RNA Biol. *6*, 508–516.

Rubio, A., Jiang, X., and Pogliano, K. (2005). Localization of translocation complex components in *Bacillus subtilis*: enrichment of the signal recognition particle receptor at early sporulation septa. J. Bacteriol. *187*, 5000–5002.

Sääf, A., Monné, M., de Gier, J.W., and von Heijne, G. (1998). Membrane topology of the 60-kDa Oxa1p homologue from *Escherichia coli*. J. Biol. Chem. *273*, 30415–30418.

Sadlish, H., Pitonzo, D., Johnson, A.E., and Skach, W.R. (2005). Sequential triage of transmembrane segments by Sec61alpha during biogenesis of a native multispanning membrane protein. Nat. Struct. Mol. Biol. *12*, 870–878.

Saito, A., Hizukuri, Y., Matsuo, E., Chiba, S., Mori, H., Nishimura, O., Ito, K., and Akiyama, Y. (2011). Post-liberation cleavage of signal peptides is catalyzed by the site-2 protease (S2P) in bacteria. Proc. Natl. Acad. Sci. U.S.A. *108*, 13740–13745.

Samuelson, J.C., Chen, M., Jiang, F., Möller, I., Wiedmann, M., Kuhn, A., Phillips, G.J., and Dalbey, R.E. (2000). YidC mediates membrane protein insertion in bacteria. Nature *406*, 637–641.

Samuelson, J.C., Jiang, F., Yi, L., Chen, M., de Gier, J.W., Kuhn, A., and Dalbey, R.E. (2001). Function of YidC for the insertion of M13 procoat protein in *Escherichia coli*: translocation of mutants that show differences in their membrane potential dependence and Sec requirement. J. Biol. Chem. *276*, 34847–34852.

Sardis, M.F., and Economou, A. (2010). SecA: a tale of two protomers. Mol. Microbiol. *76*, 1070–1081.

Sauer, R.T., and Baker, T.A. (2011). AAA+ proteases: ATP-fuelled machines of protein destruction. Annu. Rev. Biochem. *80*, 587–612.

Schaffitzel, C., Oswald, M., Berger, I., Ishikawa, T., Abrahams, J.P., Koerten, H.K., Koning, R.I., and Ban, N. (2006). Structure of the *E. coli* signal recognition particle bound to a translating ribosome. Nature *444*, 503–506.

Schiebel, E., Driessen, A.J., Hartl, F.U., and Wickner, W. (1991). Delta mu H+ and ATP function at different steps of the catalytic cycle of preprotein translocase. Cell *64*, 927–939.

Schmitz, U., James, T.L., Lukavsky, P., and Walter, P. (1999). Structure of the most conserved internal loop in SRP RNA. Nat. Struct. Biol. *6*, 634–638.

Scotti, P.A., Valent, Q.A., Manting, E.H., Urbanus, M.L., Driessen, A.J., Oudega, B., and Luirink, J. (1999). SecA is not required for signal recognition particle-mediated targeting and initial membrane insertion of a nascent inner membrane protein. J. Biol. Chem. *274*, 29883–29888.

Scotti, P.A., Urbanus, M.L., Brunner, J., de Gier, J.W., von Heijne, G., van der Does, C., Driessen, A.J., Oudega, B., and Luirink, J. (2000). YidC, the *Escherichia coli* homologue of mitochondrial Oxa1p, is a component of the Sec translocase. EMBO J. *19*, 542–549.

Serek, J., Bauer-Manz, G., Struhalla, G., van den Berg, L., Kiefer, D., Dalbey, R., and Kuhn, A. (2004). *Escherichia coli* YidC is a membrane insertase for Sec-independent proteins. EMBO J. *23*, 294–301.

Shan, S.O., Chandrasekar, S., and Walter, P. (2007). Conformational changes in the GTPase modules of the signal reception particle and its receptor drive initiation of protein translocation. J. Cell. Biol. *178*, 611–620.

Shen, K., and Shan, S.O. (2010). Transient tether between the SRP RNA and SRP receptor ensures efficient cargo delivery during cotranslational protein targeting. Proc. Natl. Acad. Sci. U.S.A. *107*, 7698–7703.

Shiomi, D., Yoshimoto, M., Homma, M., and Kawagishi, I. (2006). Helical distribution of the bacterial chemoreceptor via colocalization with the Sec protein translocation machinery. Mol. Microbiol. *60*, 894–906.

Stjepanovic, G., Kapp, K., Bange, G., Graf, C., Parlitz, R., Wild, K., Mayer, M.P., and Sinning, I. (2011). Lipids trigger a conformational switch regulating signal recognition particle (SRP)-mediated protein targeting. J. Biol. Chem. *286*, 23489–23497.

Szabady, R.L., Peterson, J.H., Skillman, K.M., and Bernstein, H.D. (2005). An unusual signal peptide facilitates late steps in the biogenesis of a bacterial autotransporter. Proc. Natl. Acad. Sci. U.S.A. *102*, 221–226.

Szyrach, G., Ott, M., Bonnefoy, N., Neupert, W., and Herrmann, J.M. (2003). Ribosome binding to the Oxa1 complex facilitates co-translational protein insertion in mitochondria. EMBO J. *22*, 6448–6457.

Talmadge, K., Stahl, S., and Gilbert, W. (1980). Eukaryotic signal sequence transports insulin antigen in *Escherichia coli*. Proc. Natl. Acad. Sci. U.S.A. *77*, 3369–3373.

Tjalsma, H., Bron, S., and van Dijl, J.M. (2003). Complementary impact of paralogous Oxa1-like proteins of *Bacillus subtilis* on post-translocational stages in protein secretion. J. Biol. Chem. *278*, 15622–15632.

Tomkiewicz, D., Nouwen, N., and Driessen, A.J. (2008). Kinetics and energetics of the translocation of maltose binding protein folding mutants. J. Mol. Biol. *377*, 83–90.

Topping, T.B., and Randall, L.L. (1994). Determination of the binding frame within a physiological ligand for the chaperone SecB. Protein Sci. *3*, 730–736.

Tsui, H.C., Keen, S.K., Sham, L.T., Wayne, K.J., and Winkler, M.E. (2011). Dynamic distribution of the SecA and SecY translocase subunits and septal localization oft he HtrA surface chaperone/protease during *Streptococcus pneumonia* D39 cell division. MBio. 2, e00202–00211.

Tsukazaki, T., Mori, H., Fukai, S., Ishitani, R., Mori, T., Dohmae, N., Perederina, A., Sugita, Y., Vassylyev, D.G., Ito, K., *et al.* (2008). Conformational transition of Sec machinery inferred from bacterial SecYE structures. Nature *455*, 988–991.

Tsukazaki, T., Mori, H., Echizen, Y., Ishitani, R., Fukai, S., Tanaka, T., Perederina, A., Vassylyev, D., Kohno, T., Maturana, A., *et al.* (2011). Structure and function of a membrane component SecDF that enhances protein export. Nature *474*, 235–238.

Tziatzios, C., Schubert, D., Lotz, M., Gundogan, D., Betz, H., Schägger, H., Haase, W., Duong, F., and Collinson, I. (2004). The bacterial protein–translocation complex: SecYEG dimers associate with one or two SecA molecules. J. Mol. Biol. *340*, 513–524.

Uchida, K., Mori, H., and Mizushima, S. (1995). Stepwise movement of preproteins in the process of translocation across the cytoplasmic membrane of *Escherichia coli*. J. Biol. Chem. *270*, 30862–30868.

Ullers, R.S., Houben, E.N., Raine, A., ten Hagen-Jongman, C.M., Ehrenberg, M., Brunner, J., Oudega, B., Harms, N., and Luirink, J. (2003). Interplay of signal recognition particle and trigger factor at L23 near the nascent chain exit site on the *Escherichia coli* ribosome. J. Cell Biol. *61*, 679–684.

Urbanus, M.L., Fröderberg, L., Drew, D., Björk, P., de Gier, J.W., Brunner, J., Oudega, B., and Luirink, J. (2002). Targeting, insertion, and localization of *Escherichia coli* YidC. J. Biol. Chem. *277*, 12718–12723.

Valent, Q.A., de Gier, J.W., von Heijne, G., Kendall, D.A., ten Hagen-Jongman, C.M., Oudega, B., and Luirink, J. (1997). Nascent membrane and presecretory proteins synthesized in *Escherichia coli* associate with signal recognition particle and trigger factor. Mol. Microbiol. *25*, 53–64.

Valent, Q.A., Scotti, P.A., High, S., de Gier, J.W., von Heijne, G., Lentzen, G., Wintermeyer, W., Oudega, B., and Luirink, J. (1998). The *Escherichia coli* SRP and SecB targeting pathways converge at the translocon. EMBO J. *17*, 2504–2512.

van Bloois, E., Jan Haan, G., de Gier, J.W., Oudega, B., and Luirink, J. (2004). F(1)F(0) ATP synthase subunit c is targeted by the SRP to YidC in the *E. coli* inner membrane. FEBS Lett. *576*, 97–100.

van Bloois, E., Nagamori, S., Koningstein, G., Ullers, R.S., Preuss, M., Oudega, B., Harms, N., Kaback, H.R., Herrmann, J.M., and Luirink, J. (2005). The Sec-independent function of *Escherichia coli* YidC is evolutionary-conserved and essential. J. Biol. Chem. *280*, 12996–13003.

van Bloois, E., Haan, G.J., de Gier, J.W., Oudega, B., and Luirink, J. (2006). Distinct requirements for translocation of the N-tail and C-tail of the *Escherichia coli* inner membrane protein CyoA. J. Biol. Chem. *281*, 10002–10009.

van Bloois, E., Dekker, H.L., Fröderberg, L., Houben, E.N., Urbanus, M.L., de Koster, C.G., de Gier, J.W., and Luirink, J. (2008). Detection of cross-links between FtsH, YidC, HflK/C suggests a linked role for these proteins in quality control upon insertion of bacterial inner membrane proteins. FEBS Lett. *582*, 1419–1424.

van Dalen, A., Killian, A., and de Kruijff, B. (1999). Delta psi stimulates membrane translocation of the C-terminal part of a signal sequence. J. Biol. Chem. *274*, 19913–19918.

Van den Berg, B., Clemons, W.M. Jr., Collinson, I., Modis, Y., Hartmann, E., Harrison, S.C., and Rapoport, T.A. (2004). X-ray structure of a protein-conducting channel. Nature *427*, 36–44.

van der Laan, M., Urbanus, M.L., Ten Hagen-Jongman, C.M., Nouwen, N., Oudega, B., Harms, N., Driessen, A.J., and Luirink, J. (2003). A conserved function of YidC in the biogenesis of respiratory chain complexes. Proc. Natl. Acad. Sci. U.S.A. *100*, 5801–5806.

van der Laan, M., Bechtluft, P., Kol, S., Nouwen, N., and Driessen, A.J. (2004a). F1F0 ATP synthase subunit c is a substrate of the novel YidC pathway for membrane protein biogenesis. J. Cell Biol. *165*, 213–222.

van der Laan, M., Nouwen, N., and Driessen, A.J. (2004b). SecYEG proteoliposomes catalyze the Deltaphi-dependent membrane insertion of FtsQ. J. Biol. Chem. *279*, 1659–1664.

van der Sluis, E.O., and Driessen, A.J. (2006). Stepwise evolution of the Sec machinery in Proteobacteria. Trends Microbiol. *14*, 105–108.

van der Sluis, E.O., Nouwen, N., and Driessen, A.J. (2002). SecY-SecY and SecY-SecG contacts revealed by site–specific crosslinking. FEBS Lett. *527*, 159–165.

van der Wolk, J.P., de Wit, J.G., and Driessen, A.J. (1997). The catalytic cycle of the *Escherichia coli* SecA ATPase comprises two distinct preprotein translocation events. EMBO J. *16*, 7297–7304.

von Heijne, G. (1990). The signal peptide J. Membr. Biol. *115*, 195–201.

Wagner, S., Pop, O.I., Haan, G.J., Baars, L., Koningstein, G., Klepsch, M.M., Genevaux, P., Luirink, J., and de Gier, J.W. (2008). Biogenesis of MalF and the MalFGK(2) maltose transport complex in *Escherichia coli* requires YidC. J. Biol. Chem. *283*, 17881–17890.

Wang, H., Na, B., Yang, H., and Tai, P.C. (2008). Additional *in vitro* and *in vivo* evidence for SecA functioning as dimers in the membrane: dissociation into monomers is not essential for protein translocation in *Escherichia coli*. J. Bacteriol. *190*, 1413–1418.

Wang, P., and Dalbey, R.E. (2011). Inserting membrane proteins: the YidC/Oxa1/Alb3 machinery in bacteria, mitochondria, and chloroplasts. Biochim. Biophys. Acta. *1808*, 866–875.

Wang, P., Kuhn, A., and Dalbey, R.E. (2010). Global change of gene expression and cell physiology in YidC-depleted *Escherichia coli*. J. Bacteriol. *192*, 2193–2209.

Weiche, B., Bürk, J., Angelini, S., Schiltz, E., Thumfart, J.O., and Koch, H.G. (2008). A cleavable N-terminal membrane anchor is involved in membrane binding of the *Escherichia coli* SRP receptor J. Mol. Biol. *377*, 761–773.

Welte, T., Kudva, R., Kuhn, P., Sturm, L., Braig, D., Müller, M., Warscheid, B., Drepper, F., and Koch, H.G. (2012). Promiscous targeting of polytopic membrane proteins to SecYEG or YidC by the *Escherichia coli* signal recognition particle. Mol. Biol. Cell *23*, 464–479.

Wickström, D., Wagner, S., Baars, L., Ytterberg, A.J., Klepsch, M., van Wijk, K.J., Luirink, J., and de Gier, J. (2011a). Consequences of depletion of the signal recognition particle in *Escherichia coli*. J. Biol. Chem. *286*, 4598–4609.

Wickström, D., Wagner, S., Simonsson, P., Pop, O., Baars, L., Ytterberg, A.J., van Wijk, K.J., Luirink, J., and de Gier, J.W. (2011b). Characterization of the consequences of YidC depletion on the inner membrane proteome of *E. coli* using 2D Blue Native/SDS-PAGE. J. Mol. Biol. *409*, 124–135.

Wilson, D.N., and Beckmann, R. (2011). The ribosomal tunnel as a functional environment for nascent polypeptide folding and translational stalling. Curr. Opin. Struct. Biol. *21*, 274–282.

Wirth, A., Jung, M., Bies, C., Frien, M., Tyedmers, J., Zimmermann, R., and Wagner, R. (2003). The Sec61p complex is a dynamic precursor activated channel. Mol. Cell *12*, 261–268.

Woolhead, C.A., McCormick, P.J., and Johnson, A.E. (2004). Nascent membrane and secretory proteins differ in FRET-detected folding far inside the ribosome and in their exposure to ribosomal proteins. Cell *116*, 725–736.

Wu, Z.C., de Keyzer, J., Kedrov, A., and Driessen, A.J.M. (2012). Competitive binding of SecA and ribosomes to the SecYEG translocon. J. Biol. Chem. *287*, 7885–7895.

Xie, K., Kiefer, D., Nagler, G., Dalbey, R.E., and Kuhn, A. (2006). Different regions of the nonconserved large periplasmic domain of *Escherichia coli* YidC are involved in the SecF interaction and membrane insertase activity. Biochemistry *45*, 13401–13408.

Xu, Z., Knafels, J.D., and Yoshino, K. (2000). Crystal structure of the bacterial protein export chaperone SecB. Nat. Struct. Biol. *7*, 1172–1177.

Yen, M.R., Harley, K.T., Tseng, Y.H., and Saier, M.H. Jr. (2001). Phylogenetic and structural analyses of the oxa1 family of protein translocases. FEMS Microbiol. Lett. *204*, 223–231.

Yi, L., Celebi, N., Chen, M., and Dalbey, R.E. (2004). Sec/SRP requirements and energetics of membrane insertion of subunits a, b, and c of the *Escherichia coli* F1F0 ATP synthase. J. Biol. Chem. *279*, 39260–39267.

Yosef, I., Bochkareva, E.S., and Bibi, E. (2010). *Escherichia coli* SRP, its protein subunit Ffh, and the Ffh M domain are able to selectively limit membrane protein expression when overexpressed. MBio. 1.

Yu, Z., Koningstein, G., Pop, A., and Luirink, J. (2008). The conserved third transmembrane segment of YidC contacts nascent *Escherichia coli* inner membrane proteins. J. Biol. Chem. *283*, 34635–34642.

Yuan, J., Phillips, G.J., and Dalbey, R.E. (2007). Isolation of cold-sensitive yidC mutants provides insights into the substrate profile of the YidC insertase and the importance of transmembrane 3 in YidC function. J. Bacteriol. *189*, 8961–8972.

Zanen, G., Antelmann, H., Meima, R., Jongbloed, J.D., Kolkman, M., Hecker, M., van Dijl, J.M., and Quax, W.J. (2006). Proteomic dissection of potential signal recognition particle dependence in protein secretion by *Bacillus subtilis*. Proteomics *6*, 3636–3648.

Zhang, D., and Shan, S.O. (2012). Translation elongation regulates selection by the signal recognition particle. J. Biol. Chem. *287*, 7652–7660.

Zhang, X., Schaffitzel, C., Ban, N., and Shan, S.O. (2009). Multiple conformational switches in a GTPase complex control co-translational protein targeting. Proc. Natl. Acad. Sci. U.S.A. *106*, 1754–1759.

Zhang, X., Rashid, R., Wang, K., and Shan, S.O. (2010). Sequential checkpoints govern substrate selection during cotranslational protein targeting. Science *328*, 757–760.

Zheng, N., and Gierasch, L.M. (1997). Domain interactions in *E. coli* SRP: stabilization of M domain by RNA is required for effective signal sequence modulation of NG domain. Mol. Cell *1*, 79–87.

Zhou, J., and Xu, Z. (2005). The structural view of bacterial translocation-specific chaperone SecB: implications for function. Mol. Microbiol. *58*, 349–357.

Zimmer, J., and Rapoport, T.A. (2009). Conformational flexibility and peptide interaction of the translocation ATPase SecA. J. Mol. Biol. *394*, 606–612.

Zimmer, J., Nam, Y., and Rapoport, T.A. (2008). Structure of a complex of the ATPase SecA and the protein-translocation channel. Nature *455*, 936–943.

Zito, C.R., and Oliver, D. (2003). Two-stage binding of SecA to the bacterial translocon regulates ribosome–translocon interaction. J. Biol. Chem. *278*, 40640–40646.

Envelope-spanning Secretion Systems in Gram-negative Bacteria

11

Matthias J. Brunner, Rémi Fronzes and Thomas C. Marlovits

Abstract

Gram-negative bacteria have a cell envelope made of two membranes separated by a thin layer of peptidoglycan. To transport macromolecules such as proteins and DNA through the entire cell envelope, several types of secretion systems are employed. This chapter focuses on the structure and function of type III and type IV secretion systems.

Type III secretion (T3S) systems are used by pathogens such as Salmonella, *Shigella* and *Yersinia*. They are composed of more than 20 different proteins, some of them present in multiple copies. Their function is to inject proteinaceous toxins, referred to as 'effectors', into the eukaryotic host cell upon intimate contact. One pathogen typically secrets many different effectors–most of them have in common that they hijack part of the host cell machinery.

Type IV secretion (T4S) systems are versatile secretion systems found in many bacteria. Typical T4S systems are made of 12 different proteins. Within the T4S family, three groups can be defined: First, T4S systems that mediate conjugative transfer of mobile genetic elements into a wide range of bacterial species but also into eukaryotic cells. Second, type IV secretion systems mediating DNA uptake or release from or into the extracellular milieu. And third, type IV secretion systems directly involved in virulence, secreting virulence factors into mammalian host cells. These are used by pathogens such as *Helicobacter pylori* or *Legionella pneumophila*.

Introduction: bacterial secretion systems

During various cellular processes, cells need to release macromolecules into their environment. Known as secretion, this process is common to all living cells. Gram-negative bacteria have a cell envelope made of two membranes separated by a thin layer of peptidoglycan. The general secretory pathway (Sec or Tat pathway) mediates the passage of most of proteins through the bacterial inner membrane into the periplasm (see Chapter 10). However, in order to transport proteins through the whole cell envelope, Gram-negative bacteria developed six different types of specialized 'molecular machines', also named secretion systems (Fig. 11.1). The architecture of these systems varies from single-component to multi–component complexes (Fig. 11.1). Type I secretion (T1S) systems are simple tripartite system mediating the passage of proteins of various sizes across the cell envelope of Gram-negative bacteria. They are formed of an ABC (ATP binding cassette) transporter or a proton-antiporter, an adaptor protein that bridges the two membranes, and an outer membrane (OM)

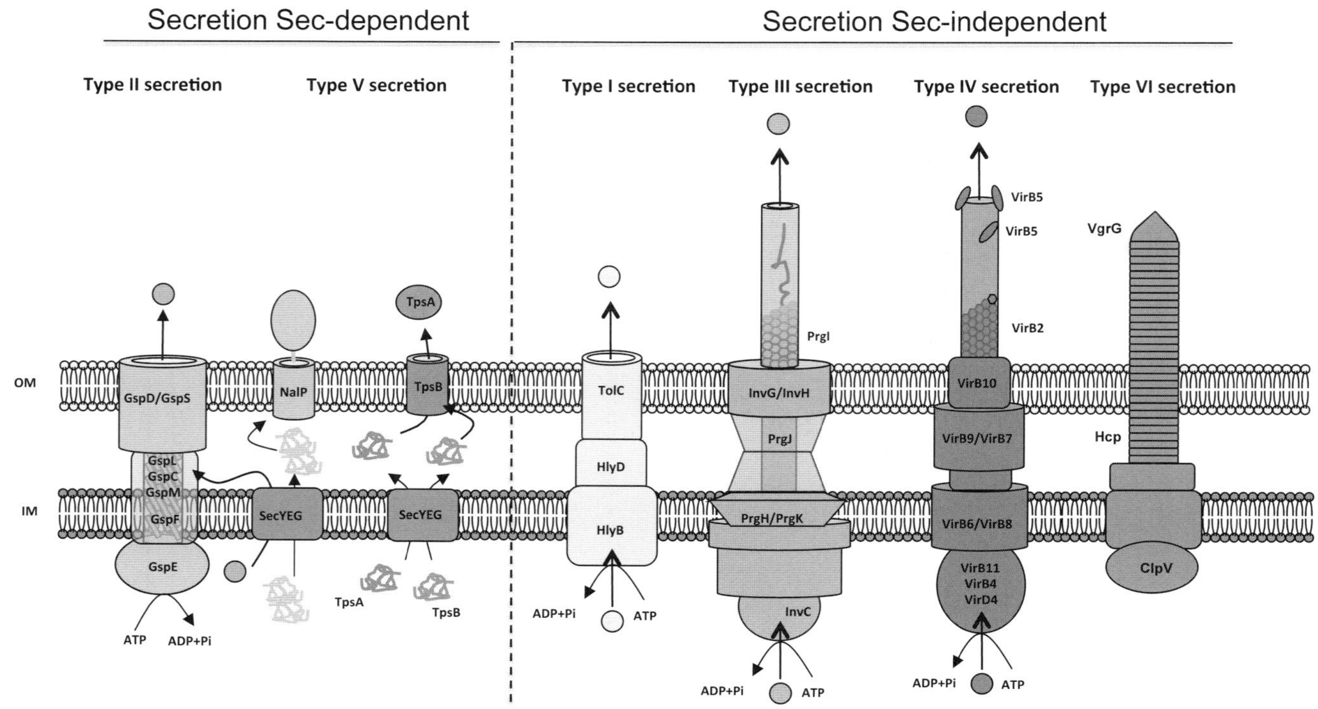

Figure 11.1 Schematic view of the secretion systems found in Gram-negative bacteria. A colour version of this figure is located in the plate section at the back of the book.

pore. They secrete their substrate in a single step without a stable periplasmic intermediate (Koronakis *et al.*, 1991). T1S systems are involved in the secretion of cytotoxins belonging to the RTX (repeats-in-toxin) protein family, cell surface layer proteins, proteases, lipases, bacteriocins, haem-acquisition proteins etc. (reviewed in Omori and Idei, 2003). Type II secretion (T2S) systems are multicomponent that use a two-step translocation mechanism. During the first step, the substrate is transported through the inner membrane via the Sec (Gold *et al.*, 2007) or the Tat translocons (Voulhoux *et al.*, 2001). Once in the periplasm, it is recognized and translocated by the T2S system through the OM. The T2S system translocon consist of 12–16 protein components (Filloux, 2004) that are found in both bacterial membranes, in the cytoplasm and periplasm. The T2S system and the type IV pilus assembly machinery are evolutionarily related (Peabody *et al.*, 2003; Craig and Li, 2008). T2S protein components are well conserved among Gram-negative bacteria, and T2S systems represent a key virulence factor in human and plant bacterial pathogens (Hales and Shuman, 1999; Iwobi *et al.*, 2003; Filloux, 2004). Type III secretion (T3S) systems, also called injectisomes, mediate a single-step secretion mechanism used by many plant and animal pathogens, including *Salmonella*, *Shigella*, *Yersinia*, enteropathogenic and enterohaemorrhagic *Escherichia coli* (EPEC, EHEC) and *Pseudomonas aeruginosa*. They deliver effector proteins into the eukaryotic host cell cytoplasm in a Sec independent manner (Cornelis and Van Gijsegem, 2000). T3S systems are evolutionarily, structurally and functionally related to bacterial flagella (Cornelis and Van Gijsegem, 2000). They are composed of more than 20 different proteins, which form a large supramolecular structure crossing the bacterial cell envelope (Cornelis and Van Gijsegem, 2000). Type IV secretion (T4S) systems are versatile secretion systems found in many bacteria. These systems play an important role in the pathogenesis of diseases including stomach ulcers (caused by *Helicobacter pylori*), Legionnaire's disease (caused by *Legionella pneumophila*) and whooping cough (caused by *Bordetella pertussis*) by mediating the release of bacterial toxins into eukaryotic cells. In addition, T4S systems that are encoded by conjugative plasmids or transposons are used to propagate genes from one bacterium to another during horizontal gene transfer events, promoting for example the spread of antibiotic resistance among pathogenic bacteria. Typical T4S systems are made of 12 proteins that form a macromolecular complex that spans the entire cell envelope. Type V secretion (T5S) systems are the simplest and most widely distributed secretion systems. They include the autotransporter (AT) secretion system and two-partner secretion (TPS) system. Over 700 pathogenic proteins with functions including auto-aggregation, adherence, invasion, cytotoxicity, serum resistance, cell-to-cell spread, and proteolysis utilize these two secretion systems to cross both inner and outer membranes during a simple two-step process (Henderson and Nataro, 2001; Mazar and Cotter, 2007). AT proteins are multidomain proteins that are exported as precursor proteins across the inner membrane via a Sec-dependent process. Subsequently, the translocator domain of the protein inserts into the outer membrane and facilitates surface localization of the passenger domain. In TPS system, a separate translocator protein mediates the secretion of effector protein through the outer membrane. Both proteins are secreted in the periplasm by the Sec translocon. Finally, type VI secretion (T6S) systems are recently discovered secretion systems that are found in several pathogens such as *Pseudomonas aeruginosa*, enteroaggregative *E. coli*, *Salmonella typhimurium*, *Vibrio cholerae* or *Yersinia pestis*. Little is know concerning the assembly and function of these secretion systems. These secretion systems are multicomponent systems that could be composed of 12–25 subunits. So far, very little is known concerning the architecture and function of these systems (for review see Cascales, 2008; Pukatzki *et al.*, 2009).

In this chapter, we chose to focus on two important envelope spanning secretion systems for which important structural and functional insights have been recently obtained. This chapter will focus on the features of the biology and structure of T3S and T4S systems, emphasizing on recent findings concerning the architecture, assembly and function of these systems.

The type III secretion system

Gram-negative pathogens such as *Yersinia, Shigella, Pseudomonas,* enteropathogenic/enterohaemorrhagic *E. coli (EPEC/EHEC)* and *Salmonella* in animals and *Erwinia, Ralstonia* and *Xanthomonas* in plants employ type III secretion systems (T3S systems) for the infection process and are required for persistence inside the host (Hueck, 1998; Galán and Wolf-Watz, 2006). Human diseases in which type III secretion (T3S) is involved range from mild, such as diarrhoea, to deadly, such as bubonic plague (Coburn *et al.*, 2007). However, not only pathogens use T3S but symbionts as well (Coombes, 2009), for instance symbionts of insects (Dale *et al.*, 2002) and legumes (Freiberg *et al.*, 1997).

T3S systems are multi-component macromolecular machineries that are usually encoded on specific pathogenicity islands (Fig. 11.2). Their function is to inject proteinaceous toxins, referred to as 'effectors', into the host cell upon intimate contact.

One pathogen typically secrets many different effectors – most of them have in common that they hijack part of the host cell machinery. For instance, *Salmonella* reprogram the cytoskeleton in order to invade host cells and thereby overcome the epithelial barrier in the gut (Haraga *et al.*, 2008). Effectors are optimized and unique for the ecological niche where pathogens co-exist with host cells. Therefore, effectors from different pathogens often vary in their biochemical function and are not necessarily evolutionary linked. Consequently, effectors are often located outside pathogenicity islands. In contrast, many other components of the T3S system, structural ones in particular, are well conserved and are usually located on specific pathogenicity islands on either the chromosome or on a virulence plasmid (Schmidt and Hensel, 2004). They were probably spread to different microorganisms by horizontal gene transfer during evolution. The structural components of the T3S system are the building blocks to form the cylindrically shaped and membrane-embedded needle complex (NC), the core of the T3S system (Kubori *et al.*, 1998; Tamano *et al.*, 2000; Sekiya *et al.*, 2001; Marlovits *et al.*, 2004; Hodgkinson *et al.*, 2009a; Schraidt *et al.*, 2010). Many components even have functional analogues or homologues in the flagellar system (Büttner, 2012) which is required for motility. Despite the different function, it is thus not surprising to observe structural similarity to some parts of the flagellar system indicating that common molecular principles are employed to build such systems even though the overall size and shape differ.

Controlled steps to build a complex molecular machine that spans three membranes

The T3S system in Gram-negative bacteria has evolved to a complex molecular machine that achieves protein translocation across three membranes – the inner and outer membrane of the bacterial cell and the plasma membrane of the eukaryotic host cell (Galán and Wolf-Watz,

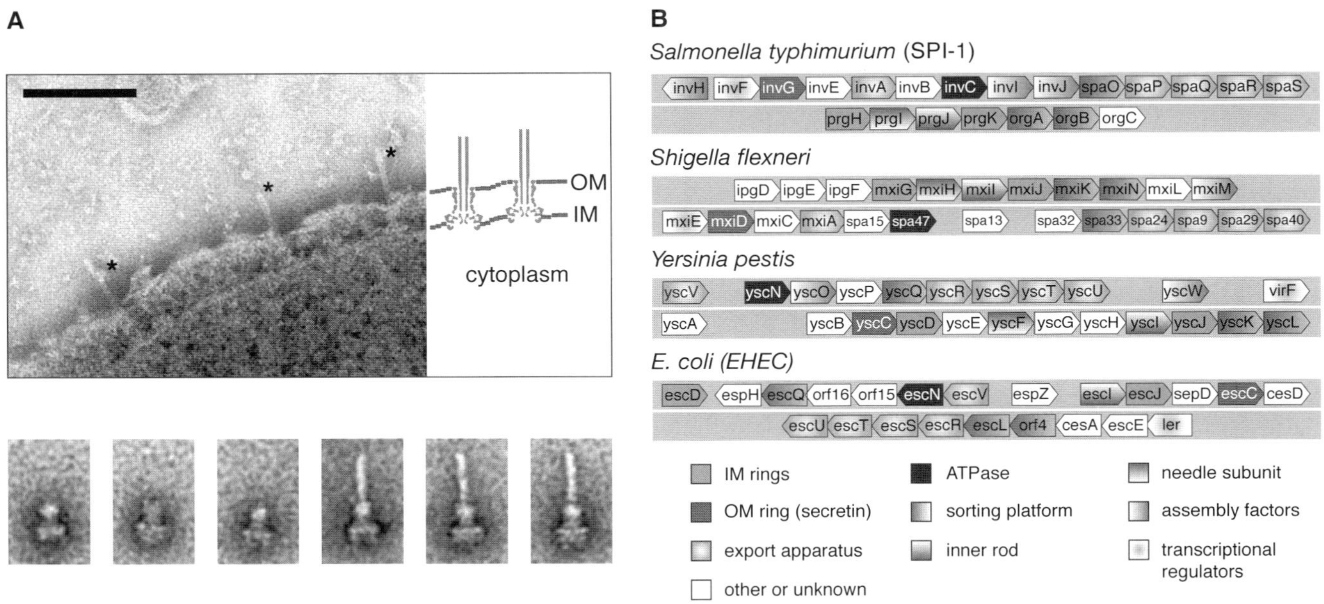

Figure 11.2 T3SSs are membrane-embedded molecular machines encoded on specific pathogenicity islands. (A) Upper panel: Electron micrograph of an osmotically shocked *S. typhimurium* cell with up-regulated SPI-1 T3SS expression; stained with phosphotungstic acid. Asterisks indicate needle complexes (NCs). Scale bar is 100nm and also applies to the lower panel. Lower panel: Electron micrographs of isolated NC basal bodies (left) and fully assembled NCs (right), stained with phosphotungstic acid. (B) Selected operons from T3SS pathogenicity islands and their gene function. Modified from (Izoré *et al.*, 2011).

2006). It consists of many components, its most prominent one being the needle complex (NC), a large hetero-oligomeric membrane protein complex with a molecular weight of about 3.5 megadaltons (Fig. 11.2). The name stems from the needle-like protrusion visible in electron micrographs of whole bacterial cells (Kubori *et al.*, 1998). These protrusions are protein filaments that engage with the host cell and are believed to serve as a conduit for the secretion substrate. The needle filament is linked to the membrane-embedded basal body, which in Gram-negative bacteria spans the inner and outer membrane (about 30 × 30nm) including the periplasmic space (Marlovits *et al.*, 2004). The basal body has a cylindrical shape, defining a central space within which the inner rod and the socket/cup are localized. The inner rod presumably connects the socket/cup with the needle filament and may help to stably anchor the filament into the basal body (Marlovits *et al.*, 2006).

The NC expressed from *Salmonella* pathogenicity island 1 (SPI-1) is composed of multiple copies of approximately ten proteins (PrgH/K/I/J, InvG, SpaP/Q/R/S, InvA) in its isolated form (Galán and Wolf-Watz, 2006). A system at this level of complexity requires defined and controlled steps during the assembly (Sukhan *et al.*, 2001; Diepold *et al.*, 2010). It is initially dependent on the cellular *sec*-machinery, in particular during the early ring-forming events of assembly. The export apparatus, a group of essential and conserved inner membrane proteins in T3S systems, plays a critical role during the initial phase of the NC assembly. It generates subcomplexes that may serve as nucleation points for the subsequent concentric ring organization of the two inner membrane rings (Samuel Wagner *et al.*, 2010a) (Fig. 11.3, step 1 (side view); see also later for alternative model). In contrast, the ring localized in the outer membrane assembles independently of the inner membrane components (Spreter *et al.*, 2009a), and extends far into the periplasm (Schraidt *et al.*, 2010) (Fig. 11.3, step 2). The outer and the inner membrane protein rings then associate to form the basal body (Fig. 11.3, step 3). At this point, the assembly only proceeds in the presence of a type III (T3)-specific ATPase (Kubori *et al.*, 2000; Sukhan *et al.*, 2001). The ATPase is required to secrete the inner rod and the needle filament proteins that become integral structural components of the NC. During this assembly phase, the septum (periplasmic gate; see Fig. 11.3, step 2), composed in part by the outer membrane protein ring, undergoes a large conformational change (Marlovits *et al.*, 2004). This allows the needle protein protomers to polymerize into a helical filamentous structure of about 25nm to 50nm length, extending into the extracellular space (Fig. 11.3, step 4). The growth termination of the needle filament is correlated with structural changes within the basal body and with a change of substrate specificity (Sukhan *et al.*, 2001; Journet *et al.*, 2003; Marlovits *et al.*, 2006).

At this stage, the T3S system becomes cognate for T3S substrates that presumably form an oligomeric cap at the tip of the extracellular needle filament (Fig. 11.3, step 5). The cap is essential for infection and may help to mediate the contact to the host cell (Lara-Tejero and Galán, 2009). Subsequently, the NC secretes translocon proteins that insert into the eukaryotic membrane and may establish a direct and stable connection between the NC and the host cell (Fig. 11.3, step 6). It is conceivable that this intimate contact is relayed back to the basal body via small conformational changes in the needle filaments and thus provides a signal for translocating effectors into the host cell cytoplasm (Cordes *et al.*, 2005; Galkin *et al.*, 2010). After the NC has docked onto the host cell, further effectors are secreted into the host cell to achieve pathogen-specific outcomes (Fig. 11.3, step 7). This process can be so fast that it is able to trigger host cell responses as soon as one minute after docking (Schlumberger *et al.*, 2005).

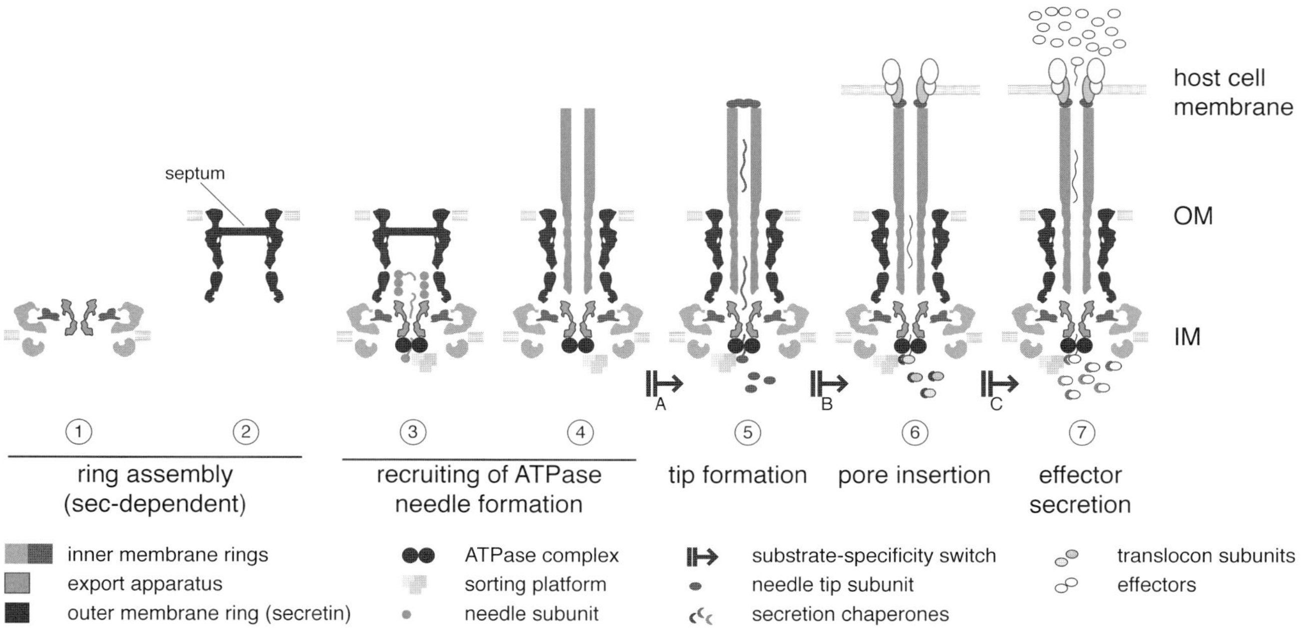

Figure 11.3 Controlled steps to build a complex molecular machine. (1) Inner and (2) outer membrane rings assemble independently of each other. The export apparatus is located inside the inner membrane ring. These steps are sec-dependent. (3) The nascent NC recruits the ATPase complex and the sorting platform. (4) The NC uses its own secretion machinery to secrete the inner rod and the needle filament. The septum opens to accommodate the nascent filament. (5) When the needle filament has reached a certain length, the complex undergoes a substrate-specificity switch. This terminates needle growth and causes a needle tip complex to form. (6) The presence of the host cell is sensed. This signal is relayed back into the bacterium through an unknown mechanism and causes further changes in substrate specificity. Pore-forming effectors insert themselves into the host cell membrane. (7) The T3SS translocates effectors into the host cell cytoplasm.

Secreting effectors with diverse functions

Effectors trigger host cell responses that promote pathogenicity or symbiosis (Cornelis, 2002; Alfano and Collmer, 2004; Schlumberger and Hardt, 2006). They are the least conserved part of the T3S system and can have, depending on the pathogen, very diverse functions such as influencing apoptosis, cytoskeleton dynamics, gene expression or vesicular trafficking (Galán, 2007).

Typically effectors employ molecular mimicry to exert their function – for instance having a specific GTPase activating protein activity without any homology to the endogenous host cell protein (Galán and Wolf-Watz, 2006) but maintaining the same chemistry during the reaction. They were produced by convergent evolution (Stebbins and Galan, 2001).

The effector secretion signal appears to be similar among many bacterial species; however, its exact nature is still not well-understood (Arnold *et al.*, 2009; Samudrala *et al.*, 2009; McDermott *et al.*, 2011). In the majority of effectors, it seems to be located within the first 30 N-terminal amino acid residues and may be largely structurally disordered. Many effectors require a specific secretion chaperone that is cognate for an N-terminal chaperone-binding domain in the effector. Bioinformatic predictions indicate that there may still be experimentally uncharacterized effectors, even in well-studied organisms such as *S. typhimurium*.

Finally, interplay with other T3S system effectors is important to create an exact outcome (Haraga *et al.*, 2008). Therefore, timing and coordination of effector secretion is necessary. This seems to be achieved, at least in part, by the cytoplasmic sorting platform (see later).

The core of the T3S system: the needle complex

With its large size and characteristic shape, the NC is well suited for single-particle reconstructions from electron microscopy. Reconstructions are available for *Shigella flexneri* (Hodgkinson *et al.*, 2009b) and *Salmonella typhimurium* (Marlovits *et al.*, 2004) (Fig. 11.4). Owing to its size, flexibility and purification difficulties the complex as a whole has not been amenable to X-ray crystallography so far. However, single domains or subunits have been solved by X-ray crystallography or NMR (Table 11.1 and Fig. 11.5). Combined with electron density maps from cryo-EM, atomic models of the inner membrane rings and the needle could be generated (Schraidt and Marlovits, 2011; Loquet *et al.*, 2012).

Several substructures of the *S. typhimurium* SPI-1 NC can be clearly distinguished in the reconstruction, supported by topological data (Schraidt *et al.*, 2010) (Fig. 11.4): The secretin ring extends far into the periplasm, making up the electron densities known as outer ring 1 (OR1), outer ring 2 (OR2) and the periplasmic neck. It consists of 15 subunits. Two concentric protein rings with 24 subunits each make up inner ring 1 (IR1), the outer protein ring extends with one domain into the cytoplasm forming the inner ring 2 (IR2). At the centre of the IR1 is the cup region, of which a substantial part is dependent on the presence of the export apparatus (Samuel Wagner *et al.*, 2010a). The needle structure has been solved in isolation as it can be treated with helical reconstruction (Cordes *et al.*, 2003; Galkin *et al.*, 2010; Fujii *et al.*, 2012).

A surprising finding in *S. typhimurium* is that the inner ring (IR) and the outer ring (OR/neck) display a non-trivial symmetry to each other: C24 and C15, respectively (Fig. 11.4). Consequently, the ring-forming base components of the NC adopt an overall three-fold (C3) symmetry indicating that the triplicate interaction of eight subunits of the IR and five subunits of the OR/neck are sufficient to generate a stable NC. The interaction is mediated by the very C-terminal end of the inner membrane protein PrgH in *Salmonella*: C-terminal

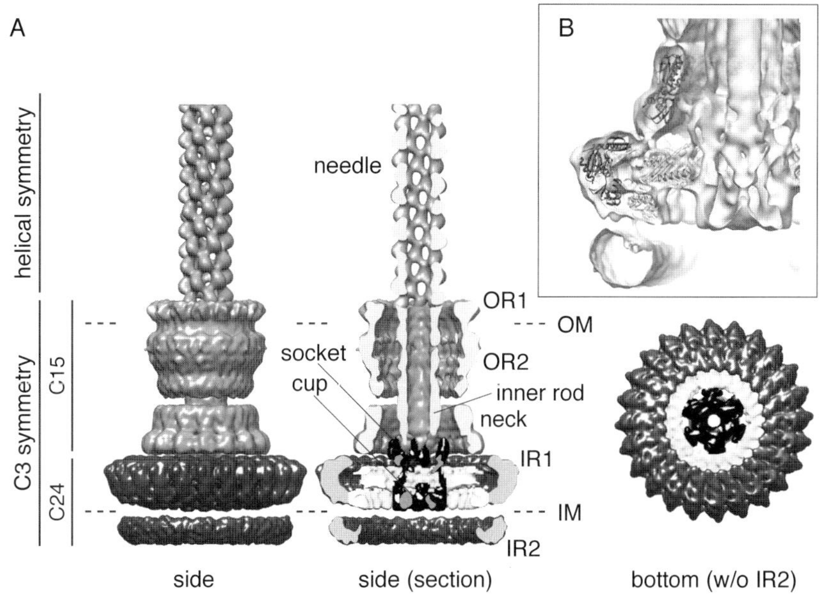

Figure 11.4 The needle complex is the core of the T3SS. (A) The NC is a large membrane protein complex (MW~3.5MDa). In *S. typhimurium* (SPI-1), the membrane rings show an overall C3 symmetry whereas the needle is helical. PrgH (dark grey) and PrgK (light grey) form the inner ring 1 (IR1), the N-terminal domain of PrgH makes up the inner ring 2 (IR2). The inner rings display a C24 symmetry. InvG (grey) forms outer rings 1 and 2 (OR1, OR2) and the neck, displaying a C15 symmetry. The export apparatus is located in the cup/socket region (black). OM, outer membrane; IM, inner membrane. Accession codes: EMD-1875 (*Salmonella* NC), EMD-1416 (*Shigella* needle filament). (B) The C-terminal domain of PrgH, a PrgK truncation and the first two domains of InvG can be docked into the electron density map. Accession codes: PDB-2Y9K, PDB-2Y9J.

truncations of four amino acids destabilize NCs (Schraidt *et al.*, 2010) and C-terminal truncations of six amino acids prevent NC formation (Sanowar *et al.*, 2010). However, the precise nature of this interaction remains unclear. Possibly, the C-terminal tail is able to adopt multiple conformations, which is required to provide stability but also flexibility during assembly, as observed previously (Marlovits *et al.*, 2004). This is consistent with crystallographic studies of the entire periplasmic domain of PrgH, as the last thirty amino acid residues could not be resolved in the crystal structure (Spreter *et al.*, 2009a).

Apart from the C-terminal tail, the periplasmic domain of PrgH is rather rigid within the complex (Marlovits *et al.*, 2004) and its precise localization within the NC could be unambiguously determined by single-particle reconstruction (Schraidt and Marlovits, 2011). Similarly, this was achieved for the periplasmic domains of the inner ring (PrgK) and the outer ring protein (InvG) for which homologous structures had been solved (EscJ and EscC, respectively) (Yip *et al.*, 2005; Spreter *et al.*, 2009b). However, the atomic structures of the cytoplasmic domain (McDowell *et al.*, 2011; Barison *et al.*, 2012) could not be unambiguously docked into the cytoplasmic IR2 due to the flexible character of this ring (unpublished results).

In contrast, single-particle reconstructions from negatively stained NC images from

Table 11.1 T3SS homologues from selected pathogens

Localization *(NC components italic)*	Organism				Characteristics	PDB
	S. typhimurium (SPI-1)	S. flexneri	*Yersinia* spp.	*E. coli* (EHEC)		
Cytoplasmic						
Sorting platform					Possibly ensures secretion order	
	SpaO	Spa33	YscQ	EscQ		
	OrgA	MxiK	YscK	Orf4		
	OrgB	MxiN	YscL	EscL		
ATPase complex	InvC	Spa47	YscN	EscN[1]	Oligomer, unfolds secretion substrates	2OBM/2OBL
IM						
IM ring						
	PrgH[1]	MxiG[1]	YscD	EscD	24-mer	3GR0/2Y9J, 2XXS
	PrgK[1]	MxiJ	YscJ	EscJ[1]	24-mer	2Y9J, 1YJ7
Export apparatus					Located at socket/cup	
	SpaP	Spa24	YscR	EscR		
	SpaR	Spa29	YscT	EscT		
	SpaS[1]	Spa40[1]	YscU[1]	EscU[1]	Self-cleavage, substrate-switching	3C01, 2VT1, 2JLI, 3BZO
	InvA[1]	MxiA	YscV	EscV		2X4A, 3LW9
	SpaQ	Spa9	YscS	EscS		

	Organism					
Localization *(NC components italic)*	*S. typhimurium* (SPI-1)	S. flexneri	*Yersinia* spp.	*E. coli* (EHEC)	Characteristics	PDB
Periplasm/OM						
Inner rod	PrgJ	MxiI	YscI	EscI		
OM ring	InvG[1]	MxiD	YscC	EscC[1]	Secretin, oligomer	2Y9K, 3GR5
	InvH	MxiM	YscW		Pilotin, undetectable in mature NC	
Extracellular						
Needle filament	PrgI[1]	MxiH[1]	YscF[1]	EscF	Polymer, hollow inside	2JOW/2X9C/ 2KV7/2LPZ, 2CA5/2V6L, 2P58
Needle tip	SipD[1]	IpaD[1]	LcrV[1]	EspA[1]	Oligomer, adaptor between needle and translocon	3NZZ, 2J0O, 1R6F, 1XOU
Host cell membrane						
Translocon					Pore insertion into host cell membrane	
	SipB	IpaB	YopB	EspD	Putative oligomer	
	SipC	IpaC	YopD	EspB	Putative oligomer	

[1] Proteins with PDB entry

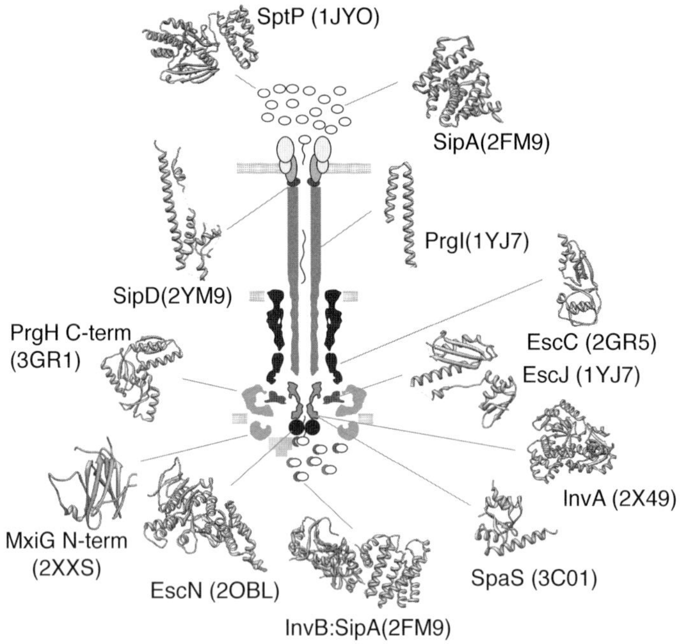

Figure 11.5 Connecting the pieces – known substructures. Selected substructures are depicted with their respective PDB ID. If not available for *S. typhimurium*, a homologue is shown.

Shigella displayed a pseudo twelve-fold symmetry throughout the entire particle (Hodgkinson *et al.*, 2009a). The larger ring associated with the inner membrane consists of 24 subunits that are not equally spaced, thus leading to 12-fold symmetry. Hodgkinson *et al.* also found 12 subunits forming the outer membrane ring, resulting in the reported 12-fold symmetry throughout the particle. It remains to be seen whether the different symmetries observed in *Salmonella* versus *Shigella* NCs can be explained by a generic inherent freedom of secretins to adopt several symmetries in rings (C_{12} to C_{15}; Linderoth *et al.*, 1997; Nouwen *et al.*, 1999; Burghout *et al.*, 2004; Schraidt and Marlovits, 2011) or due to methodological differences, such as natively vitrified versus stained sample material.

A closer look at the components

Cytosolic and inner membrane parts

Five conserved integral inner membrane proteins have been collectively called the export apparatus (Galán and Wolf-Watz, 2006). Its name stems from genetic studies which have shown that individual inhibition of their expression abolish T3-secretion (Jones and Macnab, 1990). In *Salmonella,* these proteins have been demonstrated to be critical in the very early assembly phase of the NC; most likely by establishing an assembly nucleus following ring formation mediated by the ring-forming proteins (*inside-out* model) (Wagner *et al.*, 2010a). Structural analysis of complexes from export apparatus mutant strains revealed the absence of the typical socket/cup structure which is localized centrally within the basal body. This indicates that these proteins are involved either directly or indirectly in the formation of the socket/cup. In an alternative *outside-in* model supported by experiments in *Yersinia* (Diepold *et al.*, 2011), the larger inner membrane ring is assembled first, and consequently filled up by the smaller inner ring and the export apparatus.

Both IR forming proteins have one predicted alpha-helical transmembrane domain (TMD). In *Salmonella*, the larger IR forming protein is PrgH (44 kDa), the smaller IR is composed of PrgK subunits (28 kDa) (Schraidt and Marlovits, 2011). The cytoplasmic N-terminal domain of PrgH has been shown to resemble forkhead-associated domains (FHA) which are often found as modules in signalling proteins specifically recognizing phosphothreonine-containing peptides. Two independent studies have recently shown that MxiG, the PrgH homologue in *Shigella flexneri*, indeed contains domains similar to FHA domains (McDowell *et al.*, 2011; Barison *et al.*, 2012). However, conclusions concerning the phosphothreonine-binding capabilities are inconsistent. Thus, it remains an open question whether threonine-specific phosphorylation is critical for T3S function.

In contrast to PrgH, the second ring-forming inner membrane protein PrgK displays an inverted topology, i.e. the N-terminal domain is located in the periplasm, and is separated from a short cytoplasmically localized C-terminal tail by a single alpha-helical TMD (Schraidt *et al.*, 2010). PrgK possesses a canonical N-terminal signal that is cleaved after targeting into the inner bacterial membrane, leaving a cysteine as the first amino acid of the processed PrgK polypeptide (Kubori *et al.*, 1998). Consequently, this cysteine is lipidated and presumably inserted into the outer leaflet of the inner bacterial membrane as indicated by the localization of the very N-terminal domain from single particle analysis. Interestingly, the TMD and the cytoplasmic tail are not always present in PrgK homologues, suggesting that these domains may have a less evolutionary conserved function in T3S systems (Yip *et al.*, 2005).

Taken together, the cytoplasmic IR2 is formed predominantly by the N-terminal domain of PrgH; and the periplasmic IR1 by both the C-terminal domain of PrgH and N-terminal domain of PrgK. But what are the determinants of ring-formation? Interestingly, despite lacking sequence identity, the IR forming proteins (EscD/PrgH and EscJ/PrgK) share a common motif that has been postulated as a ring-building motif (Spreter *et al.*, 2009b). This motif is wedge-shaped, with two alpha-helices folded against a beta-sheet. Moreover, the same motif is also present in the first domain of the secretin, the outer membrane protein which forms the OR. As such a motif has also been found in other proteins that do not form oligomeric rings (Valverde *et al.*, 2008; Korotkov *et al.*, 2009), it may be required but not be sufficient for ring assembly.

Type III secretion is dependent on a cytoplasmically localized ATPase (Galán and Wolf-Watz, 2006). However, it is unclear how it engages with the NC and with its substrates. From the crystal structure of the catalytic domain it became clear that it displays similarity to canonical ATPases and it has been suggested to act as a hexamer similar to F1-ATPases (Zarivach *et al.*, 2007). Functionally, the T3-specific ATPase binds and hydrolyses ATP and may thus provide the energy to dissociate secretion substrates from their bound chaperones and unfold substrates for delivery (Akeda and Galán, 2005). Possibly, the proton motive force is involved in the secretion process as well (Galán, 2008; Minamino and Namba, 2008).

Recently, three cytoplasmic proteins (SpaO/OrgA/OrgB in *Salmonella*) have been identified to form the 'sorting platform', a high-molecular weight complex critical for

T3-function (Lara-Tejero *et al.*, 2011). It can assemble in absence of the needle complex and serves as a binding platform for chaperone–effector complexes. It may ensure the correct order of substrate secretion and is well conserved among T3S systems. Currently, however, no structural details are known that could help to understand the molecular mechanistics underlying substrate loading.

Outer membrane parts and their plasticity

The OR is composed of members of the secretin family that forms rings in the outer membrane of Gram-negative bacteria (Korotkov *et al.*, 2011). The structure of the first two domains of a T3S system secretin from EPEC is known (Spreter *et al.*, 2009b). They share a structural motif that has been implicated as a ring-building motif with IR components (see previous section). A small lipoprotein, the pilotin, is required for the correct localization of the secretin ring (Daefler and Russel, 1998). The T3S system secretin ring is able to perform a substantial conformational change, opening its septum to accommodate the needle filament (Marlovits *et al.*, 2004) (Fig. 11.4, step 3/4). Interestingly, this conformational change is reversible, as removal of the needle filament from NCs *in vitro* results in a closed septum again (Schraidt *et al.*, 2010).

Extracellular parts

The helical needle filament is about 25–50 nm (Mota *et al.*, 2005; Marlovits *et al.*, 2006) long, has an outer diameter of about 10–13 nm and is hollow inside, with an inner diameter of about 3 nm. It is believed to be the secretion tunnel that bridges the extracellular space during substrate translocation. The small diameter is consistent with the idea that substrates are translocated in an unfolded state (Akeda and Galán, 2005). The needle filament consists of more than one hundred protomers of a single protein (PrgI in *Salmonella*; <10 kDa) (Broz *et al.*, 2007) and displays helical symmetry (Cordes *et al.*, 2003; Galkin *et al.*, 2010; Fujii *et al.*, 2012). Although the T3-specific ATPase is required for needle protomer transport, needle filament polymerization is ATP independent and most likely occurs at the distal end of the needle (Poyraz *et al.*, 2010). How the polymerization of the filament to a certain length, suitable for making functional NCs, is regulated (Mota *et al.*, 2005) is still under debate: In one model, as shown in *Salmonella typhimurium*, completion of the inner rod causes a substrate switch and leads the cessation of needle growth (Marlovits *et al.*, 2006). In another, as shown in *Yersinia enterocolitica*, a molecular ruler determines the length of the filament (Journet *et al.*, 2003; Wagner *et al.*, 2010b).

The structures of needle subunits from *Salmonella* and *Shigella* have been solved by NMR and X-ray crystallography (Deane *et al.*, 2006; Wang *et al.*, 2007; Poyraz *et al.*, 2010). The needle protein is made up by two extended helices, connected by a turn (Deane *et al.*, 2006). It partially refolds into a beta-strand conformation upon filament formation (Poyraz *et al.*, 2010). The structure of the whole filament from *Shigella* could be solved by helical reconstruction (Fujii *et al.*, 2012). The analysis of images from *Salmonella* needles revealed that the filament is structurally heterogeneous (in contrast to the almost crystalline flagellum), which is consistent with the idea that the needle filament could signal the host cell attachment to the bacterium and thus initiate protein translocation into eukaryotic cells (Galkin *et al.*, 2010). An atomic model of the filament was generated with constraints from solid state NMR (Loquet *et al.*, 2012).

The needle is capped distally by a tip made up of a few tip protein subunits. The tip could be first visualized in *Yersinia* (Broz *et al.*, 2007) and is considerably larger than in *Shigella* (Sani *et al.*, 2007). Structures of the tip subunit have been solved (Derewenda *et al.*, 2004; Johnson *et al.*, 2007; Chatterjee *et al.*, 2011). Common structural features of tip proteins are a central coiled coil and mixed alpha/beta domain (Chatterjee *et al.*, 2011). It has been suggested that the tip consists of five (Broz *et al.*, 2007; Lunelli *et al.*, 2011) or six (Chatterjee *et al.*, 2011) subunits. Whereas the *Yersinia* tip protein has a secretion chaperone, all other known structures are supposed to be self-chaperoned by an additional domain present in the protein (Matson and Nilles, 2001; Johnson *et al.*, 2007). Interestingly, the tips of *Salmonella* and *Shigella* have binding sites for bile salts (Stensrud *et al.*, 2008; Wang *et al.*, 2010), which may be used by the pathogens to detect changes in intestinal environments and thus time secretion accordingly.

The tip may serve as an adaptor to the translocon pore that is formed by two proteins in multiple copies. Like pore-forming toxins, they exist in a soluble form (Hayward *et al.*, 2000) and have the ability to insert into host cell membranes (Håkansson *et al.*, 1996; Neyt and Cornelis, 1999). They share a common T3S chaperone for correct targeting and preventing premature aggregation inside the bacterial cell (Ménard *et al.*, 1994). Probably due to their propensity to aggregate, structural information of the translocon proteins is very limited. The fully assembled translocon complex may adopt a size of about 500–700 kDa in its oligomeric form, in *Yersinia* (Montagner *et al.*, 2011).

Some translocon subunits possess effector domains in parts that are exposed to the host cell cytoplasm, for instance SipC from *Salmonella* has actin bundling activity which could be necessary in early stages of invasion (Hayward and Koronakis, 1999). In *Salmonella*, the tip protein is present on the bacterial surface before contact to host cells whereas the translocon proteins can only be detected shortly after incubation with host cells (Lara-Tejero and Galán, 2009). Whether the translocon inserts first and then attaches to the tip or vice versa is not known; neither is its involvement in host cell sensing.

Small molecules blocking secretion and other medical applications

Considering emerging antibiotic resistance, inhibition of T3S offers an attractive target for the development of novel antibiotics (Baron, 2010). Furthermore, these antibiotics could allow specific targeting of the pathogen while leaving the intestinal flora largely intact and, by merely preventing infection instead of killing the microbes, would decrease selective pressure for the development of resistance (Clatworthy *et al.*, 2007).

Several screens of small molecule libraries for T3S inhibition were performed, for instance assaying the influence on T3S system gene expression (Kauppi *et al.*, 2003) or secretion function (Felise *et al.*, 2008). At least three structurally unrelated compounds are known to inhibit T3S: benzimidazole, salicylidene acylhydrazide and thiazolidinone. The first inhibits transcription factors involved in T3S regulation (Garrity-Ryan *et al.*, 2010) and the second probably targets several enzymes, effectively preventing T3S by perturbing normal metabolism (Wang *et al.*, 2011). Only the last – thiazolidinone – has been implicated in directly affecting assembly of the T3S system (Felise *et al.*, 2008). It inhibits type II and type IV secretion as well, which indicates that the secretin – a common component in all three systems – could be its target. Developing these candidates into novel antibiotics will require more academic and pharmaceutical research. A more comprehensive review of the current state of research is given in (Izoré *et al.*, 2011).

Other medical applications of T3S research are the development of novel T3S system subunit vaccines (Quenee and Schneewind, 2009) and the development of artificial protein delivery systems for vaccination or therapy (Galen *et al.*, 2009; Moreno *et al.*, 2010). In the first, T3S system components such as translocon subunits are recombinantly produced and tested for the immunogenicity. This is limited to the generation of vaccines against the respective pathogens whose T3S system subunit is used as an antigen. A much broader and less conventional application is the latter – using the T3S system as a genetically engineered tool to vaccinate or modulate the immune system of the patient as a therapy for cancer and other diseases. In brief, the antigen or therapeutic protein is genetically fused to a T3S signal and expressed in attenuated pathogens with a T3S system. The live vector vaccine is then administered to the patient. The advantage of this could be that the pathogen usually elicits a strong innate immunity response, which acts as an adjuvant for the heterologous antigen. Many pre-trials were performed in animal models (Moreno *et al.*, 2010). Small clinical trials were performed with live *Salmonella* vaccines (Galen *et al.*, 2009), for instance against HIV (Kotton *et al.*, 2006). Future work will show whether these novel therapies turn out to be useful in the clinic.

Prospects

Substantial progress was made in understanding structure and function of both single components and the needle complex as a whole in the last years. However, some significant areas remain poorly understood. Most notably, they are:

First, how does substrate sorting and unfolding work? What constitutes the actual substrate signal and can additional effectors be found by prediction from the bacterial genomes? One prerequisite to answer these questions would be a structure of the ATPase (and the sorting platform) attached to the NC.

Second, how is the host cell presence detected and how is this signal transduced into the bacterial cell? Is the needle involved in this process?

Third, how does the translocon attach to the needle tip? A structure of the pore alone or in complex with the needle would greatly increase the understanding of this process.

Finally, more progress should be expected in medical applications of knowledge about the T3S system, for instance a T3S inhibitor brought to the clinic or the successful application of the T3S system as an engineered tool to prevent or cure disease.

The type IV secretion system

Type IV secretion (T4S) systems are versatile secretion systems found in many Gram-positive and Gram-negative bacteria. Within the T4S family, three groups of secretion apparatus can be defined (Fig. 11.6). First, in both Gram-positive and Gram-negative bacteria, T4S systems mediate conjugative transfer of mobile elements (plasmids or transposons) into a wide range of bacterial species but also into fungi, plants or mammalian cells (Bundock *et al.*, 1995; Grohmann *et al.*, 2003; Lawley *et al.*, 2003). This transfer requires close contacts between two cells before transfer. Importantly, conjugation promotes bacterial genome plasticity and bacterial adaptive response to changes in their environment. In particular, conjugation contributes to the spread of antibiotic resistance genes among pathogenic bacteria, leading to the emergence of multidrug resistant pathogenic strains in healthcare settings. Secondly, type IV secretion systems mediate DNA uptake or release from/in the

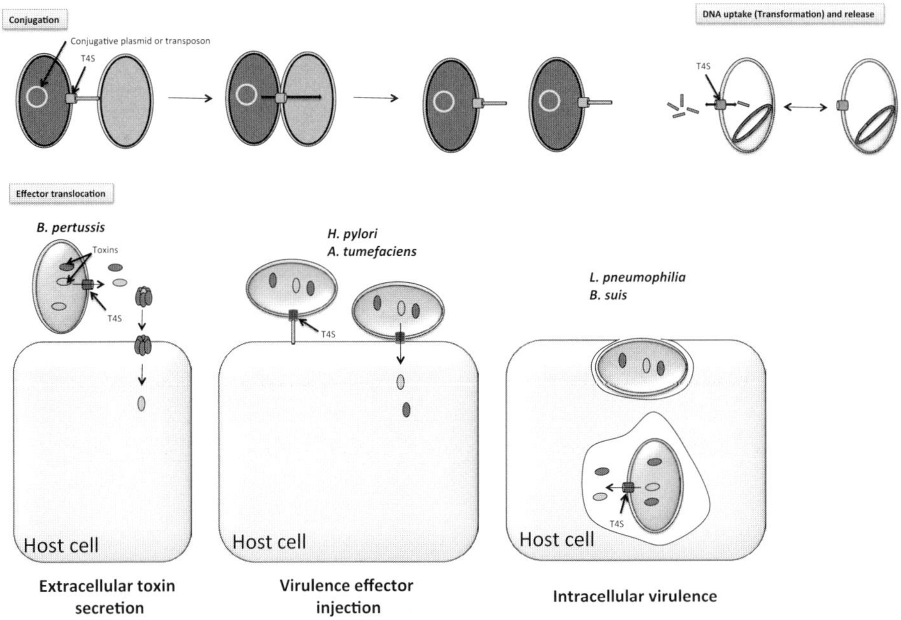

Figure 11.6 Schematic representation of roles of type IV secretion in bacteria. The two functional groups of T4S systems are shown. Protein translocation T4S systems deliver protein substrates to eukaryotic cells and are directly involved in virulence of many pathogenic Gram-negative bacteria. DNA transfer T4S systems include conjugative T4S systems, which deliver plasmids or transposons from donor to recipient bacteria in Gram-negative and Gram-positive bacteria, and DNA uptake (transformation) or release T4S systems, which mediate the exchange of DNA with extracellular milieu. They also include the *A. tumefaciens* VirB/D T4S system responsible for transferring a portion (the T-DNA) of a plasmid (the Ti plasmid) into plant.

extracellular milieu, further promoting genetic exchange between bacteria. This group comprises two systems involved in natural transformation (DNA uptake) in *Helicobacter pylori* and *Campylobacter jejuni* and a DNA release system in *Neisseria gonorrhoeae.* Finally, 'effector translocation' systems are type IV secretion systems directly involved in virulence. In pathogenic Gram-negative bacteria such as *Helicobacter pylori, Brucella suis* or *Legionella pneumophila,* T4S systems mediate the injection of virulence proteins into mammalian host cells. In *Agrobacterium tumefaciens,* the VirB/D T4SS system, delivers oncogenic DNAs into plants. *Bordetella pertussis* uses a T4S system to secrete the pertussis toxin in the extracellular milieu. T4S effectors trigger a wide range of signal transduction cascades inside the host cell promoting bacterial survival and proliferation.

Despite the wide diversity of their substrate and purpose, all the T4S systems are clearly evolutionary related (Lessl *et al.,* 1992; Lessl and Lanka, 1994). They all share the same basic components and probably assemble and function in the same manner. The genes encoding the T4S components are usually arranged in a single or few operons. The comparison of these gene clusters indicates that many T4S systems found in Gram-negative bacteria are similar to the *A. tumefaciens* VirB/D T4S system, which defines the paradigm for T4S. This system is composed of 12 proteins named VirB1-VirB11 and VirD4. In this chapter, we will name the proteins according to the *A. tumefaciens* nomenclature. T4S systems that have additional or missing components seem to have retained a basic VirB/D-like subcomplex. For example, the F-plasmid encoded T4S system consist of eight VirB-like proteins and of other components found only in F-like plasmids to perform specialized activities such as F-pilus retraction (Lawley *et al.*, 2003). The *Legionella pneumophila* Dot-ICM T4S system also contains some VirB-like components, which form the functional core of the secretion apparatus (Segal *et al.*, 1999; Vincent *et al.*, 2006). In Gram-positive bacteria, only two VirB/D components are conserved in T4S conjugative systems. In these bacteria, the T4S translocation apparatus seems to have a completely different architecture (Segal *et al.*, 1999; Grohmann *et al.*, 2003; Vincent *et al.*, 2006; Chen *et al.*, 2008a)

Assembly of the T4S apparatus

Taking into account subunit topologies, structures, functions and the network of interactions between them, it is possible to propose a model for the assembly of a functional VirB/D T4S apparatus. An inactive translocation pore made of VirB6, VirB7, VirB8, VirB9 and VirB10 would first form across the cell envelope. Then VirB3, VirB4 and pilus-associated proteins (VirB2 and VirB5) could be recruited to form a quiescent T4S machine. Finally, the ATPases VirB11 and VirD4 could finally activate the complex to either assemble the pilus or secrete the substrate.

The translocation pore assembly is likely to start with the core complex formation across the cell envelope. This complex made of VirB7, VirB9 and VirB10 assembles and multimerizes spontaneously without energy requirements (Fronzes *et al.*, 2009b; Jakubowski *et al.*, 2009). It was recently shown that VirB7 lipidation, which is essential for secretion (Fernandez *et al.*, 1996), is not required for core complex assembly (Fronzes *et al.*, 2009b). The core complex inserts and assembles first in the inner membrane. VirB7 lipidation is then needed for its insertion in the outer membrane. Both VirB6 and VirB8 have strong functional and biochemical interactions with VirB7/VirB9 and/or with VirB10. When expressed on its own with VirB7, VirB9 and VirB10, VirB8 does not make stable contacts with the core complex (Fronzes *et al.*, 2009b). The presence of VirB6 and of VirB4 could be necessary to stabilize VirB8 (Yuan *et al.*, 2005), to form the inner membrane pore and to plug it with the core complex. The assembly of the translocation pore across the cell envelope is likely to be a concerted process during which its components would gather and multimerize at the same time.

Then, VirB2, VirB3, VirB4 and VirB5 (Yuan *et al.*, 2005; Wallden *et al.*, 2012) would be recruited to form a VirB2–VirB10 complex. Studies exploring the network of interaction between the VirB/D components indicate that VirB2, VirB3 and VirB5 proteins interact peripherically with the translocation pore. It was recently shown that virB4 is able to form a stable complex with the core complex (Wallden *et al.*, 2012).

T4S apparatus is generally depicted as a single protein complex that is used for substrate secretion and T4S pilus biogenesis. However, accumulating evidences support the idea of a bifurcation in the T4S assembly once the VirB2–VirB10 complex is assembled. It appears that VirD4 is completely dispensable for pilus biogenesis while its presence is compulsory for substrate secretion. In addition, mutations in the pilin or some T4SS components such as VirB1, VirB6, VirB9 and VirB11 (Berger and Christie, 1994; Eisenbrandt *et al.*, 2000; Lai *et al.*, 2000; Sagulenko *et al.*, 2001a; Jakubowski *et al.*, 2004) can uncouple pilus biogenesis and substrate secretion. It is therefore very likely that the VirB2–VirB10 complex could then assemble differently and specialize its function to assemble the T4S pilus or to secrete the substrate. It appears that VirB3, VirB4, VirB5 and VirB8 are directly involved in pilus biogenesis in *A. tumefaciens*. A sequence of stabilizing effects and interaction has been proposed concerning pilus assembly (Yuan *et al.*, 2005): VirB4 stabilizes virB3 and virB8 protein. Then VirB8 directly interacts with virB5 in the periplasm. These interactions induce virB2–virB5 interaction and pilus formation with some contribution of virB6 and virB9. At that stage, VirB11 needs to be recruited by the T4S apparatus. This could be achieved through its interaction with VirB4. In addition, ATP binding on VirB11 could cause conformational changes that would promote its membrane binding and contacts with other T4S components. VirB4 ATPase activity is dispensable for pilus biogenesis, indicating that VirB11 would solely provide the energy used to stimulate pilus growth. Finally, it has been recently proposed that a cleaved version of virB1 secreted extracellularly (C-terminal domain of virB1 called virB1*) could promote T-pilus formation in *A. tumefaciens* (Zupan *et al.*, 2007).

Concerning the assembly of the secretion channel, it is important to note that, even if the pilus formation is not necessary for secretion to occur, the assembly of the pilus components VirB2 and VirB5 with the translocation pore is required. VirB4 and VirB10 would then recruit the ATPases VirB11 and VirD4. At that stage, the secretion channel biogenesis is complete. ATP energy and substrate binding on the CP would then activate the channel and secretion will take place.

Substrate processing and spatial positioning

During conjugation, the substrate transfer from the donor to the recipient cell occurs in three distinctive steps. First, the DNA transfer and replication proteins (DTR proteins) come together to process DNA substrate and prime it for transfer. Then, DNA is recruited by a membrane-bound substrate receptor called 'coupling protein' which transfers it to a cognate T4S transfer machine. Finally, the T4S transmembrane complex or 'mating pore' translocates the DNA across the cell envelope into the recipient bacterial cell.

During this process, only one of the two DNA strands (called T-strand) from the conjugative plasmid or transposon is transferred. The creation of a multiprotein–DNA complex called relaxosome is required to generate a transfer-competent single-stranded DNA (ssDNA). This complex comprises a relaxase/helicase protein, a transcriptional regulator, a host-encoded integration factor and the T-strand. To be processed and then transferred, DNA must carry a special sequence called origin of transfer (oriT) that will specifically be recognized by the relaxase/helicase protein. The N-terminal domain of this protein carries the relaxase activity and cut the T-strand within this sequence. A catalytic tyrosine will remain attached on the 5′ end of the T-strand. The ATP dependent helicase activity carried by the C-terminal domain of the relaxase/helicase protein then unwinds the double-stranded DNA (dsDNA) and displaces processively the T-strand from its 5′ to 3′. In the same time, the other strand is used as template by a host encoded DNA polymerase III to regenerate the initial double-stranded DNA using a rolling circle mechanism. The structures of domains of the proteins involved in the relaxosome formation were determined and are reviewed elsewhere (Gomis-Ruth and Coll, 2006). In Gram-positive bacteria, the initiation of conjugative transfer by the formation of the relaxosome is likely to follow the same mechanism (Grohmann *et al.*, 2003).

In *A. tumefaciens*, the natural substrate of the VirB/D T4S system is a DNA molecule of ~12 kB (T-DNA). Its transfer into plants induces the production of plant tumours. The VirD2 relaxase cuts within oriT-like origin of transfer border sequences in the T-DNA when included in a relaxomal complex with VirD1, VirC1 and VirC2 proteins. VirC1 is a ParA-like ATPase that binds to a sequence called 'overdrive' which is adjacent to the right border repeat of the T-DNA. It stimulates relaxosome assembly by nucleating the relaxosome assembly. VirC1 would first form a complex with VirC2 and VirD1. After binding the T-DNA right border repeats, this ternary complex would recruit the relaxase VirD2 to complete the relaxosome formation and generate the T-strand (Atmakuri *et al.*, 2007). The VirD2/T-strand complex (T-complex) would accumulate in the cytoplasm. Interestingly, several studies showed that VirB/D T4S transfer complex components locate at cell poles (see below). Recent findings indicate that VirC1, VirC2 and VirD1 also locate at the cell poles independently of each other or of the VirB/D proteins. It was shown that VirC1 could actively recruit the T-complex at the cell pole independently of the other processing proteins. Moreover, VirC1 was shown to interact with the coupling protein VirD4. Therefore, VirC1 could also help the spatial positioning of the substrate during or after its processing by the relaxosome close to or at the entrance of the T4S transfer complex. Such a spatial positioning of the relaxosome and coupling protein at the same discrete site in the cell and the substrate docking on the coupling protein by interaction with the relaxosome proteins was also recently demonstrated in the *Enterococcus faecalis* pCF10 conjugative system (Chen *et al.*, 2008a).

Remarkably, the *A. tumefaciens* T4S system is also able to transfer purely proteic substrates (Vergunst *et al.*, 2000). The transfer of the virulence factors VirF, VirE2, VirE3 or VirD5 into plant cells can be demonstrated using reporter-based assays (Vergunst *et al.*, 2000). As more T4S effectors are being identified in many bacteria, it is now possible to draw common requirements for the translocation of proteic substrates to through the T4S translocation machine. Firstly, the presence of a C-terminal signal is required for the recognition and transfer of all the T4S effectors. In *A. tumefaciens*, this C-terminal transport signal appeared to be a hydrophilic positively charged sequence (Vergunst *et al.*, 2005). Similarly, a 20-amino-acids C-terminal hydrophobic signal was shown to be necessary and sufficient for translocation of substrates by the *L. pneumophila* Dot/ICM T4S system (Nagai *et al.*, 2005). However, this signal could not be used in *A. tumefaciens* for transfer indicating that different T4S substrates display different features in their C-terminal signal peptides in order to be recognized by their cognate secretion system. Interestingly, *Bartonella henselae* substrates harbour a C-terminal positive signal peptide (Schulein *et al.*, 2005) but a proximal delivery domain required for efficient transfer was also identified. This domain was also found in many conjugative relaxases and the coupling proteins of T4S systems translocating effectors containing such delivery domain form a distinct phylogenetic cluster. This implies that these conjugative relaxases and coupling protein may have co-evolved. It is enticing to speculate that the coupling protein could then be responsible of the sorting of the T4S effectors at the entry of the translocation machine and could recruit protein substrates as well as nucleoprotein complexes substrates to the T4S system. This hypothesis is consistent with observations showing that the *A. tumefaciens* substrate VirE2 interact directly with the coupling protein VirD4 at cell poles (Atmakuri *et al.*, 2003) Moreover, the *A. tumefaciens* relaxase VirD2 could be transferred through the T4S system in absence of the T-DNA, indicating that the relaxase component could provide the translocation signal for the nucleo–protein complex (Vergunst *et al.*, 2005)

An appealing question concerning the priming of the T4S effectors is whether or not they travel through the T4S system folded or unfolded and if chaperones are required to maintain them in a partially unfolded state prior transfer. For example, in type III secretion systems, effector proteins are binding specific cytosolic chaperones (Stebbins and Galan, 2001, 2003) and an ATPase (InvC in *Salmonella enterica*) plays an important role in substrate recognition and transfer as it induce unfolding of and chaperone release from the cognate secreted protein (Akeda and Galán, 2005). Interestingly, it was shown in *A. tumefaciens* that proteins such as green fluorescent protein with the appropriate T4S C-terminal signal peptide could not be transferred into plants (Vergunst *et al.*, 2005), indicating that the folding of the proteins could influence the transfer ability of substrates. However, little evidence for chaperone involved in T4S was provided so far. Some authors suggested that CagF could be a chaperone-like protein interacting with the T4S Cag substrate (CagA) that would be required in substrate recognition by the T4S translocation complex (Couturier *et al.*, 2006; Pattis *et al.*, 2007).

Spatial positioning of T4S machines is a recent discovery and there are only few examples of spatially distributed T4S systems in the literature. This is particularly clear in *A. tumefaciens*, which interacts polarly with its target cell (plant cell). In these bacteria, some VirB proteins such as VirB3, VirB4, VirB8, VirB11 and VirD4 could locate at a cell pole independently of the other VirB proteins (Kumar *et al.*, 2000; Kumar and Das, 2002; Judd *et al.*, 2005a,b). Interestingly, the polar localization of the remaining proteins (VirB1, VirB5, VirB6, VirB7, VirB9 and VirB10) requires the presence of VirB8, which seems to be a nucleating factor for the polar positioning of the T4S machine. It was shown that the polar localization is energy independent as the ATPases virB4 and VirB11 are not required. In particular, it is worthy of note that co-expression of VirB7-VirB10 proteins is sufficient for their cell pole localization through VirB8 polar localization, confirmed that these proteins could assemble into a T4S core complex independently of the other components. It is generally though that conjugative T4S systems are not spatially localized as conjugative pili are generally found in apparently random positions at the cell surface. However, *E. coli* cells carrying the conjugative plasmid R27 display distinct foci of VirB4-like TrhC protein at the cell surface. In addition, it was recently shown in the Gram-positive bacteria *E. faecalis* that VirD4-like PcfC and relaxosome proteins (PcfF and PcfG) localize in defined foci at the cell periphery independently of each other (Chen *et al.*, 2008b).

This spatial positioning of the T4S components provides multiple advantages for the bacteria. This process may indeed help for T4S complex assembly, increasing the local concentration of the different subunits. In addition, the co-localization of the T4S translocation machine and its substrate may increase the secretion efficiency

Mechanism of T4S substrate translocation

The molecular details of the substrate translocation mechanism through the T4S apparatus are still unclear. However, substantial progress towards the understanding of this mechanism has been made. Insights in the substrate translocation pathway have been obtained thanks to the development of an assay (termed TrIP) in *A. tumefaciens* (Cascales and Christie, 2004b). It greatly helped to identify the different steps along the substrate translocation pathway. The *A. tumefaciens* VirB/D substrate makes sequential contacts with VirD4, VirB11, VirB6, VirB8, VirB9 and VirB2. The other T4S components do not interact with the substrate but are essential for the substrate transfer during different steps. This pathway and the energetic requirements of each step are discussed below.

Substrate binding to the CP (VirD4)

The TrIP assay showed that the CP, VirD4 is the first T4S component to contact the substrate. As mentioned above and consistent with these data, many evidence show that the coupling proteins act as substrate receptors at the entrance of the T4S apparatus. Many reports show that the coupling proteins are able to interact with the T4S substrate in absence of the other T4S components (see above). Interestingly, several studies also demonstrated that VirD4 is still able to interact with the T4S substrate (T-DNA or VirE2) when its ATPase activity is abolished by mutating the Walker A motif (Atmakuri *et al.*, 2003; Atmakuri *et al.*, 2004). Therefore, neither ATP hydrolysis or other T4S components are required for VirD4 to interact with the substrate.

Substrate transfer to VirB11

The TrIP assay shows that VirD4 then transfers the T4S substrate to VirB11. Interestingly, this transfer could not be detected when VirB7 was deleted, in agreement with the essential role of this protein during the early stages of the T4S apparatus biogenesis. In addition, Walker A mutants of both VirD4 and VirB11 were able to mediate this substrate translocation step. This indicates that this translocation step does not require energy and is only mediated by the interaction between the two ATPases.

Substrate transfer through the inner-membrane pore

The inner membrane components VirB6 and VirB8 are then contacting the substrate. VirB6 and VirB8 appear to function coordinately, as deletion of each protein blocks substrate translocation. Therefore, both components presumably form a pore that could allow the substrate to cross the inner membrane. The VirB6 central loop (P2 loop) could form major parts of this secretion pore (Jakubowski *et al.*, 2009). This translocation step requires each of the T4S ATPases VirB4, VirB11 and VirD4 to be present and active. This suggests that these proteins actively mediate substrate translocation across the inner membrane. It is also interesting to note that, in addition to the energetic components, the expression of VirB6, VirB7, VirB8 and either VirB9 or VirB10 is sufficient to mediate this translocation step *in vivo*. This indicates that the inner membrane pore assembly and its interaction with the ATPases could be possible even if the core complex VirB7-VirB9-VirB10 is not properly assembled (Atmakuri *et al.*, 2004).

Transfer through the outer membrane pore

During last translocation step the T4S substrate is therefore transferred across the periplasm to the outer membrane associated proteins VirB2 and VirB9. The TrIP studies indicated that VirB3, VirB5 and VirB10 are essential for substrate trafficking through the periplasm. The exact role of VirB3 and VirB5 during this process is unknown. VirB10 is a central piece of the T4S machine. It is both a structural scaffold bridging the inner- and outer membrane components (Fronzes *et al.*, 2009b; Jakubowski *et al.*, 2009) but acts also as an energy sensor/transducer during secretion. VirB10 N-terminal TMS stably insert the protein in the inner membrane. It extends via its proline-rich region (PRR) in the periplasm allowing VirB10 C-terminal domain to contact VirB7/VirB9 heterodimer and insert in the outer membrane

(Chandran *et al.*, 2009; Fronzes *et al.*, 2009b; Jakubowski *et al.*, 2009). At the inner membrane, the protein is in direct contact with the inner membrane pore component VirB8 and with the ATPases VirD4 and VirB4. Importantly, VirB10 senses ATP utilization by VirB11 and VirD4 and, in turn undergo a conformational change necessary for substrate transfer across the periplasm to the outer membrane components VirB9 and VirB2. VirB10 would presumably act as a regulator of the outer membrane pore assembly or opening. It was proposed this energy sensing would occur through VirB10 interaction with VirD4 through TM–TM interactions. However, it was shown VirB10 TMS is not the only part of VirB10 N-terminal segment to be involved in that process (Jakubowski *et al.*, 2009).

Once the outer membrane secretion pore is assembled or opened, the substrate can reach the extracellular milieu. An interaction between the pilin VirB2 and the T4S substrate does not indicate that the substrate is transferred through the T4S pilus. This question is still highly controversial. The isolation of the uncoupling mutants described above and the fact that VirB2 is essential for substrate transfer suggests instead that the pilin monomer is a component of both the translocation channel and the pilus. Interestingly, in *A. tumefaciens*, the pilin VirB2 is still directly interacting with the sample is uncoupled mutants. 'Non-piliated' pilin could directly participate to the assembly of a functional translocation channel inside the periplasm forming the 'base' of the pilus. This structure could be sufficient for transfer activity of most of T4S secretion systems. Pilus formation could then be optional for transfer activity and depend on pilin expression level and/or mediation by other type IV secretion proteins. Interestingly, VirD4 is dispensable for pilus production. Therefore, VirB10 energy sensing mechanism is specifically required for secretion. However, pilus biogenesis requires an energetic contribution by VirB11 and the presence of VirB4. As VirB4 and VirB10 also interact stably, VirB10 could be activated differently during pilus biogenesis. 'uncoupling' mutations recently identified in VirB10 suggest that VirB10 could indeed differentially regulate secretion and pilus biogenesis (Jakubowski *et al.*, 2009). It is also possible that the energy necessary for pilus biogenesis is directly transferred from VirB4/VirB11 ATPases to VirB8 and that VirB10 is not actively involved in this process.

Type IV secretion systems architecture and role of each subassembly

Molecular assembly: bits and pieces

The cytoplasmic ATPases

All the bacterial secretion systems that do not rely on the general secretory pathway to cross the inner membrane have their own cytoplasmic ATPases. T4S systems generally have three dedicated ATPases that form the power unit of the translocation apparatus. In *A. tumefaciens,* these ATPases, named VirD4 or coupling protein, VirB11 and VirB4, are essential for secretion and T4S pilus biogenesis.

Three ATPases VirD4. VirB4 and VirB11, which are essential components of the secretion system, interact with each other. Several reports describe interactions of each ATPase with the two others (Rain *et al.*, 2001; Ward *et al.*, 2002; Atmakuri *et al.*, 2004; Malek *et al.*, 2004; Terradot *et al.*, 2004). Therefore, they are likely to form a complex energizing substrate transport through the translocation apparatus. The exact architecture of this complex and the exact role of each ATPase during the secretion process or T4S pilus biogenesis are still unclear. Remarkably, VirB4 and VirD4 interact directly with VirB10 (Gilmour *et al.*, 2001;

Llosa *et al.*, 2003; Atmakuri *et al.*, 2004; Malek *et al.*, 2004; Terradot *et al.*, 2004). Moreover, the quaternary complex VirB10-VirB4-VirB11-VirD4 can be precipitated (Atmakuri *et al.*, 2004). The VirB10–VirD4 interaction has been studied in details. The interacting domains in both proteins was mapped in their N-terminal part that contain periplasmic and inner membrane TMSs. This suggests that the VirB10–VirD4 interaction occurs close to and/or in the bacterial inner membrane (Gilmour *et al.*, 2001; Llosa *et al.*, 2003; Jakubowski *et al.*, 2009). Interestingly, it was proposed that VirB10 would act as an ATP energy sensor within the T4S apparatus. Upon ATP hydrolysis by VirD4 and VirB11, VirB10 would be energized and undergo a conformational change necessary for substrate transfer through the T4S system (Cascales and Christie, 2004a; Jakubowski *et al.*, 2009). In addition, some mutations in *A. tumefaciens* VirB11 termed 'uncoupling mutations' arrested T-pilus biogenesis without abolishing substrate transfer (Sagulenko *et al.*, 2001a). Intriguingly, *B. suis* VirB4 Walker A mutants are still able to produce pili (Yuan *et al.*, 2005) but the presence of the protein is required for this process to occur, indicating that VirB4 may contribute structurally but not actively to pilus production. The T4S ATPases would then not only play a role in substrate priming and recognition (as mentioned previously) but also trigger conformational changes involved in T4S apparatus assembly and substrate transfer.

Coupling protein

Coupling proteins (CPs) are ubiquitous members of the conjugative T4S systems found in Gram-negative and Gram-positive bacteria. They are present in most other T4S systems. However, some effector translocation systems such T4S systems in *Bordetella pertussis* or *Brucella* spp. use a CPs independent mechanism for substrate recruitment and secretion (O'Callaghan *et al.*, 1999; Burns, 2003; Seubert *et al.*, 2003).

CPs are NTP-binding proteins which contain nucleotide binding motifs (Walker A and B) (Moncalian *et al.*, 1999; Gomis-Ruth *et al.*, 2001; Schroder and Lanka, 2003). Even if mutations in these motifs indicate that they are essential for the secretion process, no NTPase activity could be detected *in vitro* till recently (Moncalian *et al.*, 1999; Schroder *et al.*, 2002). It appeared that the CP involved in conjugation TrwB displays a DNA-dependent ATPase activity (Tato *et al.*, 2005). SsDNA and DsDNA are indeed necessary to promote trwB oligomerization required for ATPase activity (Tato *et al.*, 2005, 2007). The conjugative CPs preferably bind ssDNA but not in a sequence-specific manner (Moncalian *et al.*, 1999; Schroder *et al.*, 2002; Schroder and Lanka, 2003).

The molecular details CPs role remain elusive. They are essential for substrate recruitment and therefore for the secretion process. Because these proteins interact directly with the T4S substrates, CPs may act as substrate receptors at the entrance of the T4S apparatus. As mentioned above, available data in the literature indicate that substrate recognition by the T4S systems would occur through specific interactions between the CPs and proteic signals harboured by their substrates. For example, CPs involved in conjugation directly and specifically interacts with the DTR components, in particular with the relaxase covalently attached to the single-stranded DNA (Pansegrau and Lanka, 1996; Llosa *et al.*, 2003; Chen *et al.*, 2008a). Interestingly, it was recently shown that TrwA, which is a tetrameric component of the R388 relaxosome interacts with the coupling protein TrwB and stimulates its DNA-dependent ATPase activity (Tato *et al.*, 2007). In *A. tumefaciens*, the coupling protein VirD4 interacts with the relaxase VirD2 attached to the T-DNA but also recognizes transfer signal peptides harboured by some other T4S substrates at their C-terminus.

Coupling proteins are membrane proteins anchored in the bacterial inner membrane by their N-terminal membrane anchor. For some CPs, this membrane anchor is required for oligomerization (Moncalian *et al.*, 1999; Hormaeche *et al.*, 2002; Schroder *et al.*, 2002) and for relaxase binding (Schroder *et al.*, 2002). The hydrosoluble ~50 kDa cytoplasmic domain of TrwB could be purified and crystallized (Gomis-Ruth *et al.*, 2001). Its crystal structure revealed a globular hexameric assembly where each subunit has an orange segment shape and is composed of two distinct domains: An all- domain facing the cytoplasm and a NTP-binding domain linked to the inner membrane by the N-terminal membrane anchor. The first domain contains seven helices while the latter domain is made of a central twisted -sheet flanked by several helices on both sides. The six protomers assemble to form a globular ring of ~110 Å in diameter and 90 Å in height with a ~20-Å-wide channel in the centre. This channel forms an 8-Å-wide constriction at the cytoplasmic pole of the molecule. Superficial binding pockets at the interface between the subunits form the NTP binding sites.

It is enticing to postulate that CPs are molecular motors that mediate ssDNA and/or protein translocation through the inner membrane. The TrwB NTP-binding domain has an archetypal Rec-A-like fold that is found in many oligomeric ATPases including FtsK helicase and the F1-ATPase. FtsK helicase is involved in the late steps of plasmid segregation in bacteria. Structural superimposition showed that FtsK NTP-binding domain is the closest structural homologue to TrwB. Each FtsK subunit is composed of three domains (, and). The crystal structure of the and domains from FtsK was solved. They form a hexameric structure with a ~30-Å-diameter central channel that is large enough to accommodate FtsK dsDNA substrate (Massey *et al.*, 2006). Electron microscopy studies showed these domain form double head to head (to) hexameric rings on DNA, which is passing through the centre of this dodecameric structure. In TrwB, the central channel would be large enough to accommodate ssDNA (Tato *et al.*, 2007). Interestingly, like both FtsK and F1-ATPases, TrwB could function as a rotary motor with sequential ATPase catalytic cycles around the hexamer that would induce conformational change in its central channel. It was shown that TrwB undergoes conformational changes in the interior central channel upon substrate binding and hydrolysis (Gomis-Ruth *et al.*, 2002).

VirB11

VirB11 belongs to a large family of NTPases found associated with type II and type IV secretion systems. Many homologues are found in Gram-negative bacteria but also rarely in Gram-positive bacteria and archaea (Cao and Saier, 2001; Planet *et al.*, 2001; Grohmann *et al.*, 2003). The NTPase activity of several VirB11 homologues (*A. tumefaciens* VirB11, *E. coli* $TrbB_{RP4}$, $traG_{pKM101}$ and $TrwD_{R388}$, HP0525 from the cag pathogenicity island of *Helicobacter pylori*) could be determined *in vitro* (Christie et al., 1989; Rivas et al., 1997; Krause *et al.*, 2000b)

VirB11 homologues are peripheral inner membrane proteins that may be in a dynamic and regulated equilibrium between the cytoplasm and the membrane. Some VirB11 homologues fractionate as cytoplasmic and peripheral inner membrane proteins (Rashkova *et al.*, 1997; Krause *et al.*, 2000b). In addition, $TrbB_{RP4}$ and HP0525 both undergo conformational changes when interacting with lipid bilayers. Some purified VirB11 homologues see their NTPase activity stimulated by the addition of certain phospholipids (Rivas *et al.*, 1997; Krause *et al.*, 2000b).

Electron microscopy visualization of $TrbB_{RP4}$, $TrwD_{R388}$ and HP0525 showed hexameric rings of ~100–120 Å (Rivas *et al.*, 1997; Krause *et al.*, 2000a,b; Savvides *et al.*, 2003). The crystal structure of the ADP-bound HP0525 revealed that each monomer consists of two domains formed by the N- and C-terminal halves of the protein. The nucleotide-binding site is at the interface between the two domains. In the hexamer, the N- and C-terminal domains (NTD and CTD) form two separate rings defining a chamber of ~50 Å in diameter, which is open on the NTD side and closed on the CTD side. The CTD adopts a RecA-like fold while the NTD is only found in HP5025. Recently, the structure of Brucella suis VirB11 (BsB11) was determined. It showed that the VirB11 monomer differs dramatically from that of Hp0525 by a large domain swap caused by the insertion of additional sequences into the linker between the NTD and CTD. The global assembly of the VirB11 hexamer is intact compared to HP5025 but this domain organization profoundly modifies the nucleotide-binding site and the interface between subunits (Hare *et al.*, 2006). Based on sequence comparisons, it is likely that most of VirB11 homologues display a similar structure.

Biochemical and structural studies provided a detailed view of the dynamics of VirB11 hexamer (Savvides *et al.*, 2003). In the absence of nucleotide, while the CTD ring is unchanged and maintains the subunit-subunit contacts in the hexameric structure, the NTDs are flexible and display various rigid-body conformations making the NTD ring asymmetric. The binding of three ATP molecules would then induce a movement of three NTDs into a rigid conformation. The concomitant hydrolysis of the three ATP molecules and binding of three other molecules in the nucleotide-free subunits locks the hexamer into a compact and symmetric structure. When all the nucleotides are hydrolysed and released, the structure is relaxed and the VirB11 hexamer returns to its nucleotide free state.

VirB4

VirB4 proteins are ubiquitous among conjugative T4S systems, as they are associated with T4S in both Gram-positive and Gram-negative bacteria. (Grohmann *et al.*, 2003). The VirB4 homologues contain conserved Walker A and B motifs, which are essential for the secretion process (Berger and Christie, 1993; Fullner *et al.*, 1994; Cook *et al.*, 1999; Rabel *et al.*, 2003; Arechaga *et al.*, 2008). Even if several attempts to detect an ATPase activity *in vitro* with several purified VirB4 homologues failed (Rabel *et al.*, 2003), two independent reports now demonstrate such activity for both *A. tumefaciens* VirB4 and *E. coli* $TrwK_{R388}$ (Shirasu *et al.*, 1994; Arechaga *et al.*, 2008)

The cellular localization, topology and oligomerization state of VirB4 are unclear. Both direct association with the inner membrane and membrane association through interactions with other T4SS subunits were proposed (Schandel *et al.*, 1992; Dang and Christie, 1997; Gilmour *et al.*, 2001; Arechaga *et al.*, 2008). VirB4 homologues have been described as containing from zero up to four predicted transmembrane segments (TMS) (Dang and Christie, 1997; Cao and Saier, 2001; Middleton *et al.*, 2005; Arechaga *et al.*, 2008).Whether or not these proteins contain a TMS remain to be elucidated. It is worthy to note that the active VirB4 homologues were obtained from the soluble fraction without the need of detergents (Shirasu *et al.*, 1994; Arechaga *et al.*, 2008). This indicates that VirB4, if directly associated with the membrane, would rather behave like a peripheral membrane protein. The use of detergents would indeed be mandatory to isolate VirB4 as an integral membrane protein. However, this observation does not exclude that spontaneously or during the secretion process, VirB4 proteins could undergo a conformational change and be fully inserted in the inner membrane by one or several TMSs. Because of the uncertainty concerning VirB4 localization, the topologies that has been proposed are consistently very different and quite contradictory, from VirB4 entirely in the cytoplasm to most of the protein located in the periplasm (Dang and Christie, 1997; Draper *et al.*, 2006; Arechaga *et al.*, 2008). Finally, adding to this complexity, VirB4 oligomerization state is also unclear. A model based on the structural similarities between *A. tumefaciens* VirB4 C-terminus and the R388 coupling protein TrwB proposed that VirB4 C-terminal domain would assemble as homo-hexameric ring (Middleton *et al.*, 2005). Interestingly, the membrane-associated inactive *A. tumefaciens* VirB4 and the *E. coli* soluble active $TrwK_{R388}$ were respectively isolated as dimers and higher order oligomers (maybe hexamers) (Dang and Christie, 1997; Arechaga *et al.*, 2008), suggesting that the oligomerization state as well as the ATPase activity would depend on the protein localization.

The translocation pore/apparatus.

The translocation apparatus allows the substrate to reach the extracellular milieu from the cytoplasm. Therefore, it crosses the whole cell envelope and defines in both bacterial membranes and the periplasm a conduit for the substrate. This apparatus is composed of VirB6, VirB7, VirB8, VirB9 and VirB10. These proteins are inner membrane proteins such as VirB6, VirB8 or VirB10, or periplasmic proteins associated with the outer membrane such as VirB7 and VirB10. Multiple interactions were described between these proteins that would form a stable substructure (Fig. 11.7) (Beaupre *et al.*, 1997; Das *et al.*, 1997; Das and Xie, 2000; Kumar and Das, 2001; Krall *et al.*, 2002; Ward *et al.*, 2002; Cascales and Christie, 2003; Jakubowski *et al.*, 2003; Fronzes *et al.*, 2009b).

The core complex

The proteins VirB7, VirB8, VirB9 and VirB10 are often referred as core proteins of the T4S apparatus. Multiple biochemical and functional interactions between these proteins have been described in the literature (for a review see Fronzes *et al.*, 2009a). Recently, the T4S core complex from the conjugative pKM101 plasmid was identified and isolated (Fronzes *et al.*, 2009b). This work showed that the core forms a 1-MDa complex that is inserted in both inner and outer membranes and is composed of the VirB7, VirB9 and VirB10 homologues in equal stoichiometry. The structure of the complex was determined at 15 Å resolution using cryo-electron microscopy and single particle analysis. It is a cylindrical object containing 14 copies of each component. It is composed of two layers (I-layer and O-layer) linked by thin stretches of density. Each layer forms a double-walled ring-like structure that defines hollow chambers inside the complex. The I-layer is anchored in the inner membrane and resembles a cup that is opened at the base by a 55 Å diameter hole. It is composed of the N-terminal domains of the VirB9 and VirB10 homologues. The O-layer consists of a main body and a narrower cap. It is inserted in the outer membrane and is composed of the VirB7 homologue and the C-terminal domains of the VirB9 and VirB10 homologues (Fig. 11.7).

The outer membrane complex that contains the entire O-layer was crystallized and its structure at 2.6 Å resolution was solved (Chandran *et al.*, 2009). It was prepared using proteolytic cleavage of the T4S system core. This 590-kDa complex contains 14 copies each of the C-terminal domain of VirB10, the C-terminal domain of VirB9 and the full-length VirB7. It is made of two parts: a main body and a cap (Fig. 11.7). The cap is made of a hydrophobic ring of 2-helix bundles defining a 32-Å channel through the outer membrane. These helices are made of the VirB10 antenna (Fig. 11.7), establishing VirB10 homologues

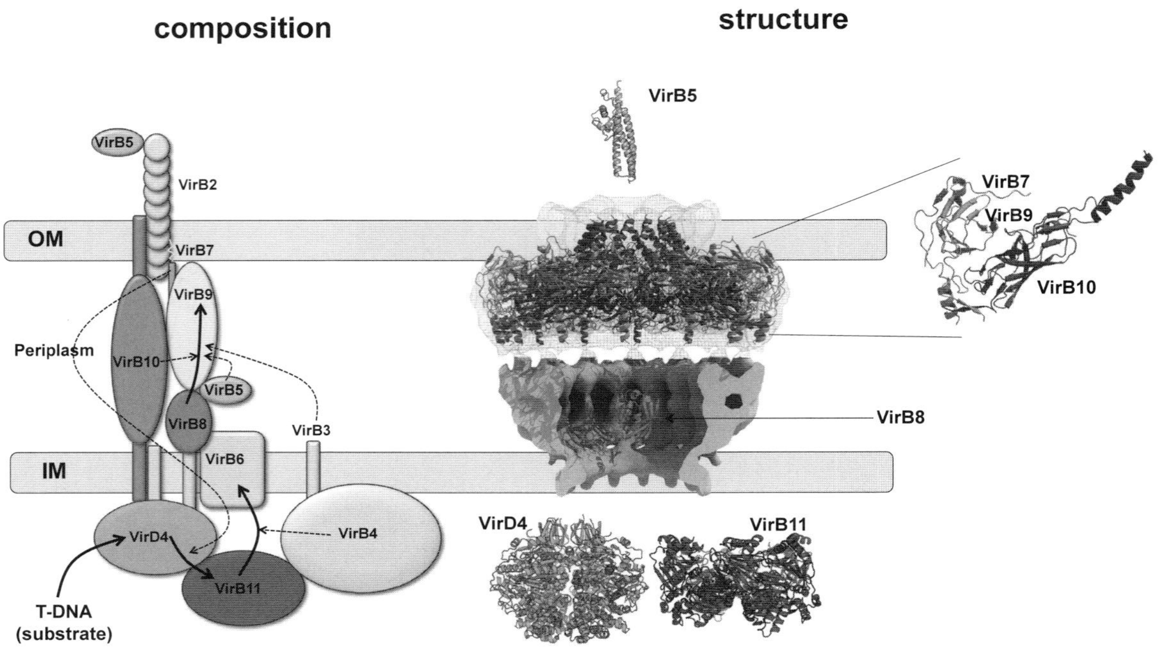

Figure 11.7 Secretion pathway and structure of the T4S system. Left: T4S system assembly and substrate pathway. The T4S components are represented according to their proposed localization. Most (but not all) of the interactions that have been confirmed biochemically (i.e. co-purification, immuno-precipitation) are indicated by the physical proximity of the schematic representations of each protein (see Chandran *et al.*, 2009, for a review of all the interactions described in the literature). Substrate (T-DNA) pathway within the T4S apparatus of *A. tumefaciens* as determined by the TrIP (Cascales and Christie, 2004b) is represented by plain arrows. The T-DNA interacts directly with VirD4, VirB11, VirB6/VirB8 and finally with VirB9/VirB2. The other T4S components influence the transfer of the substrate at different stages as represented by dotted arrows. Right: T4S system component structures. All the structures determined to date are represented at the same scale. The atomic structures are shown in ribbon representation. They include full-length VirB5, VirB8 periplasmic domain, full-length VirB11, VirD4 soluble domain and the T4S outer membrane complex (made of full-length VirB7, and the VirB9 C-terminal domain and VirB10 C-terminal domain). The structure of the core complex determined using cryo-electron microscopy and single particle analysis is rendered as a cut-out volume. The core complex is composed of the full-length VirB7, VirB9, and VirB10 proteins. A colour version of this figure is located in the plate section at the back of the book.

as outer membrane channel proteins for T4S systems. This result was unexpected, as VirB10 homologues have never been suspected to form the outer membrane channel of T4S systems. Indeed, VirB10 homologues are inserted in the inner membrane by a trans-membrane helix at the N-terminus of the protein (Jakubowski *et al.*, 2009). Thus, VirB10 is the only protein known to insert in both inner and outer membranes of Gram-negative bacteria. The main body, which has 172Å in diameter, is made of the rest of VirB10CT, VirB9CT, and VirB7. When the complex is viewed from the extracellular milieu, VirB10CT forms an inner ring surrounded by the VirB9–VirB7 complex. VirB7 forms spokes radially crossing the entire assembly (Fig. 11.7). VirB10CT forms the inner wall of the structure (Fig. X).

All outer membrane structures previously determined were made of one protein. A unique feature of the T4S system outer membrane complex is that it is made of three proteins that are essential for complex assembly and channel formation (Fronzes *et al.*, 2009b). The heterotrimer formed by VirB10CT, VirB9CT and VirB7 is shown in Fig. 11.7. VirB10CT and VirB9CT interact directly. VirB7 interacts with VirB9CT not with VirB10CT. Interestingly, the core structure of the VirB7/VirB9CT complex bound to VirB10CT is similar to the previously determined NMR structure of the isolated VirB7/VirB9CT complex (Bayliss *et al.*, 2007). VirB9CT adopts a Beta-sandwich fold around which VirB7 winds (Fig. 11.7) (Bayliss *et al.*, 2007; Chandran *et al.*, 2009). However, in the outer membrane complex, VirB7 almost entirely wraps around VirB9CT, forming an interface with VirB9CT that is much more extensive than the one observed in the NMR structure. Similarly, the structure of the isolated form of the C-terminal domain of ComB10 (ComB10CT), a VirB10 homologue from *H. pylori*, was previously solved and has a similar structure to the VirB10CT domain present in the outer membrane complex (Terradot *et al.*, 2005). In ComB10, this domain is made of a modified beta-barrel completed by a helix lying on its side (α1-helix) and a flexible helix–turn–helix antenna of 70 Å in length projecting off its top. However, in the outer membrane complex, VirB10CT differs from ComB10CT in that the flanking helix of ComB10CT is missing in VirB10CT. In addition, the N-terminus of VirB10CT forms an extended N-terminal arm (or lever arm) that projects out to contact three consecutive neighbouring heterotrimers in the tetradecamer.

The inner membrane channel/pore assembly

The I-layer of the T4S complex, made of the N-terminal domains of VirB9 and VirB10 is inserted in the inner membrane via VirB10 N-terminal TMS. However, it is very unlikely that the 14 VirB10 TMSs would form the inner membrane pore on their own. Among the T4S subunits, VirB6 and VirB8 are better candidates to form this pore. These two proteins are also inserted in the inner membrane but, unlike VirB10, they are directly contacting the substrate during secretion in *A. tumefaciens* (Cascales and Christie, 2003).

VirB8 subunits is a bitopic inner membrane protein composed of an N-terminal TMS and a large periplasmic C-terminal domain (Das and Xie, 1998; Buhrdorf *et al.*, 2003; Terradot *et al.*, 2005). This protein is thought to play a key role during pilus biogenesis (Yuan *et al.*, 2005), substrate secretion (Cascales and Christie, 2004a) and as nucleator during T4S system assembly and spatial positioning (Kumar *et al.*, 2000; Judd *et al.*, 2005a). Crystal structures of the periplasmic domains of the *B. suis* and *A. tumefaciens* VirB8 were solved (Terradot *et al.*, 2005; Bailey *et al.*, 2006). The two homologous domains display similar globular structures that comprise an extended beta-sheet flanked with five -helices (Terradot *et al.*, 2005; Bailey *et al.*, 2006) (Fig. 11.7).

VirB6 is polytopic inner membrane proteins that is essential for substrate secretion through the inner membrane (Cascales and Christie, 2003). Interestingly, *H. pylori* ComB6 and *A. tumefaciens* VirB6 share the same topology with a periplasmic N-terminus, five transmembrane segments, and a cytoplasmic C-terminal domain (Jakubowski *et al.*, 2004; Karnholz *et al.*, 2006). They also comprise a central region composed of a large periplasmic loop (P2 loop), which, in *A. tumefaciens*, mediates the interaction of VirB6 with the substrate (T-DNA). The other parts of the proteins are either essential for substrate transfer from VirB6 to VirB8 in periplasm or to VirB2 and VirB9 at the outer membrane (Jakubowski *et al.*, 2004) but do not contact the T-DNA. These data indicate that VirB6 would be central component of the inner membrane channel. In addition, the presence of VirB6 is required for the stabilization of pilus components such as VirB7 and VirB5 and is also required for pilus biogenesis (Krall *et al.*, 2002; Jakubowski *et al.*, 2003).

Both VirB6 and VirB8 interact with the outer membrane components VirB7/VirB9 (Das and Xie, 2000; Kumar *et al.*, 2000; Kumar and Das, 2001; Krall *et al.*, 2002; Jakubowski *et al.*, 2003). Surprisingly, no interaction has been observed between VirB6 and VirB8 suggesting that the two proteins are not in direct contact. Because VirB6 is a polytopic integral membrane protein, technical issues may have prevented the detection of such interactions. VirB8 was also shown to interact with VirB1, VirB4, VirB5 and VirB11 (Rambow-Larsen and Weiss, 2002; Ward *et al.*, 2002; Malek *et al.*, 2004; Hoppner *et al.*, 2005; Yuan *et al.*, 2005). It is tempting to propose that the T4S inner membrane pore would sit in the 55 Å opening at the base of the core complex and fill the I-layer cavity.

The pilus (and other T4S surface structures)

In Gram-negative bacteria, most of the T4S system gene clusters encode a pilin homologue. T4S pilin expression is essential for secretion (Haase *et al.*, 1995; Christie, 1997) and T4S related pili are associated with the expression of some T4S systems (Fullner *et al.*, 1996; Lawley *et al.*, 2003; Tanaka *et al.*, 2003). Pilus formation could help to establish a stable and specific contact between cells before substrate transfer. Some conjugative pili such as F-pili could retract by depolymerization, bringing donor and recipient cells into close contact before DNA transfer (Lawley *et al.*, 2003; Clarke *et al.*, 2008). In addition, most of the T4S pili are tubular structures and it was proposed that they could serve as a duct for T4S substrates between donor and recipient/host cells. Whether or not the T4S substrate passes through the hollow lumen of the T4S pilus remains controversial but recent observations show that DNA can be transferred between cells that are not in direct contact during conjugation (Babic *et al.*, 2008) and that DNA could be detected within the F-pilus structure (Shu *et al.*, 2008). However, it is interesting to note that pilus production and substrate secretion capacity in some T4S systems are clearly distinct events. Some mutations in the pilin (Eisenbrandt *et al.*, 2000) or some T4SS components (In *A. tumefaciens* VirB6, VirB9, VirB11) (Jakubowski *et al.*, 2004) can uncouple pilus biogenesis and substrate secretion, suggesting that pilus biogenesis is not required for secretion.

T4S pilins

T4S pre-pilins are poorly conserved. However, T4S pilins contain one or two hydrophobic patches in their sequences. In addition, T4S pre-pilin are first targeted to the periplasm by a long signal peptide that is later cleaved by a dedicated protease (the F-type pilin traAf or the P-type pilin TrbCrp4 both have their signal peptide cleaved by lepB (Majdalani and

Ippen-Ihler, 1996; Haase and Lanka, 1997). The additional maturation steps depend on the pilin type and can be separated in two classes. Most of the F-like pilins are N-acetylated and inserted in the inner membrane (Moore *et al.*, 1993). Inner membrane chaperone-like proteins could be essential for a correct insertion and accumulation of these pilins in the inner membrane (Lu *et al.*, 2002). On the other hand, the P-like conjugative pilins (RP4 TrbC but also probably pKM101 traM, R388 trwL, R6K PilX2) and several other pilins (*Agrobacterium tumefaciens* virB2 but possibly pilins from *Bordetella pertussis* toxin operon ptlA and *Brucella suis* and *Brucella abortus* virB2 homologues) are cyclic peptides (Eisenbrandt *et al.*, 1999; Kalkum *et al.*, 2002). Generally, the pre-pilin undergo a first maturation event where a C-terminal signal peptide is cleaved by an unidentified host-encoded protease. Then, a N-terminal signal peptide is removed by lepB. Finally, the plasmid encoded serine protease is responsible of pilin cyclization (traF protease in the case of the pilin trbC cyclization). Like F-pilins, cyclized pilins are likely to be inserted in the inner membrane while they are processed (Kalkum *et al.*, 2002).

In *A. tumefaciens*, VirB5 is a minor T-pilus component in addition of the structural pilin virB2 (Lai and Kado, 1998; Schmidt-Eisenlohr *et al.*, 1999). In addition, VirB7 co-purifies with the T-pilus (Sagulenko *et al.*, 2001b). Homologues of these two proteins are found in many conjugative (pKM101, RP4, R388 or F systems) and effector translocation T4S systems of other pathogenic bacteria such as *Brucella suis* or *Bartonella henselae*). The crystal structure of virB5 homologue of pKM101 conjugative plasmid (called traC) was solved (Yeo *et al.*, 2003) and comprises a three-long-helices bundle flanked by a smaller globular part (Fig. 11.7). It has been recently been shown that *A. tumefaciens* virB5 localizes at the tip of the T-pilus (Aly and Baron, 2007). Interestingly, it was recently showed in *H. pylori* that CagL is a pilus protein that would act as an adhesin on the pilus, interacting with and activating integrins at the surface of host cells (Kwok *et al.*, 2007). It was proposed that CagL would be the *H. pylori* homologue (Backert *et al.*, 2008).

Pilus morphology and structure

The morphology of the conjugative pili have been classified in two major groups: incF-like (also called F-like) pili (conjugative pili produced by inc-F, -H, -T and -J systems) and incP-like (also named P-like) pili (conjugative pili produced by inc-P, -N, -W systems) (Bradley, 1980; Lawley *et al.*, 2003). IncF-like pili are long (2–20 µm) and flexible appendages with 8–9 nm in diameter. IncP-like pili are short (<1 µm) and rigid rods with 8–12 nm in diameter (Bradley, 1980). The morphology of the pilus from non-conjugative T4S systems is rarely known or cannot be classified into these two groups. For example, *A. tumefaciens* T-pili are variable in length and flexible with 10 nm in diameter (Eisenbrandt *et al.*, 1999). *H. pylori* bacteria harbour 100–200 nm sheathed appendages at the tip of which CagA (*H. pylori* T4SS substrate) can be found (Tanaka *et al.*, 2003; Kwok *et al.*, 2007).

The structure of the F-pilus was examined using cryo-electron microscopy and single particle methods (Wang *et al.*, 2009). The tubular structure had a central lumen of ~30 Å. tubular and exhibited a different symmetry (C_4). Such a lumen diameter would be large enough to accommodate ssDNA. However, the relaxase traI, which is attached to this DNA has a molecular mass of ~200 kDa. Unless it is unfolded, it is very unlikely that it would be able to travel inside the pilus lumen. From these new data, two different subunit packing within the pilus were deduced. One is a stack of pilin rings of C4 symmetry separated by a 12.8 Å space. The other is a one-start helical symmetry with an axial rise of ~3.5 Å per subunit and a pitch of ~12.2 Å. These two packing seem to coexist within the pilus structure.

Prospects

Over the years, structures of isolated components (soluble components or domains) from the T4S apparatus have been obtained. It was only recently that an understanding of how these components assemble came to light. The structures of the T4S core complex and outer membrane complex provided the first insights on T4S system assembly. In addition, they revealed the molecular details of an outer membrane structure of unprecedented size and complexity. The next steps will be to solve the structures of other or larger subassemblies of T4S systems. Indeed, the architecture of the pore embedded in the inner membrane and its interactions with the core complex remain elusive. Another question is how the core complex associates with the T4S pilin and T4S pilus. Finally, very little is known concerning the details of the interactions of cytoplasmic ATPases with themselves and with the rest of the T4S apparatus. Together with *in vivo* functional studies, future structural work should provide the missing clues leading to a comprehensive elucidation of T4S mechanism.

Acknowledgements

We would like to thank L. Königsmaier for providing a micrograph for Fig. 11.2A and helpful discussion. MJB is supported by the FWF (DK-plus 'Structure and Interaction of Biological Macromolecules'). Research in the Marlovits laboratory is supported by the ZIT (Centre for Innovation and Technology, City of Vienna, Austria, Grant: CMCN-V Centre for Molecular and Cellular Nanostructure Vienna), by the Research Institute of Molecular Biotechnology of the Austrian Academy of Sciences (IMBA) and by the Research Institute of Molecular Pathology (IMP). Research in the Fronzes lab is supported by Institut Pasteur and CNRS (Centre National de la Recherche Scientifique).

References

Akeda, Y., and Galán, J.E. (2005). Chaperone release and unfolding of substrates in type III secretion. Nature *437*, 911–915.

Alfano, J.R., and Collmer, A. (2004). Type III secretion system effector proteins: double agents in bacterial disease and plant defense. Annu. Rev. Phytopathol. *42*, 385–414.

Aly, K.A., and Baron, C. (2007). The VirB5 protein localizes to the T-pilus tips in *Agrobacterium tumefaciens*. Microbiology *153*, 3766–3775.

Arechaga, I., Pena, A., Zunzunegui, S., del Carmen Fernandez-Alonso, M., Rivas, G., and de la Cruz, F. (2008). ATPase activity and oligomeric state of TrwK, the VirB4 homologue of the plasmid R388 type IV secretion system. J. Bacteriol. *190*, 5472–5479.

Arnold, R., Brandmaier, S., Kleine, F., Tischler, P., Heinz, E., Behrens, S., Niinikoski, A., Mewes, H.-W., Horn, M., and Rattei, T. (2009). Sequence-based prediction of type III secreted proteins. PLoS Pathog. *5*, e1000376.

Atmakuri, K., Cascales, E., Burton, O.T., Banta, L.M., and Christie, P.J. (2007). Agrobacterium ParA/MinD-like VirC1 spatially coordinates early conjugative DNA transfer reactions. EMBO J. *26*, 2540–2551.

Atmakuri, K., Cascales, E., and Christie, P.J. (2004). Energetic components VirD4, VirB11 and VirB4 mediate early DNA transfer reactions required for bacterial type IV secretion. Mol. Microbiol. *54*, 1199–1211.

Atmakuri, K., Ding, Z., and Christie, P.J. (2003). VirE2, a type IV secretion substrate, interacts with the VirD4 transfer protein at cell poles of *Agrobacterium tumefaciens*. Mol. Microbiol. *49*, 1699–1713.

Babic, A., Lindner, A.B., Vulic, M., Stewart, E.J., and Radman, M. (2008). Direct visualization of horizontal gene transfer. Science *319*, 1533–1536.

Backert, S., Fronzes, R., and Waksman, G. (2008). VirB2 and VirB5 proteins: specialized adhesins in bacterial type-IV secretion systems? Trends Microbiol. *16*, 409–413.

Bailey, S., Ward, D., Middleton, R., Grossmann, J.G., and Zambryski, P.C. (2006). *Agrobacterium tumefaciens* VirB8 structure reveals potential protein–protein interaction sites. Proc. Natl. Acad. Sci. U.S.A. *103*, 2582–2587.

Barison, N., Lambers, J., Hurwitz, R., and Kolbe, M. (2012). Interaction of MxiG with the cytosolic complex of the type III secretion system controls Shigella virulence. FASEB J...

Baron, C. (2010). Antivirulence drugs to target bacterial secretion systems. Curr. Opin. Microbiol. *13*, 100–105.

Bayliss, R., Harris, R., Coutte, L., Monier, A., Fronzes, R., Christie, P.J., Driscoll, P.C., and Waksman, G. (2007). NMR structure of a complex between the VirB9/VirB7 interaction domains of the pKM101 type IV secretion system. Proc. Natl. Acad. Sci. U.S.A. *104*, 1673–1678.

Beaupre, C.E., Bohne, J., Dale, E.M., and Binns, A.N. (1997). Interactions between VirB9 and VirB10 membrane proteins involved in movement of DNA from *Agrobacterium tumefaciens* into plant cells. J. Bacteriol. *179*, 78–89.

Berger, B.R., and Christie, P.J. (1993). The *Agrobacterium tumefaciens* virB4 gene product is an essential virulence protein requiring an intact nucleoside triphosphate-binding domain. J. Bacteriol. *175*, 1723–1734.

Berger, B.R., and Christie, P.J. (1994). Genetic complementation analysis of the *Agrobacterium tumefaciens* virB operon: virB2 through virB11 are essential virulence genes. J. Bacteriol. *176*, 3646–3660.

Bradley, D.E. (1980). Morphological and serological relationships of conjugative pili. Plasmid *4*, 155–169.

Broz, P., Mueller, C.A., Müller, S.A., Philippsen, A., Sorg, I., Engel, A., and Cornelis, G.R. (2007). Function and molecular architecture of the Yersinia injectisome tip complex. Mol. Microbiol. *65*, 1311–1320.

Buhrdorf, R., Forster, C., Haas, R., and Fischer, W. (2003). Topological analysis of a putative virB8 homologue essential for the cag type IV secretion system in *Helicobacter pylori*. Int. J. Med. Microbiol. *293*, 213–217.

Bundock, P., den Dulk-Ras, A., Beijersbergen, A., and Hooykaas, P.J. (1995). Trans-kingdom T-DNA transfer from *Agrobacterium tumefaciens* to *Saccharomyces cerevisiae*. EMBO J. *14*, 3206–3214.

Burghout, P., van Boxtel, R., Van Gelder, P., Ringler, P., Müller, S.A., Tommassen, J., and Koster, M. (2004). Structure and electrophysiological properties of the YscC secretin from the type III secretion system of Yersinia enterocolitica. J. Bacteriol. *186*, 4645–4654.

Burns, D.L. (2003). Type IV transporters of pathogenic bacteria. Curr. Opin. Microbiol. *6*, 29–34.

Büttner, D. (2012). Protein export according to schedule: architecture, assembly, and regulation of type III secretion systems from plant- and animal-pathogenic bacteria. Microbiol. Mol. Biol. Rev. *76*, 262–310.

Cao, T.B., and Saier, M.H. Jr. (2001). Conjugal type IV macromolecular transfer systems of Gram-negative bacteria: organismal distribution, structural constraints and evolutionary conclusions. Microbiology *147*, 3201–3214.

Cascales, E. (2008). The type VI secretion toolkit. EMBO Rep. *9*, 735–741.

Cascales, E., and Christie, P.J. (2003). The versatile bacterial type IV secretion systems. Nat. Rev. Microbiol. *1*, 137–149.

Cascales, E., and Christie, P.J. (2004a). Agrobacterium VirB10, an ATP energy sensor required for type IV secretion. Proc. Natl. Acad. Sci. U.S.A. *101*, 17228–17233.

Cascales, E., and Christie, P.J. (2004b). Definition of a bacterial type IV secretion pathway for a DNA substrate. Science *304*, 1170–1173.

Chandran, V., Fronzes, R., Duquerroy, S., Cronin, N., Navaza, J., and Waksman, G. (2009). Structure of the outer membrane complex of a type IV secretion system. Nature *462*, 1011–1015.

Chatterjee, S., Zhong, D., Nordhues, B.A., Battaile, K.P., Lovell, S., and De Guzman, R.N. (2011). The crystal structures of the Salmonella type III secretion system tip protein SipD in complex with deoxycholate and chenodeoxycholate. Protein Sci. *20*, 75–86.

Chen, Y., Zhang, X., Manias, D., Yeo, H.J., Dunny, G.M., and Christie, P.J. (2008a). *Enterococcus faecalis* PcfC, a spatially localized substrate receptor for type IV secretion of the pCF10 transfer intermediate. J. Bacteriol. *190*, 3632–3645.

Chen, Y., Zhang, X., Manias, D., Yeo, H.J., Dunny, G.M., and Christie, P.J. (2008b). *Enterococcus faecalis* PcfC, a spatially localized substrate receptor for type IV secretion of the pCF10 transfer intermediate. J. Bacteriol. *190*, 3632–3645.

Christie, P.J. (1997). *Agrobacterium tumefaciens* T-complex transport apparatus: a paradigm for a new family of multifunctional transporters in eubacteria. J. Bacteriol. *179*, 3085–3094.

Christie, P.J., Ward, J.E. Jr., Gordon, M.P., and Nester, E.W. (1989). A gene required for transfer of T-DNA to plants encodes an ATPase with autophosphorylating activity. Proc. Natl. Acad. Sci. U.S.A. *86*, 9677–9681.

Clarke, M., Maddera, L., Harris, R.L., and Silverman, P.M. (2008). F-pili dynamics by live-cell imaging. Proc. Natl. Acad. Sci. U.S.A. *105*, 17978–17981.

Clatworthy, A.E., Pierson, E., and Hung, D.T. (2007). Targeting virulence: a new paradigm for antimicrobial therapy. Nat. Chem. Biol. *3*, 541–548.

Coburn, B., Sekirov, I., and Finlay, B.B. (2007). Type III secretion systems and disease. Clin. Microbiol. Rev. *20*, 535–549.

Cook, D.M., Farizo, K.M., and Burns, D.L. (1999). Identification and characterization of PtlC, an essential component of the pertussis toxin secretion system. Infect. Immun. *67*, 754–759.

Coombes, B.K. (2009). Type III secretion systems in symbiotic adaptation of pathogenic and non-pathogenic bacteria. Trends Microbiol. *17*, 89–94.

Cordes, F.S., Komoriya, K., Larquet, E., Yang, S., Egelman, E.H., Blocker, A., and Lea, S.M. (2003). Helical structure of the needle of the type III secretion system of *Shigella flexneri*. J. Biol. Chem. *278*, 17103–17107.

Cordes, F.S., Daniell, S., Kenjale, R., Saurya, S., Picking, W.L., Picking, W.D., Booy, F., Lea, S.M., and Blocker, A. (2005). Helical packing of needles from functionally altered Shigella type III secretion systems. J. Mol. Biol. *354*, 206–211.

Cornelis, G.R. (2002). The *Yersinia* Ysc-Yop 'type III' weaponry. Nat. Rev. Mol. Cell Biol. *3*, 742–752.

Cornelis, G.R., and Van Gijsegem, F. (2000). Assembly and function of type III secretory systems. Annu. Rev. Microbiol. *54*, 735–774.

Couturier, M.R., Tasca, E., Montecucco, C., and Stein, M. (2006). Interaction with CagF is required for translocation of CagA into the host via the *Helicobacter pylori* type IV secretion system. Infect. Immun. *74*, 273–281.

Craig, L., and Li, J. (2008). Type IV pili: paradoxes in form and function. Curr. Opin. Struct. Biol. *18*, 267–277.

Daefler, S., and Russel, M. (1998). The Salmonella typhimurium InvH protein is an outer membrane lipoprotein required for the proper localization of InvG. Mol. Microbiol. *28*, 1367–1380.

Dale, C., Plague, G.R., Wang, B., Ochman, H., and Moran, N.A. (2002). Type III secretion systems and the evolution of mutualistic endosymbiosis. Proc. Natl. Acad. Sci. U.S.A. *99*, 12397–12402.

Dang, T.A., and Christie, P.J. (1997). The VirB4 ATPase of *Agrobacterium tumefaciens* is a cytoplasmic membrane protein exposed at the periplasmic surface. J. Bacteriol. *179*, 453–462.

Das, A., and Xie, Y.H. (1998). Construction of transposon Tn3phoA: its application in defining the membrane topology of the *Agrobacterium tumefaciens* DNA transfer proteins. Mol. Microbiol. *27*, 405–414.

Das, A., and Xie, Y.H. (2000). The *Agrobacterium* T-DNA transport pore proteins VirB8, VirB9, and VirB10 interact with one another. J. Bacteriol. *182*, 758–763.

Das, A., Anderson, L.B., and Xie, Y.H. (1997). Delineation of the interaction domains of *Agrobacterium tumefaciens* VirB7 and VirB9 by use of the yeast two-hybrid assay. J. Bacteriol. *179*, 3404–3409.

Deane, J.E., Cordes, F.S., Roversi, P., Johnson, S., Kenjale, R., Picking, W.D., Picking, W.L., Lea, S.M., and Blocker, A. (2006). Expression, purification, crystallization and preliminary crystallographic analysis of MxiH, a subunit of the *Shigella flexneri* type III secretion system needle. Acta Crystallogr. Sect. F Struct. Biol. Cryst. Commun. *62*, 302–305.

Derewenda, U., Mateja, A., Devedjiev, Y., Routzahn, K.M., Evdokimov, A.G., Derewenda, Z.S., and Waugh, D.S. (2004). The structure of *Yersinia pestis* V-antigen, an essential virulence factor and mediator of immunity against plague. Structure *12*, 301–306.

Diepold, A., Amstutz, M., Abel, S., Sorg, I., Jenal, U., and Cornelis, G.R. (2010). Deciphering the assembly of the Yersinia type III secretion injectisome. EMBO J. *29*, 1928–1940.

Diepold, A., Wiesand, U., and Cornelis, G.R. (2011). The assembly of the export apparatus (YscR,S,T,U,V) of the Yersinia type III secretion apparatus occurs independently of other structural components and involves the formation of an YscV oligomer. Mol. Microbiol.

Draper, O., Middleton, R., Doucleff, M., and Zambryski, P.C. (2006). Topology of the VirB4 C terminus in the *Agrobacterium tumefaciens* VirB/D4 type IV secretion system. J. Biol. Chem. *281*, 37628–37635.

Eisenbrandt, R., Kalkum, M., Lai, E.M., Lurz, R., Kado, C.I., and Lanka, E. (1999). Conjugative pili of IncP plasmids, and the Ti plasmid T pilus are composed of cyclic subunits. J. Biol. Chem. *274*, 22548–22555.

Eisenbrandt, R., Kalkum, M., Lurz, R., and Lanka, E. (2000). Maturation of IncP pilin precursors resembles the catalytic Dyad-like mechanism of leader peptidases. J. Bacteriol. *182*, 6751–6761.

Felise, H.B., Nguyen, H.V., Pfuetzner, R.A., Barry, K.C., Jackson, S.R., Blanc, M.-P., Bronstein, P.A., Kline, T., and Miller, S.I. (2008). An inhibitor of gram-negative bacterial virulence protein secretion. Cell Host Microbe *4*, 325–336.

Filloux, A. (2004). The underlying mechanisms of type II protein secretion. Biochim. Biophys. Acta *1694*, 163–179.

Freiberg, C., Fellay, R., Bairoch, A., Broughton, W.J., Rosenthal, A., and Perret, X. (1997). Molecular basis of symbiosis between Rhizobium and legumes. Nature *387*, 394–401.

Fronzes, R., Christie, P.J., and Waksman, G. (2009a). The structural biology of type IV secretion systems. Nat. Rev. Microbiol. *7*, 703–714.

Fronzes, R., Schafer, E., Wang, L., Saibil, H.R., Orlova, E.V., and Waksman, G. (2009b). Structure of a type IV secretion system core complex. Science *323*, 266–268.

Fujii, T., Cheung, M., Blanco, A., Kato, T., Blocker, A.J., and Namba, K. (2012). Structure of a type III secretion needle at 7-A resolution provides insights into its assembly and signaling mechanisms. Proc. Natl. Acad. Sci. U.S.A..

Fullner, K.J., Stephens, K.M., and Nester, E.W. (1994). An essential virulence protein of *Agrobacterium tumefaciens*, VirB4, requires an intact mononucleotide binding domain to function in transfer of T-DNA. Mol. Gen. Genet. *245*, 704–715.

Fullner, K.J., Lara, J.C., and Nester, E.W. (1996). Pilus assembly by Agrobacterium T-DNA transfer genes. Science *273*, 1107–1109.

Galán, J.E. (2007). SnapShot: effector proteins of type III secretion systems. Cell *130*, 192.

Galán, J.E. (2008). Energizing type III secretion machines: what is the fuel? Nat. Struct. Mol. Biol. 127–128.

Galán, J.E., and Wolf-Watz, H. (2006). Protein delivery into eukaryotic cells by type III secretion machines. Nature *444*, 567–573.

Galen, J.E., Pasetti, M.F., Tennant, S., Ruiz-Olvera, P., Sztein, M.B., and Levine, M.M. (2009). *Salmonella enterica* serovar Typhi live vector vaccines finally come of age. Immunol. Cell Biol. *87*, 400–412.

Galkin, V.E., Schmied, W.H., Schraidt, O., Marlovits, T.C., and Egelman, E.H. (2010). The structure of the Salmonella typhimurium type III secretion system needle shows divergence from the flagellar system. J. Mol. Biol. *396*, 1392–1397.

Garrity-Ryan, L.K., Kim, O.K., Balada-Llasat, J.-M., Bartlett, V.J., Verma, A.K., Fisher, M.L., Castillo, C., Songsungthong, W., Tanaka, S.K., Levy, S.B., *et al.* (2010). Small molecule inhibitors of LcrF, a *Yersinia pseudotuberculosis* transcription factor, attenuate virulence and limit infection in a murine pneumonia model. Infect. Immun. *78*, 4683–4690.

Gilmour, M.W., Lawley, T.D., Rooker, M.M., Newnham, P.J., and Taylor, D.E. (2001). Cellular location and temperature-dependent assembly of IncHI1 plasmid R27-encoded TrhC-associated conjugative transfer protein complexes. Mol. Microbiol. *42*, 705–715.

Gold, V.A., Duong, F., and Collinson, I. (2007). Structure and function of the bacterial Sec translocon. Mol. Membr. Biol. *24*, 387–394.

Gomis-Ruth, F.X., and Coll, M. (2006). Cut and move: protein machinery for DNA processing in bacterial conjugation. Curr. Opin. Struct. Biol. *16*, 744–752.

Gomis-Ruth, F.X., Moncalian, G., Perez-Luque, R., Gonzalez, A., Cabezon, E., de la Cruz, F., and Coll, M. (2001). The bacterial conjugation protein TrwB resembles ring helicases and F1-ATPase. Nature *409*, 637–641.

Gomis-Ruth, F.X., Moncalian, G., de la Cruz, F., and Coll, M. (2002). Conjugative plasmid protein TrwB, an integral membrane type IV secretion system coupling protein. Detailed structural features and mapping of the active site cleft. J. Biol. Chem. *277*, 7556–7566.

Grohmann, E., Muth, G., and Espinosa, M. (2003). Conjugative plasmid transfer in gram-positive bacteria. Microbiol. Mol. Biol. Rev. *67*, 277–301.

Haase, J., and Lanka, E. (1997). A specific protease encoded by the conjugative DNA transfer systems of IncP and Ti plasmids is essential for pilus synthesis. J. Bacteriol. *179*, 5728–5735.

Haase, J., Lurz, R., Grahn, A.M., Bamford, D.H., and Lanka, E. (1995). Bacterial conjugation mediated by plasmid RP4: RSF1010 mobilization, donor-specific phage propagation, and pilus production require the same Tra2 core components of a proposed DNA transport complex. J. Bacteriol. *177*, 4779–4791.

Håkansson, S., Schesser, K., Persson, C., Galyov, E.E., Rosqvist, R., Homble, F., and Wolf-Watz, H. (1996). The YopB protein of *Yersinia pseudotuberculosis* is essential for the translocation of Yop effector proteins across the target cell plasma membrane and displays a contact-dependent membrane disrupting activity. EMBO J. *15*, 5812–5823.

Hales, L.M., and Shuman, H.A. (1999). Legionella pneumophila contains a type II general secretion pathway required for growth in amoebae as well as for secretion of the Msp protease. Infect. Immun. *67*, 3662–3666.

Haraga, A., Ohlson, M.B., and Miller, S.I. (2008). Salmonellae interplay with host cells. Nat. Rev. Microbiol. *6*, 53–66.

Hare, S., Bayliss, R., Baron, C., and Waksman, G. (2006). A large domain swap in the VirB11 ATPase of Brucella suis leaves the hexameric assembly intact. J. Mol. Biol. *360*, 56–66.

Hayward, R.D., and Koronakis, V. (1999). Direct nucleation and bundling of actin by the SipC protein of invasive Salmonella. EMBO J. *18*, 4926–4934.

Hayward, R.D., McGhie, E.J., and Koronakis, V. (2000). Membrane fusion activity of purified SipB, a Salmonella surface protein essential for mammalian cell invasion. Mol. Microbiol. *37*, 727–739.

Henderson, I.R., and Nataro, J.P. (2001). Virulence functions of autotransporter proteins. Infect. Immun. *69*, 1231–1243.

Hodgkinson, J.L., Horsley, A., Stabat, D., Simon, M., Johnson, S., da Fonseca, P.C.A., Morris, E.P., Wall, J.S., Lea, S.M., and Blocker, A.J. (2009). Three-dimensional reconstruction of the Shigella T3SS transmembrane regions reveals 12-fold symmetry and novel features throughout. Nat. Struct. Mol. Biol. *16*, 477–485.

Hoppner, C., Carle, A., Sivanesan, D., Hoeppner, S., and Baron, C. (2005). The putative lytic transglycosylase VirB1 from Brucella suis interacts with the type IV secretion system core components VirB8, VirB9 and VirB11. Microbiology *151*, 3469–3482.

Hormaeche, I., Alkorta, I., Moro, F., Valpuesta, J.M., Goni, F.M., and De La Cruz, F. (2002). Purification and properties of TrwB, a hexameric, ATP-binding integral membrane protein essential for R388 plasmid conjugation. J. Biol. Chem. *277*, 46456–46462.

Hueck, C.J. (1998). Type III protein secretion systems in bacterial pathogens of animals and plants. Microbiol. Mol. Biol. Rev. *62*, 379–433.

Iwobi, A., Heesemann, J., Garcia, E., Igwe, E., Noelting, C., and Rakin, A. (2003). Novel virulence-associated type II secretion system unique to high-pathogenicity Yersinia enterocolitica. Infect. Immun. *71*, 1872–1879.

Izoré, T., Job, V., and Dessen, A. (2011). Biogenesis, regulation, and targeting of the type III secretion system. Structure *19*, 603–612.

Jakubowski, S.J., Krishnamoorthy, V., and Christie, P.J. (2003). *Agrobacterium tumefaciens* VirB6 protein participates in formation of VirB7 and VirB9 complexes required for type IV secretion. J. Bacteriol. *185*, 2867–2878.

Jakubowski, S.J., Krishnamoorthy, V., Cascales, E., and Christie, P.J. (2004). *Agrobacterium tumefaciens* VirB6 domains direct the ordered export of a DNA substrate through a type IV secretion System. J. Mol. Biol. *341*, 961–977.

Jakubowski, S.J., Kerr, J.E., Garza, I., Krishnamoorthy, V., Bayliss, R., Waksman, G., and Christie, P.J. (2009). Agrobacterium VirB10 domain requirements for type IV secretion and T pilus biogenesis. Mol. Microbiol. *71*, 779–794.

Johnson, S., Roversi, P., Espina, M., Olive, A., Deane, J.E., Birket, S., Field, T., Picking, W.D., Blocker, A.J., Galyov, E.E., *et al.* (2007). Self-chaperoning of the type III secretion system needle tip proteins IpaD and BipD. J. Biol. Chem. *282*, 4035–4044.

Jones, C.J., and Macnab, R.M. (1990). Flagellar assembly in Salmonella typhimurium: analysis with temperature-sensitive mutants. J. Bacteriol. *172*, 1327–1339.

Journet, L., Agrain, C., Broz, P., and Cornelis, G.R. (2003). The needle length of bacterial injectisomes is determined by a molecular ruler. Science *302*, 1757–1760.

Judd, P.K., Kumar, R.B., and Das, A. (2005a). Spatial location and requirements for the assembly of the *Agrobacterium tumefaciens* type IV secretion apparatus. Proc. Natl. Acad. Sci. U.S.A. *102*, 11498–11503.

Judd, P.K., Kumar, R.B., and Das, A. (2005b). The type IV secretion apparatus protein VirB6 of *Agrobacterium tumefaciens* localizes to a cell pole. Mol. Microbiol. *55*, 115–124.

Kalkum, M., Eisenbrandt, R., Lurz, R., and Lanka, E. (2002). Tying rings for sex. Trends Microbiol. *10*, 382–387.

Karnholz, A., Hoefler, C., Odenbreit, S., Fischer, W., Hofreuter, D., and Haas, R. (2006). Functional and topological characterization of novel components of the comB DNA transformation competence system in *Helicobacter pylori*. J. Bacteriol. *188*, 882–893.

Kauppi, A.M., Nordfelth, R., Uvell, H., Wolf-Watz, H., and Elofsson, M. (2003). Targeting bacterial virulence: inhibitors of type III secretion in Yersinia. Chem. Biol. *10*, 241–249.

Koronakis, V., Hughes, C., and Koronakis, E. (1991). Energetically distinct early and late stages of HlyB/HlyD-dependent secretion across both *Escherichia coli* membranes. EMBO J. *10*, 3263–3272.

Korotkov, K.V., Pardon, E., Steyaert, J., and Hol, W.G.J. (2009). Crystal structure of the N-terminal domain of the secretin GspD from ETEC determined with the assistance of a nanobody. Structure *17*, 255–265.

Korotkov, K.V., Gonen, T., and Hol, W.G.J. (2011). Secretins: dynamic channels for protein transport across membranes. Trends Biochem. Sci. *36*, 433–443.

Kotton, C.N., Lankowski, A.J., Scott, N., Sisul, D., Chen, L.-M., Raschke, K., Borders, G., Boaz, M., Spentzou, A., Galán, J.E., *et al.* (2006). Safety and immunogenicity of attenuated *Salmonella enterica* serovar Typhimurium delivering an HIV-1 Gag antigen via the Salmonella Type III secretion system. Vaccine *24*, 6216–6224.

Krall, L., Wiedemann, U., Unsin, G., Weiss, S., Domke, N., and Baron, C. (2002). Detergent extraction identifies different VirB protein subassemblies of the type IV secretion machinery in the membranes of *Agrobacterium tumefaciens*. Proc. Natl. Acad. Sci. U.S.A. *99*, 11405–11410.

Krause, S., Barcena, M., Pansegrau, W., Lurz, R., Carazo, J.M., and Lanka, E. (2000a). Sequence-related protein export NTPases encoded by the conjugative transfer region of RP4 and by the cag pathogenicity island of *Helicobacter pylori* share similar hexameric ring structures. Proc. Natl. Acad. Sci. U.S.A. *97*, 3067–3072.

Krause, S., Pansegrau, W., Lurz, R., de la Cruz, F., and Lanka, E. (2000b). Enzymology of type IV macromolecule secretion systems: the conjugative transfer regions of plasmids RP4 and R388 and the cag pathogenicity island of *Helicobacter pylori* encode structurally and functionally related nucleoside triphosphate hydrolases. J. Bacteriol. *182*, 2761–2770.

Kubori, T., Matsushima, Y., Nakamura, D., Uralil, J., Lara-Tejero, M., Sukhan, A., Galan, J.E., and Aizawa, S.I. (1998). Supramolecular structure of the Salmonella typhimurium type III protein secretion system. Science *280*, 602–605.

Kubori, T., Sukhan, A., Aizawa, S.I., and Galan, J.E. (2000). Molecular characterization and assembly of the needle complex of the Salmonella typhimurium type III protein secretion system. Proc. Natl. Acad. Sci. U.S.A. *97*, 10225–10230.

Kumar, R.B., and Das, A. (2001). Functional analysis of the *Agrobacterium tumefaciens* T-DNA transport pore protein VirB8. J. Bacteriol. *183*, 3636–3641.

Kumar, R.B., and Das, A. (2002). Polar location and functional domains of the *Agrobacterium tumefaciens* DNA transfer protein VirD4. Mol. Microbiol. *43*, 1523–1532.

Kumar, R.B., Xie, Y.H., and Das, A. (2000). Subcellular localization of the *Agrobacterium tumefaciens* T-DNA transport pore proteins: VirB8 is essential for the assembly of the transport pore. Mol. Microbiol. *36*, 608–617.

Kwok, T., Zabler, D., Urman, S., Rohde, M., Hartig, R., Wessler, S., Misselwitz, R., Berger, J., Sewald, N., Konig, W., *et al.* (2007). Helicobacter exploits integrin for type IV secretion and kinase activation. Nature *449*, 862–866.

Lai, E.M., and Kado, C.I. (1998). Processed VirB2 is the major subunit of the promiscuous pilus of *Agrobacterium tumefaciens*. J. Bacteriol. *180*, 2711–2717.

Lai, E.M., Chesnokova, O., Banta, L.M., and Kado, C.I. (2000). Genetic and environmental factors affecting T-pilin export and T-pilus biogenesis in relation to flagellation of *Agrobacterium tumefaciens*. J. Bacteriol. *182*, 3705–3716.

Lara-Tejero, M., and Galán, J.E. (2009). *Salmonella enterica* serovar typhimurium pathogenicity island 1-encoded type III secretion system translocases mediate intimate attachment to nonphagocytic cells. Infect. Immun. *77*, 2635–2642.

Lara-Tejero, M., Kato, J., Wagner, S., Liu, X., and Galán, J.E. (2011). A sorting platform determines the order of protein secretion in bacterial type III systems. Science *331*, 1188–1191.

Lawley, T.D., Klimke, W.A., Gubbins, M.J., and Frost, L.S. (2003). F factor conjugation is a true type IV secretion system. FEMS Microbiol. Lett. *224*, 1–15.

Lessl, M., and Lanka, E. (1994). Common mechanisms in bacterial conjugation and Ti-mediated T-DNA transfer to plant cells. Cell *77*, 321–324.

Lessl, M., Balzer, D., Pansegrau, W., and Lanka, E. (1992). Sequence similarities between the RP4 Tra2 and the Ti VirB region strongly support the conjugation model for T-DNA transfer. J. Biol. Chem. *267*, 20471–20480.

Linderoth, N.A., Simon, M.N., and Russel, M. (1997). The filamentous phage pIV multimer visualized by scanning transmission electron microscopy. Science *278*, 1635–1638.

Llosa, M., Zunzunegui, S., and de la Cruz, F. (2003). Conjugative coupling proteins interact with cognate and heterologous VirB10-like proteins while exhibiting specificity for cognate relaxosomes. Proc. Natl. Acad. Sci. U.S.A. *100,* 10465–10470.

Loquet, A., Sgourakis, N.G., Gupta, R., Giller, K., Riedel, D., Goosmann, C., Griesinger, C., Kolbe, M., Baker, D., Becker, S., *et al.* (2012). Atomic model of the type III secretion system needle. Nature *486,* 276–279.

Lu, J., Manchak, J., Klimke, W., Davidson, C., Firth, N., Skurray, R.A., and Frost, L.S. (2002). Analysis and characterization of the IncFV plasmid pED208 transfer region. Plasmid *48,* 24–37.

Lunelli, M., Hurwitz, R., Lambers, J., and Kolbe, M. (2011). Crystal structure of PrgI-SipD: insight into a secretion competent state of the type three secretion system needle tip and its interaction with host ligands. PLoS Pathog. *7,* e1002163.

Majdalani, N., and Ippen-Ihler, K. (1996). Membrane insertion of the F-pilin subunit is Sec independent but requires leader peptidase B and the proton motive force. J. Bacteriol. *178,* 3742–3747.

Malek, J.A., Wierzbowski, J.M., Tao, W., Bosak, S.A., Saranga, D.J., Doucette-Stamm, L., Smith, D.R., McEwan, P.J., and McKernan, K.J. (2004). Protein interaction mapping on a functional shotgun sequence of *Rickettsia sibirica*. Nucleic Acids Res. *32,* 1059–1064.

Marlovits, T.C., Kubori, T., Sukhan, A., Thomas, D.R., Galán, J.E., and Unger, V.M. (2004). Structural insights into the assembly of the type III secretion needle complex. Science *306,* 1040–1042.

Marlovits, T.C., Kubori, T., Lara-Tejero, M., Thomas, D., Unger, V.M., and Galán, J.E. (2006). Assembly of the inner rod determines needle length in the type III secretion injectisome. Nature *441,* 637–640.

Massey, T.H., Mercogliano, C.P., Yates, J., Sherratt, D.J., and Lowe, J. (2006). Double-stranded DNA translocation: structure and mechanism of hexameric FtsK. Mol. Cell *23,* 457–469.

Matson, J.S., and Nilles, M.L. (2001). LcrG–LcrV interaction is required for control of Yops secretion in *Yersinia pestis*. J. Bacteriol. *183,* 5082–5091.

Mazar, J., and Cotter, P.A. (2007). New insight into the molecular mechanisms of two-partner secretion. Trends Microbiol. *15,* 508–515.

McDermott, J.E., Corrigan, A., Peterson, E., Oehmen, C., Niemann, G., Cambronne, E.D., Sharp, D., Adkins, J.N., Samudrala, R., and Heffron, F. (2011). Computational prediction of type III and IV secreted effectors in gram-negative bacteria. Infect. Immun. *79,* 23–32.

McDowell, M.A., Johnson, S., Deane, J.E., Cheung, M., Roehrich, A.D., Blocker, A.J., McDonnell, J.M., and Lea, S.M. (2011). Structural and functional studies on the N-terminal domain of the Shigella type III secretion protein MxiG. J. Biol. Chem. *286,* 30606–30614.

Ménard, R., Sansonetti, P., Parsot, C., and Vasselon, T. (1994). Extracellular association and cytoplasmic partitioning of the IpaB and IpaC invasins of *S. flexneri*. Cell *79,* 515–525.

Middleton, R., Sjolander, K., Krishnamurthy, N., Foley, J., and Zambryski, P. (2005). Predicted hexameric structure of the Agrobacterium VirB4 C terminus suggests VirB4 acts as a docking site during type IV secretion. Proc. Natl. Acad. Sci. U.S.A. *102,* 1685–1690.

Minamino, T., and Namba, K. (2008). Distinct roles of the FliI ATPase and proton motive force in bacterial flagellar protein export. Nature *451,* 485–488.

Moncalian, G., Cabezon, E., Alkorta, I., Valle, M., Moro, F., Valpuesta, J.M., Goni, F.M., and de La Cruz, F. (1999). Characterization of ATP and DNA binding activities of TrwB, the coupling protein essential in plasmid R388 conjugation. J. Biol. Chem. *274,* 36117–36124.

Montagner, C., Arquint, C., and Cornelis, G.R. (2011). Translocators YopB and YopD from *Yersinia* form a multimeric integral membrane complex in eukaryotic cell membranes. J. Bacteriol..

Moore, D., Hamilton, C.M., Maneewannakul, K., Mintz, Y., Frost, L.S., and Ippen-Ihler, K. (1993). The *Escherichia coli* K-12 F plasmid gene traX is required for acetylation of F pilin. J. Bacteriol. *175,* 1375–1383.

Moreno, M., Kramer, M.G., Yim, L., and Chabalgoity, J.A. (2010). Salmonella as live trojan horse for vaccine development and cancer gene therapy. Curr. Gene Ther. *10,* 56–76.

Mota, L.J., Journet, L., Sorg, I., Agrain, C., and Cornelis, G.R. (2005). Bacterial injectisomes: needle length does matter. Science *307,* 1278.

Nagai, H., Cambronne, E.D., Kagan, J.C., Amor, J.C., Kahn, R.A., and Roy, C.R. (2005). A C-terminal translocation signal required for Dot/Icm-dependent delivery of the Legionella RalF protein to host cells. Proc. Natl. Acad. Sci. U.S.A. *102,* 826–831.

Neyt, C., and Cornelis, G.R. (1999). Insertion of a Yop translocation pore into the macrophage plasma membrane by Yersinia enterocolitica: requirement for translocators YopB and YopD, but not LcrG. Mol. Microbiol. 33, 971–981.

Nouwen, N., Ranson, N., Saibil, H., Wolpensinger, B., Engel, A., Ghazi, A., and Pugsley, A.P. (1999). Secretin PulD: association with pilot PulS, structure, and ion-conducting channel formation. Proc. Natl. Acad. Sci. U.S.A. *96*, 8173–8177.

O'Callaghan, D., Cazevieille, C., Allardet-Servent, A., Boschiroli, M.L., Bourg, G., Foulongne, V., Frutos, P., Kulakov, Y., and Ramuz, M. (1999). A homologue of the *Agrobacterium tumefaciens* VirB and *Bordetella pertussis* Ptl type IV secretion systems is essential for intracellular survival of *Brucella suis*. Mol. Microbiol. *33*, 1210–1220.

Omori, K., and Idei, A. (2003). Gram-negative bacterial ATP-binding cassette protein exporter family and diverse secretory proteins. J. Biosci. Bioeng. *95*, 1–12.

Pansegrau, W., and Lanka, E. (1996). Mechanisms of initiation and termination reactions in conjugative DNA processing. Independence of tight substrate binding and catalytic activity of relaxase (TraI) of IncPalpha plasmid RP4. J. Biol. Chem. *271*, 13068–13076.

Pattis, I., Weiss, E., Laugks, R., Haas, R., and Fischer, W. (2007). The *Helicobacter pylori* CagF protein is a type IV secretion chaperone-like molecule that binds close to the C-terminal secretion signal of the CagA effector protein. Microbiology *153*, 2896–2909.

Peabody, C.R., Chung, Y.J., Yen, M.R., Vidal-Ingigliardi, D., Pugsley, A.P., and Saier, M.H. Jr. (2003). Type II protein secretion and its relationship to bacterial type IV pili and archaeal flagella. Microbiology *149*, 3051–3072.

Planet, P.J., Kachlany, S.C., DeSalle, R., and Figurski, D.H. (2001). Phylogeny of genes for secretion NTPases: identification of the widespread tadA subfamily and development of a diagnostic key for gene classification. Proc. Natl. Acad. Sci. U.S.A. *98*, 2503–2508.

Poyraz, O., Schmidt, H., Seidel, K., Delissen, F., Ader, C., Tenenboim, H., Goosmann, C., Laube, B., Thünemann, A.F., Zychlinsky, A., *et al.* (2010). Protein refolding is required for assembly of the type three secretion needle. Nat. Struct. Mol. Biol. *17*, 788–792.

Pukatzki, S., McAuley, S.B., and Miyata, S.T. (2009). The type VI secretion system: translocation of effectors and effector-domains. Curr. Opin. Microbiol. *12*, 11–17.

Quenee, L.E., and Schneewind, O. (2009). Plague vaccines and the molecular basis of immunity against *Yersinia pestis*. Hum. Vaccin. *5*, 817–823.

Rabel, C., Grahn, A.M., Lurz, R., and Lanka, E. (2003). The VirB4 family of proposed traffic nucleoside triphosphatases: common motifs in plasmid RP4 TrbE are essential for conjugation and phage adsorption. J. Bacteriol. *185*, 1045–1058.

Rain, J.C., Selig, L., De Reuse, H., Battaglia, V., Reverdy, C., Simon, S., Lenzen, G., Petel, F., Wojcik, J., Schachter, V., *et al.* (2001). The protein–protein interaction map of *Helicobacter pylori*. Nature *409*, 211–215.

Rambow-Larsen, A.A., and Weiss, A.A. (2002). The PtlE protein of *Bordetella pertussis* has peptidoglycanase activity required for Ptl-mediated pertussis toxin secretion. J. Bacteriol. *184*, 2863–2869.

Rashkova, S., Spudich, G.M., and Christie, P.J. (1997). Characterization of membrane and protein interaction determinants of the *Agrobacterium tumefaciens* VirB11 ATPase. J. Bacteriol. *179*, 583–591.

Rivas, S., Bolland, S., Cabezon, E., Goni, F.M., and de la Cruz, F. (1997). TrwD, a protein encoded by the IncW plasmid R388, displays an ATP hydrolase activity essential for bacterial conjugation. J. Biol. Chem. *272*, 25583–25590.

Sagulenko, E., Sagulenko, V., Chen, J., and Christie, P.J. (2001a). Role of *Agrobacterium* VirB11 ATPase in T-pilus assembly and substrate selection. J. Bacteriol. *183*, 5813–5825.

Sagulenko, V., Sagulenko, E., Jakubowski, S., Spudich, E., and Christie, P.J. (2001b). VirB7 lipoprotein is exocellular and associates with the *Agrobacterium tumefaciens* T pilus. J. Bacteriol. *183*, 3642–3651.

Samudrala, R., Heffron, F., and McDermott, J.E. (2009). Accurate prediction of secreted substrates and identification of a conserved putative secretion signal for type III secretion systems. PLoS Pathog. *5*, e1000375.

Sani, M., Botteaux, A., Parsot, C., Sansonetti, P., Boekema, E.J., and Allaoui, A. (2007). IpaD is localized at the tip of the *Shigella flexneri* type III secretion apparatus. Biochim. Biophys. Acta *1770*, 307–311.

Sanowar, S., Singh, P., Pfuetzner, R.A., André, I., Zheng, H., Spreter, T., Strynadka, N.C.J., Gonen, T., Baker, D., Goodlett, D.R., *et al.* (2010). Interactions of the transmembrane polymeric rings of the *Salmonella enterica* serovar Typhimurium type III secretion system. MBio. *1*.

Savvides, S.N., Yeo, H.J., Beck, M.R., Blaesing, F., Lurz, R., Lanka, E., Buhrdorf, R., Fischer, W., Haas, R., and Waksman, G. (2003). VirB11 ATPases are dynamic hexameric assemblies: new insights into bacterial type IV secretion. EMBO J. *22*, 1969–1980.

Schandel, K.A., Muller, M.M., and Webster, R.E. (1992). Localization of TraC, a protein involved in assembly of the F conjugative pilus. J. Bacteriol. *174*, 3800–3806.

Schlumberger, M.C., and Hardt, W.-D. (2006). Salmonella type III secretion effectors: pulling the host cell's strings. Curr. Opin. Microbiol. *9*, 46–54.

Schlumberger, M.C., Müller, A.J., Ehrbar, K., Winnen, B., Duss, I., Stecher, B., and Hardt, W.-D. (2005). Real-time imaging of type III secretion: Salmonella SipA injection into host cells. Proc. Natl. Acad. Sci. U.S.A. *102*, 12548–12553.

Schmidt, H., and Hensel, M. (2004). Pathogenicity islands in bacterial pathogenesis. Clin. Microbiol. Rev. *17*, 14–56.

Schmidt-Eisenlohr, H., Domke, N., and Baron, C. (1999). TraC of IncN plasmid pKM101 associates with membranes and extracellular high-molecular-weight structures in *Escherichia coli*. J. Bacteriol. *181*, 5563–5571.

Schraidt, O., Lefebre, M.D., Brunner, M.J., Schmied, W.H., Schmidt, A., Radics, J., Mechtler, K., Galán, J.E., and Marlovits, T.C. (2010). Topology and organization of the Salmonella typhimurium type III secretion needle complex components. PLoS Pathog. *6*, e1000824.

Schraidt, O., and Marlovits, T.C. (2011). Three-dimensional model of Salmonella's needle complex at subnanometer resolution. Science *331*, 1192–1195.

Schroder, G., and Lanka, E. (2003). TraG-like proteins of type IV secretion systems: functional dissection of the multiple activities of TraG (RP4) and TrwB (R388). J. Bacteriol. *185*, 4371–4381.

Schroder, G., Krause, S., Zechner, E.L., Traxler, B., Yeo, H.J., Lurz, R., Waksman, G., and Lanka, E. (2002). TraG-like proteins of DNA transfer systems and of the *Helicobacter pylori* type IV secretion system: inner membrane gate for exported substrates? J. Bacteriol. *184*, 2767–2779.

Schulein, R., Guye, P., Rhomberg, T.A., Schmid, M.C., Schroder, G., Vergunst, A.C., Carena, I., and Dehio, C. (2005). A bipartite signal mediates the transfer of type IV secretion substrates of Bartonella henselae into human cells. Proc. Natl. Acad. Sci. U.S.A. *102*, 856–861.

Segal, G., Russo, J.J., and Shuman, H.A. (1999). Relationships between a new type IV secretion system and the icm/dot virulence system of Legionella pneumophila. Mol. Microbiol. *34*, 799–809.

Sekiya, K., Ohishi, M., Ogino, T., Tamano, K., Sasakawa, C., and Abe, A. (2001). Supermolecular structure of the enteropathogenic *Escherichia coli* type III secretion system and its direct interaction with the EspA-sheath-like Structure Proc. Natl. Acad. Sci. U.S.A. *98*, 11638–11643.

Seubert, A., Hiestand, R., de la Cruz, F., and Dehio, C. (2003). A bacterial conjugation machinery recruited for pathogenesis. Mol. Microbiol. *49*, 1253–1266.

Shirasu, K., Koukolikova-Nicola, Z., Hohn, B., and Kado, C.I. (1994). An inner-membrane-associated virulence protein essential for T-DNA transfer from *Agrobacterium tumefaciens* to plants exhibits ATPase activity and similarities to conjugative transfer genes. Mol. Microbiol. *11*, 581–588.

Shu, A.C., Wu, C.C., Chen, Y.Y., Peng, H.L., Chang, H.Y., and Yew, T.R. (2008). Evidence of DNA transfer through F-pilus channels during *Escherichia coli* conjugation. Langmuir *24*, 6796–6802.

Spreter, T., Yip, C.K., Sanowar, S., André, I., Kimbrough, T.G., Vuckovic, M., Pfuetzner, R.A., Deng, W., Yu, A.C., Finlay, B.B., *et al.* (2009). A conserved structural motif mediates formation of the periplasmic rings in the type III secretion system. Nat. Struct. Mol. Biol. *16*, 468–476.

Stebbins, C.E., and Galan, J.E. (2001a). Maintenance of an unfolded polypeptide by a cognate chaperone in bacterial type III secretion. Nature *414*, 77–81.

Stebbins, C.E., and Galan, J.E. (2001b). Structural mimicry in bacterial virulence. Nature *412*, 701–705.

Stebbins, C.E., and Galan, J.E. (2003). Priming virulence factors for delivery into the host. Nat. Rev. Mol. Cell Biol. *4*, 738–743.

Stensrud, K.F., Adam, P.R., La Mar, C.D., Olive, A.J., Lushington, G.H., Sudharsan, R., Shelton, N.L., Givens, R.S., Picking, W.L., and Picking, W.D. (2008). Deoxycholate interacts with IpaD of *Shigella flexneri* in inducing the recruitment of IpaB to the type III secretion apparatus needle tip. J. Biol. Chem. *283*, 18646–18654.

Sukhan, A., Kubori, T., Wilson, J., and Galan, J.E. (2001). Genetic analysis of assembly of the *Salmonella enterica* serovar Typhimurium type III secretion-associated needle complex. J. Bacteriol. *183*, 1159–1167.

Tamano, K., Aizawa, S., Katayama, E., Nonaka, T., Imajoh-Ohmi, S., Kuwae, A., Nagai, S., and Sasakawa, C. (2000). Supramolecular structure of the Shigella type III secretion machinery: the needle part is changeable in length and essential for delivery of effectors. EMBO J. *19*, 3876–3887.

Tanaka, J., Suzuki, T., Mimuro, H., and Sasakawa, C. (2003). Structural definition on the surface of *Helicobacter pylori* type IV secretion apparatus. Cell Microbiol. *5*, 395–404.

Tato, I., Zunzunegui, S., de la Cruz, F., and Cabezon, E. (2005). TrwB, the coupling protein involved in DNA transport during bacterial conjugation, is a DNA-dependent ATPase. Proc. Natl. Acad. Sci. U.S.A. *102*, 8156–8161.

Tato, I., Matilla, I., Arechaga, I., Zunzunegui, S., de la Cruz, F., and Cabezon, E. (2007). The ATPase activity of the DNA transporter TrwB is modulated by protein TrwA: implications for a common assembly mechanism of DNA translocating motors. J. Biol. Chem. *282*, 25569–25576.

Terradot, L., Durnell, N., Li, M., Ory, J., Labigne, A., Legrain, P., Colland, F., and Waksman, G. (2004). Biochemical characterization of protein complexes from the *Helicobacter pylori* protein interaction map: strategies for complex formation and evidence for novel interactions within type IV secretion systems. Mol. Cell Proteomics *3*, 809–819.

Terradot, L., Bayliss, R., Oomen, C., Leonard, G.A., Baron, C., and Waksman, G. (2005). Structures of two core subunits of the bacterial type IV secretion system, VirB8 from Brucella suis and ComB10 from *Helicobacter pylori*. Proc. Natl. Acad. Sci. U.S.A. *102*, 4596–4601.

Valverde, R., Edwards, L., and Regan, L. (2008). Structure and function of KH domains. FEBS J. *275*, 2712–2726.

Vergunst, A.C., Schrammeijer, B., den Dulk-Ras, A., de Vlaam, C.M., Regensburg-Tuink, T.J., and Hooykaas, P.J. (2000). VirB/D4-dependent protein translocation from *Agrobacterium* into plant cells. Science *290*, 979–982.

Vergunst, A.C., van Lier, M.C., den Dulk-Ras, A., Stuve, T.A., Ouwehand, A., and Hooykaas, P.J. (2005). Positive charge is an important feature of the C-terminal transport signal of the VirB/D4-translocated proteins of *Agrobacterium*. Proc. Natl. Acad. Sci. U.S.A. *102*, 832–837.

Vincent, C.D., Friedman, J.R., Jeong, K.C., Buford, E.C., Miller, J.L., and Vogel, J.P. (2006). Identification of the core transmembrane complex of the Legionella Dot/Icm type IV secretion system. Mol. Microbiol. *62*, 1278–1291.

Voulhoux, R., Ball, G., Ize, B., Vasil, M.L., Lazdunski, A., Wu, L.F., and Filloux, A. (2001). Involvement of the twin-arginine translocation system in protein secretion via the type II pathway. EMBO J. *20*, 6735–6741.

Wagner, S., Königsmaier, L., Lara-Tejero, M., Lefebre, M., Marlovits, T.C., and Galán, J.E. (2010a). Organization and coordinated assembly of the type III secretion export apparatus. Proc. Natl. Acad. Sci. U.S.A. *107*, 17745–17750.

Wagner, S., Stenta, M., Metzger, L.C., Dal Peraro, M., and Cornelis, G.R. (2010b). Length control of the injectisome needle requires only one molecule of Yop secretion protein P (YscP). Proc. Natl. Acad. Sci. U.S.A. *107*, 13860–13865.

Wallden, K., Williams, R., Yan, J., Lian, P.W., Wang, L., Thalassinos, K., Orlova, E.V., and Waksman, G. (2012). Structure of the VirB4 ATPase, alone and bound to the core complex of a type IV secretion system. Proc. Natl. Acad. Sci. U.S.A. *109*, 11348–11353.

Wang, D., Zetterström, C.E., Gabrielsen, M., Beckham, K.S.H., Tree, J.J., Macdonald, S.E., Byron, O., Mitchell, T.J., Gally, D.L., Herzyk, P., *et al.* (2011). Identification of bacterial target proteins for the salicylidene acylhydrazide class of virulence-blocking compounds. J. Biol. Chem. *286*, 29922–29931.

Wang, Y., Nordhues, B., Zhong, D., and De Guzman, R.N. (2010). NMR characterization of the interaction of the Salmonella type III secretion system protein SipD and bile salts. Biochemistry

Wang, Y., Ouellette, A.N., Egan, C.W., Rathinavelan, T., Im, W., and De Guzman, R.N. (2007). Differences in the electrostatic surfaces of the type III secretion needle proteins PrgI, BsaL, and MxiH. J. Mol. Biol. *371*, 1304–1314.

Wang, Y.A., Yu, X., Silverman, P.M., Harris, R.L., and Egelman, E.H. (2009). The structure of F-pili. J. Mol. Biol. *385*, 22–29.

Ward, D.V., Draper, O., Zupan, J.R., and Zambryski, P.C. (2002). Peptide linkage mapping of the *Agrobacterium tumefaciens* vir-encoded type IV secretion system reveals protein subassemblies. Proc. Natl. Acad. Sci. U.S.A. *99*, 11493–11500.

Yeo, H.J., Yuan, Q., Beck, M.R., Baron, C., and Waksman, G. (2003). Structural and functional characterization of the VirB5 protein from the type IV secretion system encoded by the conjugative plasmid pKM101. Proc. Natl. Acad. Sci. U.S.A. *100*, 15947–15952.

Yip, C.K., Kimbrough, T.G., Felise, H.B., Vuckovic, M., Thomas, N.A., Pfuetzner, R.A., Frey, E.A., Finlay, B.B., Miller, S.I., and Strynadka, N.C.J. (2005). Structural characterization of the molecular platform for type III secretion system assembly. Nature *435*, 702–707.

Yuan, Q., Carle, A., Gao, C., Sivanesan, D., Aly, K.A., Hoppner, C., Krall, L., Domke, N., and Baron, C. (2005). Identification of the VirB4-VirB8-VirB5-VirB2 pilus assembly sequence of type IV secretion systems. J. Biol. Chem. *280*, 26349–26359.
Zarivach, R., Vuckovic, M., Deng, W., Finlay, B.B., and Strynadka, N.C.J. (2007). Structural analysis of a prototypical ATPase from the type III secretion system. Nat. Struct. Mol. Biol. *14*, 131–137.
Zupan, J., Hackworth, C.A., Aguilar, J., Ward, D., and Zambryski, P. (2007). VirB1* promotes T-pilus formation in the vir-Type IV secretion system of *Agrobacterium tumefaciens*. J. Bacteriol. *189*, 6551–6563.

Part IV

Signal Transduction Across Bacterial Membranes

Signalling Mechanisms in Prokaryotes

12

Mariano Martinez, Pedro M. Alzari and Gwénaëlle André-Leroux

Abstract

Adaptation to an environmental stress is essential for cell survival in all organisms. In regard to this, the plasmatic membrane plays a fundamental role as it is the first barrier that separates the hostile environment from the cytoplasmic content, where all living reactions occur. Most of the machinery dedicated to sense upstream the external changes and to transduce the signals through the cytoplasmic membrane – upon which signals trigger a committed response downstream – is located at this special interface. This chapter is dedicated to describing the two major signalling systems found in prokaryotes: the so-called two-component systems and the eukaryotic-like serine/threonine/tyrosine protein kinases. We will focus on the basic aspects of these sensing machineries, their function, catalytic mechanism and regulation, and will extend, when possible, to more sophisticated features, such as the structural basis of the signal sensing and transduction *per se*, based on the most recent investigations.

Introduction

Normal cellular physiology requires metabolic responses to external and internal cues. It is especially true for bacteria that crucially need to sense and respond quickly to rapidly changing and hostile environments in order to survive. Signalling processes in bacteria usually arise at the membrane, as it is the cellular barrier that is in direct contact with the surroundings. Most of the signalling proteins are thus recruited or inserted in the membrane. Owing to its intrinsic complexity, a variety of phenomena can modulate signalling transduction. Among these, neighbourhood composition heterogeneities, local concentration enhancements and mechanical cooperativity are of primary impact in complex bacterial membranes (Groves and Kuriyan, 2010). Once perceived, signals need to be decoded and transduced into a proper physiological response. As in eukaryotic cells, bacteria use reversible protein phosphorylation as a general mechanism to transmit environmental and intracellular signals (Fig. 12.1). Indeed, protein modification by covalent phosphorylation is likely the most adaptable and versatile regulation mechanism of essential biological processes in the three kingdoms of life (Wurgler-Murphy *et al.*, 2004). Although it represents a small post-translational modification, protein phosphorylation can substantially alter the activity, the shape or the oligomerization state of its target. Phosphotransfer reactions allow the information to be directly transduced into a specific response, amplified following ramified pathways, or integrated with additional information from others sources.

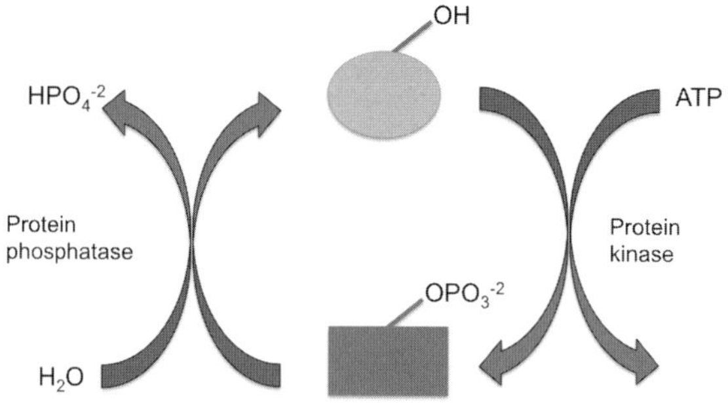

Figure 12.1 Phosphorylation conversion of proteins using ATP. Dephosphorylation via hydrolytic reaction.

It is clear that signal transduction requires specialized protein machineries shaped throughout the natural evolution process to fit the specific need of each organism. Among them, two-component systems (TCSs) and Ser/Thr/Tyr protein kinases (STPKs) are of special importance owing to their wide distribution and functional diversity in bacteria, and will be the main subject of this chapter. TCSs shuttle chemically labile phosphoryl groups between histidine (His) and aspartic acid (Asp) residues, and are mainly used by microbes to decode the rapid environmental changes into a proper adaptive response. TCSs are widespread and exist not only in nearly all prokaryotes and many archaea but also in eukaryotes such as plants, fungi and yeast. These systems are involved in sensing a wide variety of stimuli ranging from physical conditions (temperature, osmolarity, light) to concentrations of specific chemicals and ligands (nutrients or chemical signals for quorum sensing and chemotaxis). On the other hand, eukaryotic-like STPKs and their cognate phosphatases are less widely represented across the prokaryotic kingdom, but they are attracting increasing attention because of their significant roles in bacterial signal transduction (Wurgler-Murphy *et al.*, 2004; Alber, 2009; Pereira *et al.*, 2011).

In the past few years, our knowledge of the structure and function of these bacterial phosphorylation systems and the underlying molecular mechanisms of signal transduction has greatly improved. The present chapter reviews some major recent advances on signal transduction processes mediated by TCSs and STPKs in bacteria.

Two-component systems

The prototypical TCS is composed of a homodimeric membrane-anchored sensor histidine kinase (HK) and a cytosolic response regulator (RR). Given the large diversity of stimuli that must be sensed, microorganisms can have a large number of highly specialized TCSs with widely different molecular mechanisms for signal recognition. Moreover, to respond to diverse environmental changes with greater efficiency, information is often integrated between different TCSs creating complex signal transduction networks (Hilbert and Piggot, 2004; Ueki and Inouye, 2006; Thanbichler, 2009).

Signal detection triggers a series of phosphotransfer reactions, which start with the ATP-dependent phosphorylation of a conserved histidine residue on the HK (autokinase activity). The phosphate group is then transferred to a conserved aspartic residue in the cognate RR (phosphotransfer activity), inducing conformational changes that in turn trigger the response (Fig. 12.2). Once the adaptative process has been established, the phosphotransfer

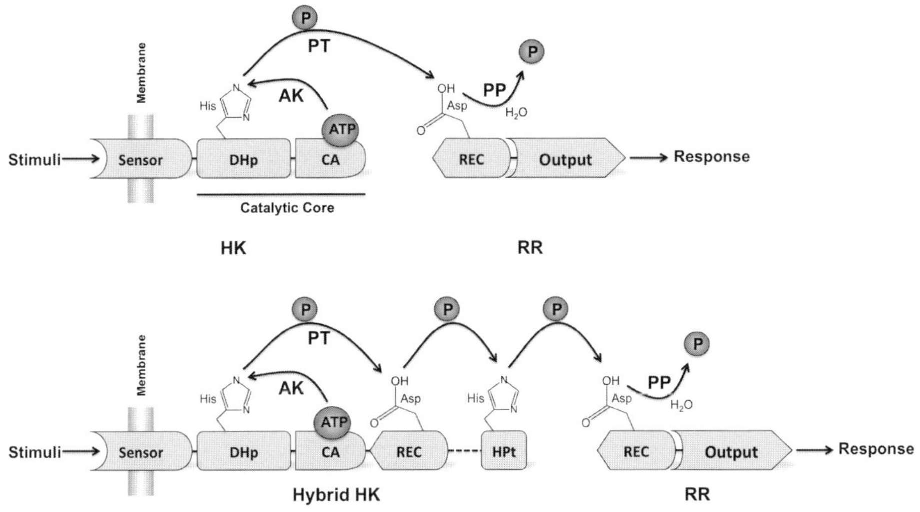

Figure 12.2 Schematic architecture of orthodox and hybrid histidine kinases and phosphoryl-transfer pathway. (A) In Orthodox Histidine kinases, the autokinase reaction (AK) involves the transfer of the γ-phosphoryl group from ATP to a conserved Histidine side chain of the HK. Then, in the phosphotransferase reaction (PT), the phosphoryl group is transferred to a conserved Aspartic residue located in the receiver (REC) domain of the RR. Phosphorylation of REC usually triggers conformational changes leading to activation of the output (OUT) domain of the RR. In the phosphatase reaction (PP), the phosphoryl group is transferred from the phospho-Asp residue to a water molecule. All three reactions require divalent metal ions, with Mg^{2+} presumably being the relevant cation *in vivo*. (B) Hybrid histidine kinases are composed of additional REC and histidine phosphotransfer (HPt) domains involved in multistep phosphotransfer events.

flow is usually stopped by HK-mediated dephosphorylation of the RR (phosphatase activity). All three catalytic activities of HKs – autokinase, phosphotransferase and phosphatase – constitute potential control points to regulate the phosphate flow and, depending on the particular system, signal perception can regulate the kinase activity (Timmen *et al.*, 2006; Fleischer *et al.*, 2007), the phosphatase activity (Brandon *et al.*, 2000), or both (Jiang *et al.*, 2000; Chamnongpol *et al.*, 2003).

Independently of the strategy employed to regulate the RR phosphorylation state, the rate of phosphorylation/dephosphorylation reactions is crucial for triggering the response, and a proper balance may be essential for cell survival. This is particularly important in cases where TCSs control complex and energetically costly cell differentiation processes. For instance, in *B. subtilis* the master regulator for entry into sporulation is the response regulator Spo0A, which directly governs the expression of over 100 genes in response to conditions of nutrient limitation (Molle *et al.*, 2003a). Interestingly, there are two categories of genes with respect to their responsiveness to the phosphorylated form of Spo0A (Spo0A-P): some are turned on/off at low doses of Spo0A-P, while others require higher doses of Spo0A-P (Fujita *et al.*, 2005). This differential profile of gene transcription allows the cell to first explore different strategies of subsistence under starvation conditions, such as biofilm formation and cannibalism; when all this fails, a higher accumulation of Spo0A-P enables a complex and irreversible process of cell differentiation that culminates in the formation of a spore, assuring cell survival under extreme environmental conditions. Being an energy-consuming process, commitment to sporulation is tightly regulated through a phosphorelay system involving several proteins, and is coordinated with other physiological functions (Higgins and Dworkin, 2012).

General architecture of histidine kinases

The exceptionally diverse family of HKs, which currently includes more than 10,000 catalogued members, can be classified into two large groups: orthodox and hybrid HKs. Most orthodox HKs are composed of two transmembrane (TM) α-helices connecting a N-terminal extracytoplasmic sensor domain to a cytoplasmic C-terminal catalytic core (Fig. 12.2). The sensor domain is highly variable in sequence and function, as it is the region of the protein that has evolved to recognize a specific signal, whereas the catalytic core is the most conserved region within the protein family. Besides this basic architecture, orthodox HKs may have a more sophisticated domain organization. Thus, some HKs contain more than one sensor domain or posses an additional transmitter domain called HAMP (domain found in histidine kinases, adenylyl cyclases, methyl binding proteins, and phosphatases). When present, this domain is located invariably between the C-terminal end of the last transmembrane segment and the catalytic core, and is thought to play an important role in relaying signals from TM regions to the kinase core (Ferris *et al.*, 2011, 2012).

Over 80% of the HKs are integral membrane proteins, some having as many as 20 transmembrane segments, while the remainders are cytosolic. These soluble HKs have been implicated in sensing intracellular stimuli such as oxygen tension (Kim *et al.*, 2010), cellular metabolism (Weiss *et al.*, 2002), or environmental stimuli that have direct access to the intracellular space by diffusion or transmission such as light.

Hybrid kinases are mainly found in higher eukaryotic organisms, although they are also present in some prokaryotes. In contrast with orthodox HKs, hybrid HKs have a distinct multidomain organization, containing multiple phosphodonor and phosphoacceptor sites within the same polypeptide chain. Thus, instead of promoting a single phosphoryl transfer, they catalyse multistep phosphotransfer reactions. The complexity of these phosphorelay systems (Fig. 12.2) admits different checkpoints for regulation, which might facilitate a better fine-tuning of the response than its orthodox counterpart. This may become particularly important in convergent or branched signalling pathways in which several different kinases affect a single response.

Sensing domains and signal perception

Sensor domains have acquired through evolution quite different mechanisms of stimulus perception and processing modes. They can be classified into three major groups based on functional aspects, mostly derived from topological and architectural organization. The largest group is comprised by HKs having a periplasmic (or extracellular)-sensing domain framed by at least two TM helices. Most of the HKs involved in sensing solutes and nutrients from the extracytoplasmic environment fall in this group. The sensor domain of these proteins frequently contains a cavity that can fit the signal molecule, promoting a conformational change that is propagated to the cytoplasm and regulates the activity of the catalytic core.

A second group is composed of HKs containing cytoplasmic sensor domains, which can be either membrane-anchored or soluble proteins. This class of sensor proteins are typically involved in detection of stimuli such as solutes or proteins monitoring the metabolic or developmental state of the cell. Finally, the third group contains HKs whose sensing mechanism is directly associated with the membrane-spanning helices. These sensor kinases can be constituted by a different number of transmembrane regions (from 2 to 20), usually connected by very short intra- or extracytoplasmic linkers. Within this group, the stimuli sensed are either membrane associated or occurs directly at the membrane interface. One of the best characterized examples is the thermosensor HK DesK from *Bacillus subtilis* (Albanesi *et al.*, 2004). Other membrane-related stimuli include mechanical properties of the cell envelope (such as swelling or mechanical stress), ion or electrochemical gradients, transport processes, and the presence of compounds that affect cell envelope integrity.

In addition to sensor HKs, in which there is a direct interaction between the signal and the kinase, another mechanism for signal sensing can use accessory proteins to perceive the stimulus and regulate HK activity. This class of signal transducers is frequently referred to as three-component systems. During the last years, the number of known members belonging to this class of proteins has steadily increased, indicating that such systems may be more common than initially thought. Just like the classification for sensor domains, the accessory proteins can be grouped into periplasmic, intramembrane or cytoplasmic, depending on where the stimulus is detected.

Structural insights into HK functional states

The conserved catalytic core of HKs is composed of two distinct structural domains: a dimerization and histidine phosphotransfer (DHp) domain, where the phospho-accepting histidine residue is located, connected *via* a fairly flexible linker to the catalytic and ATP-binding (CA) domain (Fig. 12.2). Early structural studies of HK fragments have shown that the dimeric DHp domain consists of two antiparallel helix hairpins that interact with each other to form a central four-helix bundle. The phosphorylatable His residue is positioned midway along the exposed face of the N-terminal proximal helix α1, which in turn extends beyond the four-helix bundle through a coiled-coil region, connecting the catalytic core with either HAMP domains, transmembrane helices or cytoplasmic sensor domains. In a similar way, the CA domain was found to display a β/α fold, which typically binds one ATP molecule with a highly exposed γ-phosphate.

Recently, the 3D structures of the entire catalytic cores of different HKs, including HK853 from *Thermotoga maritima* (Marina *et al.*, 2005), the thermosensor DesK from *B. subtilis* (Albanesi *et al.*, 2009) and KinB from *Geobacillus stearothermophilus* (Bick *et al.*, 2009), provided important insights into the molecular mechanisms of phosphoryl transfer. In particular, structural studies of DesK revealed three clearly distinct conformations for the catalytic core, which were assigned to the different catalytic states of the protein (Albanesi *et al.*, 2009). In one conformation, an extensive binding interface between the CA and DHp domains is stabilized by the formation of a parallel two-helix coiled-coil (Fig. 12.3A). This conformation corresponds to a kinase-incompetent state, as both the phospho-accepting His and the CA-bound nucleotide are not oriented properly for the autophosphorylation reaction to occur. Although the interface between CA and DHp has been identified as a weakly conserved motif in HKs, mutations in this region aimed at disrupting the interaction abrogate phosphatase activity in EnvZ without diminishing its kinase activity (Hsing *et al.*, 1998). Moreover, disruption of the interface in HK853 accelerates autokinase activity, whereas this activity is blocked upon interface stabilization through disulfide bond formation (Marina *et al.*, 2005). Finally, a mutant of DesK that stabilizes this conformation was

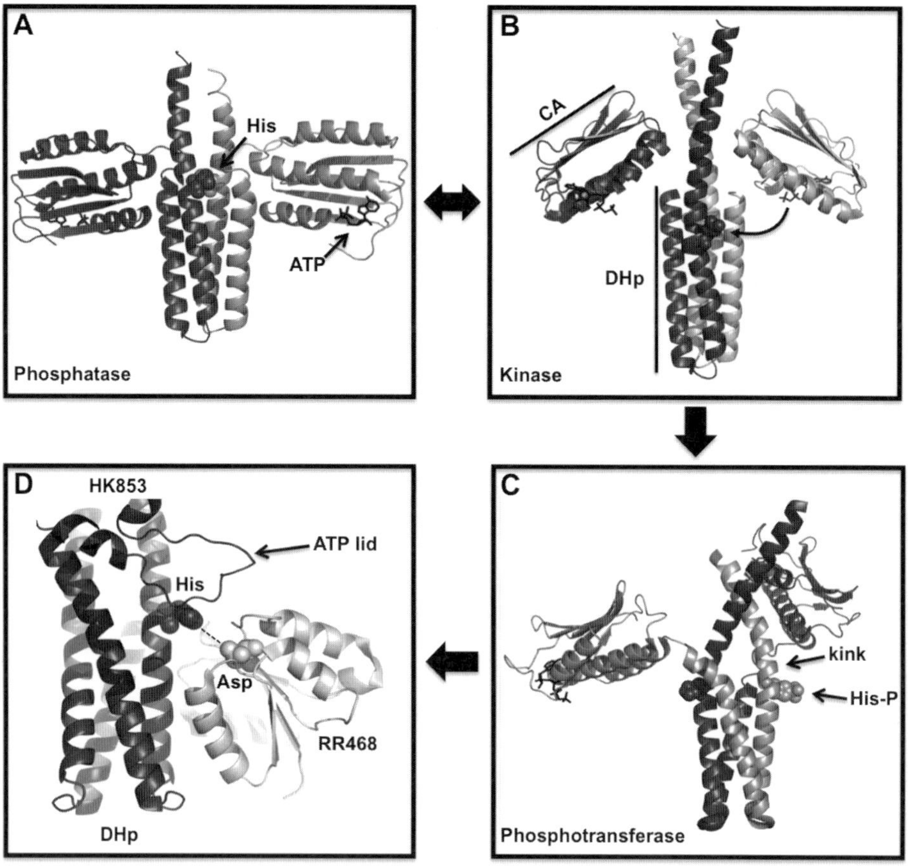

Figure 12.3 Structures of DesK and HK853 catalytic core representing each step of the phosphotransfer pathway. (A) Structure of DesK in the phosphatase competent state (PDB ID 3EHH). Light and dark grey depicts each protomer within the dimer. Arrows point out the phosphorylatable histidine located in the DHp domain and the bound ATP molecule in one of the CA domains. (B) Structure of DesK in the kinase competent state (PDB ID 3GIE). After signal detection by the transmembrane sensor domain, a conformational change disrupting the interaction between the CA and DHp domains allows histidine phosphorylation. (C) Structure of DesK in the phosphotransfer competent state (PDB ID 3GIG). For clarity, the structure was rotated about 90 degrees clockwise around the vertical axis. Arrows indicate the phosphorylated histidine and the bending of helix αl. (D) Structure of HK853 in complex with its cognate response regulator RR468 (PDB entry 3DGE). HK853 is coloured in dark grey and RR468 in light grey. For clarity, only one molecule of RR468 is shown in the figure. This close view of the complex shows that the phospho-accepting His of HK853 is aligned with the phosphorylation site of RR468 and close to RR side chains that catalyse phosphotransfer. The arrow indicates the ATP lid of CA domain that makes peripheral contacts with other mobile regions of RR.

found to produce a stable Mg^{2+}-dependent complex with its cognate response regulator DesR and to be able to carry out the dephosphorylation reaction (Albanesi *et al.*, 2009). Altogether this evidence strongly suggests that such conformation corresponds to the kinase-off state of the crystallized HKs. In the case of DesK, such conformation is compatible with a phosphatase-competent state, in which the autophosphorylation reaction is impaired but the protein is still able to promote dephosphorylation of its cognate response regulator.

In contrast, another crystal form of DesK showed a different protein conformation with highly mobile DHp and CA domains (Fig. 12.3B). This conformation is consistent with a kinase competent state, since a simple pivotal motion of the CA domain can bring the phospho-accepting histidine of the DHp in one protomer into close alignment with the γ-phosphate of a CA-bound ATP molecule in the second protomer. This rotational movement is likely assisted by complementary binding surfaces on the DHp and CA domains, as predicted by docking calculations and further supported by trapping of an intermediate state with structure-based engineered cysteine mutants (Trajtenberg *et al.*, 2010). Finally, information regarding the phosphotransferase-competent state comes from the crystal structure of autophosphorylated DesK (Albanesi *et al.*, 2009). In this structure the phosphoryl group attached to one protomer induced a markedly asymmetric conformation of the dimer (Fig. 12.3C). This functional state is further characterized by a more pronounced bending of DHp helix α1 at the site of phosphorylation, the main role of which seems to be the generation of a new interaction surface for the cognate regulator to allow complex formation and phosphoryl transfer.

The most detailed experimental evidence of the phosphotransferase reaction between the HK and RR proteins comes from the landmark crystallographic analysis of the HK *T. maritima* HK853 in complex with its response regulator RR468 (Fig. 12.3D) (Casino *et al.*, 2009). The complex is composed of two RR molecules symmetrically bound to the catalytic core of the HK and their analysis reveals that each RR binds a target region of the DHp domain just below the phosphodonor histidine residue. In agreement with this model, a similar mode of interaction was observed by NMR studies between the cytoplasmic end of the DHp domain and the RR in the EnvZ/OmpR TCS of *Escherichia coli* (Tomomori *et al.*, 1999) and in crystal structures of the ThkA–TrrA (Yamada *et al.*, 2009) and Spo0B-Spo0F (Zapf *et al.*, 2000) complexes obtained at low resolution. Furthermore, it has also been shown that this interface region determines the specificity of interaction between an HK and its cognate RR, as it has elegantly been demonstrated in experiments where a HK was 'rewired' to interact with a different RR by exchanging the cytosolic end of the DHp domain with that from another HK (Capra *et al.*, 2010).

The structure of the HK853–RR468 complex further revealed that RR interacts not only with the DHp domain, but also with peripheral regions of both HK including the DHp–CA linker, as well as with the CA domain. These contacts involve mobile regions of both proteins (β3–α3 loop of the RR and the ATP lid of the CA domain), which change their conformations according to the different functional states. Based on these findings it was proposed that the RR–CA interaction might provide a mechanism for shutting off the autokinase activity when the HK is operating either as a phosphotransferase or a phosphatase, possibly conferring a reactional specificity to the complex (Casino *et al.*, 2009).

Signal transduction mechanisms

It is clear that the different activities of the catalytic core are modulated by the input signals detected by the sensor domain. However the underlying mechanism(s) of such regulation remain elusive since for most sensor kinases, the relevant ligands have not been identified, while for those kinases whose ligands are known, structural data comparing ligand-free and ligand bound states of the kinases have not, in most cases, been available. Furthermore, little is known about the mechanism of signal transduction across the membrane or between cytoplasmic sensor domains and the catalytic core. Regardless of how stimuli are perceived by sensor domains, the propagation of the signal towards the catalytic core could involve a piston stroke movement, a tilting or rotation of the transmembrane helices, or a combination of these movements. The scarce evidence in this field indicates that signal-dependent movements are subtle and almost all relevant observations coming from structural and biochemical analysis of isolated periplasmic and cytoplasmic sensor domains are compatible with either a modest piston displacement (Falke and Erbse, 2009; Hall *et al.*, 2011) or a rotation of the signalling helices (Kwon *et al.*, 2003).

The chemotaxis aspartate receptor Tar, which complexes with the CheA kinase, has long served as a model for transmembrane signal transduction (Falke and Hazelbauer, 2001). The crystal structures of the Tar periplasmic domain with and without a bound aspartate ligand reveal subtle conformational changes, which promotes a slight asymmetry that shuts off the associated CheA kinase (Milburn *et al.*, 1991; Yeh *et al.*, 1996). A similar mechanism has been proposed for CitA (Sevvana *et al.*, 2008) and NarX (Cheung and Hendrickson, 2009), in which ligand binding to their sensor domains results in a piston-like sliding motion between terminal helices that is propagated into the cytoplasmic kinase domain.

Some recent studies of isolated sensor proteins emphasized the importance of structural plasticity of these domains, to allow the transmission of a conformational signal upon stimulus detection. For instance, structural studies of the cytoplasmic FixL sensor in various signalling states shows five different dimeric arrangements along a conserved interface (Ayers and Moffat, 2008). It is thought that distortions in FixL sensor domain after ligand binding cause quaternary changes that can be transmitted along the dimer interface towards the catalytic core. The key role of conformational symmetry and structural plasticity on signal transduction mechanisms was further emphasized by recent studies of the bacteriophytochrome photosensory core domains (Yang *et al.*, 2008) and the LuxPQ signalling complex (Neiditch *et al.*, 2006). The latter is a periplasmic protein that binds a quorum-sensing signal, AI-2, (Chen *et al.*, 2002) and controls the kinase activity of LuxQ, an integral membrane HK (Bassler *et al.*, 1994). Ligand binding to LuxPQ breaks the symmetry of the homodimer, stabilizing an asymmetrical conformational state that switches off kinase activity (Neiditch *et al.*, 2006).

It appears indeed that signal transduction mechanisms may be largely based on signal-dependent changes in quaternary structure symmetry and dynamics of transmembrane helices and their associated homodimeric HKs. Consistent with these notions, the comparison between the (symmetrical) phosphatase-competent and the (asymmetrical) kinase-competent conformations of DesK reveals a composite movement in the DHp domain, which combines a rotational shift through a concerted 'cogwheel' mechanism (Fig. 12.4) and a shearing movement that increases the tilt angle between the two helical hairpins in the homodimer (Albanesi *et al.*, 2009). A similar rotational movement has also been noticed by comparing the structure of HK853 alone and in complex with its RR (Casino *et al.*, 2009). In this way, signal-induced rotational movements of the DHp helices can directly regulate the HK catalytic activities, by modulating the binding affinity to sequester the CA domain in a kinase-inactive conformation, or by modifying the surface around the phosphorylatable His, poised to interact with the RR in the phosphotransfer and phosphatase reactions. It should be noted that signal transduction in the form of helical rotations are congruent with mechanistic models derived from the study of upstream elements

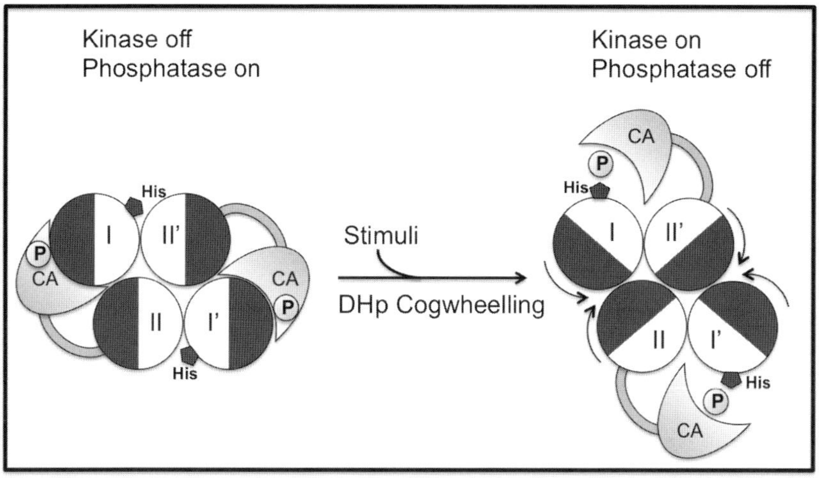

Figure 12.4 Scheme of DHp structural rearrangements that regulate interdomain interaction and catalytic activity. The scheme is a top-down view of the catalytic core in two functional states. Circles depict the four-helix bundle of DHp domain. In the kinase off/phosphatase on state (left), phospho-accepting histidine residues are buried within the bundle and CA domains are sequestered impeding phosphotransfer reaction. After signal perception, a cogwheeling rotation movement of DHp helixes release CA domains and expose phospho-accepting histidine residues (right) making possible the assembly of the kinase active site.

in different signal transducing systems. In particular, the parallel coiled-coil HAMP domain, which is present in the membrane-connecting region of many different HKs, was shown to relay the input signal through helical rotations compatible with the structural changes described above for DesK (Hulko *et al.*, 2006). Also, a combined helical rotation and tilting was involved in signal transduction by the bacterial sensory rhodopsin phototactic receptor (Moukhametzianov *et al.*, 2006) and analogous rotational movements were recently proposed as a general mechanism of signal transduction through helical coiled-coil linkers (Moglich *et al.*, 2009).

Overall structures and functional diversity of response regulators

In a typical TCS, the RR lies at the end of the phosphorylation pathway prompting the specific output response of the system, usually a modification of the transcriptional profile of the cell. Like HKs, RRs have a modular architecture that typically consists of a conserved N-terminal regulatory or receiver (REC) domain linked to a more variable C-terminal output (or effector) domain (Fig. 12.2), which in many cases is a DNA-binding domain. The REC domain catalyses phosphoryl transfer from the phospho-His of the HK to a conserved Asp residue located within the REC domain itself, and in turn controls the activity of the effector domain. Several experimental structures of REC domains revealed a conserved α/β fold composed of a central five-stranded parallel β-sheet surrounded by five amphipathic α-helices (Gao and Stock, 2009). A small set of highly conserved residues play important roles in catalysis and signal propagation. These residues define the active site and comprise a cluster of acidic residues including the phosphorylatable aspartic acid at the C-terminal end of $\beta3$, two additional acidic residues in the $\beta1$–$\alpha1$ loop that coordinate an essential Mg^{2+}

required for both phosphotransfer and phosphate hydrolysis, and a Lys residue in the β5–α5 loop that forms a salt bridge with the phosphate in the activated domain. Other highly conserved residues lie on a diagonal path, which extends away from the active site and is involved in propagation of long-range conformational changes upon RR phosphorylation (Gao and Stock, 2009). These changes may significantly modify the molecular surface of the REC domain, and therefore modulate its interaction with either the cognate HK (Casino *et al.*, 2009) or the effector domain for activity regulation (Gao and Stock, 2010).

The REC domain controls the output response by either stimulating of inhibiting the activity of the effector domain. This regulation is mainly based on the differential molecular properties of the REC surface as a function of its phosphorylation state, a simple mechanism for regulation that offers a high versatility to fit the specific needs of each particular system. Surface changes can be translated into many different strategies of effector domain activation, which are characteristic for every RR and, except for a few cases, are not conserved even within a same family of RRs. Despite this variability, it is possible to find some common patterns of activation, such as for instance phosphorylation-induced dimerization or higher-order oligomerization. Triggered by surface remodelling, this quaternary structure rearrangement can bring together the effector domains to assemble a functional active site, such as a DNA-binding motif or an enzymatic domain (Wassmann *et al.*, 2007). Other modes of oligomerization may involve heterodomain formation, as for the transcriptional factor NtrC, in which phosphorylation promotes interaction of the REC domain with the effector domain of an adjacent NtrC molecule, resulting in a ring-like protein assembly with ATPase activity (De Carlo *et al.*, 2006). In addition to this interdomain regulation, many RRs are involved in phosphorylation-modulated interactions with various different macromolecular targets. Possible RR partners not only include HKs, but also auxiliary phosphatases, effector domains, other regulatory domains, and possibly other components of the transcriptional machinery. For instance, phosphorylation of the single REC domain protein CheY regulates protein–protein interaction with the flagellar motor protein FliM to promote a switch in the rotational direction (Dyer *et al.*, 2009; Sarkar *et al.*, 2010).

Eukaryotic-like Ser/Thr/Tyr protein kinases and phosphatases

For a long time, molecular signal transduction mechanisms based on reversible Ser/Thr/Tyr protein phosphorylation had been associated primarily with eukaryotic cells. This paradigm was challenged over 20 years ago when several studies reported bacterial phosphorylation on tyrosine and serine/threonine residues. In the 1980s, Cortay and coworkers demonstrated that STPKs in *Escherichia coli* phosphorylate tyrosine residues (Cortay *et al.*, 1986). A few years later, phosphorylation by STPKs was shown in *Mycobacterium tuberculosis (Mtb)* (Chow *et al.*, 1994) and *Myxococcus xanthus* (Frasch and Dworkin, 1996). In parallel, genes coding for STPKs and Ser/Thr protein phosphatases (STPPs) were being identified in various bacterial strains, such as *Myxococcus xanthus* (Munoz-Dorado *et al.*, 1991), *Streptomyces species* (Matsumoto *et al.*, 1994) and *Mtb* (Peirs *et al.*, 1997). These initial observations were subsequently confirmed by bacterial genome sequencings. For instance, the *Mtb* genome sequencing unveiled the presence of as many genes coding for eukaryotic-like STPKs than those coding for TCSs (Cole *et al.*, 1998). However, the number and distribution of STPKs remains highly heterogeneous among bacterial genomes (Bentley *et al.*, 2002; Pereira *et al.*, 2011; Galperin and Koonin, 2010), and it has been suggested that the abundance of STPK genes might be associated with bacteria having complex life cycles (sporulation, latency) as well as a strong capacity to adapt to a changing environment (Hirakata *et al.*, 1998).

Physiological roles of STPKs

The presence of numerous eukaryotic-like STPK genes was largely confirmed in sequenced bacterial genomes (Galperin and Koonin, 2010). However, the elucidation of the physiological roles of these kinases has been much more difficult to achieve, due to functional redundancy and substrate promiscuity, and are only recently starting to be disclosed. In a similar way, extensive phosphoproteomic studies revealed that Ser/Thr/Tyr protein phosphorylation is widespread in bacteria (Macek *et al.*, 2007, 2008; Soufi *et al.*, 2008; Lin *et al.*, 2009; Ravichandran *et al.*, 2009; Sun *et al.*, 2010; Schmidl *et al.*, 2010; Parker *et al.*, 2010; Prisic *et al.*, 2010), but their actual physiological relevance remains largely unknown (Kobir *et al.*, 2011). A comprehensive review of the current knowledge on the physiological roles of STPKs in bacteria has been recently published (Pereira *et al.*, 2011); we describe below some selected examples to illustrate their wide functional relevance.

Recent studies suggest that STPKs and STPPs control important physiological functions such as cell growth, cell division, persistence or reactivation, stress response and virulence (Alber, 2009; Molle and Kremer, 2010). Thus, *Myxococcus xanthus* employs a large number of eukaryotic-like STPKs to adapt to changes in the environment as well as to communicate with other members to coordinate motion and behaviour (Perez *et al.*, 2008). The *Mtb* genome has eleven genes coding for STPKs (PknA-L), and two of them (PknA and PknB) regulate essential metabolic processes (growth and cell division) in response to environmental signals (Kang *et al.*, 2005; Fernandez *et al.*, 2006). The transcription of *pknA* and *pknB* increases upon infection of macrophages (Av-Gay *et al.*, 1999; Singh *et al.*, 2006) whilst *pknB* expression is down regulated in a model of *Mtb* persistence, during nutrient starvation (Betts *et al.*, 2002). Overexpression of *pknB* in *M. smegmatis* and *M. bovis* BCG shows a phenotype with elongated cell morphology, suggesting dysfunctional cell wall synthesis and cell division processes (Kang *et al.*, 2005). Along the same line, FtsZ, a cell division protein equivalent to eukaryotic tubulin, was identified as a substrate for PknA, PknB, PknL and for the phosphatase PPp, suggesting a role of these proteins in regulating cell division in *Mtb* (Thakur and Chakraborti, 2006; Schultz *et al.*, 2009). Moreover, phosphorylation of *Mtb* FtsZ by PknA inhibits its GTPase activity (Thakur and Chakraborti, 2006). *PknA* and *pknB* genes share the same operon close to the chromosomal origin of replication, which comprises seven genes that are involved in cell division and signal transduction pathways, and is conserved in *Actinomycetales* (Kang *et al.*, 2005; Fernandez *et al.*, 2006).

Recently, *Mtb* PknH was shown to regulate the growth of bacilli in a mouse model and in response to nitric oxide (NO) stress *in vitro*. Moreover, deletion of *pknH* was shown to cause a hypervirulent phenotype when this mutant of *Mtb* strain was used to infect BALB/c mice (Papavinasasundaram *et al.*, 2005). Endogenous substrates have been identified for PknH: EmbR (Molle *et al.*, 2003b), DacB1 and Rv0681 (Zheng *et al.*, 2007) but their exact physiological implication remains unclear. In addition, substrates and downstream effectors of PknH signalling in response to NO stimulus are yet to be identified. Another mycobacterial kinase, PknG, has been reported to act as a secreted virulence factor to promote *Mtb* survival within human macrophages (Walburger *et al.*, 2004). Interestingly, this same kinase acts as a regulator of central metabolism in *Mtb* and *Corynebacterium glutamicum*. PknG was found to control the cellular pool of glutamate by phosphorylating the GarA regulator, a fork-head-associated (FHA) protein that interacts with – and modulates the activity of – metabolic enzymes at the crossroad between glutamate metabolism and energy production through the tricarboxylic acid (TCA) cycle (O'Hare *et al.*, 2008; Nott *et al.*, 2009; Chao *et al.*, 2010; Wagner *et al.*, 2011).

Bacterial STPKs were also proposed to be involved in persistence and reactivation, or in the virulence of pathogenic bacteria. In *Bacillus subtilis*, the STPK PrkC (the orthologue of *Mtb* PknB) was shown to mediate spore reactivation in response to peptidoglycan fragments (Shah *et al.*, 2008). Growing bacteria release muropeptides from the cell wall into the milieu, and these molecules are detected by the extracellular signal sensing domains of PrkC to promote germination of dormant *B. subtilis* spores (Shah *et al.*, 2008; Shah and Dworkin, 2010). PtpkA, a *Pseudomonas aeruginosa* STPK, was shown to be responsible for the virulence of acute and chronic cystic fibrosis in humans. PtpkA phosphorylates a Thr residue of Fha1, an FHA-containing protein, which regulates assembly of the type VI secretion system, and has an impact on the virulence of clinical isolates (Mougous *et al.*, 2007; Hsu *et al.*, 2009). It was also reported that Stk, a STPK of the Gram-positive species *Streptococcus pyogenes*, that causes myositis and pharyngitis in human, is required to induce disease in a murine myositis model of infection. Transcriptional data show that Stk activates genes of virulence factors, including proteins involved in the metabolism of α-glucans, fatty acid biosynthesis or cell wall synthesis. This was further confirmed by the phenotype of a Stk deletion mutant (Bugrysheva *et al.*, 2011). Similar work in *Enterococcus faecalis* has also shown a similar role for the homologous STPK PrkC (Kristich *et al.*, 2007).

The few examples described above highlight the involvement of STPKs in numerous signalling pathways of physiological functions, ranging from cell wall remodelling during growth and/or division to regulation of central metabolism such as TCA cycle. In addition, since the phosphorylation of Ser, Thr and Tyr residues is ubiquitous in eukaryotes, many bacterial pathogens may secrete specific kinases to modulate host signalling pathways. Pathogenic bacteria have thus developed strategies to use STPKs in order to colonize ecological niches, survive to the hostile macrophage environment or interfere with the human immune response (Cozzone, 2005; Greenstein *et al.*, 2005; Kulasekara and Miller, 2007; Chao *et al.*, 2010). Considering the important role of STPKs in host–pathogen interactions, these enzymes are interesting targets for therapeutic applications and are intensively studied in order to develop specific inhibitors (Wehenkel *et al.*, 2006, 2008; Lougheed *et al.*, 2011).

Overall structures of STPK catalytic domains

Extensive genetic, biochemical and structural data established that prokaryotic STPKs are subject to similar mechanisms of substrate recognition and regulation than their eukaryotic counterparts (Greenstein *et al.*, 2005; Pereira *et al.*, 2011). Based on sequence homology, the kinase domain (KD) of prokaryotic Ser/Thr and Tyr kinases belong to the eukaryotic STP/Y protein kinase superfamily (Hanks and Hunter, 1995). It should be noted here that many bacterial tyrosine kinases belong to a different enzyme family, the so called BY family (Grangeasse *et al.*, 2007; Grangeasse *et al.*, 2010); these enzymes are not involved in bacterial signalling and fall beyond the scope of this review. The KDs of different eukaryotic-like kinases adopt strikingly similar conformations in its active form ('on' state). By contrast, structural studies of STPKs in inactive states have revealed a remarkable plasticity of the protein core, which can adopt very distinct 'off' conformations since these are not subject to the functional constraints that the active state must satisfy (Huse and Kuriyan, 2002). This

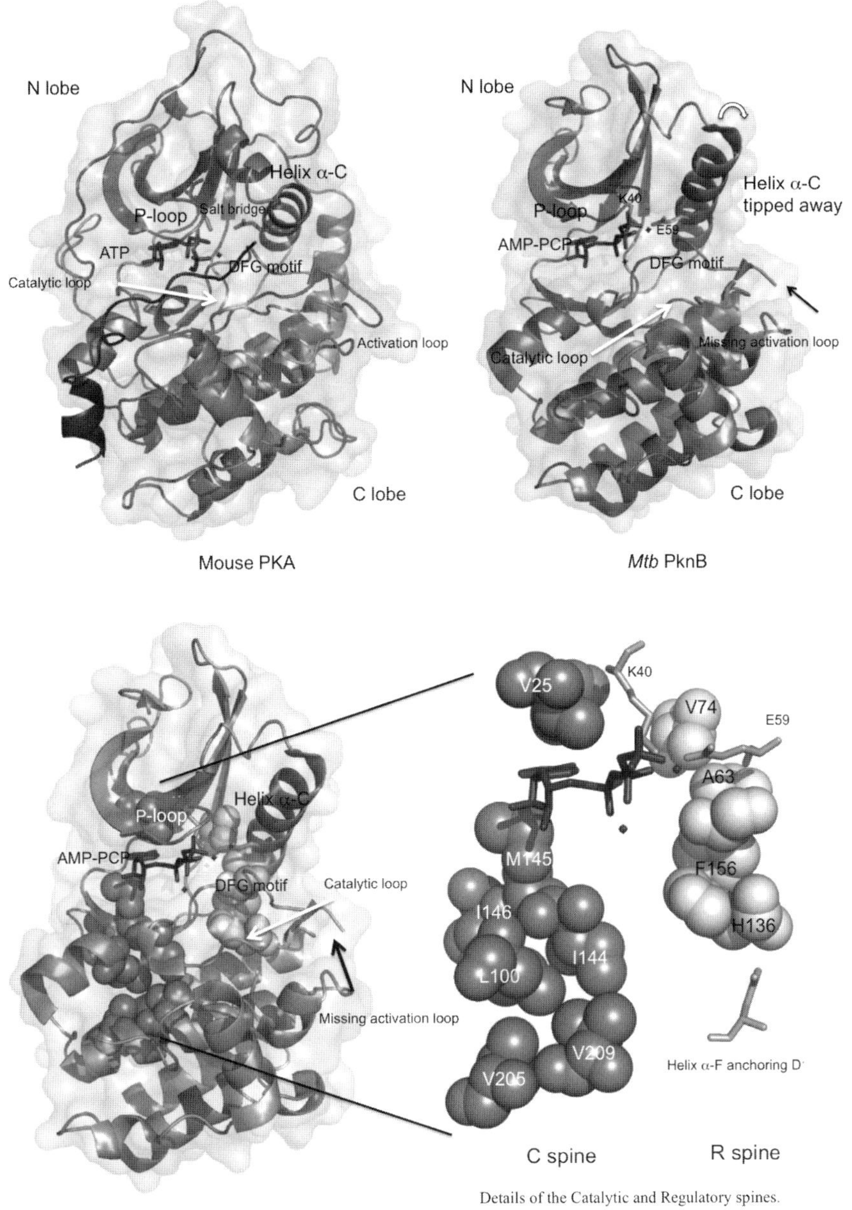

Figure 12.5 (A) Active conformation depicted by key features in Mouse PKA and *Mtb* PknB. (B) Key features of the 'on state' of *Mtb* PknB. (C) Back-to-back dimer of PknB. (D) Integral membrane PknB: KD and PASTA domains.

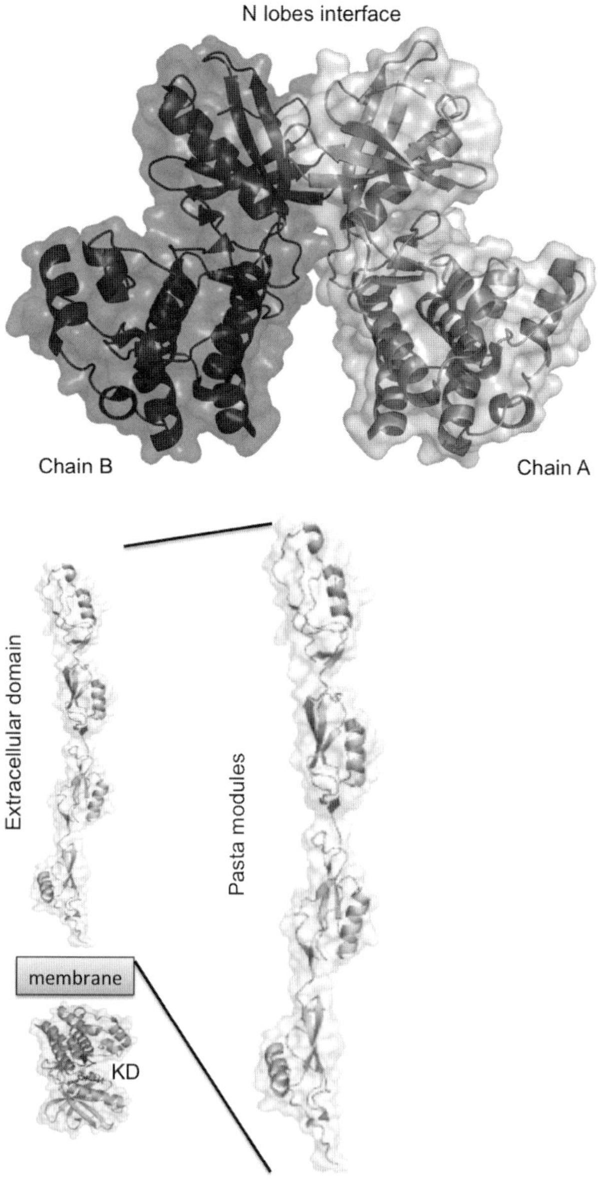

Figure 12.5 (continued)

means that the exquisite control of kinase activity can be mediated by multiple allosteric mechanisms that converge to the conserved active KD conformation (Fig. 12.5A).

The overall structure of KD folds into a two-lobed core, where the smaller N-terminal lobe (N lobe) is composed of a five-stranded β sheet and one prominent α helix called helix α-C, and the larger C-terminal lobe (C lobe) is predominantly helical (Knighton *et al.*, 1991a,b). ATP is bound in a deep cleft between the two lobes and sits beneath a highly conserved loop connecting strands β1 and β2. This phosphate-binding loop (P-loop) displays a conserved glycine-rich sequence motif, which binds the ATP phosphates through backbone interactions (Knighton *et al.*, 1991a,b; Huse and Kuriyan, 2002; Nolen *et al.*, 2004). The first crystal structure of a prokaryotic kinase, *Mtb* PknB (Ortiz-Lombardia *et al.*, 2003; Young *et al.*, 2003), revealed a remarkable similarity of bacterial STPK protein folding and catalytic machinery with its eukaryotic homologues (Fig. 12.5A and B). In particular, the PknB structures reveal several features known to be associated with an active kinase conformation. These features include: nucleotide binding between the N and C lobes; autophosphorylation of the KD activation loop; formation of an ion pair between two universally conserved residues, Lys40 (from the β3 strand) and Glu59 (from helix α-C), required to properly position the ATP phosphates for catalysis; engagement of key structural elements (P-loop, catalytic loop, DFG motif, helix α-C) into a catalytically competent conformation; and proper assembly of the catalytic and regulatory spines (Kornev *et al.*, 2008; Kornev and Taylor, 2010; Taylor and Kornev, 2011) (Fig. 12.5A and B). A similar active conformation has been also observed for *Mtb* PknE (Gay *et al.*, 2006; Scherr *et al.*, 2007), but apparently not for *Mtb* PknG (Gay *et al.*, 2006; Scherr *et al.*, 2007), the structure of which presents some unusual features incompatible with an 'on' kinase state, and whose activation might require the action of external (as yet unknown) regulatory factor(s).

Regulation mechanisms of kinase activity

Extensive studies of eukaryotic systems revealed that STPKs can be activated by a wide diversity of molecular mechanisms (Johnson *et al.*, 1996; Huse and Kuriyan, 2002). Phosphorylation of one or more residues in the KD activation loop, a centrally located segment flanked by the DFG and APE sequence motifs, is usually required for full kinase activity, and this represents one of the most common mechanisms of kinase regulation in eukaryotic STPKs (Nolen *et al.*, 2004). The phosphorylated activation loop can then stabilize the active kinase state, as shown by the crystal structures of PKA (Knighton *et al.*, 1991a,b), and/or relieve an auto-inhibition mechanism by which the unphosphorylated loop could block the catalytic groove (Huse and Kuriyan, 2002; Pirruccello *et al.*, 2006). Furthermore, an active kinase can be switched off by phosphatase-mediated dephosphorylation of the activation loop. In terms of signal transduction, this mechanism provides a convenient strategy to propagate an external signal across the membrane, because extracellular ligands could induce oligomerization of the sensor domain, which in turn brings together the corresponding intracellular kinase domains, thus promoting kinase activation. Phosphorylation of the activation loop was shown to regulate kinase activity of *Mtb* PknB (Boitel *et al.*, 2003; Ortiz-Lombardia *et al.*, 2003; Young *et al.*, 2003) and other mycobacterial STPKs (Boitel *et al.*, 2003; Duran *et al.*, 2005). Unlike a typical active STPK, however, the activation loops of PknB (Boitel *et al.*, 2003; Ortiz-Lombardia *et al.*, 2003; Young *et al.*, 2003) and PknE (Gay *et al.*, 2006; Scherr *et al.*, 2007) were disordered, suggesting that phosphorylation may not stabilize a unique active structure or, alternatively, that binding to an unidentified cofactor or substrate(s) may be required to fold the activation loop.

STPK activation by kinase domain dimerization was also observed in mycobacterial enzymes (Greenstein *et al.*, 2007; Lombana *et al.*, 2010) and in *P. aeruginosa* PpkA (Mougous *et al.*, 2007; Hsu *et al.*, 2009). In *Mtb*, both PknB and PknE were seen to form a 'back-to-back' dimer through a conserved interface in its N-lobe (Ortiz-Lombardia *et al.*, 2003; Young *et al.*, 2003; Gay *et al.*, 2006; Wehenkel *et al.*, 2006; Pereira *et al.*, 2011) (Fig.

12.5C). Recent work from Tom Alber team further showed that monomeric PknB mutants can be found in different conformational states, whereas formation of the 'back-to-back' dimer stabilizes the active kinase conformation for autophosphorylation (Greenstein *et al.*, 2007; Lombana *et al.*, 2010). Interestingly, this dimerization mode also mediates allosteric activation of human PKR, a RNA-dependent protein kinase (Dar *et al.*, 2005; Dey *et al.*, 2005), suggesting that this is an ancient, widely distributed mechanism of STPK regulation (Greenstein *et al.*, 2007; Alber, 2009; Lombana *et al.*, 2010).

Signal sensing domains

Many bacterial STPKs are receptor-like (transmembrane) enzymes composed of a cytoplasmic kinase domain linked through a transmembrane region to an extracellular – putative signal sensing – domain. The best characterized of these kinase-associated sensor modules is the PASTA (penicillin and Ser/Thr kinase associated) domain, which consists of a 70-residues module and is usually found in tandem of several repeat units. *Mtb* PknB has four such PASTA domains (Fig. 12.5D) in a fully extended arrangement (Barthe *et al.*, 2010), and a similar elongated conformation is present in many PknB homologues from *Mycobacteria* and *Firmicutes*, such as *B. subtilis* PrkC (Barthe *et al.*, 2010; Paracuellos *et al.*, 2010) and *S. aureus* PrkC (Paracuellos *et al.*, 2010; Ruggiero *et al.*, 2011). As mentioned before, muropeptide signalling through PASTA domains in *B. subtilis* is able to germinate dormant spores (Shah *et al.*, 2008; Shah and Dworkin, 2010; Squeglia *et al.*, 2011). Noteworthy, the Rpf (resuscitation-promoting factor) protein in *Micrococcus luteus* has a muralytic activity capable of degrading peptidoglycan and thus of stimulating bacteria to exit from dormancy (Keep *et al.*, 2006). In agreement with these data, the PASTA domains of *Mtb* PknB were also reported to bind specific peptidoglycan fragments (Mir *et al.*, 2011), although a possible role of the PknB PASTA domains in the reactivation of latent *Mtb* cells remains to be confirmed

Another type of extracellular sensor domain is that of PknD, which forms a rigid, symmetrical six-bladed β-propeller with a flexible tether to the transmembrane domain (Fig. 12.6) (Good *et al.*, 2004). The authors suggest that this domain could bind a multivalent signal molecule that could act by changing the quaternary structure of the intracellular kinase domain, but no specific ligand is currently known. In eukaryotic cells, β-propeller proteins tend to oligomerize to form larger assemblies, and some of these have been implicated in the pathogenesis of a variety of diseases (Pons *et al.*, 2003; Chen *et al.*, 2011). Interestingly, it has been recently shown that the PknD β-propeller is sufficient to trigger *Mtb* invasion of brain endothelia (Be *et al.*, 2012). This process is tissue-specific, as it was not observed in lung tissues, and suggests a crucial role of the PknD sensor domain in central nervous system (CNS) tuberculosis, a devastating and poorly known form of the disease primarily affecting young children. However, it remains unclear whether this β-propeller is a *bona fide* sensor domain, as well as the mechanisms by which the putative signal could be propagated across the membrane.

Protein phosphatases

In contrast to TCSs, which shuttle labile phosphoryl groups between His and Asp residues, phosphorylation of Ser/Thr/Tyr residues is chemically stable and requires the action of protein phosphatases to ensure reversibility of signalling cascades (Wehenkel *et al.*, 2008; Pereira *et al.*, 2011). Homologues of different eukaryotic families of protein phosphatases

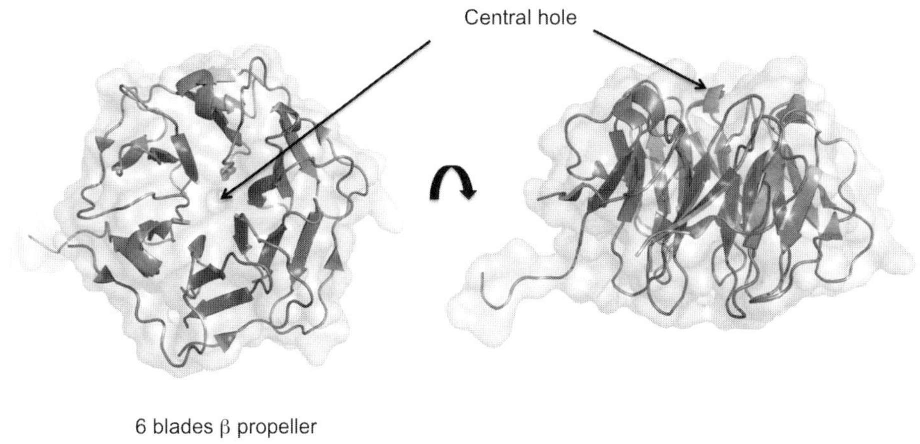

Figure 12.6 PknD extracellular domain.

are present in archaea and bacteria, including conventional and low molecular weight Tyr phosphatases, phosphoprotein phosphatases (PPPs) and metal-dependent protein phosphatases (PPMs). However, their actual roles in prokaryotic signalling and virulence are still poorly understood (Pereira *et al.*, 2011), and therefore bacterial protein phosphatases will only be briefly reviewed here.

Eukaryotic-like tyrosine phosphatases are used by many bacterial pathogens as virulence factors. For example, StpP from *Salmonella typhimurium* inhibits host MAPK (mitogen-activated protein kinase) (Lin *et al.*, 2003), YopH from pathogenic yersiniae inhibits phagocytosis by interfering with the cytoskeletal machinery of host cells (Yuan *et al.*, 2005; Chao *et al.*, 2010). In *Mtb*, two Tyr phosphatases (PtpA and PtpB) are also thought to interfere with host signalling in infected macrophages (Wehenkel *et al.*, 2006; Chao *et al.*, 2010), because inactivation of the *ptpB* gene compromises the survival of bacteria in activated macrophages (Singh *et al.*, 2003) and PtpA interferes with phagocytosis mechanisms and actin polymerization inside macrophages (Castandet *et al.*, 2005). The crystal structures of both *Mtb* PtpA and PtpB in complex with inhibitors (Greenstein *et al.*, 2005; Madhurantakam *et al.*, 2005), as well as that more recent of *S. aureus* PtpA (Vega *et al.*, 2011), confirmed their close overall similarity to eukaryotic PTPs (Fig. 12.7A and B), although unique structural features in the prokaryotic active sites can be exploited to design selective inhibitors (Grundner *et al.*, 2007).

Eukaryotic-like Ser/Thr protein phosphatases (STPPs) are metallohydrolases that contains a dinuclear metal centre essential for catalytic activity. Although eukaryotic STPPs have been classified in two subfamilies, PPPs and PPMs (Cohen and Cohen, 1989; Barford *et al.*, 1998; Jackson and Denu, 2001), all STPPs share a common structure and catalytic mechanism. The crystal structures of various bacterial (PPM) STPPs have been determined, including the catalytic domain of membrane-anchored *Mtb* PstP (Pullen *et al.*, 2004; Wehenkel *et al.*, 2007), *Mycobacterium smegmatis* MspP (Bellinzoni *et al.*, 2007), *Streptococcus agalactiae* SaSTP (Rantanen *et al.*, 2007), and *Thermosynechococcus elongatus* tPphA (Schlicker *et al.*, 2008). All of these were found to display a very similar overall fold and dinuclear-binding site to that observed for human PP2Cα (Das *et al.*, 1996; Pereira *et al.*,

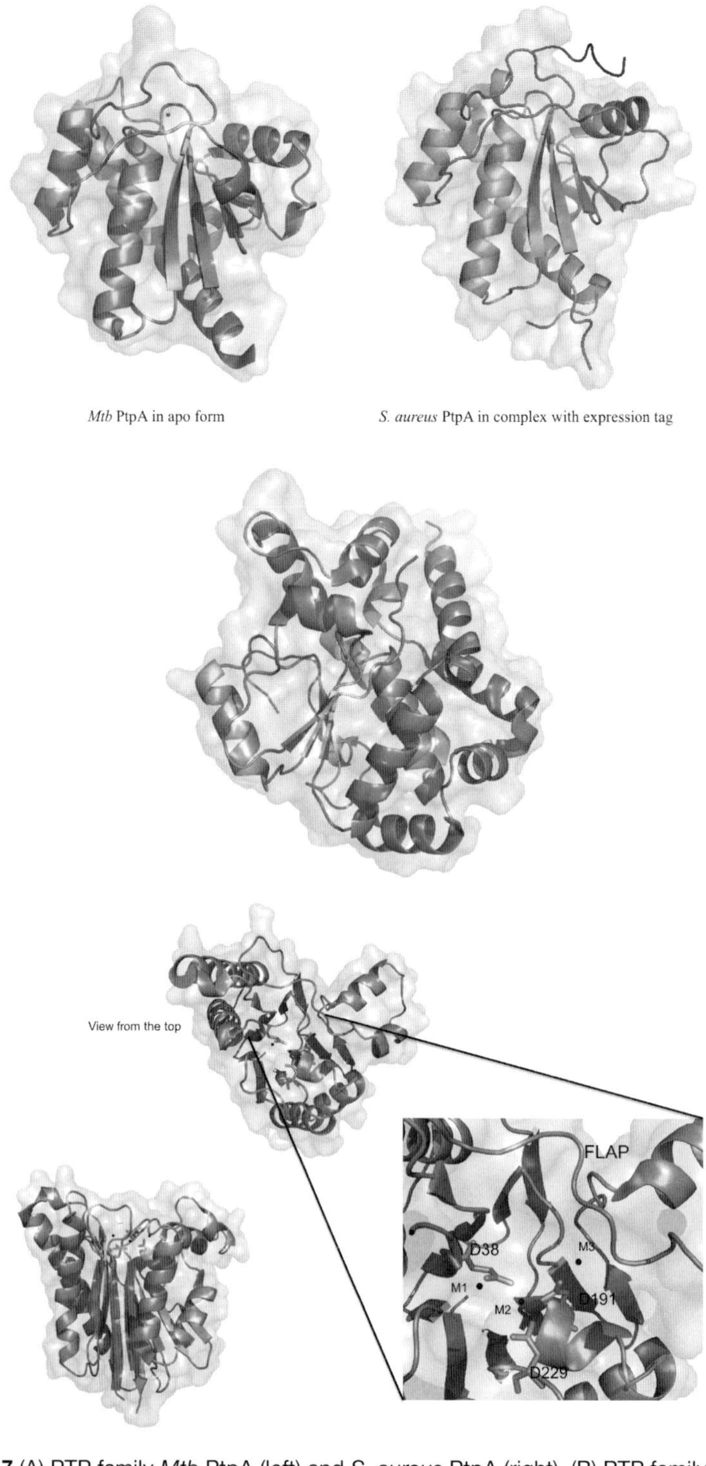

Figure 12.7 (A) PTP family *Mtb* PtpA (left) and *S. aureus* PtpA (right). (B) PTP family *Mtb* PtpB. (C) PPP family *Mtb* PstP and details on binding of showing three metal ions.

2011) (Fig. 12.7C). However, the structures of microbial STPPs revealed some significant differences, in particular supporting an associative mechanism of catalysis (Bellinzoni *et al.*, 2007; Wehenkel *et al.*, 2007) and the presence of a third metal-binding site in the active site. This third metal ion, which might be a general feature within this enzyme family (Schlicker *et al.*, 2008; Su *et al.*, 2011), is not directly involved in catalysis but modulates the conformation of the flap region (Pullen *et al.*, 2004; Bellinzoni *et al.*, 2007; Wehenkel *et al.*, 2007, 2008) important for substrate binding and catalysis (Schlicker *et al.*, 2008; Su *et al.*, 2011).

Concluding remarks

More than 20 years ago, two-component systems were firmly recognized as a fundamental principle in bacterial signal perception and transduction (Nixon *et al.*, 1986), and at about the same time the initial observations were being made of eukaryotic-like Ser/Thr/Tyr phosphorylation in bacteria (Cortay *et al.*, 1986; Munoz-Dorado *et al.*, 1991). Since then significant progress has been accomplished in our understanding of the proteins involved, in particular TCSs and STPKs, and their important roles in bacterial physiology. However, the mechanistic basis underlying these roles is still far from clear. Experimental data and molecular models for signal detection and signal transduction across the membrane remain scarce and largely restricted to a few well-characterized examples, and most of the evidence for specific kinase/phosphatase substrates is based on *in vitro* phosphorylation experiments that may not reflect the *in vivo* situation. To further enhance complexity, evidence of crosstalk between TCSs and STPKs have been recently reported in *M. xanthus* (Lux and Shi, 2005) and *Mtb* (Chao *et al.*, 2010). Hopefully, current research efforts will allow us in the near future to better understand bacterial signalling in mechanistic detail, as well as the interactions of the different systems with each other and their relevance for cell physiology.

Acknowledgements

We are deeply grateful to N. De Val, M. N. Lisa and A. Wehenkel for their careful reading and comments of the chapter.

References

Albanesi, D., Mansilla, M.a.C., and Mendoza, D.D. (2004). The membrane fluidity sensor DesK of *Bacillus subtilis* controls the signal decay of its cognate response regulator. Society *186*, 2655–2663.

Albanesi, D., Martin, M., Trajtenberg, F., Mansilla, M.C., Haouz, A., Alzari, P.M., de Mendoza, D., and Buschiazzo, A. (2009). Structural plasticity and catalysis regulation of a thermosensor histidine kinase. Proc. Natl. Acad. Sci. U.S.A. *106*, 16185–16190.

Alber, T. (2009). Signaling mechanisms of the *Mycobacterium tuberculosis* receptor Ser/Thr protein kinases. Curr. Opin. Struct. Biol. *19*, 650–657.

Av-Gay, Y., Jamil, S., and Drews, S.J. (1999). Expression and characterization of the *Mycobacterium tuberculosis* serine/threonine protein kinase PknB. Infect. Immun. *67*, 5676–5682.

Ayers, R.A., and Moffat, K. (2008). Changes in quaternary structure in the signaling mechanisms of PAS domains. Biochemistry *47*, 12078–12086.

Barford, D., Das, A.K., and Egloff, M.P. (1998). The structure and mechanism of protein phosphatases: insights into catalysis and regulation. Annu. Rev. Biophys. Biomol. Struct. *27*, 133–164.

Barthe, P., Mukamolova, G.V., Roumestand, C., and Cohen-Gonsaud, M. (2010). The structure of PknB extracellular PASTA domain from mycobacterium tuberculosis suggests a ligand-dependent kinase activation. Structure *18*, 606–615.

Bassler, B.L., Wright, M., and Silverman, M.R. (1994). Multiple signalling systems controlling expression of luminescence in Vibrio harveyi: sequence and function of genes encoding a second sensory pathway. Mol. Microbiol. *13*, 273–286.

Be, N.A., Bishai, W.R., and Jain, S.K. (2012). Role of *Mycobacterium tuberculosis* pknD in the pathogenesis of central nervous system tuberculosis. BMC Microbiol. *12*, 7.

Bellinzoni, M., Wehenkel, A., Shepard, W., and Alzari, P.M. (2007). Insights into the catalytic mechanism of PPM Ser/Thr phosphatases from the atomic resolution structures of a mycobacterial enzyme. Structure *15*, 863–872.

Bentley, S.D., Chater, K.F., Cerdeno-Tarraga, A.M., Challis, G.L., Thomson, N.R., James, K.D., Harris, D.E., Quail, M.A., Kieser, H., Harper, D., *et al.* (2002). Complete genome sequence of the model actinomycete *Streptomyces coelicolor* A3(2). Nature *417*, 141–147.

Betts, J.C., Lukey, P.T., Robb, L.C., McAdam, R.A., and Duncan, K. (2002). Evaluation of a nutrient starvation model of *Mycobacterium tuberculosis* persistence by gene and protein expression profiling. Mol. Microbiol. *43*, 717–731.

Bick, M.J., Lamour, V., Rajashankar, K.R., Gordiyenko, Y., Robinson, C.V., and Darst, S.A. (2009). How to switch off a histidine kinase: crystal structure of Geobacillus stearothermophilus KinB with the inhibitor Sda. J. Mol. Biol. *386*, 163–177.

Boitel, B., Ortiz-Lombardia, M., Duran, R., Pompeo, F., Cole, S.T., Cervenansky, C., and Alzari, P.M. (2003). PknB kinase activity is regulated by phosphorylation in two Thr residues and dephosphorylation by PstP, the cognate phospho-Ser/Thr phosphatase, in *Mycobacterium tuberculosis*. Mol. Microbiol. *49*, 1493–1508.

Brandon, L., Dorus, S., Epstein, W., Altendorf, K., and Jung, K. (2000). Modulation of KdpD phosphatase implicated in the physiological expression of the kdp ATPase of *Escherichia coli*. Mol. Microbiol. *38*, 1086–1092.

Bugrysheva, J., Froehlich, B.J., Freiberg, J.A., and Scott, J.R. (2011). Serine/threonine protein kinase Stk is required for virulence, stress response, and penicillin tolerance in Streptococcus pyogenes. Infect. Immun. *79*, 4201–4209.

Capra, E.J., Perchuk, B.S., Lubin, E.A., Ashenberg, O., Skerker, J.M., and Laub, M.T. (2010). Systematic dissection and trajectory-scanning mutagenesis of the molecular interface that ensures specificity of two-component signaling pathways. PLoS Genet. *6*, e1001220.

Casino, P., Rubio, V., and Marina, A. (2009). Structural insight into partner specificity and phosphoryl transfer in two-component signal transduction. Cell *139*, 325–336.

Castandet, J., Prost, J.F., Peyron, P., Astarie-Dequeker, C., Anes, E., Cozzone, A.J., Griffiths, G., and Maridonneau-Parini, I. (2005). Tyrosine phosphatase MptpA of *Mycobacterium tuberculosis* inhibits phagocytosis and increases actin polymerization in macrophages. Res. Microbiol. *156*, 1005–1013.

Chamnongpol, S., Cromie, M., and Groisman, E.A. (2003). Mg2+ sensing by the Mg2+ sensor PhoQ of *Salmonella enterica*. J. Mol. Biol. *325*, 795–807.

Chao, J., Wong, D., Zheng, X., Poirier, V., Bach, H., Hmama, Z., and Av-Gay, Y. (2010). Protein kinase and phosphatase signaling in *Mycobacterium tuberculosis* physiology and pathogenesis. Biochim. Biophys. Acta *1804*, 620–627.

Chen, C.K., Chan, N.L., and Wang, A.H. (2011). The many blades of the beta-propeller proteins: conserved but versatile. Trends Biochem. Sci. *36*, 553–561.

Chen, X., Schauder, S., Potier, N., Van Dorsselaer, A., Pelczer, I., Bassler, B.L., and Hughson, F.M. (2002). Structural identification of a bacterial quorum-sensing signal containing boron. Nature *415*, 545–549.

Cheung, J., and Hendrickson, W.A. (2009). Structural analysis of ligand stimulation of the histidine kinase NarX. Structure *17*, 190–201.

Chow, K., Ng, D., Stokes, R., and Johnson, P. (1994). Protein tyrosine phosphorylation in *Mycobacterium tuberculosis*. FEMS Microbiol. Lett. *124*, 203–207.

Cohen, P., and Cohen, P.T. (1989). Protein phosphatases come of age. J. Biol. Chem. *264*, 21435–21438.

Cole, S.T., Brosch, R., Parkhill, J., Garnier, T., Churcher, C., Harris, D., Gordon, S.V., Eiglmeier, K., Gas, S., Barry, C.E. 3rd, *et al.* (1998). Deciphering the biology of *Mycobacterium tuberculosis* from the complete genome sequence. Nature *393*, 537–544.

Cortay, J.C., Duclos, B., and Cozzone, A.J. (1986). Phosphorylation of an *Escherichia coli* protein at tyrosine. J. Mol. Biol. *187*, 305–308.

Cozzone, A.J. (2005). Role of protein phosphorylation on serine/threonine and tyrosine in the virulence of bacterial pathogens. J. Mol. Microbiol. Biotechnol. *9*, 198–213.

Dar, A.C., Dever, T.E., and Sicheri, F. (2005). Higher-order substrate recognition of eIF2alpha by the RNA-dependent protein kinase PKR. Cell *122*, 887–900.

Das, A.K., Helps, N.R., Cohen, P.T., and Barford, D. (1996). Crystal structure of the protein serine/threonine phosphatase 2C at 2.0 A resolution. EMBO J. *15*, 6798–6809.

De Carlo, S., Chen, B., Hoover, T.R., Kondrashkina, E., Nogales, E., and Nixon, B.T. (2006). The structural basis for regulated assembly and function of the transcriptional activator NtrC. Genes Dev. *20,* 1485–1495.

Dey, M., Cao, C., Dar, A.C., Tamura, T., Ozato, K., Sicheri, F., and Dever, T.E. (2005). Mechanistic link between PKR dimerization, autophosphorylation, and eIF2alpha substrate recognition. Cell *122,* 901–913.

Duran, R., Villarino, A., Bellinzoni, M., Wehenkel, A., Fernandez, P., Boitel, B., Cole, S.T., Alzari, P.M., and Cervenansky, C. (2005). Conserved autophosphorylation pattern in activation loops and juxtamembrane regions of *Mycobacterium tuberculosis* Ser/Thr protein kinases. Biochem. Biophys. Res. Commun. *333,* 858–867.

Dyer, C.M., Vartanian, A.S., Zhou, H., and Dahlquist, F.W. (2009). A molecular mechanism of bacterial flagellar motor switching. J. Mol. Biol. *388,* 71–84.

Falke, J.J., and Erbse, A.H. (2009). The piston rises again. Structure *17,* 1149–1151.

Falke, J.J., and Hazelbauer, G.L. (2001). Transmembrane signaling in bacterial chemoreceptors. Trends Biochem. Sci. *26,* 257–265.

Fernandez, P., Saint-Joanis, B., Barilone, N., Jackson, M., Gicquel, B., Cole, S.T., and Alzari, P.M. (2006). The Ser/Thr protein kinase PknB is essential for sustaining mycobacterial growth. J. Bacteriol. *188,* 7778–7784.

Ferris, H.U., Dunin-Horkawicz, S., Mondejar, L.G., Hulko, M., Hantke, K., Martin, J., Schultz, J.E., Zeth, K., Lupas, A.N., and Coles, M. (2011). The mechanisms of HAMP-mediated signaling in transmembrane receptors. Structure *19,* 378–385.

Ferris, H.U., Dunin-Horkawicz, S., Hornig, N., Hulko, M., Martin, J., Schultz, J.E., Zeth, K., Lupas, A.N., and Coles, M. (2012). Mechanism of regulation of receptor histidine kinases. Structure *20,* 56–66.

Fleischer, R., Heermann, R., Jung, K., and Hunke, S. (2007). Purification, reconstitution, and characterization of the CpxRAP envelope stress system of *Escherichia coli.* J. Biol. Chem. *282,* 8583–8593.

Frasch, S.C., and Dworkin, M. (1996). Tyrosine phosphorylation in *Myxococcus xanthus,* a multicellular prokaryote. J. Bacteriol. *178,* 4084–4088.

Fujita, M., Gonzalez-Pastor, J.E., and Losick, R. (2005). High- and low-threshold genes in the Spo0A regulon of *Bacillus subtilis.* J. Bacteriol. *187,* 1357–1368.

Galperin, M.Y., and Koonin, E.V. (2010). From complete genome sequence to 'complete' understanding? Trends Biotechnol. *28,* 398–406.

Gao, R., and Stock, A.M. (2009). Biological insights from structures of two-component proteins. Annu. Rev. Microbiol. *63,* 133–154.

Gao, R., and Stock, A.M. (2010). Molecular strategies for phosphorylation-mediated regulation of response regulator activity. Curr. Opin. Microbiol. *13,* 160–167.

Gay, L.M., Ng, H.L., and Alber, T. (2006). A conserved dimer and global conformational changes in the structure of apo-PknE Ser/Thr protein kinase from *Mycobacterium tuberculosis.* J. Mol. Biol. *360,* 409–420.

Good, M.C., Greenstein, A.E., Young, T.A., Ng, H.L., and Alber, T. (2004). Sensor domain of the *Mycobacterium tuberculosis* receptor Ser/Thr protein kinase, PknD, forms a highly symmetric beta propeller. J. Mol. Biol. *339,* 459–469.

Grangeasse, C., Cozzone, A.J., Deutscher, J., and Mijakovic, I. (2007). Tyrosine phosphorylation: an emerging regulatory device of bacterial physiology. Trends Biochem. Sci. *32,* 86–94.

Grangeasse, C., Terreux, R., and Nessler, S. (2010). Bacterial tyrosine-kinases: structure–function analysis and therapeutic potential. Biochim. Biophys. Acta *1804,* 628–634.

Greenstein, A.E., Grundner, C., Echols, N., Gay, L.M., Lombana, T.N., Miecskowski, C.A., Pullen, K.E., Sung, P.Y., and Alber, T. (2005). Structure/function studies of Ser/Thr and Tyr protein phosphorylation in *Mycobacterium tuberculosis.* J. Mol. Microbiol. Biotechnol. *9,* 167–181.

Greenstein, A.E., Echols, N., Lombana, T.N., King, D.S., and Alber, T. (2007). Allosteric activation by dimerization of the PknD receptor Ser/Thr protein kinase from *Mycobacterium tuberculosis.* J. Biol. Chem. *282,* 11427–11435.

Groves, J.T., and Kuriyan, J. (2010). Molecular mechanisms in signal transduction at the membrane. Nat. Struct. Mol. Biol. *17,* 659–665.

Grundner, C., Perrin, D., Hooft van Huijsduijnen, R., Swinnen, D., Gonzalez, J., Gee, C.L., Wells, T.N., and Alber, T. (2007). Structural basis for selective inhibition of *Mycobacterium tuberculosis* protein tyrosine phosphatase PtpB. Structure *15,* 499–509.

Hall, B.A., Armitage, J.P., and Sansom, M.S. (2011). Transmembrane helix dynamics of bacterial chemoreceptors supports a piston model of signalling. PLoS Comput. Biol. *7,* e1002204.

Hanks, S.K., and Hunter, T. (1995). Protein kinases 6. The eukaryotic protein kinase superfamily: kinase (catalytic) domain structure and classification. FASEB J. *9*, 576–596.

Higgins, D., and Dworkin, J. (2012). Recent progress in *Bacillus subtilis* sporulation. FEMS Microbiol. Rev. *36*, 131–148.

Hilbert, D.W., and Piggot, P.J. (2004). Compartmentalization of gene expression during *Bacillus subtilis* spore formation. Microbiol. Mol. Biol. Rev. *68*, 234–262.

Hirakata, T., Kieser, H., Hopwood, D., Urabe, H., and Ogarawa, H. (1998). Putative protein serine/threonine kinase genes are located in several positions on the chromosome of *Streptomyces coelicolor* A3(2). FEMS Microbiol. Lett. *159*, 1–5.

Hsing, W., Russo, F.D., Bernd, K.K., and Silhavy, T.J. (1998). Mutations that alter the kinase and phosphatase activities of the two-component sensor EnvZ. J. Bacteriol. *180*, 4538–4546.

Hsu, F., Schwarz, S., and Mougous, J.D. (2009). TagR promotes PpkA-catalysed type VI secretion activation in *Pseudomonas aeruginosa*. Mol. Microbiol. *72*, 1111–1125.

Hulko, M., Berndt, F., Gruber, M., Linder, J.U., Truffault, V., Schultz, A., Martin, J., Schultz, J.E., Lupas, A.N., and Coles, M. (2006). The HAMP domain structure implies helix rotation in transmembrane signaling. Cell *126*, 929–940.

Huse, M., and Kuriyan, J. (2002). The conformational plasticity of protein kinases. Cell *109*, 275–282.

Jackson, M.D., and Denu, J.M. (2001). Molecular reactions of protein phosphatases – insights from structure and chemistry. Chem. Rev. *101*, 2313–2340.

Jiang, P., Atkinson, M.R., Srisawat, C., Sun, Q., and Ninfa, A.J. (2000). Functional dissection of the dimerization and enzymatic activities of *Escherichia coli* nitrogen regulator II and their regulation by the PII protein. Biochemistry *39*, 13433–13449.

Johnson, L.N., Noble, M.E., and Owen, D.J. (1996). Active and inactive protein kinases: structural basis for regulation. Cell *85*, 149–158.

Kang, C.M., Abbott, D.W., Park, S.T., Dascher, C.C., Cantley, L.C., and Husson, R.N. (2005). The *Mycobacterium tuberculosis* serine/threonine kinases PknA and PknB: substrate identification and regulation of cell shape. Genes Dev. *19*, 1692–1704.

Keep, N.H., Ward, J.M., Cohen-Gonsaud, M., and Henderson, B. (2006). Wake up! Peptidoglycan lysis and bacterial non-growth states. Trends Microbiol. *14*, 271–276.

Kim, M.-J., Park, K.-J., Ko, I.-J., Kim, Y.M., and Oh, J.-I. (2010). Different roles of DosS and DosT in the hypoxic adaptation of Mycobacteria. J. Bacteriol. *192*, 4868–4875.

Knighton, D.R., Zheng, J.H., Ten Eyck, L.F., Ashford, V.A., Xuong, N.H., Taylor, S.S., and Sowadski, J.M. (1991a). Crystal structure of the catalytic subunit of cyclic adenosine monophosphate-dependent protein kinase. Science *253*, 407–414.

Knighton, D.R., Zheng, J.H., Ten Eyck, L.F., Xuong, N.H., Taylor, S.S., and Sowadski, J.M. (1991b). Structure of a peptide inhibitor bound to the catalytic subunit of cyclic adenosine monophosphate-dependent protein kinase. Science *253*, 414–420.

Kobir, A., Shi, L., Boskovic, A., Grangeasse, C., Franjevic, D., and Mijakovic, I. (2011). Protein phosphorylation in bacterial signal transduction. Biochim. Biophys. Acta *1810*, 989–994.

Kornev, A.P., and Taylor, S.S. (2010). Defining the conserved internal architecture of a protein kinase. Biochim. Biophys. Acta *1804*, 440–444.

Kornev, A.P., Taylor, S.S., and Ten Eyck, L.F. (2008). A helix scaffold for the assembly of active protein kinases. Proc. Natl. Acad. Sci. U.S.A. *105*, 14377–14382.

Kristich, C.J., Wells, C.L., and Dunny, G.M. (2007). A eukaryotic-type Ser/Thr kinase in *Enterococcus faecalis* mediates antimicrobial resistance and intestinal persistence. Proc. Natl. Acad. Sci. U.S.A. *104*, 3508–3513.

Kulasekara, H.D., and Miller, S.I. (2007). Threonine phosphorylation times bacterial secretion. Nat. Cell Biol. *9*, 734–736.

Kwon, O., Georgellis, D., and Lin, E.C. (2003). Rotational on–off switching of a hybrid membrane sensor kinase Tar-ArcB in *Escherichia coli*. J. Biol. Chem. *278*, 13192–13195.

Lin, M.H., Hsu, T.L., Lin, S.Y., Pan, Y.J., Jan, J.T., Wang, J.T., Khoo, K.H., and Wu, S.H. (2009). Phosphoproteomics of Klebsiella pneumoniae NTUH-K2044 reveals a tight link between tyrosine phosphorylation and virulence. Mol. Cell Proteomics *8*, 2613–2623.

Lin, S.L., Le, T.X., and Cowen, D.S. (2003). SptP, a Salmonella typhimurium type III-secreted protein, inhibits the mitogen-activated protein kinase pathway by inhibiting Raf activation. Cell Microbiol. *5*, 267–275.

Lombana, T.N., Echols, N., Good, M.C., Thomsen, N.D., Ng, H.L., Greenstein, A.E., Falick, A.M., King, D.S., and Alber, T. (2010). Allosteric activation mechanism of the *Mycobacterium tuberculosis* receptor Ser/Thr protein kinase, PknB. Structure *18*, 1667–1677.

Lougheed, K.E., Osborne, S.A., Saxty, B., Whalley, D., Chapman, T., Bouloc, N., Chugh, J., Nott, T.J., Patel, D., Spivey, V.L., *et al.* (2011). Effective inhibitors of the essential kinase PknB and their potential as anti-mycobacterial agents. Tuberculosis (Edinb) *91*, 277–286.

Lux, R., and Shi, W. (2005). A novel bacterial signalling system with a combination of a Ser/Thr kinase cascade and a His/Asp two-component system. Mol. Microbiol. *58*, 345–348.

Macek, B., Gnad, F., Soufi, B., Kumar, C., Olsen, J.V., Mijakovic, I., and Mann, M. (2008). Phosphoproteome analysis of *E. coli* reveals evolutionary conservation of bacterial Ser/Thr/Tyr phosphorylation. Mol. Cell Proteomics *7*, 299–307.

Macek, B., Mijakovic, I., Olsen, J.V., Gnad, F., Kumar, C., Jensen, P.R., and Mann, M. (2007). The serine/threonine/tyrosine phosphoproteome of the model bacterium *Bacillus subtilis*. Mol. Cell Proteomics *6*, 697–707.

Madhurantakam, C., Rajakumara, E., Mazumdar, P.A., Saha, B., Mitra, D., Wiker, H.G., Sankaranarayanan, R., and Das, A.K. (2005). Crystal structure of low-molecular-weight protein tyrosine phosphatase from *Mycobacterium tuberculosis* at 1.9-A resolution. J. Bacteriol. *187*, 2175–2181.

Marina, A., Waldburger, C.D., and Hendrickson, W.A. (2005). Structure of the entire cytoplasmic portion of a sensor histidine-kinase protein. EMBO J. *24*, 4247–4259.

Matsumoto, A., Hong, S.K., Ishizuka, H., Horinouchi, S., and Beppu, T. (1994). Phosphorylation of the AfsR protein involved in secondary metabolism in Streptomyces species by a eukaryotic-type protein kinase. Gene *146*, 47–56.

Milburn, M.V., Prive, G.G., Milligan, D.L., Scott, W.G., Yeh, J., Jancarik, J., Koshland, D.E. Jr., and Kim, S.H. (1991). Three-dimensional structures of the ligand-binding domain of the bacterial aspartate receptor with and without a ligand. Science *254*, 1342–1347.

Mir, M., Asong, J., Li, X., Cardot, J., Boons, G.J., and Husson, R.N. (2011). The extracytoplasmic domain of the *Mycobacterium tuberculosis* Ser/Thr kinase PknB binds specific muropeptides and is required for PknB localization. PLoS Pathog *7*, e1002182.

Moglich, A., Ayers, R.A., and Moffat, K. (2009). Design and signaling mechanism of light-regulated histidine kinases. J. Mol. Biol. *385*, 1433–1444.

Molle, V., and Kremer, L. (2010). Division and cell envelope regulation by Ser/Thr phosphorylation: mycobacterium shows the way. Mol. Microbiol. *75*, 1064–1077.

Molle, V., Fujita, M., Jensen, S.T., Eichenberger, P., Gonzalez-Pastor, J.E., Liu, J.S., and Losick, R. (2003a). The Spo0A regulon of *Bacillus subtilis*. Mol. Microbiol. *50*, 1683–1701.

Molle, V., Kremer, L., Girard-Blanc, C., Besra, G.S., Cozzone, A.J., and Prost, J.F. (2003b). An FHA phosphoprotein recognition domain mediates protein EmbR phosphorylation by PknH, a Ser/Thr protein kinase from *Mycobacterium tuberculosis*. Biochemistry *42*, 15300–15309.

Mougous, J.D., Gifford, C.A., Ramsdell, T.L., and Mekalanos, J.J. (2007). Threonine phosphorylation post-translationally regulates protein secretion in *Pseudomonas aeruginosa*. Nat. Cell Biol. *9*, 797–803.

Moukhametzianov, R., Klare, J.P., Efremov, R., Baeken, C., Goppner, A., Labahn, J., Engelhard, M., Buldt, G., and Gordeliy, V.I. (2006). Development of the signal in sensory rhodopsin and its transfer to the cognate transducer. Nature *440*, 115–119.

Munoz-Dorado, J., Inouye, S., and Inouye, M. (1991). A gene encoding a protein serine/threonine kinase is required for normal development of *M. xanthus*, a gram-negative bacterium. Cell *67*, 995–1006.

Neiditch, M.B., Federle, M.J., Pompeani, A.J., Kelly, R.C., Swem, D.L., Jeffrey, P.D., Bassler, B.L., and Hughson, F.M. (2006). Ligand-induced asymmetry in histidine sensor kinase complex regulates quorum sensing. Cell *126*, 1095–1108.

Nixon, B.T., Ronson, C.W., and Ausubel, F.M. (1986). Two-component regulatory systems responsive to environmental stimuli share strongly conserved domains with the nitrogen assimilation regulatory genes ntrB and ntrC. Proc. Natl. Acad. Sci. U.S.A. *83*, 7850–7854.

Nolen, B., Taylor, S., and Ghosh, G. (2004). Regulation of protein kinases; controlling activity through activation segment conformation. Mol. Cell *15*, 661–675.

Nott, T.J., Kelly, G., Stach, L., Li, J., Westcott, S., Patel, D., Hunt, D.M., Howell, S., Buxton, R.S., O'Hare, H.M., *et al.* (2009). An intramolecular switch regulates phosphoindependent FHA domain interactions in *Mycobacterium tuberculosis*. Sci. Signal *2*, ra12.

O'Hare, H.M., Duran, R., Cervenansky, C., Bellinzoni, M., Wehenkel, A.M., Pritsch, O., Obal, G., Baumgartner, J., Vialaret, J., Johnsson, K., *et al.* (2008). Regulation of glutamate metabolism by protein kinases in mycobacteria. Mol. Microbiol. *70*, 1408–1423.

Ortiz-Lombardia, M., Pompeo, F., Boitel, B., and Alzari, P.M. (2003). Crystal structure of the catalytic domain of the PknB serine/threonine kinase from *Mycobacterium tuberculosis*. J. Biol. Chem. *278*, 13094–13100.

Papavinasasundaram, K.G., Chan, B., Chung, J.H., Colston, M.J., Davis, E.O., and Av-Gay, Y. (2005). Deletion of the *Mycobacterium tuberculosis* pknH gene confers a higher bacillary load during the chronic phase of infection in BALB/c mice. J. Bacteriol. *187*, 5751–5760.

Paracuellos, P., Ballandras, A., Robert, X., Kahn, R., Herve, M., Mengin-Lecreulx, D., Cozzone, A.J., Duclos, B., and Gouet, P. (2010). The extended conformation of the 2.9-A crystal structure of the three-PASTA domain of a Ser/Thr kinase from the human pathogen *Staphylococcus aureus*. J. Mol. Biol. *404*, 847–858.

Parker, J.L., Jones, A.M., Serazetdinova, L., Saalbach, G., Bibb, M.J., and Naldrett, M.J. (2010). Analysis of the phosphoproteome of the multicellular bacterium *Streptomyces coelicolor* A3(2) by protein/peptide fractionation, phosphopeptide enrichment and high-accuracy mass spectrometry. Proteomics *10*, 2486–2497.

Peirs, P., De Wit, L., Braibant, M., Huygen, K., and Content, J. (1997). A serine/threonine protein kinase from *Mycobacterium tuberculosis*. Eur. J. Biochem. *244*, 604–612.

Pereira, S.F., Goss, L., and Dworkin, J. (2011). Eukaryote-like serine/threonine kinases and phosphatases in bacteria. Microbiol. Mol. Biol. Rev. *75*, 192–212.

Perez, J., Castaneda-Garcia, A., Jenke-Kodama, H., Muller, R., and Munoz-Dorado, J. (2008). Eukaryotic-like protein kinases in the prokaryotes and the myxobacterial kinome. Proc. Natl. Acad. Sci. U.S.A. *105*, 15950–15955.

Pirruccello, M., Sondermann, H., Pelton, J.G., Pellicena, P., Hoelz, A., Chernoff, J., Wemmer, D.E., and Kuriyan, J. (2006). A dimeric kinase assembly underlying autophosphorylation in the p21 activated kinases. J. Mol. Biol. *361*, 312–326.

Pons, T., Gomez, R., Chinea, G., and Valencia, A. (2003). Beta-propellers: associated functions and their role in human diseases. Curr. Med. Chem. *10*, 505–524.

Prisic, S., Dankwa, S., Schwartz, D., Chou, M.F., Locasale, J.W., Kang, C.M., Bemis, G., Church, G.M., Steen, H., and Husson, R.N. (2010). Extensive phosphorylation with overlapping specificity by *Mycobacterium tuberculosis* serine/threonine protein kinases. Proc. Natl. Acad. Sci. U.S.A. *107*, 7521–7526.

Pullen, K.E., Ng, H.L., Sung, P.Y., Good, M.C., Smith, S.M., and Alber, T. (2004). An alternate conformation and a third metal in PstP/Ppp, the *M. tuberculosis* PP2C-Family Ser/Thr protein phosphatase. Structure *12*, 1947–1954.

Rantanen, M.K., Lehtio, L., Rajagopal, L., Rubens, C.E., and Goldman, A. (2007). Structure of *Streptococcus agalactiae* serine/threonine phosphatase. The subdomain conformation is coupled to the binding of a third metal ion. FEBS J. *274*, 3128–3137.

Ravichandran, A., Sugiyama, N., Tomita, M., Swarup, S., and Ishihama, Y. (2009). Ser/Thr/Tyr phosphoproteome analysis of pathogenic and non-pathogenic Pseudomonas species. Proteomics *9*, 2764–2775.

Ruggiero, A., Squeglia, F., Marasco, D., Marchetti, R., Molinaro, A., and Berisio, R. (2011). X-ray structural studies of the entire extracellular region of the serine/threonine kinase PrkC from *Staphylococcus aureus*. Biochem. J. *435*, 33–41.

Sarkar, M.K., Paul, K., and Blair, D. (2010). Chemotaxis signaling protein CheY binds to the rotor protein FliN to control the direction of flagellar rotation in *Escherichia coli*. Proc. Natl. Acad. Sci. U.S.A. *107*, 9370–9375.

Scherr, N., Honnappa, S., Kunz, G., Mueller, P., Jayachandran, R., Winkler, F., Pieters, J., and Steinmetz, M.O. (2007). Structural basis for the specific inhibition of protein kinase G, a virulence factor of *Mycobacterium tuberculosis*. Proc. Natl. Acad. Sci. U.S.A. *104*, 12151–12156.

Schlicker, C., Fokina, O., Kloft, N., Grune, T., Becker, S., Sheldrick, G.M., and Forchhammer, K. (2008). Structural analysis of the PP2C phosphatase tPphA from Thermosynechococcus elongatus: a flexible flap subdomain controls access to the catalytic site. J. Mol. Biol. *376*, 570–581.

Schmidl, S.R., Gronau, K., Pietack, N., Hecker, M., Becher, D., and Stulke, J. (2010). The phosphoproteome of the minimal bacterium Mycoplasma pneumoniae: analysis of the complete known Ser/Thr kinome suggests the existence of novel kinases. Mol. Cell Proteomics *9*, 1228–1242.

Schultz, C., Niebisch, A., Schwaiger, A., Viets, U., Metzger, S., Bramkamp, M., and Bott, M. (2009). Genetic and biochemical analysis of the serine/threonine protein kinases PknA, PknB, PknG and PknL of *Corynebacterium glutamicum*: evidence for non-essentiality and for phosphorylation of OdhI and FtsZ by multiple kinases. Mol. Microbiol. *74*, 724–741.

Sevvana, M., Vijayan, V., Zweckstetter, M., Reinelt, S., Madden, D.R., Herbst-Irmer, R., Sheldrick, G.M., Bott, M., Griesinger, C., and Becker, S. (2008). A ligand-induced switch in the periplasmic domain of sensor histidine kinase CitA. J. Mol. Biol. *377*, 512–523.

Shah, I.M., and Dworkin, J. (2010). Induction and regulation of a secreted peptidoglycan hydrolase by a membrane Ser/Thr kinase that detects muropeptides. Mol. Microbiol. *75*, 1232–1243.

Shah, I.M., Laaberki, M.H., Popham, D.L., and Dworkin, J. (2008). A eukaryotic-like Ser/Thr kinase signals bacteria to exit dormancy in response to peptidoglycan fragments. Cell *135*, 486–496.

Singh, A., Singh, Y., Pine, R., Shi, L., Chandra, R., and Drlica, K. (2006). Protein kinase I of *Mycobacterium tuberculosis*: cellular localization and expression during infection of macrophage-like cells. Tuberculosis (Edinb) *86*, 28–33.

Singh, R., Rao, V., Shakila, H., Gupta, R., Khera, A., Dhar, N., Singh, A., Koul, A., Singh, Y., Naseema, M., *et al.* (2003). Disruption of mptpB impairs the ability of *Mycobacterium tuberculosis* to survive in guinea pigs. Mol. Microbiol. *50*, 751–762.

Soufi, B., Gnad, F., Jensen, P.R., Petranovic, D., Mann, M., Mijakovic, I., and Macek, B. (2008). The Ser/Thr/Tyr phosphoproteome of *Lactococcus lactis* IL1403 reveals multiply phosphorylated proteins. Proteomics *8*, 3486–3493.

Squeglia, F., Marchetti, R., Ruggiero, A., Lanzetta, R., Marasco, D., Dworkin, J., Petoukhov, M., Molinaro, A., Berisio, R., and Silipo, A. (2011). Chemical basis of peptidoglycan discrimination by PrkC, a key kinase involved in bacterial resuscitation from dormancy. J. Am. Chem. Soc. *133*, 20676–20679.

Su, J., Schlicker, C., and Forchhammer, K. (2011). A third metal is required for catalytic activity of the signal-transducing protein phosphatase M tPphA. J. Biol. Chem. *286*, 13481–13488.

Sun, X., Ge, F., Xiao, C.L., Yin, X.F., Ge, R., Zhang, L.H., and He, Q.Y. (2010). Phosphoproteomic analysis reveals the multiple roles of phosphorylation in pathogenic bacterium *Streptococcus pneumoniae*. J. Proteome Res. *9*, 275–282.

Taylor, S.S., and Kornev, A.P. (2011). Protein kinases: evolution of dynamic regulatory proteins. Trends Biochem. Sci. *36*, 65–77.

Thakur, M., and Chakraborti, P.K. (2006). GTPase activity of mycobacterial FtsZ is impaired due to its transphosphorylation by the eukaryotic-type Ser/Thr kinase, PknA. J. Biol. Chem. *281*, 40107–40113.

Thanbichler, M. (2009). Spatial regulation in *Caulobacter crescentus*. Curr. Opin. Microbiol. *12*, 715–721.

Timmen, M., Bassler, B.L., and Jung, K. (2006). AI-1 influences the kinase activity but not the phosphatase activity of LuxN of Vibrio harveyi. J. Biol. Chem. *281*, 24398–24404.

Tomomori, C., Tanaka, T., Dutta, R., Park, H., Saha, S.K., Zhu, Y., Ishima, R., Liu, D., Tong, K.I., Kurokawa, H., *et al.* (1999). Solution structure of the homodimeric core domain of *Escherichia coli* histidine kinase EnvZ. Nat. Struct. Biol. *6*, 729–734.

Trajtenberg, F., Grana, M., Ruetalo, N., Botti, H., and Buschiazzo, A. (2010). Structural and enzymatic insights into the ATP binding and autophosphorylation mechanism of a sensor histidine kinase. J. Biol. Chem. *285*, 24892–24903.

Ueki, T., and Inouye, S. (2006). A novel regulation on developmental gene expression of fruiting body formation in Myxobacteria. Appl. Microbiol. Biotechnol. *72*, 21–29.

Vega, C., Chou, S., Engel, K., Harrell, M.E., Rajagopal, L., and Grundner, C. (2011). Structure and substrate recognition of the *Staphylococcus aureus* protein tyrosine phosphatase PtpA. J. Mol. Biol. *413*, 24–31.

Wagner, T., Bellinzoni, M., Wehenkel, A., O'Hare, H.M., and Alzari, P.M. (2011). Functional plasticity and allosteric regulation of alpha-ketoglutarate decarboxylase in central mycobacterial metabolism. Chem. Biol. *18*, 1011–1020.

Walburger, A., Koul, A., Ferrari, G., Nguyen, L., Prescianotto-Baschong, C., Huygen, K., Klebl, B., Thompson, C., Bacher, G., and Pieters, J. (2004). Protein kinase G from pathogenic mycobacteria promotes survival within macrophages. Science *304*, 1800–1804.

Wassmann, P., Chan, C., Paul, R., Beck, A., Heerklotz, H., Jenal, U., and Schirmer, T. (2007). Structure of BeF3- -modified response regulator PleD: implications for diguanylate cyclase activation, catalysis, and feedback inhibition. Structure *15*, 915–927.

Wehenkel, A., Fernandez, P., Bellinzoni, M., Catherinot, V., Barilone, N., Labesse, G., Jackson, M., and Alzari, P.M. (2006). The structure of PknB in complex with mitoxantrone, an ATP-competitive inhibitor, suggests a mode of protein kinase regulation in mycobacteria. FEBS Lett. *580*, 3018–3022.

Wehenkel, A., Bellinzoni, M., Schaeffer, F., Villarino, A., and Alzari, P.M. (2007). Structural and binding studies of the three-metal center in two mycobacterial PPM Ser/Thr protein phosphatases. J. Mol. Biol. *374*, 890–898.

Wehenkel, A., Bellinzoni, M., Grana, M., Duran, R., Villarino, A., Fernandez, P., Andre-Leroux, G., England, P., Takiff, H., Cervenansky, C., *et al.* (2008). Mycobacterial Ser/Thr protein kinases and phosphatases: physiological roles and therapeutic potential. Biochim. Biophys. Acta *1784*, 193–202.

Weiss, V., Kramer, G., Dunnebier, T., and Flotho, A. (2002). Mechanism of regulation of the bifunctional histidine kinase NtrB in *Escherichia coli*. J. Mol. Microbiol. Biotechnol. *4*, 229–233.

Wurgler-Murphy, S.M., King, D.M., and Kennelly, P.J. (2004). The phosphorylation site database: a guide to the serine-, threonine-, and/or tyrosine-phosphorylated proteins in prokaryotic organisms. Proteomics *4*, 1562–1570.

Yamada, S., Sugimoto, H., Kobayashi, M., Ohno, A., Nakamura, H., and Shiro, Y. (2009). Structure of PAS-linked histidine kinase and the response regulator complex. Structure *17*, 1333–1344.

Yang, X., Kuk, J., and Moffat, K. (2008). Crystal structure of *Pseudomonas aeruginosa* bacteriophytochrome: photoconversion and signal transduction. Proc. Natl. Acad. Sci. U.S.A. *105*, 14715–14720.

Yeh, J.I., Biemann, H.P., Prive, G.G., Pandit, J., Koshland, D.E. Jr., and Kim, S.H. (1996). High-resolution structures of the ligand binding domain of the wild-type bacterial aspartate receptor. J. Mol. Biol. *262*, 186–201.

Young, T.A., Delagoutte, B., Endrizzi, J.A., Falick, A.M., and Alber, T. (2003). Structure of *Mycobacterium tuberculosis* PknB supports a universal activation mechanism for Ser/Thr protein kinases. Nat. Struct. Biol. *10*, 168–174.

Yuan, M., Deleuil, F., and Fallman, M. (2005). Interaction between the Yersinia tyrosine phosphatase YopH and its macrophage substrate, Fyn-binding protein, Fyb. J. Mol. Microbiol. Biotechnol. *9*, 214–223.

Zapf, J., Sen, U., Madhusudan, Hoch., J.A., and Varughese, K.I. (2000). A transient interaction between two phosphorelay proteins trapped in a crystal lattice reveals the mechanism of molecular recognition and phosphotransfer in signal transduction. Structure *8*, 851–862.

Zheng, X., Papavinasasundaram, K.G., and Av-Gay, Y. (2007). Novel substrates of *Mycobacterium tuberculosis* PknH Ser/Thr kinase. Biochem. Biophys. Res. Commun. *355*, 162–168.

Part V

Bacterial Membranes in Pathogenesis

Outer Membrane-embedded and -associated Proteins and their Role in Adhesion and Pathogenesis

13

Vincent van Dam, Virginie Roussel-Jazédé, Jesús Arenas, Martine P. Bos and Jan Tommassen

Abstract

The cell envelope of Gram-negative bacteria is composed of two membranes, which are separated by the periplasm containing a layer of peptidoglycan. The outer membrane is in contact with the environment. It contains a myriad of integral and associated proteins that are involved in adhesion to biotic and abiotic surfaces and, in the case of pathogens, in virulence. To reach the cell surface, these proteins have to cross the entire cell envelope, which is accomplished via various secretion pathways. *Neisseria meningitidis*, a commensal that lives in the nasopharynx but occasionally causes sepsis and/or meningitis, expresses a plethora of these virulence factors including type IV pili, proteins secreted via the type V secretion pathway, and cell surface-exposed lipoproteins. Here, we discuss the biogenesis and the function of such virulence factors with emphasis on those produced by *N. meningitidis*.

Introduction

The cell envelope of Gram-negative bacteria consists of two membranes, the inner or cytoplasmic membrane and the outer membrane. The membranes are separated by the periplasm containing a cell wall composed of peptidoglycan. In most Gram-negative bacteria, the outer membrane is an asymmetrical bilayer with phospholipids present exclusively in the inner leaflet and an outer leaflet composed of lipopolysaccharides (LPS, a.k.a. endotoxin) (Bos *et al.*, 2007). It is impermeable to hydrophobic and large hydrophilic compounds and functions as a protective barrier preventing the access of toxic compounds from the environment to the cell interior.

In contrast to the integral proteins usually found in biological membranes, which span the membrane in the form of α-helices composed of hydrophobic amino acids, integral outer membrane proteins (OMPs) generally form β-barrel structures (Bos *et al.*, 2007). They consist of an even number of antiparallel membrane-spanning amphipathic β-strands, which are connected by short turns at the periplasmic side and long loops at the surface-exposed side of the membrane. Most OMPs are involved in nutrient acquisition. The most abundant OMPs are the porins, which form large water-filled channels allowing the passage of hydrophilic solutes with molecular weights up to ~600 Da by passive diffusion. Besides integral OMPs, the outer membrane contains also lipoproteins, which are usually anchored to the membrane only via an N-terminal *N*-acyl-diacylglycerylcysteine moiety with the proteinaceous part extending either in the periplasm or in the external medium (Bos *et al.*, 2007). They are involved in a variety of functions including the biogenesis and the maintenance of the integrity of the outer membrane.

Gram-negative bacteria are capable of secreting proteins across the entire cell envelope to the cell surface or into the extracellular milieu. For this purpose, they have developed a large variety of secretion pathways (Gerlach and Hensel, 2007). Some of these pathways involve two separate steps for the translocation of the substrates across the inner and the outer membrane. In other systems, the substrates are transported in a single step without a periplasmic intermediate across the entire cell envelope into the extracellular milieu or directly into the cytosol of a eukaryotic target cell. The secretion machineries involved can be fairly simple consisting of only one or a few proteins, or they can be very complex consisting of up to 25 different proteins (Gerlach and Hensel, 2007).

Many surface-associated and secreted proteins of pathogenic bacteria function as virulence factors. Among the secreted proteins are cytotoxins, which destroy host tissues, and hydrolytic enzymes, which digest large macromolecules to allow for the uptake of the resulting residues as nutrients. Cell surface-associated proteins include adhesins, with which the bacteria adhere to biotic and abiotic surfaces, and proteins that help the bacteria to escape from the defence mechanisms of the host.

In this review, we will discuss outer membrane-embedded and surface-associated proteins that play a role in the interaction of bacteria with a eukaryotic host. Collectively, Gram-negative bacteria have developed a large variety of such proteins, which cannot be covered within the space limitations of this review. In the examples discussed, the virulence factors produced by *Neisseria meningitidis* and its close relative *Neisseria gonorrhoeae* are emphasized. *N. meningitidis* is a Gram-negative diplococcus that lives in the nasopharynx but occasionally causes sepsis and/or meningitis. Meningococcal disease has a high morbidity and mortality and a universally protective vaccine is very much wanted. Among the cell surface-exposed virulence factors produced are pili, proteins secreted by the type V secretion pathway, lipoproteins, and integral OMPs. We will discuss the role of these virulence factors and of related virulence factors from other bacteria in the interaction with the host. We will also discuss what is known about the biogenesis of these proteins, except for the biogenesis of integral OMPs, for which readers should refer to Chapter 3.

Pili

Pili constitute an important class of adhesive structures used by bacteria to adhere to host-cell surfaces. Pili are hair-like filamentous surface appendages (Fig. 13.1) that are composed of several hundreds or even thousands of subunits, named pilins. They extend from the cell surface, thereby allowing initial interactions with host cells from a distance. Apart from having adhesive properties, pili can play a role in bacterial colonization and pathogenesis via DNA transfer, biofilm formation, cell aggregation, host-cell invasion, and twitching motility. In addition, pili are potent inducers of the host's immune response. They have been studied extensively in Gram-negative bacteria but have recently also been discovered in Gram-positive bacteria, such as *Corynebacterium diphtheria* (Ton-That and Schneewind, 2003) and various *Streptococci* (Telford *et al.*, 2006). In Gram-negative bacteria, pili can be classified based on their assembly pathways into four different groups: (i) pili assembled

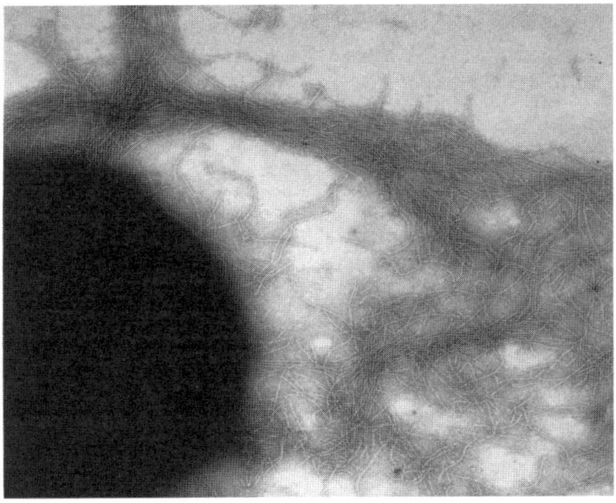

Figure 13.1 Transmission electron micrograph showing numerous type IV pili protruding from the surface of *N. gonorrhoeae.* Magnification: 135,000×.

via the chaperone-usher pathway, (ii) the CS1 pilus family assembled via the alternative chaperone-usher pathway, (iii) the curli pili assembled by the nucleation/precipitation pathway, and (iv) the type IV pili (T4P), which are also expressed in *N. meningitidis* and *N. gonorrhoeae,* and will be most extensively discussed below.

Pili assembled via the chaperone-usher pathway

In the chaperone-usher pathway (Waksman and Hultgren, 2009), pilins are transported into the periplasm via the Sec protein-translocation machinery. In the periplasm, a specific chaperone is bound, which prevents premature assembly of the subunits. Subsequently, the pilin/chaperone complex is delivered to an outer membrane-embedded usher protein that serves as a platform for pilus assembly. This usher forms a pore in the outer membrane that allows for the transport of pilins that are thereby incorporated into the growing pilus. After the delivery of the subunit, the chaperone is recycled into the periplasm.

Type I pili form an example of pili that are assembled via this pathway. They are found throughout the family of *Enterobacteriaceae* and are the most prevalent pilus structures in uropathogenic *Escherichia coli* (UPEC). The type I pilus organelle is encoded by the *fim* gene cluster and consists of a 1–2 μm long helical rod formed mainly by 500–3000 copies of the major subunit FimA. Type I pili contribute to the adhesion of UPEC to host cells in the urinary tract by exposing the adhesin FimH at the tip of the pilus (Jones *et al.*, 1995; Connell *et al.*, 1996; Hahn *et al.*, 2002). FimH specifically interacts with mannosylated glycoprotein receptors that are expressed by several types of host cells, such as bladder and kidney epithelial cells, mast cells, and macrophages (Zhou *et al.*, 2001). Adherence of UPEC to epithelial cells of the urinary tract can lead to urinary tract infections.

CS1 pili

CS1 pili are assembled in a similar way as the chaperone-usher pili; however, the major pilin CooA lacks any significant homology to other pilins (Perez-Casal *et al.*, 1990). Moreover, only four different structural and accessory proteins are required for the assembly of functional pili (Froehlich *et al.*, 1994). CS1 pili play a role in adhesion to intestinal cells and are present in enterotoxigenic *E. coli* (ETEC) (Gaastra and Svennerholm, 1996; Wolf, 1997), which are a major cause of human diarrheal disease, and in some other species, such as *Salmonella enterica* serovar Typhi (Folkeson *et al.*, 1999), which causes typhoid fever.

Curli

Curli are also found in enteric bacteria such as *E. coli* and *Salmonella* (Olsén *et al.*, 1989). Curli are assembled by the nucleation/precipitation pathway (Barnhart and Chapman, 2006). In this pathway, soluble curli subunits CsgA and CsgB are thought to be secreted into the extracellular environment assisted by the outer membrane lipoprotein CsgG. On the cell surface, the major subunit CsgA is then nucleated into a fibre by CsgB (Hammar *et al.*, 1996; Bian and Normark, 1997). Curli fibres show resemblance to eukaryotic amyloid fibres and are produced via a highly regulated process (Olsén *et al.*, 1993).

The role of curli in pathogenesis is not clear. It has been described that curli bind to several proteins, among which several coagulation factors (Herwald *et al.*, 1998) and fibronectin, an extracellular matrix protein (Olsén *et al.*, 1989). Because of the role these proteins play in blood agglutination, curli-mediated binding of bacteria to these proteins may lead to delay of blood clotting, which could facilitate bacterial spreading.

Type IV pili

T4P are found in a wide variety of Gram-negative bacteria, including *E. coli*, *S. enterica*, *Pseudomonas* spp., and *Neisseria* spp. (Fig. 13.1), but also in some Gram-positive bacteria. Generally, T4P are flexible fibres of 5–8 nm in width and several μm in length (Craig and Li, 2008). Individual fibres often interact to form bundles that are characteristic for T4P. T4P consist of polymers of pilins and, in some cases, an adhesive minor subunit can be found at the tip. The pilus subunits are variable in sequence and length among different species and are synthesized as precursors (prepilins). Prepilins possess a hydrophilic leader peptide that ends with a glycine and is cleaved by a specific prepilin peptidase.

Two subtypes of T4P are recognized based on the length of the mature protein and the leader peptide. Type IVa pilins are synthesized with short leader peptides of less than 10 amino acids (aa) and have a typical length of ~150–160 aa. The leader peptides of type IVb pilins are longer and also the mature pilins are generally longer (180–200 aa). However, the type IVb family includes also the fimbrial low-molecular-weight protein (Flp) prepilin subfamily. Members of this subfamily have a very short mature pilin (50–80 aa) (Kachlany *et al.*, 2001). The impact of these differences might be that T4P are assembled via different mechanisms. Indeed, conflicting data on the assembly of type IVa and type IVb pili have been proposed to point to different mechanisms (Pelicic, 2008).

Biogenesis, structure and function of type IV pili

T4P assembly involves at least 12 conserved proteins. Several of these proteins are referred to as core proteins because they are required in all T4P systems. These core proteins include (i) the major pilin, (ii) a specific inner membrane-embedded prepilin peptidase, (iii) an ATPase that powers traffic of pilins during assembly, (iv) an integral inner membrane protein that recruits the traffic ATPase to the pilus-assembly site, and (v) an integral OMP, the secretin, which is required for the emergence of T4P on the bacterial cell surface. Apart from these core proteins, other non-core proteins exist that are system or species specific.

The T4P-assembly apparatus is a macromolecular machine linking the cytoplasm with the extracellular environment. The molecular assembly mechanism must therefore be highly coordinated and is not fully understood. However, the role of several components has been elucidated. The nomenclature of these components differs between bacterial species; we will discuss here the genes and proteins involved with the neisserial type IVa pilus.

The prepilins are encoded by the *pilE* gene. Their initial transport across the inner membrane is performed by the Sec translocase. After translocation, prepilins are folded and stabilized by the formation of an intramolecular disulfide bond, a process mediated by the oxidoreductase DsbA (Peek and Taylor, 1992; Tinsley *et al.*, 2004). The prepilin peptidase PilD removes the N-terminal leader peptide (Strom *et al.*, 1993). PilD also processes several accessory components, often referred to as pseudopilins or minor subunits.

T4P are assembled at the inner membrane (Wolfgang *et al.*, 2000; Carbonelle *et al.*, 2005), where PilG exists as a transmembrane tetramer (Collins *et al.*, 2007) that allows contact with periplasmic and cytoplasmic proteins, and the hexameric ATPase PilF powers polymerization of the subunits (Morand *et al.*, 2004). The growing pilus crosses the periplasm including the peptidoglycan layer to emerge from the outer membrane via the secretin PilQ (Wolfgang *et al.*, 2000; Carbonelle *et al.*, 2005). PilQ is among the most abundant OMPs in *Neisseria* (Newhall *et al.*, 1980) and exists in large multimeric complexes (Collins *et al.*, 2001).

Extended pili can be retracted from the cell surface. Retraction requires another hexameric ATPase, PilT, and is regulated by the PilC proteins (Morand *et al.*, 2004). When pili are retracted, the pilus is disassembled and pilin subunits are drawn back into the inner membrane. This extrusion/retraction mechanism allows phage infection, DNA uptake and twitching motility (Burrows, 2005). Pilus retraction also mediates movement of bacteria towards the host cell surface where intimate attachment can occur (Pujol *et al.*, 1999; Merz *et al.*, 2000).

The pseudopilins have been proposed to play a role in pili dynamics by promoting extension by polymerization of the pilus. Some of these proteins, such as PilX and PilV, are thought to insert into the pilus. PilX is a key protein involved in the formation of pili aggregates and in adhesion (Hélaine *et al.*, 2005). PilV has been identified as a factor that triggers reorganization of the host-cell plasma membrane (Mikaty *et al.*, 2009). In addition, pseudopilins PilH-L were shown to be of crucial influence in T4P dynamics and function (Winther-Larsen *et al.*, 2005).

In *N. meningitidis*, PilE is prone to extensive antigenic variation mediated by chromosomal rearrangements. In this process, silent partial pilin genes (referred to as *pilS*) are recombined via a RecA-dependent mechanism into the active *pilE* expression locus (Haas and Meyer, 1986; Andrews and Gojobori, 2004). Variation of PilE is also caused by post-translational modifications, such as O-linked glycosylation, O-linked phosphoethanolamine, phosphocholine, or phosphoglycerol modification, and phosphorylation (Virji *et al.*, 1996b; Marceau and Nassif, 1999; Hegge *et al.*, 2004; Chamot-Rooke *et al.*, 2007). These post-translational modifications probably influence pilus structure and function. Recently, glycosylation of gonococcal pilins was suggested to be involved in mediating cervical infection by binding the I-domain region of complement receptor 3 (Jennings *et al.*, 2011). In addition, posttranslational addition of phosphoglycerol onto T4P in *N. meningitidis* was shown to specifically cause release of T4P-dependent contacts between bacteria (Chamot-Rooke *et al.*, 2011). This regulated detachment process allowed bacteria to migrate across the epithelium and to new colonization sites.

Involvement of type IV pili in neisserial pathogenesis

T4P are crucial virulence factors involved in the initial interaction of *N. meningitidis* with host cells. Adhesion of encapsulated meningococci to human epithelial and endothelial cells appeared to be strictly dependent on T4P (Nassif *et al.*, 1994). In non-encapsulated strains, which are usually non-virulent, these pili are not essential for host cell interactions. In those bacteria, adhesion can also be mediated by other adhesins, such as the Opa and Opc proteins (Virji *et al.*, 1995a), which will be discussed later in this chapter. Apparently, the capsule forms a physical barrier that can only be crossed by the long T4P adhesin structure. The efficiency of pili-mediated attachment is dependent on the structure of the pilus as well as on the type of host cell (Virji *et al.*, 1992, 1995b).

Adhesion of meningococci to human cells can be divided into two steps, which both require T4P, i.e. (i) the initial interaction of single diplococci with a host cell and (ii) the formation of intimate contact via non-pilus structures. In the first step, an important role for PilC has been proposed because this protein is located in the pilus fibre and has been suggested to possess a cell-binding domain (Rudel *et al.*, 1995a). Originally, two *pilC* genes were discovered in gonococci encoding the proteins PilC1 and PilC2 (Jonsson *et al.*, 1991). Expression of these proteins is prone to phase variation in both gonococci (Jonsson *et al.*, 1991) and meningococci (Rytkönen *et al.*, 2004), and strains depleted for PilC show impaired pilus expression and lose their natural transformation competency (Jonsson *et al.*, 1991; Rudel *et al.*, 1995b). Initially, only PilC1 was shown to be required for adhesion in *N. meningitidis*, while PilC2 was only described to be involved in pilus assembly (Nassif *et al.*, 1994; Morand *et al.*, 2001). However, recent findings indicated that also PilC2 is capable of mediating adhesion (Morand *et al.*, 2009). The two different meningococcal PilC variants, however, triggered different cellular responses, and unlike PilC1, which enables attachment to many epithelial and endothelial cell types, PilC2 only mediated adhesion to one specific cell line (Morand *et al.*, 2009).

The human membrane co-factor protein CD46 has been suggested to act as a receptor for meningococcal pili, but this role is under debate. CD46 is a glycoprotein that also serves as a receptor for other pathogens and functions in the human cell as an important inactivator of cell-attached complement factors C3b and C4b (Riley-Vargas *et al.*, 2004). Observations that point to a role of CD46 as a receptor for T4P include experiments in which transgenic mice expressing human CD46 appeared susceptible to meningococcal infection, while *N. meningitidis* is naturally a strictly human pathogen (Johansson *et al.*, 2003). Also, a direct interaction between pilin-expressing gonococci and purified recombinant CD46 was observed (Källström *et al.*, 1997). Moreover, T4P of gonococci and meningococci were shown to be required in the induction of presenilin/γ-secretase-mediated processing of CD46, which could lead to modulation of cellular responses (Weyand *et al.*, 2010). There are also arguments against a role for CD46 as pilin receptor. CD46-independent binding of gonococci or purified PilC to human epithelial cells suggested the presence of another receptor (Kirchner *et al.*, 2005). Other receptors that have been proposed include the human complement regulator C4BP to which the N-terminal part of PilC was found to bind (Blom *et al.*, 2001) and the I-domain of integrins, which was shown to interact with gonococcal pili (Edwards and Apicella, 2005). Recently, a β_2-adrenoreceptor has been identified as a receptor for the meningococcal pili (Coureuil *et al.*, 2010). Binding activates recruitment of the Par3/Par6/PKζ polarity complex in brain endothelial cells (Coureuil *et al.*, 2009). This complex plays a key role in eukaryotic cell polarity and in the formation of intercellular junctions. Pili-mediated recruitment of this complex causes opening of the intercellular junctions of the brain–endothelial interface and allows *N. meningitidis* to cross the blood–brain barrier and invade the meninges.

During proliferation, pili-mediated adhesion leads to triggering of the host's signalling pathways such as the abovementioned β-arrestin signalling pathway. In this way, *N. meningitidis* induces local elongation of microvilli towards the bacteria, which can lead to engulfment and internalization of a small portion of the bacteria (Merz *et al.*, 1996; Pujol *et al.*, 1997). The formation of these microvilli results from the organization of specific molecular complexes, the so-called cortical plaques, underneath bacterial microcolonies (Merz *et al.*, 1999; Eugène *et al.*, 2002). The formation of cortical plaques involves dense actin polymerization and accumulation of membrane-associated proteins, such as Intercellular adhesion molecule-1 (ICAM-1), CD44 and epidermal growth factor receptor, underneath bacterial aggregations (Merz *et al.*, 1999). These modifications of the host cell are important for *N. meningitidis* to resist various fluid flows and mechanical stress in the extracellular surroundings in the nasopharynx but also to resist detachment from the blood vessel wall caused by blood flow (Mairey *et al.*, 2006).

Type V secreted proteins

Numerous proteins secreted via a type V secretion pathway play a role in bacterial adhesion and pathogenesis. The type V secretion pathway consists of three main categories of two-step secretion systems (Fig. 13.2), i.e. the classical autotransporter (AT) pathway (type Va), the two-partner secretion (TPS) pathway (type Vb), and the so-called oligomeric coiled-coil autotransporters (Oca or type Vc). In addition, a new system, called type Vd, was recently discovered, and the class of invasin/intimin proteins, although not generally categorized as a type V secretion pathway, follows a pathway that appears rather similar.

Type Va: the classical autotransporters

Structure and secretion mechanism

The AT pathway seems to be a simple two-step secretion pathway (Fig. 13.2a). ATs are modular proteins consisting of an N-terminal signal sequence for transport across the inner membrane, a C-terminal translocator domain for transport across the outer membrane, and, in between, the secreted passenger domain. The protein is synthesized in the cytoplasm and crosses the inner membrane via the Sec machinery (Brandon *et al.*, 2003; Sijbrandi *et al.*, 2003). It was long assumed that the translocator domain subsequently folds and inserts into the outer membrane to form a pore that allows the passenger domain to be secreted (Pohlner *et al.*, 1987). The structures of most of the passengers are β-helical (Otto *et al.*, 2005), and it was suggested that their folding at the cell surface would provide the energy needed for the translocation across the outer membrane. Also the Donnan potential across the outer membrane could provide an energy source for passenger translocation as was suggested recently in planar lipid bilayer experiments, which showed the stable translocation of a passenger fragment across the bilayer after switching the transmembrane potential (Roussel-Jazédé *et al.*, 2011).

The postulated role of the translocator domain has been challenged by recent findings on the secretion mechanism. The structure of the translocator domain of NalP of *N. meningitidis*

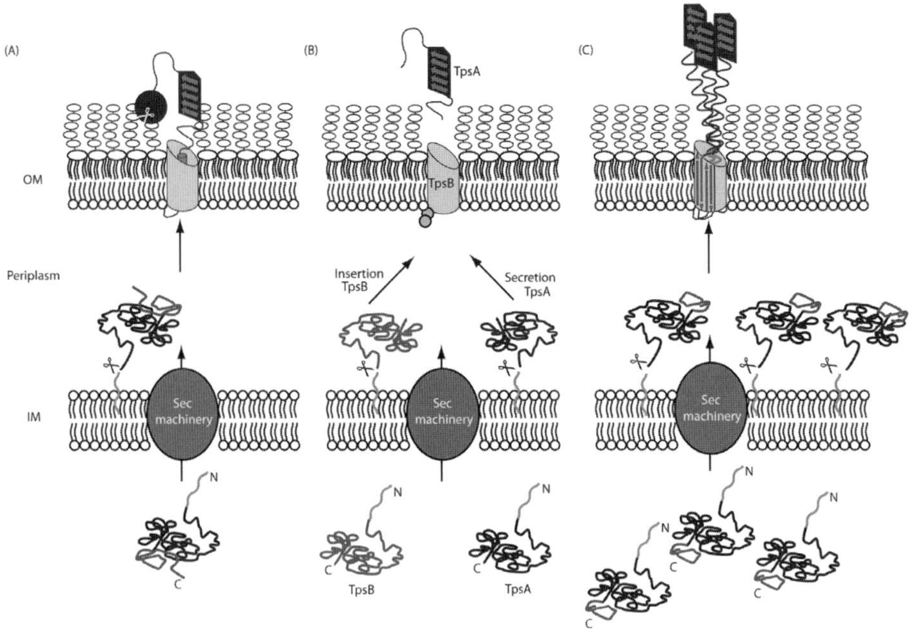

Figure 13.2 Schematic representation of the three main type V secretion pathways. In all cases, the secreted protein is first transported across the inner membrane via the Sec machinery after which the signal sequence is cleaved off at the periplasmic face of this membrane by leader peptidase (scissors). (A) The type Va or classical AT pathway. The C-terminal translocator domain inserts into the outer membrane as a 12-stranded β-barrel that may form a pore to mediate the translocation of the covalently attached N-terminal passenger domain to the cell surface. An alternative model for the secretion of the passenger involving the Bam machinery is discussed in the text. After translocation across the outer membrane, the passenger may be cleaved off autocatalytically by an endogenous protease domain (sphere with scissors) or by other proteases. (B) The type Vb or TPS pathway. TpsB is inserted into the outer membrane as a β-barrel with two POTRA domains (spheres) exposed at the periplasmic side. The N-terminal TPS domain of TpsA is recognized by the POTRA domains of TpsB, and TpsA is subsequently translocated across the outer membrane through a pore formed by the β-barrel domain of TpsB. (C) The type Vc pathway. This pathway is similar to the classical AT pathway except that translocator domain is in this case much smaller. Consequently, three protomers are required to form a 12-stranded β-barrel that is needed to expose three passenger domains at the cell surface.

was solved few years ago and showed a 12-stranded β-barrel with a pore size being too narrow to allow for the passage of partially folded domains. Therefore, an alternative model was proposed (Oomen *et al.*, 2004) that suggests an important role for the Bam complex in the process. The Bam complex, of which Omp85 (BamA) of *N. meningitidis* was the first identified component, is required for the assembly of integral OMPs (Voulhoux *et al.*, 2003). It was demonstrated to be required also for autotransporter secretion (Voulhoux *et al.*, 2003; Jain and Goldberg, 2007). The Bam complex likely assists the insertion of the translocator domain into the outer membrane, but, during that process, it might also provide the conduit that allows for the transport of the passenger across the membrane. Thus, the translocator domain would function primarily as a recognition signal for the Bam complex, while the passenger domain would be transported through a pore formed by the Bam complex, rather than through the pore formed by the translocator domain itself. Various cross-linking studies with trapped translocation intermediates support an important role for the Bam complex in autotransporter secretion (Ieva and Bernstein, 2009; Sauri *et al.*, 2009; Ieva *et al.*, 2011). Recently, a new protein transport system consisting of the Omp85-family member TamA in the outer membrane and an inner membrane protein, designated TamB, was implicated in autotransporter secretion in *E. coli* (Selkrig *et al.*, 2012). These components, however, were not identified in the cross-linking studies cited above and their precise contribution in the autotransporter secretion mechanism remains to be elucidated.

Functions of classical autotransporters in *N. meningitidis*

Based upon their common architecture, ATs can fairly easily be recognized within bacterial genome sequences. Their functions can be quite diverse (Henderson and Nataro, 2001); in this section, we will describe only those of meningococcal ATs. Several meningococcal ATs are serine proteases. Serine proteases are ubiquitously distributed among prokaryotes and eukaryotes. They have been grouped into six clans. Among the meningococcal ATs, two clans can be found, SA (chymotrypsin) and SB (subtilisin).

The chymotrypsin clan, as its name indicates, shares a common fold with chymotrypsin. The catalytic triad is conserved and the order is always His/Asp/Ser. Meningococcal ATs that carry this protease domain are IgA protease (Vitovski and Sayers, 2007), App (Hadi *et al.*, 2001), and AusI (van Ulsen *et al.*, 2006). IgA protease of *N. gonorrhoeae* was the first described AT (Pohlner *et al.*, 1987). Subsequently, homologous IgA proteases were identified in other pathogenic bacteria, including *N. meningitidis* and *Haemophilus influenzae*. The precise biological function of IgA proteases is not completely clear. They are able to cleave human IgA by targeting a sequence in the hinge region, thereby producing intact Fabα and Fcα fragments (Mansa and Kilian, 1986), but the role of this cleavage during infection is not completely understood. The biological role of the other ATs, App and AusI, is poorly understood. However, App, a homologue of Hap from *H. influenzae*, and AusI were shown to mediate adhesion to epithelial cells (Serruto *et al.*, 2003; Turner *et al.*, 2006). The protease domain of all the proteins mentioned above is needed for their autocatalytic release from the cell surface and has probably additional functions during infection.

In the subtilisin-like proteases, the order of the catalytic triad residues within the primary sequence is Asp/His/Ser. The AT NalP of *N. meningitidis* is a member of this subfamily. Its protease activity is involved in the autocatalytic release of the passenger from the cell surface (van Ulsen *et al.*, 2003). In addition, NalP functions in the processing of other ATs, i.e. IgA protease, App, and AusI (van Ulsen *et al.*, 2003, 2006). Different forms of these NalP substrates are released dependent on whether they are processed by NalP or autocatalytically. NalP-mediated proteolytic cleavage results in the release of an extended passenger with an extra domain, called the α-peptide, relative to the passenger that is released upon autocatalytic cleavage. It was also demonstrated that NalP is responsible for cleavage of two surface-exposed lipoproteins, the LbpB component of the bipartite lactoferrin receptor (Roussel-Jazédé *et al.*, 2010) and the heparin-binding protein NhbA (Serruto *et al.*, 2010). The release of LbpB reduced the complement-mediated killing of the bacteria when incubated with an LbpB-specific bactericidal antiserum and complement. Since antibodies directed against LbpB are found in convalescent patient sera, the release of an immunogenic protein as LbpB may represent a novel means for *N. meningitidis* to escape the human immune response (Roussel-Jazédé *et al.*, 2010). Expression of NalP is phase-variable (Martin *et al.*, 2004), suggesting that NalP may be under immunological pressure.

Type Vb: the two-partner secretion system

Structure and secretion mechanism

The TPS pathway is used for the secretion of very large proteins (generally over 1000 residues). The prototypical TPS system is that of the filamentous haemagglutinin (FHA) of *Bordetella pertussis* (Jacob-Dubuisson *et al.*, 2001). A TPS system is composed of two proteins, the secreted protein and an outer membrane-embedded transport protein, which are generically called TpsA and TpsB, respectively (Fig. 13.2b). TPS systems were classified as type Vb, because they were supposed to be 'uncoupled' ATs, i.e. they contain an independently expressed transporter instead of a transporter fused to the secreted protein. However, there are major differences between the TPS and AT secretion routes. For example, the TpsBs are rather conserved members of the Omp85 superfamily and are markedly different from the translocator domains of ATs, both in sequence and in barrel size.

The TpsA proteins are translocated across the inner membrane via the Sec machinery into the periplasm, where they interact with the cognate outer membrane-embedded TpsB transporter (Fig. 13.2b). TpsB proteins consist of two polypeptide transport-associated (POTRA) domains that extend into the periplasm and a C-terminal domain that is inserted as a β-barrel in the outer membrane forming a channel, through which the TpsA protein is presumably translocated to the cell surface (Jacob-Dubuisson *et al.*, 2001; Clantin *et al.*, 2007). TpsA proteins contain a ~110-residues long N-terminal sequence, the TPS domain, which is essential for interaction with the POTRA domains of the cognate TpsB (Hodak *et al.*, 2006; Clantin *et al.*, 2007). The POTRA and TPS domains are highly conserved between different species; however, small sequence divergence may result in specificity of the interactions allowing for the presence of different TpsA/TpsB systems in a given genome (van Ulsen *et al.*, 2008). After or during translocation across the outer membrane, the exoprotein folds at the cell surface. A recently established *in vitro* translocation assay suggested that protein folding at the trans side of the membrane might be the driving force for protein translocation and also confirmed that TpsB is the only membrane protein needed for TpsA translocation (Fan *et al.*, 2012). At the cell surface, TpsA may be cleaved by specific extracellular proteases (Coutte *et al.*, 2001). Subsequently, it can either remain associated with the cell surface or be released into the extracellular milieu.

TPS systems in *N. meningitidis*

Based upon sequence differences in their TPS domain, up to three different TPS systems can be distinguished within a single meningococcal strain, and also, several copies of each system can be present in one strain (van Ulsen and Tommassen, 2006). When the distribution of these systems was investigated, one system, system 1, appeared to be ubiquitous, while the other two systems were significantly more prevalent among isolates of hyperinvasive lineages than among isolates of poorly invasive strains (van Ulsen *et al.*, 2008). Acquisition of additional TPS systems and duplication of TPS systems, which is accompanied by large genomic rearrangements, may be mediated by the transposase-like elements that are often found encoded in the genetic islands that contain these systems (van Ulsen and Tommassen, 2006; J. Arenas and J. Tommassen, unpublished).

The *tpsA* and *tpsB* genes are usually organized in an operon with the *tpsB* gene at the promoter-proximal end. In *N. meningitidis*, several shorter, *tpsA*-related open-reading frames (ORFs), designated *tpsC* cassettes, are usually located downstream of the *tpsA* gene with intervening ORFs located in between these *tpsC* cassettes (Klee *et al.*, 2000; van Ulsen and Tommassen, 2006). Fig. 13.3 illustrates the genetic organization of a TPS system in one *N. meningitidis* strain. The *tpsC*s share sequences of variable length within their 5′ region with *tpsA* but differ at their 3′ extreme (van Ulsen and Tommassen, 2006). Via homologous recombination at these shared sequences, a *tpsC* can be moved into the *tpsA* expression locus resulting in alteration of the C-terminal region of TpsA, but the frequency of such recombination events appears low (J. Arenas and J. Tommassen, unpublished).

In *N. meningitidis*, TpsA has been reported to be proteolytically processed, but its release into the extracellular medium seems to be strain dependent (Schmitt *et al.*, 2007; van Ulsen *et al.*, 2008; Neil and Apicella, 2009). Therefore, it seems that the processing of TpsA does not regulate its release into the medium, which is also the case for FHA of *B. pertussis*, which is processed but remains at the bacterial cell surface.

The role of type Vb secreted proteins in meningococcal pathogenesis

TpsA of system 1 (TpsA1) appears of considerable relevance for pathogenesis, as evidenced, for example, by its attribution to haemolytic activity (Talà *et al.*, 2008). Also, disruption of the *tpsA1* and/or *tpsB1* genes revealed a function of the protein in adherence to some epithelial cell lines (Schmitt *et al.*, 2007). This adherence function, which was inhibited by the presence of capsule and LPS structures, is in agreement with adhesion properties of TpsAs of *H. influenzae* and *Moraxella catarrhalis* (Avadhanula *et al.*, 2006; Plamondon *et al.*, 2007). Additionally, an essential role of meningococcal TpsA1 in intracellular survival and escape from infected cells has been reported (Talà *et al.*, 2008). TpsA1 was also shown to have a role at later stages of biofilm formation (Neil and Apicella, 2009) possibly by mediating cell-to-cell contact and autoaggregation. This has also been reported for TpsA of *Xanthomonas axonopodis* pv. *citri*, a phytopathogen responsible for citrus cancer (Gottig *et al.*, 2009).

Up to date, no function has been reported for the meningococcal TPS systems 2 and 3. Evolutionary analysis revealed that those systems, more than system 1, are related with TpsAs from other bacteria (J. Arenas and J. Tommassen, unpublished). Thus, the ubiquitous nature of system 1 and the additional presence of systems 2 and 3 in highly invasive strains

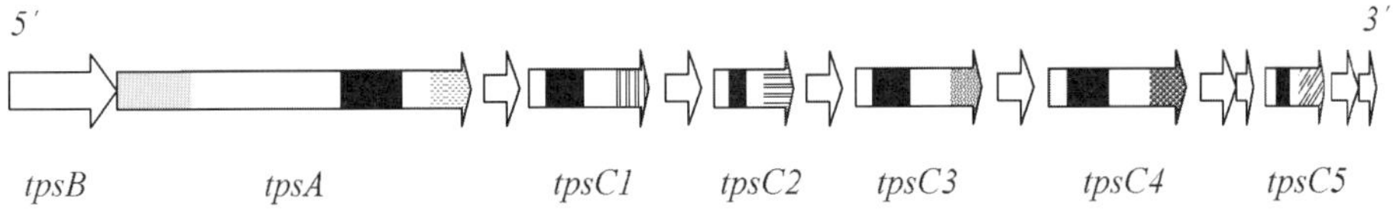

Figure 13.3 Genetic organization of the TPS system of *N. meningitidis* strain FAM18. The genome of FAM18 contains a single TPS island formed by a *tpsB* gene (NMC0443), one *tpsA* gene (NMC0444), and five *tpsC* cassettes interspersed with additional ORFs (NMC0446 to NMC0456). The *tpsA* gene contains a sequence at the 5′ extreme that codes for a TPS domain (coloured grey), which is required for recognition by TpsB for translocation through the outer membrane. The *tpsC* cassettes do not encode a TPS domain. The 5′ ends of the *tpsC*s share sequence similarity over a region of variable-length with a part of *tpsA* (indicated in black), while their 3′ ends are very divergent (variably coloured). The intervening ORFs located in between the *tpsC* cassettes are not coloured for clarity. The length of the various genes is not to scale.

(van Ulsen *et al.*, 2008) could point to an essential role of system 1 during colonization and supplemental virulence functions for systems 2 and 3 in determined genetic backgrounds.

Interestingly, it has been reported that certain *E. coli* isolates are able to contact other bacteria that compete for nutrients in the same niche and inhibit their growth through the TPS system (Aoki *et al.*, 2005), a mechanism termed contact-dependent growth inhibition (CDI). The C-terminal region of TpsA and the immediately downstream ORF play important roles in this mechanism (Hayes *et al.*, 2010). In the proposed model, cell-to-cell contact is required for growth inhibition. For that, TpsA binds to BamA (Omp85) on target cells. After binding, the C-terminal region of TpsA is cleaved and introduced into the cytoplasm of the target cell where CDI is developed. The protein encoded by the ORF immediately downstream of TpsA functions as an immunity protein and prevents CDI in the producing strain by counteracting the growth-inhibitory signal (Aoki *et al.*, 2010). It was suggested that the CDI mechanism of TpsAs is conserved among various bacteria (Aoki *et al.*, 2010), a hypothesis that was recently confirmed amongst others in *Burkholderia pseudomallei* (Nikolakakis *et al.*, 2012) and *N. meningitidis* (J. Arenas and J. Tommassen, unpublished). In this respect, it is interesting to note that, as a consequence of recombination of *tpsC* into *tpsA*, the altered C-terminus of TpsA may confer different CDI effects and alter the CDI in a defined population.

Type Vc: the trimeric autotransporters

Structure and secretion mechanism

The domain organization of Ocas is similar to that of classical ATs, i.e. a signal sequence, a passenger, and a C-terminal translocator domain. However, the latter domain is much shorter than in the case of classical ATs, but they show some sequence similarity. While the translocator domains of classical ATs form monomeric 12-stranded β-barrels (Oomen *et al.*, 2004), the protomers of the trimeric ATs each contribute a four-stranded β-sheet to form together a 12-stranded β-barrel with similar dimensions as that of the classical ATs (Barnard *et al.*, 2007) (Fig. 13.2C). In contrast to classical ATs, none of the studied Ocas appears to be cleaved at the cell surface and released into the extracellular milieu. Thus far, all characterized members of this subfamily have been found to possess adhesive activity, in most cases mediating bacterial adherence to eukaryotic cells or extracellular matrix proteins and in some cases resulting in binding of circulating host factors, such as immunoglobulins or complement components (Linke *et al.*, 2006; Lyskowski *et al.*, 2011).

The role of trimeric autotransporters in meningococcal adhesion

N. meningitidis has two members of the Oca family, NhhA (*Neisseria* Hia homologue A) and NadA (*Neisseria* adhesion A), which both possess adhesin-like properties. NadA was found to be expressed in approximately 50% of *N. meningitidis* strains isolated from patients but only in 5% of strains from healthy individuals (Comanducci *et al.*, 2004). The expression of NadA is phase variable being controlled by slipped-strand mispairing at a tetranucleotide repeat (TAAA) upstream of the *nadA* promoter (Martin *et al.*, 2003). It was shown that NadA might have an important role in mucosal colonization by *N. meningitidis*. Expression of NadA in *E. coli* enhanced bacterial association with Chang epithelial cells, and a *nadA* knockout mutant of *N. meningitidis* showed decreased cell adhesion and invasion compared to the wild-type strain (Capecchi *et al.*, 2005). In addition, specific receptors for NadA were observed in monocytes, macrophages, and monocyte-derived dendritic cells. It was also shown recently that NadA is able to bind β1-integrins (Nägele *et al.*, 2011). For all these reasons, NadA likely is an important factor in host–pathogen interaction during neisserial infection.

NhhA possesses sequence similarity with the Oca's Hia and Hsf of *H. influenzae*. The passenger domains of Hia and Hsf are characterized by the presence of internal repeats and distinct cell-binding sites (Laarmann *et al.*, 2002; Cotter *et al.*, 2005). The amino acid sequence of NhhA shows a similar organization of its passenger domain, but the number of repeats is lower, rendering its length considerably shorter than those of Hia and Hsf (Scarselli *et al.*, 2001). More important, two cell-binding domains called BD1 and BD2 in Hia (Meng *et al.*, 2008) are missing in NhhA. Inactivation of the *nhhA* gene in *N. meningitidis* strain MC58 significantly reduced the adhesion capacity compared to the wild-type strain. Binding of recombinant NhhA to heparan sulfate and laminin was demonstrated suggesting that NhhA is an adhesin that is able to bind extracellular matrix components (Scarselli *et al.*, 2006). Also, NhhA plays an important role in evading the host's innate immune response. It was shown that NhhA inhibits phagocytosis by inducing macrophage apoptosis (Sjölinder *et al.*, 2012). In addition, by binding activated vitronectin, NhhA protects bacteria against complement-mediated killing (Griffiths *et al.*, 2011). These immune evasion capacities of NhhA enhance both nasal colonization and the development of sepsis *in vivo* (Sjölinder *et al.*, 2008).

Type Vd secretion system

Recently, a novel class of type V secretion systems has been proposed. This proposition was based on the discovery of a novel secreted protein of *P. aeruginosa*, PlpD (Salacha *et al.*, 2010). The protein was identified as a lipolytic enzyme of the bacterial patatin-like protein family because its N-terminal domain contains the four conserved domains that characterize this protein family and that are required for lipase activity (Salacha *et al.*, 2010). Similar to the classical ATs, PlpD is synthesized as a single molecule composed of a signal peptide, a secreted domain, and a C-terminal transporter domain. However, in contrast to the 12-stranded β-barrels formed by the translocator domains of ATs, the transporter domain of PlpD consists of a POTRA motif and a 16-stranded β-barrel, similar to those of the TpsB transporters of the type Vb secretion systems. The unique organization of a TpsB-like component associated with a passenger domain led to the proposal of a new class of type V secretion systems, type Vd. PlpD exhibits lipase activity *in vitro* but its function is not precisely understood (Salacha *et al.*, 2010). Patatins are common in plants and are considered as tuber storage proteins that show hydrolase activity (Shewry, 2003). Earlier described patatin homologues are substrates for the type III or type IV secretion systems (Finck-Babançon *et al.*, 1997; VanRheenen *et al.*, 2006). The major cytotoxin of *P. aeruginosa*, ExoU, is a patatin homologue that is delivered via the type III secretion system directly into the cytosol of a eukaryotic cell (Finck-Babançon *et al.*, 1997), where it induces rapid cell death. Together, more than 200 orthologues of PlpD have been found in various Gram-negative bacteria (Salacha *et al.*, 2010), but not *Neisseria*. Apart for patatin domains, also other protein domains were found to be fused to a TpsB-like domain (Arnold *et al.*, 2010). Probably, these proteins are also secreted via a type Vd mechanism.

The intimin/invasin family

Proteins of the intimin/invasin family (Int/Inv family) have been suggested to be secreted in a similar manner as ATs but their sequential organization is different (Touze *et al.*, 2004). Juxtaposed to the signal sequence, required for transport across the inner membrane, is first a hydrophilic α-helical domain, the α-domain, which is sometimes decorated with a peptidoglycan-binding LysM domain (Bateman and Bycroft, 2000; Tsai *et al.*, 2010). C-terminal of this α-domain is a highly conserved β-barrel domain, followed by a second α-helical domain, called α′-domain. A surface-localized passenger domain can be found at the C-terminus (Wentzel *et al.*, 2001).

A model for the secretion of Int/Inv proteins has been proposed (Tsai *et al.*, 2010) in which the protein, after its translocation across the inner membrane, is bound by general periplasmic chaperones and delivered to the Bam complex. The Bam complex then enables incorporation of the β-barrel into the outer membrane. Evidence for this has been found (Bodelón *et al.*, 2009). Similar as suggested for the C-terminal β-barrel of ATs, the N-terminal β-barrel domain of Int/Inv proteins would function as a pore through which the passenger domain is transported to the cell surface.

Recent bioinformatic studies revealed that Int/Inv proteins, although widespread among the bacterial kingdom, are predominantly associated with γ-proteobacteria (Tsai *et al.*, 2010). They are not found in *Neisseria*. Int/Inv proteins from β-, γ-, and ε-proteobacteria and *Chlamydia* possess characteristic C-terminal passenger domains with C-type lectin domains. Int/Inv proteins of α-proteobacteria, cyanobacteria and chlorobi only possess predicted β-barrel domains (Tsai *et al.*, 2010). The function of these proteins is completely unknown.

The function of the intimins from enteropathogenic and enterohaemorrhagic *E. coli* (EPEC and EHEC, respectively) and the invasins from *Yersinia* spp. is well established. The β-barrel domain of these proteins anchors the C-terminal passenger in the outer membrane. The passenger is rod shaped and consists of four N-terminal immunoglobulin-like domains and a C-terminal C-type lectin domain. Residues in the last immunoglobulin and the C-type lectin domains of invasin are crucial for binding β_1-integrins (Isberg and Leong, 1990; Leong *et al.*, 1995; Saltman *et al.*, 1996). Upon binding, invasin clusters and activates the β_1-integrins. This promotes uptake of bacteria into the host cells by activation of autophagy (Deuretzbacher *et al.*, 2009). This step is important in initial infection and enables pathogenic *Yersinia* in the penetration of the host mucosa and subsequent traverse of the Peyer's patches (Isberg *et al.*, 1987; Isberg and Barnes, 2002).

Intimin was first described as a protein associated with formation of attaching and effacing intestinal lesions that can lead to diarrhoea and kidney damage (Jerse *et al.*, 1990). It binds a receptor on the epithelial cell membrane, the translocated intimin receptor (Tir) (Frankel *et al.*, 1998; Hartland *et al.*, 1999), which is produced by the bacteria and injected into the host cell by a type III secretion system (Kenny *et al.*, 1997). The extracellular exposed domain of Tir binds the C-type lectin domain of intimin (Batchelor *et al.*, 2000) leading to intimate attachment of the bacteria to the host cell. Intimin can oligomerize, and its binding to Tir results in clustering of Tir (Luo *et al.*, 2000; Touze *et al.*, 2004) and a cascade of events that ultimately leads to reorganization of the host cytoskeleton and the formation of actin-rich pedestal-like structures (Hayward *et al.*, 2006). These pedestals firmly anchor the bacterium to the surface of the host cell (Goosney *et al.*, 1999). Their role in pathogenesis *in vivo*, however, has not yet been established.

Besides Tir, intimin binds other, host-encoded receptors. Like invasin, intimin has affinity for β_1-integrins on the host-cell surface. This interaction was not required for intimin-mediated adhesion of EPEC to cultured epithelial cells (Liu *et al.*, 1999). However, it was shown that intimin, probably by interacting with β_1-integrin, could act as an immune modulator in mice in a *Citrobacter rodentium* infection, inducing massive colonic hyperplasia *in vivo* (Higgins *et al.*, 1999).

Also several intimin subtypes have been reported to bind nucleolin that is localized to the host-cell plasma membrane (Sinclair and O'Brien, 2002). This occurs probably also via the C-type lectin domain (Reece *et al.*, 2002), and Tir and nucleolin appear to compete for intimin during bacterial adhesion. The role of intimin–nucleolin interaction is not exactly known but is likely involved in the initial stages of adhesion (Sinclair and O'Brien, 2002).

Cell-surface exposed lipoproteins

Biogenesis of lipoproteins

Lipoproteins are usually not directly involved in bacterial adhesion to host cells but, nevertheless, they can play a pivotal role in pathogenesis. The maturation and transport of lipoproteins is illustrated in Fig. 13.4. Lipoproteins are initially synthesized as preprolipoproteins possessing an N-terminal signal sequence for export via the Sec machinery. The signal sequences of lipoproteins are special because of a conserved C-terminal region, the lipobox, which has a consensus sequence [LVI][ASTVI][GAS]C. The cysteine, which is the first residue after the signal sequence, is crucial because a diacylglycerol moiety derived from phosphatidylglycerol is covalently attached to its thiol group by the enzyme lipoprotein diacylglyceryl transferase (Lgt). Subsequently, the signal sequence of the prolipoprotein is cleaved by the lipoprotein signal peptidase Lsp, after which the α-amino group of the N-terminal cysteine is additionally acylated by the enzyme phospholipid:apolipoprotein transacylase Lnt (Sankaran and Wu, 1994).

After their maturation, lipoproteins can remain attached to the inner membrane or be transported to the outer membrane. Sorting of inner and outer membrane-destined lipoproteins depends on the identity of the amino acid residue adjacent to the lipidated cysteine; this is called the '+2 rule'. Inner membrane lipoproteins usually contain an aspartate residue at this position. A different amino acid at this position results in localization to the outer membrane (Yamaguchi *et al.*, 1988). This +2 rule was established in *E. coli*, but, particularly in other bacteria, other residues near the N-terminus may play a predominant role in sorting (Narita and Tokuda, 2007).

Lipoproteins are targeted to the periplasmic side of the outer membrane by the Lol system, which consists of a transmembrane protein complex LolCDE, a periplasmic chaperone LolA, and an outer membrane receptor LolB (Fig. 13.4) (for a review, see Tokuda and Matsuyama, 2004). The LolCDE complex is an ABC transporter, where LolC and LolE are integral membrane proteins with mutual sequence similarity and LolD is an associated ATPase. It provides the energy for extrusion of the lipoprotein from the inner membrane by ATP hydrolysis. The lipoprotein is transported through the periplasm as a soluble complex with LolA and is subsequently transferred to the receptor LolB, which mediates its insertion into the inner leaflet of the outer membrane (Fig. 13.4). For further reading on lipoprotein biogenesis, the reader is referred to Chapter 4.

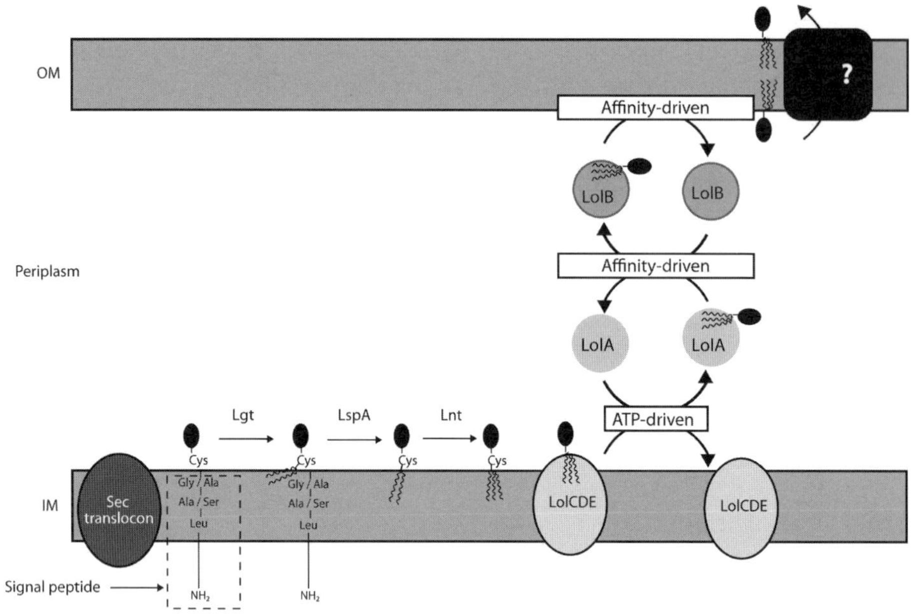

Figure 13.4 Biogenesis of lipoproteins. Lipoproteins are synthesized as precursors with an N-terminal signal sequence that contains at its C-terminal end a consensus sequence Leu-(Ala/Ser)-(Gly-Ala)-Cys, called a lipobox. They are translocated across the inner membrane by the Sec machinery. Maturation of the precursors is catalysed by three well-conserved enzymes, Lgt, LspA and Lnt, which are required for the addition of a diacylglycerol moiety, the removal of the signal sequence, and the N-acylation of the cysteine at position +1, respectively. Transport to the outer membrane requires the inner membrane-embedded ABC transporter LolCDE, the periplasmic chaperone LolA, and the outer membrane-associated receptor LolB, which itself is a lipoprotein. Release of the lipoproteins from the inner membrane is driven by ATP hydrolysis, whereas the subsequent steps, the transfer from LolA to LolB and from LolB into the inner leaflet of the outer membrane are affinity driven. How lipoproteins are transported to the cell surface is unknown.

Analysis of the available *N. meningitidis* genome sequences revealed that these strains contain homologues of LolA-D, but a LolE homologue is absent (Kovacs-Simon *et al.*, 2011). Presumably, whilst LolC and LolE of *E. coli* form a heterodimer, the LolC homologue of *N. meningitidis* is functional as a homodimer.

In *E. coli*, most, if not all outer membrane lipoproteins face the periplasm, but in other Gram-negative bacteria, lipoproteins can also be exposed to the cell surface. Apart from pullulanase of *Klebsiella oxytoca*, which is secreted via a type II secretion mechanism (Pugsley, 1993) and ATs such as NalP from *N. meningitidis*, hardly anything is known about the process by which lipoproteins are transported across the outer membrane, but release from the inner membrane and translocation across the outer membrane are likely determined by separate N-terminal residues (Schulze *et al.*, 2010). Studies in *Borrelia burgdorferi* showed that translocation of the surface-exposed lipoprotein OspA across the outer membrane is initiated at the C-terminus. This suggests a mechanism by which the lipid anchor is inserted in the outer membrane on the periplasmic side, followed by translocation of the protein to the cell surface (Schulze *et al.*, 2010). The latter step requires an unfolded conformation indicating that periplasmic chaperones may be needed to prevent premature folding. This observation and the fact that outer membrane translocation of surface-exposed lipoproteins takes place in the absence of ATP and the proton-motive force seem to suggest that the process is driven by ongoing extracellular folding and assembly (Schulze *et al.*, 2010).

Role of lipoproteins in pathogenesis

Lipoproteins can play various roles in bacterial pathogenesis in several bacterial species. For instance, outer membrane lipoproteins of *B. burgdorferi* were shown to interact with human brain microvascular endothelial cells, and it was proposed that two of these proteins, Vsp1 and Vsp2, are involved in adhesion to the host cell surface (Gandhi *et al.*, 2010). Lipoproteins can also be indirectly involved in virulence by functioning as essential components of bacterial transport systems or as chaperones. An example of a lipoprotein that indirectly influences meningococcal pathogenesis is DsbA. This protein belongs to the family of oxidoreductases that catalyse disulfide-bond formation in the periplasm between cysteine residues in exported proteins. Such disulfide bonds are needed for the stability and/or activity of these proteins. The activity of DsbA has been shown to be required for adhesion, invasion and intracellular survival in several pathogenic species, including EPEC (Zhang and Donnenberg, 1996), *Vibrio cholerae* (Peek and Taylor, 1992), and *Shigella flexneri* (Watarai *et al.*, 1995). In *N. meningitidis*, three DsbA proteins are present, one soluble periplasmic protein and two inner membrane-associated lipoproteins. The presence of at least one of the two lipoproteins is required for the formation of functional T4P (Tinsley *et al.*, 2004).

Lipoproteins are potent inducers of host inflammatory responses. The lipid tail is thought to play an important role in their recognition by the host immune system. For instance, *Staphylococcus* strains that are deficient in the maturation of preprolipoproteins fail to elicit an adequate immune response, suggesting a role for lipoproteins in pathogen recognition and innate immunity (Stoll *et al.*, 2005; Bubeck Wardenburg *et al.*, 2006).

Cell surface-exposed neisserial lipoproteins

Searching the available *N. meningitidis* genome sequences for signal sequences ending with a lipobox suggested the presence of 72 and 82 putative lipoproteins in serogroup A strain Z2491 (Juncker *et al.*, 2003) and serogroup B strain MC58 (J. Grijpstra and J. Tommassen, unpublished), respectively. Some of these lipoproteins are cell-surface exposed and play important roles in meningococcal pathogenesis.

Neisserial lipoproteins and iron uptake

Some lipoproteins are involved in iron uptake. Iron is an essential element for bacterial growth and free iron is scarce in the human body. Transferrins form a family of proteins that bind free iron in the blood and body fluids. Their function is to deliver iron to cells, to reduce the concentration of toxic free iron, and to deplete microbial invaders from this essential nutrient. *N. meningitidis* has developed ways to strip iron from transferrin and other human iron-binding proteins. The surface-exposed lipoprotein transferrin-binding protein (Tbp) B forms together with the integral OMP TbpA a transferrin receptor (Legrain *et al.*, 1993). After transport of the stripped iron across the outer membrane, the ferric-binding protein A (FbpA) transports the iron across the periplasm and initiates the process of uptake into the cytoplasm. Similar transferrin receptors, composed of TbpA and TbpB homologues, have been found in various other pathogens, including *N. gonorrhoeae*, *H. influenzae*, *Actinobacillus pleuropneumoniae*, and *M. catarrhalis* (Gray-Owen and Schryvers, 1996). Both proteins are only expressed under iron limitation and are essential for iron acquisition from human transferrin, since mutants deficient in TbpA or TbpB were incapable of utilizing transferrin for growth (Irwin *et al.*, 1993). In *N. gonorrhoeae*, however, mutants lacking TbpB can still grow on transferrin, but they have lower iron-uptake efficiency (Anderson *et al.*, 1994). TbpA and TbpB form a complex (Boulton *et al.*, 1998), but both proteins can independently bind specifically human transferrin (Pintor *et al.*, 1998). In contrast to TbpA, TbpB is capable of discriminating between iron-loaded and iron-free transferrin and preferably binds the iron-bound form (Cornelissen and Sparling, 1996). Recent structural analysis (Noinaj *et al.*, 2012) revealed that TbpA and TbpB can bind transferrin simultaneously, thereby positioning one of the iron-binding sites in transferrin directly above the plug that closes the channel in TbpA. A helix finger in one of the cell surface-exposed loops of TbpA inserts in the iron-binding cleft of transferrin, thereby releasing the bound iron. The iron is then transported through the channel in TbpA after removal of the plug at the expense of energy provided by the proton gradient across the inner membrane, which is coupled to the receptor via the TonB complex (Noinaj *et al.*, 2012).

The lactoferrin-binding protein (Lbp) B of *N. meningitidis* is a lipoprotein that, together with the integral OMP LbpA, forms a bipartite receptor for human lactoferrin (Prinz *et al.*, 1999). Lactoferrin is an iron-binding protein that is found in phagocytes and body fluids, such as milk, mucus, and tears. The LbpA and LbpB proteins show sequence similarity with TbpA and TbpB, respectively (Pettersson *et al.*, 1994), and the mechanistic of these receptors is likely very similar. In contrast to LbpA, LbpB is not essential for iron acquisition from lactoferrin although it was shown to be capable of binding lactoferrin *in vitro* (Pettersson *et al.*, 1998). Antibodies against LbpB were found in convalescent-patient sera, implicating that the protein is expressed during infection (Pettersson *et al.*, 2006). One strategy of *N. meningitidis* against LbpB-directed immune responses is that the protein is released from the cell surface by the proteolytic AT NalP (Roussel-Jazédé *et al.*, 2010). Because LbpB is dispensable for iron uptake via lactoferrin, it may very well perform (an) alternative function(s) in *N. meningitidis*. One possibility is that it might antagonize the antimicrobial activity of lactoferrin. Iron-depleted lactoferrin has been shown to release LPS from the outer membrane of enteric bacteria, which is bactericidal (Ellison *et al.*, 1988, 1990). It has been shown that both membrane-bound and released LbpB can bind lactoferrin and it is a possibility that LbpB prevents the antimicrobial activity of lactoferrin by scavenging it away from the outer membrane surface (Roussel-Jazédé *et al.*, 2010). Furthermore, lactoferrin was shown to proteolytically cleave the ATs Hap and IgA1 protease of *H. influenzae* (Hendrixson *et al.*, 2003). These ATs, which are thought to be important for colonization of the nasopharynx, are homologues of App and IgA protease of *N. meningitidis*. Thus, binding of lactoferrin to LbpB may assist in colonization by preventing proteolysis of these ATs (Roussel-Jazédé *et al.*, 2010).

Factor H-binding protein

The surface-exposed lipoprotein factor H (fH)-binding protein (fHbp) plays an important role in helping the bacteria to evade the alternative complement pathway (Madico *et al.*, 2006; Schneider *et al.*, 2006). This pathway is activated when complement factors recognize pathogen surfaces and leads to the death of the microbes. Activation of the complement cascade on host cell surfaces is inhibited by binding of fH to charged carbohydrates. *N.*

meningitidis is able to bind fH via fHbp by using protein–protein interactions (Schneider *et al.*, 2009), thereby limiting complement activation at its cell surface. In line with this, expression of fHbp, even at low levels, was found to be important for bacterial survival in human blood and serum (Seib *et al.*, 2009). It was described that oxygen limitation stimulates fHbp expression, suggesting that fHbp also plays role in meningococcal survival in environments lacking oxygen, such as the submucosa and intracellular environments (Oriente *et al.*, 2010). An additional role of fHbp in mediating resistance to the human immune system is that it protects the bacteria against the cationic antimicrobial peptide LL-37. LL-37 targets negatively charged bacterial membrane surfaces and is believed to cause membrane destabilization and permeabilization. In addition, it modulates the innate immune response (Oren *et al.*, 1999; Scott *et al.*, 2002). It has been suggested that fHbp-mediated protection against LL-37 involves direct electrostatic interaction between the molecules (Seib *et al.*, 2009).

Because of its immunomodulatory abilities, fHbp is a favourite candidate antigen for the development of a new vaccine against *N. meningitidis* (Madico *et al.*, 2006). Antibodies directed against fHbp would be able to kill bacteria in two ways. First, the antibody–antigen interaction induces the classical pathway of the human complement system. Secondly, antibodies binding to fHbp may prevent binding of fH to the bacterial cell surface, thereby increasing the susceptibility of the bacteria to the alternative complement pathway.

Neisserial heparin-binding antigen

Also the neisserial heparin-binding antigen (NhbA) was described to play a role in protection of *N. meningitidis* in humans. NhbA possesses an arginine-rich region that was shown to bind heparin *in vitro* and binding of heparin was suggested to be correlated with increased survival of *N. meningitidis* in human serum (Serruto *et al.*, 2010). Various virulence factors from many different bacteria have been described to bind heparin (Rostand and Esko, 1997). Heparin interacts with several inhibitors of the complement system such as fH, and it is possible that bacteria binding heparin recruit these factors, thereby escaping from the host immune system.

Integral OMPs

Apart from cell-surface-exposed proteins, also integral OMPs have been implicated in bacterial adhesion and pathogenesis. In the case of *N. meningitidis*, a family of integral OMPs known as the opacity-associated proteins (Opa and Opc) plays a major role in adhesion. Also, the most abundant OMPs in *Neisseria*, the porins, have been shown to play a role in pathogenesis. Opa, Opc and porins are examples of integral OMPs with a β-barrel structure. These structures contain extracellular loops, which can facilitate interaction of the bacteria with the host.

Opacity-associated proteins

Multiple *opa* genes are present in the genomes of the *Neisseriaceae*. The Opa protein repertoire is much greater in gonococci, which contain eleven different *opa* genes, than in meningococci, which contain three or four *opa* loci. The *opa* genes are constitutively transcribed, but Opa protein expression is subject to phase variation owing to the presence of pentameric repeat sequences (5′-CTCTT-3′) located within the signal-sequence-encoding part of the genes. The number of repeats varies due to slipped-strand mispairing during DNA replication resulting in high-frequency on/off switching of Opa protein expression (Stern *et al.*, 1986). Opa proteins consist of a conserved framework of an eight-stranded β-barrel, which exposes one conserved, one semi-variable and two highly variable loops at the cell surface (Malorny *et al.*, 1998). Opa proteins facilitate interaction with various host cells, such as neutrophils, epithelial and endothelial cells. They recognize two types of receptors: the heparan sulfate proteoglycans (HSPG) (Chen *et al.*, 1995; van Putten and Paul, 1995) and, as illustrated in Fig. 13.5, the CEACAM glycoprotein family (Chen and Gotschlich, 1996; Virji *et al.*, 1996a; Gray-Owen *et al.*, 1997a,b). Each Opa protein recognizes a specific set of receptor molecules (Bos *et al.*, 1997a; Chen *et al.*, 1997; Popp *et al.*, 1999). The specificity of the interaction is dictated by either one or both of the hyper-variable loops, suggesting that these extracellular loops closely interact (Bos *et al.*, 2002). Furthermore, the HSPG-binding Opa proteins can recruit heparin-binding host molecules, such as vitronectin, to the bacterial cell surface via a molecular bridge formed by heparin-like molecules. This mechanism allows for vitronectin-mediated uptake of bacteria into host cells expressing vitronectin receptors (Duensing and van Putten, 1998) or for sequestering chemokines, thereby affecting chemotactic gradients, which attract immune cells (Duensing *et al.*, 1999).

The large Opa protein repertoire available to the bacteria might serve to mediate interactions with different cell types encountered during different stages of infection or to evade the immune system. The interactions of Opa proteins with the immune system are complex. Opa proteins are immunogenic in humans as anti-Opa antibodies are generated following both invasive meningococcal disease and vaccination (Poolman *et al.*, 1983; Rosenqvist *et al.*, 1995). Opa proteins were shown to protect gonococci against killing by limited amounts of human serum, i.e. circumstances that may be prevalent on mucosal surfaces (Bos *et al.*, 1997b). It was shown that the interaction of Opa proteins with CEACAM1 on $CD4^+$ T lymphocytes inhibited activation and proliferation of these T cells (Boulton and Gray-Owen, 2002), although in another study no such inhibitory effect was found (Youssef *et al.*, 2009). Furthermore, Opa has been demonstrated to inhibit antibody production from human B

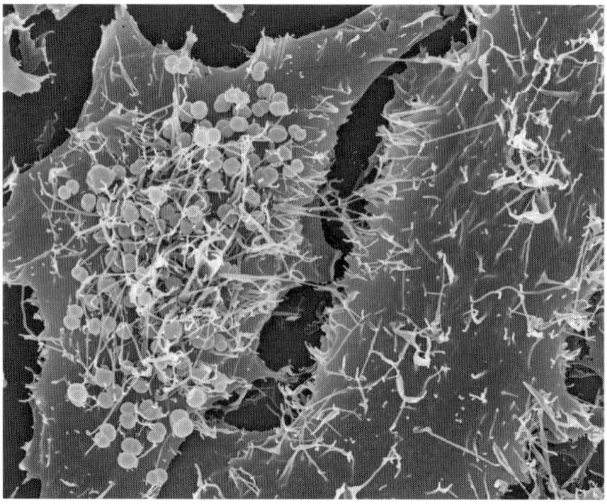

Figure 13.5 Scanning electron micrograph of a mixed cell culture of HeLa-CEACAM1 positive and negative cells infected with Opa-expressing *N. gonorrhoeae*. Only the CEACAM1-positive cell on the left is recognized by the bacteria. Magnification: 50,000×.

lymphocytes (Pantelic *et al.*, 2005). Such results obviously raise the question whether inclusion of Opa proteins in neisserial vaccines should be avoided. Unfortunately, lack of suitable animal models and obvious limitations of human studies make it difficult to extrapolate these results to the *in vivo* situation. However, a strong indication for an important role of Opa proteins in gonococcal infections is the finding that Opa^+ bacteria are recovered during natural gonococcal infection (James and Swanson, 1978) and following inoculation of humans with Opa^- bacteria (Swanson *et al.*, 1988; Jerse *et al.*, 1994).

The NspA protein is a distant member of the Opa family. Like the Opa proteins, it forms an eight-stranded β-barrel and it shows sequence similarity to the Opa proteins, but its cell-surface-exposed loops are considerably shorter (Vandeputte-Rutten *et al.*, 2003). Unlike the *opa* genes, expression of *nspA* is not prone to phase variation by slipped-strand mispairing. Recently, it was demonstrated that NspA, like fHbp, inhibits complement activation via the alternative pathway by binding fH (Lewis *et al.*, 2010).

The Opc protein is functionally similar to the Opa proteins in many aspects (Virji *et al.*, 1993), but its crystal structure revealed that it is structurally different, in that it consists of a 10-stranded β-barrel with five surface-exposed loops (Prince *et al.*, 2002). Opc expression levels are also variable, but this is caused by slipped-strand mispairing in a poly-C tract in the *opc* promoter. Gonococcal Opc levels are generally much lower than those in meningococci, but the *opc* gene is completely absent in certain clonal lineages of *N. meningitidis* (Zhu *et al.*, 1999). Opc resembles the class of heparin-binding Opa proteins because it also binds to heparin-like molecules and HSPG receptors on human epithelial cells (de Vries *et al.*, 1998). Both vitronectin and fibronectin were shown to promote entry of Opc^+ bacteria into human host cells (Virji *et al.*, 1994; Unkmeir *et al.*, 2002). Vitronectin can bind to Opc via heparin bridges or via a direct interaction (Sa E Cunha *et al.*, 2010).

Porins

Porins form narrow, water-filled pores whose main function is to exchange nutrients and other small solutes over the outer membrane of Gram-negative bacteria. Meningococci and gonococci stably express either one of two variants of PorB, called class 2 and class 3 in *N. meningitidis* and PorB1A and PorB1B in *N. gonorrhoeae*. In addition, meningococci phase-variably express another porin, PorA. The crystal structure of PorB revealed a 16-stranded β-barrel conformation similar to most other known bacterial porin structures (Tanabe *et al.*, 2010). Besides their role in nutrient acquisition, a range of other functions have been attributed to the neisserial porins. The PorB1A protein confers invasion of *N. gonorrhoeae* into epithelial cells (van Putten *et al.*, 1998). Gonococcal porins were shown to protect the bacteria from the bactericidal effects of human serum by binding the complement regulatory proteins fH and C4b-binding protein (Ram *et al.*, 1998; Ram *et al.*, 2001). The former interaction can also create a molecular bridge through the simultaneous binding of fH to PorB on bacteria and to complement receptor 3 on cervical epithelial cells (Agarwal *et al.*, 2010). Binding of fH to porin appears to be species specific as it was not observed in *N. meningitidis* (Madico *et al.*, 2007).

Conflicting data have been reported on the effect of neisserial PorB proteins on apoptosis with both pro-apoptotic effects reported for the gonococcal porin PorB1B (Müller *et al.*, 1999) and anti-apoptotic effects for meningococcal PorB (Massari *et al.*, 2000). PorB can act as an adjuvant in immune responses through activation of Toll-like receptor 2 (Wetzler, 2010).

Conclusions and future prospects

In this chapter, we reviewed outer membrane-embedded and surface-associated proteins and their role in the interaction of bacteria, in particular *N. meningitidis*, with the host during pathogenesis. Many of the proteins discussed are involved in direct intimate adherence of the bacterium with host cells either during colonization or infection. The large number of different proteins that are considered as adhesins raises questions about their exact role. Why would there be so many proteins involved in adherence to host cells?

It is possible that different proteins assist during different stages in neisserial pathogenesis. T4P are only strictly required for attachment to host epithelia when *N. meningitidis* is capsulated. It has been proposed that pilus-mediated initial attachment of *N. meningitidis* to cells induces expression of the regulatory protein CrgA, which in turn mediates a negative feedback regulation of the expression of the *pilC1*, *pilE*, and the capsule-related *sia* genes (Deghmane *et al.*, 2002). Consequently, pili and capsule are lost, thereby unmasking proteins that are involved in intimate adhesion. These may include type V-secreted proteins or membrane-embedded OMPs such as the opacity-associated proteins and the porins.

Besides this, redundancy of functions appears to be associated with the meningococcal life style. Forced by the selection pressure of the immune system, the meningococcus is able to switch on and off the expression of most proteins that are exposed at the cell surface. As important functions involved in the interaction with the host may not be lost, the bacteria should have alternative proteins that are able to take over the function of a protein whose synthesis is switched off. Most of these proteins may have other functions in addition to mediating intimate attachment. Pili, for example, are also required for bacterial movement across epithelial surfaces, DNA uptake, and host cell invasion. This multipurpose nature also holds true for many other proteins. For instance, the type Vb secreted TpsA has been suggested to function in adherence to host cells but also in intracellular survival and in bacterial competition. Porins can, apart from their well-described function as channel for small nutrients and solutes, also play a role in adhesion and invasion. Lipoproteins such as TbpB and LbpB appear to be designed to scavenge iron under iron-limited conditions, but they could also play a role in evading the host defence system by binding antimicrobial compounds. The role of the lipoprotein fHbp in the evasion of host defences is also not restricted to the binding of a single ligand. The ability of OMPs to participate in several pathogenesis-related processes points to a versatile and efficient nature of these proteins. From a bacterial perspective, it is highly effective when a limited set of surface-exposed proteins performs a vast array of functions in bacterial infection and can take over the functions of other proteins when their synthesis is switched off.

It must also be realized that pathogenesis is a rare event in the meningococcal life cycle. Usually *N. meningitidis* harmlessly colonizes the human nasopharynx and only in exceptional cases this is followed by infection. This might implicate that proteins involved in pathogenesis primarily display functions that are related with the commensal life style and survival on the nasopharyngeal epithelium. The involvement of these proteins in pathogenesis might therefore have to be considered as side-effects of otherwise rather innocuous proteins.

Another aspect that complicates elusion and interpretation of the role of OMPs in the case of *N. meningitidis* is that almost all studies have been performed in *in vitro* systems. Because *N. meningitidis* pathogenesis is restricted to humans, experimental systems are limited. Therefore, all *in vitro* studies have to be extrapolated to the *in vivo* situation, which could possibly lead to misinterpretations.

The OMPs and surface-exposed proteins that have been discussed in this chapter cover only a part of the total spectrum of such proteins, but even the role of most of these proteins in colonization and pathogenesis is far from resolved. Particularly, the knowledge of the functions of the type V-secreted proteins appears fragmentary. Elucidating their exact functions and mechanisms requires extensive research. In particular, the role of these proteins in the interaction with epithelial and host immune cells requires further attention. Identification of the molecular receptors and ligands for bacterial surface proteins and the host cell signalling pathways they induce, thus renders a wide field open for new discoveries.

Acknowledgements

Jan Grijpstra is acknowledged for bioinformatic analysis and Fred Hayes and Dave Dorward (Rocky Mountain Laboratories, NIAID, NIH, USA) for electron microscopy. Research in the authors' laboratory is supported by grants from the Netherlands Organization for Health Research and Development (ZonMW) and the Division for Chemical Sciences (CW) from the Netherlands Organization for Scientific Research (NWO), and from GlaxoSmithKline biologicals s.a.

References

Agarwal, S., Ram, S., Ngampasutadol, J., Gulati, S., Zipfel, P.F., and Rice, P.A. (2010). Factor H facilitates adherence of *Neisseria gonorrhoeae* to complement receptor 3 on eukaryotic cells. J. Immunol. *185*, 4344–4353.

Anderson, J.E., Sparling, P.F., and Cornelissen, C.N. (1994). Gonococcal transferrin-binding protein 2 facilitates but is not essential for transferrin utilization. J. Bacteriol. *176*, 3162–3170.

Andrews, T.D., and Gojobori, T. (2004). Strong positive selection and recombination drive the antigenic variation of the PilE protein of the human pathogen *Neisseria meningitidis*. Genetics *166*, 25–32.

Aoki, S.K., Diner, E.J., De Roodenbeke, C.T., Burgess, B.R., Poole, S.J., Braaten, B.A., Jones, A.M., Webb, J.S., Hayes, C.S., Cotter, P.A., *et al.* (2010). A widespread family of polymorphic contact-dependent toxin delivery systems in bacteria. Nature *468*, 439–442.

Aoki, S.K., Pamma, R., Hernday, A.D., Bickham, J.E., Braaten, B.A., and Low, D.A. (2005). Contact-dependent inhibition of growth in *Escherichia coli*. Science *309*, 1245–1248.

Arnold, T., Zeth, K., and Linke, D. (2010). Omp85 from the thermophilic cyanobacterium *Thermosynechococcus elongatus* differs from proteobacterial Omp85 in structure and domain composition. J. Biol. Chem. *285*, 18003–18015.

Avadhanula, V., Rodriguez, C.A., Ulett, G.C., Bakaletz, L.O., and Adderson, E.E. (2006). Nontypeable *Haemophilus influenzae* adheres to intercellular adhesion molecule 1 (ICAM-1) on respiratory epithelial cells and upregulates ICAM-1 expression. Infect. Immun. *74*, 830–838.

Barnhart, M.M., and Chapman, M.R. (2006). Curli biogenesis and function. Annu. Rev. Microbiol. *60*, 131–147.

Barnard, T.J., Dautin, N., Lukacik, P., Bernstein, H.D., and Buchanan, S.K. (2007). Autotransporter structure reveals intra-barrel cleavage followed by conformational changes. Nat. Struct. Mol. Biol. *14*, 1214–1220.

Batchelor, M., Prasannan, S., Daniell, S., Reece, S., Connerton, I., Bloomberg, G., Dougan, G., Frankel, G., and Matthews, S. (2000). Structural basis for recognition of the translocated intimin receptor (Tir) by intimin from enteropathogenic *Escherichia coli*. EMBO J. *19*, 2452–2464.

Bateman, A., and Bycroft, M. (2000). The structure of a LysM domain from *E. coli* membrane-bound lytic murein transglycosylase D (MltD). J. Mol. Biol. *299*, 1113–1119.

Bian, Z., and Normark, S. (1997). Nucleator function of CsgB for the assembly of adhesive surface organelles in *Escherichia coli*. EMBO J. *16*, 5827–5836.

Blom, A.M., Rytkönen, A., Vasquez, P., Lindahl, G., Dahlbäck, B., and Jonsson, A.B. (2001). A novel interaction between type IV pili of *Neisseria gonorrhoeae* and the human complement regulator C4B-binding protein. J. Immunol. *166*, 6764–6770.

Bodelón, G., Marín, E., and Fernández, L.A. (2009). Role of periplasmic chaperones and BamA (YaeT/Omp85) in folding and secretion of intimin from enteropathogenic *Escherichia coli* strains. J. Bacteriol. *191*, 5169–5179.

Bos, M.P., Grunert, F., and Belland, R.J. (1997a). Differential recognition of members of the carcinoembryonic antigen family by Opa variants of *Neisseria gonorrhoeae*. Infect. Immun. *65*, 2353–2361.

Bos, M.P., Hogan, D., and Belland, R.J. (1997b). Selection of Opa$^+$ *Neisseria gonorrhoeae* by limited availability of normal human serum. Infect. Immun. *65*, 645–650.

Bos, M.P., Kao, D., Hogan, D.M., Grant, C.C., and Belland, R.J. (2002). Carcinoembryonic antigen family receptor recognition by gonococcal Opa proteins requires distinct combinations of hypervariable Opa protein domains. Infect. Immun. *70*, 1715–1723.

Bos, M.P., Robert, V., and Tommassen, J. (2007). Biogenesis of the gram-negative bacterial outer membrane. Annu. Rev. Microbiol. *61*, 191–214.

Boulton, I.C., and Gray-Owen, S.D. (2002). Neisserial binding to CEACAM1 arrests the activation and proliferation of CD4$^+$ T lymphocytes. Nat. Immunol. *3*, 229–236.

Boulton, I.C., Gorringe, A.R., Allison, N., Robinson, A., Gorinsky, B., Joannou, C.L., and Evans, R.W. (1998). Transferrin-binding protein B isolated from *Neisseria meningitidis* discriminates between apo and diferric human transferrin. Biochem. J. *334*, 269–273.

Brandon, L.D., Goehring, N., Janakiraman, A., Yan, A.W., Wu, T., Beckwith, J., and Goldberg, M.B. (2003). IcsA, a polarly localized autotransporter with an atypical signal peptide, uses the Sec apparatus for secretion, although the Sec apparatus is circumferentially distributed. Mol. Microbiol. *50*, 45–60.

Bubeck Wardenburg, J., Williams, W.A., and Missiakas, D. (2006). Host defenses against *Staphylococcus aureus* infection require recognition of bacterial lipoproteins. Proc. Natl. Acad. Sci. U.S.A. *103*, 13831–13836.

Burrows, L.L. (2005). Weapons of mass retraction. Mol. Microbiol. *57*, 878–888.

Capecchi, B., Adu-Bobie, J., Di Marcello, F., Ciucchi, L., Masignani, V., Taddei, A., Rappuoli, R., Pizza, M., and Arico, B. (2005). *Neisseria meningitidis* NadA is a new invasin which promotes bacterial adhesion to and penetration into human epithelial cells. Mol. Microbiol. *55*, 687–698.

Carbonnelle, E., Hélaine, S., Prouvensier, L., Nassif, X., and Pelicic, V. (2005). Type IV pilus biogenesis in *Neisseria meningitidis*: PilW is involved in a step occurring after pilus assembly, essential for fibre stability and function. Mol. Microbiol. *55*, 54–64.

Chamot-Rooke, J., Rousseau, B., Lanternier, F., Mikaty, G., Mairey, E., Malosse, C., Bouchoux, G., Pelicic, V., Camoin, L., Nassif, X., *et al.* (2007). Alternative *Neisseria* spp. type IV pilin glycosylation with a glyceramido acetamido trideoxyhexose residue. Proc. Natl. Acad. Sci. U.S.A. *104*, 14783–14788.

Chamot-Rooke, J., Mikaty, G., Malosse, C., Soyer, M., Dumont, A., Gault, J., Imhaus, A.F., Martin, P., Trellet, M., Clary, G., *et al.* (2011). Posttranslational modification of pili upon cell contact triggers *N. meningitidis* dissemination. Science *331*, 778–782.

Chen, T., and Gotschlich, E.C. (1996). CGM1a antigen of neutrophils, a receptor of gonococcal opacity proteins. Proc. Natl. Acad. Sci. U.S.A. *93*, 14851–14856.

Chen, T., Belland, R.J., Wilson, J., and Swanson, J. (1995). Adherence of pilus$^-$ Opa$^+$ gonococci to epithelial cells *in vitro* involves heparan sulfate. J. Exp. Med. *182*, 511–517.

Chen, T., Grunert, F., Medina-Marino, A., and Gotschlich, E.C. (1997). Several carcinoembryonic antigens (CD66) serve as receptors for gonococcal opacity proteins. J. Exp. Med. *185*, 1557–1564.

Clantin, B., Delattre, A.S., Rucktooa, P., Saint, N., Meli, A.C., Locht, C., Jacob-Dubuisson, F., and Villeret, V. (2007). Structure of the membrane protein FhaC: a member of the Omp85-TpsB transporter superfamily. Science *317*, 957–961.

Collins, R.F., Davidsen, L., Derrick, J.P., Ford, R.C., and Tønjum, T. (2001). Analysis of the PilQ secretin from *Neisseria meningitidis* by transmission electron microscopy reveals a dodecameric quaternary structure J. Bacteriol. *183*, 3825–3832.

Collins, R.F., Saleem, M., and Derrick, J.P. (2007). Purification and three-dimensional electron microscopy structure of the *Neisseria meningitidis* type IV pilus biogenesis protein PilG. J. Bacteriol. *189*, 6389–6396.

Comanducci, M., Bambini, S., Caugant, D.A., Mora, M., Brunelli, B., Capecchi, B., Ciucchi, L., Rappuoli, R., and Pizza, M. (2004). NadA diversity and carriage in *Neisseria meningitidis*. Infect. Immun. *72*, 4217–4223.

Connell, I., Agace, W., Klemm, P., Schembri, M., Mărild, S., and Svanborg, C. (1996). Type 1 fimbrial expression enhances *Escherichia coli* virulence for the urinary tract. Proc. Natl. Acad. Sci. U.S.A. *93*, 9827–9832.

Cornelissen, C.N., and Sparling, P.F. (1996). Binding and surface exposure characteristics of the gonococcal transferrin receptor are dependent on both transferrin-binding proteins. J. Bacteriol. *178*, 1437–1444.

Cotter, S.E., Yeo, H.J., Juehne, T., and St Geme, J.W. 3rd. (2005). Architecture and adhesive activity of the *Haemophilus influenzae* Hsf adhesin. J. Bacteriol. *187*, 4656–4664.

Coureuil, M., Mikaty, G., Miller, F., Lécuyer, H., Bernard, C., Bourdoulous, S., Duménil, G., Mège, R.M., Weksler, B.B., Romero, I.A., *et al.* (2009). Meningococcal type IV pili recruit the polarity complex to cross the brain endothelium. Science *325*, 83–87.

Coureuil, M., Lécuyer, H., Scott, M.G.H., Boularan, C., Enslen, H., Soyer, M., Mikaty, G., Bourdoulous, S., Nassif, X., and Marullo, S. (2010). Meningococcus hijacks a β2-adrenoceptor/β-arrestin pathway to cross brain microvasculature endothelium. Cell *143*, 1149–1160.

Coutte, L., Antoine, R., Drobecq, H., Locht, C., and Jacob-Dubuisson, F. (2001). Subtilisin-like autotransporter serves as maturation protease in a bacterial secretion pathway. EMBO J. *20*, 5040–5048.

Craig, L., and Li, J. (2008). Type IV pili: paradoxes in form and function. Curr. Opin. Struct. Biol. *18*, 267–277.

De Vries, F.P., Cole, R., Dankert, J., Frosch, M., and van Putten, J.P.M. (1998). *Neisseria meningitidis* producing the Opc adhesin binds epithelial cell proteoglycan receptors. Mol. Microbiol. *27*, 1203–1212.

Deghmane, A.E., Giorgini, D., Larribe, M., Alonso, J.M., and Taha, M.K. (2002). Down-regulation of pili and capsule of *Neisseria meningitidis* upon contact with epithelial cells is mediated by CrgA regulatory protein. Mol. Microbiol. *43*, 1555–1564.

Deuretzbacher, A., Czymmeck, N., Reimer, R., Trülzsch, K., Gaus, K., Hohenberg, H., Heesemann, J., Aepfelbacher, M., and Ruckdeschel, K. (2009). β1 integrin-dependent engulfment of *Yersinia enterocolitica* by macrophages is coupled to the activation of autophagy and suppressed by type III protein secretion. J. Immunol. *183*, 5847–5860.

Duensing, T.D., and van Putten, J.P.M. (1998). Vitronectin binds to the gonococcal adhesin OpaA through a glycosaminoglycan molecular bridge. Biochem. J. *334*, 133–139.

Duensing, T.D., Wing, J.S., and van Putten, J.P.M. (1999). Sulfated polysaccharide-directed recruitment of mammalian host proteins: a novel strategy in microbial pathogenesis. Infect. Immun. *67*, 4463–4468.

Edwards, J.L., and Apicella, M.A. (2005). I-domain-containing integrins serve as pilus receptors for *Neisseria gonorrhoeae* adherence to human epithelial cells. Cell. Microbiol. *7*, 1197–1211.

Ellison, R.T. 3rd, Giehl, T.J., and LaForce, F.M. (1988). Damage of the outer membrane of enteric gram-negative bacteria by lactoferrin and transferrin. Infect. Immun. *56*, 2774–2781.

Ellison, R.T. 3rd, LaForce, F.M., Giehl, T.J., Boose, D.S., and Dunn, B.E. (1990). Lactoferrin and transferrin damage of the Gram-negative outer membrane is modulated by Ca^{2+} and Mg^{2+}. J. Gen. Microbiol. *136*, 1437–1446.

Eugène, E., Hoffmann, I., Pujol, C., Couraud, P.O., Bourdoulous, S., and Nassif, X. (2002). Microvilli-like structures are associated with the internalization of virulent capsulated *Neisseria meningitidis* into vascular endothelial cells. J. Cell. Sci. *115*, 1231–1241.

Fan, E., Fiedler, S., Jacob-Dubuisson, F., and Müller, M. (2012). Two-partner secretion of gram-negative bacteria: a single β-barrel protein enables transport across the outer membrane. J. Biol. Chem. *287*, 2591–2599.

Finck-Babançon, V., Goranson, J., Zhu, L., Sawa, T., Wiener-Kronish, J.P., Fleiszig, S.M.J., Wu, C., Mende-Mueller, L., and Frank, D.W. (1997). ExoU expression by *Pseudomonas aeruginosa* correlates with acute cytotoxicity and epithelial injury. Mol. Microbiol. *25*, 547–557.

Folkesson, A., Advani, A., Sukupolvi, S., Pfeifer, J.D., Normark, S., and Löfdahl, S. (1999). Multiple insertions of fimbrial operons correlate with the evolution of *Salmonella* serovars responsible for human disease. Mol. Microbiol. *33*, 612–622.

Frankel, G., Phillips, A.D., Novakova, M., Batchelor, M., Hicks, S., and Dougan, G. (1998). Generation of *Escherichia coli* intimin derivatives with differing biological activities using site-directed mutagenesis of the intimin C-terminus domain. Mol. Microbiol. *29*, 559–570.

Froehlich, B.J., Karakashian, A., Melsen, L.R., Wakefield, J.C., and Scott, J.R. (1994). CooC and CooD are required for assembly of CS1 pili. Mol. Microbiol. *12*, 387–401.

Gaastra, W., and Svennerholm, A.M. (1996). Colonization factors of human enterotoxigenic *Escherichia coli* (ETEC). Trends Microbiol. *4*, 444–452.

Gandhi, G., Londoño, D., Whetstine, C.R., Sethi, N., Kim, K.S., Zückert, W.R., and Cadavid, D. (2010). Interaction of variable bacterial outer membrane lipoproteins with brain endothelium. PLoS One *5*, e13257.

Gerlach, R.G., and Hensel, M. (2007). Protein secretion systems and adhesins: the molecular armory of Gram-negative pathogens. Int. J. Med. Microbiol. *297*, 401–415.

Goosney, D.L., de Grado, M., and Finlay, B.B. (1999). Putting *E. coli* on a pedestal: a unique system to study signal transduction and the actin cytoskeleton. Trends Cell Biol. *19,* 11–14.

Gottig, N., Garavaglia, B.S., Garofalo, C.G., Orellano, E.G., and Ottado, J. (2009). A filamentous hemagglutinin-like protein of *Xanthomonas axonopodis* pv. citri, the phytopathogen responsible for citrus canker, is involved in bacterial virulence. PLoS One *4,* e4358.

Gray-Owen, S.D., and Schryvers, A.B. (1996). Bacterial transferrin and lactoferrin receptors. Trends Microbiol. *4,* 185–191.

Gray-Owen, S.D., Dehio, C., Haude, A., Grunert, F., and Meyer, T.F. (1997a). CD66 carcinoembryonic antigens mediate interactions between Opa-expressing *Neisseria gonorrhoeae* and human polymorphonuclear phagocytes. EMBO J. *16,* 3435–3445.

Gray-Owen, S.D., Lorenzen, D.R., Haude, A., Meyer, T.F., and Dehio, C. (1997b). Differential Opa specificities for CD66 receptors influence tissue interactions and cellular response to *Neisseria gonorrhoeae.* Mol. Microbiol. *26,* 971–980.

Griffiths, N.J., Hill, D.J., Borodina, E., Sessions, R.B., Devos, N.I., Feron, C.M., Poolman, J.T., and Virji, M. (2011). Meningococcal surface fibril (Msf) binds to activated vitronectin and inhibits the terminal complement pathway to increase serum resistance. Mol. Microbiol. *82,* 1129–1149.

Haas, R., and Meyer, T.F. (1986). The repertoire of silent pilus genes in *Neisseria gonorrhoeae*: evidence for gene conversion. Cell *44,* 107–115.

Hadi, H.A., Wooldridge, K.G., Robinson, K., and Ala'Aldeen, D.A.A. (2001). Identification and characterization of App: an immunogenic autotransporter protein of *Neisseria meningitidis.* Mol. Microbiol. *41,* 611–623.

Hahn, E., Wild, P., Hermanns, U., Sebbel, P., Glockshuber, R., Häner, M., Taschner, N., Burkhard, P., Aebi, U., and Müller, S.A. (2002). Exploring the 3D molecular architecture of *Escherichia coli* type 1 pili. J. Mol. Biol. *323,* 845–857.

Hammar, M., Bian, Z., and Normark, S. (1996). Nucleator-dependent intercellular assembly of adhesive curli organelles in *Escherichia coli.* Proc. Natl. Acad. Sci. U.S.A. *93,* 6562–6566.

Hartland, E.L., Batchelor, M., Delahay, R.M., Hale, C., Matthews, S., Dougan, G., Knutton, S., Connerton, I., and Frankel, G. (1999). Binding of intimin from enteropathogenic *Escherichia coli* to Tir and to host cells. Mol. Microbiol. *32,* 151–158.

Hayes, C.S., Aoki, S.K., and Low, D.A. (2010). Bacterial contact-dependent delivery systems. Annu. Rev. Genet., *44,* 71–90.

Hayward, R.D., Leong, J.M., Koronakis, V., and Campellone, K.G. (2006). Exploiting pathogenic *Escherichia coli* to model transmembrane receptor signalling. Nat. Rev. Microbiol. *4,* 358–370.

Hegge, F.T., Hitchen, P.G., Aas, F.E., Kristiansen, H., Løvold, C., Egge-Jacobsen, W., Panico, M., Leong, W.Y., Bull, V., Virji, M., *et al.* (2004). Unique modifications with phosphocholine and phosphoethanolamine define alternate antigenic forms of *Neisseria gonorrhoeae* type IV pili. Proc. Natl. Acad. Sci. U.S.A. *101,* 10798–10803.

Hélaine, S., Carbonnelle, E., Prouvensier, L., Beretti, J.L., Nassif, X., and Pelicic, V. (2005). PilX, a pilus-associated protein essential for bacterial aggregation, is a key to pilus-facilitated attachment of *Neisseria meningitidis* to human cells. Mol. Microbiol. *55,* 65–77.

Henderson, I.R., and Nataro, J.P. (2001). Virulence functions of autotransporter proteins. Infect. Immun. *69,* 1231–1243.

Hendrixson, D.R., Qiu, J., Shewry, S.C., Fink, D.L., Petty, S., Baker, E.N., Plaut, A.G., and St Geme, J.W. 3rd. (2003). Human milk lactoferrin is a serine protease that cleaves *Haemophilus* surface proteins at arginine-rich sites. Mol. Microbiol. *47,* 607–617.

Herwald, H., Mörgelin, M., Olsén, A., Rhen, M., Dahlbäck, B., Müller-Esterl, W., and Björck, L. (1998). Activation of the contact-phase system on bacterial surfaces – a clue to serious complications in infectious diseases. Nat. Med. *4,* 298–302.

Higgins, L.M., Frankel, G., Connerton, I., Gonçalves, N.S., Dougan, G., and MacDonald, T.T. (1999). Role of bacterial intimin in colonic hyperplasia and inflammation. Science *285,* 588–591.

Hodak, H., Clantin, B., Willery, E., Villeret, V., Locht, C., and Jacob-Dubuisson, F. (2006). Secretion signal of the filamentous haemagglutinin, a model two-partner secretion substrate. Mol. Microbiol. *61,* 368–382.

Ieva, R., and Bernstein, H.D. (2009). Interaction of an autotransporter passenger domain with BamA during its translocation across the bacterial outer membrane. Proc. Natl. Acad. Sci. U.S.A. *106,* 19120–19125.

Ieva, R., Tian, P., Peterson, J.H., and Bernstein, H.D. (2011). Sequential and spatially restricted interactions of assembly factors with an autotransporter β domain. Proc. Natl. Acad. Sci. U.S.A. *108,* E383-E391.

Irwin, S.W., Averil, N., Cheng, C.Y., and Schryvers, A.B. (1993). Preparation and analysis of isogenic mutants in the transferrin receptor protein genes, *tbpA* and *tbpB*, from *Neisseria meningitidis*. Mol. Microbiol. *8*, 1125–1133.

Isberg, R.R., and Barnes, P. (2002). Dancing with the host; flow-dependent bacterial adhesion. Cell *110*, 1–4.

Isberg, R.R., and Leong, J.M. (1990). Multiple β_1 chain integrins are receptors for invasin, a protein that promotes bacterial penetration into mammalian cells. Cell *60*, 861–871.

Isberg, R.R., Voorhis, D.L., and Falkow, S. (1987). Identification of invasin: a protein that allows enteric bacteria to penetrate cultured mammalian cells. Cell *50*, 769–778.

Jacob-Dubuisson, F., Locht, C., and Antoine, R. (2001). Two-partner secretion in Gram-negative bacteria: a thrifty, specific pathway for large virulence proteins. Mol. Microbiol. *40*, 306–313.

Jain, S., and Goldberg, M.B. (2007). Requirement for YaeT in the outer membrane assembly of autotransporter proteins. J. Bacteriol. *189*, 5393–5398.

James, J.F., and Swanson, J. (1978). Studies on gonococcus infection. XIII. Occurrence of color/opacity colonial variants in clinical cultures. Infect. Immun. *19*, 332–340.

Jennings, M.P., Jen, F.E., Roddam, L.F., Apicella, M.A., and Edwards, J.L. (2011). *Neisseria gonorrhoeae* pilin glycan contributes to CR3 activation during challenge of primary cervical epithelial cells. Cell. Microbiol. *13*, 885–896.

Jerse, A.E., Yu, J., Tall, B.D., and Kaper, J.B. (1990). A genetic locus of enteropathogenic *Escherichia coli* necessary for the production of attaching and effacing lesions on tissue culture cells. Proc. Natl. Acad. Sci. U.S.A. *87*, 7839–7843.

Jerse, A.E., Cohen, M.S., Drown, P.M., Whicker, L.G., Isbey, S.F., Seifert, H.S., and Cannon, J.G. (1994). Multiple gonococcal opacity proteins are expressed during experimental urethral infection in the male. J. Exp. Med. *179*, 911–920.

Johansson, L., Rytkönen, A., Bergman, P., Albiger, B., Källström, H., Hökfelt, T., Agerberth, B., Cattaneo, R., and Jonsson, A.B. (2003). CD46 in meningococcal disease. Science *301*, 373–375.

Jones, C.H., Pinkner, J.S., Roth, R., Heuser, J., Nicholes, A.V., Abraham, S.N., and Hultgren, S.J. (1995). FimH adhesin of type 1 pili is assembled into a fibrillar tip structure in the *Enterobacteriaceae*. Proc. Natl. Acad. Sci. U.S.A. *92*, 2081–2085.

Jonsson, A.B., Nyberg, G., and Normark, S. (1991). Phase variation of gonococcal pili by frameshift mutation in *pilC*, a novel gene for pilus assembly. EMBO J. *10*, 477–488.

Juncker, A.S., Willenbrock, H., Von Heijne, G., Brunak, S., Nielsen, H., and Krogh, A. (2003). Prediction of lipoprotein signal peptides in Gram-negative bacteria. Protein Sci. *12*, 1652–1662.

Kachlany, S.C., Planet, P.J., Desalle, R., Fine, D.H., Figurski, D.H., and Kaplan, J.B. (2001). Flp-1, the first representative of a new pilin gene subfamily, is required for non-specific adherence of *Actinobacillus actinomycetemcomitans*. Mol. Microbiol. *40*, 542–554.

Källström, H., Liszewski, M.K., Atkinson, J.P., and Jonsson, A.B. (1997). Membrane cofactor protein (MCP or CD46) is a cellular pilus receptor for pathogenic *Neisseria*. Mol. Microbiol. *25*, 639–647.

Kenny, B., DeVinney, R., Stein, M., Reinscheid, D.J., Frey, E.A., and Finlay, B.B. (1997). Enteropathogenic *E. coli* (EPEC) transfers its receptor for intimate adherence into mammalian cells. Cell *91*, 511–520.

Kirchner, M., Heuer, D., and Meyer, T.F. (2005). CD46-independent binding of neisserial type IV pili and the major pilus adhesin, PilC, to human epithelial cells. Infect. Immun. *73*, 3072–3082.

Klee, S.R., Nassif, X., Kusecek, B., Merker, P., Beretti, J.L., Achtman, M., and Tinsley, C.R. (2000). Molecular and biological analysis of eight genetic islands that distinguish *Neisseria meningitidis* from the closely related pathogen *Neisseria gonorrhoeae*. Infect. Immun. *68*, 2082–2095.

Kovacs-Simon, A., Titball, R.W., and Michell, S.L. (2011). Lipoproteins of bacterial pathogens. Infect. Immun. *79*, 548–561.

Laarmann, S., Cutter, D., Juehne, T., Barenkamp, S.J., and St Geme, J.W. 3rd. (2002). The *Haemophilus influenzae* Hia autotransporter harbours two adhesive pockets that reside in the passenger domain and recognize the same host cell receptor. Mol. Microbiol. *46*, 731–743.

Legrain, M., Mazarin, V., Irwin, S.W., Bouchon, B., Quentin-Millet, M.J., Jacobs, E., and Schryvers, A.B. (1993). Cloning and characterization of *Neisseria meningitidis* genes encoding the transferrin-binding proteins Tbp1 and Tbp2. Gene *130*, 73–80.

Leong, J.M., Morrissey, P.E., Marra, A., and Isberg, R.R. (1995). An aspartate residue of the *Yersinia pseudotuberculosis* invasin protein that is critical for integrin binding. EMBO J. *14*, 422–431.

Lewis, L.A., Ngampasutadol, J., Wallace, R., Reid, J.E., Vogel, U., and Ram, S. (2010). The meningococcal vaccine candidate neisserial surface protein A (NspA) binds to factor H and enhances meningococcal resistance to complement. PLoS Pathog. *29*, e1001027.

Linke, D., Riess, T., Autenrieth, I.B., Lupas, A., and Kempf, V.A. (2006). Trimeric autotransporter adhesins: variable structure, common function. Trends Microbiol. *14*, 264–270.

Liu, H., Magoun, L., and Leong, J.M. (1999). β_1-chain integrins are not essential for intimin-mediated host cell attachment and enteropathogenic *Escherichia coli*-induced actin condensation. Infect. Immun. *67*, 2045–2049.

Luo, Y., Frey, E.A., Pfuetzner, R.A., Creagh, A.L., Knoechel, D.G., Haynes, C.A., Finlay, B.B., and Strynadka, N.C.J. (2000). Crystal structure of enteropathogenic *Escherichia coli* intimin–receptor complex. Nature *405*, 1073–1077.

Lyskowski, A., Leo, J.C., and Goldman, A. (2011). Structure and biology of trimeric autotransporter adhesins. Adv. Exp. Med. Biol. *715*, 143–158.

Madico, G., Welsch, J.A., Lewis, L.A., McNaughton, A., Perlman, D.H., Costello, C.E., Ngampasutadol, J., Vogel, U., Granoff, D.M., and Ram, S. (2006). The meningococcal vaccine candidate GNA1870 binds the complement regulatory protein factor H and enhances serum resistance. J. Immunol. *177*, 501–510.

Madico, G., Ngampasutadol, J., Gulati, S., Vogel, U., Rice, P.A., and Ram, S. (2007). Factor H binding and function in sialylated pathogenic neisseriae is influenced by gonococcal, but not meningococcal, porin. J. Immunol. *178*, 4489–4497.

Mairey, E., Genovesio, A., Donnadieu, E., Bernard, C., Jaubert, F., Pinard, E., Seylaz, J., Olivo-Marin, J.C., Nassif, X., and Duménil, G. (2006). Cerebral microcirculation shear stress levels determine *Neisseria meningitidis* attachment sites along the blood–brain barrier. J. Exp. Med. *203*, 1939–1950.

Malorny, B., Morelli, G., Kusecek, B., Kolberg, J., and Achtman, M. (1998). Sequence diversity, predicted two-dimensional protein structure, and epitope mapping of neisserial Opa proteins. J. Bacteriol. *180*, 1323–1330.

Mansa, B., and Kilian, M. (1986). Retained antigen-binding activity of Fab α-fragments of human monoclonal immunoglobulin A1 (IgA1) cleaved by IgA1 protease. Infect. Immun. *52*, 171–174.

Marceau, M., and Nassif, X. (1999). Role of glycosylation at Ser63 in production of soluble pilin in pathogenic *Neisseria*. J. Bacteriol. *181*, 656–661.

Martin, P., van de Ven, T., Mouchel, N., Jeffries, A.C., Hood, D.W., and Moxon, E.R. (2003). Experimentally revised repertoire of putative contingency loci in *Neisseria meningitidis* strain MC58: evidence for a novel mechanism of phase variation. Mol. Microbiol. *50*, 245–257.

Martin, P., Sun, L., Hood, D.W., and Moxon, E.R. (2004). Involvement of genes of genome maintenance in the regulation of phase variation frequencies in *Neisseria meningitidis*. Microbiology *150*, 3001–3012.

Massari, P., Ho, Y., and Wetzler, L.M. (2000). *Neisseria meningitidis* porin PorB interacts with mitochondria and protects cells from apoptosis. Proc. Natl. Acad. Sci. U.S.A. *97*, 9070–9075.

Meng, G., St Geme, J.W. 3rd, and Waksman, G. (2008). Repetitive architecture of the *Haemophilus influenzae* Hia trimeric autotransporter. J. Mol. Biol. *384*, 824–836.

Merz, A.J., Rifenbery, D.B., Arvidson, C.G., and So, M. (1996). Traversal of a polarized epithelium by pathogenic *Neisseriae*: facilitation by type IV pili and maintenance of epithelial barrier function. Mol. Med. *2*, 745–754.

Merz, A.J., Enns, C.A., and So, M. (1999). Type IV pili of pathogenic *Neisseriae* elicit cortical plaque formation in epithelial cells. Mol. Microbiol. *32*, 1316–1332.

Merz, A.J., So, M., and Sheetz, M.P. (2000). Pilus retraction powers bacterial twitching motility. Nature *407*, 98–102.

Mikaty, G., Soyer, M., Mairey, E., Henry, N., Dyer, D., Forest, K.T., Morand, P., Guadagnini, S., Prévost, M.C., Nassif, X., *et al.* (2009). Extracellular bacterial pathogen induces host cell surface reorganization to resist shear stress. PLoS Pathog. *5*, e1000314.

Morand, P.C., Tattevin, P., Eugène, E., Beretti, J.L., and Nassif, X. (2001). The adhesive property of the type IV pilus-associated component PilC1 of pathogenic *Neisseria* is supported by the conformational structure of the N-terminal part of the molecule. Mol. Microbiol. *40*, 846–856.

Morand, P.C., Bille, E., Morelle, S., Eugène, E., Beretti, J.L., Wolfgang, M., Meyer, T.F., Koomey, M., and Nassif, X. (2004). Type IV pilus retraction in pathogenic *Neisseria* is regulated by the PilC proteins. EMBO J. *23*, 2009–2017.

Morand, P.C., Drab, M., Rajalingam, K., Nassif, X., and Meyer, T.F. (2009). *Neisseria meningitidis* differentially controls host cell motility through PilC1 and PilC2 components of type IV pili. PLoS One *4*, e6834.

Müller, A., Günther, D., Düx, F., Naumann, M., Meyer, T.F., and Rudel, T. (1999). Neisserial porin (PorB) causes rapid calcium influx in target cells and induces apoptosis by the activation of cysteine proteases. EMBO J. *18*, 339–352.

Nägele, V., Heesemann, J., Schielke, S., Jiménez-Soto, L.F., Kurzai, O., and Ackermann, N. (2011). *Neisseria meningitidis* adhesin NadA targets β1 integrins: functional similarity to *Yersinia* invasin. J. Biol. Chem. *286*, 20536–20546.

Narita, S., and Tokuda, H. (2007). Amino acids at positions 3 and 4 determine the membrane specificity of *Pseudomonas aeruginosa* lipoproteins. J. Biol. Chem. *282*, 13372–13378.

Nassif, X., Beretti, J.L., Lowy, J., Stenberg, P., O'Gaora, P., Pfeifer, J., Normark, S., and So, M. (1994). Roles of pilin and PilC in adhesion of *Neisseria meningitidis* to human epithelial and endothelial cells. Proc. Natl. Acad. Sci. U.S.A. *91*, 3769–3773.

Neil, R.B., and Apicella, M.A. (2009). Clinical and laboratory evidence for *Neisseria meningitidis* biofilms. Future Microbiol. *4*, 555–563.

Newhall, W.J., Wilde, C.E. 3rd, Sawyer, W.D., and Haak, R.A. (1980). High-molecular-weight antigenic protein complex in the outer membrane of *Neisseria gonorrhoeae*. Infect. Immun. *27*, 475–482.

Nikolakakis, K., Amber, S., Wilbur, J.S., Diner, E.J., Aoki, S.K., Poole, S.J., Tuanyok, A., Keim, P.S., Peacock, S., Hayes, C.S., *et al.* (2012). The toxin/immunity network of *Burkholderia pseudomallei* contact-dependent growth inhibition (CDI) systems. Mol. Microbiol. *84*, 516–529.

Noinaj, N., Easley, N.C., Mizuno, N., Gumbart, J., Boura, E., Steere, A.N., Zak, O., Aisen, P., Tajkhorshid, E., Evans, R.W., *et al.* (2012). Structural basis for iron piracy by pathogenic *Neisseria*. Nature *483*, 53–58.

Olsén, A., Jonsson, A., and Normark, S. (1989). Fibronectin binding mediated by a novel class of surface organelles on *Escherichia coli*. Nature *338*, 652–655.

Olsén, A., Arnqvist, A., Hammar, M., and Normark, S. (1993). Environmental regulation of curli production in *Escherichia coli*. Infect. Agents Dis. *2*, 272–274.

Oomen, C.J., van Ulsen, P., Van Gelder, P., Feijen, M., Tommassen, J., and Gros, P. (2004). Structure of the translocator domain of a bacterial autotransporter. EMBO J. *23*, 1257–1266.

Oren, Z., Lerman, J.C., Gudmundsson, G.H., Agerberth, B., and Shai, Y. (1999). Structure and organization of the human antimicrobial peptide LL-37 in phospholipid membranes: relevance to the molecular basis for its non-cell-selective activity. Biochem. J. *341*, 501–513.

Oriente, F., Scarlato, V., and Delany, I. (2010). Expression of factor H binding protein of meningococcus responds to oxygen limitation through a dedicated FNR-regulated promoter. J. Bacteriol. *192*, 691–701.

Otto, B.R., Sijbrandi, R., Luirink, J., Oudega, B., Heddle, J.G., Mizutani, K., Park, S.Y., and Tame, J.R. (2005). Crystal structure of hemoglobin protease, a heme binding autotransporter protein from pathogenic *Escherichia coli*. J. Biol. Chem. *280*, 17339–17345.

Pantelic, M., Kim, Y.J., Bolland, S., Chen, I., Shively, J., and Chen, T. (2005). *Neisseria gonorrhoeae* kills carcinoembryonic antigen-related cellular adhesion molecule 1 (CD66a)-expressing human B cells and inhibits antibody production. Infect. Immun. *73*, 4171–4179.

Peek, J.A., and Taylor, R.K. (1992). Characterization of a periplasmic thiol:disulfide interchange protein required for the functional maturation of secreted virulence factors of *Vibrio cholerae*. Proc. Natl. Acad. Sci. U.S.A. *89*, 6210–6214.

Pelicic, V. (2008). Type IV pili: e pluribus unum? Mol. Microbiol. *68*, 827–837.

Perez-Casal, J., Swartley, J.S., and Scott, J.R. (1990). Gene encoding the major subunit of CS1 pili of human enterotoxigenic *Escherichia coli*. Infect. Immun. *58*, 3594–3600.

Pettersson, A., Klarenbeek, V., van Deurzen, J., Poolman, J.T., and Tommassen, J. (1994). Molecular characterization of the structural gene for the lactoferrin receptor of the meningococcal strain H44/76. Microb. Pathog. *17*, 395–408.

Pettersson, A., Prinz, T., Umar, A., van der Biezen, J., and Tommassen, J. (1998). Molecular characterization of LbpB, the second lactoferrin-binding protein of *Neisseria meningitidis*. Mol. Microbiol. *27*, 599–610.

Pettersson, A., Kortekaas, J., Weynants, V.E., Voet, P., Poolman, J.T., Bos, M.P., and Tommassen, J. (2006). Vaccine potential of the *Neisseria meningitidis* lactoferrin-binding proteins LbpA and LbpB. Vaccine *24*, 3545–3557.

Pintor, M., Gómez, J.A., Ferrón, L., Ferreirós, C.M., and Criado, M.T. (1998). Analysis of TbpA and TbpB functionality in defective mutants of *Neisseria meningitidis*. J. Med. Microbiol. *47*, 757–760.

Plamondon, P., Luke, N.R., and Campagnari, A.A. (2007). Identification of a novel two-partner secretion locus in *Moraxella catarrhalis*. Infect. Immun. *75*, 2929–2936.

Pohlner, J., Halter, R., Beyreuther, K., and Meyer, T.F. (1987). Gene structure and extracellular secretion of *Neisseria gonorrhoeae* IgA protease. Nature *325*, 458–462.

Poolman, J.T., Hopman, C.T., and Zanen, H.C. (1983). Immunogenicity of meningococcal antigens as detected in patient sera. Infect. Immun. *40*, 398–406.

Popp, A., Dehio, C., Grunert, F., Meyer, T.F., and Gray-Owen, S.D. (1999). Molecular analysis of neisserial Opa protein interactions with the CEA family of receptors: identification of determinants contributing to the differential specificities of binding. Cell. Microbiol. *1*, 169–181.

Prince, S.M., Achtman, M., and Derrick, J.P. (2002). Crystal structure of the OpcA integral membrane adhesin from *Neisseria meningitidis*. Proc. Natl. Acad. Sci. U.S.A. *99*, 3417–3421.

Prinz, T., Meyer, M., Pettersson, A., and Tommassen, J. (1999). Structural characterization of the lactoferrin receptor from *Neisseria meningitidis*. J. Bacteriol. *181*, 4417–4419.

Pugsley, A.P. (1993). The complete general secretory pathway in gram-negative bacteria. Microbiol. Rev. *57*, 50–108.

Pujol, C., Eugène, E., de Saint Martin, L., and Nassif, X. (1997). Interaction of *Neisseria meningitidis* with a polarized monolayer of epithelial cells. Infect. Immun. *65*, 4836–4842.

Pujol, C., Eugène, E., Marceau, M., and Nassif, X. (1999). The meningococcal PilT protein is required for induction of intimate attachment to epithelial cells following pilus-mediated adhesion. Proc. Natl. Acad. Sci. U.S.A. *96*, 4017–4022.

Ram, S., McQuillen, D.P., Gulati, S., Elkins, C., Pangburn, M.K., and Rice, P.A. (1998). Binding of complement factor H to loop 5 of porin protein 1A: a molecular mechanism of serum resistance of nonsialylated *Neisseria gonorrhoeae*. J. Exp. Med. *188*, 671–680.

Ram, S., Cullinane, M., Blom, A.M., Gulati, S., McQuillen, D.P., Monks, B.G., O'Connell, C., Boden, R., Elkins, C., Pangburn, M.K., *et al.* (2001). Binding of C4b-binding protein to porin: a molecular mechanism of serum resistance of *Neisseria gonorrhoeae*. J. Exp. Med. *193*, 281–295.

Reece, S., Simmons, C.P., Fitzhenry, R.J., Ghaem-Maghami, M., Mundy, R., Hale, C., Matthews, S., Dougan, G., Phillips, A.D., and Frankel, G. (2002). Tyrosine residues at the immunoglobulin-C-type lectin interdomain boundary of intimin are not involved in Tir-binding but implicated in colonisation of the host. Microbes Infect. *4*, 1389–1399.

Riley-Vargas, R.C., Gill, D.B., Kemper, C., Liszewski, M.K., and Atkinson, J.P. (2004). CD46: expanding beyond complement regulation. Trends Immunol. *25*, 496–503.

Rosenqvist, E., Høiby, E.A., Wedege, E., Bryn, K., Kolberg, J., Klem, A., Rønnild, E., Bjune, G., and Nøkleby, H. (1995). Human antibody responses to meningococcal outer membrane antigens after three doses of the Norwegian group B meningococcal vaccine. Infect. Immun. *63*, 4642–4652.

Rostand, K.S., and Esko, J.D. (1997). Microbial adherence to and invasion through proteoglycans. Infect. Immun. *65*, 1–8.

Roussel-Jazédé, V., Jongerius, I., Bos, M.P., Tommassen, J., and van Ulsen, P. (2010). NalP-mediated proteolytic release of lactoferrin-binding protein B from the meningococcal cell surface. Infect. Immun. *78*, 3083–3089.

Roussel-Jazédé, V., Van Gelder, P., Sijbrandi, R., Rutten, L., Otto, B.R., Luirink, J., Gros, P., Tommassen, J., and van Ulsen, P. (2011). Channel properties of the translocator domain of the autotransporter Hbp of *Escherichia coli*. Mol. Membr. Biol. *28*, 157–169.

Rudel, T., Scheurerpflug, I., and Meyer, T.F. (1995a). *Neisseria* PilC protein identified as type-4 pilus tip-located adhesin. Nature *373*, 357–359.

Rudel, T., Facius, D., Barten, R., Scheuerpflug, I., Nonnenmacher, E., and Meyer, T.F. (1995b). Role of pili and the phase-variable PilC protein in natural competence for transformation of *Neisseria gonorrhoeae*. Proc. Natl. Acad. Sci. U.S.A. *92*, 7986–7990.

Rytkönen, A., Albiger, B., Hansson-Palo, P., Källström, H., Olcén, P., Fredlund, H., and Jonsson, A.B. (2004). *Neisseria meningitidis* undergoes PilC phase variation and PilE sequence variation during invasive disease. J. Infect. Dis. *189*, 402–409.

Sa E Cunha, C., Griffiths, N.J., and Virji, M. (2010). *Neisseria meningitidis* Opc invasin binds to the sulfated tyrosines of activated vitronectin to attach to and invade human brain endothelial cells. PLoS Pathog. *6*, e1000911.

Salacha, R., Kovačić, F., Brochier-Armanet, C., Wilhelm, S., Tommassen, J., Filloux, A., Voulhoux, R., and Bleves, S. (2010). The *Pseudomonas aeruginosa* patatin-like protein PlpD is the archetype of a novel Type V secretion system. Environ. Microbiol. *12*, 1498–1512.

Saltman, L.H., Lu, Y., Zaharias, E.M., and Isberg, R.R. (1996). A region of the *Yersinia pseudotuberculosis* invasin protein that contributes to high affinity binding to integrin receptors. J. Biol. Chem. *271*, 23438–23444.

Sankaran, K., and Wu, H.C. (1994). Lipid modification of bacterial prolipoprotein. Transfer of diacylglyceryl moiety from phosphatidylglycerol. J. Biol. Chem. *269*, 19701–19706.

Sauri, A., Soprova, Z., Wickström, D., de Gier, J.W., Van der Schors, R.C., Smit, A.B., Jong, W.S.P., and Luirink, J. (2009). The Bam (Omp85) complex is involved in secretion of the autotransporter haemoglobin protease. Microbiology *155*, 3982–3991.

Scarselli, M., Rappuoli, R., and Scarlato, V. (2001). A common conserved amino acid motif module shared by bacterial and intercellular adhesins: bacterial adherence mimicking cell–cell recognition? Microbiology *147*, 250–252.

Scarselli, M., Serruto, D., Montanari, P., Capecchi, B., Adu-Bobie, J., Veggi, D., Rappuoli, R., Pizza, M., and Aricò, B. (2006). *Neisseria meningitidis* NhhA is a multifunctional trimeric autotransporter adhesin. Mol. Microbiol. *61*, 631–644.

Schmitt, C., Turner, D., Boesl, M., Abele, M., Frosch, M., and Kurzai, O. (2007). A functional two-partner secretion system contributes to adhesion of *Neisseria meningitidis* to epithelial cells. J. Bacteriol. *189*, 7968–7976.

Schneider, M.C., Exley, R.M., Chan, H., Feavers, I., Kang, Y.H., Sim, R.B., and Tang, C.M. (2006). Functional significance of factor H binding to *Neisseria meningitidis*. J. Immunol. *176*, 7566–7575.

Schneider, M.C., Prosser, B.E., Caesar, J.J., Kugelberg, E., Li, S., Zhang, Q., Quoraishi, S., Lovett, J.E., Deane, J.E., Sim, R.B., *et al.* (2009). *Neisseria meningitidis* recruits factor H using protein mimicry of host carbohydrates. Nature *458*, 890–893.

Schulze, R.J., Chen, S., Kumru, O.S., and Zückert, W.R. (2010). Translocation of *Borrelia burgdorferi* surface lipoprotein OspA through the outer membrane requires an unfolded conformation and can initiate at the C-terminus. Mol. Microbiol. *76*, 1266–1278.

Scott, M.G., Davidson, D.J., Gold, M.R., Bowdish, D., and Hancock, R.E.W. (2002). The human antimicrobial peptide LL-37 is a multifunctional modulator of innate immune responses. J. Immunol. *169*, 3883–3891.

Seib, K.L., Serruto, D., Oriente, F., Delany, I., Adu-Bobie, J., Veggi, D., Aricò, B., Rappuoli, R., and Pizza, M. (2009). Factor H-binding protein is important for meningococcal survival in human whole blood and serum and in the presence of the antimicrobial peptide LL-37. Infect. Immun. *77*, 292–299.

Selkrig, J., Mosbahi, K., Webb, C.T., Beloussof, M.J., Perry, A.J., Well, T.J., Morris, F., Leyton, D.I., Totsika, M., Phan, M.D., *et al.* (2012). Discovery of an archetypical protein transport system in bacterial outer membranes. Nat. Struct. Mol. Biol. *19*, 506–510.

Serruto, D., Adu-Bobie, J., Scarselli, M., Veggi, D., Pizza, M., Rappuoli, R., and Aricò, B. (2003). *Neisseria meningitidis* App, a new adhesin with autocatalytic serine protease activity. Mol. Microbiol. *48*, 323–334.

Serruto, D., Spadafina, T., Ciucchi, L., Lewis, L.A., Ram, S., Tontini, M., Santini, L., Biolchi, A., Seib, K.L., Giuliani, M.M., *et al.* (2010). *Neisseria meningitidis* GNA2132, a heparin-binding protein that induces protective immunity in humans. Proc. Natl. Acad. Sci. U.S.A. *107*, 3770–3775.

Shewry, P.R. (2003). Tuber storage proteins. Ann. Bot. *91*, 755–769.

Sijbrandi, R., Urbanus, M.L., ten Hagen-Jongman, C.M., Bernstein, H.D., Oudega, B., Otto, B.R., and Luirink, J. (2003). Signal recognition particle (SRP)-mediated targeting and Sec-dependent translocation of an extracellular *Escherichia coli* protein. J. Biol. Chem. *278*, 4654–4659.

Sinclair, J.F., and O'Brien, A.D. (2002). Cell surface-localized nucleolin is a eukaryotic receptor for the adhesin intimin-γ of enterohemorrhagic *Escherichia coli* O157:H7. J. Biol. Chem. *277*, 2876–2885.

Sjölinder, H., Eriksson, J., Maudsdotter, L., Aro, H., and Jonsson, A.B. (2008). Meningococcal outer membrane protein NhhA is essential for colonization and disease by preventing phagocytosis and complement attack. Infect. Immun. *76*, 5412–5420.

Sjölinder, M., Altenbacher, G., Hagner, M., Sun, W., Schedin-Weiss, S., and Sjölinder, H. (2012). Meningococcal outer membrane protein NhhA triggers apoptosis in macrophages. PLoS One *7*, e29586.

Stern, A., Brown, M., Nickel, P., and Meyer, T.F. (1986). Opacity genes in *Neisseria gonorrhoeae*: control of phase and antigenic variation. Cell *47*, 61–71.

Stoll, H., Dengjel, J., Nerz, C., and Götz, F. (2005). *Staphylococcus aureus* deficient in lipidation of prelipoproteins is attenuated in growth and immune activation. Infect. Immun. *73*, 2411–2423.

Strom, M.S., Nunn, D.N., and Lory, S. (1993). A single bifunctional enzyme, PilD, catalyzes cleavage and N-methylation of proteins belonging to the type IV pilin family. Proc. Natl. Acad. Sci. U.S.A. *90*, 2404–2408.

Swanson, J., Barrera, O., Sola, J., and Boslego, J. (1988). Expression of outer membrane protein II by gonococci in experimental gonorrhea. J. Exp. Med. *168*, 2121–2129.

Talà, A., Progida, C., De Stefano, M., Cogli, L., Spinosa, M.R., Bucci, C., and Alifano, P. (2008). The HrpB-HrpA two-partner secretion system is essential for intracellular survival of *Neisseria meningitidis*. Cell. Microbiol. *10*, 2461–2482.

Tanabe, M., Nimigean, C.M., and Iverson, T.M. (2010). Structural basis for solute transport, nucleotide regulation, and immunological recognition of *Neisseria meningitidis* PorB. Proc. Natl. Acad. Sci. U.S.A. *107*, 6811–6816.

Telford, J.L., Barocchi, M.A., Margarit, I., Rappuoli, R., and Grandi, G. (2006). Pili in gram-positive pathogens. Nat. Rev. Microbiol. *4*, 509–519.

Tinsley, C.R., Voulhoux, R., Beretti, J.L., Tommassen, J., and Nassif, X. (2004). Three homologues, including two membrane-bound proteins, of the disulfide oxidoreductase DsbA in *Neisseria meningitidis*: effects on bacterial growth and biogenesis of functional type IV pili. J. Biol. Chem. *279*, 27078–27087.

Tokuda, H., and Matsuyama, S.I. (2004). Sorting of lipoproteins to the outer membrane in *E. coli*. Biochim. Biophys. Acta *1693*, 5–13.

Ton-That, H., and Schneewind, O. (2003). Assembly of pili on the surface of *Corynebacterium diphtheriae*. Mol. Microbiol. *50*, 1429–1438.

Touze, T., Hayward, R.D., Eswaran, J., Leong, J.M., and Koronakis, V. (2004). Self-association of EPEC intimin mediated by the β-barrel-containing anchor domain: a role in clustering of the Tir receptor. Mol. Microbiol. *51*, 73–87.

Tsai, J.C., Ming-Ren, Y., Castillo, R., Leyton, D.L., Henderson, I.R., and Saier, M.H. Jr. (2010). The bacterial intimins and invasins: a large and novel family of secreted proteins. PLoS One *5*, e14403.

Turner, D.P., Marietou, A.G., Johnston, L., Ho, K.K., Rogers, A.J., Wooldridge, K.G., and Ala'Aldeen, D.A.A. (2006). Characterization of MspA, an immunogenic autotransporter protein that mediates adhesion to epithelial and endothelial cells in *Neisseria meningitidis*. Infect. Immun. *74*, 2957–2964.

Unkmeir, A., Latsch, K., Dietrich, G., Wintermeyer, E., Schinke, B., Schwender, S., Kim, K.S., Eigenthaler, M., and Frosch, M. (2002). Fibronectin mediates Opc-dependent internalization of *Neisseria meningitidis* in human brain microvascular endothelial cells. Mol. Microbiol. *46*, 933–946.

Vandeputte-Rutten, L., Bos, M.P., Tommassen, J., and Gros, P. (2003). Crystal structure of neisserial surface protein A (NspA), a conserved outer membrane protein with vaccine potential. J. Biol. Chem. *278*, 24825–24830.

Van Putten, J.P.M., and Paul, S.M. (1995). Binding of syndecan-like cell surface proteoglycan receptors is required for *Neisseria gonorrhoeae* entry into human mucosal cells. EMBO J. *14*, 2144–2154.

Van Putten, J.P.M., Duensing, T.D., and Carlson, J. (1998). Gonococcal invasion of epithelial cells driven by P.IA, a bacterial ion channel with GTP binding properties. J. Exp. Med. *188*, 941–952.

VanRheenen, S.M., Luo, Z.-Q., O'Connor, T., and Isberg, R.R. (2006). Members of a *Legionella pneumophila* family of proteins with ExoU (phospholipase A) active sites are translocated to target cells. Infect. Immun. *74*, 3597–3606.

Van Ulsen, P., and Tommassen, J. (2006). Protein secretion and secreted proteins in pathogenic *Neisseriaceae*. FEMS Microbiol. Rev. *30*, 292–319.

Van Ulsen, P., van Alphen, L., ten Hove, J., Fransen, F., van der Ley, P., and Tommassen, J. (2003). A Neisserial autotransporter NalP modulating the processing of other autotransporters. Mol. Microbiol. *50*, 1017–1030.

Van Ulsen, P., Adler, B., Fassler, P., Gilbert, M., van Schilfgaarde, M., van der Ley, P., van Alphen, L., and Tommassen, J. (2006). A novel phase-variable autotransporter serine protease, AusI, of *Neisseria meningitidis*. Microbes Infect. *8*, 2088–2097.

Van Ulsen, P., Rutten, L., Feller, M., Tommassen, J., and van der Ende, A. (2008). Two-partner secretion systems of *Neisseria meningitidis* associated with invasive clonal complexes. Infect. Immun. *76*, 4649–4658.

Virji, M., Alexandrescu, C., Ferguson, D.J., Saunders, J.R., and Moxon, E.R. (1992). Variations in the expression of pili: the effect on adherence of *Neisseria meningitidis* to human epithelial and endothelial cells. Mol. Microbiol. *6*, 1271–1279.

Virji, M., Makepeace, K., Ferguson, D.J., Achtman, M., and Moxon, E.R. (1993). Meningococcal Opa and Opc proteins: their role in colonization and invasion of human epithelial and endothelial cells. Mol. Microbiol. *10*, 499–510.

Virji, M., Makepeace, K., and Moxon, E.R. (1994). Distinct mechanisms of interactions of Opc-expressing meningococci at apical and basolateral surfaces of human endothelial cells; the role of integrins in apical interactions. Mol. Microbiol. *14*, 173–184.

Virji, M., Makepeace, K., Peak, I.R., Ferguson, D.J., Jennings, M.P., and Moxon, E.R. (1995a). Opc- and pilus-dependent interactions of meningococci with human endothelial cells: molecular mechanisms and modulation by surface polysaccharides. Mol. Microbiol. *18*, 741–754.

Virji, M., Makepeace, K., Peak, I., Payne, G., Saunders, J.R., Ferguson, D.J., and Moxon, E.R. (1995b). Functional implications of the expression of PilC proteins in meningococci. Mol. Microbiol. *16*, 1087–1097.

Virji, M., Makepeace, K., Ferguson, D.J., and Watt, S.M. (1996a). Carcinoembryonic antigens (CD66) on epithelial cells and neutrophils are receptors for Opa proteins of pathogenic neisseriae. Mol. Microbiol. *22*, 941–950.

Virji, M., Stimson, E., Makepeace, K., Dell, A., Morris, H.R., Payne, G., Saunders, J.R., and Moxon, E.R. (1996b). Posttranslational modifications of meningococcal pili. Identification of a common trisaccharide substitution on variant pilins of strain C311. Ann. N.Y. Acad. Sci. *797*, 53–64.

Vitovski, S., and Sayers, J.R. (2007). Relaxed cleavage specificity of an immunoglobulin A1 protease from *Neisseria meningitidis*. Infect. Immun. *75*, 2875–2885.

Voulhoux, R., Bos, M.P., Geurtsen, J., Mols, M., and Tommassen, J. (2003). Role of a highly conserved bacterial protein in outer membrane protein assembly. Science *299*, 262–265.

Waksman, G., and Hultgren, S.J. (2009). Structural biology of the chaperone-usher pathway of pilus biogenesis. Nat. Rev. Microbiol. *7*, 765–774.

Watarai, M., Tobe, T., Yoshikawa, M., and Sasakawa, C. (1995). Disulfide oxidoreductase activity of *Shigella flexneri* is required for release of Ipa proteins and invasion of epithelial cells. Proc. Natl. Acad. Sci. U.S.A. *92*, 4927–4931.

Wentzel, A., Christmann, A., Adams, T., and Kolmar, H. (2001). Display of passenger proteins on the surface of *Escherichia coli* K-12 by the enterohemorrhagic *E. coli* intimin EaeA. J. Bacteriol. *183*, 7273–7284.

Wetzler, L.M. (2010). Innate immune function of the neisserial porins and the relationship to vaccine adjuvant activity. Future Microbiol. *5*, 749–758.

Weyand, N.J., Calton, C.M., Higashi, D.L., Kanack, K.J., and So, M. (2010). Presenilin/γ-secretase cleaves CD46 in response to *Neisseria* infection. J. Immunol. *184*, 694–701.

Winther-Larsen, H.C., Wolfgang, M., Dunham, S., van Putten, J.P.M., Dorward, D., Løvold, C., Aas, F.E., and Koomey, M. (2005). A conserved set of pilin-like molecules controls type IV pilus dynamics and organelle-associated functions in *Neisseria gonorrhoeae*. Mol. Microbiol. *56*, 903–917.

Wolf, M.K. (1997). Occurrence, distribution, and associations of O and H serogroups, colonization factor antigens, and toxins of enterotoxigenic *Escherichia coli*. Clin. Microbiol. Rev. *10*, 569–584.

Wolfgang, M., van Putten, J.P.M., Hayes, S.F., Dorward, D., and Koomey, M. (2000). Components and dynamics of fiber formation define a ubiquitous biogenesis pathway for bacterial pili. EMBO J. *19*, 6408–6418.

Yamaguchi, K., Yu, F., and Inouye, M. (1988). A single amino acid determinant of the membrane localization of lipoproteins in *E. coli*. Cell 53, 423–432.

Youssef, A.R., van der Flier, M., Estevão, S., Hartwig, N.G., van der Ley, P., and Virji, M. (2009). Opa^+ and Opa^- isolates of *Neisseria meningitidis* and *Neisseria gonorrhoeae* induce sustained proliferative responses in human $CD4^+$ T cells. Infect. Immun. *77*, 5170–5180.

Zhang, H.Z., and Donnenberg, M.S. (1996). DsbA is required for stability of the type IV pilin of enteropathogenic *Escherichia coli*. Mol. Microbiol. *21*, 787–797.

Zhou, G., Mo, W.J., Sebbel, P., Min, G., Neubert, T.A., Glockshuber, R., Wu, X.R., Sun, T.T., and Kong, X.P. (2001). Uroplakin Ia is the urothelial receptor for uropathogenic *Escherichia coli*: evidence from *in vitro* FimH binding. J. Cell. Sci. *114*, 4095–4103.

Zhu, P., Morelli, G., and Achtman, M. (1999). The *opcA* and Ψ*opcB* regions in *Neisseria*: genes, pseudogenes, deletions, insertion elements and DNA islands. Mol. Microbiol. 33, 635–650.

Bacterial Membranes as Drug Targets

14

Alvin Lo, Gaetano Castaldo and Han Remaut

Abstract

Bacterial membranes play a pivotal role in maintaining cell integrity, in chemical energy generation and in the interplay between bacteria and their environment. Given their indispensable functions in bacterial physiology and metabolism, it comes as no surprise that bacterial membranes and membrane-associated biosynthetic and metabolic pathways form important targets for antibacterial compounds. Bacterial membranes and membrane-associated proteins and processes have formed the targets of natural antibiotics as well as promising starting points for the search and design of new synthetic antibacterials. In this chapter, we provide a review of existing examples of membrane-targeting antibiotics as well as an overview of the pathways that are currently under investigation for chemical knock-down.

Introduction

The cell membrane forms an encasing barrier that segregates the bacterial intracellular space from the extracellular environment. At this frontier resides a battery of membrane proteins of various functions accounting for up to 75% of the membrane's mass (Guidotti, 1972) and approximately 30% of all cellular proteins (Arinaminpathy *et al.*, 2009). Their abundance reflects the importance of membrane proteins in facilitating an array of vital cellular processes including signal transduction, ion conductivity, translocation of metabolites and virulence-associated molecules (both protein and nucleic acids), and cell adhesion. In addition, membrane proteins are pivotal components in energy conversion and in the construction and maintenance of cell integrity (von Heijne, 2007). The ability of bacteria to tightly regulate the synthesis of membrane proteins and fine-tune their biochemical and biophysical properties accordingly allow them to persist in a wide range of environments. Like soluble proteins, membrane proteins often form larger complexes to carry out their biological functions. A greater knowledge of the composition, assembly and turnover of these protein complexes is crucial in deciphering their regulation and functions and identify routes for chemical intervention.

The increasing prevalence of drug-resistance and multidrug-resistant bacteria (Taubes, 2008) and the lack of new classes of clinically effective antibiotics discovered in over 40 years (Clatworthy *et al.*, 2007) signal the urgent need to better understand the molecular mechanisms for the development of resistance, as well as to explore for new antibacterial therapies. The current approaches to antimicrobial drug discovery are shifting away from conventional combinatorial chemistry and the empirical high-throughput screening of natural products and chemical libraries, towards more structure-guided methods by utilizing the available protein structural information to identify and design new ligands (Fernandes, 2006; Simmons *et al.*, 2010). The increasing wealth of knowledge gained from the advances and concerted effort of structural genomics and computation has fuelled the use of structure-based drug design strategies as an indispensable tool in medicinal chemistry (Simmons *et al.*, 2010). The amalgamation of structure-guided methods, *in silico* screening and combinatorial chemistry holds great promise for the future landscape of drug design.

This chapter examines the discovery and progress of various approaches undertaken in the development of antimicrobial drugs targeting the bacterial cell membrane integrity and synthesis, signal transduction pathways and secretion pathways (Table 14.1). This chapter also discusses the challenges and considerations that should be taken into account for future development of these approaches.

Perturbing membrane synthesis and physiology

Fatty acid biosynthesis inhibitors

Membrane lipid homeostasis is essential and bacteria adjust their membrane lipid composition through biosynthesis and modification of fatty acids to match cell division and environmental requirements (Zhang and Rock, 2008). The crucial role of fatty acid biosynthesis in the maintenance of membrane homeostasis is of a tremendous interest in the development of new antimicrobial drugs (Heath and Rock, 2004). Bacteria employ type II fatty acid biosynthesis (FASII) to produce the fatty acid components of phospholipids through different enzymatic reactions catalysed by discrete monofunctional proteins (Fig. 14.1). This is in contrast to the mammalian counterpart type I fatty acid biosynthesis (FASI), where all stepwise reactions are carried out by a multifunctional polypeptide (White *et al.*, 2005). The first step in FASII pathway is carried out by the acetyl-CoA carboxylase (ACC) to generate malonyl-CoA that is subsequently used for the biosynthesis of the acyl chain. The growing fatty acid undergoes successive downstream reactions such as reduction and dehydration carried out by fatty acid biosynthetic (Fab) enzymes (Lai and Cronan, 2003).

The potential of the FASII pathway as target for the development of novel antibiotics has been fuelled by the discovery that two widely used anti-mycobacterial compounds, *triclosan* and *isoniazid*, exert their activities by inhibiting enoyl-acyl carrier protein reductase (ER; FabI) in type II fatty acid synthesis (Fig. 14.1) (Lu and Tonge, 2008). Isoniazid is a pro-drug broadly used in the first-line treatment of tuberculosis. In Mycobacteria, isoniazid is coupled to NADH via the catalase-peroxidase KatG to form isonicotinic acyl-NADH (Fig. 14.2), which serves as a competitive inhibitor to the enoyl-acyl carrier protein reductase InhA, leading to the inhibition of mycolic acid synthesis (Rozwarski *et al.*, 1998). Triclosan and *triclocarban* (trichloro-carbanilide) (Fig. 14.2) are bacteriostatics commonly used in cosmetic products, household disinfectants and toothpaste (Perencevich *et al.*, 2001). They bind and stabilize ER-NAD$^+$ enzyme cofactor complexes, thereby inhibiting the FASII pathway (Levy *et al.*, 1999). As exemplified by these two classes of antibacterials, ER forms an attractive target for antibiotic development because of its essential role in fatty acid synthesis, the high degree of structural conservation in bacteria and distinct sequence and structure to the mammalian counterpart. These results have prompted the structure-based drug design

Table 14.1 Summary of antimicrobial agents and lead molecules targeting bacterial membranes and membrane-associated processes

Pathway	Source/chemical composition	Target/mode of action	*In vivo/ex vivo* relevance	Reference
Fatty acid biosynthesis				
Type II fatty acid biosynthesis (FASII)	Triclosan, isoniazid, triclocarban, cephalochromin, aquastatin A, vinaxanthone, kalimantacin and batumin	Enoyl-ACP reductase (ER) FabI	Gram-positive bacteria, MRSA, Mycobacteria	Kamigiri *et al.* (1996), Smirnov *et al.* (2000), Zheng *et al.* (2007), Lu and Tonge (2008), Zheng *et al.* (2009), Mattheus *et al.* (2010)
	Aquastatin A	Enoyl-ACP reductase (ER) isoform FabK	Gram-positive bacteria, MRSA, Mycobacteria	Kwon *et al.* (2009)
	Thiolactomycin and phomallenic acids A, B and C	β-Ketoacyl-carrier protein synthase III (KSIII) FabH and FaB/FabF isoforms	MRSA, *Haemophilus influenzae* and *Bacillus subtilis*	Noto *et al.* (1982), Young *et al.* (2006)
	Cerulenin	β-Ketoacyl-carrier protein synthase III (KSIII) FabH and FabB/FabF isoforms		D'Agnolo *et al.* (1973), Young *et al.* (2006)
	Platensimycin	β-Ketoacyl-carrier protein synthase III (KSIII) FabF isoform	Gram-positive bacteria	Wang *et al.* (2006)
	Secnidazole derivatives and Vinylogous carbamates	β-Ketoacyl-carrier protein synthase III (KSIII) FabH	Broad-spectrum activity	Li *et al.* (2011), Zhang *et al.* (2011)
	Moiramide B, Andrimid	Acetyl-CoA carboxylase (ACC)	Broad-spectrum activity	Fredenhagen *et al.* (1987), Needham *et al.* (1994), Freiberg *et al.* (2004), Freiberg *et al.* (2006)
Membrane lipid biosynthesis	Acyl-phosphate mimetics	Glycerol-3-phosphate acyltransferase PlsX/PlsY system	Gram-positive bacteria	Grimes *et al.* (2008)
Lipoproteins processing and sorting	Globomycin (SF1902-A1) and its derivatives, MAC13243	LpsA pathway of lipoprotein processing	Gram-negative bacteria, *Mycobacterium tuberculosis*	Inukai *et al.* (1978), Dev *et al.* (1985), Pathania *et al.* (2009)
Mycolic acid pathway	Cerulenin, Thiolactomycin and its derivatives	β-Ketoacyl-carrier protein synthases KasA and KasB involved in the mycobacterial FASII elongation steps	Mycobacteria	Kremer *et al.* (2000), Schaeffer *et al.* (2001), Senior *et al.* (2003), Senior *et al.* (2004), Kamal *et al.* (2005), Kim *et al.* (2006)

Table 14.1 (continued)

Pathway	Source/chemical composition	Target/mode of action	*In vivo/ex vivo* relevance	Reference
	Isoxyl and thiacetazone	B-hydroxyacyl dehydratases Had AB and Had BC involved in *M. tubercolosis* FA SII	*Mycobacterium tubercolosis*	Grzegorzewicz *et al.* (2012)
Membrane integrity				
	Histatins	Damaging bacterial membrane	Broad spectrum activity	MacKay *et al.* (1984a,b), De Smet and Contreras (2005)
	Defensins	Damaging bacterial membrane	Broad spectrum activity	Lehrer (2004), Ganz (2005)
	Cathelicidin	Damaging bacterial membrane	Broad spectrum activity	Turner *et al.* (1998), Hancock and Diamond (2000), Sorensen and Borregaard (2005)
	Bacteriocins (colicin)	Damaging bacterial membrane	Broad-spectrum activity	Cotter *et al.* (2005), Willey and van der Donk (2007)
	Lantibiotics (nisin A)	Damaging bacterial membrane	Broad spectrum activity, MRSA and VRE	Hsu *et al.* (2004), Cotter *et al.* (2005), Breukink and de Kruijff(2006), Hasper *et al.* (2006), Brand *et al.* (2010), Nishie *et al.* (2012)
	Anionic and cationic lipopeptides (daptomycin and polymyxins)	Damaging bacterial membrane	Broad spectrum activity	Silverman *et al.* (2003), Jung *et al.* (2004), Baltz *et al.* (2005), Straus and Hancock (2006), Vaara (2008), Nation and Li (2009), Vaara (2010)
	Phospholipase A2	Phosphatidylglycerol hydrolysis	Gram-positive bacteria	Murakami *et al.* (1997), Foreman-Wykert *et al.* (1999)

Table 14.1 (continued)

Pathway	Source/chemical composition	Target/mode of action	*In vivo/ex vivo* relevance	Reference
Bacterial importers				
	Albomycin: ferrichrome–thioribosyl-pyrimidine conjugate Salmycin: ferrioxamine–aminosaccharide conjugate	Trojan horse siderophore – antimicrobial conjugates that hijack siderophore import for uptake of the conjugated antibiotic moiety Antibiotic moiety targets t-RNA synthetases.	Albomycin: Gram-negative bacteria Salmycin: *Staphylococcus aureus*, *Streptococcus pneumoniae*	Gause (1955), Fiedler *et al*. (1985), Vértesy *et al*. (1995), Pramanik *et al*. (2007), Braun *et al*. (2009)
	Rifamycin CGP 4832	DNA-dependent RNA polymerase inhibitor imported through FhuA	Gram-negative bacteria	Puglsey *et al*. (1987), Ferguson *et al*. (2001
	Microcin C7	Aspartyl t-RNA synthetase inhibitor imported via Yej oligopeptide importer	*Escherichia coli*	Guijarro *et al*. (1995), Novikova *et al*. (2007)
	Agrocin 84	LeuRS t-RNA synthetase inhibitor internalized via opine permease	*Agrobacterium tumefaciens*	Kim *et al*. (1997), Reader *et al*. (2005)
Signal transduction				
Histidine kinases	Radicicol (macrocyclic lactone compound)	Binds PhoQ sensor histidine kinase, preventing autophosphorylation		Guarnieri *et al*. (2008)
Histidine kinases receptor domain	LED209 (*N*-phenyl-4 [[(phenylamino) thioxomethyl] amino]benzene-sulfonamide hydrate)	Interfere with the binding of ligands to QseC sensor histidine kinase, preventing its activation	Effective in reducing the pathogenicity of *Salmonella typhimurium* and *Francisella tularensis* but not enterohaemorrhagic *E. coli* in animal models	Rasko *et al*. (2008)
Secretion systems				
Resistance–nodulation–division (RND) efflux pumps	$Co(NH_3)_6^{3+}$	Binds directly to the electronegative periplasmic entrance of TolC-like OM channel protein, impeding its conductivity	*In vivo* potency not assessed. Toxic to cell at micromolar concentration	Andersen *et al.* (2002), Higgins *et al.* (2004)

Table 14.1 (continued)

Pathway	Source/chemical composition	Target/mode of action	*In vivo/ex vivo* relevance	Reference
	PAβN (phenylalanyl-arginyl-β-naphthylamide)	Competitively binds MexB, RND inner membrane transporter, and is preferentially extruded resulting intracellular accumulation of cognate antibiotic	Effective against RND transporters in *Pseudomonas aeruginosa*, *Klebsiella pneumoniae*, and *Salmonella enterica*	Renau *et al.* (1999), Chevalier *et al.* (2000), Lomovskaya *et al.* (2001), Baucheron *et al.* (2002), Hasdemir *et al.* (2004)
Type II secretion systems (T2SSs)	Virstatin (4-[*N*-(1,8-naphthalimide)]-n-butyric acid)	Transcriptional down-regulation of T2SS effectors	Confers *in vivo* protection in infant mice against colonization of *Vibrio cholerae*	Hung *et al.* (2005), Shakhnovich *et al.* (2007), Chatterjee *et al.* (2011)
Type III secretion systems (T3SSs)	INPs salicylidene acylhydrazides	Inhibit T3S machinery, molecular target unknown	Broad-spectrum activity*: Yersinia*, *Chlamydia*, *Salmonella and Shigella*	Kauppi *et al.* (2003), Muschiol *et al.* (2006), Bailey *et al.* (2007), Hudson *et al.* (2007), Muschiol *et al.* (2009), Veenendaal *et al.* (2009)
	2-Imino-5-arylidene thiazolidinone	Possible targeting related outer membrane component of T2SSs and T3SSs	Broad spectrum activity: *Yersinia*, *Salmonella*, as well as inhibition of T2SS of *P. aeruginosa* and *Francisella novicida*	Felise *et al.* (2008)
Type IV secretion systems (T4SSs)	CHIR-1 (4-methyl-2-[4-(trifluoromethoxy) phenyl]–1,2,4-thiadiazolidine-3,5-dione)	Inhibits the trafficking of cytotoxin-associated antigen, CagA, from *H. pylori* into host cells	Impaired colonization of mouse gastric mucosa by CHIR-1-treated *Helicobacter pylori*	Hilleringmann *et al.* (2006)
	2-hydroxy-N′-{[5-(2-nitrophenyl)–2-furyl]methylene} benzohydrazide	Interrupts the dimerization of VirB8, inhibiting transcription of *virB* operon	Attenuated T4SS-dependent intercellular survival of *Brucella abortus* in macrophages	Paschos *et al.* (2011)
Chaperone-usher pathways	Bicyclic 2-pyridones pilicides	Prevent the recruitment of chaperone/subunit complexes to usher	UPEC type 1 pili mediated biofilm formation and adherence to bladder epithelial cells	Pinkner *et al.* (2006)

Table 14.1 (continued)

Pathway	Source/chemical composition	Target/mode of action	*In vivo/ex vivo* relevance	Reference
	Receptor analogues for fimbrial adhesins Examples: alkyl- and aryl-mannosides	Receptor analogues for type 1 pilus adhesin FimH. Competitive inhibition of type 1 binding to high-mannan glycoproteins.	Effective in treating chronic UTIs and preventing reinfection of UPEC when orally administered in murine model	Han *et al.* (2010), Cusumano *et al.* (2011)

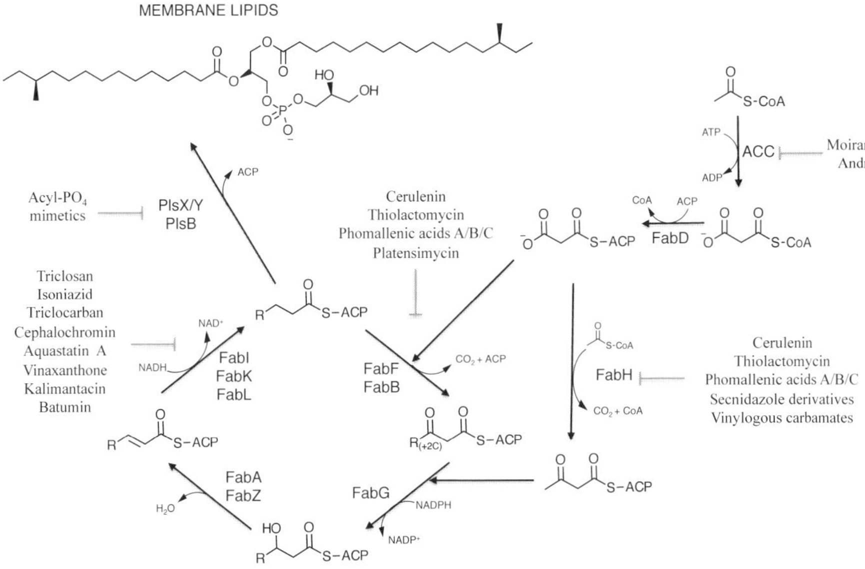

Figure 14.1 Schematic diagram of Type II fatty acid biosynthesis (FASII) and steps that are subject to antibacterial targeting. Acetyl-CoA carboxylase (ACC) is responsible for the carboxylation of Acetyl-CoA, resulting in the production of malonate successively transacylated by the malonyl-CoA:ACP transacylase (FabD). The initial condensation step is carried out by the β-ketoacyl-ACP syntase III (FabH) with the homologues FabF/FabB responsible for the elongation steps. Acetoacetyl-ACP and growing fatty acid chain undergo successive reduction steps carried out by the β-ketoacyl-ACP reductase (FabG), dehydratation by the β-hydroxyacyl-ACP dehydratase (FabA and FabZ) and final reduction by the the enoyl-ACP reductase (FabI, FabL and FabK isoforms). Fatty acids are finally incorporated into membrane lipids by PLsB or PLsX/Y. Known FASII inhibiting antibacterials are indicated alongside their molecular targets.

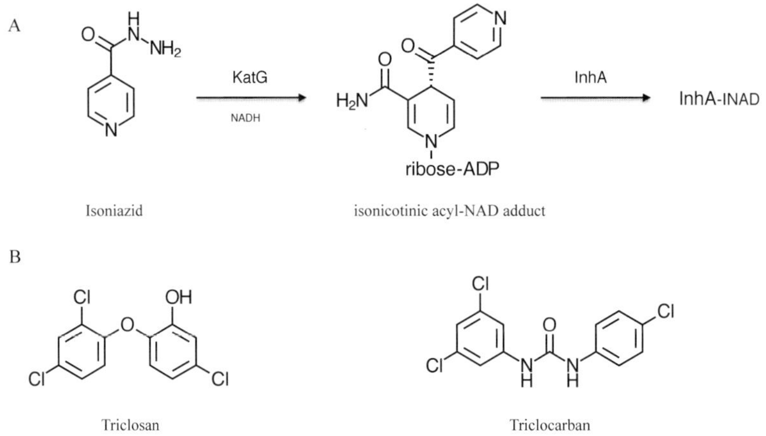

Figure 14.2 Mechanism of antibacterial compound Isoniazid and structure of compounds Triclosan and Triclocarban. (A) Isoniazid is an anti-mycobacterial pro-drug that inhibits the enoyl-acyl reductase (ER) InhA (homologous to *E. coli* FabI). Isoniazid is converted to a covalent adduct by the catalase-peroxidase enzyme KatG in the presence of NAD^+, resulting in a isonicotinic acyl–NAD adduct that competitively inhibits InhA. (B) Triclosan and Triclocarban interact non-covalently with the enzyme–cofactor complex ER-NAD^+ resulting in a rearrangment of an amino acid loop in proximity of the binding site.

for the development of novel FabI inhibitors against *Staphylococcus aureus* infections (Bogdanovich *et al.*, 2007; Karlowsky *et al.*, 2009). Kim and coworkers have identified different natural products that possess high inhibitory activity towards FabI (Fig. 14.1) (Zheng *et al.*, 2007, 2009). *Cephalochromin* was isolated from an unidentified fungal isolate and exhibited bactericidal activity with a MIC values between 2–8 µg/ml against Gram-positive bacteria including methicillin and quinolone-resistant *S. aureus* (MRSA and QRSA respectively), *S. epidermidis* and *B. subtilis,* whilst it displayed no antimicrobial effect against Gram-negative bacteria such as *E. coli* and *P. aeruginosa* (Zheng *et al.*, 2007). In the course of their screening process, the same laboratory has also identified the FabI inhibitor *vinaxanthone* from a fermented extract of *Penicillium* sp. F131. This compound selectively inhibited FabI from different Gram-positive bacteria including MRSA with an IC_{50} of 0.9 µg/ml (Zheng *et al.*, 2009). Furthermore, *aquastatin A,* a natural product isolated from *Sporothrix* sp. FN611, has shown potent *S. aureus* FabI inhibitory effect with an IC_{50} of 3.2 µM and with MIC values between 16 and 32 µg/ml for *S. aureus* and MRSA (Kwon *et al.*, 2009). In addition, it has also been shown to be active against the FabK isoform in *Streptococcus pneumoniae* (Kwon *et al.*, 2009). *Kalimantacin* and *batumin* are two polyketide antibiotics isolated from the fermentation broth of *Alcaligenes* sp. YL-02632S and *Pseudomonas batumici* respectively (Kamigiri *et al.*, 1996; Smirnov *et al.*, 2000). Both polyketides possess anti-MRSA activity with a MIC of 0.05 µg/ml due to FabI inhibition (Mattheus *et al.*, 2010). More recently, new potent FabI and FabK inhibitors have been reported by screening a synthetic library of compounds based on the triclosan scaffold and the natural products isolated from microbial sources (Gerusz *et al.*, 2012; Kwon *et al.*, 2013).

β-Ketoacyl-acyl carrier protein synthases (β-ketoacyl-ACP synthase or KAS) catalyse the condensation of malonyl-ACP with the growing fatty acid chain and serve in the initiation and elongation steps of Type II fatty acid synthesis (Fig. 14.1) (Lai and Cronan, 2003).

Many bacteria including *E. coli* possess three KAS enzymes, KASI (FabB) is important for elongation of unsaturated fatty acids, KASII (FabF) regulates synthesis of temperature-induced fatty acids, and KASIII (FabH) catalyses the initial condensation step of the fatty acid synthesis and is essential for bacterial survival (Lai and Cronan, 2003). FabH is highly conserved in many human pathogens and has been a promising target for the development of novel antibiotics (Li *et al.*, 2011). Natural products such as *cerulenin* and *thiolactomycin,* produced by *Cephalosporium caerulens* and *Nocardia* sp. respectively, were found to be inhibitors of FabH as well as FabB and FabF (D'Agnolo *et al.*, 1973; Noto *et al.*, 1982). More recently, a library of over 250,000 natural product extracts from actinomyces and fungi was screened resulting in the identification of the FabH and FabB/FabF inhibitors *phomallenic acids A, B* and *C* as effective antibacterial compounds against MRSA, *Haemophilus influenzae* and *Bacillus subtilis.* Phomallenic acid C exhibited the most potent overall antibacterial activity, with MIC against *S. aureus* of 3.9 µg/ml and IC_{50} of 0.77 µg/ml (Young *et al.*, 2006). A combination of target-based whole cell and biochemical assays allowed systematic screening of 250,000 natural product extracts and led to the identification of the small molecule inhibitor *platensimycin* from *Streptomyces platensis.* This compound showed good membrane permeability and potent FabF inhibitory activity with IC_{50} values of 48 nM and160 nM for *S. aureus* and *E. coli* respectively (Wang *et al.*, 2006). Potent FabH inhibitors have also been identified amongst the derivatives of *secnidazole,* an antibiotic effective in the treatment of amoebiasis, giardiasis, trichomoniasis and bacterial vaginosis (Gillis and Wiseman, 1996). Screening of a library of novel cinnamic acid secnidazole ester derivatives has been performed against two Gram-negative bacteria, *E. coli* and *P. aeruginosa* and two Gram-positive strains, *B. subtilis* and *S. aureus* in order to identify FabH inhibitors. The most effective compound exhibited MIC between 1.56 and 6.25 µg/ml with an inhibitory activity against *E. coli* FabH with IC_{50} of 2.5 µM (Zhang *et al.*, 2011). Li *et al.* have carried out a similar combination of screening assays as above against a library of novel *vinylogous carbamates* developed by structure-based approaches. Two compounds with an IC_{50} between 2.6 and 3.3 µM have been identified as the most potent *E. coli* FabH inhibitors (Li *et al.*, 2011). Recently, Zhu and coworkers have designed and synthesized libraries of Schiff bases and acylhydrazone derivatives identifying novel promising *E. coli* FabH inhibitors (Wang *et al.*, 2012; Zhou *et al.*, 2013).

The broad structural conservation of the essential metabolic enzyme ACC amongst bacteria and its dissimilarity when compared to its mammalian counterpart represents an attractive target for novel antimicrobials (Fig. 14.1) (Polyak *et al.*, 2012). ACC is selectively inhibited by *moiramide B* and *andrimid,* two natural pseudopeptidic pyrrolidinedione produced by marine isolate of the bacterium *Pseudomonas fluorescens* (Fredenhagen *et al.*, 1987; Needham *et al.*, 1994; Freiberg *et al.*, 2004). Moiramide B is a competitive inhibitor of the natural substrate, malonyl-CoA, with a K_i value of 5 nM (Freiberg *et al.*, 2004). Although these antibiotics showed broad-spectrum activity, the IC_{50} for the Gram-positive bacteria was markedly increased by a factor of 10 compared with that of the Gram-negative bacteria. In addition, isolation of resistant strains of *S. aureus* and *B. subtilis* displayed mutations in the conserved motif of ACC (Freiberg *et al.*, 2004). These results have led to the rational design of novel moiramide B derivatives that have exhibited improved activities against Gram-positive bacteria and a MIC_{90} of 0.1 µg/ml against different *S. aureus* clinical isolates (Freiberg *et al.*, 2006). Recently, Gramajo and coworkers reported the discovery of a small molecule inhibitor targeting the carboxyltransferase domain of the *M. tuberculosis* ACC with an interesting inhibitory activity (K_i 8 μM) (Kurth *et al.*, 2009). Antimicrobial properties of natural products that target ACC have also stimulated the structure-based design of small molecule inhibitors against this essential component of FASII pathway (Polyak *et al.*, 2012). Despite of these successes, to date there are no reported ACC inhibitors in clinical use as antibiotic (Polyak *et al.*, 2012).

Mycolic acids are a class of α-alkyl β-hydroxy fatty acids that is present in the cell wall of all mycobacteria (see Chapter 5). They are typically found as esters of the complex polysaccharide arabinogalactan that replace some of the muramic acid residues in the mycobacteria peptidoglycan structure (Brennan and Nikaido, 1995). This peptidoglycan complex is essential for the viability of *M. tuberculosis* and other mycobacteria (Dover *et al.*, 2004). Mycolic acids are also bound to trehalose to form the trehalose dimycolate (TDM) (Brennan and Nikaido, 1995). These fatty acids confer important biophysical characteristics to mycobacteria including chemical resistance, low permeability to hydrophobic antibiotics (Glickman *et al.*, 2001), virulence (Dubnau *et al.*, 2000; Glickman *et al.*, 2000), biofilm formation (Ojha *et al.*, 2005) and persistence (Daffe and Draper, 1998; Glickman *et al.*, 2000). Mycolic acids are produced by both type I and type II FAS. Type I FAS resembles the eukaryotic multifunctional FAS and is responsible for the *de novo* synthesis of fatty acids from acetyl-CoA. Type II FAS is similar to those found in bacteria (Fig. 14.1) and act successively to elongate and modify the growing acyl chain (Takayama *et al.*, 2005). The enzymes involved in mycolic acid biosynthesis are essential for mycobacterial survival therefore they represent an attractive antibacterial drug target (Bhatt *et al.*, 2007). KasA and KasB are β-ketoacyl-ACP synthases involved in the mycobacterial FASII elongation steps and are homologues of the *E. coli* FabB and FabF (Fig. 14.1) (Kremer *et al.*, 2000). KasA and KasB are effectively inhibited by the natural products cerulenin and thiolactomycin (Kremer *et al.*, 2000; Schaeffer *et al.*, 2001). Thiolactomycin derivatives have been reported to be more potent with improved pharmacokinetics properties (Kremer *et al.*, 2000; Kamal *et al.*, 2005; Kim *et al.*, 2006). Their development however has been hampered by their complex chemical synthetic routes (Bhatt *et al.*, 2007). Structure-based approach has fuelled novel thiolactomycin derivatives with increased potency *in vitro* against mtFabH but they failed *in vivo* (Senior *et al.*, 2003, 2004). *Nitroimidazole PA-824* and *OPC67683* are mycolic acid biosynthesis inhibitors highly effective both *in vitro* and *in vivo* currently under phase II clinical trials for the treatment of multidrug- resistant-TB (Shi and Sugawara, 2010; Cole and Riccardi, 2011). Recently, it has been discovered that the two antimycobacterial prodrugs isoxyl and thiacetazone block mycolic acid biosynthesis by targeting the β-hydroxyacyl dehydratases HadAB and HadBC involved in *M. tuberculosis* FAS II (Grzegorzewicz *et al.*, 2012).

Most of the bacterial fatty acid biosynthesis studies have been carried out in *E. coli*. In Gram-negative bacteria FASII pathway is essential for cell growth albeit their ability to uptake and incorporate exogenous fatty acids into their membrane lipids (Cronan and Rock, 2008). The essentiality of FASII stems from their indispensable requirement of β-hydroxy fatty acids to assemble the lipid A core structure of lipopolysaccharides embedded into the outer membrane (Raetz *et al.*, 2007). The *E. coli* paradigm for FASII pathway has been used to support the potential of such biosynthetic pathway as target for new broad-spectrum antimicrobials. The applicability of this strategy is however in question as members of *Bacillus/Lactobacillus/Streptococcus* possess an additional isoform of the essential ER FabI, FabK, conferring higher resistance to triclosan (Marrakchi *et al.*, 2003). More recently Brinster *et al.* (2009) reported that in *Streptococcus agalactiae* FASII pathway is not essential if the culture is supplemented with exogenous fatty acid or human serum. In a contradicting report, Koul and coworkers were unable to revert growth of *S. aureus* treated with triclosan by addition of fatty acids reaffirming the essentiality of FASII for this bacterium (Balemans *et al.*, 2010). Brinster *et al.* in response have instead showed the ability of a *fabI* knock-out strain of *S. aureus* to incorporate extracellular fatty acids (Brinster *et al.*, 2010).

A greater knowledge of the diversity of fatty acid biosynthesis between the model organism *E. coli* and Gram-positive bacteria is required and would certainly be of benefit for the use of FASII as target for an effective antimicrobial therapy. Regarding the formation of phosphatidic acid, a key intermediate in membrane lipid biosynthesis, *E. coli* utilizes the glycerol-3-phosphate acyltransferase PlsB (Heath *et al.*, 1994). However, majority of bacteria including many Gram-positive pathogens, lack of PlsB and rely on the PlsX/PlsY system to produce an acyl-phosphate intermediate that is not found in *E. coli* and mammalians (Lu *et al.*, 2006). This finding has led to the structure-based development of acyl-phosphate mimetics as PlsX/PlsY inhibitors (Grimes *et al.*, 2008).

Despite contradicting reports on the essentiality of the FASII pathway in Gram-positive bacteria, this biosynthetic pathway remains an attractive target for antimicrobial therapy in order to stay ahead in arms race against drug resistance in pathogens.

Inhibitors of lipoprotein biogenesis

Bacterial lipoproteins are a class of proteins modified at their N-terminal Cys residue with diacyl-glycerol that derives from membrane phospholipids (see Chapter 4). The lipid moiety serves to anchor lipoproteins to the membrane to exert their biological functions at the membrane–aqueous interface (Okuda and Tokuda, 2011). Lipopeptides are synthesized as precursors in the cytoplasm with a signal peptide important for their sorting and translocation to the periplasmic side of the cytoplasmic (inner) membrane (Tokuda and Matsuyama, 2004). The signal peptide is successively cleaved off by the lipoprotein signal peptidase LspA (or signal peptidase II). In diderm bacteria (Gram-negative bacteria and mycobacteria) lipoprotein processing and sorting pathways are essential for growth and/or virulence (Tokuda, 2009). On the contrary, in monoderm bacteria (Gram-positive bacteria) unprocessed lipoprotein precursors are usually still active, with the hydrophobic signal peptide functionally similar to the diacyl-glycerol moiety (Venema *et al.*, 2003). Given the absence of LspA homologues in mammalians, the lipoprotein processing pathway has drawn much attention as a promising target for novel antibiotics towards diderm bacteria. *Globomycin* or *SF1902-A1* is a cyclic peptide antibiotic produced by *Streptomyces* that inhibits LpsA-mediated signal removal during lipoprotein processing in monoderm and diderm bacteria (Inukai *et al.*, 1978; Dev *et al.*, 1985). Inhibition of the lipoprotein processing leads to the accumulation of precursors in the cytoplasmic membrane and ultimately cell death (Dev *et al.*, 1985). Globomycin exhibits bactericidal effect only towards Gram-negative bacteria due the essentiality of the LpsA pathway (Kiho *et al.*, 2004). A library screen of globomycin analogues revealed a long alkyl side chain derivative conferred the most potent antimicrobial activity, with a MIC value of 1.56 μg/ml against *E. coli*. Unexpectedly the analogue also showed a moderate activity against MRSA (MIC 12.5 μg/ml) suggesting the possibility that globomycin, or its derivatives, might have additional targets (Kiho *et al.*, 2004). In addition, globomycin is also active against *M. tuberculosis* although via a LspA-independent mechanism (Banaiee *et al.*, 2007). More recently, a cell-based small molecule screening has resulted in the discovery of the compound *MAC13243* that is active against *P. aeruginosa* by inhibition of the lipoprotein-specific carrier LolA (Pathania *et al.*, 2009).

Cationic antimicrobial peptides (AMPs)

Cationic antimicrobial peptides (AMPs) are widespread in nature. In animals and plants they play a fundamental role in innate immune defence against different ranges of pathogens from bacteria to viruses (Devine and Hancock, 2002). They show broad-spectrum killing activity and rapid action, generally within minutes *in vitro*. In addition, only few bacterial species are resistant to these compounds (Straus and Hancock, 2006). Available evidences indicate that AMPs exert their activity by perturbing the permeability of bacterial cell membranes (Huang, 2000). The observed selectivity for prokaryotic membranes stems from the electrostatic attraction between the cationic AMPs and the negatively charged bacterial membranes, compared to the zwitterionic nature of phospholipids in the outer leaflet of mammalian membranes. Consequently, AMPs have become novel promising candidates as drug for an antimicrobial therapy (Bolla *et al.*, 2011).

AMPs exhibit great diversity in amino acid composition, size (mostly between 12 and 100 amino acids) and structural build-up, including β-sheets, α-helices, loops and extended structures (Fig. 14.3) (Hancock, 1997; Hancock, 2001; Jenssen *et al.*, 2006). They are positively charged with a net charge between +2 to +9 owing to the high content of arginine and lysine and possess up to 50% of hydrophobic residues. A common structural feature of AMPs is their folding in amphipathic molecules upon interaction and insertion into membranes (Hancock, 2005). The interaction of AMPs to the lipid matrix of the membranes is stepwise, encompassing a rapid membrane association driven by electrostatic forces, followed by a slower membrane insertion step. Bacterial membranes are known to be rich in negatively charged phospholipids such as phosphatidylglycerol, cardiolipin or phosphatidylserine, as well as the zwitterionic phosphatidylethanolamine (Verkleij *et al.*, 1973). On the other hand, the eukaryotic membranes are predominantly composed of phospholipids with a neutral net charge: phosphatidylethanolamine, phosphatidylcholine or sphingomyelin (Yeaman and Yount, 2003). In addition, compared to that of prokaryotic membranes, eukaryotic membranes contain cholesterol, which stabilizes the lipid bilayer and thereby further impairing the AMPs damaging activity (Matsuzaki, 1999). In Gram-negative bacteria the strong association of AMPs to the outer membrane leads to displacement of Mg^{2+} and Ca^{2+} that stabilize the LPS by interacting with the phosphate groups. Removal or displacement of these bivalent cations causes destabilization of the bilayer, promoting the uptake of more AMPs molecules in a process termed 'self-promoted uptake' (Hancock, 1997).

AMPs are thought to be predominantly unstructured in solution and fold in their secondary structures (β-sheets, α-helices) upon binding to the membranes. Despite differences in the secondary structures, AMPs adopt a common amphipathic cylindrical shape, with one side hydrophobic and one side hydrophilic (Fig. 14.3). This configuration allows the AMPs to adopt a parallel orientation to the membrane and ready to be incorporated at the bilayer hydrophilic–hydrophobic interface, leading to the thinning of the membrane (Straus and Hancock, 2006). The binding of AMPs to the bacterial membrane is a key step for the antimicrobial activity (Hancock and Rozek, 2002); however, the precise mechanism of membrane damaging is still unclear. Three possible models of action, shown in Fig. 14.3, have been proposed to explain the permeabilizing process: the formation of toroidal-pores, barrel-stave pores and carpet-like patches.

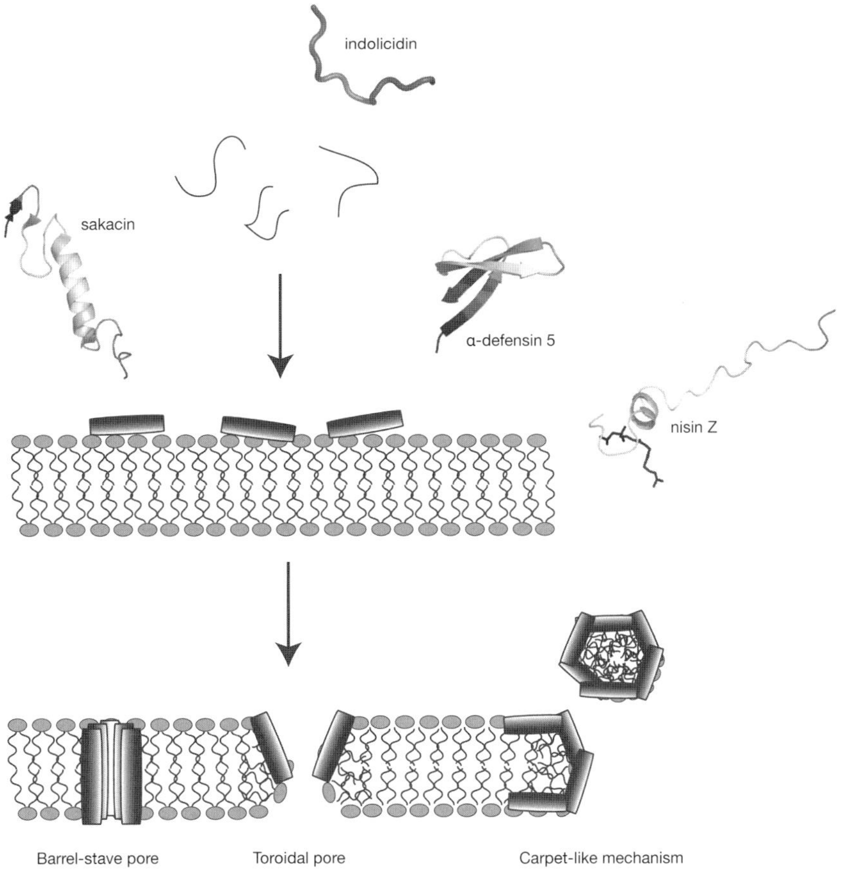

Figure 14.3 Membrane-active antimicrobial peptides (AMPs) and their proposed modes of action. Membrane-active antimicrobial peptides are short secreted peptides that fold into pore-forming amphipathic molecules upon contact with the lipid bilayer, AMPs share an amphipathic nature but are otherwise structurally diverse, including α-helical (example shown: sakacin, PDB: 1OG7), β-sheet (example shown: α-defensin 5, PDB: 1ZMQ), coiled (examples shown: indolicidin, PDB 1G89; nisin Z shown in complex with lipid II, PDB 1WCO) and mixed α–β structures (not shown). Three models have been proposed for membrane disruption: the formation of barrel-stave pores, toroidal pores or a carpet-like (detergent-like) mechanism.

In the toroidal model (Matsuzaki *et al.*, 1995), the AMPs inserts perpendicularly to the membrane, with the hydrophobic regions binding to the lipid tails and the hydrophilic parts interacting with the phospholipid head groups. These results in a tilting of the lipid molecules that leads to the bending of the membrane and the formation of aqueous pores formed by the lipid head groups and the hydrophilic face of the AMPs (Yeaman and Yount, 2003; Fig. 14.3). Similarly to the toroidal model, in the barrel-stave model the AMPs reorient perpendicularly to the membrane forming an aqueous pore or channel of variable sizes. The peptides are positioned as staves in a barrel-shaped configuration with the hydrophobic regions interacting with the lipid core of the membrane and hydrophilic regions facing each other (Yeaman and Yount, 2003). In the carpet model (Pouny *et al.*, 1992), AMPs bind parallel to the membrane, covering its surface in a carpet-like fashion. This association destabilizes the membrane and results in micelle formation and bilayer dissolution (Fig. 14.3). Despite the differences of the three models, the bacterial cells are killed by damaging the membrane that leads ultimately to its depolarization, dissolution, variation in lipids composition and interference with intracellular biosynthetic pathways (Straus and Hancock, 2006).

Host defence peptides (HDPs)

Considering the *in vivo* role of antimicrobial peptides in the innate immunity, some authors prefer the term host defence peptides (HDPs) to AMPs (Hancock and Devine, 2004). Three important groups of human HDPs have been characterized and represent promising leads for the development of new drugs: *defensins, histatins* and *cathelicidin* (De Smet and Contreras, 2005). HDPs are gene-encoded and produced as preproteins that become active through one or two proteolytic steps. Generally small in size, between 10 and 50 amino acids, these cationic proteins have an amphiphilic residue distribution but show some different structural features (Tossi and Sandri, 2002). HDPs have a similar mode of action to AMPs and possess a broad-spectrum killing activity against bacteria, fungi and enveloped viruses (White *et al.*, 1995; Bals, 2000).

Mammalian defensins are non-glycosylated cationic peptides rich in arginine and contain six cysteine residues that are involved in three disulfide bonds (Lehrer, 2004; Ganz, 2005). Human defensins have considerable sequence variation but they share a similar tertiary structure with a triple-stranded β-sheet and a β-hairpin loop that contains the cationic charges (Fig. 14.3). Generally they are classified as either α- or β-defensins based on the position of their disulfide bonds (Ganz, 2005).

Cathelicidins are derived from preproteins with a conserved signal sequence and a proregion homologous to the cathepsin L inhibitor, cathelin, but with a variable C-terminal domain (Hancock and Diamond, 2000; Sorensen and Borregaard, 2005). In humans, the cathelicidin named LL-37 derives from the proteolytic cleavage of the CAP18 C-terminal domain. It has a bactericidal activity against both Gram-positive and Gram-negative bacteria (Turner *et al.*, 1998) and is expressed in different cell lines including leucocytes and epithelial cells of skin and gastrointestinal and respiratory tracts (Kosciuczuk *et al.*, 2012).

Histatins are a family of HDPs rich in histidine with molecular mass between 3 and 4 kDa present in human saliva (MacKay *et al.*, 1984b). They were first described in 1970 (Holbrook and Molan, 1973) and play an important role in maintaining a healthy oral cavity due to their potent bactericidal activity (MacKay *et al.*, 1984a). In the histatin family, histatin 1, 3 and 5 have been extensively studied: they form amphipathic α-helices with seven histidines residues (Oppenheim *et al.*, 1988; Castagnola *et al.*, 2004). Histatin 5 possesses the strongest killing activity towards bacteria and fungi and represents the most interesting peptide for the development of new antibacterials (Tsai and Bobek, 1997; De Smet and Contreras, 2005). In addition, the 12-amino-acid fragment of histatin 5 has been successfully used in a mouth rinse and gel formulations in clinical trials with positive results in safety and activity against plaque and gingivitis (De Smet and Contreras, 2005).

Class I and II bacteriocins are ribosomally synthesized cationic antimicrobial peptides produced by bacteria to compete against other bacteria usually of same or closely related genus or species (Cotter *et al.*, 2005). Bacteriocins are produced as peptide precursors with an N-terminal signal sequence and a C-terminal proregion. The leader peptide keeps the molecule inactive and, therefore, protects the producing bacteria from their own antimicrobial drug (Willey and van der Donk, 2007). *Lantibiotics* are class I bacteriocins that contain the thioether amino acid lanthionine (Cotter *et al.*, 2005). This modified amino acid forms multiple ring structures and confers to lantibiotics improved physical and chemical stability (Sahl *et al.*, 1995; Bierbaum *et al.*, 1996). Bacteriocins possess membrane-permeabilizing activity similar to other HDPs and AMPs but a single bacteriocin might have different modes of action to kill the target bacteria (Hasper *et al.*, 2006). Different lantibiotics, including *nisin*, interact with the bactoprenol precursor of the cell wall, Lipid II forming pores in the bacterial membrane (Fig. 14.3) (Hsu *et al.*, 2004; Breukink and de Kruijff, 2006). Although bacteriocins confer potent bactericidal activity, resistance has been observed and is related to variation in the cell membrane fluidity and surface charge (Vadyvaloo *et al.*, 2004). Bacteriocins represent promising antimicrobial compounds given their characteristics of being safe for humans, their chemical and physical stability, and potent killing activity. Nisin A is one of the few examples of commercially available bacteriocins currently used as food-preservative (Nishie *et al.*, 2012). In addition, nisin and other lantibiotics have also been shown to be active against multi-drug resistant pathogens such as MRSA and vancomycin-resistant enterococci (VRE) (Brand *et al.*, 2010).

The development of antimicrobial peptides has been largely hampered in the past by the high production cost of the conventional solid-phase synthesis methods, and therefore limiting the accessibility to these peptides. The introduction of inexpensive high-throughput synthesis and screening of thousands of peptide variants on a cellulose support and in combination with a highly sensitive antimicrobial assays has accelerated the identification of an eight amino acid peptide conferring broad spectrum inhibitory activity (Hilpert *et al.*, 2005). The design of peptidomimetic compounds based on natural peptide templates such as *magainins* (Porter *et al.*, 2000) or in particular HDP *protegrin 1* (Shankaramma *et al.*, 2002; Robinson *et al.*, 2005) has also drawn considerable attention over the last decade due to potential of such molecules to mitigate lability to proteases and therefore assuming a more favourable pharmacokinetics compared to that of their natural templates. Recently, after several chemical optimization cycles of the previously identified HDP protegrin 1 mimetics (Shankaramma *et al.*, 2002; Robinson *et al.*, 2005), a lead compound conferring narrow spectrum of activity against *Pseudomonas* spp. with nanomolar range potency was obtained (Srinivas *et al.*, 2010). Subsequent biochemical and genetic interrogations revealed that the mimetic confers a non-lytic mechanism of action and identified LptD, an outer membrane protein important in LPS transport, as its molecular target. Its *in vivo* efficacy in murine septicaemia model was also reported (Srinivas *et al.*, 2010).

Lipopeptide antibiotics (daptomycin and polymyxins)

AMPs are not the only membrane active antibiotics. Lipopeptide antibiotics are highly active against bacteria as well as some fungi (Baltz *et al.*, 2005). They contain a lipid tail covalently attached to the N-terminus of a linear or cyclic peptide sequence either with a net positive or negative charge. Despite the difference in the net charge of the peptide moiety, both cationic and anionic lipopeptide antibiotics bind to bacterial membranes. They act by inserting their lipid tail into the bilayer and successively damaging the membrane through mechanisms similar to those proposed for AMPs although the precise mechanism is still unclear (Straus and Hancock, 2006).

All lipopeptide antibiotics share the tendency to oligomerize with formation of micelles. *Daptomycin*, an anionic lipopeptide antibiotic in the presence of equimolar amount of divalent ions, such as Ca^{2+}, triggers oligomerization and renders the drug more amphiphilic allowing the transition to its active form (Fig. 14.4A) (Jung *et al.*, 2004). Although oligomerization is a key property, in close proximity to the membrane lipopeptide micelles would dissociate in order to bind to the membrane. Their ability to interact with the bacterial bilayer via their lipid moiety is essential for their antimicrobial activity. It has been shown that the addition of a hydrophobic tail at an appropriate length increases the binding affinity of lipopeptides to the membrane regardless of the net charge of the peptide moiety. Moreover, conjugation of a lipid moiety to a cationic lipopeptide promotes the formation of secondary structures important for their association to the membrane. This exemplified by the addition of lipid tail might increase lipopeptide antimicrobial properties by enhancing the ability to form secondary structures upon interaction with bacterial membranes (Epand, 1997; Straus and Hancock, 2006).

For anionic lipopeptides divalent ions such as Ca^{2+} play an essential role both for the formation of micelles and the interaction with the membranes (Andrew *et al.*, 1987). The interaction of anionic lipopeptide daptomycin with the membrane is thought to be mediated by calcium, through binding with the negatively charged phosphatidylglycerol headgroups, allowing the insertion of its lipid moiety into the bilayer. This interaction results in the

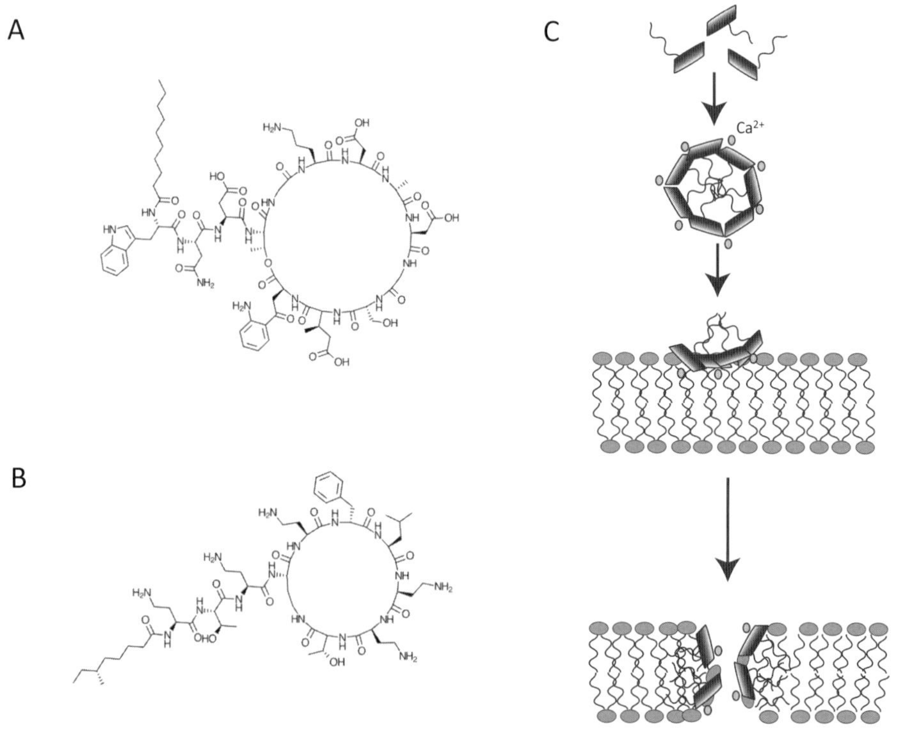

Figure 14.4 Lipopeptide antimicrobials. Figure shows the chemical structures of (A) Daptomycin (B) Polymyxin B and (C) the proposed mode of action of daptomycin. In the presence of a molar equivalent of Ca^{2+}, the lipopeptide oligomerizes into a micellar structure. In proximity of the bacterial membrane, the multimer dissociates and daptomycin binds to the membrane in a process facilitated by Ca^{2+}. Finally the drug is incorporated into the bilayer causing damages leading to leakages and ultimately to cell death.

formation of a channel for K^+ efflux from the cell leading to loss of membrane potential and ultimately to cell death (Silverman *et al.*, 2003). The proposed mechanism of action of daptomycin is illustrated in Fig. 14.4C.

Polymyxins are examples of cationic lipopeptides (Fig. 14.4B). They are a class of pentacationic cyclic lipodecapeptides with bactericidal activity due to their strong binding to the anionic bacterial outer membrane (Vaara, 2010). Discovered in late 1940s they were abandoned between 1970 and 1980 because of their nephrotoxicity. The rise of resistant bacterial pathogens has prompted renewed interest in polymyxins as antimicrobial agents and they have recently been used as a treatment of last resort for infections caused by MDR Gram-negative bacteria (Nation and Li, 2009). To overcome toxicity and recently observed resistance, the development of new polymyxin derivatives is urgently required. The positive charges in the cyclic core of polymyxins have been implicated as a cause of nephrotoxicity via a similar mechanism as aminoglycosides (Vaara, 2010). New polymyxin analogues carrying only three positive charges in the cyclic peptide core instead of the classical five of polymyxin B have been synthesized (Vaara *et al.*, 2008). Further pharmacokinetic studies will explain whether these new analogues might be less toxic whilst retaining the same bacterial effect of their ancestors, polymyxin B and E.

AMPs and lipopeptides have garnered great interest in modern medicine, predominantly as a result of their ability to target the essential structure of the membrane bilayer and to act selectively against bacteria. These membrane-damaging agents could also potentially interfere with different bacterial targets including proteins involved in cell division, energy transfer and macromolecular biosynthesis (Hancock, 2001). Given their unique mechanism of action and selectivity, AMPs and lipopeptides represent novel and promising antimicrobial compounds. Furthermore, the resistance to drugs that target the membrane bilayer is limited and less frequent for those that have a unique target. For daptomycin, the resistance is multifactorial and in *S. aureus* has been observed through overexpression of MprF, a protein that adds L-lysine to the phosphatidylglycerol diminishing the net negative charge of the bacterial membrane essential for the antimicrobial activity (Peschel *et al.*, 2001; Andra *et al.*, 2011).

Phospholipases

Mammalian group II secreted phospholipase A2 (gIIa $sPLA_2$) is an enzyme secreted by a variety of cells involved in the inflammation process and has a role in the innate immune defence (Murakami *et al.*, 1997). PLA_2 possesses antimicrobial activity and acts synergistically with other antibacterial proteins and peptides. This enzyme selectively hydrolyses the anionic phosphatidylglycerol of the bacterial cell membranes whilst showing negligible activity against zwitterionic phospholipids such as phosphatidylcholine that is usually found in the eukaryotic membranes. The affinity for bacterial membranes with their negative net charge is due to the cationic nature of PLA_2 (Scott *et al.*, 1994) a characteristic shared with AMPs and lipopeptides discussed above. Bactericidal activity of PLA_2 depends on the accessibility of the phospholipid bilayer and the ability to get across the anionic murein wall. Peptidoglycan can show different degrees of cross-linking and that would influence the antimicrobial activity of the PLA_2 depending on the species of bacteria especially amongst Gram-positive bacteria (Foreman-Wykert *et al.*, 1999). It has been reported that preliminary treatment of bacteria either with lysostaphin, a specific murein protease, or lysozyme promotes bactericidal effect against bacteria that are usually resistant to PLA_2 (Buckland *et al.*, 2000).

PLA_2 is most effective against Gram-positive bacteria as the outer membrane of Gram-negative bacteria provides an extra barrier to the cell membrane. Bactericidal/permeability-increasing proteins (BPI) are additional permeabilizing factors initially isolated from polymorphonuclear leucocytes that might act synergically with PLA_2 (Elsbach, 1998). BPI is rich in basic amino acids, mainly lysine, present in the amino terminal half of the protein and binds with high affinity to LPS due to electrostatic interactions. The complex formation between BPI and LPS in live bacteria led to an immediate growth inhibition (in't Veld *et al.*, 1988) The potent bactericidal activity of BPI *in vitro* has encouraged its use as antibiotic in clinical trials with promising results (Elsbach, 1998).

Hijacking bacterial import systems

The bacterial cell envelope forms a formidable barrier for the penetration of foreign substances and antimicrobials, and thereby provides a significant contribution to resistance (see also Chapter 9). Nutrient uptake faces the same barriers, and bacteria have evolved a multitude of specific importers to facilitate membrane translocation. In the last two decades, it has become evident that a number of natural antibiotic substances, as well as viruses, have exploited these import systems for their translocation across the bacterial envelope.

A commonly used port of entry is siderophore importers, iron acquisition systems that utilize a small high-affinity iron chelator to import ferric iron into the periplasm and/or cytoplasm (Roosenberg *et al.*, 2000; Mollmann *et al.*, 2009). *Albomycin*, a ferrichrome–thioribosyl–pyrimidine conjugate (Gause, 1955; Fiedler *et al.*, 1985) and *salmycin*, a ferrioxamine–aminosaccharide conjugate (Vértesy *et al.*, 1995; Pramanik and Braun, 2006), are two well-studied naturally occurring siderophore–antimicrobial conjugates produced by *Actinomyces subtropicus* and *Streptomyces violaceus* respectively. Both antibiotics are Trojan horse compounds that internalize an aminoacyl-tRNA synthetase targeting antimicrobial moiety by virtue of its covalent coupling to a siderophore-like moiety that is specifically imported by the bacterial siderophore importers. Albomycin exhibits marked antimicrobial activity against a range of Gram-negative bacteria, where it uses the FhuA-TonB and FhuD-FhuB-FhuC systems to translocate across the outer and inner membrane, respectively (Braun *et al.*, 2009). Salmycin is equally active against a number of Gram-positive bacteria, including *Streptococcus pneumonia* and *Staphylococcus aureus*, where it uses a FhuD-FhuB-FhuC ABC transporter system (Braun *et al.*, 2009). Albomycin has also been shown to be effective in murine infection model (Pramanik *et al.*, 2007) and salmycin on other the hand is rather unstable *in vivo* and requires shorter intervals between administration for effectiveness (Braun *et al.*, 2009). A sufficient amount of albomycin either isolated from natural source or chemically derived is required for it to be useful antibiotic therapy (Braun *et al.*, 2009). Besides these natural siderophore conjugates, the *rifamycin derivative CGP 4832* has a 200-fold increased uptake compared to Rifampicin in *E. coli* and *Salmonella* species through binding and import via the FhuA- TonB system (Pugsley *et al.*, 1987; Ferguson *et al.*, 2001). Similarly, *microcin C7* forms a Trojan Horse inhibitor that conjugates a non-hydrolysable aspartyl-adenosine moiety that targets aspartyl t-RNA synthetase to a hexapeptide that is being internalized via the YejABEF oligopeptide transport system (Guijarro *et al.*, 1995; Novikova *et al.*, 2007). Another example of a Trojan Horse antibiotic that utilizes bacterial importers is *agrocin 84*, a LeuRS t-RNA synthetase inhibitor targeting *Agrobacterium tumefaciens* (Reader *et al.*, 2005). The toxic group of agrocin 84 is linked to a D-glucofuran phosphate moiety that forms an agrocinopine mimic and is specifically imported by *Agrobacteria* through the opine permease encoded by the Ti plasmid (Kim and Farrand, 1997).

Lead by the example of the natural siderophore conjugates, several siderophore analogues have been produced synthetically to improve production yield and for optimal drug delivery (Schnabelrauch *et al.*, 2000; Mies *et al.*, 2008). The antimicrobial moiety to be conjugated with the iron chelator requires careful considerations to ensure that the coupling will not impede the antimicrobial activity as observed with sulfonamides (Zahner *et al.*, 1977) and the broad-spectrum action of the antimicrobial moiety coupled is important to maximize its utilization. Furthermore, the ease of delivering the drug to its site of action is a crucial factor. This is evident with β-lactams as they are required to be shuttled to as far as the periplasm to elicit effect (Heinisch *et al.*, 2002; Wittmann *et al.*, 2002). A marked improvement in *in vitro* efficacy of siderophore–β-lactam conjugates was observed against *P. aeruginosa, Stenotrophomonas maltophilia* and a number of Enterobacteria (Wittmann *et al.*, 2002; Mollmann *et al.*, 2009; Ji *et al.*, 2012). A number of selected siderophore–β-lactam conjugates have also been shown to be effective *in vivo* murine protection model against *P. aeruginosa* (Wittmann *et al.*, 2002) and more recently in rat soft-tissue infection model with multidrug- resistance *Acinetobacter baumannii* (Russo *et al.*, 2011).

Jamming signal transduction

Bacteria are constantly subjected to a myriad of environmental fluctuations, antimicrobials and host responses. The remarkable ability of the bacteria to sift through a diverse combination of environmental cues and assaults in their microenvironment and to accurately respond by orchestrating gene expression pivotal for adaptation to these prevailing conditions is the key to their survival and persistence. Two-component signal transduction systems (TCSTSs; reviewed in Chapter 12) are the major means by which bacteria monitor environmental signals and in response regulate an array of physiological traits including virulence, drug-resistance, nutrient uptake, biofilm and quorum sensing (QS) (Stock *et al.*, 2000; Mascher *et al.*, 2006).

TCSTSs have been the most well studied signalling pathways in bacteria since their first discovery more than 20 years ago (Ninfa and Magasanik, 1986). The prototypical TCSTSs comprise of a transmembrane histidine kinase (HK) and a cytoplasmic response regulator (RR) (Fig. 14.5). The HKs constitute a divergent N-terminal sensor domain that recognizes specific stimulus. Upon stimulation (through ligand-binding or physical perturbation), the SHK undergoes ATP-dependant autophosphorylation at a highly conserved histidine residue in the transmitter domain. The phosphoryl group is then transferred to a conserved aspartate residue in the receiver domain of the RR, inducing a conformational change that activates the RR effector domain to elicit binding to the upstream regulatory region of its target genes (Fig. 14.5). Given the complexity of environmental signals, bacteria have evolved to possess multiple TCSTSs that each recognizes and responds to specific stimuli including peptides, sugars, antimicrobials and QS signals. To coordinate responses to diverse stimuli, cross-phosphorylation can occur between different TCSTSs forming a large and intricate cascade of regulatory networks (Eguchi and Utsumi, 2008; Mitrophanov and Groisman, 2008). This notion is highlighted in a recent genome-wide transcriptional study

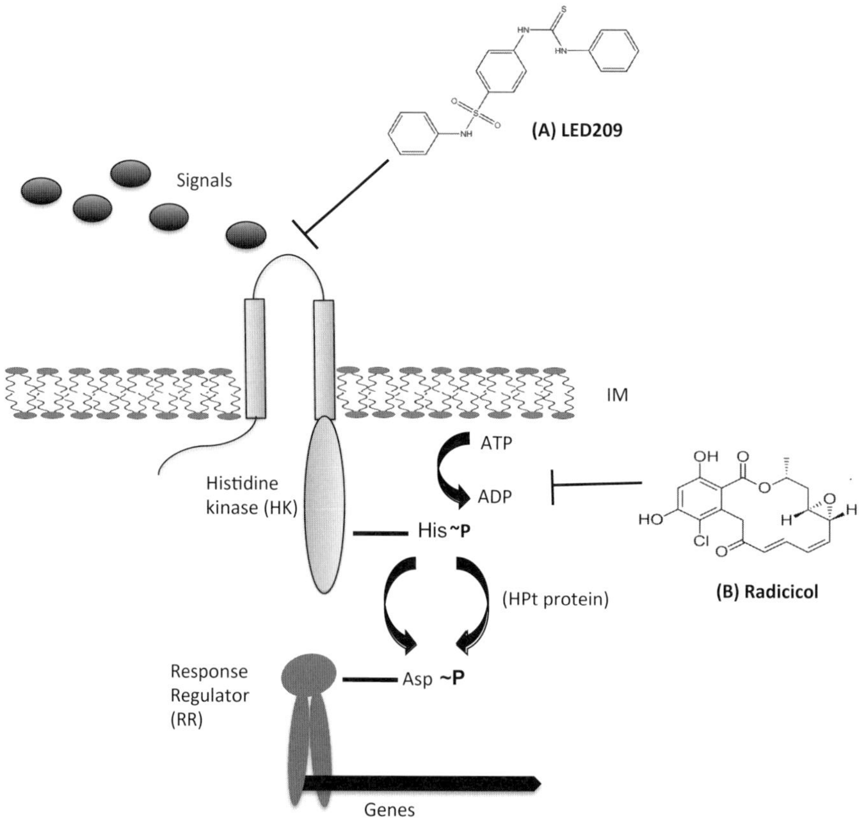

Figure 14.5 Schematic representation of a typical two-component signal transduction system. Upon stimulation by specific signal(s), the sensor histidine kinase (HK) undergoes ATP-dependent autophosphorylation at a highly conserved histidine residue in the transmitter domain. The phosphoryl-group is then transferred to a conserved aspartate residue (directly or via a soluble HPt protein) in the receiver domain of the response regulator (RR), inducing a conformational change that activate the RR effector domain to elicit binding to the upstream regulatory regions of its target genes, thus modulating their expression. (A) LED209 inhibits binding of AI-3, adrenaline and noradrenaline to the QseC (quorum sensing HK) and concurrently abrogates QseC autophosphorylation and expression of virulence genes under the control of QseC. (B) *Radicicol* competitively binds the HK ATP-binding pocket of PhoQ (virulence sensor HK), inhibiting its auto-kinase activity and downstream signalling. Abbreviations; IM, inner membrane; OM, outer membrane; P, phosphoryl-group.

showing that a single HK, FimS (FimS-FimR), can alter the expression levels of up to 10% of the bacterial transcriptome (Lo *et al.*, 2010).

TCSTSs have several features that underscore their potential as either broad-spectrum or pathogen-specific antivirulence targets for antimicrobial therapy. First, TCSTSs are widely distributed among bacteria, fungi, and plant kingdoms, however their presence in the animal kingdom has not been reported and thus potentiates selective toxicity. Second, despite their functional diversity, HKs and RRs from different bacterial species share conserved amino acid sequence motifs and structural homologies in particular within their active domains (Gao and Stock, 2009). Thus, given the role of TCSTS as a common denominator in bacterial signalling networks that regulate an array of physiological processes including virulence and antibiotic resistance, TCSTSs inhibitors that chemically interfere with the conserved active domains could be developed, conferring a broad-spectrum antibacterial activity. Alternatively, the selectively targeting of the receptor domains of non-essential TCSTSs that are involved in regulating virulence phenotypes including antibiotic resistance, biofilm formation, adherence to and invasion of host cells, provides promising anti-virulence approaches. The use of conventional antibiotics that primarily target the essential cellular processes for bacterial growth imposes strong selective pressure and accelerates development of drug resistance (Werner *et al.*, 2008). In light of this, much of the recent antimicrobial therapy developments have been geared towards targeting virulence factors rather than bacterial growth in anticipation that such strategies engender a gentler selective pressure on bacteria to develop drug resistance (Clatworthy *et al.*, 2007; Cegelski *et al.*, 2008; Rasko and Sperandio, 2010).

An example of broad-acting inhibitors of the HK kinase domain is found in the PhoQ-PhoP system. PhoQ-PhoP is one of the most extensively studied TCSTSs involved in pathogenesis in a number of Gram-negative bacterial species including *Salmonella typhimurium, Shigella flexneri, Yersinia pestis* and *Pseudomonas aeruginosa* (Miller *et al.*, 1989; Moss *et al.*, 2000; Oyston *et al.*, 2000; Gooderham *et al.*, 2009). PhoQ sensor is known to monitor extracellular Mg^{2+} concentration, antimicrobial peptides and acidic environment (Garcia Vescovi *et al.*, 1996; Bader *et al.*, 2005; Prost *et al.*, 2007). In *S. typhimurium*, the inactivation of either *phoQ* or *phoR*, renders the bacteria avirulent, exhibiting defects in cell invasion and a reduced ability to survive within macrophages (Galan and Curtiss, 1989; Miller *et al.*, 1989; Behlau and Miller, 1993). The ATP-binding domains of HKs and the GHL (gyrase, Hsp90 and MutL) family of proteins share structural conservation (Dutta and Inouye, 2000) and the GHL inhibitor *Radicicol* (an Hsp90 inhibitor) was found to bind directly to residues within ATP-binding pocket of *S. typhimurium* PhoQ, inhibiting its auto-kinase activity albeit with low affinity (K_d of 715 ± 78 μM) (Guarnieri *et al.*, 2008). In a more recent study, four small compound inhibitors targeting the *S. flexneri* PhoQ have been discovered through a high-throughput virtual screening of the ATP-binding pocket (Cai *et al.*, 2011). These compounds bound to *S. flexneri* PhoQ with dissociation constants ranging from 4.5 to 10.6 μM. Three of these were equally effective in inhibiting cell invasion of *S. flexneri* or *S. typhimurium*, thus suggesting that these compounds provide good starting points for development of broad-spectrum inhibitors targeting multiple HKs. Although these compounds have been shown to interact with *S. flexneri* PhoQ through enzymatic assays, whether their auto-kinase inhibition activity is due to specific binding to the PhoQ catalytic domain remains to be experimentally established.

To date, the ligands that trigger most HKs autophosphorylation are largely unknown. It is thought that these ligand–receptor interactions can be exploited as conserved targets for drug intervention. QseC sensor HK is known to respond to the QS molecule AI-3 and/or human hormones adrenaline and/or noradrenaline (Clarke *et al.*, 2006) and initiating an intricate phosphorylation cascade that modulates expression of virulence genes (Clarke and Sperandio, 2005; Clarke *et al.*, 2006; Njoroge and Sperandio, 2011). QseC homologues have been detected in at least 25 pathogenic bacteria of human and plant, and the inactivation of *qseC* has been shown to result in reduced virulence of enterohaemorrhagic *E. coli* (EHEC), *S. typhimurium* and *F. tularensis* in animal infection models (Clarke *et al.*, 2006; Weiss *et al.*, 2007; Bearson and Bearson, 2008; Moreira *et al.*, 2010). A small compound inhibitor, *LED209* (*N*-phenyl-4 {[(phenylamino)thioxomethyl]amino}benzene-sulfonamide hydrate), that has been identified through a high throughput screen of chemical library is able to interfere with the binding of AI-3, adrenaline and noradrenaline to QseC and concurrently abrogates QseC autophosphorylation and expression of virulence genes under the control of QseC (Rasko *et al.*, 2008). LED209 has also been shown to be effective in reducing the pathogenicity of *S. typhimurium* and *F. tularensis* but not EHEC in animal models (Rasko *et al.*, 2008). LED209 is a promising drug that provides proof of principle and illustrates the feasibility of development of anti-virulence intervention targeting the conserved bacterial signalling mechanism. More recently, Ng *et al.* (2013) identified board-spectrum inhibitors of virulence in *Vibrio* species including *V. cholera*, *V. harveyi* and *V. parahaemolyticus*. These inhibitors target the ATP hydrolysis domain of LuxO, a highly conserved quorum-sensing response regulator of the NtrC-family (Ng *et al.*, 2013).

In Gram-positive bacteria, QS systems typically constitute of posttranslationally modified secreted autoinducing peptides (AIPs) that interact with the HK sensor of TCSTS. AgrC is the HK of the TCSTS (AgrC-AgrA) in *S. aureus* that responds to the increasing concentration of AIPs as a function of cell density and mediate the expression of a vast array of invasive virulence factors including secreted toxins, lipases and proteases (George and Muir, 2007). *S. aureus* is the most common cause of nosocomial infections that imposes serious burden on health care resources. Several strategies of disrupting the AIPs interaction with AgrC have been reported over the years. Synthetic AIP thiolactone inhibitors that interfere with the binding of cognate AIPs to AgrC have been developed (Mayville *et al.*, 1999; Lyon *et al.*, 2000). Apolipoprotein B is a major structural protein of lipoproteins that has been shown to sequester AIP1 thereby preventing its interaction with AgrC and subsequent signalling through ArgC (Peterson *et al.*, 2008). The use of antibodies that have been raised against AIP in *S. aureus* resulted in reduction of expression of virulence determinants modulated by *agr* system and treatment with the antibodies reduces skin lesion in mice challenged with *S. aureus* (Park *et al.*, 2007). However, contrasting studies have implicated the role of *agr* system in repressing the expression of crucial biofilm formation determinants and have shown that dysfunction of the *agr* system caused persistence bacteraemia in *S. aureus* (Vuong *et al.*, 2000; Fowler *et al.*, 2004). These observations show that caution is needed in developing QS inhibitor approaches and further emphasize the need for a defined knowledge of the genetic and molecular interplay between QS signals and the regulatory networks in the organism in order to pinpoint quintessential factors to be targeted.

To date no TCSTSs inhibitors have undergone extensive clinical study. It remains to be seen whether TCSTS inhibitors can live up to their promised potentials, further research and careful consideration are warranted.

Blocking the secretion pathways

Bacteria have evolved an array of secretion systems or machineries to transport diverse virulence effector molecules across the cell membranes into the extracellular milieu as well as directly into the host cell. These apparatus are pivotal for bacterial pathogenesis and their inactivation would render the pathogen avirulent and more susceptible to host immune surveillance, thus making them attractive targets for anti-virulence drugs development. To date, these secretion systems have been divided into seven distinctive groups, types I through VII (T1SS–T7SS), and two additional groups comprising the chaperone-usher and curli biogenesis pathways (Tseng *et al.*, 2009; Rego *et al.*, 2010). Effector molecules are translocated across the bacterial envelope either in a single step, bypassing the periplasm (including type I, type III or type IV systems; see Chapter 11) or a two-step process with an intermediate passage of the substrates through the periplasm (type II, type V, and the chaperone-usher and curli biogenesis pathways) (Tseng *et al.*, 2009; Rego *et al.*, 2010). In the latter systems, the substrates are secreted into the periplasm via the SEC or twin-arginine (Tat) pathways and are subsequently exported across the outer membrane using a pathway-specific transporter in the outer membrane. To formulate drugs that specifically target these machineries, the following inhibitory mechanisms can be of consideration: blockage of the outer membrane transport protein, inhibition of the periplasmic adaptor proteins, interference of the apparatus assembly or deprivation of the energy transfer within the mechanism.

Type I secretion systems (T1SSs) and resistance-nodulation-division (RND) efflux pumps

T1SSs are known to orchestrate the secretion of virulence polypeptides including toxins, lipases and adhesins of varying sizes ranging from 10 kDa to 900 kDa (Delepelaire, 2004). Multi-drug resistance (MDR) efflux pumps on the other hand as the name describes confer resistance to antibiotics in clinically relevant pathogens by actively pumping out xeno-compounds (Li and Nikaido, 2004; Piddock, 2006). The MDR phenotype has frequently been associated with the overexpression of resistance–nodulation–cell division (RND)-type efflux pumps (Poole, 2005; Blair and Piddock, 2009). T1SSs and RND-efflux pumps in Gram-negative bacteria are organized as tripartite systems constituting of an outer membrane channel-forming protein, a periplasmic-spanning adaptor (membrane fusion protein) and an inner membrane transporter (Fig. 14.6) (Delepelaire, 2004; Blair and Piddock, 2009). The outer membrane channel-forming proteins of the TolC family are key constituents of both T1SSs and RND efflux pumps, facilitating pathogenesis and drug resistance in a number of clinically relevant bacteria such as *Legionella pneumophila, S. typhimurium,* and *F. tularensis* by exporting a range of unrelated substrates including toxins, proteases and antibiotics (Buckley *et al.*, 2006; Gil *et al.*, 2006 Ferhat *et al.*, 2009). Given the TolC-like channel dependent nature of T1SSs and RND-efflux pumps and their crucial roles in secretion of virulence factors and in drug resistance, they will be discussed in parallel as attractive targets for antimicrobial drug development.

To date, the development of drugs targeting T1SSs that are not involved in MDR is lacking compared to that of the NDR-efflux pumps. Several trivalent cations, in particular $Co(NH_3)_6^{3+}$, have been shown to be effective in impeding the conductivity of TolC-like channels in *E. coli* by directly binding (with low nanomolar K_d values) to the electronegative periplasmic entrance that is widely conserved throughout the TolC family proteins (Fig. 14.6) (Andersen *et al.*, 2002; Higgins *et al.*, 2004). However, the potency of this compound *in vivo* could not be assessed due to its cytotoxicity to cells at micromolar concentration. Nevertheless, a better understanding of its mechanism of inhibition may provide a basis for rational design strategy to develop a less toxic and broad-spectrum inhibitor. *Phenylalanyl-arginyl-β-naphthylamide* (PAβN) compound is known to inhibit MexB (OprM-MexA-MexB) in *P. aeruginosa* (Renau *et al.*, 1999; Lomovskaya *et al.*, 2001) and a few other related RND transporters in *K. pneumoniae*, and *S. enterica* (Chevalier *et al.*, 2000; Baucheron *et al.*, 2002; Hasdemir *et al.*, 2004). PAβN is a competitive inhibitor that is preferentially extruded by MexB resulting in intercellular accumulation of its cognate

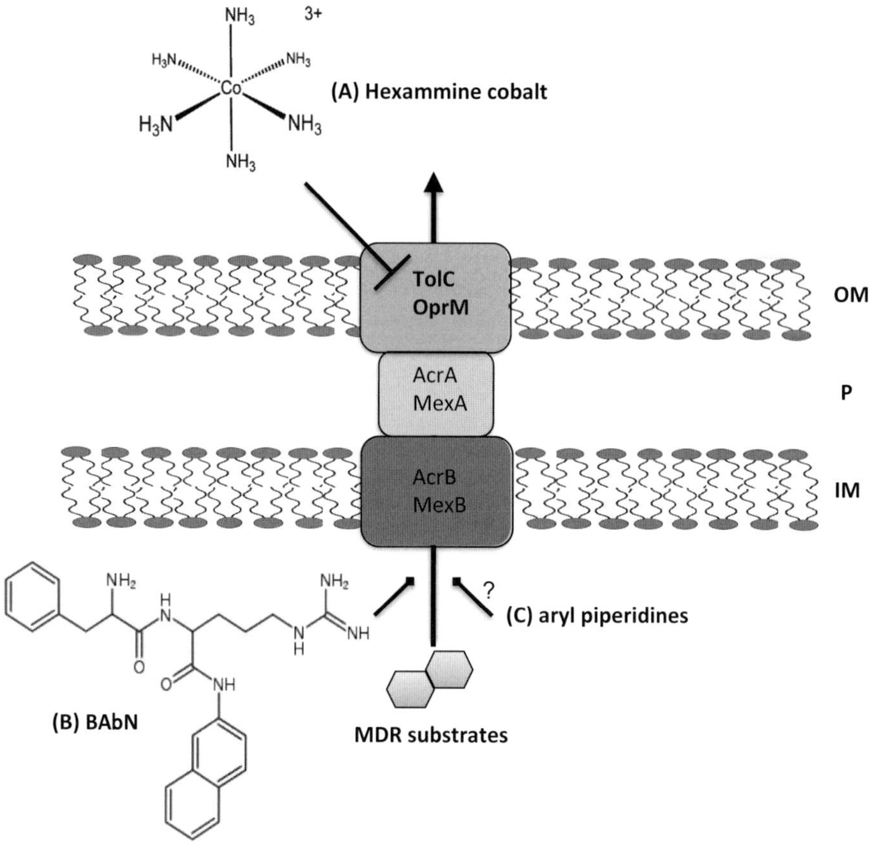

Figure 14.6 Inhibiting multidrug transport. The extrusion of antibiotic across the cell envelope is illustrated for the TolC-AcrA-AcrB (*E. coli*) or OprM-MexA-MexB (*P. aeruginosa*) systems. The arrows indicate the direction of flow of molecules (MDR substrates) through the machineries. (A) Hexamine cobalt ($Co(NH_3)_6^{3+}$) blocks substrate extrusion at the level of the TolC-like outer membrane channel (B) phenylalanyl-arginyl-β-naphthylamide blocks substrate extrusion through competitive binding to the IM transporter (C) A similar mechanism is believed to be responsible for the observed inhibition by aryl piperidines such as 1-(1-naphthyl)-piperazine and 1-(1-naphthylmethyl)-piperazine (not shown). Abbreviations: IM, inner membrane; OM, outer membrane; P, periplasm.

antibiotic substrate and subsequent restoration of antibiotic sensitivity as lower concentration of antibiotics is required for toxicity (Fig. 14.6). Despite of the broad-spectrum activity of PAβN, it has been reported that this molecule at high concentration can affect membrane integrity (Lomovskaya *et al.*, 2001).

Several *arylpiperazines* including 1-(1-naphthyl)-piperazine and 1-(1-naphthylmethyl)-piperazine (NPM) have also been shown to significantly enhance *Enterobacteriaceae* sensitivity to antibiotics (Bohnert and Kern, 2005; Schumacher *et al.*, 2006). Although the exact mechanism of NPM inhibition is not yet known, the ability of NPM to reverse the extrusion of levofloxacin in an AcrAB-overproducing *E. coli* strain in a similar manner to PAβN suggests inhibition of efflux pump (Bohnert and Kern, 2005). Recent studies have also shown that licensed drugs other than antimicrobial agents for use in human confer synergistic activity when co-administered with antibiotics against resistant strains; these include chlorpromazine and trimethoprim (Bailey *et al.*, 2008; Piddock *et al.*, 2010).

Type II secretion systems (T2SSs) and Type IV pili

The distribution of T2SSs is restricted to *Proteobacteria*, where they are important for secretion of virulence factors in human pathogens including enterotoxigenic *E. coli* (ETEC), *L. pneumophila* and *Vibrio cholerae* (Filloux, 2004; Cianciotto, 2005). Virulence factors that are exported through the T2SS include the heat labile toxin of ETEC, cholera toxin of *V. cholerae* and exotoxin A of *P. aeruginosa* (Hirst and Holmgren, 1987; Wick *et al.*, 1990 Tauschek *et al.*, 2002). The translocation of substrates across the OM by the T2SS is much more complex than the T1SS, involving 12 to 15 accessory proteins (Filloux, 2004; Johnson *et al.*, 2006). T2SS and type IV pili assembly share many accessory components that are not only structurally related but also functionally complementary (Sauvonnet *et al.*, 2000; Vignon *et al.*, 2003; Ayers *et al.*, 2010). Type IV pili have been shown to be involved a number of virulence traits including surface adhesion, biofilm formation and DNA uptake (Craig *et al.*, 2004; Nudleman and Kaiser, 2004) and thus making it an additional attractive drug target. Despite a lack of development of drugs targeting the T2SS and Type IV pili to date, a small molecule compound *virstatin*, 4-[*N*-(1,8-naphthalimide)]-*n*-butyric acid that do not directly interfere with the type II secretion machinery itself, but rather inhibits the dimerization of the transcriptional regulator ToxT, thereby abrogating the expression of cholera toxin and also the toxin co-regulated pilus of *V. cholerae* (Hung *et al.*, 2005; Shakhnovich *et al.*, 2007) was identified. Virstatin confers *in vivo* protection in infant mice against colonization of *V. cholerae* (Hung *et al.*, 2005). More recently, virstatin is also shown to interact with *V. cholerae* accessory cholera enterotoxin resulting in partial unfolding of this toxin (Chatterjee *et al.*, 2011).

Type III secretion systems (T3SSs)

T3SSs are the key virulence determinants that are prevalent in a large number of pathogenic Gram-negative bacteria of human, animals and plants (Mota and Cornelis, 2005; Moraes *et al.*, 2008). These systems serve as conduits for direct transportation of virulence-effector molecules into the cytosol of targeted host cells, where these molecules dysregulate an array of host cell functions including immune and signalling responses (Galan and Wolf-Watz, 2006). The interspecies conservation both structurally and functionally of most of the core components in the T3SS (Rosqvist *et al.*, 1995; Cornelis and Van Gijsegem, 2000; Lilic *et al.*, 2006) potentiate the development of broad-spectrum T3SS inhibitors that are effective independent of the nature of the effector molecules that they translocate.

Three different inhibitors, the *salicylidene acylhydrazides*, the *salicylanilides* and the *sulfonylaminobenzanilides* were first identified through cell-based assays by measuring the expression of T3SS-related genes with a transcriptional reporter system in *Y. pseudotuberculosis* (Kauppi *et al.*, 2003). These compounds were reported to have no effect on growth and each confers different modes of inhibition. It was thought that the salicylidene acylhydrazides and likely the sulfonylaminobenzanilides directly interfere with the secretion mechanism and the salicylanilides likely to target the transcriptional regulation of T3SS (Kauppi *et al.*, 2003). Subsequent detailed study of the derivatives of the salicylidene acylhydrazides (INPs) showed that the most active derivative directly targets the secretion machinery and inhibits *Y. pseudotuberculosis* virulence in an *in vivo* HeLa cell model (Nordfelth *et al.*, 2005). These molecules are also active against *Chlamydia trachomatis, C. pneumoniae, S. typhimurium and S. flexneri* (Muschiol *et al.*, 2006; Bailey *et al.*, 2007; Hudson *et al.*, 2007; Muschiol *et al.*, 2009; Veenendaal *et al.*, 2009).

In a separate study, chemical screens for inhibitors of T3SS-dependent translocation of a phospholipase in *S. typhimurium* resulted in the identification of *2-imino-5-arylidene thiazolidinone*. This molecule also confers broad-spectrum activity against T3SSs of *Yersinia* spp. and *S. typhimurium* as well as inhibition of T2SS of *P. aeruginosa* and *Francisella novicida* (Felise *et al.*, 2008). Despite the identification of numerous inhibitors against T3SS, their specific molecular targets and the precise mechanism of inhibition remain largely unknown. Recently, instead of random whole cell-based screening of chemical libraries, a rational and targeted approach of using computational chemistry was employed to select for compounds that would inhibit the T3SS ATPase activity in *Y. pestis* by targeting its catalytic domain (Swietnicki *et al.*, 2011). The ATPase is the conserved and essential for T3SS function and therefore virulence in many pathogens. The active compounds identified are effective in preventing the translocation of YopE effector by attenuated *Y. pestis* into culture medium and mammalian cells. They are also effective against a homologous ATPase of *Burkholderia mallei* with similar potency. These compounds are reportedly different to the salicylidene acylhydrazides, the salicylanilides, the sulfonylaminobenzanilides and the thiazolidinones (Swietnicki *et al.*, 2011). The rational design approach is promising in terms of narrowing down the molecular target of the inhibitors and this is important for downstream structure–activity relationship optimization.

Type IV secretion systems (T4SSs)

T4SSs are multi-component membrane-spanning translocators that are widely distributed and essential in many important pathogenic Gram-negative bacteria including *A. tumefaciens, Bordetella pertussis, Legionella pneumophila, Bartonella* species and *H. pylori* (Baron, 2005; Backert and Meyer, 2006; Baron and Coombes, 2007). These systems share a common evolutionary origin with the bacterial conjugation machineries (Lawley *et al.*, 2003; Christie, 2004). T4SSs are known to translocate virulence factors of protein(s), DNA or protein–DNA complexes out of bacterial cells into recipient cells altering host physiological processes or facilitating the introgression of antibiotic resistance and virulence genes (Cascales and Christie, 2003; Christie *et al.*, 2005). Their indispensable roles in pathogenesis underscore their potential importance as drug targets.

The VirB/VirD4 of *A. tumefaciens* has served a model system for the study of T4SS (Christie, 1997). This system constitute of 12 secretion components (VirB1 to VirB11 and VirD4) that assemble into a transmembrane complex (Fronzes *et al.*, 2009). The VirB11 subunits are structurally homologous to a large family of ATPases that typically involved in translocation of macromolecules including T2SSs and type IV pili assembly systems. These subunits are ubiquitous in Gram-negative bacteria and crucial for the assembly of the multi-protein secretion complexes (Cao and Saier, 2001; Planet *et al.*, 2001; Christie *et al.*, 2005). A HTS of compound libraries against Caga (a VirB11 homologue in *H. pylori*) ATPase activity has identified three active compounds, two of which are structurally related compounds of thiadiazolidine-3,5-dione class (Hilleringmann *et al.*, 2006). One of the compounds, 4-methyl-2-[4-(trifluoromethoxy)phenyl]-1,2,4-thiadiazolidine-3,5-dione (CHIR-1), was shown to inhibit the trafficking of cytotoxin-associated antigen, CagA from

H. pylori into host cells and also impaired the ability of *H. pylori* to colonize mice gastric mucosal. The exact mode of inhibition and the molecular target(s) for these compounds are however unknown. Given the cross-species conservation of the ATPase subunits structurally, it will be of a great interest to assess the effectiveness of these compounds other pathogenic species.

More recently, a *salicylidene acylhydrazide* derivative, 2-hydroxy-N′-{[5-(2-nitrophenyl)–2-furyl]methylene}benzohydrazide that is structurally related to the ones that are active against the T3SSs (as described above) was identified in a *Brucella abortus* whole cell-based HTS assay. This compound has been shown to interrupt the dimerization of VirB8, which is pivotal for T4SS function and the transcription of *virB* operon. At phenotypic level, this compound greatly reduced the T4SS-dependent intercellular survival of *B. abortus* in macrophages (Paschos *et al.*, 2011). Similarly to the active compounds against VirB11, the actual mode of action and molecular target(s) of this salicylidene acylhydrazide remain unclear. Nevertheless, all these potential inhibitors do provide a promising starting point for further molecular tweaking to achieve a more ideal inhibitor.

A high-throughput *in vivo* conjugation assay screen of a library of bacterial and fungal extracts revealed two atypical unsaturated fatty acids, linoleic and oleic acids (*cis*-unsaturated C_{18} fatty acids), inhibit the shuttling of plasmid R388 between *E. coli* strains tested (Fernandez-Lopez *et al.*, 2005). The questions whether these fatty acids directly inhibit T4SS or interfere with the cell membrane dynamics remain to be answered. The identification of their molecular target(s) is warranted for further chemical optimization to obtain ideal inhibitors.

Chaperone-usher (CU) pathways

To date, CU pathways are one of the best-characterized assembly systems utilized by a myriad of Gram-negative pathogens (including *E. coli*, *Salmonella* spp., *Yersinia* spp., *Pseudomonas* spp. and *Klebsiella* spp.) to assemble adhesive organelles that are crucial for biofilm formation, attachment and invasion of host cells (Nuccio and Baumler, 2007; Fronzes *et al.*, 2008; Zav'yalov *et al.*, 2009). P pili and type 1 pili of uropathogenic strains of *E. coli* (UPEC) are the prototypical CU pathway assemblies that facilitate the adherence and colonization of the host kidney (Roberts *et al.*, 1994; Dodson *et al.*, 2001) and bladder (Langermann *et al.*, 1997; Mulvey *et al.*, 1998), respectively. P pili assembly constitutes of several pilus subunits; PapA, E, F, G, H, and K and FimA, F, G, and H in type 1 pili. PapA and FimA are the major subunits; PapF/E/K and FimF/G form the linker subunits of P pili and type 1 pili respectively (Fig. 14.7). The adhesin subunits PapG and FimH dictate the binding specificity of P pili and type 1 pili to galabiose on kidney cells and mannose on bladder cells respectively. During the assembly process, chaperone PapD/FimC stabilize the folding of pilus subunits as they emerge from the SEC translocon into the periplasm and shuttle them to the PapC/FimD usher, the outer membrane assembly platform where they are incorporated in the growing fibre (Fig. 14.7) (Sauer *et al.*, 2004; Waksman and Hultgren, 2009). Receptor analogues have successfully been employed to inhibit pilus-dependent adhesion; the use of alkyl- and aryl-mannosides for the inhibition of type 1 pilus adhesin FimH is currently one of the most active lines of research (Fig. 14.7A) (Klein *et al.*, 2010; Cusumano *et al.*, 2011). Besides targeting the adhesin, the conservation of structural and biochemical mechanisms of CU pilus assembly, and the prevalence CU pathways in pathogenic species, suggest the attractiveness of these systems as drug targets of broad-spectrum potential.

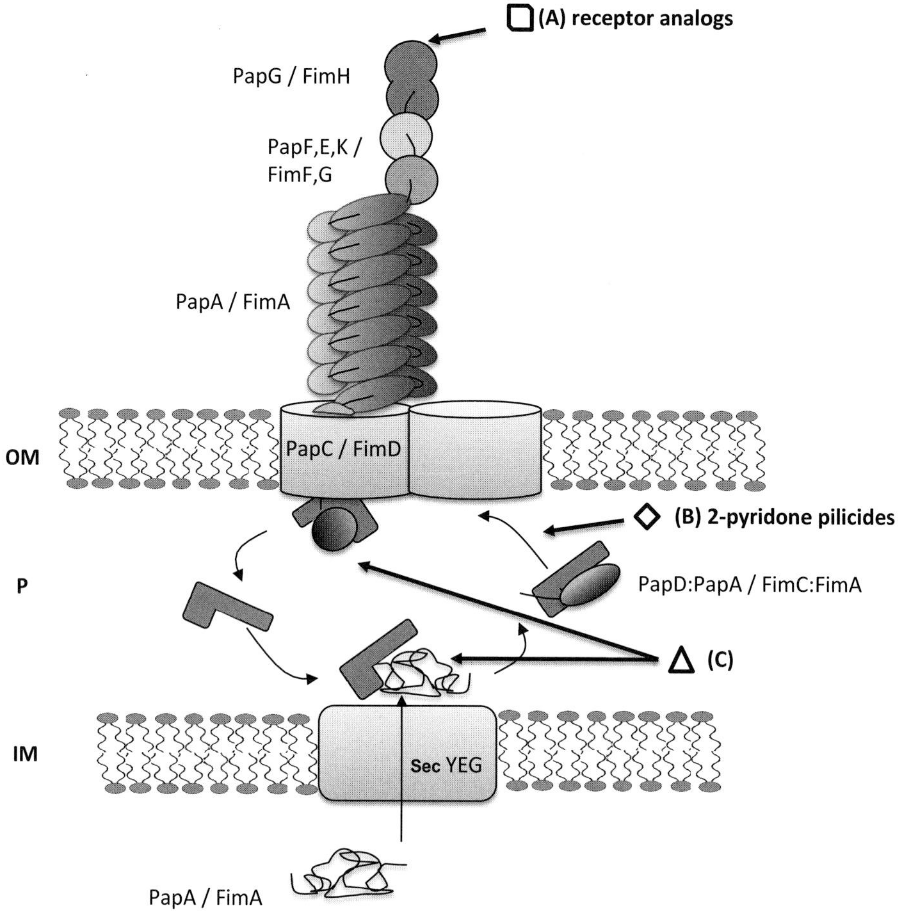

Figure 14.7 Chemical knockout of chaperone/usher pili and pilus assembly. Schematic representation of pilus assembly by the chaperone-usher pathway, the type 1 and P pilus model systems are shown. Subunits entering the periplasmic space (P) are bound by the periplasmic chaperone (PapD/FimC) and shuttled to the outer membrane usher (PapC/FimD) where they are assembled into non-covalent polymers and translocated to the bacterial surface. Sites of known and proposed chemical interference with pilus function and assembly encompass: (A) competitive inhibition of adhesin function by means of receptor analogues (e.g. mannosides in case of type 1 pili) (B) 2-pyridone pilicides block pilus biogenesis by competitive inhibition of chaperone-subunit recruitment to the usher, thereby arresting new subunit incorporations at the base of the growing pilus. (C) subunit folding by the chaperone and subunit polymerization at the usher have been proposed as target for pilus biogenesis inhibition. Abbreviations; IM, inner membrane; OM, outer membrane.

A wealth of structural data of P pili and type 1 pili and their subunits prompted the structure-based design of peptidomimetic inhibitors termed *pilicides* (bicyclic 2-pyridones) that were initially designed to interact with the highly conserved Arg and Lys pair in the interdomain cleft of the chaperone (Arg8 and Lys112 in the P pilus chaperone PapD), crucial for chaperone–subunit association (Svensson *et al.*, 2001). However, subsequent studies revealed that the pilicides actually bind to a hydrophobic patch on the chaperone that coincides with the molecular surface responsible for the interaction with the N-terminal domain of the usher, the outer membrane protein responsible for the polymerization and translocation of the nascent pilus across the outer membrane (Hedenstrom *et al.*, 2005; Pinkner *et al.*, 2006). By blocking the usher-binding site on the chaperone, pilicides inhibit the new delivery of subunits to the growing fibre and stop pilus biogenesis (Fig. 14.7B). Pilicides have been shown to inhibit the assembly of both P pili and type 1 pili (Pinkner *et al.*, 2006). Pilicides confer a dose-dependent inhibitory effect on P pili and type 1 pili mediated haemagglutination, type 1 pili dependent-biofilm formation and adherence to bladder epithelial cells (Pinkner *et al.*, 2006). Unpublished data from our laboratory show that the subunit polymerization step can form an additional target in pilus biogenesis inhibition (Fig. 14.7C). The recent advances and novel strategies in the chemical attenuation of chaperone-usher pili have been reviewed in greater detail elsewhere (Lo *et al.*, 2013).

Concluding remarks

The avid development of drugs targeting the bacterial cell membrane integrity and synthesis, signal transduction pathways and secretion pathways holds promising potential to combating bacterial pathogenesis. The increasing threat of multidrug-resistance bacteria and the lack of new clinically relevant antibiotics in the industrial pipeline has prompted the departure from conventional combinatorial chemistry and empirical high-throughput screening of natural products and of chemical libraries to the more structure-guided methods by utilizing the ever-increasing wealth of protein structural information to identify and design new ligands for chemical attenuation. The potential antimicrobials discovered from these methods provide promising starting points and their future developments require better understanding of their mode of action and identifying their molecular targets for downstream structure activity relationship optimization to improve their pharmacokinetics, potency and efficacy *in vivo*.

References

Andersen, C., Koronakis, E., Bokma, E., Eswaran, J., Humphreys, D., Hughes, C., and Koronakis, V. (2002). Transition to the open state of the TolC periplasmic tunnel entrance. Proc. Natl. Acad. Sci. U.S.A. *99*, 11103–11108.

Andra, J., Goldmann, T., Ernst, C.M., Peschel, A., and Gutsmann, T. (2011). Multiple peptide resistance factor (MprF)-mediated Resistance of *Staphylococcus aureus* against antimicrobial peptides coincides with a modulated peptide interaction with artificial membranes comprising lysyl-phosphatidylglycerol. J. Biol. Chem. *286*, 18692–18700.

Andrew, J.H., Wale, M.C., Wale, L.J., and Greenwood, D. (1987). The effect of cultural conditions on the activity of LY146032 against staphylococci and streptococci. J. Antimicrob. Chemother. *20*, 213–221.

Arinaminpathy, Y., Khurana, E., Engelman, D.M., and Gerstein, M.B. (2009). Computational analysis of membrane proteins: the largest class of drug targets. Drug Discov. Today *14*, 1130–1135.

Ayers, M., Howell, P.L., and Burrows, L.L. (2012). Architecture of the type II secretion and type IV pilus machineries. Future Microbiol. *5*, 1203–1218.

Backert, S., and Meyer, T.F. (2006). Type IV secretion systems and their effectors in bacterial pathogenesis. Curr. Opin. Microbiol. *9*, 207–217.

Bader, M.W., Sanowar, S., Daley, M.E., Schneider, A.R., Cho, U., Xu, W., Klevit, R.E., Le Moual, H., and Miller, S.I. (2005). Recognition of antimicrobial peptides by a bacterial sensor kinase. Cell *122*, 461–472.

Bailey, A.M., Paulsen, I.T., and Piddock, L.J. (2008). RamA confers multidrug resistance in *Salmonella enterica* via increased expression of acrB, which is inhibited by chlorpromazine. Antimicrob. Agents Chemother. *52*, 3604–3611.

Bailey, L., Gylfe, A., Sundin, C., Muschiol, S., Elofsson, M., Nordstrom, P., Henriques-Normark, B., Lugert, R., Waldenstrom, A., Wolf-Watz, H., *et al.* (2007). Small molecule inhibitors of type III secretion in Yersinia block the *Chlamydia pneumoniae* infection cycle. FEBS Lett. *581*, 587–595.

Balemans, W., Lounis, N., Gilissen, R., Guillemont, J., Simmen, K., Andries, K., and Koul, A. (2010). Essentiality of FASII pathway for *Staphylococcus aureus*. Nature *463*, E3; discussion E4.

Bals, R. (2000). Epithelial antimicrobial peptides in host defense against infection. Respir. Res. *1*, 141–150.

Baltz, R.H., Miao, V., and Wrigley, S.K. (2005). Natural products to drugs: daptomycin and related lipopeptide antibiotics. Nat. Prod. Rep. *22*, 717–741.

Banaiee, N., Jacobs, W.R., and Ernst, J.D. (2007). LspA-independent action of globomycin on *Mycobacterium tuberculosis*. J. Antimicrob. Chemother. *60*, 414–416.

Baron, C. (2005). From bioremediation to biowarfare: on the impact and mechanism of type IV secretion systems. FEMS Microbiol. Lett. *253*, 163–170.

Baron, C., and Coombes, B. (2007). Targeting bacterial secretion systems: benefits of disarmament in the microcosm. Infect. Disord. Drug Targets *7*, 19–27.

Baucheron, S., Imberechts, H., Chaslus-Dancla, E., and Cloeckaert, A. (2002). The AcrB multidrug transporter plays a major role in high-level fluoroquinolone resistance in *Salmonella enterica* serovar typhimurium phage type DT204. Microb. Drug Resist. *8*, 281–289.

Bearson, B.L., and Bearson, S.M. (2008). The role of the QseC quorum-sensing sensor kinase in colonization and norepinephrine-enhanced motility of *Salmonella enterica* serovar Typhimurium. Microb. Pathog. *44*, 271–278.

Behlau, I., and Miller, S.I. (1993). A PhoP-repressed gene promotes Salmonella typhimurium invasion of epithelial cells. J. Bacteriol. *175*, 4475–4484.

Bhatt, A., Molle, V., Besra, G.S., Jacobs, W.R. Jr., and Kremer, L. (2007). The *Mycobacterium tuberculosis* FAS-II condensing enzymes: their role in mycolic acid biosynthesis, acid-fastness, pathogenesis and in future drug development. Mol. Microbiol. *64*, 1442–1454.

Bienemann, A.S., Lee, Y.B., Howarth, J., and Uney, J.B. (2008). Hsp70 suppresses apoptosis in sympathetic neurones by preventing the activation of c-Jun. J. Neurochem. *104*, 271–278.

Bierbaum, G., Szekat, C., Josten, M., Heidrich, C., Kempter, C., Jung, G., and Sahl, H.G. (1996). Engineering of a novel thioether bridge and role of modified residues in the lantibiotic Pep5. Appl. Environ. Microbiol. *62*, 385–392.

Blair, J.M., and Piddock, L.J. (2009). Structure, function and inhibition of RND efflux pumps in Gram-negative bacteria: an update. Curr. Opin. Microbiol. *12*, 512–519.

Bogdanovich, T., Clark, C., Kosowska-Shick, K., Dewasse, B., McGhee, P., and Appelbaum, P.C. (2007). Antistaphylococcal activity of CG400549, a new experimental FabI inhibitor, compared with that of other agents. Antimicrob. Agents Chemother. *51*, 4191–4195.

Bohnert, J.A., and Kern, W.V. (2005). Selected arylpiperazines are capable of reversing multidrug resistance in *Escherichia coli* overexpressing RND efflux pumps. Antimicrob. Agents Chemother. *49*, 849–852.

Bolla, J.M., Alibert-Franco, S., Handzlik, J., Chevalier, J., Mahamoud, A., Boyer, G., Kiec-Kononowicz, K., and Pages, J.M. (2011). Strategies for bypassing the membrane barrier in multidrug resistant Gram-negative bacteria. FEBS Lett. *585*, 1682–1690.

Brand, A.M., de Kwaadsteniet, M., and Dicks, L.M. (2010). The ability of nisin F to control *Staphylococcus aureus* infection in the peritoneal cavity, as studied in mice. Lett. Appl. Microbiol. *51*, 645–649.

Braun, V., Pramanik, A., Gwinner, T., Koberle, M., and Bohn, E. (2009). Sideromycins: tools and antibiotics. Biometals *22*, 3–13.

Brennan, P.J., and Nikaido, H. (1995). The envelope of mycobacteria. Annu. Rev. Biochem. *64*, 29–63.

Breukink, E., and de Kruijff, B. (2006). Lipid II as a target for antibiotics. Nat. Rev. Drug Discov. *5*, 321–332.

Brinster, S., Lamberet, G., Staels, B., Trieu-Cuot, P., Gruss, A., and Poyart, C. (2009). Type II fatty acid synthesis is not a suitable antibiotic target for Gram-positive pathogens. Nature *458*, 83–86.

Brinster, S., Lamberet, G., Staels, B., Trieu-Cuot, P., Gruss, A., and Poyart, C. (2010). Brinster *et al.* reply. Nature *463*, E4.

Buckland, A.G., Heeley, E.L., and Wilton, D.C. (2000). Bacterial cell membrane hydrolysis by secreted phospholipases A(2): a major physiological role of human group IIa sPLA(2) involving both bacterial cell wall penetration and interfacial catalysis. Biochim. Biophys. Acta *1484*, 195–206.

Buckley, A.M., Webber, M.A., Cooles, S., Randall, L.P., La Ragione, R.M., Woodward, M.J., and Piddock, L.J. (2006). The AcrAB-TolC efflux system of *Salmonella enterica* serovar Typhimurium plays a role in pathogenesis. Cell Microbiol. *8*, 847–856.

Cai, X., Zhang, J., Chen, M., Wu, Y., Wang, X., Chen, J., Shen, X., Qu, D., and Jiang, H. (2011). The effect of the potential PhoQ histidine kinase inhibitors on *Shigella flexneri* virulence. PLoS One *6*, e23100.

Cao, T.B., and Saier, M.H. Jr. (2001). Conjugal type IV macromolecular transfer systems of Gram-negative bacteria: organismal distribution, structural constraints and evolutionary conclusions. Microbiology *147*, 3201–3214.

Cascales, E., and Christie, P.J. (2003). The versatile bacterial type IV secretion systems. Nat. Rev. Microbiol. *1*, 137–149.

Castagnola, M., Inzitari, R., Rossetti, D.V., Olmi, C., Cabras, T., Piras, V., Nicolussi, P., Sanna, M.T., Pellegrini, M., Giardina, B., *et al.* (2004). A cascade of 24 histatins (histatin 3 fragments) in human saliva. Suggestions for a pre-secretory sequential cleavage pathway. J. Biol. Chem. *279*, 41436–41443.

Cegelski, L., Marshall, G.R., Eldridge, G.R., and Hultgren, S.J. (2008). The biology and future prospects of antivirulence therapies. Nat. Rev. Microbiol. *6*, 17–27.

Chatterjee, T., Mukherjee, D., Dey, S., Pal, A., Hoque, K.M., and Chakrabarti, P. (2011). Accessory cholera enterotoxin, Ace, from *Vibrio cholerae*: structure, unfolding, and virstatin binding. Biochemistry *50*, 2962–2972.

Chevalier, J., Pages, J.M., Eyraud, A., and Mallea, M. (2000). Membrane permeability modifications are involved in antibiotic resistance in Klebsiella pneumoniae. Biochem. Biophys. Res. Commun. *274*, 496–499.

Christie, P.J. (1997). *Agrobacterium tumefaciens* T-complex transport apparatus: a paradigm for a new family of multifunctional transporters in eubacteria. J. Bacteriol. *179*, 3085–3094.

Christie, P.J. (2004). Type IV secretion: the Agrobacterium VirB/D4 and related conjugation systems. Biochim. Biophys. Acta *1694*, 219–234.

Christie, P.J., Atmakuri, K., Krishnamoorthy, V., Jakubowski, S., and Cascales, E. (2005). Biogenesis, architecture, and function of bacterial type IV secretion systems. Annu. Rev. Microbiol. *59*, 451–485.

Cianciotto, N.P. (2005). Type II secretion: a protein secretion system for all seasons. Trends Microbiol. *13*, 581–588.

Clarke, M.B., and Sperandio, V. (2005). Transcriptional regulation of flhDC by QseBC and sigma (FliA) in enterohaemorrhagic *Escherichia coli*. Mol. Microbiol. *57*, 1734–1749.

Clarke, M.B., Hughes, D.T., Zhu, C., Boedeker, E.C., and Sperandio, V. (2006). The QseC sensor kinase: a bacterial adrenergic receptor. Proc. Natl. Acad. Sci. U.S.A. *103*, 10420–10425.

Clatworthy, A.E., Pierson, E., and Hung, D.T. (2007). Targeting virulence: a new paradigm for antimicrobial therapy. Nat. Chem. Biol. *3*, 541–548.

Cole, S.T., and Riccardi, G. (2011). New tuberculosis drugs on the horizon. Curr. Opin. Microbiol. *14*, 570–576.

Cornelis, G.R., and Van Gijsegem, F. (2000). Assembly and function of type III secretory systems. Annu. Rev. Microbiol. *54*, 735–774.

Cotter, P.D., Hill, C., and Ross, R.P. (2005). Bacteriocins: developing innate immunity for food. Nat. Rev. Microbiol. *3*, 777–788.

Cronan, J.E., and Rock, C.O. (2008). Biosynthesis of membrane lipids. In Eco-Sal-*Escherichia coli* and salmonella typhimurium. Cell Mol. Biol., Chapter 3.6.4.

Cusumano, C.K., Pinkner, J.S., Han, Z., Greene, S.E., Ford, B.A., Crowley, J.R., Henderson, J.P., Janetka, J.W., and Hultgren, S.J. (2011). Treatment and prevention of urinary tract infection with orally active FimH inhibitors. Sci. Transl. Med. *3*, 109ra115.

D'Agnolo, G., Rosenfeld, I.S., Awaya, J., Omura, S., and Vagelos, P.R. (1973). Inhibition of fatty acid synthesis by the antibiotic cerulenin. Specific inactivation of beta-ketoacyl-acyl carrier protein synthetase. Biochim. Biophys. Acta *326*, 155–156.

Daffe, M., and Draper, P. (1998). The envelope layers of mycobacteria with reference to their pathogenicity. Adv. Microb. Physiol. *39*, 131–203.

De Smet, K., and Contreras, R. (2005). Human antimicrobial peptides: defensins, cathelicidins and histatins. Biotechnol. Lett. *27*, 1337–1347.

Delepelaire, P. (2004). Type I secretion in gram-negative bacteria. Biochim. Biophys. Acta *1694*, 149–161.

Dev, I.K., Harvey, R.J., and Ray, P.H. (1985). Inhibition of prolipoprotein signal peptidase by globomycin. J. Biol. Chem. *260*, 5891–5894.

Devine, D.A., and Hancock, R.E. (2002). Cationic peptides: distribution and mechanisms of resistance. Curr. Pharm. Des. *8*, 703–714.

Dodson, K.W., Pinkner, J.S., Rose, T., Magnusson, G., Hultgren, S.J., and Waksman, G. (2001). Structural basis of the interaction of the pyelonephritic *E. coli* adhesin to its human kidney receptor. Cell *105*, 733–743.

Dover, L.G., Cerdeno-Tarraga, A.M., Pallen, M.J., Parkhill, J., and Besra, G.S. (2004). Comparative cell wall core biosynthesis in the mycolated pathogens, *Mycobacterium tuberculosis* and Corynebacterium diphtheriae. FEMS Microbiol. Rev. *28*, 225–250.

Dubnau, E., Chan, J., Raynaud, C., Mohan, V.P., Laneelle, M.A., Yu, K., Quemard, A., Smith, I., and Daffe, M. (2000). Oxygenated mycolic acids are necessary for virulence of *Mycobacterium tuberculosis* in mice. Mol. Microbiol. *36*, 630–637.

Dutta, R., and Inouye, M. (2000). GHKL, an emergent ATPase/kinase superfamily. Trends Biochem. Sci. *25*, 24–28.

Eguchi, Y., and Utsumi, R. (2008). Introduction to bacterial signal transduction networks. Adv. Exp. Med. Biol. *631*, 1–6.

Elsbach, P. (1998). The bactericidal/permeability-increasing protein (BPI) in antibacterial host defense. J. Leukoc. Biol. *64*, 14–18.

Epand, R.M. (1997). Biophysical studies of lipopeptide–membrane interactions. Biopolymers *43*, 15–24.

Felise, H.B., Nguyen, H.V., Pfuetzner, R.A., Barry, K.C., Jackson, S.R., Blanc, M.P., Bronstein, P.A., Kline, T., and Miller, S.I. (2008). An inhibitor of gram-negative bacterial virulence protein secretion. Cell Host Microbe *4*, 325–336.

Ferguson, A.D., Kodding, J., Walker, G., Bos, C., Coulton, J.W., Diederichs, K., Braun, V., and Welte, W. (2001). Active transport of an antibiotic rifamycin derivative by the outer-membrane protein FhuA. Structure *9*, 707–716.

Ferhat, M., Atlan, D., Vianney, A., Lazzaroni, J.C., Doublet, P., and Gilbert, C. (2009). The TolC protein of Legionella pneumophila plays a major role in multi-drug resistance and the early steps of host invasion. PLoS One *4*, e7732.

Fernandes, P. (2006). Antibacterial discovery and development – the failure of success? Nat. Biotechnol. *24*, 1497–1503.

Fernandez-Lopez, R., Machon, C., Longshaw, C.M., Martin, S., Molin, S., Zechner, E.L., Espinosa, M., Lanka, E., and de la Cruz, F. (2005). Unsaturated fatty acids are inhibitors of bacterial conjugation. Microbiology *151*, 3517–3526.

Festini, F., Sperotto, S., and Neri, S. (2007). The safety of drug therapies: strategies and methods for nurses. Assist. Inferm. Ric. *26*, 165–180.

Fiedler, H.P., Walz, F., and Zähner, H. (1985). Albomycin: studies on fermentation, isolation, and quantitative determination. Appl. Microbiol. Biotechnol. *21*, 341–347.

Fields, P.I., Groisman, E.A., and Heffron, F. (1989). A Salmonella locus that controls resistance to microbicidal proteins from phagocytic cells. Science *243*, 1059–1062.

Filloux, A. (2004). The underlying mechanisms of type II protein secretion. Biochim. Biophys. Acta *1694*, 163–179.

Foreman-Wykert, A.K., Weinrauch, Y., Elsbach, P., and Weiss, J. (1999). Cell-wall determinants of the bactericidal action of group IIA phospholipase A2 against Gram-positive bacteria. J. Clin. Invest. *103*, 715–721.

Fowler, V.G. Jr., Sakoulas, G., McIntyre, L.M., Meka, V.G., Arbeit, R.D., Cabell, C.H., Stryjewski, M.E., Eliopoulos, G.M., Reller, L.B., Corey, G.R., *et al.* (2004). Persistent bacteremia due to methicillin-resistant *Staphylococcus aureus* infection is associated with agr dysfunction and low-level *in vitro* resistance to thrombin-induced platelet microbicidal protein. J. Infect. Dis. *190*, 1140–1149.

Fredenhagen, A., Tamura, S.Y., Kenny, P.T.M., Komura, H., Naya, K., Nakanishi, K., Nishiyama, K., Sugiura, M., and Kita, H. (1987). Andrimid, a new peptide antibiotic produced by an intracellular bacterial symbiont isolated from a brown planthopper. J. Am. Chem. Soc. *109*, 4409–4411.

Freiberg, C., Brunner, N.A., Schiffer, G., Lampe, T., Pohlmann, J., Brands, M., Raabe, M., Habich, D., and Ziegelbauer, K. (2004). Identification and characterization of the first class of potent bacterial acetyl-CoA carboxylase inhibitors with antibacterial activity. J. Biol. Chem. *279*, 26066–26073.

Freiberg, C., Pohlmann, J., Nell, P.G., Endermann, R., Schuhmacher, J., Newton, B., Otteneder, M., Lampe, T., Habich, D., and Ziegelbauer, K. (2006). Novel bacterial acetyl coenzyme A carboxylase inhibitors with antibiotic efficacy *in vivo*. Antimicrob. Agents Chemother. *50*, 2707–2712.

Freire-Moran, L., Aronsson, B., Manz, C., Gyssens, I.C., So, A.D., Monnet, D.L., and Cars, O. (2011). Critical shortage of new antibiotics in development against multidrug-resistant bacteria-Time to react is now. Drug Resist. Updat. *14*, 118–124.

Fronzes, R., Remaut, H., and Waksman, G. (2008). Architectures and biogenesis of non-flagellar protein appendages in Gram-negative bacteria. EMBO J. *27*, 2271–2280.

Fronzes, R., Christie, P.J., and Waksman, G. (2009). The structural biology of type IV secretion systems. Nat. Rev. Microbiol. *7*, 703–714.

Galan, J.E., and Curtiss, R. 3rd (1989). Virulence and vaccine potential of phoP mutants of Salmonella typhimurium. Microb. Pathog. *6*, 433–443.

Galan, J.E., and Wolf-Watz, H. (2006). Protein delivery into eukaryotic cells by type III secretion machines. Nature *444*, 567–573.

Galatoire, O., Touitou, V., Heran, F., Amar, N., Jacomet, P.V., Gheck, L., Berete-Coulibaly, R., Benchekroun, S., and Morax, S. (2007). High-resolution magnetic resonance imaging of the upper eyelid: correlation with the position of the skin crease in the upper eyelid. Orbit *26*, 165–171.

Ganesh, A., Kim, A., Casale, P., and Cucchiaro, G. (2007). Low-dose intrathecal morphine for postoperative analgesia in children. Anesth. Analg. *104*, 271–276.

Ganz, T. (2005). Defensins and other antimicrobial peptides: a historical perspective and an update. Comb. Chem. High Throughput Screen. *8*, 209–217.

Gao, R., and Stock, A.M. (2009). Biological insights from structures of two-component proteins. Annu. Rev. Microbiol. *63*, 133–154.

Garcia, B., Penarrocha, M., Marti, E., Gay-Escodad, C., and von Arx, T. (2007). Pain and swelling after periapical surgery related to oral hygiene and smoking. Oral Surg. Oral Med. Oral Pathol. Oral Radiol. Endod. *104*, 271–276.

Garcia Vescovi, E., Soncini, F.C., and Groisman, E.A. (1996). Mg2+ as an extracellular signal: environmental regulation of Salmonella virulence. Cell *84*, 165–174.

Gause, G.H. (1955). Recent studies on albomycin, a new antibiotic. BMJ *12*, 1177.

Geibel, S., and Waksman, G. (2011). Crystallography and electron microscopy of chaperone/usher pilus systems. Adv. Exp. Med. Biol. *715*, 159–174.

George, E.A., and Muir, T.W. (2007). Molecular mechanisms of agr quorum sensing in virulent staphylococci. Chembiochem. *8*, 847–855.

Gerusz, V., Denis, A., Faivre, F., Bonvin, Y., Oxoby, M., Briet, S., LeFralliec, G., Oliveira, C., Desroy, N., Raymond, C., *et al.* (2012). From triclosan toward the clinic: discovery of nonbiocidal, potent FabI inhibitors for the treatment of resistant bacteria. J. Med. Chem. *55*, 9914–9928.

Gil, H., Platz, G.J., Forestal, C.A., Monfett, M., Bakshi, C.S., Sellati, T.J., Furie, M.B., Benach, J.L., and Thanassi, D.G. (2006). Deletion of TolC orthologs in *Francisella tularensis* identifies roles in multidrug resistance and virulence. Proc. Natl. Acad. Sci. U.S.A. *103*, 12897–12902.

Gillis, J.C., and Wiseman, L.R. (1996). Secnidazole. A review of its antimicrobial activity, pharmacokinetic properties and therapeutic use in the management of protozoal infections and bacterial vaginosis. Drugs *51*, 621–638.

Glickman, M.S., Cahill, S.M., and Jacobs, W.R. Jr. (2001). The *Mycobacterium tuberculosis* cmaA2 gene encodes a mycolic acid trans-cyclopropane synthetase. J. Biol. Chem. *276*, 2228–2233.

Glickman, M.S., Cox, J.S., and Jacobs, W.R. Jr. (2000). A novel mycolic acid cyclopropane synthetase is required for cording, persistence, and virulence of *Mycobacterium tuberculosis*. Mol. Cell *5*, 717–727.

Gooderham, W.J., Gellatly, S.L., Sanschagrin, F., McPhee, J.B., Bains, M., Cosseau, C., Levesque, R.C., and Hancock, R.E. (2009). The sensor kinase PhoQ mediates virulence in *Pseudomonas aeruginosa*. Microbiology *155*, 699–711.

Grimes, K.D., Lu, Y.J., Zhang, Y.M., Luna, V.A., Hurdle, J.G., Carson, E.I., Qi, J., Kudrimoti, S., Rock, C.O., and Lee, R.E. (2008). Novel acyl phosphate mimics that target PlsY, an essential acyltransferase in gram-positive bacteria. ChemMedChem 3, 1936–1945.

Grzegorzewicz, A.E., Kordulákovaá, J., Jones, V., Born, S.E., Belardinelli, J.M., Vaquié, A., Gundi, V.A., Madacki, J., Siama, N., Laval, F., *et al.* (2012). A common mechanism of inhibition of the *Mycobacterium tuberculosis* mycolic acid biosynthetic pathway by isoxyl and thiacetazone. J. Biol. Chem. *287*, 38434–38441.

Guarnieri, M.T., Zhang, L., Shen, J., and Zhao, R. (2008). The Hsp90 inhibitor radicicol interacts with the ATP-binding pocket of bacterial sensor kinase PhoQ. J. Mol. Biol. *379*, 82–93.

Guidotti, G. (1972). Membrane proteins. Annu. Rev. Biochem. *41*, 731–752.

Han, Z., Pinkner, J.S., Ford, B., Obermann, R., Nolan, W., Wildman, S.A., Hobbs, D., Ellenberger, T., Cusumano, C.K., Hultgren, S.J., *et al.* (2010). Structure-based drug design and optimization of mannoside bacterial FimH antagonists. J. Med. Chem. 53, 4779–4792.

Hancock, R.E. (1997). Peptide antibiotics. Lancet *349*, 418–422.

Hancock, R.E. (2001). Cationic peptides: effectors in innate immunity and novel antimicrobials. Lancet. Infect. Dis. *1*, 156–164.

Hancock, R.E. (2005). Mechanisms of action of newer antibiotics for Gram-positive pathogens. Lancet. Infect. Dis. *5*, 209–218.

Hancock, R.E., and Devine, D.R. (2004). Mammalian host defences peptides (Cambridge University Press, Cambridge), pp. 1–4.

Hancock, R.E., and Diamond, G. (2000). The role of cationic antimicrobial peptides in innate host defences. Trends Microbiol. *8*, 402–410.

Hancock, R.E., and Rozek, A. (2002). Role of membranes in the activities of antimicrobial cationic peptides. FEMS Microbiol. Lett. *206*, 143–149.

Hasdemir, U.O., Chevalier, J., Nordmann, P., and Pages, J.M. (2004). Detection and prevalence of active drug efflux mechanism in various multidrug-resistant Klebsiella pneumoniae strains from Turkey. J. Clin. Microbiol. *42*, 2701–2706.

Hasper, H.E., Kramer, N.E., Smith, J.L., Hillman, J.D., Zachariah, C., Kuipers, O.P., de Kruijff, B., and Breukink, E. (2006). An alternative bactericidal mechanism of action for lantibiotic peptides that target lipid II. Science *313*, 1636–1637.

Heath, R.J., Jackowski, S., and Rock, C.O. (1994). Guanosine tetraphosphate inhibition of fatty acid and phospholipid synthesis in *Escherichia coli* is relieved by overexpression of glycerol-3-phosphate acyltransferase (plsB). J. Biol. Chem. *269*, 26584–26590.

Heath, R.J., and Rock, C.O. (2004). Fatty acid biosynthesis as a target for novel antibacterials. Curr. Opin. Investig. Drugs *5*, 146–153.

Hedenstrom, M., Emtenas, H., Pemberton, N., Aberg, V., Hultgren, S.J., Pinkner, J.S., Tegman, V., Almqvist, F., Sethson, I., and Kihlberg, J. (2005). NMR studies of interactions between periplasmic chaperones from uropathogenic *E. coli* and pilicides that interfere with chaperone function and pilus assembly. Org. Biomol. Chem. *3*, 4193–4200.

Heinisch, L., Wittmann, S., Stoiber, T., Berg, A., Ankel-Fuchs, D., and Mollmann, U. (2002). Highly antibacterial active aminoacyl penicillin conjugates with acylated bis-catecholate siderophores based on secondary diamino acids and related compounds. J. Med. Chem. *45*, 3032–3040.

Higgins, M.K., Eswaran, J., Edwards, P., Schertler, G.F., Hughes, C., and Koronakis, V. (2004). Structure of the ligand-blocked periplasmic entrance of the bacterial multidrug efflux protein TolC. J. Mol. Biol. *342*, 697–702.

Hilleringmann, M., Pansegrau, W., Doyle, M., Kaufman, S., MacKichan, M.L., Gianfaldoni, C., Ruggiero, P., and Covacci, A. (2006). Inhibitors of *Helicobacter pylori* ATPase Cagalpha block CagA transport and cag virulence. Microbiology *152*, 2919–2930.

Hilpert, K., Volkmer-Engert, R., Walter, T., and Hancock, R.E. (2005). High-throughput generation of small antibacterial peptides with improved activity. Nat. Biotechnol. *23*, 1008–1012.

Hirst, T.R., and Holmgren, J. (1987). Conformation of protein secreted across bacterial outer membranes: a study of enterotoxin translocation from *Vibrio cholerae*. Proc. Natl. Acad. Sci. U.S.A. *84*, 7418–7422.

Holbrook, I.B., and Molan, P.C. (1973). A further study of the factors enhancing glycolysis in human saliva. Arch. Oral Biol. *18*, 1275–1282.

Horvath, I., Weise, C.F., Andersson, E.K., Chorell, E., Sellstedt, M., Bengtsson, C., Olofsson, A., Hultgren, S.J., Chapman, M., Wolf-Watz, M., *et al.* (2012). Mechanisms of protein oligomerization: inhibitor of functional amyloids templates alpha-synuclein fibrillation. J. Am. Chem. Soc. *134*, 3439–3444.

Hsu, S.T., Breukink, E., Tischenko, E., Lutters, M.A., de Kruijff, B., Kaptein, R., Bonvin, A.M., and van Nuland, N.A. (2004). The nisin-lipid II complex reveals a pyrophosphate cage that provides a blueprint for novel antibiotics. Nat. Struct. Mol. Biol. *11*, 963–967.

Huang, H.W. (2000). Action of antimicrobial peptides: two-state model. Biochemistry *39*, 8347–8352.

Hudson, D.L., Layton, A.N., Field, T.R., Bowen, A.J., Wolf-Watz, H., Elofsson, M., Stevens, M.P., and Galyov, E.E. (2007). Inhibition of type III secretion in *Salmonella enterica* serovar Typhimurium by small-molecule inhibitors. Antimicrob. Agents Chemother. *51*, 2631–2635.

Hung, D.T., Shakhnovich, E.A., Pierson, E., and Mekalanos, J.J. (2005). Small-molecule inhibitor of *Vibrio cholerae* virulence and intestinal colonization. Science *310*, 670–674.

in't Veld, G., Mannion, B., Weiss, J., and Elsbach, P. (1988). Effects of the bactericidal/permeability-increasing protein of polymorphonuclear leukocytes on isolated bacterial cytoplasmic membrane vesicles. Infect. Immun. *56*, 1203–1208.

Inukai, M., Takeuchi, M., Shimizu, K., and Arai, M. (1978). Mechanism of action of globomycin. J. Antibiot. (Tokyo) *31*, 1203–1205.

Jenssen, H., Hamill, P., and Hancock, R.E. (2006). Peptide antimicrobial agents. Clin Microbiol. Rev. *19*, 491–511.

Ji, C., Miller, P.A., and Miller, M.J. (2012). Iron transport-mediated drug delivery: practical syntheses and *in vitro* antibacterial studies of tris-catecholate siderophore-aminopenicillin conjugates reveals selectively potent antipseudomonal activity. J. Am. Chem. Soc. *134*, 9898–9901.

Johnson, T.L., Abendroth, J., Hol, W.G., and Sandkvist, M. (2006). Type II secretion: from structure to function. FEMS Microbiol. Lett. *255*, 175–186.

Jung, D., Rozek, A., Okon, M., and Hancock, R.E. (2004). Structural transitions as determinants of the action of the calcium-dependent antibiotic daptomycin. Chem. Biol. *11*, 949–957.

Kamal, A., Shaik, A.A., Sinha, R., Yadav, J.S., and Arora, S.K. (2005). Antitubercular agents. Part 2: new thiolactomycin analogues active against *Mycobacterium tuberculosis*. Bioorg. Med. Chem. Lett. *15*, 1927–1929.

Kamigiri, K., Suzuki, Y., Shibazaki, M., Morioka, M., Suzuki, K., Tokunaga, T., Setiawan, B., and Rantiatmodjo, R.M. (1996). Kalimantacins A, B and C, novel antibiotics from Alcaligenes sp. YL-02632S. I. Taxonomy, fermentation, isolation and biological properties. J. Antibiot. (Tokyo) *49*, 136–139.

Karlowsky, J.A., Kaplan, N., Hafkin, B., Hoban, D.J., and Zhanel, G.G. (2009). AFN-1252, a FabI inhibitor, demonstrates a Staphylococcus-specific spectrum of activity. Antimicrob. Agents Chemother. *53*, 3544–3548.

Kauppi, A.M., Nordfelth, R., Uvell, H., Wolf-Watz, H., and Elofsson, M. (2003). Targeting bacterial virulence: inhibitors of type III secretion in Yersinia. Chem. Biol. *10*, 241–249.

Keyser, P., Elofsson, M., Rosell, S., and Wolf-Watz, H. (2008). Virulence blockers as alternatives to antibiotics: type III secretion inhibitors against Gram-negative bacteria. J. Intern. Med. *264*, 17–29.

Kiho, T., Nakayama, M., Yasuda, K., Miyakoshi, S., Inukai, M., and Kogen, H. (2004). Structure–activity relationships of globomycin analogues as antibiotics. Bioorg. Med. Chem. *12*, 337–361.

Kim, H., and Farrand, S.K. (1997). Characterization of the acc operon from the nopaline-type Ti plasmid pTiC58, which encodes utilization of agrocinopines A and B and susceptibility to agrocin 84. J. Bacteriol. *179*, 7559–7572.

Kim, P., Zhang, Y.M., Shenoy, G., Nguyen, Q.A., Boshoff, H.I., Manjunatha, U.H., Goodwin, M.B., Lonsdale, J., Price, A.C., Miller, D.J., *et al.* (2006). Structure–activity relationships at the 5-position of thiolactomycin: an intact (5R)-isoprene unit is required for activity against the condensing enzymes from *Mycobacterium tuberculosis* and *Escherichia coli*. J. Med. Chem. *49*, 159–171.

Klein, T., Abgottspon, D., Wittwer, M., Rabbani, S., Herold, J., Jiang, X., Kleeb, S., Luthi, C., Scharenberg, M., Bezencon, J., *et al.* (2010). FimH antagonists for the oral treatment of urinary tract infections: from design and synthesis to *in vitro* and *in vivo* evaluation. J. Med. Chem. *53*, 8627–8641.

Kosciuczuk, E.M., Lisowski, P., Jarczak, J., Strzalkowska, N., Jozwik, A., Horbanczuk, J., Krzyzewski, J., Zwierzchowski, L., and Bagnicka, E. (2012). Cathelicidins: family of antimicrobial peptides. A review. Mol. Biol. Rep. *39*, 10957–10970.

Kremer, L., Douglas, J.D., Baulard, A.R., Morehouse, C., Guy, M.R., Alland, D., Dover, L.G., Lakey, J.H., Jacobs, W.R. Jr., Brennan, P.J., *et al.* (2000). Thiolactomycin and related analogues as novel anti-mycobacterial agents targeting KasA and KasB condensing enzymes in *Mycobacterium tuberculosis*. J. Biol. Chem. *275*, 16857–16864.

Kurth, D.G., Gago, G.M., de la Iglesia, A., Bazet Lyonnet, B., Lin, T.W., Morbidoni, H.R., Tsai, S.C., and Gramajo, H. (2009). ACCase 6 is the essential acetyl-CoA carboxylase involved in fatty acid and mycolic acid biosynthesis in mycobacteria. Microbiology *155*, 2664–2675.

Kwon, Y.J., Fang, Y., Xu, G.H., and Kim, W.G. (2009). Aquastatin A, a new inhibitor of enoyl-acyl carrier protein reductase from Sporothrix sp. FN611. Biol. Pharm. Bull. *32*, 2061–2064.

Kwon, Y.J., Sohn, M.J., Oh, T., Cho, S.N., Kim, C.J., and Kim, W.G. (2013). Panosialins, inhibitors of enoyl-ACP reductase from *Streptomyces* sp. AN1761. J. Microbiol. Biotechnol. *23*, 184–188.

Lai, C.Y., and Cronan, J.E. (2003). Beta-ketoacyl-acyl carrier protein synthase III (FabH) is essential for bacterial fatty acid synthesis. J. Biol. Chem. *278*, 51494–51503.

Lawley, T.D., Klimke, W.A., Gubbins, M.J., and Frost, L.S. (2003). F factor conjugation is a true type IV secretion system. FEMS Microbiol. Lett. *224*, 1–15.

Lehrer, R.I. (2004). Primate defensins. Nat. Rev. Microbiol. *2*, 727–738.

Levy, C.W., Roujeinikova, A., Sedelnikova, S., Baker, P.J., Stuitje, A.R., Slabas, A.R., Rice, D.W., and Rafferty, J.B. (1999). Molecular basis of triclosan activity. Nature *398*, 383–384.

Li, H.Q., Luo, Y., and Zhu, H.L. (2011). Discovery of vinylogous carbamates as a novel class of beta-ketoacyl-acyl carrier protein synthase III (FabH) inhibitors. Bioorg. Med. Chem. *19*, 4454–4459.

Lilic, M., Vujanac, M., and Stebbins, C.E. (2006). A common structural motif in the binding of virulence factors to bacterial secretion chaperones. Mol. Cell *21*, 653–664.

Lo, A., Seers, C., Dashper, S., Butler, C., Walker, G., Walsh, K., Catmull, D., Hoffmann, B., Cleal, S., Lissel, P., *et al.* (2010). FimR and FimS: biofilm formation and gene expression in Porphyromonas gingivalis. J. Bacteriol. *192*, 1332–1343.

Lo, A.W., Moonens, K., and Remaut, H. (2013). Chemical attenuation of pilus function and assembly in Gram-negative bacteria. Curr. Opin. Microbiol. *16*, 85–92.

Lomovskaya, O., Warren, M.S., Lee, A., Galazzo, J., Fronko, R., Lee, M., Blais, J., Cho, D., Chamberland, S., Renau, T., *et al.* (2001). Identification and characterization of inhibitors of multidrug resistance efflux pumps in *Pseudomonas aeruginosa*: novel agents for combination therapy. Antimicrob. Agents Chemother. *45*, 105–116.

Lu, H., and Tonge, P.J. (2008). Inhibitors of FabI, an enzyme drug target in the bacterial fatty acid biosynthesis pathway. Acc. Chem. Res. *41*, 11–20.

Lu, Y.J., Zhang, Y.M., Grimes, K.D., Qi, J., Lee, R.E., and Rock, C.O. (2006). Acyl-phosphates initiate membrane phospholipid synthesis in Gram-positive pathogens. Mol. Cell *23*, 765–772.

Lutticken, D., Segers, R.P., and Visser, N. (2007). Veterinary vaccines for public health and prevention of viral and bacterial zoonotic diseases. Rev. Sci. Tech. *26*, 165–177.

Lyon, G.J., Mayville, P., Muir, T.W., and Novick, R.P. (2000). Rational design of a global inhibitor of the virulence response in *Staphylococcus aureus,* based in part on localization of the site of inhibition to the receptor-histidine kinase, AgrC. Proc. Natl. Acad. Sci. U.S.A. *97*, 13330–13335.

MacKay, B.J., Denepitiya, L., Iacono, V.J., Krost, S.B., and Pollock, J.J. (1984a). Growth-inhibitory and bactericidal effects of human parotid salivary histidine-rich polypeptides on Streptococcus mutans. Infect. Immun. *44*, 695–701.

MacKay, B.J., Pollock, J.J., Iacono, V.J., and Baum, B.J. (1984b). Isolation of milligram quantities of a group of histidine-rich polypeptides from human parotid saliva. Infect. Immun. *44*, 688–694.

Markman, M. (2007). The objectivity of reports describing the use of novel diagnostic tests in the management of ovarian cancer: a critique. Gynecol. Oncol. *104*, 271–272.

Marrakchi, H., Dewolf, W.E. Jr., Quinn, C., West, J., Polizzi, B.J., So, C.Y., Holmes, D.J., Reed, S.L., Heath, R.J., Payne, D.J., *et al.* (2003). Characterization of *Streptococcus pneumoniae* enoyl-(acyl-carrier protein) reductase (FabK). Biochem. J. *370*, 1055–1062.

Mascher, T., Helmann, J.D., and Unden, G. (2006). Stimulus perception in bacterial signal-transducing histidine kinases. Microbiol. Mol. Biol. Rev. *70*, 910–938.

Matsuzaki, K. (1999). Why and how are peptide–lipid interactions utilized for self-defense? Magainins and tachyplesins as archetypes. Biochim. Biophys. Acta *1462*, 1–10.

Matsuzaki, K., Murase, O., and Miyajima, K. (1995). Kinetics of pore formation by an antimicrobial peptide, magainin 2, in phospholipid bilayers. Biochemistry *34*, 12553–12559.

Mattheus, W., Masschelein, J., Gao, L.J., Herdewijn, P., Landuyt, B., Volckaert, G., and Lavigne, R. (2010). The kalimantacin/batumin biosynthesis operon encodes a self-resistance isoform of the FabI bacterial target. Chem. Biol. *17*, 1067–1071.

Mayville, P., Ji, G., Beavis, R., Yang, H., Goger, M., Novick, R.P., and Muir, T.W. (1999). Structure-activity analysis of synthetic autoinducing thiolactone peptides from *Staphylococcus aureus* responsible for virulence. Proc. Natl. Acad. Sci. U.S.A. *96*, 1218–1223.

Messoudi, A., Fnini, S., Labsaili, N., Ghrib, S., Rafai, M., and Largab, A. (2007). Giant cell tumors of the tendon sheath of the hand: 32 cases. Chir. Main. *26*, 165–169.

Mies, K.A., Gebhardt, P., Mollmann, U., and Crumbliss, A.L. (2008). Synthesis, siderophore activity and iron(III) chelation chemistry of a novel mono-hydroxamate, bis-catecholate siderophore mimic: N(alpha),-N(epsilon)-Bis[2,3-dihydroxybenzoyl]-l-lysyl-(gamma-N-methyl-N-hydroxya mido)-L-glutamic acid. J. Inorg. Biochem. *102*, 850–861.

Miller, S.I., Kukral, A.M., and Mekalanos, J.J. (1989). A two-component regulatory system (phoP phoQ) controls Salmonella typhimurium virulence. Proc. Natl. Acad. Sci. U.S.A. *86*, 5054–5058.

Mitrophanov, A.Y., and Groisman, E.A. (2008). Signal integration in bacterial two-component regulatory systems. Genes Dev. *22*, 2601–2611.

Mollmann, U., Heinisch, L., Bauernfeind, A., Kohler, T., and Ankel-Fuchs, D. (2009). Siderophores as drug delivery agents: application of the 'Trojan Horse' strategy. Biometals *22*, 615–624.

Moraes, T.F., Spreter, T., and Strynadka, N.C. (2008). Piecing together the type III injectisome of bacterial pathogens. Curr. Opin. Struct. Biol. *18*, 258–266.

Moreira, C.G., Weinshenker, D., and Sperandio, V. (2010). QseC mediates *Salmonella enterica* serovar typhimurium virulence *in vitro* and *in vivo.* Infect. Immun. *78*, 914–926.

Moss, J.E., Fisher, P.E., Vick, B., Groisman, E.A., and Zychlinsky, A. (2000). The regulatory protein PhoP controls susceptibility to the host inflammatory response in *Shigella flexneri*. Cell Microbiol. *2*, 443–452.

Mota, L.J., and Cornelis, G.R. (2005). The bacterial injection kit: type III secretion systems. Ann. Med. *37*, 234–249.

Mulvey, M.A., Lopez-Boado, Y.S., Wilson, C.L., Roth, R., Parks, W.C., Heuser, J., and Hultgren, S.J. (1998). Induction and evasion of host defenses by type 1-piliated uropathogenic *Escherichia coli*. Science *282*, 1494–1497.

Murakami, M., Nakatani, Y., Atsumi, G., Inoue, K., and Kudo, I. (1997). Regulatory functions of phospholipase A2. Crit. Rev. Immunol. *17*, 225–283.

Muschiol, S., Bailey, L., Gylfe, A., Sundin, C., Hultenby, K., Bergstrom, S., Elofsson, M., Wolf-Watz, H., Normark, S., and Henriques-Normark, B. (2006). A small-molecule inhibitor of type III secretion inhibits different stages of the infectious cycle of Chlamydia trachomatis. Proc. Natl. Acad. Sci. U.S.A. *103*, 14566–14571.

Muschiol, S., Normark, S., Henriques-Normark, B., and Subtil, A. (2009). Small molecule inhibitors of the Yersinia type III secretion system impair the development of Chlamydia after entry into host cells. BMC Microbiol. *9*, 75.

Nation, R.L., and Li, J. (2009). Colistin in the 21st century. Curr. Opin. Infect. Dis. *22*, 535–543.

Needham, J., Kelly, M.T., Ishige, M., and Andersen, R.J. (1994). Andrimin and moiramides A-C metabolites produced in culture by a marine isolate of the bacterium *Pseudomonas fluorescens*: structure elucidation and biosynthesis. J. Org. Chem. *59*, 2058–2063.

Ng, W.L., Perez, L., Cong, J., Semmelhack, M.F., and Bassler, B.L. (2012). Broad spectrum pro-quorum-sensing molecules as inhibitors of virulence in vibrios. PLoS Pathog. *8*, e1002767.

Ninfa, A.J., and Magasanik, B. (1986). Covalent modification of the glnG product, NRI, by the glnL product, NRII, regulates the transcription of the glnALG operon in *Escherichia coli*. Proc. Natl. Acad. Sci. U.S.A. *83*, 5909–5913.

Nishie, M., Nagao, J., and Sonomoto, K. (2012). Antibacterial peptides 'bacteriocins': an overview of their diverse characteristics and applications. Biocontrol Sci. *17*, 1–16.

Njoroge, J., and Sperandio, V. (2012). Enterohemorrhagic *Escherichia coli* virulence regulation by the two bacterial adrenergic kinases QseC and QseE. Infect. Immun. *80*, 688–703.

Nordfelth, R., Kauppi, A.M., Norberg, H.A., Wolf-Watz, H., and Elofsson, M. (2005). Small-molecule inhibitors specifically targeting type III secretion. Infect. Immun. *73*, 3104–3114.

Noto, T., Miyakawa, S., Oishi, H., Endo, H., and Okazaki, H. (1982). Thiolactomycin, a new antibiotic. III. *In vitro* antibacterial activity. J. Antibiot. (Tokyo) *35*, 401–410.

Nuccio, S.P., and Baumler, A.J. (2007). Evolution of the chaperone/usher assembly pathway: fimbrial classification goes Greek. Microbiol. Mol. Biol. Rev. *71*, 551–575.

Nudleman, E., and Kaiser, D. (2004). Pulling together with type IV pili. J. Mol. Microbiol. Biotechnol. *7*, 52–62.

Ojha, A., Anand, M., Bhatt, A., Kremer, L., Jacobs, W.R. Jr., and Hatfull, G.F. (2005). GroEL1: a dedicated chaperone involved in mycolic acid biosynthesis during biofilm formation in mycobacteria. Cell *123*, 861–873.

Okuda, S., and Tokuda, H. (2011). Lipoprotein sorting in bacteria. Annu. Rev. Microbiol. *65*, 239–259.

Oppenheim, F.G., Xu, T., McMillian, F.M., Levitz, S.M., Diamond, R.D., Offner, G.D., and Troxler, R.F. (1988). Histatins, a novel family of histidine-rich proteins in human parotid secretion. Isolation, characterization, primary structure, and fungistatic effects on Candida albicans. J. Biol. Chem. *263*, 7472–7477.

Oyston, P.C., Dorrell, N., Williams, K., Li, S.R., Green, M., Titball, R.W., and Wren, B.W. (2000). The response regulator PhoP is important for survival under conditions of macrophage-induced stress and virulence in *Yersinia pestis*. Infect. Immun. *68*, 3419–3425.

Park, J., Jagasia, R., Kaufmann, G.F., Mathison, J.C., Ruiz, D.I., Moss, J.A., Meijler, M.M., Ulevitch, R.J., and Janda, K.D. (2007). Infection control by antibody disruption of bacterial quorum sensing signaling. Chem. Biol. *14*, 1119–1127.

Paschos, A., den Hartigh, A., Smith, M.A., Atluri, V.L., Sivanesan, D., Tsolis, R.M., and Baron, C. (2011). An *in vivo* high-throughput screening approach targeting the type IV secretion system component VirB8 identified inhibitors of *Brucella abortus* 2308 proliferation. Infect. Immun. *79*, 1033–1043.

Pathania, R., Zlitni, S., Barker, C., Das, R., Gerritsma, D.A., Lebert, J., Awuah, E., Melacini, G., Capretta, F.A., and Brown, E.D. (2009). Chemical genomics in *Escherichia coli* identifies an inhibitor of bacterial lipoprotein targeting. Nat. Chem. Biol. *5*, 849–856.

Perencevich, E.N., Wong, M.T., and Harris, A.D. (2001). National and regional assessment of the antibacterial soap market: a step toward determining the impact of prevalent antibacterial soaps. Am. J. Infect. Control *29*, 281–283.

Peschel, A., Jack, R.W., Otto, M., Collins, L.V., Staubitz, P., Nicholson, G., Kalbacher, H., Nieuwenhuizen, W.F., Jung, G., Tarkowski, A., *et al.* (2001). *Staphylococcus aureus* resistance to human defensins and evasion of neutrophil killing via the novel virulence factor MprF is based on modification of membrane lipids with l-lysine. J. Exp. Med. *193*, 1067–1076.

Peterson, M.M., Mack, J.L., Hall, P.R., Alsup, A.A., Alexander, S.M., Sully, E.K., Sawires, Y.S., Cheung, A.L., Otto, M., and Gresham, H.D. (2008). Apolipoprotein B Is an innate barrier against invasive *Staphylococcus aureus* infection. Cell Host Microbe *4*, 555–566.

Piddock, L.J. (2006). Multidrug-resistance efflux pumps – not just for resistance. Nat. Rev. Microbiol. *4*, 629–636.

Piddock, L.J., Garvey, M.I., Rahman, M.M., and Gibbons, S. (2010). Natural and synthetic compounds such as trimethoprim behave as inhibitors of efflux in Gram-negative bacteria. J. Antimicrob. Chemother. *65*, 1215–1223.

Pinkner, J.S., Remaut, H., Buelens, F., Miller, E., Aberg, V., Pemberton, N., Hedenstrom, M., Larsson, A., Seed, P., Waksman, G., *et al.* (2006). Rationally designed small compounds inhibit pilus biogenesis in uropathogenic bacteria. Proc. Natl. Acad. Sci. U.S.A. *103*, 17897–17902.

Planet, P.J., Kachlany, S.C., DeSalle, R., and Figurski, D.H. (2001). Phylogeny of genes for secretion NTPases: identification of the widespread tadA subfamily and development of a diagnostic key for gene classification. Proc. Natl. Acad. Sci. U.S.A. *98*, 2503–2508.

Polyak, S.W., Abell, A.D., Wilce, M.C., Zhang, L., and Booker, G.W. (2012). Structure, function and selective inhibition of bacterial acetyl-coa carboxylase. Appl. Microbiol. Biotechnol. *93*, 983–992.

Poole, K. (2005). Efflux-mediated antimicrobial resistance. J. Antimicrob. Chemother. *56*, 20–51.

Porter, E.A., Wang, X., Lee, H.S., Weisblum, B., and Gellman, S.H. (2000). Non-haemolytic beta-amino-acid oligomers. Nature *404*, 565.

Pouny, Y., Rapaport, D., Mor, A., Nicolas, P., and Shai, Y. (1992). Interaction of antimicrobial dermaseptin and its fluorescently labeled analogues with phospholipid membranes. Biochemistry *31*, 12416–12423.

Pramanik, A., and Braun, V. (2006). Albomycin uptake via a ferric hydroxamate transport system of *Streptococcus pneumoniae* R6. J. Bacteriol. *188*, 3878–3886.

Pramanik, A., Stroeher, U.H., Krejci, J., Standish, A.J., Bohn, E., Paton, J.C., Autenrieth, I.B., and Braun, V. (2007). Albomycin is an effective antibiotic, as exemplified with Yersinia enterocolitica and *Streptococcus pneumoniae*. Int. J. Med. Microbiol. *297*, 459–469.

Prost, L.R., Daley, M.E., Le Sage, V., Bader, M.W., Le Moual, H., Klevit, R.E., and Miller, S.I. (2007). Activation of the bacterial sensor kinase PhoQ by acidic pH. Mol. Cell *26*, 165–174.

Pugsley, A.P., Zimmerman, W., and Wehrli, W. (1987). Highly efficient uptake of a rifamycin derivative via the FhuA-TonB-dependent uptake route in *Escherichia coli*. J. Gen. Microbiol. *133*, 3505–3511.

Raetz, C.R., Reynolds, C.M., Trent, M.S., and Bishop, R.E. (2007). Lipid A modification systems in gram-negative bacteria. Annu. Rev. Biochem. *76*, 295–329.

Rasko, D.A., and Sperandio, V. (2010). Anti-virulence strategies to combat bacteria-mediated disease. Nat. Rev. Drug Discov. *9*, 117–128.

Rasko, D.A., Moreira, C.G., Li de, R., Reading, N.C., Ritchie, J.M., Waldor, M.K., Williams, N., Taussig, R., Wei, S., Roth, M., *et al.* (2008). Targeting QseC signaling and virulence for antibiotic development. Science *321*, 1078–1080.

Reader, J.S., Ordoukhanian, P.T., Kim, J.G., de Crecy-Lagard, V., Hwang, I., Farrand, S., and Schimmel, P. (2005). Major biocontrol of plant tumors targets tRNA synthetase. Science *309*, 1533.

Rego, A.T., Chandran, V., and Waksman, G. (2010). Two-step and one-step secretion mechanisms in Gram-negative bacteria: contrasting the type IV secretion system and the chaperone-usher pathway of pilus biogenesis. Biochem. J. *425*, 475–488.

Renau, T.E., Leger, R., Flamme, E.M., Sangalang, J., She, M.W., Yen, R., Gannon, C.L., Griffith, D., Chamberland, S., Lomovskaya, O., *et al.* (1999). Inhibitors of efflux pumps in *Pseudomonas aeruginosa* potentiate the activity of the fluoroquinolone antibacterial levofloxacin. J. Med. Chem. *42*, 4928–4931.

Roberts, J.A., Marklund, B.I., Ilver, D., Haslam, D., Kaack, M.B., Baskin, G., Louis, M., Mollby, R., Winberg, J., and Normark, S. (1994). The Gal(alpha 1–4)Gal-specific tip adhesin of *Escherichia coli* P-fimbriae is needed for pyelonephritis to occur in the normal urinary tract. Proc. Natl. Acad. Sci. U.S.A. *91*, 11889–11893.

Robinson, J.A., Shankaramma, S.C., Jetter, P., Kienzl, U., Schwendener, R.A., Vrijbloed, J.W., and Obrecht, D. (2005). Properties and structure–activity studies of cyclic beta-hairpin peptidomimetics based on the cationic antimicrobial peptide protegrin I. Bioorg. Med. Chem. *13*, 2055–2064.

Roosenberg, J.M. 2nd, Lin, Y.M., Lu, Y., and Miller, M.J. (2000). Studies and syntheses of siderophores, microbial iron chelators, and analogs as potential drug delivery agents. Curr. Med. Chem. *7*, 159–197.

Rosqvist, R., Hakansson, S., Forsberg, A., and Wolf-Watz, H. (1995). Functional conservation of the secretion and translocation machinery for virulence proteins of yersiniae, salmonellae and shigellae. EMBO J. *14*, 4187–4195.

Rozwarski, D.A., Grant, G.A., Barton, D.H., Jacobs, W.R. Jr., and Sacchettini, J.C. (1998). Modification of the NADH of the isoniazid target (InhA) from *Mycobacterium tuberculosis*. Science *279*, 98–102.

Russo, T.A., Page, M.G., Beanan, J.M., Olson, R., Hujer, A.M., Hujer, K.M., Jacobs, M., Bajaksouzian, S., Endimiani, A., and Bonomo, R.A. (2011). *In vivo* and *in vitro* activity of the siderophore monosulfactam BAL30072 against Acinetobacter baumannii. J. Antimicrob. Chemother. *66*, 867–873.

Sahl, H.G., Jack, R.W., and Bierbaum, G. (1995). Biosynthesis and biological activities of lantibiotics with unique post-translational modifications. Eur. J. Biochem. *230*, 827–853.

Sauer, F.G., Remaut, H., Hultgren, S.J., and Waksman, G. (2004). Fiber assembly by the chaperone-usher pathway. Biochim. Biophys. Acta *1694*, 259–267.

Sauvonnet, N., Gounon, P., and Pugsley, A.P. (2000). PpdD type IV pilin of *Escherichia coli* K-12 can Be assembled into pili in *Pseudomonas aeruginosa*. J. Bacteriol. *182*, 848–854.

Schaeffer, M.L., Agnihotri, G., Volker, C., Kallender, H., Brennan, P.J., and Lonsdale, J.T. (2001). Purification and biochemical characterization of the *Mycobacterium tuberculosis* beta-ketoacyl-acyl carrier protein synthases KasA and KasB. J. Biol. Chem. *276*, 47029–47037.

Schnabelrauch, M., Wittmann, S., Rahn, K., Mollmann, U., Reissbrodt, R., and Heinisch, L. (2000). New synthetic catecholate-type siderophores based on amino acids and dipeptides. Biometals *13*, 333–348.

Schumacher, A., Steinke, P., Bohnert, J.A., Akova, M., Jonas, D., and Kern, W.V. (2006). Effect of 1-(1-naphthylmethyl)-piperazine, a novel putative efflux pump inhibitor, on antimicrobial drug susceptibility in clinical isolates of Enterobacteriaceae other than *Escherichia coli*. J. Antimicrob. Chemother. *57*, 344–348.

Scott, D.L., Mandel, A.M., Sigler, P.B., and Honig, B. (1994). The electrostatic basis for the interfacial binding of secretory phospholipases A2. Biophys. J. *67*, 493–504.

Senior, S.J., Illarionov, P.A., Gurcha, S.S., Campbell, I.B., Schaeffer, M.L., Minnikin, D.E., and Besra, G.S. (2003). Biphenyl-based analogues of thiolactomycin, active against *Mycobacterium tuberculosis* mtFabH fatty acid condensing enzyme. Bioorg. Med. Chem. Lett. *13*, 3685–3688.

Senior, S.J., Illarionov, P.A., Gurcha, S.S., Campbell, I.B., Schaeffer, M.L., Minnikin, D.E., and Besra, G.S. (2004). Acetylene-based analogues of thiolactomycin, active against *Mycobacterium tuberculosis* mtFabH fatty acid condensing enzyme. Bioorg. Med. Chem. Lett. *14*, 373–376.

Shakhnovich, E.A., Hung, D.T., Pierson, E., Lee, K., and Mekalanos, J.J. (2007). Virstatin inhibits dimerization of the transcriptional activator ToxT. Proc. Natl. Acad. Sci. U.S.A. *104*, 2372–2377.

Shankaramma, S.C., Athanassiou, Z., Zerbe, O., Moehle, K., Mouton, C., Bernardini, F., Vrijbloed, J.W., Obrecht, D., and Robinson, J.A. (2002). Macrocyclic hairpin mimetics of the cationic antimicrobial peptide protegrin I: a new family of broad-spectrum antibiotics. Chembiochem. *3*, 1126–1133.

Shi, R., and Sugawara, I. (2010). Development of new anti-tuberculosis drug candidates. Tohoku J. Exp. Med. *221*, 97–106.

Silverman, J.A., Perlmutter, N.G., and Shapiro, H.M. (2003). Correlation of daptomycin bactericidal activity and membrane depolarization in *Staphylococcus aureus*. Antimicrob. Agents Chemother. *47*, 2538–2544.

Simmons, K.J., Chopra, I., and Fishwick, C.W. (2010). Structure-based discovery of antibacterial drugs. Nat. Rev. Microbiol. *8*, 501–510.

Smirnov, V.V., Churkina, L.N., Perepnikhatka, V.I., Mukvich, N.S., Garagulia, A.D., Kiprianova, E.A., Kravets, A.N., and Dovzhenko, S.A. (2000). Isolation of highly active strain producing the antistaphylococcal antibiotic batumin. Prikl. Biokhim. Mikrobiol. *36*, 55–58.

Sorensen, O.E., and Borregaard, N. (2005). Cathelicidins – nature's attempt at combinatorial chemistry. Comb. Chem. High Throughput Screen. *8*, 273–280.

Srinivas, N., Jetter, P., Ueberbacher, B.J., Werneburg, M., Zerbe, K., Steinmann, J., Van der Meijden, B., Bernardini, F., Lederer, A., Dias, R.L., *et al.* (2010). Peptidomimetic antibiotics target outer-membrane biogenesis in *Pseudomonas aeruginosa*. Science *327*, 1010–1013.

Stock, A.M., Robinson, V.L., and Goudreau, P.N. (2000). Two-component signal transduction. Annu. Rev. Biochem. *69*, 183–215.

Straus, S.K., and Hancock, R.E. (2006). Mode of action of the new antibiotic for Gram-positive pathogens daptomycin: comparison with cationic antimicrobial peptides and lipopeptides. Biochim. Biophys. Acta *1758*, 1215–1223.

Svensson, A., Larsson, A., Emtenas, H., Hedenstrom, M., Fex, T., Hultgren, S.J., Pinkner, J.S., Almqvist, F., and Kihlberg, J. (2001). Design and evaluation of pilicides: potential novel antibacterial agents directed against uropathogenic *Escherichia coli*. Chembiochem. *2*, 915–918.

Swietnicki, W., Carmany, D., Retford, M., Guelta, M., Dorsey, R., Bozue, J., Lee, M.S., and Olson, M.A. (2011). Identification of small-molecule inhibitors of *Yersinia pestis* Type III secretion system YscN ATPase. PLoS One *6*, e19716.

Takayama, K., Wang, C., and Besra, G.S. (2005). Pathway to synthesis and processing of mycolic acids in *Mycobacterium tuberculosis*. Clin Microbiol. Rev. *18*, 81–101.

Taubes, G. (2008). The bacteria fight back. Science *321*, 356–361.

Tauschek, M., Gorrell, R.J., Strugnell, R.A., and Robins-Browne, R.M. (2002). Identification of a protein secretory pathway for the secretion of heat-labile enterotoxin by an enterotoxigenic strain of *Escherichia coli*. Proc. Natl. Acad. Sci. U.S.A. *99*, 7066–7071.

Tokuda, H. (2009). Biogenesis of outer membranes in Gram-negative bacteria. Biosci. Biotechnol. Biochem. *73*, 465–473.

Tokuda, H., and Matsuyama, S. (2004). Sorting of lipoproteins to the outer membrane in *E. coli*. Biochim. Biophys. Acta *1693*, 5–13.

Tossi, A., and Sandri, L. (2002). Molecular diversity in gene-encoded, cationic antimicrobial polypeptides. Curr. Pharm. Des. *8*, 743–761.

Tsai, H., and Bobek, L.A. (1997). Human salivary histatin-5 exerts potent fungicidal activity against Cryptococcus neoformans. Biochim. Biophys. Acta *1336*, 367–369.

Tseng, T.T., Tyler, B.M., and Setubal, J.C. (2009). Protein secretion systems in bacterial–host associations, and their description in the Gene Ontology. BMC Microbiol. *9* (Suppl 1), S2.

Turner, J., Cho, Y., Dinh, N.N., Waring, A.J., and Lehrer, R.I. (1998). Activities of LL-37, a cathelin-associated antimicrobial peptide of human neutrophils. Antimicrob. Agents Chemother. *42*, 2206–2214.

Vaara, M. (2010). Polymyxins and their novel derivatives. Curr. Opin. Microbiol. *13*, 574–581.

Vaara, M., Fox, J., Loidl, G., Siikanen, O., Apajalahti, J., Hansen, F., Frimodt-Moller, N., Nagai, J., Takano, M., and Vaara, T. (2008). Novel polymyxin derivatives carrying only three positive charges are effective antibacterial agents. Antimicrob. Agents Chemother. *52*, 3229–3236.

Vadyvaloo, V., Arous, S., Gravesen, A., Hechard, Y., Chauhan-Haubrock, R., Hastings, J.W., and Rautenbach, M. (2004). Cell-surface alterations in class IIa bacteriocin-resistant *Listeria monocytogenes* strains. Microbiology *150*, 3025–3033.

Veenendaal, A.K., Sundin, C., and Blocker, A.J. (2009). Small-molecule type III secretion system inhibitors block assembly of the Shigella type III secreton. J. Bacteriol. *191*, 563–570.

Venema, R., Tjalsma, H., van Dijl, J.M., de Jong, A., Leenhouts, K., Buist, G., and Venema, G. (2003). Active lipoprotein precursors in the Gram-positive eubacterium *Lactococcus lactis*. J. Biol. Chem. *278*, 14739–14746.

Verkleij, A.J., Zwaal, R.F., Roelofsen, B., Comfurius, P., Kastelijn, D., and van Deenen, L.L. (1973). The asymmetric distribution of phospholipids in the human red cell membrane. A combined study using phospholipases and freeze-etch electron microscopy. Biochim. Biophys. Acta *323*, 178–193.

Vértesy, L., Arentz, W., Fehlhaber, H.W., and Kogler, H. (1995). Salmycin A-D, Antibiotika aus *Streptomyces violaceus*, DSM8286, mit Siderophor-Aminoglycosid-Struktur. Helv. Chim. Acta *78*, 46–60.

Vignon, G., Kohler, R., Larquet, E., Giroux, S., Prevost, M.C., Roux, P., and Pugsley, A.P. (2003). Type IV-like pili formed by the type II secreton: specificity, composition, bundling, polar localization, and surface presentation of peptides. J. Bacteriol. *185*, 3416–3428.

von Heijne, G. (2007). The membrane protein universe: what's out there and why bother? J. Intern. Med. *261*, 543–557.

Vuong, C., Saenz, H.L., Gotz, F., and Otto, M. (2000). Impact of the agr quorum-sensing system on adherence to polystyrene in *Staphylococcus aureus*. J. Infect. Dis. *182*, 1688–1693.

Wagner, W., Vogel, M., and Goebel, W. (1983). Transport of hemolysin across the outer membrane of *Escherichia coli* requires two functions. J. Bacteriol. *154*, 200–210.

Waksman, G., and Hultgren, S.J. (2009). Structural biology of the chaperone-usher pathway of pilus biogenesis. Nat. Rev. Microbiol. *7*, 765–774.

Wandersman, C., and Delepelaire, P. (1990). TolC, an *Escherichia coli* outer membrane protein required for hemolysin secretion. Proc. Natl. Acad. Sci. U.S.A. *87*, 4776–4780.

Wang, J., Soisson, S.M., Young, K., Shoop, W., Kodali, S., Galgoci, A., Painter, R., Parthasarathy, G., Tang, Y.S., Cummings, R., *et al.* (2006). Platensimycin is a selective FabF inhibitor with potent antibiotic properties. Nature *441*, 358–361.

Weiss, D.S., Brotcke, A., Henry, T., Margolis, J.J., Chan, K., and Monack, D.M. (2007). *In vivo* negative selection screen identifies genes required for Francisella virulence. Proc. Natl. Acad. Sci. U.S.A. *104*, 6037–6042.

Werner, G., Strommenger, B., and Witte, W. (2008). Acquired vancomycin resistance in clinically relevant pathogens. Future Microbiol. *3*, 547–562.

White, S.H., Wimley, W.C., and Selsted, M.E. (1995). Structure, function, and membrane integration of defensins. Curr. Opin. Struct. Biol. *5*, 521–527.

White, S.W., Zheng, J., Zhang, Y.M., and Rock (2005). The structural biology of type II fatty acid biosynthesis. Annu. Rev. Biochem. *74*, 791–831.

Wick, M.J., Hamood, A.N., and Iglewski, B.H. (1990). Analysis of the structure–function relationship of *Pseudomonas aeruginosa* exotoxin A. Mol. Microbiol. *4*, 527–535.

Willey, J.M., and van der Donk, W.A. (2007). Lantibiotics: peptides of diverse structure and function. Annu. Rev. Microbiol. *61*, 477–501.

Wittmann, S., Schnabelrauch, M., Scherlitz-Hofmann, I., Mollmann, U., Ankel-Fuchs, D., and Heinisch, L. (2002). New synthetic siderophores and their beta-lactam conjugates based on diamino acids and dipeptides. Bioorg. Med. Chem. *10*, 1659–1670.

Yeaman, M.R., and Yount, N.Y. (2003). Mechanisms of antimicrobial peptide action and resistance. Pharmacol. Rev. *55*, 27–55.

Young, K., Jayasuriya, H., Ondeyka, J.G., Herath, K., Zhang, C., Kodali, S., Galgoci, A., Painter, R., Brown-Driver, V., Yamamoto, R., *et al.* (2006). Discovery of FabH/FabF inhibitors from natural products. Antimicrob. Agents Chemother. *50*, 519–526.

Zahner, H., Diddens, H., Keller-Schierlein, W., and Nageli, H.-U. (1977). Some experiments with semisynthetic sideromycins. Jpn. J. Antibiot. *30*, S201.

Zav'yalov, V., Zavialov, A., Zav'yalova, G., and Korpela, T. (2009). Adhesive organelles of Gram-negative pathogens assembled with the classical chaperone/usher machinery: structure and function from a clinical standpoint. FEMS Microbiol. Rev. *34*, 317–378.

Zhang, H.J., Zhu, D.D., Li, Z.L., Sun, J., and Zhu, H.L. (2011). Synthesis, molecular modeling and biological evaluation of beta-ketoacyl-acyl carrier protein synthase III (FabH) as novel antibacterial agents. Bioorg. Med. Chem. *19*, 4513–4519.

Zhang, Y.M., and Rock, C.O. (2008). Membrane lipid homeostasis in bacteria. Nat. Rev. Microbiol. *6*, 222–233.

Zheng, C., Han, L., Yap, C.W., Xie, B., and Chen, Y. (2006). Progress and problems in the exploration of therapeutic targets. Drug Discov. Today *11*, 412–420.

Zheng, C.J., Sohn, M.J., and Kim, W.G. (2009). Vinaxanthone, a new FabI inhibitor from Penicillium sp. J. Antimicrob. Chemother. *63*, 949–953.

Zheng, C.J., Sohn, M.J., Lee, S., Hong, Y.S., Kwak, J.H., and Kim, W.G. (2007). Cephalochromin, a FabI-directed antibacterial of microbial origin. Biochem. Biophys. Res. Commun. *362*, 1107–1112.

Zhou, Y., Du, Q.R., Sun, J., Li, J.R., Fang, F., Li, D.D., Qian, Y., Gong, H.B., Zhao, J., and Zhu, H.L. (2013). Novel Schiff-base-derived FabH inhibitors with dioxygenated rings as antibiotic agents. ChemMedChem *8*, 433–441.

Index

Notes: As the subject of this book is bacterial membranes, entries have been kept to a minimum under this term. Users are advised to look under more specific terms. Page references in *italics* refer to figures and those in **bold** to tables. Abbreviations used are as follows: LPS, lipopolysaccharide; OMP, outer membrane protein; SRP, signal recognition pathway.

M

N

O

P

T

V

W

Y

Z

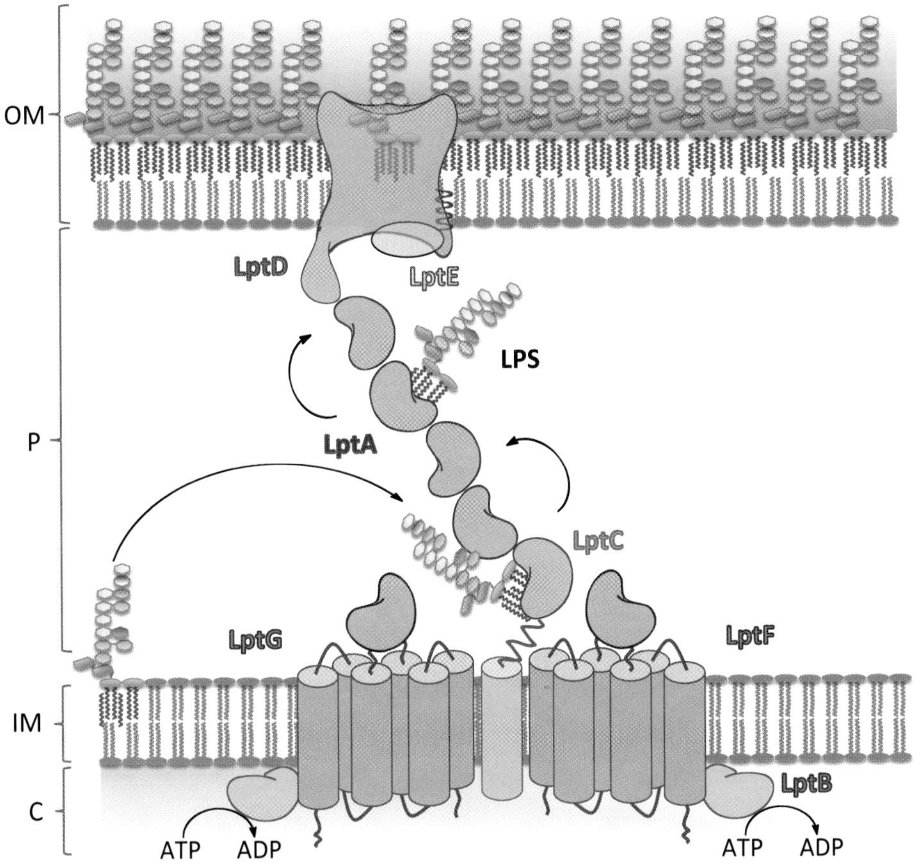

Figure 2.5 LPS transport through the cell envelope. The flipped LPS molecule is extracted from the IM by the ABC transporter LptBCFG. According to the 'trans–envelope complex' model LptA, LptE and LptD constitute a multiprotein complex with LptBCFG which spans the cell envelope by bridging IM and OM components (see text for details). The IM and OM components are indicated as in Figure 2.2. The name of the specific Lpt components is indicated; the bean shape indicates the OstA_N domains present in different proteins of the complex (see Figure 2.6 and text for details).

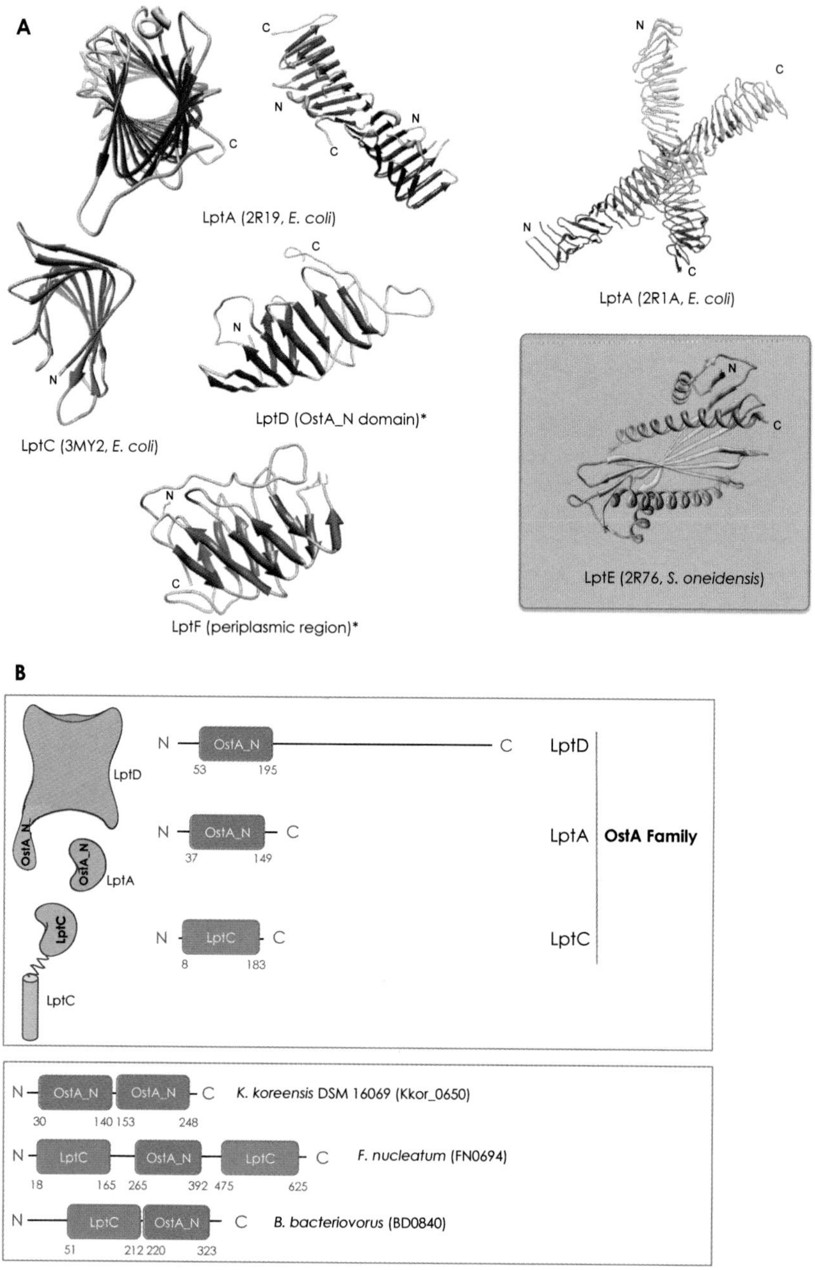

Figure 2.6 Crystal structures and structure prediction of Lpt proteins. (A) The crystal structure of LptA, LptC and LptE are reported together with the respective PDB codes. LptA structures obtained in the presence (PDB: 2R1A) or in the absence of LPS (PDB: 2R19) are shown. The structure of LptE has been solved from three different *E. coli* orthologues, only the structure of LptE from *Shewanella oneidensis* is shown. Structure predictions are marked with an asterisk: OstA_N domain of *E. coli* LptD and periplasmic region of *E. coli* LptF. (B) Upper panel: conserved functional domains in Lpt proteins. The OstA domain family includes the OstA_N signature in both LptD and LptA and LptC signature in LptC. Lower panel: domains organization of PF03968 (OstA_N) and PF036835 (LptC) in selected genomes.

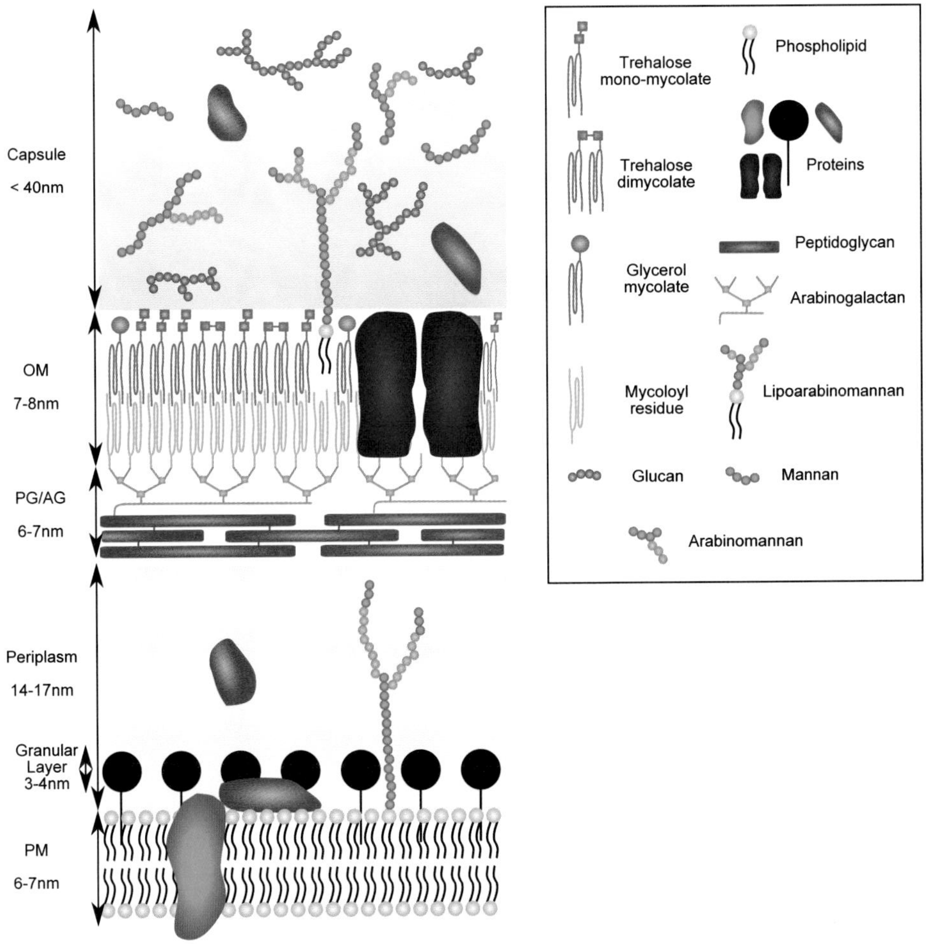

Figure 5.2 Model of the cell envelope and the capsule of mycobacteria. The cell envelope consists of (i) the plasma membrane, (ii) the periplasmic space, which contains the granular layer, (iii) the cell wall skeleton, which is made of peptidoglycan, arabinogalactan and mycoloyl residues, and (iv) the outer membrane.

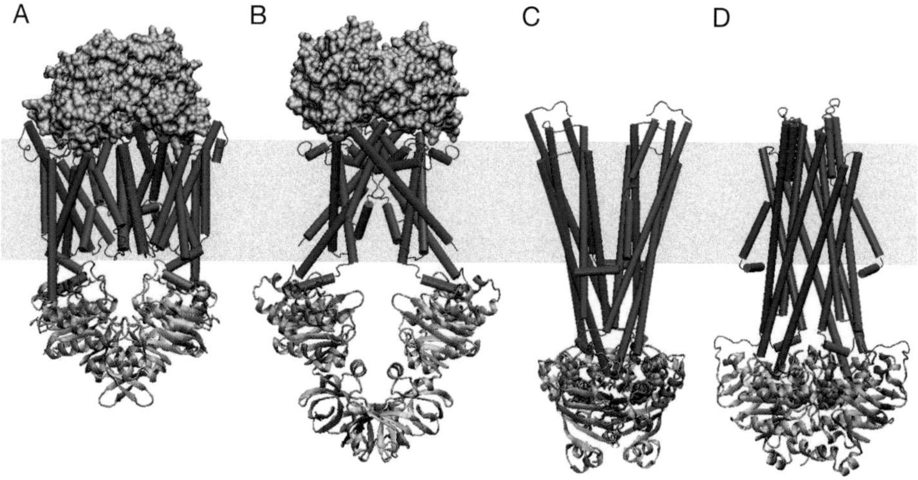

Figure 7.1 Prokaryotic ABC transporter architectures. The TMDs (blue and purple) of the binding-protein dependent ABC importers have two general conformation types: (A) that of the densely packed 10-TM helix HI1470/1 (2NQ2.pdb), with only subtle tilt changes in the central pore lining helices (red) throughout the transport cycle; and (B) the inverted tepee conformation of the *M. acetivorans* ModBC (with BP docked) (2ONK.pdb), in which the TMDs undergo large-scale structural rearrangements induced by NBD (orange and gold) dimerization and separation during the transport cycle. The two ABC exporters crystallized, Sav1866 and murine MDR1A, depict TMD helical domain swapping and the same NBD:TMD interface, but differ in the conformation of the TMDs, with the nucleotide-free MDR1A adopting an inverted tepee TMD configuration (not shown). (C) Side view of Sav1866 (2HYD.pdb) depicting domain swapping in the TMDs, which form divergent open wings in the nucleotide-bound state. (D) Front view of Sav1866 depicting the two coupling helices (short, almost horizontal purple and blue cylinders just above the NBDs). (Reprinted with permission from Jones *et al.*, 2009.)

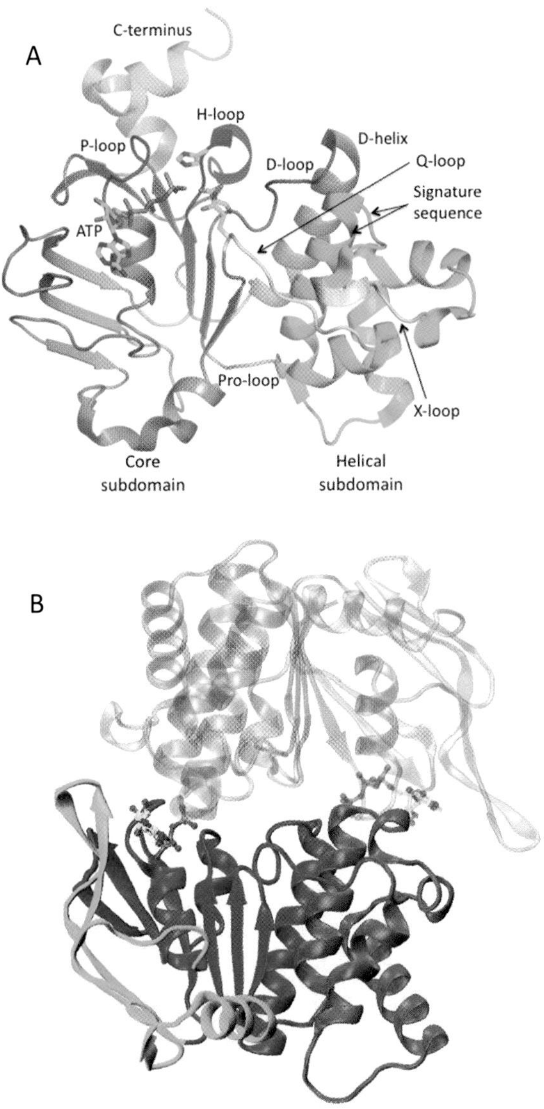

Figure 7.2 NBD monomer and dimer showing the major conserved motifs and residues. (A) NBD monomer in ribbon representation with ATP in stick format. The core subdomain (CSD) is coloured grey and the helical subdomain (HSD) is green. The Walker A (P-loop) motif is coloured magenta. The catalytic glutamate (in stick form) is depicted at the start of the D-loop (dark orange) followed immediately by the D-helix (light orange). The Q-loop is in yellow with the glutamine side chain in stick form. The H-loop (histidine side chain in stick form) is in dark blue; and the Pro-loop is in light blue at the bottom of the figure connecting the two subdomains and leading into the signature sequence (vertical green helix and loop adjacent to the orange D-helix). The C-terminus is coloured cyan. The short orange helix and loop in the centre of the HSD is the X-loop. (B) The ATP-binding cassette dimer. Ribbon diagram of the MJ0796 ABC ATPase dimer (1L2T.pdb). One monomer is coloured with the ATP-binding core subdomain blue and the antiparallel β subdomain green, together comprising lobe I. The α-helical subdomain or lobe II is coloured red. ATP is shown in ball-and-stick form with carbon in yellow, nitrogen in blue, oxygen in red and phosphorus in tan. The opposite monomer and ATP are shown in ghost representation. (Reprinted with permission from Jones *et al.*, 2009.).

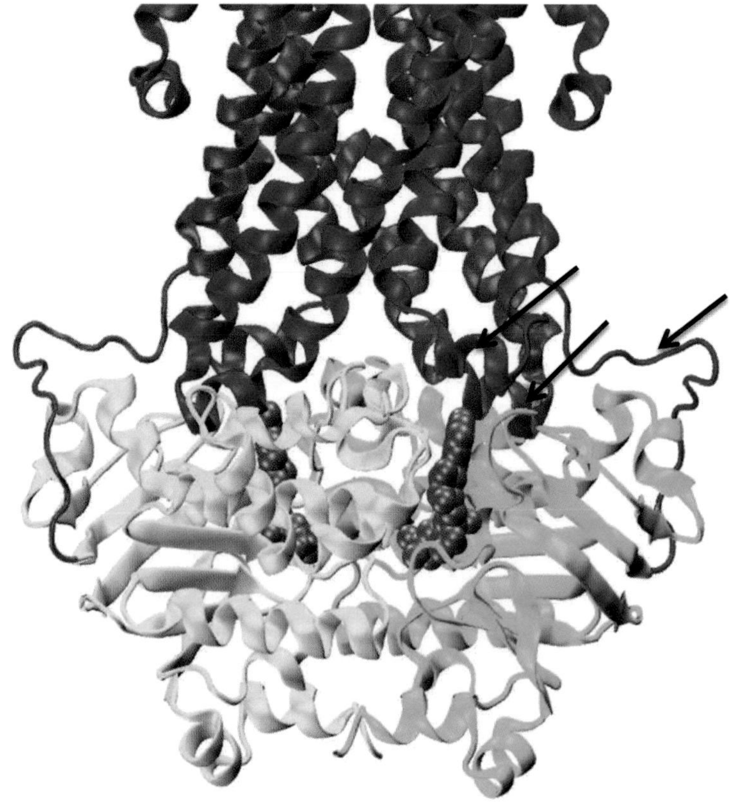

Figure 7.3 The cross-protomer interaction of Sav1866. The intracellular portion of a Sav1866 homodimer is represented in cartoon fashion; the two Sav1866 molecules are coloured yellow and red, and blue and green. Bound nucleotide is rendered in grey space-filling representation. The cross-protomer ('domain swapping') interaction is illustrated by the intracellular loops of one TMD (blue) interacting primarily with the NBD of the opposite protomer (yellow). Five helices in each protomer of Sav1866 provide three main contact points to the NBD. The extension to TM6 has a direct covalent linkage into the NBD itself, whereas the linker regions (ICL1 and ICL2) between TM helices 2 and 3 and between TM helices 4 and 5 provide non-covalent interactions with sites on the NBD. These three regions are arrowed in the figure. (Reprinted with permission from Kerr *et al.*, 2010.)

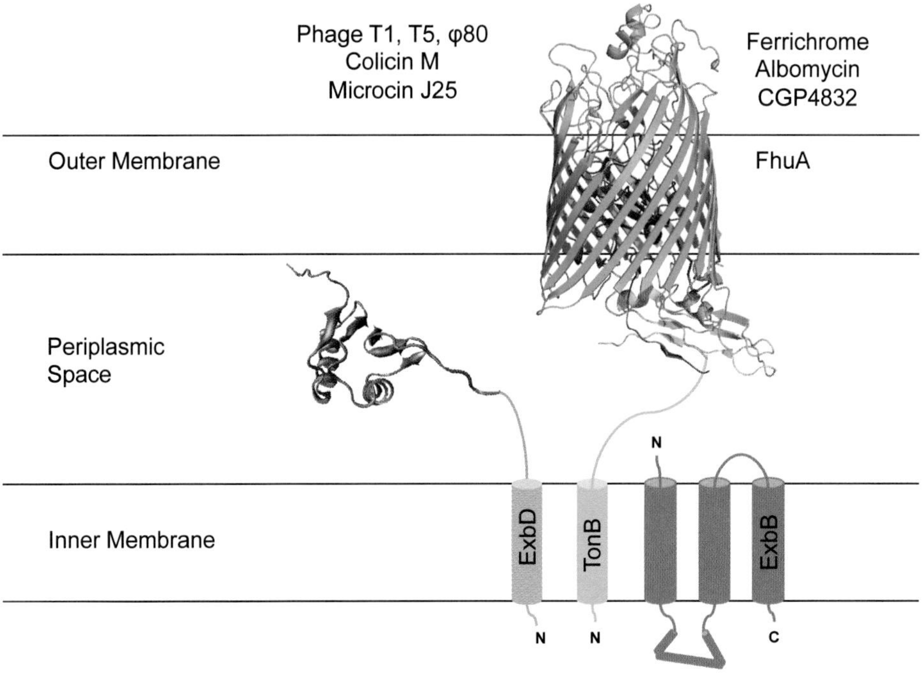

Figure 8.1 Proteins involved in ferrichrome transport in *Escherichia coli.* Shown are the crystal structure of the FhuA protein (Ferguson *et al.*, 1998; Locher *et al.*, 1998) with the ligands that use FhuA to enter cells; the crystal structure of a C-terminal portion of the TonB protein bound to the TonB box of FhuA (Pawelek *et al.*, 2006); the NMR structure of a C-terminal portion of ExbD (residues 64–133; Garcia-Herrero *et al.*, 2007); and the transmembrane topology of TonB, ExbB, and ExbD. The TonB, ExbB, and ExbD proteins form a complex of at least six ExbB proteins and one ExbD protein (Pramanik *et al.*, 2010, 2011; see also Higgs *et al.*, 2002). The structure of ExbB is not known.

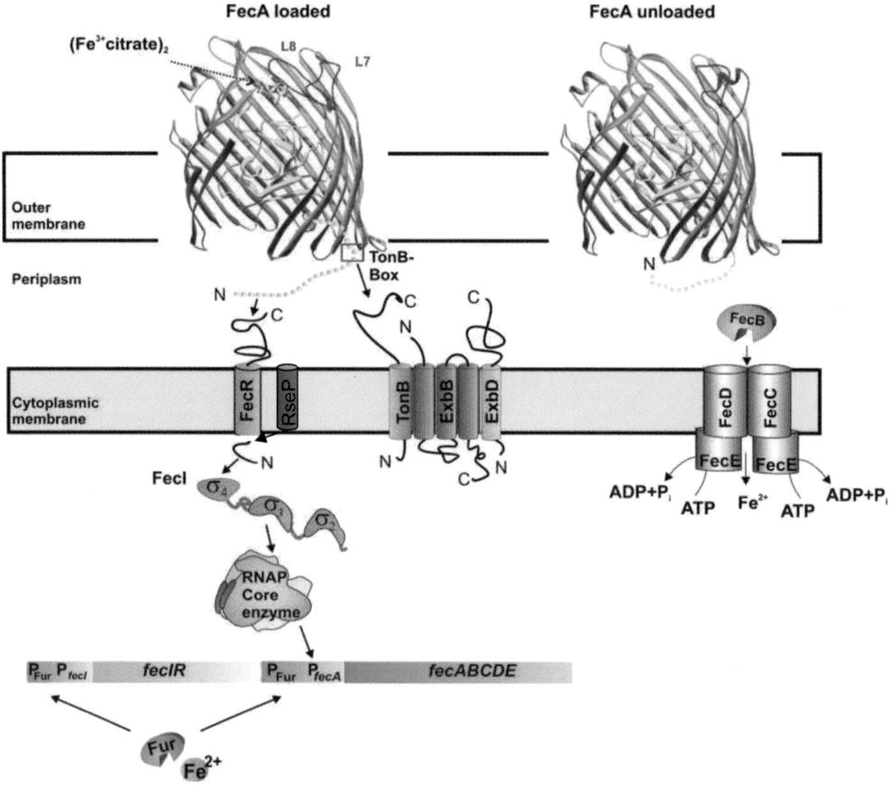

Figure 8.2 Transenvelope regulation of the *fec* genes (left side) and transport of ferric citrate (right side). The N-terminal region of FecA (yellow dashed line) is not seen in the crystal structure but has been determined by NMR. TonB–ExbB–ExbD are required for both regulation and transport. A signal elicited by binding of diferric dicitrate to FecA is transmitted across the OM to FecR, which transmits the signal across the cytoplasmic membrane to the ECF sigma factor FecI. It is predicted that the signal changes the structure of FecR such that it is cleaved by the RseP protease and the resulting cytoplasmic FecR fragment binds to FecI which then binds to the *fecABCDE* promoter. Transcription of the *fecIRA* regulatory genes and the *fecABCDE* transport genes is repressed by the Fur repressor protein loaded with Fe^{2+}. Under iron-limiting growth conditions, Fe^{2+} dissociates from Fur and transcription of *fecIR* and in turn *fecABCDE* is initiated. After transport of diferric dicitrate into the periplasm, iron is transported across the cytoplasmic membrane by the ABC transporter FecBCDE.

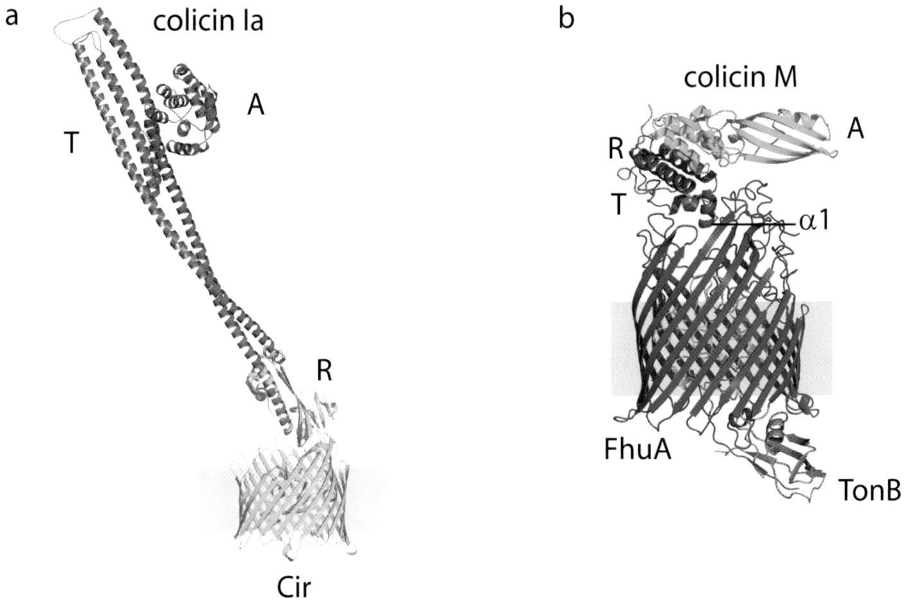

Figure 8.3 (a) Crystal structure of the Cir OM transporter with bound colicin Ia. The crystal structure of Cir with the colicin Ia receptor binding domain (designated R) has been determined (Buchanan *et al.*, 2007). The translocation domain (T) and the activity domain (A) are shown. The proteins are drawn to scale. Cir has a height of 40 Å, and colicin Ia is 210 Å long. (Courtesy of Susan Buchanan, NIH, Bethesda, Maryland, USA). (b) Crystal structure of colicin M modelled on the crystal structure of the FhuA OM transporter such that the α1 helix of colicin M contacts FhuA in its surface pocket. Colour code: green, colicin M activity domain (A); magenta, receptor binding domain (R); orange, translocation domain (T); blue, FhuA; yellow, bound C-terminal TonB fragment. For the sake of clarity the drawing scale of (b) is larger than that of (a).

a

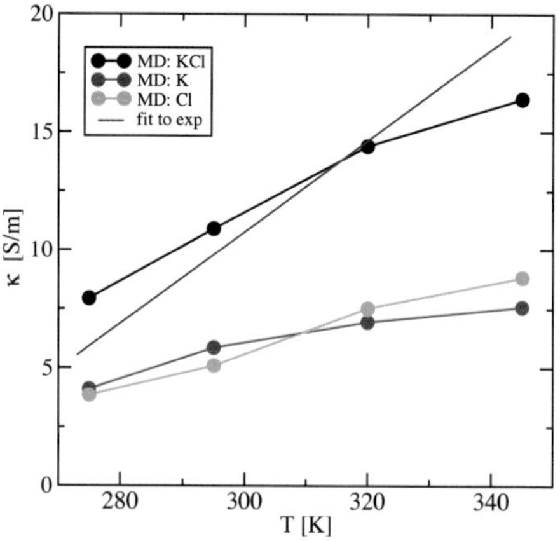

b

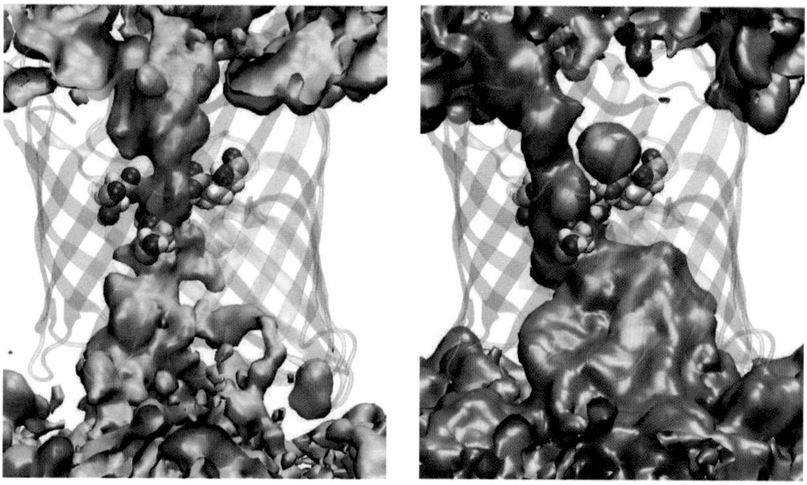

Figure 9.3 (a) Temperature-dependent conductivity of KCl in bulk water. Shown are the experimental findings compared to MD simulations. In the simulations one also gets the individual currents of the two ion types. For details see Pezeshki *et al.* (2009). (b) Iso-density surfaces of the Cl– (left) and K+ ion densities (right) in an OmpF channel averaged over the full trajectory. The important residues in the constriction zone are shown as well. For details see Pezeshki *et al.* (2009).

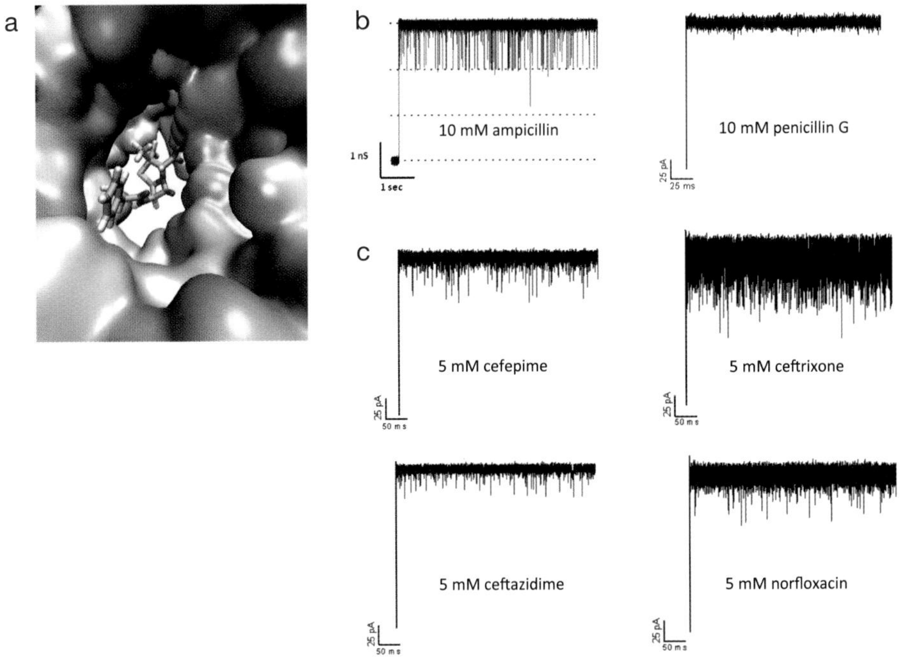

Figure 9.4 (a) Schematic representation of antibiotic translocation through the constriction zone of a single OmpF channel. (b) Typical tracks of ion currents through single trimeric OmpF channels reconstituted into planar lipid membranes in the presence of zwitterionic and anionic penicillins. Membrane bathing solutions contained 1 M NaCl (pH 5.0) and 5.7mM of the indicated antibiotic, and the applied voltage was 100mV (adopted from Danelon *et al.*, 2006). (c) Typical tracks of ion conductance through single trimeric OmpF channels reconstituted into miniaturized free standing bilayer setup in the presence of cephalosporins and fluoroquinolones. The membrane was formed from DPhPC; membrane bathing solutions contained 1 M KCl, pH6 and few millimolar antibiotics. Applied voltage was –50mV (adapted from Mahendran *et al.*, 2010).

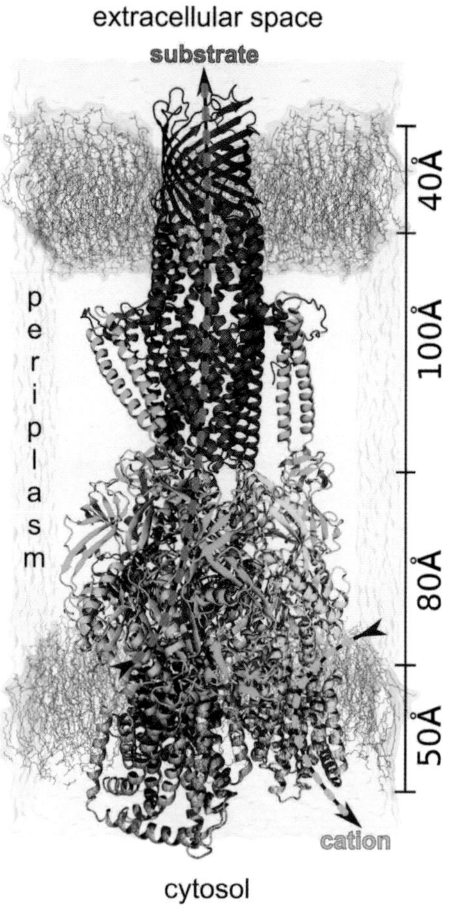

Figure 9.5 The tripartite efflux system ArcAB-TolC spanning two membranes according to the model of Symmons *et al.* (2009).

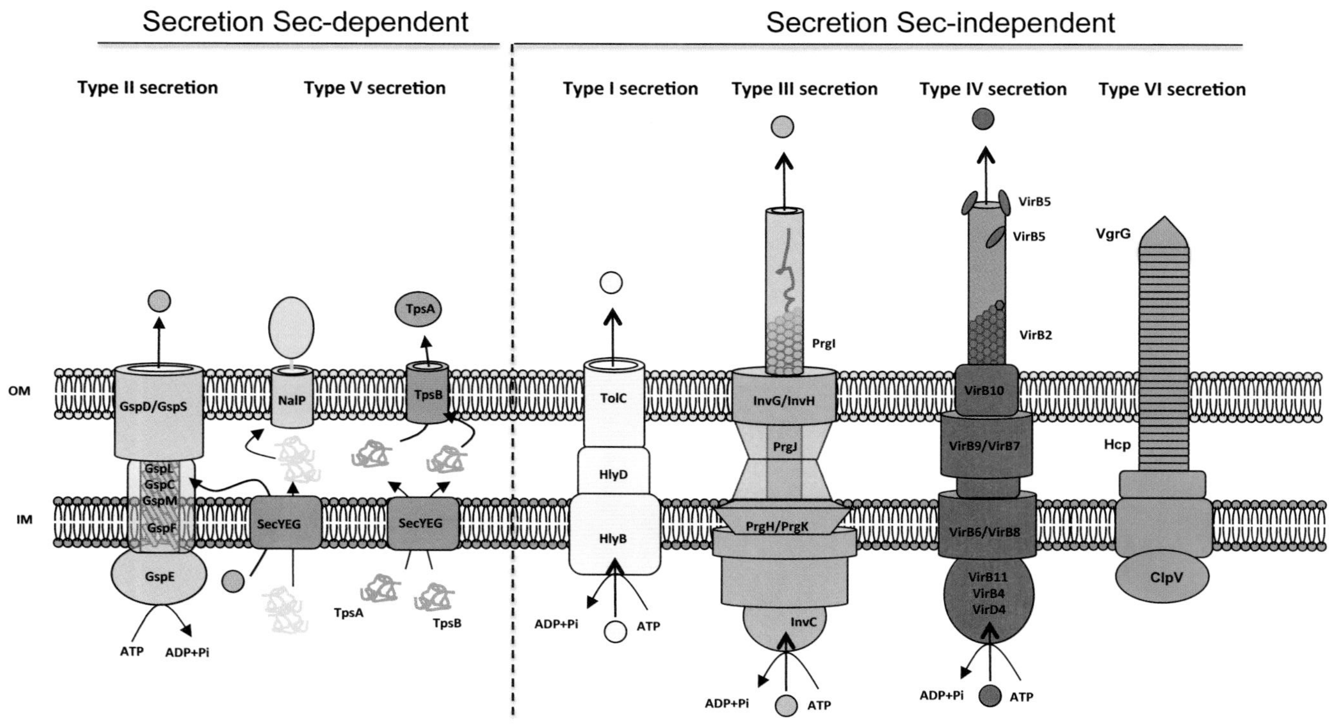

Figure 11.1 Schematic view of the secretion systems found in Gram-negative bacteria.

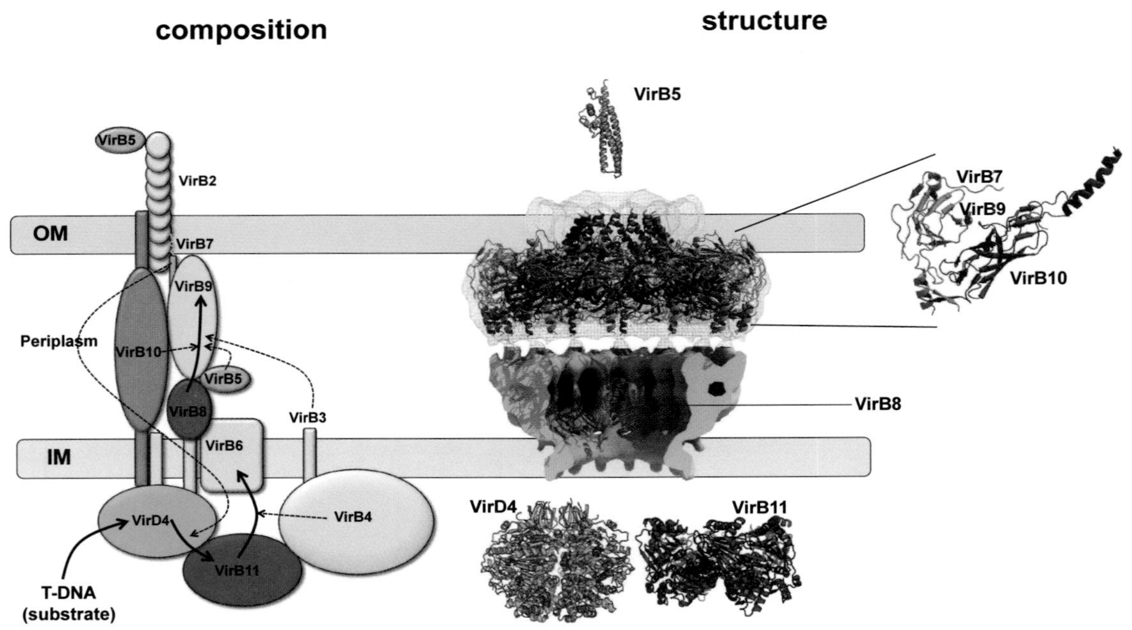

Figure 11.7 Secretion pathway and structure of the T4S system. Left: T4S system assembly and substrate pathway. The T4S components are represented according to their proposed localization. Most (but not all) of the interactions that have been confirmed biochemically (i.e. co-purification, immuno-precipitation) are indicated by the physical proximity of the schematic representations of each protein (see Chandran *et al.*, 2009, for a review of all the interactions described in the literature). Substrate (T-DNA) pathway within the T4S apparatus of *A. tumefaciens* as determined by the TrIP (Cascales and Christie, 2004b) is represented by plain arrows. The T-DNA interacts directly with VirD4, VirB11, VirB6/VirB8 and finally with VirB9/VirB2. The other T4S components influence the transfer of the substrate at different stages as represented by dotted arrows. Right: T4S system component structures. All the structures determined to date are represented at the same scale. The atomic structures are shown in ribbon representation. They include full-length VirB5, VirB8 periplasmic domain, full-length VirB11, VirD4 soluble domain and the T4S outer membrane complex (made of full-length VirB7, and the VirB9 C-terminal domain and VirB10 C-terminal domain). The structure of the core complex determined using cryo-electron microscopy and single particle analysis is rendered as a cut-out volume. The core complex is composed of the full-length VirB7, VirB9, and VirB10 proteins.